Evolutionary History of the "Robust" Australopithecines

Evolutionary History of the "Robust" Australopithecines

Edited by
Frederick E. Grine

LONDON AND NEW YORK

First published 1988 by Transaction Publishers

Published 2017 by Routledge
2 Park Square, Milton Park, Abingdon, Oxon OX14 4RN
711 Third Avenue, New York, NY 10017, USA

Routledge is an imprint of the Taylor & Francis Group, an informa business

Copyright © 1988 by Taylor & Francis.

All rights reserved. No part of this book may be reprinted or reproduced or utilised in any form or by any electronic, mechanical, or other means, now known or hereafter invented, including photocopying and recording, or in any information storage or retrieval system, without permission in writing from the publishers.

Notice:
Product or corporate names may be trademarks or registered trademarks, and are used only for identification and explanation without intent to infringe.

Library of Congress Catalog Number: 2007024721

Library of Congress Cataloging-in-Publication Data

Evolutionary history of the "robust" australopithecines : a concise history of the major figures, ideas, and schools of sociological thought / [edited by] Frederick E. Grine.
    p. cm.
Originally printed: New York : A. de Gruyter, 1988, in series: Foundations of human behavior.
Includes bibliographical references and index.
ISBN 978-0-202-36137-6 (alk. paper)
    1. Australopithecines--Congresses. 2. Paranthropus--Congresses. I. Grine, Frederick E., 1952-

GN283.E93 2007
569.9'3--dc22

2007024721

ISBN 13: 978-0-202-36137-6 (pbk)

Dedicated To

My Parents
Edward and Phyllis

and

My Wife
Elizabeth

For Their Unfailing Encouragement,
Support and Forbearance

Dedicated also to
ROBERT BROOM

On the Half Century Since He First Encountered
the Face of *Paranthropus robustus*
(Tuesday, June 14, 1938)
And for His Recognition of the Significance of
This Novel Hominid Taxon

Dr. Robert Broom, FRS

(1866–1951)

Contents

Foreword by *F.C. Howell*	xi
Preface	xvii
Acknowledgments	xix
List of Contributors	xx

I. Studies of the Craniodental Evidence

1 Enamel Thickness and Development in *Australopithecus* and *Paranthropus*   *3*
   Frederick E. Grine and Lawrence B. Martin

2 Growth of Teeth and Development of the Dentition in *Paranthropus*   *43*
   M. Christopher Dean

3 Implications of *In Vivo* Experiments for Interpreting the Functional Significance of "Robust" Australopithecine Jaws   *55*
   William L. Hylander

4 Enlarged Occipital/Marginal Sinuses and Emissary Foramina: Their Significance in Hominid Evolution   *85*
   Dean Falk

5 "Robust" Australopithecine Brain Endocasts: Some Preliminary Observations   *97*
   Ralph L. Holloway

6 Growth Processes in the Cranial Base of Hominoids and Their Bearing on Morphological Similarities that Exist in the Cranial Base of *Homo* and *Paranthropus*   *107*
   M. Christopher Dean

II. Studies of the Postcranial Evidence

7 New Estimates of Body Size in Australopithecines   *115*
   William L. Jungers

8 Associated Cranial and Postcranial Bones of *Australopithecus boisei*   *127*
   Hannah M. Grausz, Richard E. Leakey, Alan C. Walker, and Carol V. Ward

9 New Estimates of Body Weight in Early Hominids and Their Significance to Encephalization and Megadontia in "Robust" Australopithecines   *133*
   Henry M. McHenry

10 New Postcranial Remains from Swartkrans and Their Bearing on the
Functional Morphology and Behavior of *Paranthropus robustus*   *149*
Randall L. Susman

III. Studies of Variation and Taxonomy

11 Variation, Sexual Dimorphism and the Taxonomy of *Australopithecus*   *175*
William H. Kimbel and Tim D. White

12 On Variation in the Masticatory System of *Australopithecus boisei*   *193*
Yoel Rak

13 Evolution of the "Robust" Australopithecines in the Omo Succession:
Evidence from Mandibular Premolar Morphology   *199*
Gen Suwa

14 New Craniodental Fossils of *Paranthropus* from the Swartkrans Formation
and Their Significance in "Robust" Australopithecine Evolution   *223*
Frederick E. Grine

IV. Studies of Evolutionary Relationships

15 The Evolution of *Australopithecus boisei*   *247*
Alan C. Walker and Richard E. Leakey

16 Implications of KNM-WT 17000 for the Evolution of "Robust"
Australopithecus   *259*
William H. Kimbel, Tim D. White and Donald C. Johanson

17 Are "Robust" Australopithecines a Monophyletic Group?   *269*
Bernard A. Wood

18 A New *Australopithecus* Cranium from Sterkfontein and Its Bearing on the
Ancestry of *Paranthropus*   *285*
Ronald J. Clarke

19 Numerous Apparently Synapomorphic Features in *Australopithecus
robustus, Australopithecus boisei* and *Homo habilis:* Support for the
Skelton-McHenry-Drawhorn Hypothesis   *293*
Phillip V. Tobias

V. Studies of Paleogeography, Paleoecology and Natural History

20 New Information from the Swartkrans Cave of Relevance to "Robust"
Australopithecines   *311*
Charles K. Brain

21 Chronology of South African Australopith Site Units   *317*
Eric Delson

22 "Robust" Hominids and Plio-Pleistocene Paleogeography of the Turkana Basin, Kenya and Ethiopia *Frank H. Brown and Craig S. Feibel*	325
23 Habitat Preference and Paleoecology of *Australopithecus boisei* in Eastern Africa *Pat Shipman and John M. Harris*	343
24 The Deep-Sea Oxygen Isotope Record, The Global Ice Sheet System and Hominid Evolution *Michael L. Prentice and George H. Denton*	383
25 Late Pliocene Climatic Events and Hominid Evolution *Elisabeth S. Vrba*	405
26 Tooth Morphology, Wear and Diet in *Australopithecus* and *Paranthropus* from Southern Africa *Richard F. Kay and Frederick E. Grine*	427
27 The Comparative Biology of "Robust" *Australopithecus:* Clues from Context *Tim D. White*	449
28 Divergence between Early Hominid Lineages: The Roles of Competition and Culture *Milford H. Wolpoff*	485
29 The Causes of "Robust" Australopithecine Extinction *Richard G. Klein*	499

VI. Summary Comments

30 Evolutionary History of the "Robust" Australopithecines: A Summary and Historical Perspective *Frederick E. Grine*	509
Specimen Index	521
Subject Index	525

Foreword

F. CLARK HOWELL

Fred Grine first discussed with me early in 1985 the appropriateness and feasibility of a workshop/conference devoted to those early hominids often collectively referred to as "robust" australopithecines. I was strongly supportive of the appropriateness of the idea, but had a certain hesitancy with respect to the feasibility given the substantial prevalence of controversy and dissent. Grine spoke to me from the perspective of my own concern with such research, and he proceeded with the necessary planning, sought to raise the necessary travel and maintenance funds for invited participants and targeted the gathering for early autumn, 1986. This date was ultimately unfeasible and the workshop eventuated in early spring, 1987, immediately prior to the annual meeting of the AAPA in New York City. It was both a very busy and heady time; the papers in this volume testify to the diversity of topics presented, and reflect, in some part, the wide range of subject matter discussed by us over a 5-day period. The success of the endeavor is reflected not only in these contributions, but also in the level, balance and candidness of discourse that was (almost) consistently maintained through the masterful efforts of both the organizer and a succession of session moderators. I believe the occasion, and the resultant volume, do in fact constitute a significant watershed in the varied efforts, by an increasing number and diversity of investigators, to focus on problems of hominid origins and to seek to elucidate, in particular, the complexities of differentiation, adaptation and phylogeny of Plio-Pleistocene Hominidae. Whereas the workshop centered on the "robust" australopithecine problem, the discussions as well as many of the published contributions concern substantially broader issues relevant, both methodologically and empirically, to the whole field of hominid evolutionary biology. Consequently, this volume is destined to have an even more significant and broader impact.

The central problems of concern at the workshop were (a) the cladogenetic of phyletic evolutionary events attending the differentiation of "robust" australopithecine lineage(s); (b) the overall morphologic pattern and distinctive character states peculiar to the lineage(s); (c) the functional correlatives of such characters; (d) the phylogenetic affinities between representatives of the "robust" lineage(s), and correspondingly the question of requisite systematics, and affinities with other, valid hominid taxa; (e) the paleoenvironmental settings and paleoecological circumstances attending the origin and evolution of "robust" australopithecines; and (f) the extinction of "robust" australopithecines. Essentially *all* participants in the workshop approached such problems from the perspective of a multiplicity of species of *Australopithecus* [minimally *A. afarensis*, *A. africanus*, *A. robustus* (including or not *A. crassidens* as a valid taxon), and *A. boisei*] and, at least, two extinct species of *Homo* (minimally, *H. habilis* and *H. erectus*). There was, however, even prior to the workshop neither consensus on the composition of the attendant hypodigms of said taxa—that is, on the necessity for the recognition of additional taxa at the specific level (*aethiopicus* in the case of *Australopithecus*, *ergaster* in the case of *Homo*) or at the generic/subgeneric level (the *Paranthropus* issue). Although addressed to a varying extent, none of those issues was fully resolved as a consequence of the workshop; nonetheless, the appropriateness of current taxonomic attributions and the need for a re-examination of supraspecific categories was expressed on repeated occasions. In my view, these constitute important and appropriate questions worthy of continued concern and investigation, both in respect to the problem foci of this workshop and the broader issues in hominid phylogenetics.

Neither the origin of Hominidae nor of the genus *Australopithecus (sensu lato)* were formally discussed at the workshop. Most participants assumed a notably empiricist stance, and presumably since so little fossil hominoid material of immediate relevance has been recovered there was a natural tendency to skirt the issue. Nonetheless, fragmetary manidbular/dental specimens from the Lukeino, Lothagam and Chemeron formations (Kenya Rift Valley)[1] are directly relevant to the issue as they are not clearly hominoid, but also largely morphometrically approach or conform with a hominid pattern. In the case of the Lothagam and Tabarin mandible fragments, and the Mabaget proximal humerus there are some demonstrably specific points of concordance with the morphology for *A. afarensis*. Unfortunately, the relevant elements requisite to an elucidation of postural and locomotor adaptation are still lacking. Moreover, specifically "robust" australopithecine features are absent.

Thus, the morphotype epitomized by *A. afarensis* must constitute (for now, at least) the basis for comparison for other, subsequent samples of the genus in the evaluation of their phylogenetic relationships. (Equally relevant, *Pan* is accordingly *not* an appropriate morphotype, however useful for outgroup comparison with Hominidae.) Fortunately, *A. afarensis* is very well represented by juveniles and adults, male and female individuals, and by literally all skeletal elements—some (mandibles and teeth) are preserved in substantial numbers, such that ranges of variability can be assessed. If its roots fall within the uppermost Miocene, as appears very probable, its best documentation is within the middle and upper Pliocene. Although there was considerable discussion, and even some argumentation among conference participants, consensus seemed to emerge that this was not only a valid taxon (contra several prior sceptics, some of whom were present), but that a single species was represented by the large Hadar Formation sample, and that the species hypodigm certainly included *both* the Laetoli and Hadar samples (as long maintained by D.C. Johanson and T.D. White, in particular). Probably, most concerned investigators would also include within the hypodigm the Maka proximal femur and Belohdelie frontal remains (middle Awash Valey, Ethiopia) of earlier Pliocene age. Cranial parts KNM-ER 2602 (Area 117) and associated lower premolar–molar teeth KNM-ER 5431 from the Tulu Bor Member, Koobi Fora Formation probably testify to the presence of this species in East Turkana. The affinities of fragmentary hominid specimens (dental, mandibular, postcranial) from the equivalent-aged (lower) Lomekwi Member of the Nachukui Formation of west Turkana have yet to be ascertained. In the lower Omo Basin, Ethiopia, this species (only) is represented in the Usno Formation (Member U-12) and the lowest members (A and B, and perhaps lower C) of the Shungura Formation. All of these occurrences are of approximately, or slightly greater than 3 Myr. This species, in my view, is still unknown in southern Africa.

The first definitive appearance of "robust" australopithecine characters (in mandibular and dental elements) was initially noted in the 2.5 Myr range in the Shungura Formation succession (F.E. Grine, F. C. Howell). This has been solidly confirmed recently by the recovery of a quite complete cranium, KNM-WT 17000 (upper Lomekwi Member), and a partial mandible with dentition, KNM-WT 16005 (Lokalalei Member), from the Nachukui Formation at an age of c.2.5-2.4 Myr. Apparently then, the roots of this taxon (or hominid sublineage) fall within the <3.0 and >2.5 Myr time span. Although some investigators would seek to maintain a taxonomic distinction (as the species, *A. aethiopicus*) for these (and somewhat younger) samples (F. E. Grine, F. C. Howell, D. C. Johanson, W. H. Kimbel, T. D. White), others (R. E. Leakey and A. C. Walker) consider that it seems unwise at present to divide the *A. boisei* lineage into separate species. The extent to which there are persistent cranial resemblances (12 symplesiomorphies) with *A. afarensis,* shared derived resemblances (12 synapomorphies) with *A. robustus* and *A. boisei,* and few shared derived resemblances (2 synapomorphies) with *A. boisei* alone, of 32 major characters chosen, has been set out by W.H. Kimbel, T. D. White and D. C. Johanson, If this cranial character list was to be expanded there would be at most only several more features shared exclusively with *A. boisei;* this would appear to be equally the case with mandibular structure and the dentition.

Nonetheless there was apparent consensus that the *A. boisei* or the *A. aethopicus/boisei* lineage

---

[1]These are the mandible body fragments KNM-TH 13500 (Tabarin, 5.0 Myr) and KNM-LT 329 (Lothagam, ? 5.5 Myr), a left M^1, KNM-LN 335 (Lukeino, 6.0–5.6 Myr), and a left proximal humerus, KNM–BC 1745 (Mabaget, <5.1 Myr).

was indeed of upper Pliocene antiquity, and that it displayed certain discernible evolutionary trends in endocranial volume and conformation, craniofacial, mandibular and dental morphology in the course of its 1.5 Myr (or greater) existence. Sexual dimorphism in body size/weight remains consistently substantial, and such size/weight estimates for *Australopithecus* taxa appear to overlap markedly both in averages and ranges (W. L Jungers, H. M. McHenry), in contrast to some previous assertions.

A species-level distinction between "robust" australopithecine taxa in two sites in southern Africa was proposed by R. Broom 40 years ago, and subsequently supported and then ultimately rejected by J. T. Robinson. Most other workers have downplayed such a distinction, between *A. robustus* (Kromdraai-B) and *A. crassidens* (Swartkrans), although several (F. E. Grine, F. C. Howell) have insisted that at least the dental morphology, and particularly that of the deciduous dentition, strongly argues for specific recognition. Important additions both to the Kromdraai (Kromdraai B East, Member 3) sample through the work of E. S. Vrba and to the Swartkrans sample by C. K. Brain certainly tend not only to support a specific distinction and a sister group relationship between these taxa, but also suggest an *A. robustus->crassidens*, ancestral-descendant relationship (F. E. Grine). Unfortunately, most macromammal evidence neither affirms nor controverts an age assessment of KBE-3 as antecedent or subsequent to SK-1; however, some microvertebrate taxa certainly tend to support a greater age for the former infilling. Interestingly, the *A. crassidens* morph, now represented by upward of some 120-125 individuals, persists seemingly unchanged through three successive members of the Swartkrans Formation and, most importantly, it is unquestionably associated with a species of *Homo*, in Members 1 and 2 (C. K. Brain, F. E. Grine). Perhaps the most recent occurrence of any "robust" australopithecine is documented in Member 3 (C. K. Brain). Clearly, the Kromdraai *A. robustus* sample, limited though it may be, is of the greatest importance in the search to decipher the relationships between the well-known *A. crassidens* and *A. boisei* species, as well as in respect to their respective ancestral morphotypes and the nature of linkages with *A. afarensis*. To be more direct, is it not appropriate to consider *A. robustus-crassidens* and *A. aethopicus-boisei* (sub)lineages in the perspective of an antecedent, *A. afarensis* morphotype?

The matter of phylogenetic affinities among australopithecine taxa has been a significant issue for over 40 years. It has become an ever more central and hotly debated issue as successive discoveries in southern Africa (as at Swartkrans) and particularly in eastern Africa have vastly enhanced the hominid fossil record. In the past 15 years the concern has assumed a prominent place in hominid evolutionary biology, with diverse efforts increasingly directed toward cladistic analyses along with an enhanced appreciation of the nature and significance of particular character states, their retention, the degree to which they are shared, and their singular appearances. Whereas monophyly or quite simple branching events were initially posited, numbers of investigators (M. C. Dean, F. E. Grine, W. H. Kimbel, H. M. McHenry, P. V. Tobias, T. D. White and B. A. Wood) have come to recognize and assess the enhanced complexity of early hominid diversity. The question of independently acquired character states (homoplasies) is now of widespread concern and a central research focus (B. A. Wood, A. C. Walker and R. E. Leakey). Probably it is fair to say that more students of the subject consider now that such parallel or convergent manifestations are distinct possibilities and perhaps also of significant magnitude in regard to cranial, mandibular and dental morphology. This is of far-reaching consequence for human evolutionary studies as a whole.

Australopithecus africanus (hypodigm Taung; Sterkfontein Member 4; Makapansgat Members 3 and 4) is clearly primitive relative to "robust" australopithecines and, of course, to genus *Homo* (a consensus view). The nature of many of these craniodental differences has been examined by Kimbel, White, Johanson, Rak, Wood, Holloway and Tobias. The phylogenetic affinities of the former species has long been a subject of central concern and patently continues to remain so. Its position as directly (phylogenetically) antecedent to *Homo*, to "robust" australopithecines alone, to "robust" australopithecines *and* to *Homo*, to *A. robustus/crassidens* only, to *A. robustus/crassidens* and *Homo,* and as representative of a wholly separate lineage has been proposed and/or supported by some at one time or another. Morphologic differences between samples comprising the hypodigm continue to be proffered (R. J. Clarke, T. D. White and W. H. Kimbel). Undoubtedly there was at the workshop less measure of consensus in respect to this taxon than on any other single such issue. In a way this is all the more startling given the very substantial, and ever increasing samples (notably that from Sterkfontein Member 4). The possibility that this sample might comprise more than a single species has been put forth by R. J. Clarke. Most (but not necessarily all) workers familiar with the sample would hesitate to affirm this interpretation. Nonetheless, there may indeed be merit in Clarke's perspective; the matter should not simply be offhandedly dismissed.

Both the age and possible duration of the Member 4 infilling at Sterkfontein are of relevance in this regard. Six members of the Sterkfonetin Formation have been formally defined, having an aggregate thickness of 27 + m, although in excess of 50m may be represented. Member 4, with four beds (A to D) almost 10m in thickness, comprises a succession of roof collapses and externally derived sediments (A), succeeded by a conelike accumulation of externally derived colluvial sediments, incorporating coarse to finer clastics (B,C) capped by thin, precipitated calcite lenses (D). The whole of this Member's accumulation was probably relatively protracted, although that at the base was both intermittent and correspondingly occasionally rapid. Vertebrates are (locally) common in B and C, from which all *A. africanus* remains are inferred or known to have derived. Geomagnetic polarity measurements through the succession indicate, not unexpectedly, disturbed conditions of deposition. An age within the upper Pliocene is generally acknowledged, and a possible range of 2.8 to 2.4/2.3 Myr (E. S. Vrba) has been specifically proposed, whereas based on a biochronology of African cercopithecoids an age of 2.5 to 2.4 Myr (= Shungura Formation Tuff D equivalent) has been inferred (E. Delson). Scant proboscidean and suid remains are consistent with such estimates. Some bovids show links with Makapansgat (e.g., the ovibovine, *Makapania broomi; Hippotragus cookei)*,while others (e.g., the presence of *Antidorcas*) do not (E. S. Vrba). An upper age might be indicated by the occurrence of the large zebrine, *Equus capensis*, which occurs first in the Turkana basin at about 2.3 Myr. Among carnivores, *Pachycrocuta* sp. is also a link with Makapansgat. However, the occurrences of *Chasmaporthetes* sp. (or spp.)[2] and *Crocuta* (or aff. *Crocuta*) are congruent with the presence of one or both earlier in the Pliocene (as at Laetoli and Hadar). At any rate, the Sterkfontein Formation demonstrably encompasses a substantial time span, and Member 4 may link more closely than usually envisioned with Makapansgat (before) and Kromdraai-B (after). The Sterkfontein locality is obviously of singular importance both for further understanding differentiation within *Australopithecus* and the roots of genus *Homo*.

I concur, for the record, with suggestions (E. Delson, E. S. Vrba) that Taung (and its type hominid) constitutes a not greatly younger occurrence of *A. africanus* relative to its occurrence in Sterkfontein Member 4.

Vast and fundamental advances have been made toward the establishment of a time scale for the late Cenozoic and the recognition of significant global paleoclimatic events. This is largely a consequence of the long-term DSDP effort in the world oceans, and the establishment of both an adequate biostratigraphy (planktonic and benthic microorganisms) and a geochronologically firm, high-resolution oxygen isotope record of paleotemperature and ice volume variations (Prentice and Denton). It has also resulted from appropriate in-depth investigations of significant stratotypes in Eurasia. The implications of major transformations in global climate and meteorological regimes has appropriately impacted paleobiologists, including in particular those concerned with changes in mammal communities, their structure and composition, both in Africa and in Eurasia. The potential significance for hominid origins and early evolution has been forcibly stressed in recent years, particularly by C. K. Brain and E. S. Vrba. As there are long, stratigraphically well-controlled and sampled successions in several eastern African basins (e.g., Awash, Omo-Turkana, Baringo and Olduvai-Laetoli) in which there is also generally firm geochronological control (K/Ar, paleomagnetics), these serve as requisite reference sections for both provincial

---

[2]Well-known in Ruscinian age (and younger) faunas in north China, it is earlier represented at Langebaanweg in the form of *"Percrocuta" australis*, and might well be at Sahabi (F. C. Howell). It is unclear whether this interesting hyenid is of African or Eurasian origin. I have a strong feeling that other carnivores, particularly canids, will prove to be of major biostratigraphic and biogeographic import.

and extra-areal comparison and correlation. Each also affords some diversity of paleogeographic and paleoenvironmental data, of first or second order, relevant to such comparisons, correlations and related inferences.

Several sorts of comparisons may be (and have been) attempted and attendant correlations sought. One sort seeks to delineate local or provincial paleoenvironmental patterns within the framework of global paleoclimatic events, while another sort seeks correlations and causal consequences (origins, speciation events/radiations, extinctions, dispersals, biotidal effects) within such an established framework. E. S. Vrba has contributed substantially in this regard, and herein poses three important, testable hypotheses relative to refugeal vs. biotidal areas, turnover pulses and climatic-initiating causes. A different sort of comparison seeks to employ particular sets of taxa of known habitat preference(s) to elucidate (local or provincial) paleoenvironments in order, for example, to illuminate preferences and potential adaptative adjustments and strategies of other, associated (extinct) taxa, including hominids; the chapter by P. Shipman and J. M. Harris is such an example. All three sorts of comparisons are useful and potentially valid endeavors in hominid paleobiology, but each has its drawbacks and limitations.

Analytical and comparative studies of these kinds are clearly capable of affording invaluable insights into a diversity of paleobiological problems. The level of stratigraphic resolution and the utilization of taxonomic diversity within groups should be maximized in such efforts. The adaptive-niche significance of other families (e.g., suids and cercopithecoids) similarly requires comparable examination and comprison with the substantial data base derived (largely) from bovid frequencies, although the amount of paleoenvironmental information that may be afforded by microvertebrates is indeed substantial. Moreover, the treatment of sedimentary environments and taphonomic circumstances attending samples still merits more refined acknowledgment, as do also ecotonal and mosaic structures of extant African situations employed in ecological characterizations. In a number of respects, and particularly in regard to the specifics of early hominid ecological niches and habitat preferences, such studies are still in a very formative state.

The organizer and editor is to be commended highly on having produced a workshop and its attendant volume that are both timely and of immense benefit to students of early hominids, in particular, and of substantive interest to all those concerned with hominid evolution, in general.

Preface

Over the past decade or so there have been quite a few large conferences at which members of the paleoanthropological community have gathered to present papers on diverse topics ranging from Miocene hominoids to Magdalenian tools. There also have been a number of smaller gatherings in which the symposium participants have given papers on more restricted themes in human evolution. Neither the larger nor the smaller conferences that have been held to date, however, have focused on the evolutionary history of that commonly neglected "side-branch" of human evolution comprised by the "robust" australopithecines.

In the whole of the field of paleoanthropology there is perhaps no more neglected (but nevertheless interesting) group of hominids than the "robust" australopithecines. The recent discovery of the "black skull" and the amount of popular press that it engendered suggests that the "robust" australopithecines are perhaps beginning to acquire a level of general interest that is certainly their due. Through them we have the opportunity to examine the origin, natural history and ultimate extinction of not just a single species but an entire radiation in the hominid fossil record; this is undoubtedly a unique situation. This group of early hominids has long held my interest and fascination, and in 1985 I began to formulate plans to organize a workshop of a small group of colleagues who also held an interest in these Plio-Pleistocene creatures.

An international workshop on The Evolutionary History of the "Robust" Australopithecines was convened at the State University of New York at Stony Brook on March 27th, 1987, and for the next 5 days a group of some 33 scientists—paleontologists, functional morphologists, anthropologists, paleoclimatologists, geologists and archaeologists–were cloistered at Danfords Inn, Port Jefferson to discuss a variety of topics related to these extinct hominids. The first 2 days of the workshop were devoted to the presentation of "formal" papers, and the succeeding 3 days were devoted to open and frank, round-table discussion.

In order to ensure that this workshop attained the stated goal of holding open and frank discussions on selected topics, it was necessary to restrict attendance to a reasonably small number of participants. With only one exception, the papers that comprise the present volume were presented at the workshop and have been revised in light of the discussions that were held during the course of that gathering. Though an invited participant, P. V. Tobias was prevented from attending the workshop due to circumstances beyond his control; the paper that he planned to deliver, however, is included here. The papers published herein thus provide a synopsis of the significant contributions to the workshop.

In organizing the workshop, and in choosing its title (being the title of the present book), it was necessary to convey meaning while avoiding the use of taxonomic nomenclature to which the participants might have objected. Some of the participants objected to the name *Australopithecus*, while others refused (and still refuse) to admit the validity of *Paranthropus*. Although I especially dislike the sobriquet, "robust" australopithecine since the adjective "robust" appears to be inappropriate at best and the name "australopithecine" is downright misleading in its taxonomic sense, the majority of the workshop participants seemed to be able to live with it. Despite its being manifestly inappropriate, the name "robust" australopithecine has gained such wide usage that it is very unlikely to fall into disuse in the near future. Even I can live with it.

The 28 papers that comprise this volume are grouped for convenience under the headings: Studies of the Craniodental Evidence, Studies of the Postcranial Evidence, Studies of Variation and Taxonomy, Studies of Evolutionary Relationships, and Studies of Paleogeography, Paleoecology and Natural History. Clearly a number of individual papers span two or more of these groupings, indicating the inter-disciplinary relationships that have developed in studies of the early hominid fossil record.

This volume is presented as an attempt to marshall and summarize some of what we have learned (or, at least, think we have learned) about the evolutionary history of the "robust" australopithecines in the 50 years since Robert Broom first encountered the visage of a new kind of ape-man from Kromdraai. New discoveries spanning the 50 some years from Kromdraai to Lomekwi have served to keep us aware that the paleontological record for hominid evolution is hardly exhausted; and because of such finds no single volume can hope to stand as a summary on the "robust" australopithecines for very long. The papers in the book act to shed new light upon some old questions, and they also act to provide new questions. The answers to those questions will hopefully bring us even closer to a fuller understanding and appreciation of the origins, evolution and ultimate demise of the "robust" australopithecines. If for no other reason than the fact that the "robust" australopithecines most likely stood as our closest relatives, a better understanding of their origin, history and demise can only serve to provide heightened appreciation of the course of human evolution itself.

Frederick E. Grine
Stony Brook, NY

Acknowledgments

From the inception of the idea to convene a workshop that would be dedicated to a discussion of the evolutionary history of the "robust" australopithecines, through the actual running of the workshop and on to the present volume a number of individuals have provided invaluable assistance, support and encouragement.

For their unfailing stimulation during countless lunch-time conversations, their sage advice and unstinting support during the seemingly endless months of preparation, and for their ability to cope with and assuage occasional displays of somewhat less than sangfroid behavior, I would like to express my gratitude to N. Creel, J. G Fleagle, W. L. Jungers, D. W. Krause, S. Larson, L. B. Martin, J. T. Stern and R. L. Susman. I am very grateful to N. Creel, J. G. Fleagle, D. W. Krause and J. T. Stern, who chaired the paper and discussion sessions at the workshop. They managed to keep us on time and on track. I am deeply indebted to M. M. Dewey for the encouragement and wise counsel that he has provided.

I would like to express my thanks to the Office of Conferences and Special Events of the State University of New York at Stony Brook, and especially A. Forkin and A. Brody, for their assistance in organizing the workshop reception and registration, and I am grateful to K. Passafiumi and the staff of Danfords Inn, Port Jefferson for their considerable efforts and attention to detail which ensured that the workshop ran according to schedule. I am grateful to the Stony Brook graduate students who gave freely of their time to provide transportation for the workshop participants and for their assistance in running the workshop; my special thanks go to D. Daegling, D. Waddle, D. Blaustein, B. Dumont, K. Fudge, N. Handler, M. Maas, R. Malenky, M. McClinton, J. Meldrum, T. Naiman, L. Shapiro, S. Strait and P. Ungar.

B. Asfaw, R. J. Clarke, R. L. Holloway, D. C. Johanson, W. H. Kimbel, G. Suwa, A. C. Walker and T. D. White made available casts of important fossil specimens during the course of the workshop, and I am grateful also to C. K. Brain, who very kindly made available during the workshop the original fossils excavated by him from the Swartkrans cave.

The workshop could not have been held, and the present volume would not have been made available had it not been for the generous support of a number of individuals and institutions. I am exceedingly grateful to the following for financial support of the international workshop on the evolutionary history of the "robust" australopithecines: The National Science Foundation, The L. S. B. Leakey Foundation, The Institute of Human Origins, The State University of New York at Stony Brook, The Holt Family Foundation, The Boise Fund, and Aldine de Gruyter.

I would also like to express my sincere thanks to L. Betti, L. Jungers and S. Nash for providing artwork for this endeavor at brief notice, and to M. Stewart and T. Manning for untiring and patient photographic asistance. I am grateful to S. Johnston, T. Leger, A. Obuck and D. Greitzer of Aldine de Gruyter for their invaluable assistance in preparing the present volume.

List of Contributors

Charles K. Brain
　Transvaal Museum
　P.O. Box 413
　Pretoria 0001
　South Africa

Frank H. Brown
　Department of Geology and
　　Geophysics
　University of Utah
　Salt Lake City, Utah 84112
　U.S.A.

Ronald J. Clarke
　31 Wescott Road
　Wokingham
　Berkshire RG11 2ER
　England

M. Christopher Dean
　Department of Anatomy and Embryology
　University College London
　London WC1E 6BT
　England

Eric Delson
　Department of Anthropology
　City University of New York;
　Department of Vertebrate Paleontology
　American Museum of Natural History
　New York, New York 10024
　U.S.A.

George H. Denton
　Institute for Quaternary Studies
　Department of Geological Sciences
　University of Maine
　Orono, Maine 04469
　U.S.A.

Dean Falk
　Department Anthropology
　State University of New York at Albany
　Albany, New York 12222
　U.S.A.

Craig S. Feibel
　Department of Geology and
　　Geophysics
　University of Utah
　Salt Lake City, Utah 84112
　U.S.A.

Hannah M. Grausz
　Department of Cell Biology and Anatomy
　Johns Hopkins University School of
　　Medicine
　Baltimore, Maryland 21205
　U.S.A.

Frederick E. Grine
　Departments of Anthropology and
　　Anatomical Sciences
　State University of New York at
　　Stony Brook
　Stony Brook, New York 11794
　U.S.A.

John M. Harris
　Division of Earth Sciences
　Natural History Museum of Los Angeles
　　County
　Los Angeles, California 90007
　U.S.A.

Ralph L. Holloway
　Department of Anthropology
　Columbia University
　New York, New York 10027
　U.S.A.

F. Clark Howell
　Department of Anthropology
　University of California
　Berkeley, California 94720
　U.S.A.

William L. Hylander
　Departments of Anatomy and Biological
　　Anthropology
　Duke University
　Durham, North Carolina 27710
　U.S.A.

Donald C. Johanson
　Institute of Human Origins
　Berkeley, California 94720
　U.S.A.

William L. Jungers
　Department of Anatomical Sciences
　School of Medicine
　State University of New York at
　　Stony Brook
　Stony Brook, New York 11794
　U.S.A.

Richard F. Kay
  Department of Anatomy and Biological
    Anthropology
  Duke University Medical Center
  Durham, North Carolina 27710
  U.S.A.
William H. Kimbel
  Institute of Human Origins
  Berkeley, California 94720
  U.S.A.
Richard G. Klein
  Department of Anthropology
  University of Chicago
  Chicago, Illinois 60637
  U.S.A.
Richard E. F. Leakey
  National Museums of Kenya
  P.O. Box 40658
  Nairobi
  Kenya
Lawrence B. Martin
  Department of Anthropology
  State University of New York at
    Stony Brook
  Stony Brook, New York 11794
  U.S.A.
Henry M. McHenry
  Department of Anthropology
  University of California
  Davis, California 95616
  U.S.A.
Michael L. Prentice
  Institute for Quaternary Studies
  Department of Geological Sciences
  University of Maine
  Orono, Maine 04469
  U.S.A.
Yoel Rak
  Department of Anatomy and
    Anthropology
  Sackler Faculty of Medicine
  Tel Aviv University
  Ramat Aviv, Tel Aviv
  Israel
Pat Shipman
  Department of Cell Biology and Anatomy
  Johns Hopkins University School of
    Medicine
  Baltimore, Maryland 21205
  U.S.A.

Randall L. Susman
  Department of Anatomical Sciences
  School of Medicine
  State University of New York at
    Stony Brook
  Stony Brook, New York 11794
  U.S.A.
Gen Suwa
  Primate Research Institute
  Kyoto University
  Inuyama City, Aichi
  Japan
Phillip V. Tobias
  Department of Anatomy
  University of the Witwatersrand
  Medical School
  Johannesburg 2193
  South Africa
Elisabeth S. Vrba
  Department of Geology and Geophysics
  Yale University
  New Haven, Connecticut 06511
  U.S.A.
Alan C. Walker
  Department of Cell Biology and Anatomy
  Johns Hopkins University School of
    Medicine
  Baltimore, Maryland 21205
  U.S.A.
Carol V. Ward
  Department of Cell Biology and Anatomy
  Johns Hopkins University School of
    Medicine
  Baltimore, Maryland 21205
  U.S.A.
Tim D. White
  Department of Anthropology
  University of California
  Berkeley, California 94720
  U.S.A.
Milford H. Wolpoff
  Department of Anthropology
  University of Michigan
  Ann Arbor, Michigan 48109
  U.S.A.
Bernard A. Wood
  Department of Anatomy and Cell Biology
  The University of Liverpool
  Liverpool L69 3X
  England

Studies of the Craniodental Evidence

I

Enamel Thickness and Development in *Australopithecus* and *Paranthropus*

FREDERICK E. GRINE AND LAWRENCE B. MARTIN

Teeth comprise a significant portion of the hominid fossil record, and for this reason they have been the subject of numerous morphometric analyses. To date, almost all studies of early hominid dentitions have concentrated on aspects of crown size and morphology, root configurations and occlusal wear. Despite the importance that has been attached to tooth enamel thickness and structure in analyses of Miocene hominoid fossils and interpretations of hominid origins (e.g., Jolly, 1970; Pilbeam, 1972; Simons, 1972, 1976; Simons and Pilbeam, 1972; Szalay, 1972; Gantt, 1977, 1982, 1983, 1986; Gantt et al., 1977; Kay, 1981; Martin, 1983, 1985), these parameters have received surprisingly little attention in studies of the Plio-Pleistocene hominid fossil record.

Robinson (1956) appears to have been the first and, until quite recently, the only worker to have published measurements of early hominid tooth enamel thickness. He recorded maximum and minimum enamel thickness measurements for six naturally fractured *Paranthropus* molars that ranged between 1.0 and 3.0 mm with a mean of 2.3 mm (Robinson, 1956: 21). All six specimens measured by him came from the Swartkrans Member 1 "Hanging Remnant" breccia. He noted that "*Australopithecus [africanus]* does not appear to differ markedly in this respect, though one would expect that the smaller teeth would have slightly thinner enamel" (Robinson, 1956: 21); although he alluded to this later (Robinson, 1963), no measurements were provided. Nevertheless, thick enamelled cheek-teeth have come to be regarded as a characteristic, if not a diagnostic feature of earlier and later hominids, with the australopithecines being regarded as very thick enamelled (e.g., Jolly, 1970; Simons, 1972, 1976; Pilbeam, 1972). This characterization, however, would appear to be related more to observations on tooth wear and the timing of dentine exposure on fossil teeth than to actual measurements of enamel thickness.

Zonneveld and Wind (1985) have recorded a maximum occlusal enamel thickness of 3.3 mm from a CT scan of a worn M^2 of *Paranthropus robustus* (TM 1517) from Kromdraai, and Sperber (1985) has published maximum and minimum measurements obtained from lateral radiographs of unworn upper and lower cheek-teeth of *Australopithecus* (Taung and Sterkfontein) and *Paranthropus* (Kromdraai and Swartkrans) specimens. Measurements of enamel thickness via radiographic techniques were evaluated by Gantt (1977), who concluded that such measurements could not be considered accurate because they varied by up to 50% from the true values obtained from thin sections of the same specimens. Sperber (1985), who recognized some of the limitations of accuracy imposed by this indirect method of measurement, concluded that *Paranthropus* molar enamel was generally thicker than that of *A. africanus*, which, in turn, outranked the thickness displayed by modern humans. These results, however, are hardly surprising in view of the fact that Swartkrans *Paranthropus* molars tend to be larger while modern human teeth tend to be smaller than those of *A. africanus* (Robinson, 1956).

Based upon measurements of broken molar teeth, Gantt (1986: 466) recorded that both *A. africanus* and southern African *Paranthropus* specimens possess "significantly" thicker enamel than any known hominoid, including modern humans. However, the sources of Gantt's data are unclear, for in one paper (1982: 97) he lists that five cheek-teeth from four individuals of *Australopithecus* (sensu lato) were examined, while in another instance (1983: 278) this sample comprised a lower molar from Swartkrans identified as *Australopithecus* sp., the Garusi M^3 of *A. afarensis* and a lower molar from Koobi Fora attributed to "*A. africanus?*". Gantt (1982, fig.

8.3; 1983, fig. 26; 1986, fig. 6) has illustrated a thickness value for *A. robustus* (presumably the Swartkrans lower molar), but as noted by Beynon and Wood (1986), the value of some 2.7 mm read from his graph is at variance with the measurements of 4.0 to 4.5 mm for *A. africanus* and *A. robustus* recorded by him (Gantt, 1985, 1986). Although Gantt illustrated a thickness value for *A. robustus* regressed against an estimated body weight, he (1983, 1986) noted, following the work by Kay (1981) and Martin (1983), that enamel thickness should be compared to tooth size in order to correctly evaluate fossil forms.

The first comparative analysis of enamel thickness in early hominid taxa, undertaken by Beynon and Wood (1986), entailed the linear measurement of occlusal, cuspal apex and lateral thicknesses on a series of naturally fractured cheek-teeth attributable to *Paranthropus boisei* and "early" *Homo (Homo habilis* and *H. erectus*). That study represents the only instance to date in which any attempt has been made to correct enamel thickness measurements for estimates of tooth size. Beynon and Wood (1986) related linear enamel measurements to the crown base area (a value that includes enamel) and a crude estimate of dentine area.

The absolute enamel thicknesses recorded by Beynon and Wood (1986) for the permanent molars of *P. boisei* were noted by them as being significantly greater than the corresponding means for the "early" *Homo* sample, and the size corrected values for cuspal apex and occlusal thickness of *P. boisei* were found to be significantly larger than the corresponding "early" *Homo* averages. Beynon and Wood (1986), however, noted that the techniques of size correction employed by them are rather crude, and that linear measurements of enamel thickness should be related to the more accurate measures employed by Martin (1983).

Studies of enamel structure in fossil hominids have been restricted to the documentation of enamel prism packing patterns (Vrba and Grine, 1978a,b; Boyde and Martin, 1984), decussation, and developmental rates (Robinson, 1956; Bromage and Dean, 1985; Beynon and Wood, 1986, 1987). Robinson (1956) noted that in southern African specimens of *Australopithecus* and *Paranthropus* the perikymata are distinctly more concentrated near the cervical than the occlusal and/or incisal margins, and that the fossils appeared to differ from modern human teeth insofar as the perikymata appeared to be more regular in the areas of highest concentration. Studies of perikymata have been used by Bromage and Dean (1985) to infer age at death for juvenile fossil hominid specimens.

The initial study of early hominid enamel structure by Vrba and Grine (1978a,b) documented the clear predominance of Pattern 3 prism packing in teeth of *Paranthropus* from Swartkrans and Kromdraai and *A. africanus* from Sterkfontein. Vrba and Grine also noted fields of subsurface Pattern 1 prisms in regions such as cuspal apices, but concluded that Pattern 3 prisms predominate in australopithecine enamel. Claims that hominids can be differentiated from apes on the basis of the presence of Pattern 3 enamel in the former and Pattern 1 enamel in the latter (Gantt *et al.*, 1977) were refuted by Vrba and Grine (1978a,b), who observed a predominance of Pattern 3 enamel in all hominoid taxa, and by Boyde and Martin (1982) on the basis of more detailed studies of developing enamel surfaces. Boyde and Martin (1984) have documented enamel prism packing patterns for two lower molars of *P. boisei* from the Shungura Formation in which Pattern 3 packing was found at all depths sampled deep to the subsurface region. They found the surface region to be composed of from 6 to 10 μm of prismless enamel with a thin layer of Pattern 1 enamel deep to this.

Beynon and Wood (1986) have recorded that Hunter-Schreger bands are both straighter and narrower in *P. boisei* than in "early" *Homo*, suggestive of a greater degree of prism decussation in the latter. They (1986, 1987) have also argued that molar crown formation time was shorter in *P. boisei* than in *Homo* on the basis of differences observed in the disposition of the Brown Striae of Retzius.

The purpose of the present study was to investigate enamel thickness and development in early hominid and modern hominoid taxa as determined from controlled sections through permanent molar crowns, and to attempt to relate these findings to those made by others on the basis of naturally fractured early hominid teeth. The other goal of this study was to relate differences in enamel thickness to modes of enamel accretion in that such data would appear to be of both phylogenetic and functional relevance.

Material

Comparative data relating to tooth enamel thickness derive for the most part from the study by Martin (1983) of sectioned molars of *Pan troglodytes, Gorilla gorilla, Pongo pygmaeus* and *Homo sapiens*. These data pertain to measurements of buccolingual (BL) sections through the apices of the mesial cusps of upper and lower molars. The numbers of specimens comprising each of these samples are recorded in Table 1.1; N refers to both number of teeth and number of individuals. In addition, enamel thickness data were recorded for five of the *H. sapiens* mandibular molars ($M_1 = 2$, $M_2 = 1$, $M_3 = 2$) that had been sectioned along a BL plane through the tips of the entoconid and hypoconid, and for three *H. sapiens* lower molars ($M_1 = 1$, $M_2 = 2$) not included in the above sample for mesiodistal (MD) sections through the tips of the protoconid and hypoconid for comparison with the values obtained for two of the fossil specimens (see below).

Six specimens representing *Australopithecus* and *Paranthropus* formed the basis of the study of enamel thickness (Table 1.2). *A. africanus* is represented by two unworn permanent maxillary molars, Stw 284 and Stw 402 (Figs. 1.1 and 1.2), obtained recently by A. R. Hughes from ex-

Table 1.1. Composition of Extant Hominoid Molar Samples

Taxon	M1	M2	M3	Total
Pan troglodytes				
Maxillary	2	3	1	6
Mandibular	3	4	1	8
Combined	5	7	2	14
Gorilla gorilla				
Maxillary	3	3	3	9
Mandibular	2	4	2	8
Combined	5	7	5	17
Pongo pygmaeus				
Maxillary	3	3	2	8
Mandibular	3	3	3	9
Combined	6	6	5	17
Homo sapiens				
Maxillary	2	2	3	7
Mandibular	2	2	2	6
Combined	4	4	5	13

Table 1.2. Fossil Hominid Molars Examined in This Study

Specimen	Tooth	Derivation
Australopithecus africanus		
Stw 284	LM^2	Sterkfontein Fm., Member 4
Stw 402	RM^1	Sterkfontein Fm., Member 4
Paranthropus robustus		
KB 5223	RM_1	Kromdraai B East Fm., Member 3
Paranthropus crassidens		
SKX 21841	RM^3	Swartkrans Fm., Member 3
Paranthropus boisei		
L 10–21	RM_1	Shungura Fm., Member E
L 398–847	LM_3	Shungura Fm., Member F

cavations of decalcified Member 4 breccia at Sterkfontein. These specimens have not yet been published. The crown of the Stw 284 M^2 measures 15.0 mm MD and 16.9 mm BL; the corresponding diameters of the Stw 402 M^1 are 13.8 mm and 14.8 mm. *P. robustus* is represented by the unworn RM_1 of KB 5223 that was obtained from *in situ* Member 3 breccia of the Kromdraai B East Formation (Vrba, 1981); it has been described in detail elsewhere (Grine, 1982). This crown (Fig. 1.3) measures 14.3 mm MD and 12.5 mm BL across the trigonid cusps. The Swartkrans "robust" australopithecine, for which Howell (1978) and Grine (1982, 1985, this volume, chapter 14) argue that the specific designation *P. crassidens* Broom, 1949 be retained, is represented by SKX 21841 (Fig. 1.4). This isolated and unworn RM^3 was obtained from *in situ* Member 3 sediments (Brain, 1982); the crown measures 15.5 mm MD and 16.6 mm BL (Grine, this volume, chapter 14).

The *P. boisei* sample consists of two incomplete permanent mandibular molars from Members E and F of the Shungura Formation. Specimen L 10-21 (Fig. 1.5), which derives from Member E, was identified as a RM_3 by Howell and Coppens (1974: 9), although it is regarded by us and by G. Suwa (pers. comm.) to be a RM_1. It preserves the metaconid and entoconid, a portion of the hypoconid that extends to (or close to) the tip, and parts of the protoconid, hypoconulid and tuberculum sextum. It evinces slight enamel facets on the mesial marginal ridge, protoconid, metaconid and hypoconid, and faint wear on the entoconid. Specimen L 398-847 (Fig. 1.6) comes from sediments related to the derived tuffaceous equivalent (F') of Tuff F, and thus from one of the earliest localities within Member F (Howell and Coppens, 1974). This fragment has been identified by Howell and Coppens (1974: 10) as the distal portion of a RM_3, although it is regarded by us as comprising the mesiobuccal moiety of a LM_3 with the protoconid and portions of the metaconid and hypoconid. Whereas L 10-21 does not appear to have been specifically identified, L 398-847 has been attributed to *P. boisei* (Howell and Coppens, 1976: 528; Coppens, 1980: 216).

Specimen Preparation and Examination

The techniques utilized in sectioning the extant hominoid molars have been detailed elsewhere by Martin (1983). These same techniques were employed with reference to the two *P. boisei* specimens except that whereas the extant hominoid molars were cut with a Buehler Isomet diamond wafering blade in peripheral configuration with a resultant 350 μm wide cut, the two *P. boisei* specimens were sectioned using a Cambridge Microslice diamond saw in annular configuration which produced a 175 μm wide cut. Specimen L 10-21 was sectioned buccolingually through the apex of the entoconid and the middle of the preserved portion of the hypoconid. Specimen L 398-847 was sectioned mesiodistally through the apices of the protoconid and hypoconid.

The molars from Sterkfontein, Swartkrans and Kromdraai were first photographed and cast, then refluxed in a chloroform/methanol mixture at 46°C for 5 days, and then transferred to methyl methacrylate (MMA) monomer and stored in the dark at between 3° and 4°C (Boyde and Tamarin, 1984). The MMA was changed at 24-hour intervals for a total of four changes to ensure that all of the chloroform/methanol had been replaced by MMA, which was then allowed to polymerize at room temperature. Each of the embedded molars was then cut along a transverse (BL) plane through the apices of the mesial cusps using a wire saw developed by the Polish Academy of Sciences with a wire thickness of 60 μm resulting in a cut some 70 μm wide. The cut faces possessed a surface flatness of about 1.0 μm, and some were finished with 0.1 μm diamond polishing to produce effectively topography free surfaces.

As noted by Martin (1983), due to the influence of obliquity of section in increasing apparent enamel thickness, the measurements obtained from two adjacent cut faces are not identical. Martin (1983) proposed that the face that maximizes the dentine component and minimizes the enamel component of the crown section should be used for computations of enamel thickness because obliquity can only act to exaggerate the enamel component. The application of these criteria means that enamel thickness data for the extant hominoid sample derive from planes within 175 μm of the ideal plane of section. The data for the *P. boisei* specimens, which were

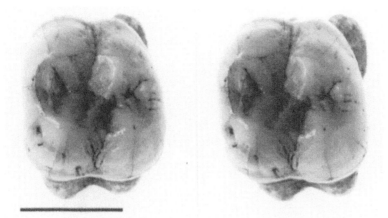

FIGURE 1.1. Occlusal stereopair of Stw 284. *Australopithecus africanus*, Sterkfontein Formation, Member 4. Scale bar = 10 mm.

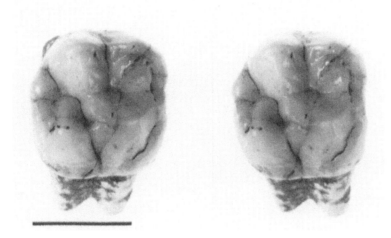

FIGURE 1.2. Occlusal stereopair of Stw 402. *Australopithecus africanus*, Sterkfontein Formation, Member 4. Scale bar = 10 mm.

FIGURE 1.3. Occlusal stereopair of KB 5223. *Paranthropus robustus*, Kromdraai B East Formation, Member 3. Scale bar = 10 mm.

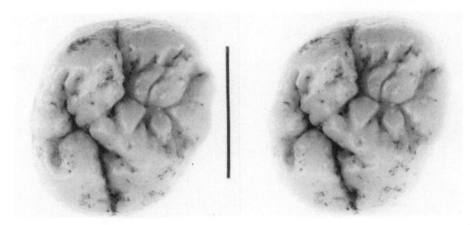

FIGURE 1.4. Occlusal stereopair of SKX 21841. *Paranthropus crassidens*, Swartkrans Formation, Member 3. Scale bar = 10 mm.

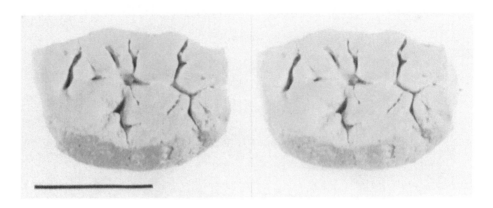

FIGURE 1.5. Occlusal stereopair of L 10-21 (cast). *Paranthropus boisei*, Shungura Formation, Member E. Scale bar = 10 mm.

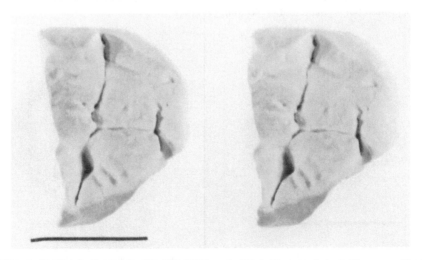

FIGURE 1.6. Occlusal stereopair of L 398-847 (cast). *Paranthropus boisei*, Shungura Formation, Member F. Scale bar = 10 mm.

sectioned and studied during 1984 and 1985 and had already been returned to Ethiopia before the South African fossils were obtained, derive from faces within 87.5 μm of the ideal plane of section. Because of the narrower cuts made through the South African fossils, enamel thickness data for these specimens derive from faces that are no more than 35 μm from the ideal plane of section in which there is no obliquity.

Each of the cut faces was photographed according to the method described by Martin (1983: 177–178), and all measurements of enamel thickness were recorded from the photographs using either a Graf-Pen sonic digitizer or a dial-equipped Mauser vernier caliper with specially sharpened ends. At least three determinations were made of each area and linear measurement, and the average was utilized in subsequent calculations.

Aspects of enamel structure and development were studied utilizing polarizing light microscopy (PLM) and scanning electron microscopy (SEM). To facilitate SEM study, the cut faces of the fossil specimens were lightly etched-up using 0.5% HCl for periods of 30 sec. and coated with carbon by vacuum evaporation.

Measurements of Enamel Thickness

The measurements of enamel thickness recorded for the extant hominoid and fossil hominid molars are depicted in Fig. 1.7 and 1.8. The majority (i.e., those designated by lower case letters a through l) are those used by Martin (1983). These measurements are defined as follows:

Measurement a. The total area of the tooth crown section delineated by the outer enamel perimeter and a straight line between the buccal and lingual cervices.

Measurement b. The area of dentine (and pulp) enclosed by the enamel–dentine junction (EDJ) and a straight line between the buccal and lingual cervices. This provides a measure of tooth crown size excluding the contribution of the enamel cap.

Measurement c. The area of the sectioned enamel cap.

Measurement e. The perimeter length of the EDJ from the buccal to the lingual cervix.

Measurements f and g. The vertical thickness of enamel of the buccal (f) and lingual (g) cuspal apices. Measured from the cuspal tips perpendicular to a line drawn tangent to the apices of the

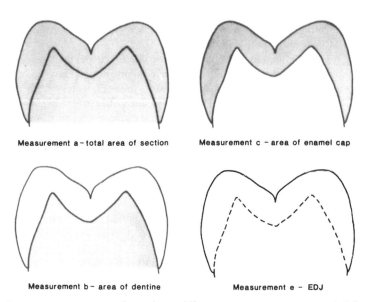

FIGURE I.7. Area measurements a, b, and c and linear measurement e recorded for modern and fossil molars. See text for definitions. Adapted from Martin (1983, fig. 4.1).

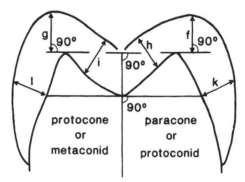

Linear measurements f–l (Martin 1983)

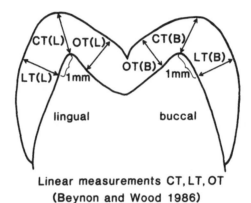

Linear measurements CT, LT, OT
(Beynon and Wood 1986)

FIGURE 1.8. Linear measurements of enamel thickness f, g, h, i, k and l as defined by Martin (1983, fig. 4.2) and CT, LT and OT as defined by Beynon and Wood (1986, fig. 1A).

dentine horns. These measurements correspond closely to those designated "A" and "B" by Gantt (1977) and Molnar and Gantt (1977).

Measurement CT. As defined by Beynon and Wood (1986), cuspal thickness is measured from the tip of the dentine horn to the tip of the cusp. While no distinction between buccal and lingual measurements was made by Beynon and Wood (1986), these diameters were recorded here as CT(B) and CT(L), respectively (Fig. 1.8). The value of linear measurement CT will be equivalent to that of f or g only in those instances in which the dentine horn and cuspal tips occupy the same sagittal plane; in all other instances the value for CT will exceed that for f or g.

Measurements h and i. The maximum linear thickness of enamel on the occlusal (= lingual) face of the buccal cusp (measurement h) and the occlusal (= buccal) face of the lingual cusp (measurement i) recorded perpendicular to the EDJ. These measurements correspond to "EA" and "EB" of Gantt (1977), and to "OT" of Beynon and Wood (1986).

Measurement OT. As defined by Beynon and Wood (1986), it is recorded at least 0.5 mm from the tip of the dentine horn. Whereas Beynon and Wood (1986) did not differentiate buccal from lingual cuspal measurements, these are designated OT(B) and OT(L), respectively, here (Fig. 1.8).

Measurements k and l. The linear thickness of enamel measured perpendicular to the EDJ on the sides of the buccal and lingual cusps from the point on the EDJ at which a line drawn parallel to one between the apices of the dentine horns and tangent to the lowest point of the EDJ between the cusps intersects the EDJ at the sides of the crown. Measurements k and l correspond respectively to "JJ" and "KK" of Gantt (1977).

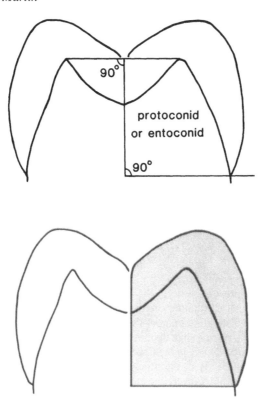

FIGURE 1.9. Determination of enamel-cap and dentine areas of individual cusps used in measurements of L 10-21 (entoconid) and L 398-847 (protoconid).

Measurement LT. As defined by Beynon and Wood (1986), this records the maximum lateral enamel thickness at least 1.0 mm cervical to the dentine horn and perpendicular to the EDJ. Because this is a measure of maximum lateral enamel thickness, LT [here designated LT(B) and LT(L) for the buccal and lingual crown sides] will likely be greater than measurements k and l of Martin (1983).

Because the two *P. boisei* molars included in this study are incomplete, measurements a, b, c and e as defined above (Fig. 1.7) could not be recorded directly. In order to evaluate these values, measurements a, b, c and e were recorded separately for BL sections through the entoconid, and BL and MD sections through the protoconid, of modern human molars. The individual "cuspal" values were determined for that region of the section enclosed by a vertical line drawn perpendicular to and at the midpoint of a line connecting the apices of the dentine horns, and by a line drawn from the cervix of the protoconid (or entoconid) perpendicular to the vertical line (Fig. 1.9).

Four indices of average or relative enamel thickness were calculated from measurements a, b, c and e for each of the specimens. The first, c/e, is an expression of the average thickness of the enamel cap over the length of the EDJ. The relative contribution of the enamel cap to the entire crown section is expressed by c/a × 100. Relative enamel thickness may be expressed by $\sqrt{c}/e \times 100$, which relates a linear value for the enamel cap to the length of the EDJ. Relative enamel thickness may be expressed also by comparing the c/e to a linear value of the dentine area (b) according to the formula $(c/e)/\sqrt{b} \times 100$. In those cases where measurements were recorded for the protoconid or entoconid only, the appropriate denominator is $\sqrt{b} \times 2$. All linear measurements of enamel thickness are compared to $\sqrt{b}$ (or to $\sqrt{b} \times 2$) as a linear value for crown size that does not contain enamel.

Enamel Thickness in Extant Large-Bodied Hominoids

Photographs of sections of representative molars of extant large-bodied hominoids, and the enamel cap profiles for these specimen sections are depicted in Figs. 1.10 and 1.11. The values for measurements a, b, c and e obtained from BL sections through the mesial cusps of the upper and lower molars of *Pan, Gorilla, Pongo* and *Homo* are recorded in Table 1.3.

In all species samples, the means for maxillary molars are larger than those for mandibular molars; the only exception pertains to measurement c in the *Gorilla* sample, and in this instance the mandibular average is but exiguously larger.

The values for the indices that express average and relative enamel thickness for these samples are recorded in Table 1.4. In *Pan, Gorilla* and *Homo* the index value means are larger for lower than for upper molars, and while this holds for the relative enamel thickness expression $(c/e)/\sqrt{b} \times 100$ in *Pongo*, the other index values in the orangutan sample are slightly larger for the maxillary molars. In most instances, however, the differences between the upper and lower molar means are not statistically significant (Table 1.4), and while no sample shows a significant difference between upper and lower molars in average enamel thickness, the mandibular molar sample of *Pan troglodytes* displays significantly larger values for both indices of relative enamel thickness and in the proportion of the enamel cap to the total crown area.

Thus, according to the present sample, the mandibular molars of *P. troglodytes* appear to have significantly thicker enamel than do their maxillary molars by estimates of relative enamel thickness. At the same time, relative enamel thickness as expressed by the index $\sqrt{c}/e \times 100$ is significantly greater in the mandibular than in the maxillary molars comprising the present *H. sapiens* sample. The differences between the upper and lower molars of the modern human sample, however, do not differ significantly in the other measures of relative enamel thickness.

While there is a slight tendency for relative enamel thickness to increase from M1 to M3 within these extant hominoid samples (Martin, 1983: appendix A), evaluations of the very small samples of teeth at each position revealed no significant differences among molars at different positions within the upper or lower arcades. In view of the differences between the maxillary and mandibular molars of *Pan* and *Homo*, however, due caution should be exercised when comparing upper and lower molars of fossil taxa.

The average enamel thickness values for *P. troglodytes* molars are significantly smaller than

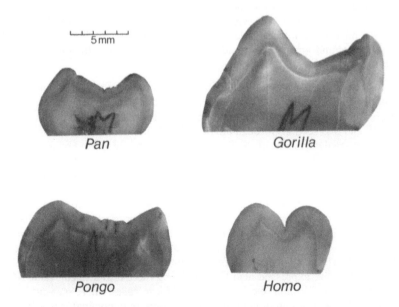

FIGURE 1.10. Photographs of BL sections through the mesial cusps of representative molars of *Pan troglodytes* (RM_2, buccal to left), *Gorilla gorilla* (LM^3, buccal to right), *Pongo pygmaeus* (LM^2, buccal to right) and *Homo sapiens* (RM_3, buccal to left).

those for the other three hominoid samples (Table 1.5). Similarly, the *Gorilla* c/e means are significantly smaller than those for humans, and although the *Gorilla* and *Pongo* means do not differ significantly for mandibular molars, the maxillary molar and combined sample means are significantly smaller in *Gorilla* (Table 1.5). The human maxillary c/e mean does not differ significantly from that for the *Pongo* sample, and although the lower molar and combined sample means differ significantly between these two taxa, the confidence limits are considerably lower than those pertaining to the other comparisons (Table 1.5). Thus, average enamel thickness tends to be significantly less in *Pan* than in *Gorilla*, significantly less in *Gorilla* than in *Pongo*, and less in *Pongo* than in *Homo*.

On the other hand, compared to the size of the tooth as a whole or the dentine contribution to crown size, enamel thickness does not differ significantly between *Pan* and *Gorilla* for either the upper or the lower molars (Table 1.5). Relative enamel thickness values of both *Pan* and *Gorilla* are significantly lower than those for *Pongo* and *Homo* (Table 1.5). While measures of relative enamel thickness do not differ significantly between *Pongo* and *Homo* for upper molars, the latter means are larger, and the *Homo* values are significantly larger than those for *Pongo* for the lower molar and combined molar samples (Table 1.5). Thus, while the African apes do not differ from one another in terms of relative enamel thickness, their enamel is significantly thinner than that of orangutans and modern humans, while human enamel tends to be relatively thicker than that of *Pongo*.

Linear measurements of enamel thickness recorded for the extant hominoid samples are recorded in Table 1.6. Comparisons of the buccal and lingual values (i.e., f vs. g, h vs. i, k vs. l) reveal that they differ significantly only in the maxillary molar sample of *Pan* with regard to cusp tip dimensions, and in the maxillary molar sample of *Pongo* with regard to the lateral dimensions. It is perhaps noteworthy, however, that in all four taxa the occlusal thickness mean for the buccal cusp of the mandibular molar (measurement h) exceeds that for the lingual cusp (measurement i), and in all four the lingual lateral thickness (measurement l) mean for the upper molars exceeds the buccal lateral (measurement k) value. The latter situation holds for all eight specimens comprising the *Pongo* sample and all seven that constitute the *Homo* upper molar sample, although only in the *Pongo* sample do the buccal and lingual lateral thickness means differ significantly. In the *Pongo*, *Homo* and *Gorilla* lower molar samples the buccal lateral thickness measurement (k) exceeds the average for lingual lateral thickness (measurement l). Enamel, therefore, tends to be thicker on the lingual face of the protocone and on the buccal face of the protoconid; these are the regions in which cingular remnants occur.

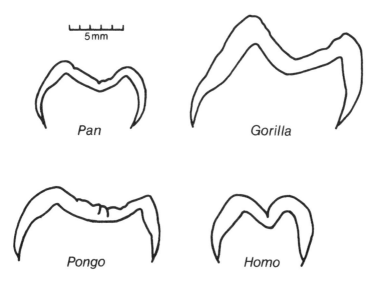

FIGURE 1.11. Enamel cap profiles of extant hominoid specimens depicted in Fig. 1.10.

Table 1.3. Values of Measurements a, b, c and e from BL Sections through the Mesial Cusps of Extant Hominoid Molars

	N	$\bar{X}$	SD	SE	Observed range
Measurement a—Total area of section					
Pan					
Maxillary	6	55.17	6.88	2.81	48.0–66.5
Mandibular	8	43.31	7.78	2.75	29.1–51.3
Combined	14	48.39	9.37	2.50	29.1–66.5
Gorilla					
Maxillary	9	105.77	16.84	5.61	80.5–136.6
Mandibular	8	95.71	10.37	3.67	84.0–114.4
Combined	17	101.04	14.68	3.56	80.5–136.6
Pongo					
Maxillary	8	70.28	4.19	1.48	64.4–77.8
Mandibular	9	63.73	5.08	1.69	54.3–71.2
Combined	17	66.81	5.65	1.37	54.3–77.8
Homo					
Maxillary	7	60.23	9.50	3.59	48.9–71.3
Mandibular	6	46.77	6.44	2.63	36.7–53.7
Combined	13	54.02	10.55	2.92	36.7–71.3
Measurement b—Dentine Area					
Pan					
Maxillary	6	43.65	7.01	2.86	34.3–54.9
Mandibular	8	31.94	5.70	2.01	22.3–37.5
Combined	14	36.96	8.52	2.28	22.3–54.9
Gorilla					
Maxillary	9	82.32	14.14	4.71	60.1–107.6
Mandibular	8	71.90	8.52	3.01	61.5–84.4
Combined	17	77.42	12.67	3.07	60.1–107.6
Pongo					
Maxillary	8	48.06	4.12	1.46	41.7–53.7
Mandibular	9	43.69	4.75	1.58	34.5–52.2
Combined	17	45.75	4.87	1.18	34.5–53.7
Homo					
Maxillary	7	38.06	4.47	1.69	31.5–45.6
Mandibular	6	26.70	4.38	1.79	18.8–31.9
Combined	13	32.82	7.26	2.01	18.8–45.6
Measurement c—Enamel cap area					
Pan					
Maxillary	6	11.52	2.01	0.82	9.1–14.0
Mandibular	8	11.38	2.26	0.80	6.8–13.8
Combined	14	11.44	2.07	0.55	6.8–14.0
Gorilla					
Maxillary	9	23.44	2.94	0.98	20.2–29.0
Mandibular	8	23.81	4.75	1.68	17.5–32.3
Combined	17	23.62	3.77	0.92	17.5–32.3
Pongo					
Maxillary	8	22.21	2.71	0.96	17.2–25.5
Mandibular	9	20.04	2.12	0.71	17.4–25.1
Combined	17	21.06	2.59	0.63	17.2–25.5
Homo					
Maxillary	7	22.17	6.97	2.64	15.2–34.3
Mandibular	6	20.07	3.80	1.55	14.5–25.1
Combined	13	21.20	5.62	1.56	14.5–34.3

Table 1.3. *Continued*

	N	$\overline{X}$	SD	SE	Observed range
Measurement e—Enamel–dentine junction length					
Pan					
Maxillary	6	20.57	1.17	0.48	19.2–22.3
Mandibular	8	17.83	1.73	0.61	14.9–19.3
Combined	14	19.00	2.03	0.54	14.9–22.3
Gorilla					
Maxillary	9	27.94	2.33	0.78	24.3–31.8
Mandibular	8	26.03	2.00	0.71	23.0–28.4
Combined	17	27.04	2.33	0.57	23.0–31.8
Pongo					
Maxillary	8	20.30	0.57	0.20	19.6–20.8
Mandibular	9	19.46	1.61	0.54	16.9–22.2
Combined	17	19.85	1.28	0.31	16.9–22.2
Homo					
Maxillary	7	18.51	1.10	0.42	16.7–19.8
Mandibular	6	15.22	1.26	0.52	12.8–16.4
Combined	13	16.99	2.05	0.57	12.8–19.8

In view of the observation that in only two instances do the buccal and lingual linear measurement means differ significantly within a sample, the averages of these two measurements can be compared to the dentine area (b) to obtain relative thickness values for the cuspal tip, occlusal and lateral enamel (Table 1.7). As was noted for the other indices of relative enamel thickness (Table 1.4), the linear values (Table 1.7) are larger for the mandibular than for the maxillary molars in all instances save for lateral thickness in the *Pongo* sample. The differences between the corresponding upper and lower molar means are significant, however, only in the *Pan* sample (Table 1.7).

Table 1.4. Index Values of Average and Relative Enamel Thickness for Molars of Extant Hominoids

	N	$\overline{X}$	SD	SE	Observed range	"*t*"	*p*
Average Enamel Thickness – (c/e)							
Pan							
Maxillary	6	0.56	0.11	0.04	0.46–0.71	1.31	NS
Mandibular	8	0.63	0.09	0.03	0.46–0.74		
Combined	14	0.60	0.10	0.03	0.46–0.74		
Gorilla							
Maxillary	9	0.84	0.06	0.02	0.75–0.93	1.33	NS
Mandibular	8	0.92	0.17	0.06	0.62–1.14		
Combined	17	0.87	0.13	0.03	0.62–1.14		
Pongo							
Maxillary	8	1.10	0.14	0.05	0.83–1.29	0.71	NS
Mandibular	9	1.05	0.15	0.05	0.89–1.36		
Combined	17	1.07	0.14	0.03	0.83–1.36		
Homo							
Maxillary	7	1.19	0.33	0.13	0.82–1.80	0.80	NS
Mandibular	6	1.32	0.24	0.10	0.92–1.63		
Combined	13	1.25	0.29	0.08	0.82–1.80		

Continued

Table 1.4. *Continued*

	N	$\bar{X}$	SD	SE	Observed range	"t"	p
Proportion of enamel cap to total area – (c/a × 100)							
Pan							
Maxillary	6	21.08	4.49	1.83	17.4–28.5	2.94	<0.02
Mandibular	8	26.20	1.86	0.66	23.3–28.6		
Combined	14	24.01	4.07	1.09	17.4–28.6		
Gorilla							
Maxillary	9	22.30	1.50	0.50	20.0–25.3	1.86	NS
Mandibular	8	24.89	3.88	1.37	17.2–29.1		
Combined	17	23.62	3.77	0.92	17.2–29.1		
Pongo							
Maxillary	8	31.65	3.68	1.30	24.3–35.9	0.05	NS
Mandibular	9	31.56	3.33	1.11	26.7–36.5		
Combined	17	31.60	3.38	0.82	24.3–36.5		
Homo							
Maxillary	7	36.21	6.31	2.38	30.0–48.1	2.00	NS
Mandibular	6	42.92	5.68	2.32	34.0–48.8		
Combined	13	39.31	6.74	1.87	30.0–48.8		
Relative Enamel thickness – ($\sqrt{c}/e$ × 100)							
Pan							
Maxillary	6	16.50	1.76	0.72	14.7–19.3	2.98	<0.02
Mandibular	8	18.86	1.21	0.43	17.5–20.9		
Combined	14	17.85	1.86	0.50	14.7–20.9		
Gorilla							
Maxillary	9	17.34	0.77	0.26	16.1–18.6	1.93	NS
Mandibular	8	18.74	2.03	0.72	14.8–21.0		
Combined	17	18.00	1.62	0.39	14.8–21.0		
Pongo							
Maxillary	8	23.19	1.68	0.59	20.0–25.6	0.05	NS
Mandibular	9	23.14	2.47	0.82	19.6–27.1		
Combined	17	23.17	2.07	0.50	19.6–27.1		
Homo							
Maxillary	7	25.14	3.07	1.16	21.0–30.7	2.44	<0.05
Mandibular	6	29.47	3.33	1.36	24.3–33.1		
Combined	13	27.14	3.79	1.05	21.0–33.1		
Relative enamel thickness – [(c/e)/$\sqrt{b}$ × 100]							
Pan							
Maxillary	6	8.60	2.12	0.86	7.0–12.1	2.95	<0.02
Mandibular	8	11.23	1.21	0.43	9.6–13.3		
Combined	14	10.10	2.09	0.56	7.0–13.3		
Gorilla							
Maxillary	9	9.29	0.79	0.26	8.3–10.8	2.07	NS
Mandibular	8	10.88	2.16	0.76	6.8–13.4		
Combined	17	10.04	1.74	0.42	6.8–13.4		
Pongo							
Maxillary	8	15.89	2.37	0.84	11.3–19.1	0.06	NS
Mandibular	9	15.97	2.76	0.92	12.3–20.5		
Combined	17	15.93	2.51	0.61	11.3–20.5		
Homo							
Maxillary	7	19.30	5.25	1.99	13.8–29.6	2.18	NS
Mandibular	6	25.90	5.66	2.31	17.4–32.3		
Combined	13	22.35	6.23	1.73	13.8–32.3		

Table 1.5. Comparisons of Average and Relative Enamel Thickness Values of Extant Hominoid Samples

	$\overline{X}$	Pan		Gorilla		Pongo	
		"t"	p	"t"	p	"t"	p
Average enamel thickness – (c/e)							
Maxillary							
Pan	0.56						
Gorilla	0.84	6.41	<0.001				
Pongo	1.10	7.79	<0.001	5.09	<0.001		
Homo	1.19	4.44	<0.001	3.15	<0.01	0.71	NS
Mandibular							
Pan	0.63						
Gorilla	0.92	4.26	<0.001				
Pongo	1.05	6.88	<0.001	1.68	NS		
Homo	1.32	7.54	<0.001	3.66	<0.005	2.70	<0.02
Combined							
Pan	0.60						
Gorilla	0.87	6.37	<0.001				
Pongo	1.07	10.53	<0.001	4.32	<0.001		
Homo	1.25	7.91	<0.001	4.82	<0.001	2.25	<0.05
Proportion of enamel cap area to total area – [c/a × 100]							
Maxillary							
Pan	21.08						
Gorilla	22.30	0.77	NS				
Pongo	31.65	4.85	<0.001	7.02	<0.001		
Homo	36.21	4.89	<0.001	6.44	<0.001	1.74	NS
Mandibular							
Pan	26.20						
Gorilla	24.89	0.86	NS				
Pongo	31.56	4.02	<0.002	3.82	<0.002		
Homo	42.92	7.87	<0.001	7.08	<0.001	4.91	<0.001
Combined							
Pan	24.01						
Gorilla	23.62	0.28	NS				
Pongo	31.60	5.68	<0.001	6.50	<0.001		
Homo	39.31	7.20	<0.001	8.11	<0.001	4.10	<0.001
Relative enamel thickness – [√c/e × 100]							
Maxillary							
Pan	16.50						
Gorilla	17.34	1.28	NS				
Pongo	23.19	7.23	<0.001	9.42	<0.001		
Homo	25.14	6.07	<0.001	7.40	<0.001	1.56	NS
Mandibular							
Pan	18.86						
Gorilla	18.74	0.14	NS				
Pongo	23.14	4.44	<0.001	3.98	<0.002		
Homo	29.47	8.40	<0.001	7.50	<0.001	4.24	<0.001
Combined							
Pan	17.85						
Gorilla	18.00	0.24	NS				
Pongo	23.17	7.45	<0.001	8.11	<0.001		
Homo	27.14	8.18	<0.001	8.97	<0.001	3.67	<0.002

Continued

Table 1.5. Continued

		Pan		Gorilla		Pongo	
	$\overline{X}$	"t"	p	"t"	p	"t"	p
Relative enamel thickness – $[(c/e)/\sqrt{b} \times 100]$							
Maxillary							
Pan	8.60						
Gorilla	9.29	0.90	NS				
Pongo	15.89	5.95	<0.001	7.90	<0.001		
Homo	19.30	4.65	<0.001	5.69	<0.001	1.66	NS
Mandibular							
Pan	11.23						
Gorilla	10.88	0.40	NS				
Pongo	15.97	4.48	<0.001	4.19	<0.001		
Homo	25.90	7.21	<0.001	6.94	<0.001	4.57	<0.001
Combined							
Pan	10.10						
Gorilla	10.04	0.09	NS				
Pongo	15.93	6.93	<0.001	7.95	<0.001		
Homo	22.35	6.96	<0.001	7.80	<0.001	3.87	<0.001

Comparisons of the means for the combined molar samples (Table 1.8) reveal that *Pan* and *Gorilla* do not differ significantly in any of the three linear measures of relative enamel thickness, but that for all such measures the African ape means are significantly lower than those for *Pongo* and *Homo*. The *Pongo* sample means are, in turn, significantly smaller than those for *Homo*.

Thus, utilizing both area and linear measurements, the African apes do not appear to differ from one another in terms of relative enamel thickness. The molar enamel of *Pan* and *Gorilla* is, however, significantly thinner than that of *Pongo* and *Homo*, while human molar enamel tends to be significantly thicker than that of the orangutan.

Table 1.6. Values of Linear Enamel Thickness Measurements f through l from BL Sections Through the Mesial Cusps of Extant Hominoid Molars

Sample	Measurement	N	$\overline{X}$	SD	SE	Observed range	"t"	p
Pan					Maxillary			
	f	5	0.49	0.07	0.03	0.41–0.59	2.63	<0.05
	g	6	0.36	0.09	0.04	0.24–0.53		
	h	6	0.60	0.13	0.05	0.40–0.73	0.50	NS
	i	6	0.57	0.07	0.03	0.50–0.70		
	k	6	0.68	0.11	0.05	0.59–0.91	1.01	NS
	l	6	0.75	0.13	0.05	0.61–0.98		
					Mandibular			
	f	6	0.58	0.18	0.08	0.41–0.83	0.39	NS
	g	8	0.55	0.11	0.04	0.41–0.74		
	h	8	0.66	0.10	0.04	0.52–0.81	0.57	NS
	i	8	0.63	0.11	0.04	0.48–0.81		
	k	8	0.72	0.16	0.06	0.48–0.94	0.53	NS
	l	8	0.76	0.14	0.05	0.57–0.98		
Gorilla					Maxillary			
	f	9	0.84	0.22	0.07	0.41–1.06	1.12	NS
	g	9	1.01	0.40	0.13	0.35–1.53		
	h	9	0.83	0.13	0.04	0.68–1.06	0.14	NS
	i	9	0.82	0.17	0.06	0.59–1.06		
	k	9	0.91	0.09	0.03	0.77–1.05	0.57	NS
	l	9	0.94	0.13	0.04	0.73–1.10		

Table 1.6. *Continued*

Sample	Measurement	N	$\bar{X}$	SD	SE	Observed range	"t"	p
					Mandibular			
	f	7	0.87	0.28	0.11	0.59–1.30	0.61	NS
	g	8	0.95	0.23	0.08	0.71–1.42		
	h	8	0.99	0.22	0.08	0.83–1.30	0.85	NS
	i	8	0.89	0.25	0.09	0.57–1.19		
	k	8	1.05	0.22	0.08	0.83–1.36	1.35	NS
	l	8	0.93	0.12	0.04	0.80–1.12		
Pongo						Maxillary		
	f	8	1.05	0.18	0.06	0.85–1.42	1.45	NS
	g	7	1.25	0.34	0.13	0.71–1.77		
	h	8	1.12	0.20	0.07	0.86–1.53	0.58	NS
	i	8	1.17	0.14	0.05	0.98–1.42		
	k	8	1.09	0.08	0.03	1.03–1.25	4.26	<0.001
	l	8	1.32	0.13	0.05	1.17–1.57		
						Mandibular		
	f	7	1.19	0.29	0.11	0.77–1.59	0.15	NS
	g	8	1.21	0.23	0.08	0.83–1.53		
	h	9	1.30	0.15	0.05	1.04–1.46	1.92	NS
	i	9	1.14	0.20	0.07	0.74–1.39		
	k	9	1.10	0.13	0.04	0.87–1.32	0.70	NS
	l	9	1.06	0.11	0.04	0.92–1.26		
Homo						Maxillary		
	f	7	1.27	0.25	0.09	0.94–1.65	0.65	NS
	g	7	1.41	0.51	0.19	0.59–2.06		
	h	7	1.19	0.34	0.13	0.88–1.89	0.20	NS
	i	7	1.23	0.41	0.15	0.87–2.06		
	k	7	1.09	0.23	0.09	0.84–1.44	1.80	NS
	l	7	1.44	0.46	0.17	0.91–2.24		
						Mandibular		
	f	6	1.35	0.37	0.15	0.77–1.77	0.40	NS
	g	6	1.44	0.41	0.17	1.00–2.18		
	h	6	1.28	0.28	0.11	0.94–1.73	0.29	NS
	i	6	1.23	0.31	0.13	0.93–1.79		
	k	6	1.29	0.31	0.13	0.94–1.79	0.57	NS
	l	6	1.21	0.15	0.06	1.10–1.51		

Table 1.7. Relative Enamel Thickness Values Determined from Linear Measurements of BL Sections Through Mesial Cusps of Extant Hominoid Molars

	N	$\bar{X}$	SD	SE	Observed range	"t"	p
Relative vertical cuspal thickness–$[(f+g)/2]/\sqrt{b}$							
Pan							
Maxillary	6	6.36	1.11	0.45	4.45–7.68	4.33	<0.001
Mandibular	8	10.55	2.15	0.76	6.70–12.54		
Combined	14	8.75	2.75	0.74	4.45–12.54		
Gorilla							
Maxillary	9	10.35	2.93	0.98	6.29–16.00	0.30	NS
Mandibular	8	10.79	3.04	1.08	7.43–16.58		
Combined	17	10.56	2.90	0.70	6.29–16.58		

Continued

Table 1.7. *Continued*

	N	$\overline{X}$	SD	SE	Observed range	"t"	p
Pongo							
Maxillary	8	16.41	4.04	1.43	10.64–23.70	0.97	NS
Mandibular	8	18.36	4.00	1.41	12.35–24.19		
Combined	16	17.38	4.01	1.00	10.64–24.19		
Homo							
Maxillary	7	21.67	5.31	2.01	14.26–30.59		
Mandibular	6	27.19	6.63	2.70	16.79–36.08		
Combined	13	24.22	6.37	1.77	14.26–36.08		
Relative Occlusal Thickness–$[(h+i)/2]/\sqrt{b}$							
Pan							
Maxillary	6	9.03	2.12	0.87	6.07–12.29	2.56	<0.05
Mandibular	8	11.46	1.44	0.51	9.11–13.14		
Combined	14	10.42	2.10	0.56	6.07–13.14		
Gorilla							
Maxillary	9	9.14	0.73	0.24	7.96–10.38	2.08	NS
Mandibular	8	11.18	2.85	1.01	7.33–15.24		
Combined	17	10.10	2.22	0.54	7.33–15.24		
Pongo							
Maxillary	8	16.65	2.69	0.95	12.55–21.93	1.45	NS
Mandibular	9	18.63	2.90	0.97	12.31–22.15		
Combined	17	17.70	2.90	0.70	12.31–22.15		
Homo							
Maxillary	7	19.64	5.87	2.22	14.55–32.24	1.53	NS
Mandibular	6	24.27	4.83	1.97	19.62–32.90		
Combined	13	21.78	5.72	1.59	14.55–32.90		
Relative Lateral Thickness–$[(k+1)/2]/\sqrt{b}$							
Pan							
Maxillary	6	11.08	1.27	0.52	9.04–12.46	2.38	<0.05
Mandibular	8	13.23	1.91	0.67	10.82–16.73		
Combined	14	12.30	1.95	0.52	9.04–16.73		
Gorilla							
Maxillary	9	10.34	1.33	0.44	8.50–12.52	1.71	NS
Mandibular	8	11.76	2.06	0.73	9.03–15.82		
Combined	17	11.01	1.81	0.44	8.50–15.82		
Pongo							
Maxillary	8	17.42	1.37	0.49	14.87–19.11	1.30	NS
Mandibular	9	16.44	1.69	0.56	14.22–19.40		
Combined	17	16.90	1.58	0.38	14.22–19.40		
Homo							
Maxillary	7	20.58	5.30	2.00	14.77–29.77	1.46	NS
Mandibular	6	24.59	4.44	1.81	18.58–31.11		
Combined	13	22.43	5.16	1.43	14.77–31.11		

Table 1.8. Comparisons of Relative Enamel Thickness Values from Linear Measurements of Extant Hominoids (Combined Maxillary and Mandibular Molars)

	$\overline{X}$	Pan		Gorilla		Pongo	
		"t"	p	"t"	p	"t"	p
Vertical cuspal thickness–$[(f+g)/2]/\sqrt{b}$							
Pan	8.75						
Gorilla	10.56	1.77	NS				
Pongo	17.38	6.77	<0.001	5.62	<0.001		
Homo	24.22	8.30	<0.001	7.87	<0.001	3.53	<0.002
Occlusal thickness–$[(h+i)/2]/\sqrt{b}$							
Pan	10.42						
Gorilla	10.10	0.41	NS				
Pongo	17.70	7.84	<0.001	8.58	<0.001		
Homo	21.78	6.95	<0.001	7.73	<0.001	2.55	<0.02
Lateral thickness–$[(k+1)/2]\sqrt{b}$							
Pan	12.30						
Gorilla	11.01	1.91	NS				
Pongo	16.90	7.26	<0.001	10.11	<0.001		
Homo	22.43	6.85	<0.001	8.50	<0.001	4.19	<0.001

Enamel Thickness in *Australopithecus* and *Paranthropus*

Photographs of sections through the early hominid molars examined in this study and the resultant enamel cap profiles of these specimens are depicted in Figs. 1.12 and 1.13. The irregular contour of the *P. robustus* (KB 5223) EDJ results from the loss of internal enamel tissue, especially in the cuspal horns, and almost all of the dentine, presumably through acidic leaching during the process of fossilization and/or initial preparation of the specimen.

The values for measurements a, b, c and e and the resultant indices of average and relative enamel thickness obtained for the complete *A. africanus* and *P. crassidens* crowns are recorded in Table 1.9. The values recorded in Table 1.9 for the *P. boisei* specimens are for a BL section through the entoconid (L 10-21) and a MD section through the protoconid (L 398-847). The values recorded in Table 1.9 for the *P. robustus* crown are those measured directly from the specimen in which the EDJ has been artifically enlarged. Thus, while the value for measurement a of the KB 5223 molar is accurate, the values for b and e are inflated, and those for measurement c and the resultant indices of average and relative enamel thickness are too low. The latter should be regarded only as minimum values. Values for our reconstruction of this crown may be found below (see Table 1.13). While both cut faces of each specimen were measured, the values recorded in Table 1.9 are for that face which maximized dentine area and thus represented the section least influenced by obliquity.

The average and relative thickness values obtained for the two *A. africanus* molars are very similar to one another. The values for the *P. crassidens* molar are larger than those for the *A. africanus* specimens, while relative enamel thickness measures for the partial *P. boisei* crowns are larger still than the *P. crassidens* values (Table 1.9). Since the measurements for the *P. boisei* specimens are for incomplete teeth—measurements for a single cusp in both cases—it is necessary to establish their comparability to measurements recorded for complete crown sections.

The L 10-21 entoconid relative thickness values [$\sqrt{c}/e = 46.86$; $(c/e)\sqrt{b} = 30.97$] are large compared to the corresponding entoconid values for modern human molars ($N = 5, \overline{X} = 39.34$, $SD = 3.50$ and $N = 5, \overline{X} = 24.24, SD = 3.80$) (Fig. 1.14). Utilizing the proportions of entoconid values to those obtained for total crown sections through the entoconid and hypoconid of human molars, the total crown estimates for the L 10-21 molar may be computed (Table 1.10). The estimated (reconstructed) values for c/e, c/a and $(c/e)/\sqrt{b}$ are very similar to those obtained for the entoconid alone, while the value for the relative thickness index $\sqrt{c}/e$ is noticeably lower.

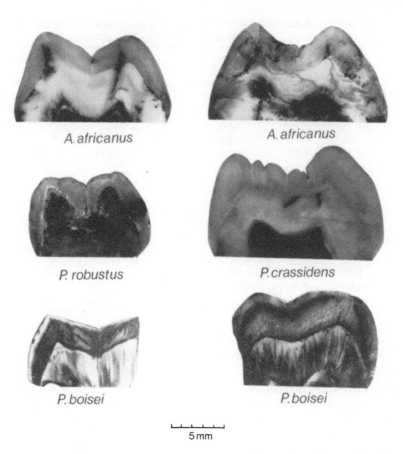

FIGURE 1.12. Photographs of sections through early hominid molars. BL sections through mesial cusps of molars of *A. africanus* (Stw 402, RM^1, buccal to left), *A. africanus* (Stw 284, LM^2, buccal to left), *P. robustus* (KB 5223, RM_1, buccal to right), and *P. crassidens* (SKX 21841, RM^3, buccal to left). BL section through entoconid and part of hypoconid of *P. boisei* (L 10-21, RM_1, buccal to right). MD section through protoconid and partial hypoconid of *P. boisei* (L 398-847, LM_3, mesial to right).

Nevertheless, all four average and relative enamel thickness values estimated for the total crown of L 10-21 are correspondingly large when compared to the measurements for modern human molars sectioned through the entoconid and hypoconid (Table 1.11).

Moreover, the means recorded for the distal cusp profile of modern human molars (Table 1.11) are slightly smaller than those obtained for the mesial cusp profiles (Tables 1.3 and 1.4), and in three of the five specimens for which both mesial (protoconid and metaconid) and distal (entoconid and hypoconid) sections were measured, the former values were slightly larger. Thus, the estimated total crown values recorded here for the L 10-21 *P. boisei* specimen would appear to be slightly conservative in comparison to the values recorded for specimens sectioned through the mesial cusps.

The relative enamel thickness values recorded from the MD section of the protoconid of L 398-847 are large by comparison to the values obtained for a modern human sample (Table 1.12; Fig. 1.15). The very similar means for MD and BL sections through the protoconid in modern humans permit an estimate to be made for the BL protoconid profile of L 398-847 (Table 1.12; Fig. 1.15). The proportionate relationships between values obtained for BL sections through the protoconid and those recorded for BL sections through both mesial cusps in human molars permits estimates for the total crown values of the L 398-847 crown (Table 1.12). As was noted for the

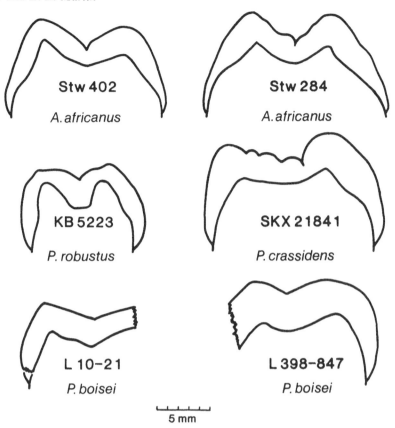

FIGURE 1.13. Enamel cap profiles of early hominid molars depicted in Fig. 1.12. The cervix of the L 10-21 entoconid is reconstructed following the trajectories of the EDJ and outer enamel surface.

Table 1.9. Measurements a, b, c and e and Indices of Average and Relative Enamel Thickness for Early Hominid Specimens

	a	b	c	e	c/e	c/a	$\sqrt{c}$/e	(c/e)/$\sqrt{b}$
BL section through mesial cusps								
Austalopithecus africanus								
Stw 284	118.67	72.37	46.30	25.59	1.81	39.02	26.57	21.27
Stw 402	98.28	59.61	38.67	21.69	1.78	39.35	28.68	23.06
Paranthropus robustus								
KB 5223[a]	78.00	46.29	31.71	19.56	1.62	40.65	28.78	23.82
Paranthropus crassidens								
SKX 21841	140.27	78.26	62.01	23.64	2.62	44.21	33.33	29.61
BL section through entoconid								
Paranthropus boisei								
L 10–21	58.62	31.28	27.35	11.16	2.45	46.66	46.86	30.97
MD section through protoconid								
P. boisei								
L 398–847	57.24	28.07	29.17	10.08	2.89	50.96	53.57	38.58

[a]Values measured from uncorrected enamel cap profile; value of a is accurate, values of b and e are inflated, values of c and resultant indices are artificially low and should be regarded as minimums only. See text for explanation and Table 1.13 for reconstructed values.

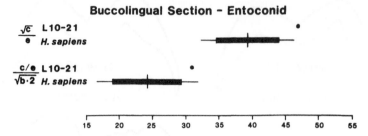

FIGURE 1.14. Comparison of relative enamel thickness values for the L 10-21 *P. boisei* specimen with those for a sample of modern humans. Vertical line = $\overline{X}$, horizontal line = $\overline{X} \pm 2\text{SD}$, horizontal bar = $\overline{X} \pm 3\text{SE}$.

L 10-21 specimen, the estimated total crown values for average and relative enamel thickness for L 398-847 are very similar to the measured single cusp values. These data suggest that measurements of enamel thickness recorded from incomplete crowns are quite accurate where (1) the section passes through the tip of the dentine horn and (2) the area of dentine below the cusp can be determined. The reconstructed crown profile through the mesial cusps of the *P. boisei* molar L 398-847 is illustrated in Fig. 1.16.

As noted above, the values recorded in Table 1.9 for the *P. robustus* (KB 5223) specimen are inaccurate owing to the fact that some of the innermost enamel has been lost. Although much of the inner enamel contour is damaged, at least three small areas of intact EDJ are clearly identifiable on the mesial posterior face of the section (Fig. 1.17). These preserved segments of the EDJ permit a reasonably accurate reconstruction of the original enamel cap profile. The average and relative enamel thickness measurements determined from the uncorrected profile (Table 1.9) are minimum values only. Nevertheless, they fall within the ranges for modern human values, and they are comparable to the values recorded for the two *A. africanus* specimens. The thickness values obtained from the corrected enamel cap profile of KB 5223 (Table 1.13) are comparable to those recorded for the most thickly enamelled human molars, the *P. crassidens* specimen and the *P. boisei* lower molar.

The average and relative enamel thickness values of the *A. africanus* and *P. crassidens* specimens (Table 1.9), the individual cuspal (Table 1.9) or estimated total crown values (Tables 1.10 and 1.12) of the *P. boisei* specimens, and the minimum (Table 1.9) and corrected values (Table 1.13) of the *P. robustus* molar are substantially greater than those for extant African apes. The

Table 1.10. Comparison of Enamel Thickness Values for BL Section of Entoconid and Total Crown in *Homo sapiens* and Estimated Values for L 10–21

	a	b	c	e	c/e	c/a	$\sqrt{c/e}$	$(c/e)/\sqrt{b}$
Proportion of entoconid to total crown values (ento/total × 100)								
N	5	5	5	5	5	5	5	5
$\overline{X}$	44.22	42.79	46.41	47.67	97.55	105.15	143.06	105.71
SD	3.14	4.05	1.93	2.21	2.68	4.13	4.69	6.61
Measured entoconid values of L 10–21								
	58.62	31.28	27.35	11.16	2.45	46.66	46.86	30.97
Estimated total crown values for L 10–21 from means								
	132.56	73.10	58.93	23.41	2.51	44.37	32.76	29.30
Estimated low and high values for L 10–21 from mean ± 2SD								
	116.08	61.47	54.41	21.42	2.38	41.14	30.74	26.04
	154.51	90.17	64.28	25.80	2.66	48.16	35.05	33.48

Table 1.11. Estimated Enamel Thickness Values for L 10-21 Compared to Values for Human Molars Sectioned Through Entoconid and Hypoconid

	a	b	c	e	c/e	c/a	$\sqrt{c}/e$	$(c/e)/\sqrt{b}$
Estimated values for L 10–21								
	132.56	73.10	58.93	23.41	2.51	44.37	32.76	29.30
Human molar values								
N	5	5	5	5	5	5	5	5
$\overline{X}$	50.02	29.68	20.34	16.44	1.24	40.77	27.56	23.05
SD	5.83	4.30	2.43	1.45	0.18	3.73	3.07	4.44
SE	2.61	1.92	1.09	0.65	0.08	1.67	1.37	1.99

fossil hominid values fall above the sample 95% confidence limits ($\overline{X} \pm 2SD$) for *Pongo* (Table 1.4). At the same time, the *A. africanus* values fall comfortably within the 95% confidence limits for the modern human sample, while the *P. crassidens* values fall at the very upper extremes of the human sample limits for maxillary molars (Fig. 1.18). The enamel thickness values for the *P. robustus* and L 10-21 *P. boisei* molars fall within the 95% confidence limits for the human lower molar sample, whereas the values for the other *P. boisei* tooth fall above these limits (Fig. 1.18).

Table 1.12. Comparison of Enamel Thickness Values for MD Section of Protoconid, BL Section of Protoconid and BL Section of Total Crown in *Homo sapiens* and Estimated Values for L 398–847

	a	b	c	e	c/e	c/a	$\sqrt{c}/e$	$(c/e)/\sqrt{b}$
Human values for MD section of protoconid								
N	3	3	3	3	3	3	3	3
$\overline{X}$	25.97	14.70	11.27	7.89	1.43	43.66	42.69	26.64
SD	3.58	2.75	0.83	0.76	0.03	2.83	2.55	3.03
Human values for BL section of protoconid								
N	6	6	6	6	6	6	6	6
$\overline{X}$	26.76	15.51	11.26	8.03	1.40	42.30	41.88	25.65
SD	4.99	3.52	2.20	0.92	0.21	5.22	4.02	4.91
Comparison of human MD and BL values (MD/BL × 100)								
	97.05	94.78	100.09	98.26	102.14	103.22	101.93	103.86
Measured MD protoconid values of L 398–847								
	57.24	28.07	29.17	10.08	2.89	50.96	53.57	38.58
Estimated BL protoconid values for L 398–847								
	58.98	29.62	29.14	10.26	2.83	49.37	52.56	37.15
Human proportions BL protoconid/Total crown values (proto/total)								
N	6	6	6	6	6	6	6	6
$\overline{X}$	54.32	57.08	50.77	52.24	97.53	93.77	136.62	91.69
SD	5.95	7.72	4.52	2.81	9.80	5.83	9.51	9.31
Estimated total BL crown values for L 398–847 from human means								
	108.58	51.89	57.40	19.64	2.90	52.65	38.47	40.52
Estimated low and high crown values for L 398–847 ($\overline{X} \pm 2SD$)								
	89.07	40.84	48.72	17.73	2.42	46.83	33.77	33.68
	139.04	71.13	69.83	22.01	3.63	60.13	44.69	50.84

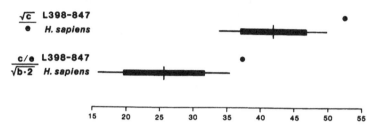

FIGURE 1.15. Comparison of relative enamel thickness values for MD section through protoconid of L 398-847 *P. boisei* specimen with those for a sample of modern humans, and comparison of estimated BL section values for L 398-847 with corresponding human values. Vertical line = $\overline{X}$, horizontal line = $\overline{X} \pm 2SD$, horizontal bar = $\overline{X} \pm 3SE$.

The *A. africanus* relative enamel thickness values fall comfortably within the 99% confidence limits for the means ($\overline{X} \pm 3SE$) of the combined upper and lower molar samples for modern *H. sapiens*, whereas the *P. robustus*, *P. crassidens* and *P. boisei* values fall well above these limits (Fig. 1.18). The values for the *P. robustus*, *P. crassidens* and L 10-21 *P. boisei* specimens fall within the upper limits of the 95% confidence intervals for the combined upper and lower molar samples for modern humans.

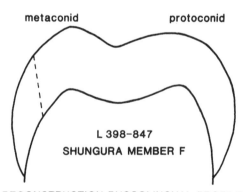

FIGURE 1.16. Reconstructed BL profile through the mesial cusps of the L 398-847 *P. boisei* M$_3$. The predicted relative enamel thickness values (Table 1.12) are compared with those obtained from measurements of the reconstruction.

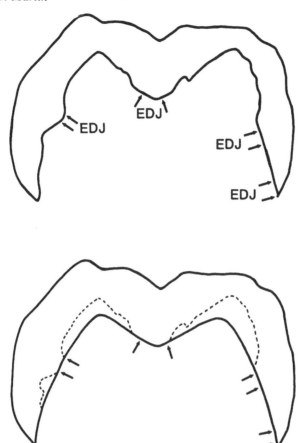

FIGURE 1.17. Drawing of actual enamel cap profile preserved by the mesial posterior face of the KB 5223 *P. robustus* M_1 (above) and the reconstructed profile (below) based on preserved regions of the enamel–dentine junction (EDJ) on the buccal, lingual and occlusal aspects.

Thus, the molars of *A. africanus* are comparable to those of modern humans, whereas the *P. robustus*, *P. crassidens*, and *P. boisei* teeth possess notably thicker enamel. Following the convention established by Martin (1983, 1985), average enamel thickness is plotted against the dentine component of tooth size for the extant and fossil hominoid samples in Fig. 1.19. Considering only the thick-enamelled modern and fossil hominid specimens, there appears to be a tendency for average enamel thickness to increase with tooth size, but compared to *H. sapiens* and *A. africanus*, the molars of *P. robustus*, *P. crassidens* and *P. boisei* possess comparatively thick enamel for their sizes. These findings support the observations made by Beynon and Wood (1986),

Table 1.13. Comparison of Enamel Thickness Measurements and Indices Obtained from Damaged Enamel Cap Profile and Reconstructed Profile for KB 5223

	a	b	c	e	c/e	c/a	$\sqrt{c}/e$	$(c/e)/\sqrt{b}$
Damaged values	78.00	46.29	31.71	19.56	1.62	40.65	28.78	23.82
Reconstructed values	78.00	41.61	36.39	18.00	2.02	46.65	33.50	31.32

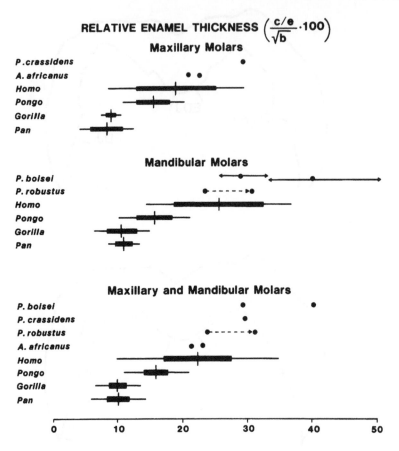

FIGURE 1.18. Comparison of relative enamel thickness values for extant hominoid samples and fossil hominid specimens. Vertical line = $\bar{X}$, horizontal line = $\bar{X} \pm$ 2SD, horizontal bar = $\bar{X} \pm$ 3SE. Lines on either side of *P. boisei* specimens indicate extreme low and high values for reconstructed estimates (see Tables 1.10 and 1.12). *P. robustus* value to left = minimum value from damaged profile, value to right = corrected value from reconstructed enamel cap profile (see Table 1.13).

whose linear measurements indicated molar enamel to be thicker in *P. boisei* than early *Homo* in relation to estimates of the dentine contribution to crown size.

Martin (1983, 1985) has demonstrated that relative enamel thickness in modern *H. sapiens* is similar to that evinced by the Miocene hominoid *Sivapithecus*, a condition that is interpreted as being primitive for the great ape and human clade. While the *A. africanus* values fall within the 95% confidence limits for the *Sivapithecus* sample (Table 1.14), the *P. robustus*, *P. crassidens* and *P. boisei* values fall well above these limits (indeed, they are at or above the upper 99% confidence limits for that sample).

The linear measurements employed by Beynon and Wood (1986) and those used by Martin (1983) are recorded for the *Australopithecus* and *Paranthropus* molars in Table 1.15. The values of measurements h and i of Martin (1983) are equivalent to OT(B) and OT(L) of Beynon and Wood (1986). The corresponding values for BL sections through the mesial cusps of modern human molars show no significant differences between buccal and lingual sides (Table 1.16), although in the maxillary molars the lingual (protocone) values tend to be larger while in the mandibular molars the buccal (protoconid) values tend to be larger. This situation holds also for the *A. africanus* and *P. crassidens* specimen values (Table 1.15).

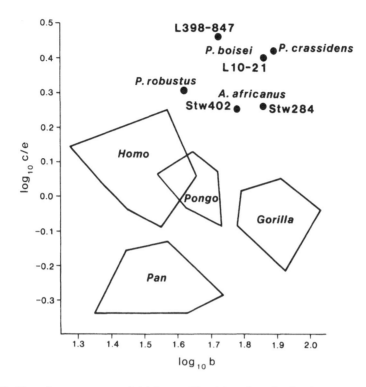

FIGURE 1.19. Plot of average enamel thickness (Y axis) against the dentine component of tooth size (X axis). The polygons encompass all specimens comprising the extant hominoid samples.

In view of the absence of significant differences between the buccal and lingual crown values in the modern human sample, relative enamel thickness values can be computed by using the averages of the corresponding measurements; this enables the upper and lower molar sample values to be compared (Table 1.17). As was noted for other measures of enamel thickness, the OT, LT and CT values for modern humans tend to be larger for mandibular than maxillary molars, although the differences are not statistically significant. The relative enamel thickness values determined from linear measurements for the early hominid molars are recorded in Table 1.18, and while the *A. africanus* and *P. crassidens* values can be compared directly to those recorded for modern human homologues (Table 1.17), the values for the *P. boisei* specimens are for single cusps only. Nevertheless, the correspondence between individual cusp and total crown values is reasonably close, indicating the accuracy of the *P. boisei* values.

The CT, OT and LT values obtained for the L 10-21 and L 398-847 *P. boisei* molars (Table 1.15) are comparable to the raw values recorded by Beynon and Wood (1986) for a sample of geochronologically younger *P. boisei* specimens from Koobi Fora and Olduvai Gorge. That is, the Shungura Formation molar values fall within the observed ranges and close to the corresponding means recorded by Beynon and Wood (1986, table 3). At the same time, the L 10-21 and L 398-847 values fall above the upper limits of an early *Homo* sample observed by them (1986, table 4).

Enamel at the cuspal tip (CT) and over the lateral aspect (LT) tends to be relatively thicker than that over the occlusal basin (OT) in human maxillary and mandibular molars (Table 1.17). This pattern of relative enamel thickness is evinced also by one of the *A. africanus* specimens (Stw 284) and the *P. crassidens* specimen, whereas in the other *A. africanus* and in both

Table 1.14. Comparison of Average and Relative Enamel Thickness Values of *Sivapithecus*, *Australopithecus* and *Paranthropus*

		c/e	c/a	$\sqrt{c/e}$	$(c/e)/\sqrt{b}$
Sivapithecus[a]					
Maxillary	N	4	4	4	4
	$\overline{X}$	1.14	36.84	25.41	19.49
	SD	0.16	3.09	1.78	2.64
Mandibular	N	5	5	5	5
	$\overline{X}$	1.21	36.00	26.43	19.95
	SD	0.22	3.27	1.94	2.77
Combined	N	9	9	9	9
	$\overline{X}$	1.18	36.37	25.98	19.75
	SD	0.18	3.02	1.83	2.55
Australopithecus africanus					
Stw 284		1.81	39.02	26.57	21.27
Stw 402		1.78	39.35	28.68	23.06
Paranthropus robustus					
KB 5223		2.02	46.65	33.50	31.32
Paranthropus crassidens					
SKX 21841		2.62	44.21	33.33	29.61
Paranthropus boisei					
L 10–21		2.51	44.37	32.76	29.30
L 398–847		2.90	52.65	38.47	40.52

[a]Data for *Sivapithecus* from Martin (1983, Appendix A).

P. boisei specimens the cuspal and occlusal enamel is relatively thicker than that over the lateral aspect of the crown (Table 1.18). Occlusal enamel is also relatively thicker than lateral enamel in the *P. robustus* specimen.

Comparison of the relative enamel thickness values obtained for the fossil hominid molars (Table 1.18) reveals that in most instances they fall within the 95% confidence limits (i.e., $\overline{X} \pm$ 2 SD) for modern human samples (Table 1.17). However, it is noteworthy that the cuspal thickness values for the *P. crassidens* and L 398-847 *P. boisei* molars, and the occlusal thickness values for both *P. boisei* molars fall above these confidence limits. The occlusal thickness value for the

Table 1.15. Values of Linear Enamel Thickness Measurements for Fossil Hominid Molars Examined in this Study

Measurements	Stw 284	Stw 402	KB 5223	SKX 21841	L 10–21	L 398–847[a]
f	2.40	1.54	—	3.33	(2.12)	3.86
g	2.82	2.46	—	3.39	2.79	3.19
h	1.86	1.98	—	2.72	(2.29)	3.45
i	2.12	2.08	2.15	2.30	2.88	3.54
k	2.11	1.64	—	2.27	—	2.50
l	2.48	1.67	—	2.93	2.29	—
CT(B)	2.43	1.75	—	3.35	(2.17)	3.86
CT(L)	2.82	2.46	—	3.42	2.94	3.19
OT(B)	1.86	1.98	—	2.72	(2.29)	3.45
OT(L)	2.12	2.08	2.15	2.30	2.88	3.54
LT(B)	2.19	1.68	1.83	2.48	—	2.70
LT(L)	2.72	1.90	1.57	3.35	2.34	—

[a]For specimen L 398–847 measurements g, i, CT(L), and LT(L) refer to hypoconid diameters, while measurements k and LT(B) refer to diameters across the mesial aspect of the protoconid.

Table 1.16. Values of Linear Enamel Thickness Measurements for BL Sections Through the Mesial Cusps of Modern Human Molars

Measurements	N	$\bar{X}$	SD	SE	Observed range	"t"	p
Maxillary molars							
f	7	1.27	0.25	0.09	0.94–1.65	0.65	NS
g	7	1.41	0.51	0.19	0.59–2.06		
h = OT(B)	7	1.19	0.34	0.13	0.88–1.89	0.20	NS
i = OT(L)	7	1.23	0.41	0.15	0.87–2.06		
k	7	1.09	0.23	0.09	0.84–1.44	1.80	NS
l	7	1.44	0.46	0.17	0.91–2.24		
CT(B)	5	1.46	0.19	0.09	1.16–1.65	1.57	NS
CT(L)	5	1.77	0.40	0.18	1.38–2.34		
LT(B)	6	1.48	0.39	0.16	0.96–1.97	1.76	NS
LT(L)	6	1.93	0.49	0.20	1.41–2.59		
Mandibular molars							
f	6	1.35	0.37	0.15	0.77–1.77	0.40	NS
g	6	1.44	0.41	0.17	1.00–2.18		
h = OT(B)	6	1.28	0.28	0.11	0.94–1.73	0.29	NS
i = OT(L)	6	1.23	0.31	0.13	0.93–1.79		
k	6	1.29	0.31	0.13	0.94–1.79	0.57	NS
l	6	1.21	0.15	0.06	1.10–1.51		
CT(B)	6	1.65	0.24	0.10	1.39–2.00	0.68	NS
CT(L)	6	1.51	0.44	0.18	1.02–2.18		
LT(B)	6	1.64	0.23	0.09	1.30–1.97	0.66	NS
LT(L)	6	1.55	0.24	0.10	1.28–1.96		

Table 1.17. Relative Enamel Thickness Values from Averaged Buccal and Lingual Linear Measurements from BL Sections Through the Mesial Cusps of Modern Human Molars

Sample	N	$\bar{X}$	SD	SE	Observed range	"t"	p
Relative cuspal thickness – $[(f+g)/2]/\sqrt{b}$							
Maxillary	7	21.67	5.31	2.01	14.26–30.59	1.67	NS
Mandibular	6	27.19	6.63	2.70	16.79–36.08		
Combined	13	24.22	6.37	1.77	14.26–36.08		
Relative cuspal thickness – $\{[CT(B)+CT(L)]/2\}/\sqrt{b}$							
Maxillary	6	26.55	4.99	2.04	20.68–32.90	1.39	NS
Mandibular	6	30.71	5.37	2.19	23.40–37.94		
Combined	12	28.63	5.40	1.56	20.68–37.94		
Relative lateral thickness – $[(h+i)/2]/\sqrt{b}$							
Maxillary	7	19.64	5.87	2.22	14.55–32.24	1.53	NS
Mandibular	6	24.27	4.83	1.97	19.62–32.90		
Combined	13	21.78	5.72	1.59	14.55–32.90		
Relative occlusal thickness – $[(k+l)/2]/\sqrt{b}$							
Maxillary	7	20.58	5.30	2.00	14.77–29.77	1.46	NS
Mandibular	6	24.59	4.44	1.81	18.58–31.11		
Combined	13	22.43	5.16	1.43	14.77–31.11		
Relative lateral thickness – $\{[LT(B)+LT(L)]/2\}/\sqrt{b}$							
Maxillary	6	28.09	6.46	2.64	19.01–37.50	0.99	NS
Mandibular	6	30.94	2.86	1.17	25.47–33.08		
Combined	12	29.52	4.99	1.44	19.01–37.50		

Table 1.18. Relative Enamel Thickness Values of Early Hominid Molars Determined from Averaged Linear Measurements

Measurements	Stw 284	Stw 402	KB 5223	SKX 21841	L 10–21	L 398–847
f+g	30.67	25.91	—	37.97	35.27	47.13
h+i	23.38	26.30	33.33	28.36	36.41	46.73
k+l	27.03	21.50	—	29.38	28.95	33.38
CT	30.91	27.33	—	38.31	37.17	47.13
LT	28.91	23.19	26.36	32.99	29.58	36.05

P. robustus molar (33.33) falls very close to the upper 95% human sample confidence limit of 33.93 (Table 1.17).

Thus, the molars of the *Paranthropus* specimens appear to display rather thick occlusal and/or cuspal enamel in comparison to *A. africanus* and modern humans (Fig. 1.20). Beynon and Wood (1986) found that size corrected occlusal and cuspal thickness values of *P. boisei* were significantly greater than the values of an early *Homo* sample, but that lateral thickness means did not differ significantly between these two samples. The present data appear to support these

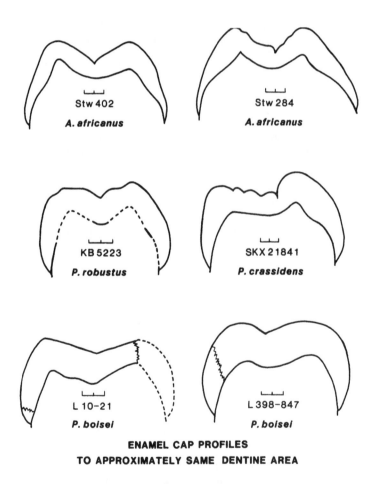

FIGURE 1.20. Enamel cap profiles of *Australopithecus* and *Paranthropus* molars drawn to approximately same dentine areas. Scale bars = 2 mm.

findings for *P. boisei* in relation to modern human as well as fossil *Homo* samples, and they indicate that *P. crassidens* shared relatively thick cuspal (but not lateral) enamel with *P. boisei*. Thus, *Paranthropus* specimens, in comparison to modern human and *A. africanus* molars, tend to possess moderate lateral enamel thickness and very thick enamel on the chewing surfaces of these teeth.

The finding that three species of early hominids, *P. robustus*, *P. crassidens* and *P. boisei*, possess thicker enamel than that found in *Homo*, *Sivapithecus* or *A. africanus* is of considerable importance in terms of defining the phylogenetic relationships among the three so-called "robust" australopithecine taxa. Following the conclusions of Martin (1985), the increase in relative enamel thickness displayed by the three *Paranthropus* species must represent a derived condition within the great ape and human clade. Thus, since *Homo* is hypothesized to retain the ancestral pattern for that clade, the derived condition evinced by the three *Paranthropus* species would appear to preclude any from being regarded as ancestral to *Homo*. Perhaps more significantly, especially in light of arguments against "robust" australopithecine monophyly (Walker and Leakey, this volume, chapter 15; Wood, this volume, chapter 17), the present findings on enamel thickness provide a derived character state by which to unite a *Paranthropus* clade. However, it has been shown previously that analysis of enamel thickness data in isolation can lead to serious errors in the interpretation of homology (e.g., the apparent sharing of thin enamel by gibbons and African apes), and that it is essential to consider the pattern of enamel accretion in the identification of synapomorphy (Martin, 1983, 1985).

Enamel Development in *Australopithecus* and *Paranthropus*

The discovery that *Australopithecus africanus* possesses thick enamel [as defined by Martin (1983, 1985)], while *Paranthropus robustus*, *P. boisei* and *P. crassidens* exhibit "hyper-thick" enamel raises important phylogenetic and adaptive questions. Is the thick enamel of *A. africanus* a retention of the ancestral condition for the great ape and human clade, and is the "hyper-thick" enamel homologous in the three species in which it has been documented? In order to provide answers to these questions, it is necessary to consider enamel accretion or development.

The amount (thickness) of enamel on a tooth is a product of several variables: (i) the average daily secretory rate of the ameloblasts; (ii) the length of the secretory life of the ameloblasts and the duration of crown formation; and (iii) the number of ameloblasts that are secretory at any one time. The average daily secretory rate of the ameloblasts has been shown to account for the differences in completed enamel thickness among modern large-bodied hominoids, whereas differences in ameloblast secretory life and duration of amelogenesis account for the difference in enamel thickness between gibbons and humans (Martin, 1983, 1985). The number of ameloblasts active at any one time has been shown to vary between deciduous and permanent teeth in large-bodied hominoids, where deciduous teeth form more enamel per unit time than permanent teeth (Dean and Wood, 1981). In relation to the findings for *Paranthropus* species, one or more of these variables could co-vary such that "hyper-thick" enamel might be produced through at least seven developmental pathways.

Discovery that the same developmental pathway was utilized in the production of "hyper-thick" enamel in more than one taxon would constitute strong evidence for the homology of change in enamel thickness, since all three parameters are known to vary within extant Hominoidea (i.e., developmental pathways do not appear to be evolutionarily constrained within this clade).

With regard to the period of enamel secretion, although little is currently known about the relationship of the duration of amelogenesis to other growth parameters, it seems reasonable to postulate that these are linked at least to the degree that shifts in the duration of enamel secretory activity would reflect other changes in overall growth strategy. As such, it would seem rather unlikely that *Paranthropus* would have had a longer period of dental development than fossil or modern representatives of *Homo* (Smith, 1986; Dean, this volume, chapter 2), although conclusive and convincing evidence is not yet available. This possibility might be addressed through the approach employed by Boyde (1963) in a study of thin sections of human premolars, where cumulative counts of the total number of increments of tooth formation were made.

The average daily secretory rate (i.e., the accretionary rate of enamel apposition) can be determined from observations of prism cross-striation intervals. Although there has been some debate in the literature over whether these features actually represent a circadian pattern of amelogenesis, experimental (Mimura, 1939) and abundant circumstantial evidence (Boyde, 1963, 1964, 1976) indicate that they are correlated with daily rhythmic perturbations in matrix secretion. Prism cross-striations can be observed directly in topography free specimens using back scattered SEM, or through the use of TSRLM, or PLM.

Martin (1985) has reported that differences in daily secretory rates among hominoid taxa appear to be correlated with the developmental packing patterns of the prisms such that Pattern 3 enamel is associated with cross-striation repeat intervals of between 5 and 7 μm, and Pattern 1 enamel is associated with repeat intervals of less than 2.5 μm. Beynon and Wood (1987) have recorded cross-striation repeat intervals for specimens of *Paranthropus boisei* that are slightly in excess of the 5 to 7 μm usually encountered in hominoids. Since Beynon and Wood obtained their measurements from regions in which segments of prisms were close to having been longitudially sectioned (= parazones) on naturally fractured surfaces, their figures should be comparable to those recorded by Martin (1985) in terms of accuracy. Nevertheless, the figures recorded by Beynon and Wood (1987) certainly do not account for the extent of the enamel thickness in *P. boisei* recorded either here or by them, and thus it is clear that an increase in daily secretory rate does not represent the principal mechanism.

Dean and Wood (1981) have shown that living hominoids form their deciduous teeth in a manner that is quite distinct from that employed in the formation of the permanent dentition; thus, greater numbers of ameloblasts are active at any one time in deciduous teeth, resulting in enamel accretion along a broader front. The conformation of the secretory front of the enamel may be analyzed in mature tissue by examination of the Brown Striae of Retzius, and variation may be expressed by comparing the lengths of individual striae from their intercept with the EDJ to the point at which they reach the outer tooth surface.

Alternatively, variation can be expressed by measuring the angles at which the striae diverge from the EDJ. It is difficult, however, to obtain accurate and repeatable results using the latter method of analysis because both the EDJ and the striae in question compose manifestly curved lines. Nevertheless, when the secretory front is longer because of the activity of greater numbers of cells at any given time, the striae make an increasingly acute angle with the EDJ such that the striae follow a course more nearly parallel to that of the EDJ. Secretion across a broad front thus results in the presence of relatively fewer striae that can be traced from the first that reaches the outer surface to the last that forms the cervix (Fig. 1.21).

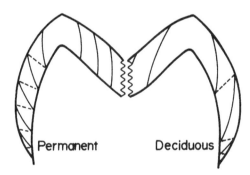

FIGURE 1.21. Schematic representation of patterns of enamel accretion in deciduous and permanent molars of modern humans and other large-bodied hominoids. The first of the Brown Striae of Retzius that is traceable from outer surface of cuspal apex to enamel dentine junction is indicated by first line drawn lateral to cuspal apex. Dotted lines indicate path followed perpendicular to EDJ from interception of Brown Striae to outer enamel surface and interception of next striation. Adapted from Dean and Wood (1981).

Evidence for the accretionary pattern displayed by each fossil hominid taxon was investigated by examination of the length and orientation of the Brown of Striae of Retzius. The cut surfaces of the embedded teeth were polished lightly with 1200-grit silicon carbide paper to remove scratches produced during sectioning, and the specimens cut with the diamond wafering blade were polished with 1 μm diamond paste on a lapping wheel. The polished faces were examined without etching by epipolarized light using a Zeiss Ultraphot with a 10x pol lens, and montages of the entire section were prepared using a 6.3x dry objective lens (Fig. 1.22). Variations in the orientation of the striae were expressed in relation to the stria that first reaches the definitive outer enamel surface (i.e., the first stria that runs from the cuspal tip at the outer tooth surface to the EDJ) (Figs. 1.21 and 1.23). Measurements of length and area were made using a Graf-Pen sonic digitizer. Given the nature of enamel thicknesses recorded for the buccal and lingual sides of maxillary and mandibular molars of extant hominoids and fossil hominids (see above), the most reliable comparisons appear to be between the buccal face of the protoconid and the lingual aspect of the protocone; thus, the measurements and comparisons reported here are restricted to these regions.

The linear and areal measurements recorded for the fossil hominid specimens are provided in Table 1.19. Several indices derived from these measurements serve to express differences in the configuration of enamel accretion in the fossil hominid sample (Table 1.20). Thus, the index t/v expresses the proportion of the EDJ length that is covered by the first increment to reach the outer tooth surface; index r/t provides an indication of the degree to which the envelope of enamel enclosed by the first increment to reach the tooth surface is skewed; index (r + s)/t compares the length of the first striae to reach the tooth surface to the length of the entire EDJ; and index (r + s)/u compares the length of the first line to reach the tooth surface to the length of the outer surface from the cuspal tip to the cervix (Fig. 1.23). Finally, index w/x expresses the proportion of lateral enamel that is included within the first incremental line to reach the outer tooth surface (Fig. 1.23).

Despite the fact that the samples for each taxon are very small, several clear patterns of difference emerge (Table 1.20). Thus, in both *A. africanus* specimens the value for the index (r + s)/v falls under 100 (84–97) indicating that the first incremental striae to reach the tooth surface is shorter than the EDJ, whereas in all three *Paranthropus* specimens the index value is substantially greater than 100% (114–132). Thus, *Paranthropus* molars display a condition in which the first incremental line is extremely elongate and, assuming constancy in ameloblast size, this indicates that *Paranthropus* had a relatively larger number of cells active at any one time than did *A. africanus*. Also, in *A. africanus*, the first increment to reach the outer tooth surface forms less than two-thirds (59–62%) of the tooth crown, while in *Paranthropus* it is responsible for almost 75% or more (74–83%) of the crown (Table 1.20, index w/x). The other indices reveal similar elongation of the first stria to reach the outer enamel surface (with a concomitant lengthening of the other striae) in *Paranthropus* specimens.

Thus, the three species of *Paranthropus* for which molar enamel has been examined appear to be characterized by the presence of "hyper-thick" enamel that is formed through the secretory activity of a relatively large number of cells at any one time. That is, *Paranthropus* permanent molars follow a developmental pattern like that of deciduous molars described by Dean and Wood (1981). This finding, which has been reported for *P. boisei* by Beynon and Wood (1986, 1987), would seem to represent a significant developmental synapomorphy linking the three species of *Paranthropus*, and distinguishing them from *A. africanus*, and from early and later species of *Homo* (Fig. 1.24).

Several points that emerge from these comparisons (Table 1.20), however, suggest differences among the *Paranthropus* species. Thus, the value of the index r/t in *P. robustus* differs from the values recorded for the other two *Paranthropus* species, and this reflects the fact that while the first stria to reach the outer tooth surface on the *P. robustus* molar is responsible for a significant portion of the crown, it tucks in rather rapidly to make a large angle of incidence with the EDJ. Thus, the cervical portion of the crown (i.e., the last portion of the crown to be formed) appears to have developed somewhat more slowly in *P. robustus* than in *P. crassidens* and *P. boisei* specimens (Fig. 1.24).

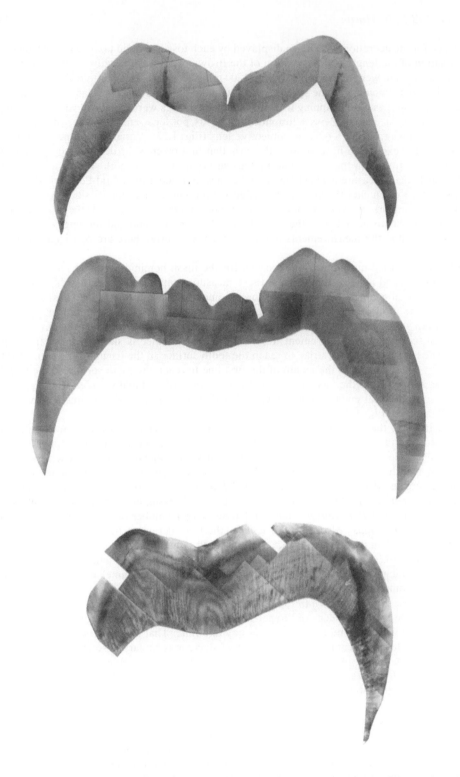

FIGURE 1.22. Montage reconstructions of early hominid enamel cap profiles as viewed via polarizing light microscopy. Top: *A. africanus*, Stw 402, RM^1, buccal to left. Middle: *P. crassidens*, SKX 21841, RM^3, buccal to left. Bottom: *P. boisei*, L 398-847, LM$_3$ mesial to right. All to same scale.

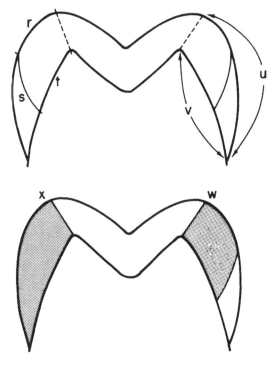

FIGURE 1.23. Top: Drawing of section through hominoid molar crown showing first incremental striation traceable from outer enamel surface to enamel–dentine junction, and the linear measurements recorded from polarizing light microscopic montages of the early hominid fossils. Bottom: Drawing of section through hominoid molar crown showing area of lateral enamel cap enclosed by first incremental line to reach the outer tooth surface relative to total lateral enamel cap.

In addition, in *P. boisei* the striae in the occlusal enamel are continuous across both cusps (Fig. 1.22), implying that the whole crown formed essentially as a single unit rather than from discrete developmental lobes. This differs from the pattern evinced by the Swartkrans molar, and it is noticeably dissimilar to the Kromdraai specimen in which perikymata are clearly visible on the occlusal surface.

Thus, there appears to be some evidence for differentiation among the *Paranthropus* species examined here, inasmuch as the *P. robustus* crown displays striae that tend to become more highly angled to the EDJ in the cervical region than do the molars of *P. crassidens* and *P. boisei*, and the occlusal enamel striae in *P. boisei* are continuous across both cusps. These differences, while less impressive perhaps than the principal difference in developmental pattern that separates *Paranthropus* from *Australopithecus* and *Homo*, may provide further evidence of species level differentiation with the genus *Paranthropus*.

Table 1.19. Values of Brown Striae, Outer Enamel Surface, and Enamel-Dentine Junction Lengths for Fossil Hominid Specimens Examined in this Study[a]

Measurements	Stw 284	Stw 402	KB 5223	SKX 21841	L 398–847[b]
r	3.54	1.56	2.96	4.40	2.98
s	6.26	6.41	3.60	7.78	7.21
t	5.10	5.25	4.09	7.74	5.74
u	11.32	10.79	8.37	14.41	12.42
v	8.84	8.22	5.77	9.72	7.72

[a]Measurements in mm. See Fig. 1.23.
[b]Values for mesial aspect of protoconid; all other values for lingual aspect of protocone or buccal aspect of protoconid. L 10–21 was not measured because the cervix is missing.

Table 1.20. Values of Indices Relating to Enamel Development Recorded for Fossil Hominid Specimens Examined in this Study

Index	Stw 284	Stw 402	KB 5223	SKX 21841	L 398–847[a]
t/v	0.58	0.64	0.71	0.80	0.74
r/t	0.23	0.30	0.72	0.57	0.52
(r+s)/u	0.66	0.74	0.78	0.85	0.82
(r+s)/v	0.84	0.97	1.14	1.25	1.32
w/x	61.40	59.00	74.30	79.00	83.30

[a]Values for mesial aspect of protoconid; all other values for lingual aspect of protocone or buccal aspect of protoconid.

It is also significant to note that prism decussation, as evidenced by the formation of Hunter-Schreger bands, is clearly seen in all species, and it is especially evident in *P. boisei* (Fig. 1.22). In *P. boisei* a very clear pattern of decussation is manifest in the occlusal enamel, in which the Hunter-Schreger bands run almost perpendicularly from the EDJ to the occlusal surface. Thus, suggestions that *P. boisei* enamel displays little decussation in comparison to other hominids (Beynon and Wood, 1986), and inferences that *Paranthropus* teeth would therefore be more prone to fracture under heavy masticatory loads would appear to be without foundation.

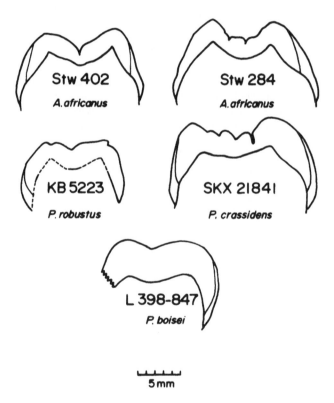

FIGURE 1.24. Tracings of cross-sections through early hominid fossil molars showing the position and configuration of the initial incremental line to course from the outer tooth surface to the EDJ. Tracing from polarizing light microscopy montages.

Conclusions

The findings of this study reveal that the permanent molar teeth of *Paranthropus* specimens that have been examined from Members E and F of the Shungura Formation *(P. boisei)*, Member 3 of the Kromdraai B East Formation *(P. robustus)* and Member 3 of the Swartkrans Formation *(P. crassidens)* possess enamel that is relatively thicker than that displayed by modern humans or by specimens of *A. africanus* from Sterkfontein. Human and *A. africanus* enamel is of the same relative thickness, and this is like the condition displayed by the Miocene hominoid *Sivapithecus*. According to previous interpretations by Martin (1983, 1985), the thick enamel found in humans and *A. africanus* represents a retention of the primitive condition for the great ape and human clade; the "hyper-thick" enamel of *Paranthropus* specimens would appear to represent a synapomorphy of that clade. The results obtained here for relative enamel thickness in extant and fossil hominids are in agreement with the finding by Beynon and Wood (1986) that *P. boisei* possessed significantly thicker enamel than fossil representatives of the genus *Homo*.

Developmental parameters have been examined in an attempt to establish homology for the "hyper-thick" enamel displayed by *Paranthropus* specimens, and in each of the three species examined here the Brown Striae of Retzius are longer and more acutely angled to the EDJ than is the case for *A. africanus*. The condition shown by *A. africanus* is similar to that displayed by modern humans and to the configuration reported for early *Homo* specimens by Beynon and Wood (1986, 1987). Thus, the permanent molar crowns of *Paranthropus* formed at a rapid rate through the amelogenic activity of relatively large numbers of cells at any given time. This pattern, which represents a retention of a deciduous mode of enamel accretion (Dean and Wood, 1981), is specialized within the Hominoidea and would appear, therefore, to establish the homology of the "hyper-thick" enamel of *Paranthropus* specimens examined here. This would seem to constitute a significant synapomorphy shared by *Paranthropus* species. Differences among the teeth of *P. boisei*, *P. robustus*, and *P. crassidens* in the angles assumed by the striae over the cervical portion of the crown and in the configuration of striae over the occlusal aspect, suggest that species level distinctions among these samples should be retained (Howell, 1978; Grine, 1982, 1985, this volume, chapter 14).

Paranthropus developed thicker molar enamel than has been documented for any other hominoid taxon. This "hyper-thick" enamel developed at a faster rate than the thick enamel on *Sivapithecus*, *Homo* and *Australopithecus africanus* molars through the secretory activity of relatively greater numbers of ameloblasts at any given time, but the molar crowns of *Paranthropus* may not necessarily have taken a longer period of time to form than those of other large-bodied hominoids. The differences in suggested developmental timing reported by Beynon and Wood (1987) indicate that *P. boisei* and early *Homo* molar crown formation rates were not terribly dissimilar; indeed, Beynon and Wood (1986, 1987) have suggested that *P. boisei* may actually have completed crown formation somewhat more quickly.

The fact that the great thickness of enamel displayed by the *Paranthropus* specimens could be formed without evidence of other concomitant major alterations in growth strategy (Dean, this volume, chapter 2) suggests that enamel thickness was increased through selective pressures relating to some functional role. This could be increased levels of wear and/or occlusal loads as, contrary to studies by Beynon and Wood (1986, 1987), the enamel of *Paranthropus*, and the occlusal enamel of *P. boisei* in particular, possesses well-developed Hunter-Schreger band formation, indicative of the presence of crack-stopping structures related to prism decussation.

Examination of enamel thickness and structure in *A. afarensis* should provide valuable information concerning its purported affinities with *Paranthropus*. Moreover, examination of the enamel on the teeth of KNM-WT 17000 should provide similar information pertaining to its membership in the *Paranthropus* clade and to its specific status within the genus (cf. Walker and Leakey, this volume, chapter 15; Kimbel *et al.*, this volume, chapter 16).

Acknowledgments

This project would have been impossible without the cooperation and encouragement of the individuals entrusted with the curatorship of the hominid fossil record. In particular, we are grateful to F. C. Howell for access to the teeth from the Shungura Formation, C. K. Brain for making available the teeth from Kromdraai and Swartkrans, and P. V. Tobias and A. R. Hughes for permission to examine the molars from Sterkfontein. We thank M. Dewey and A. Boyde for their support and encouragement and for permitting us access to critical laboratory facilities, and we are indebted to D. Colflesh for his time and expertise in obtaining the PLM images. We thank K. Jones and G. Shidlovsky, Brookhaven National Laboratory, for access to the AMR 1400.

We have benefited greatly from discussions with many colleagues, but in particular we would like to acknowledge the input of A. Boyde and M. C. Dean. We are grateful also to the participants in the workshop for fruitful discussion about "robust" australopithecine monophyly, evolution and adaptation.

This work was supported by grants from the L. S. B. Leakey Foundation (FEG and LM), a SUNY Faculty Grant-in-Aid (LM), and by a SUNY Research Development Grant (LM and FEG).

References

Beynon, A. D. and Wood, B. A. (1986). Variation in enamel thickness and structure in East African hominids. *Amer. J. Phys. Anthropol., 70*: 177–193.

Beynon, A. D. and Wood, B. A. (1987). Patterns and rates of enamel growth in the molar teeth of early hominids. *Nature, 326*: 493–496.

Boyde, A. (1963). Estimation of age at death of young human skeletal remains from incremental lines in dental enamel. *Proc. 3rd Intern. Congr. Forensic Immunol., Med., Pathol., Toxicol. London. Excerpta Med. Intern. Congr. Ser., 80*: 36–46.

Boyde, A. (1964). The structure and development of mammalian tooth enamel. Ph.D. Thesis, University of London.

Boyde, A. (1976). Amelogenesis and the development of teeth. In B. Cohen and I. R. H. Kramer (eds.), *Scientific foundations of dentistry*, pp. 335–352. Heinemann, London.

Boyde, A. and Martin, L. B. (1982). Enamel microstructure determination in hominoid and cercopithecoid primates. *Anat. Embryol., 165*: 193–212.

Boyde, A. and Martin, L. B. (1984). A non-destructive survey of prism packing patterns in primate enamels. In R. W. Fearnhead and S. Suga (eds.), *Tooth enamel IV*, pp. 417–421. Elsevier, New York.

Boyde, A. and Tamarin, A. (1984). Improvement to critical point drying technique for SEM. *Scanning, 6*: 30–35.

Brain, C. K. (1982). The Swartkrans site: stratigraphy of the fossil hominids and a reconstruction of the environment of early *Homo*. *Pretirage 1er Congr. Intern. Paleontol. Hum.*, Nice, vol. 2, pp. 676–706.

Bromage, T. G. and Dean, M. C. (1985). Re-evaluation of the age at death of immature fossil hominids. *Nature, 317*: 525–527.

Coppens, Y. (1980). The differences between *Australopithecus* and *Homo;* preliminary conclusions from the Omo research expeditions. In L-K Konigsson (ed.), *Current argument on early Man*, pp. 207–225. Pergamon, New York.

Dean, M. C. and Wood, B. A. (1981). Developing pongid dentition and its use for ageing individual crania in comparative cross-sectional growth studies. *Folia Primatol., 36*: 111–127.

Gantt, D. G. (1977). Enamel of primate teeth: its thickness and structure with reference to functional and phyletic implications. Ph.D. Thesis, Washington University, St. Louis, MO.

Gantt, D. G. (1982). Neogene hominoid evolution: a tooth's inside view. *In* B. Kurten (ed.), *Teeth: form, function and evolution*, pp. 93–108. Columbia University Press, New York.
Gantt, D. G. (1983). The enamel of Neogene hominoids: structural and phyletic implications. *In* R. L. Ciochon and R. S. Corruccini (eds.), *New interpretations of ape and human ancestry*, pp. 249–298. Plenum, New York.
Gantt, D. G. (1985). Enamel thickness and human evolution. *Amer. J. Phys. Anthropol., 66*: 171.
Gantt, D. G. (1986). Enamel thickness and ultrastructure in hominoids: with reference to form, function and phylogeny. *In* D. R. Swindler and J. Erwin (eds.), *Comparative primate biology*, vol. 1: *Systematics, evolution and anatomy*, pp. 453–475. Alan R. Liss, New York.
Gantt, D. G., Pilbeam, D. R. and Steward, G. (1977). Hominoid enamel prism patterns. *Science, 198*: 1155–1157.
Grine, F. E. (1982). A new juvenile hominid (Mammalia: Primates) from Member 3, Kromdraai Formation, Transvaal, South Africa. *Ann. Tvl. Mus., 33*: 165–239.
Grine, F. E. (1985). Australopithecine evolution: the deciduous dental evidence. *In* E. Delson (ed.), *Ancestors: the hard evidence*, pp. 153–167. Alan R. Liss, New York.
Howell, F. C. (1978). Hominidae. *In* V. J. Maglio and H. B. S. Cooke (eds.), *Evolution of African mammals*, pp. 154–248. Harvard University Press, Cambridge, MA.
Howell, F. C. and Coppens, Y. (1974). Inventory of remains of Hominidae from Pliocene/Pleistocene formations of the lower Omo Basin, Ethiopia (1967–1972). *Amer. J. Phys. Anthropol., 40*: 1–16.
Howell, F. C. and Coppens, Y. (1976). An overview of Hominidae from the Omo Succession, Ethiopia. *In* Y. Coppens, F. C. Howell, G. Ll. Isaac and R. E. F. Leakey (eds.), *Earliest Man and environments in the Lake Rudolf Basin*, pp. 522–532. University of Chicago Press, Chicago, IL.
Jolly, C. J. (1970). The seed-eaters: a new model of hominid differentiation based on a baboon analogy. *Man, 5*: 5–26.
Kay, R. F. (1981). The nut-crackers: a new theory of the adaptations of the Ramapithecinae. *Amer. J. Phys. Anthropol., 55*: 141–152.
Martin, L. B. (1983). The relationships of the later Miocene Hominoidea. Ph.D. Thesis, University of London.
Martin, L. B. (1985). Significance of enamel thickness in hominoid evolution. *Nature, 314*: 260–263.
Mimura, F. (1939). The periodicity of growth lines seen in enamel. *Kobyo-shi, 13*: 454–455.
Molnar, S. and Gantt, D. G. (1977). Functional implications of primate enamel thickness. *Amer. J. Phys. Anthropol., 46*: 447–454.
Pilbeam, D. R. (1972). *The ascent of Man*. Macmillan, New York.
Robinson, J. T. (1956). The dentition of the Australopithecinae. *Mem. Tvl. Mus., 9*: 1–179.
Robinson, J. T. (1963). Adaptive radiation in the australopithecines and the origin of Man. *In* F. C. Howell and F. Bourliere (eds.), *African ecology and human evolution*, pp. 385–416. Aldine, Chicago, IL.
Simons, E. L. (1972). *Primate evolution: an introduction to Man's place in Nature*. Macmillan, New York.
Simons, E. L. (1976). The nature of the transition in the dental mechanism from pongids to hominids. *J. Hum. Evol., 5*: 511–528.
Simons, E. L. and Pilbeam, D. R. (1972). Hominoid paleoprimatology. *In* R. Tuttle (ed.), *The functional and evolutionary biology of primates*, pp. 36–62. Aldine, Chicago, IL.
Smith, B. H. (1986). Dental development in *Australopithecus* and early *Homo*. *Nature, 323*: 327–330.
Sperber, G. H. (1985). Comparative primate dental enamel thickness: a radiodontological study. *In* P. V. Tobias (ed.), *Hominid evolution: past, present and future*, pp. 443–454. Alan R. Liss, New York.
Szalay, F. S. (1972). Hunting–scavenging protohominids: a model for hominid origins. *Man, 10*: 420–429.

Vrba, E. S. (1981). The Kromdraai australopithecine site revisited in 1980: recent investigations and results. *Ann. Tvl. Mus., 33*: 17–60.

Vrba, E. S. and Grine, F. E. (1978a). Analysis of South African australopithecine enamel prism patterns. *Proc. Electron Microsc. Soc. Sthn. Afr., 8*: 125–126.

Vrba, E. S. and Grine, F. E. (1978b). Australopithecine enamel prism patterns. *Science, 202*: 890–892.

Zonneveld, F. W. and Wind, J. (1985). High-resolution computed tomography of fossil hominid skulls: a new method and some results. *In* P. V. Tobias (ed.), *Hominid evolution: past, present and future,* pp. 427–436. Alan R. Liss, New York.

Growth of Teeth and Development of the Dentition in *Paranthropus*

2

M. CHRISTOPHER DEAN

Hard tissues are unique in that they preserve the secretory rhythms of the cells that form them. Dental hard tissues are especially unique because, for the most part, they are not turned over during the life of an individual. Enamel preserves the best incremental record of its formation because it mineralizes very quickly after secretion.

Experiments using pigs and dogs have confirmed a circadian rhythm during both enamel (Mimura, 1939) and dentine formation (Yilmaz et al., 1977), and there is overwhelming circumstantial evidence that points to a daily rhythmic slowing of matrix secretion during the formation of enamel in modern humans (Asper, 1916; Komai, 1942; Boyde, 1963, 1964, 1976). Enamel prisms are formed with alternating constrictions and varicosities along their lengths which reflect this daily rhythm. Enamel cross-striations are another manifestation of this rhythm visible when ground sections of teeth are viewed with light microscopy. Regular density variations in atomic number contrast along the length of the enamel prisms have the same periodicity as cross-striations (Boyde, 1976; Boyde and Jones, 1983) and so demonstrate that cross-striations are not likely to be optical artifacts. Scanning electron microscopy (SEM) of replicas of fractured surfaces of fossil hominid teeth occasionally reveal alternating enamel prism varicosities and constrictions (Fig. 2.1), and when these can be counted over the whole of an axially fractured surface it is possible to estimate the time of enamel formation in these teeth (Beynon and Dean, 1987). When more than one tooth from a single individual can be sectioned for light microscopy, counts of cross-striations can be continued from one growing tooth to another and used to estimate an age at death (Boyde, 1963, 1964).

Other incremental features are easily seen with light microscopy but rarely with SEM. These are Brown Striae of Retzius, and they are commonly 7, 8 or 9 cross-striations apart in primates (Gysi, 1939; Fukuhara, 1959; Newman and Poole, 1974; Bullion, 1986, 1987; Beynon and Reid, 1987; Dean 1987a) but it seems each individual (at least in modern and archeological populations of humans), demonstrates a constant periodicity for all teeth (Bullion, 1987). This fact, together with data from other studies that have demonstrated that the pattern of prominent Striae of Retzius can be matched in both contralateral teeth (Gustafson, 1959; Gustafson and Gustafson, 1967) as well as in teeth from different parts of the same mouth (Fujita, 1939), strongly suggests an underlying systemic cause for Striae of Retzius. No clear cause has been identified but many other similar rhythms have been documented in modern humans (see Dean, 1987a, for a review of the literature).

Tooth Growth and Development in *Paranthropus*

Beynon and Dean (1987) have made use of enamel varicosities that can be seen on replicas of fractured surfaces of teeth of *Paranthropus boisei* and Striae of Retzius that can be seen on the same original specimen when it is viewed in polarized light under alcohol. Indeed, it is likely that in *P. boisei,* where there is little decussation of the enamel prisms (Dean, 1987a) and where natural fractures of enamel occur for great distances along prisms, better estimates of crown formation time can be obtained by counting daily varicosities on fractured surfaces than could

FIGURE 2.1. SEM of an epoxy replica of specimen KNM-ER 733D, an upper P4, showing enamel prisms about 6 μm wide with alternating varicosities and constrictions along their lengths. In this field the repeat interval between varicosities is between 5 and 6 μm. The original magnification of the micrograph was 500 times.

be made from using sections of the original teeth. A time for the period of enamel formation in a premolar tooth of *P. boisei* (KNM-ER 733D) of 2.4 years maximum has been calculated by counting both the daily and near weekly incremental features in the enamel. This time for premolar crown formation matches estimates for molar crown formation times in *P. boisei* closely (Beynon and Wood, 1987).

This important finding reveals that not only had the premolars in *Paranthropus* become molarized in their morphology but also more than likely in their crown formation times, since premolars in great apes and modern humans take longer to form (between 3.3 and 4.75 years) than molars (between 2.2 and 3.0 years). The mechanism by which rapid crown formation is achieved in *Paranthropus* involves many more ameloblasts in tooth formation at any one time than are active in *Homo sapiens* (Dean and Wood 1981; Beynon and Wood 1987; Grine and Martin, this volume, chapter 1). An important consequence of this is that in *P. boisei* the bulk of the outline of the crowns of molars and premolars is formed quickly giving the impression on radiographs that the tooth crown is nearly completely formed long before it really is. In fact the last half of crown formation time may involve only 10% of the area of enamel in a cross-section of the crown (Beynon and Wood, 1987). This means that the conventional stages of crown formation used in human radiographic studies must be interpreted with caution in studies of fossil hominids.

In addition to providing a time for premolar crown formation, the study of Beynon and Dean (1987) demonstrated that the mean daily incremental rate of enamel formation in the cervical part of this premolar tooth (KNM-ER 733D) was divisible into the mean rate measured between Retzius lines by 6.98. Further confirmation for this relationship is evident in the molar tooth of this specimen (KNM-ER 733A), as 7 varicosities can also be counted between Retzius lines in

FIGURE 2.2. Oblique view of the buccal surface of an epoxy replica of specimen KNM-ER 1804, an upper P4, in a maxillary fragment attributed to *Paranthropus boisei*. The view demonstrates the fault planes associated with the striae of Retzius. Their continuity with perikymata at the surface of the tooth is easily seen in this oblique orientation. Each "step," that represents the thickness of enamel between two adjacent Retzius lines, is between 30 and 40 μm deep. The original magnification of the micrograph was 50 times.

places on a replica of the specimen where both varicosities and Retzius lines can be identified together using SEM (Dean, 1987a). This strongly suggests that *P. boisei* was unlikely to have differed significantly from modern humans in this important repeat interval and that a similar range of values is also to be expected within the sample of fossil teeth attributed to *Paranthropus*.

When fossil teeth are not naturally fractured, other nondestructive methods must usually be employed to estimate periods of enamel formation. Perikymata, which represent the surface manifestations of Striae of Retzius (Gysi, 1939), can be seen on many fossil hominid teeth (Fig. 2.2) and these have been used to calculate incisor crown formation times (Bromage and Dean, 1985; Dean *et al.*, 1986). Incisor enamel thickness is more uniform than enamel in other teeth and incisors also begin to form closer to birth than other teeth that might otherwise be used to estimate an age at death. A combination of methods used to estimate crown formation times can now be used to cross-check times for other tooth crowns formed over the same period of time. Beynon and Wood (1987) have estimated from daily increments in fractured molar teeth of *P. boisei* that crown formation time, in their opinion, was rapid (about 2.2 to 2.5 years). Dean (1987b) has confirmed that this agrees well with estimates of age at death for juvenile specimens of *P. boisei* (made by counting perikymata on incisors) where the M1 is also just completed in the same specimens. The central incisor crowns and the first permanent molar crowns can be seen from radiographs to have completed their formation together in KNM-ER 812, KNM-ER 1477, KNM-ER 1820 and probably also SK 3978 (Dean, 1987b; Skinner and Sperber, 1982). When allowance for small amounts of root formation are made on some of these teeth and the fact that

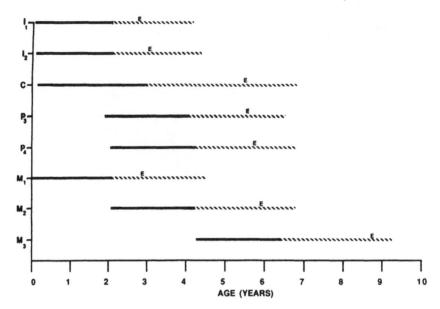

FIGURE 2.3. Preliminary chart depicting the dental developmental sequence of the combined upper and lower permanent dentition in *Paranthropus*. The chart is based upon mean times of crown formation for incisors, molars and premolars published elsewhere (see text) and the radiographic appearance of many immature specimens illustrated in Skinner and Sperber (1982), Mann (1975), and Dean (1987b). Information about the timing of premolar and canine crown completion and root initiation relative to incisor root completion has been derived from specimen SKX 162 and descriptions of this specimen have also been used to compile the chart [see text and Grine (this volume, chapter 14)].

the molars begin to form at or just before birth and incisors a month or so afterward, the agreement of these ages in cross-checked crown formation times is remarkable.

Reliable crown formation times are also invaluable for studies that concentrate on the sequence or pattern of dental development. Figure 2.3 represents the probable sequence of dental developmental events in *Paranthropus*. It is likely that some differences exist between South and East African species of *Paranthropus,* but these are not yet properly documented and Figure 2.3 is intended as a preliminary working model. Mean known values for incisor, premolar and molar crown formation times (Bromage and Dean, 1985; Beynon and Wood, 1987; Beynon and Dean, 1987) have been used to construct this chart. Root formation times are estimates based on the appearance of roots and the coinciding stages of tooth formation of all the permanent teeth in *Paranthropus* illustrated in or documented by Mann (1975), Skinner and Sperber (1982), Smith (1986) and Dean (1987b). Both crown and root formation times in the canine are estimates, but a new juvenile maxilla attributed to *Paranthropus* from Swartkrans (SKX 162) and two incisors associated with it (SKX 308 and SKX 1788) (see Grine, this volume, chapter 14) provide important information about this. The specimen demonstrates that the upper central incisors, although missing, had completed root formation (the empty sockets being 16.5 mm deep). At the same time there was 6.5 mm of root on the developing canine root and 3 or 4 mm of root on the developing premolar roots. This specimen also has been used to construct part of the chart in Fig. 2.3.

It is clear from Fig. 2.3 that this sequence of developing teeth does in fact provide some data that support a rapid rate of dental development and by implication general growth and development in *Paranthropus* (Smith, 1986). It would not, for example, be usual to find a normal human child with first permanent molars in occlusion and first premolars only 1 year or so into their formation period. Human first premolars are some 4 years into their development when the first permanent molars come into occlusion.

Some recent studies of dental development in early hominids have emphasized advanced permanent incisor formation or eruption relative to the first permanent molar in *Paranthropus* and early and modern *Homo* as compared to *Australopithecus* or extant great apes (Dean, 1985, 1987b; Smith, 1986; Bromage, 1987). It more than likely follows that teeth that complete their crowns and begin to develop roots at the same time will begin to erupt and emerge into the mouth close to each other, within the time span of the total growth period. It is, however, clear that calcification sequences of teeth rather than eruption sequences are a more objective way of documenting different patterns of tooth development (Garn and Lewis, 1963). Grine (1987) has questioned a clear distinction in incisor and first permanent molar eruption sequence between (and within) groups of early hominids and has also correctly drawn attention to the developing but unerupted permanent incisors of SK 61. Nevertheless, Figs. 2.4 and 2.5, which illustrate the calcification stages of the mandibular first permanent molars and lower central incisors in several juvenile hominids (described in Mann, 1975; Skinner and Sperber, 1982; Dean, 1985; and Dean 1987b), emphasize a distinction between *Australopithecus* and the great apes, on the one hand, and *Homo* and *Paranthropus*, on the other. In the future, new data about the growth of fossil hominid teeth will reduce the need to refer to patterns of dental development as "*Homo*-like" or "great apelike", and it is clear even now that while many early hominid taxa share patterns in common with other taxa, they are, in other ways, distinct in their own right.

These studies of incisor and first permanent molar calcification (or eruption) sequence do not provide any information about rates of general growth and development. They point only to shifts in rate of formation that have occurred between two tooth types. As yet there are no good data available about molar crown formation times in *Australopithecus* or *Paranthropus* in South Africa but it appears likely that greatly shortened incisor crown formation periods are the major underlying cause of the calcification sequence observed in *Paranthropus*. The eruption sequence of teeth is, futhermore, intimately associated with the growth of the jaws and the space available for permanent teeth to emerge into the mouth. The growth mechanisms that underlie the flat retruded face in *Paranthropus* and *Homo* differ from those that contribute to the more prognathic face of great apes and *Australopithecus*. Dean (1987c) has postulated that a combination of smaller anterior teeth and an anterior dental arch that is likely to be more spacious earlier in the growth period may well be one important contributory factor underlying the early emergence of incisors in *Paranthropus*.

Some Possible Taxonomic Implications

A knowledge of crown formation times can be used to provide an age at death for certain juvenile hominid specimens (Bromage and Dean, 1985; Dean *et al.*, 1986; Dean, 1987b). This means that reliable comparisons can now be made between, for example, the size of the mandible at different stages of the growth period. (In reality, because of the paucity of other material, mandibles are probably the only bone for which it is possible to present fossil hominid growth data.) Cross-sectional area of the mandible is an excellent measure of size. With some careful reconstruction to restore a more reliable outline to the mandibular corpus in a few individuals where bone has been eroded or damaged, area measurements can be made from contact prints of sections cut through high resolution silicone molds of the mandibular corpora of eight South and East African specimens of *Paranthropus*. However, because the size and shape of the juvenile mandible changes rapidly during the growth period (and is also markedly different at M1 and dm2), it is better to use a mean value of the areas at the mesial aspect of M1 and the mesial aspect of dm2 as a measure of overall size. Figure 2.6 illustrates sections of these mandibles arranged according to age. One specimen, KNM-ER 1820, is twice the size of two other East African specimens (KNM-ER 1477 and KNM-ER 812), that are of near identical age (Dean, 1987b). Indeed, this individual exceeds all other known juvenile hominids attributed to *Australopithecus, Paranthropus* or early *Homo* in its mean mandibular cross-sectional area at M1

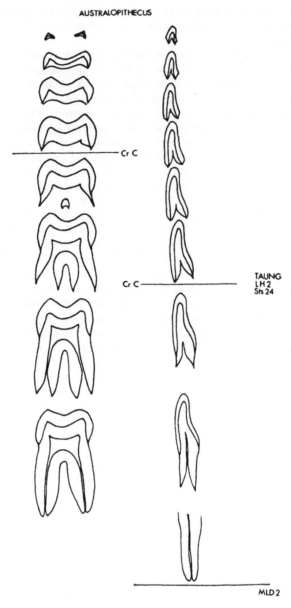

FIGURE 2.4. Diagram illustrating the coinciding calcification stages of mandibular first permanent molars and lower central incisors in great apes and *Australopithecus*. The approximate stages of development of certain key specimens are indicated. Maxillary specimens are not included in the diagram but follow the same pattern (e. g., LH-3, LH-6, LH-21, Taung, Sts 24, AL 333-86).

and dm2 (Dean and Brett, in prep.). Sexual dimorphism is unknown in hominoids as young as 2.5 to 3 years of age. In fact, it cannot be detected in gorilla mandibles until adolescence and even the adult male and female ranges overlap considerably. Neither does the range of cross-sectional areas in an aged series of gorilla mandibles include individuals that are twice the size of others of a similar age until adolescence (Dean and Brett, in prep.). (Interestingly, it has also been noted that KNM-ER 1820 stands out as having the largest talonid area of the M1 in a study

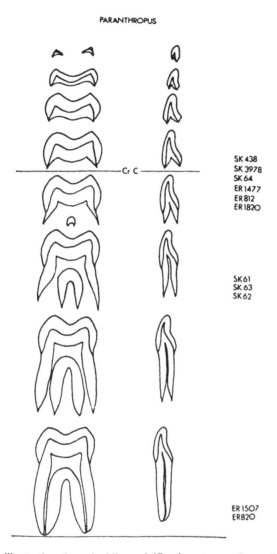

FIGURE 2.5. Diagram illustrating the coinciding calcification stages of mandibular first permanent molars and lower central incisors in *Paranthropus* and early *Homo*. The approximate stages of development and probable order of dental maturity of key specimens are indicated on the right hand side of the diagram.

of cusp areas in *P. boisei*) (Wood *et al.*, 1983). While these data set KNM-ER 1820 apart from other specimens of *Paranthropus*, it is clear that there are very few juvenile specimens from East Africa with which to make comparisons and that great caution about the possible implications of one very large specimen is warranted. Nevertheless, these data for juvenile hominid mandibles suggest that similar measurements for adult specimens might usefully be reexamined.

Walker and Leakey (this volume, chapter 15) noted that in East Africa both large and small mandibles attributed to *Paranthropus* are present throughout the whole geological time span of the genus, and so are not indicative of an increase in size through time. This is confirmed by the geochronological ages presented by Brown and Feibel (this volume, chapter 21) for hominid mandibles in East Africa. It is also generally accepted that the larger crania such as KNM-WT 17000, KNM-ER 406 and OH 5 are compatible in size with specimens such as KNM-ER 729 and

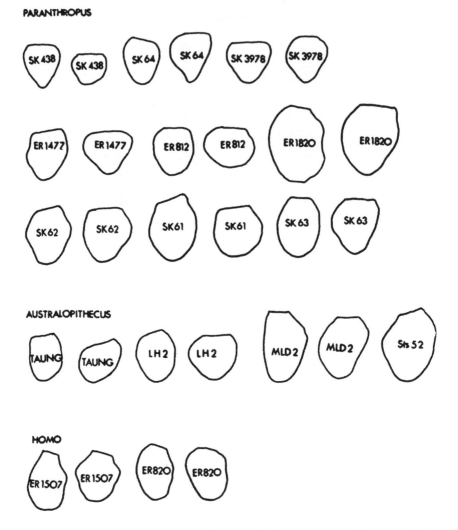

FIGURE 2.6. Mandibular cross-sectional outlines, made through the mesial aspect of the M1 and the mesial aspect of dm2 (excluding supra-alveolar tooth crowns), of eight specimens attributed to *Paranthropus*. Five specimens attributed to *Australopithecus* and two attributed to early *Homo* are included for comparison. Both sections are illustrated for each specimen, the left hand one of each pair being at dm2 or P4 and the right hand one of the pair being at M1. Only one section at M1 is illustrated for specimen Sts 52, which is sub-adult. Specimen KNM-ER 1820 has a mean cross-sectional area of 512 mm^2 and is twice the size of either KNM-ER 812 or KNM-ER 1477 (256 and 255 mm^2, respectively), which are of nearly identical age (Dean, 1987b).

KNM-ER 3230 (Walker *et al.*, 1986). Walker and Leakey (this volume, chapter 15) have also noted that smaller mandibles are now being recovered that are likely to be compatible in size with smaller crania such as KNM-ER 732 and KNM-ER 407. However, Chamberlain and Wood (1985) have presented data for mandibular cross-sectional area in a large sample of adult specimens attributed to *Paranthropus* and have identified four extremely large "male" specimens (Omo 7, KNM-ER 818, KNM-ER 1468, and KNM-ER 1469) that far exceed KNM-ER 729 and KNM-ER 3230 in size. This suggests that the crania associated with these mandibles would have been even bigger than KNM-WT 17000, KNM-ER 406 or OH 5. If this proves to be so, then these crania would be unreasonably large "males" to match with "female" crania such as KNM-ER 407 or KNM-ER 732. While it is generally accepted, and highly likely, that a sexually dimorphic species of *Paranthropus* existed in East Africa, an alternative hypothesis which is compatible with the data presented here for both juvenile and adult mandibles, would be that there existed alongside this dimorphic species, another larger species of *Paranthropus* of which KNM-ER 1820 is a juvenile representative. Should KNM-WT 17000 and other specimens such as the Omo 18-1967-18 mandible prove on morphological grounds to be distinct from *P. boisei*, then this would probably indicate the case for a third species of *Paranthropus* in East Africa.

Conclusions

With the notable exception of Taung (and descriptions of deciduous teeth), juvenile specimens have not until recently figured prominently in studies of early fossil hominids. Comparative studies of tooth growth and dental development are changing this. Many of the features of both individual tooth growth and general dental growth and development in *Paranthropus* appear to be occurring more quickly than in modern humans and also somewhat faster than is usual for modern great apes (although not necessarily faster than in fossil apes). It is particularly notable that premolars seem to be forming their crowns in the same period of time that molars do and that incisor crowns are forming relatively quickly in *Paranthropus*. Information about the likely age at death of juvenile specimens means that they can be incorporated into studies of hominid phylogeny more reliably than before. In particular, data about juvenile mandibular cross-sectional areas presented here suggest that it would be wise to re-examine the premise that all variation in size among the East African specimens of *Paranthropus* can be attributed solely to sexual dimorphism and or an increase in size through time.

Acknowledgments

I am grateful to the Governments of Kenya and Tanzania and to the Governors and Director of the National Museums of Kenya and the Transvaal Museum, South Africa for granting me permission to study fossil hominid material in their care. I am especially grateful to Fred Grine for inviting me to participate in the "robust" australopithecine workshop at Stony Brook, New York. This study was made possible by grants from the Wellcome Trust, The Nuffield Foundation, The Leakey Trust, The Boise Fund and The Royal Society. I am particularly grateful to Fred Brett for time and help in preparing the mandibular data presented here.

References

Asper, von. H. (1916). Uber die "Braune Retzinusgsche Parallelstreifung" im Schmelz der menschlichen Zahne. *Schweiz Vierteljahrschr. Zahnheilkunde, 26*: 275–214.
Beynon, A. D. and Dean, M. C. (1987). Crown formation time of a fossil hominid premolar tooth. *Arch Oral Biol., 32*: 773–780.
Beynon, A. D. and Reid, D. J. (1987). Relations between perikymata counts and crown formation times in the human permanent dentition. *J. Dent. Res., 66*: 889.
Beynon, A. D. and Wood, B. A. (1987). Patterns and rates of enamel growth in the molar teeth of early hominids. *Nature, 326*: 493–496.
Boyde, A. (1963). Estimation of age at death of young human skeletal remains from incremental lines in dental enamel. *Proc. 3rd Intern. Congr. Forensic Immunol., Med., Pathol., Toxicol., London. Excerpta Med. Intern. Congr. Ser., 80*: 36–46.

Boyde, A. (1964). The structure and development of mammalian enamel. Ph.D. Thesis, University of London.
Boyde, A. (1976). Amelogenesis and the development of teeth. *In* B. Cohen and I. R. H. Kramer (eds.), *Scientific foundations of dentistry*, pp. 335–352. Heinemann, London.
Boyde, A. and Jones, S. J. (1983). Backscattered electron imaging of dental tissues. *Anat. Embryol., 168*: 211–226.
Bromage, T. G. (1987). The biological and chronological maturation of early hominids. *J. Hum. Evol., 16*: 257–272.
Bromage, T. G. and Dean, M. C. (1985). Re-evaluation of the age at death of Plio-Pleistocene fossil hominids. *Nature, 317*: 525–528.
Bullion, S. K. (1986). Information from teeth on the growth and developmental history of individuals. *In* E. Cruwys and R. A. Foley (eds.), *Teeth and anthropology*, pp. 133–136. B.A.R. International series 291, Cambridge University Press.
Bullion, S. K. (1987). The incremental structures within tooth enamel and their application in archeology. PhD. Thesis, University of Lancaster.
Chamberlain, A. and Wood, B. A. (1985). A reappraisal of variation in hominid mandibular corpus dimensions. *Amer. J. Phys. Anthropol., 66*: 399–405.
Dean, M. C. (1985). The eruption pattern of the permanent incisors and first permanent molars in *Australopithecus (Paranthropus) robustus*. *Amer. J. Phys. Anthropol., 67*: 251–259.
Dean, M. C. (1987a), Growth layers and incremental markings in hard tissues; a review of the literature and some preliminary observations about enamel structure in *Paranthropus boisei*. *J. Hum. Evol., 16*: 157–172.
Dean, M. C. (1987b). The dental developmental status of six juvenile fossil hominids from East Africa. *J. Hum. Evol., 16*: 197–213.
Dean, M. C. (1987c). Rates of tooth growth and dental development in hominoids. *Primate Eye, 32*: 6–7.
Dean, M. C. and Wood, B. A. (1981). Developing pongid dentition and its use for aging individual crania in comparative cross-sectional growth studies. *Folia Primatol., 36*: 111–127.
Dean, M. C., Stringer, C. B. and Bromage, T. G. (1986). A new age at death for the Neanderthal child from Devil's Tower, Gibraltar and the implications for studies of general growth and development in Neanderthals. *Amer. J. Phys. Anthropol., 70*: 301–309.
Fujita, T. (1939). Neue Feststellungen uber Retzius'schen Parallelstreifung des Zahnschmelzes. *Anat. Anz., 87*: 350–355.
Fukuhara, T. (1959). Comparative anatomical studies of the growth lines in the enamel of mammalian teeth. *Acta Anat. Nipp., 34*: 322–332.
Garn, S. M. and Lewis, A. B. (1963). Phylogenetic and intra-specific variations in tooth sequence polymorphisms. *In* D. R. Brothwell (ed.), *Dental anthropology*, pp. 53–73. Pergamon, New York.
Grine, F. E. (1987). On the eruption pattern of the incisors and first permanent molars in *Paranthropus*. *Amer. J. Phys. Anthropol., 72*: 352–359.
Gustafson, A-G. (1959). A morphological investigation of certain variations in the structure and mineralization of human dental enamel. *Odont. Tidskr., 67*: 361–472.
Gustafson, G. and Gustafson, A.-G. (1967). Microanatomy and histochemistry of enamel. *In* A. E. E. Miles (ed.), *Structural and chemical organisation of teeth*, vol. 2, pp. 135–162. Academic Press, New York.
Gysi, A. (1939). Metabolism in adult enamel. *Dent. Dig., 37*: 661–668.
Komai, S. (1942). A study of parallel lines in the enamel rods of *Homo*. *Kobyo-shi, 16*: 280–292.
Mann, A. E. (1975). Some paleodemographic aspects of the South African australopithecines. University of Pennsylvania Publications in Anthropology, Philadelphia, PA.
Mimura, F. (1939). The periodicity of growth lines seen in enamel. *Kobyo-shi, 13*: 454–455.
Newman, H. N. and Poole, D. F. G. (1974). Observations with scanning and transmission electron microscopy on the structure of human surface enamel. *Arch. Oral Biol., 19*: 1135–1143.
Skinner, M. A. and Sperber, G. H. (1982). *Atlas of radiographs of early man*. A. Liss, New York.

Smith, H. B. (1986). Dental development in *Australopithecus* and early *Homo*. *Nature, 323*: 327–330.
Walker, A., Leakey, R. E., Harris, J. M. and Brown, F. H. (1986). 2.5-Myr *Australopithecus boisei* from west of Lake Turkana, Kenya. *Nature, 322*: 517–522.
Wood, B. A., Abbott, S. A. and Graham, S. H. (1983). Analysis of the dental morphology of Plio-Pleistocene hominids. *J. Anat., 137*: 287–314.
Yilmaz, S., Newman, H. N. and Poole, D. F. G. (1977). Diurnal periodicity of von Ebner growth lines in pig dentine. *Arch. Oral Biol., 22*: 511–513.

Implications of *In Vivo* Experiments for Interpretating the Functional Significance of "Robust" Australopithecine Jaws

3

WILLIAM L. HYLANDER

There is a general consensus that the craniofacial region of "robust" australopithecines was especially adapted for the generation and dissipation of powerful masticatory forces (e.g., Robinson, 1956; Leakey, 1959; Tobias, 1967; Crompton and Hiiemae, 1969; Jolly, 1970; Pilbeam, 1972; Wolpoff, 1973; Simons, 1972; DuBrul, 1977; Hylander, 1979a; Grine, 1981, 1984; Rak, 1983, this volume, chapter 12; Kay and Grine, this volume, chapter 26). Recent experimental work on muscle-activity patterns, loading patterns, and jaw movements in primates provides a better indication as to the possible nature of these masticatory muscle and reaction forces. This chapter will focus on the implications of these experimental data for interpreting the morphology of "robust" australopithecine jaws.

This paper is based on two major assumptions: First, if the "robust" australopithecine face was indeed adapted for the generation and dissipation of powerful forces, then the size and shape of the mandibular corpus and symphysis reflect a structural adaptation to resist stress generated during the power stroke of unilateral postcanine biting or mastication. It is unlikely that these regions were especially adapted to resist stress generated during incisor or canine biting because these teeth are greatly reduced in size, and presumably in functional importance (cf., Jolly, 1970; Hylander, 1975). For similar reasons, it is unlikely that the enlarged dimensions of the symphysis and corpus were due to the requirements of enclosing the mandibular canine root (cf., Chamberlain and Wood, 1985). Moreover, there is no reason to suspect that these regions were especially adapted to counter stress during other behaviors such as jaw opening, swallowing, respiration, licking, etc. (cf. Hylander, 1984). The second assumption is that an *in vivo* stress analysis of the mandible of *Macaca fascicularis* (Hylander, 1979b,c, 1981, 1984 1985; Hylander and Crompton, 1986; Hylander *et al.*, 1987) is a useful model for understanding how the australopithecine mandible was stressed during the power stroke of mastication and unilateral isometric biting. This assumption is based on the overall similarities of the anatomy and physiology of the chewing apparatus of all catarrhine primates (cf. Hiiemae, 1978).

Patterns of stress and strain for both the mandibular symphysis and corpus in *M. fascicularis* will be briefly reviewed prior to discussing the implications of these data. Based on the first working assumption this review will be limited to only those stress patterns associated with the power stroke of mastication and unilateral isometric molar biting. Although several loading regimes occur along the corpus and symphysis during these periods, not all of these regimes will be discussed. Instead, attention will be focused upon various bending and twisting regimes because it is under these conditions that maximum stress concentrations ordinarily occur and structural adaptations will be required for mandibles routinely exposed to elevated levels of stress (Hylander, 1979a, 1985; Bouvier and Hylander, 1981.)[1]

---

[1]An obvious exception is that there can also be structural adaptations to large local stress concentrations due to direct shear or axial loads. For example, thick enamel among higher primates may be a structural adaptation to counter powerful loads associated with the bite force (Kay, 1981). Unfortunately there are no *in vivo* data available that deal with these important local effects on teeth or bone.

Patterns of Stress along the Mandibular Corpus During Mastication and Unilateral Biting

The Working Side

In vivo experimental stress analysis data demonstrate that the working-side mandibular corpus in the molar region is primarily twisted about its long axis, directly sheared dorsoventrally, and slightly bent in both the parasagittal and transverse planes during the power stroke of both mastication and unilateral molar biting (Hylander, 1979b, 1981; Hylander *et al.*, 1987). Of these stress regimes, twisting (or torsion) of the mandibular corpus about its long axis is apparently the most important for causing the occurrence of large stress concentrations.

Twisting occurs because the muscle and reaction forces do not all act through the long axis of the mandibular corpus. That is, because the resultant adductor muscle force lies lateral to the mandibular corpus, there is a tendency to evert the lower border of the mandible and invert the alveolar process in the molar region during both mastication and unilateral biting (Fig. 3.1A). Moreover, the position of the bite point and the direction of the bite force also influence the twisting pattern (Hylander, 1979a,b; Demes *et al.*, 1984).

Twisting of the mandibular corpus about its long axis results in maximum torsional stress along its outer (periosteal) surface. Instead of being evenly distributed, the torsional stress is maximal along the outermost aspect of the corpus that is closest to the twisting axis of neutrality. Torsional stress is maximal and concentrated along the periosteal surface of the medial and lateral aspects of the corpus because the transverse thickness of the corpus is less than its vertical depth (Fig. 3.1B). The ability of the mandibular corpus to counter this stress is directly proportional to its polar moment of inertia (Arges and Palmer, 1963), and the most effective way to increase this polar moment is to deposit additional cortical bone along the periosteal surface in the area of maximum torsional stress, i.e., create a transversely thicker corpus (Hylander, 1979a,b).

The Balancing Side

There are major differences in patterns of stress between mastication and unilateral biting for the balancing side. Stress analysis data demonstrate that the balancing-side mandibular corpus in the molar region is primarily twisted about its long axis, directly sheared dorsoventrally, and

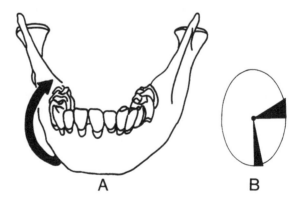

FIGURE 3.1. Twisting of the mandibular corpus. The arrow in (A) indicates the manner in which the resultant muscle force of the jaw adductors twists the mandibular corpus. The darkly shaded areas in (B) indicate the distribution of torsional stress along an oval section. Torsional stress is absent along the twisting axis of neutrality (the black dot), and it is maximal along the outer surface of the section closest to the twisting axis of neutrality.

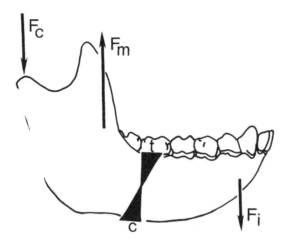

FIGURE 3.2. Bending of the balancing-side mandibular corpus in the parasagittal plane. F_c, vertical component of the condylar reaction force, F_m, vertical component of the jaw adductor muscle force; F_i, vertical component of the internal force transmitted across the symphysis from balancing to working side; t, tensile bending stress; and c, compressive bending stress. The darkly shaded areas indicate the distribution of bending stress if the corpus was a symmetrical beam. This bending stress is maximal along the upper and lower borders of the mandibular corpus and it is absent along the neutral axis which is located approximately midway between these two borders.

powerfully bent in the parasagittal plane during the power stroke of both mastication and unilateral molar biting (Hylander, 1979b, 1981). The mandibular corpus is also bent laterally in the transverse plane ("wishboning") during mastication but not (or much less) during unilateral biting (Hylander, 1984; Hylander et al., 1987). These twisting, parasagittal bending, and wishboning loading regimes are important for causing relatively large stress concentrations in the mandibular corpus, although the strain data and theoretical considerations suggest that bending in the parasagittal plane is the most important for the balancing-side corpus (Hylander, 1979a,b). The above data do not support the assumption made by Demes et al. (1984) and Wolff (1984) that medial transverse bending of the mandibular corpus is a critical stress regime during the power stroke.

Similar to the situation on the working side, twisting along the balancing side occurs because the resultant adductor muscle force lies lateral to the mandibular corpus; thus, there is a tendency to evert the lower border of the mandible and invert the alveolar process in the molar region (Fig. 3.1A). Those forces that tend to bend the balancing-side mandibular corpus in the parasagittal plane during both mastication and unilateral biting are indicated in Fig. 3.2. The tendency for the balancing-side mandibular corpus to experience wishboning during mastication (Fig. 3.3) is due, at least in part, to the persistent activity of the balancing-side deep masseter and the simultaneous cessation of activity of the balancing-side superficial masseter and medial pterygoid muscles during the terminal portion of the power stroke (Hylander et al., 1987).

Unlike the conditions during twisting (Fig. 3.1B), parasagittal bending stress is concentrated along the upper and lower borders of the mandibular corpus in the molar region (Fig. 3.2). This bending stress is absent along the bending axis of neutrality, which is close to the region that experiences maximum twisting or torsional stress. The ability of the mandibular corpus to counter parasagittal bending stress is directly proportional to its second moment of inertia, and the most effective way to increase the second moment of inertia for this type of bending is to deposit additional cortical bone along the periosteal surface of the upper and lower borders of the mandibular corpus, i.e., create a vertically deeper mandible (Hylander 1979a,b). In contrast, wishboning bending stress is concentrated along the medial and lateral aspects of the mandibular corpus along the same regions that experience maximum torsional stress (Fig. 3.3). The most effective way to increase the second moment of inertia so as to counter an increase in bending

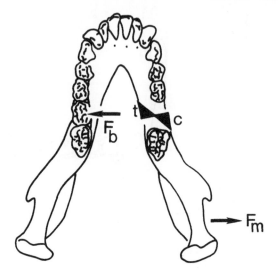

FIGURE 3.3. Bending of the balancing-side mandibular corpus in the transverse plane (wishboning). F_b, transverse component to the bite force; F_m, transverse component to the balancing-side jaw adductor muscle force; t, tensile bending stress; and c, compressive bending stress. The darkly shaded areas indicate the distribution of bending stress if the corpus was a symmetrical beam. Note that bending stress is maximal along the medial and lateral borders of the mandibular corpus and it is absent along the neutral axis which is located approximately midway between these two borders.

stress due to wishboning is to deposit additional bone along the periosteal surfaces of the medial and lateral aspects of the mandibular corpus (Smith, 1983; Kelley and Pilbeam, 1986), i.e., create a transversely thicker mandible. In summary, a vertically deep mandibular corpus is an efficient solution to counter parasagittal bending while a transversely thick mandibular corpus is an efficient solution to counter both twisting about its long axis and wishboning.

Patterns of Stress Along the Mandibular Symphysis During Mastication and Unilateral Biting

Patterns of symphyseal stress differ between the power stroke of mastication and unilateral biting. Stress analysis data indicate that during mastication the symphysis experiences dorsoventral shear, bending due to twisting of the corpora about their long axes, and lateral transverse bending (wishboning) (Hylander, 1984, 1985). Wishboning of the symphysis is probably the most important regime for causing the occurrence of large stress concentrations during mastication even though the horizontally aligned external forces causing it are small relative to the vertically aligned forces (Hylander et al., 1987). In contrast, wishboning of the symphysis is not a consistently dominant stress pattern during unilateral biting. Apparently the most important stress regimes during unilateral biting are twisting about a transverse axis and bending due to twisting of the mandibular corpora about their long axes (cf., Hylander, 1984, 1985).

Bending of the symphysis due to twisting of the mandibular corpora about their long axes (Figs. 3.1A and 3.4) results in maximum bending stress along its upper and lower borders. Twisting of the symphysis about a transverse axis, which occurs when the moments about the symphysis in the lateral projection are unequal (Fig. 3.5), results in torsional stress distributed similarly to torsional stress in the mandibular corpus, i.e., it is concentrated along the periosteal surface that is closest to the twisting axis of neutrality (Fig. 3.1B). Wishboning of the symphysis, which

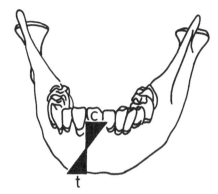

FIGURE 3.4. Bending of the mandibular symphysis due to twisiting of the mandibular corpus about is long axis (Fig. 3.1A). The darkly shaded areas indicate the distribution of bending stress if the symphysis was a symmetrical beam. Note that bending stress is maximal along the upper and lower borders of the mandibular symphysis. c, compressive bending stress and t, tensile bending stress.

occurs for the same reasons that the mandibular corpus is bent laterally in the transverse plane (Figs. 3.3 and 3.6), results in much more stress along the lingual or concave surface of the symphysis than there is along the labial or convex surface because the mandible functions as a sharply curved beam at this time (Hylander, 1984, 1985) (Fig. 3.6).

The ability of the mandibular symphysis to counter all of the above twisting and bending stresses is directly proportional to either its polar moment of inertia (for twisting) or its various second moments of inertia (for bending). The most effective way to increase the polar moment

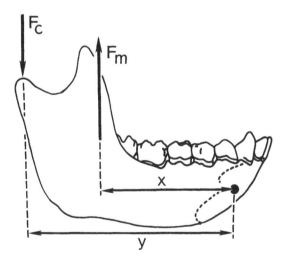

FIGURE 3.5. Twisting of the mandibular symphysis about a transverse axis (the black dot). F_c, vertical component of balancing-side condylar reaction force; F_m, vertical component of balancing-side jaw adductor muscle force; x, moment arm associated with F_m; y, moment arm associated with F_c. There are two moments acting about the center of the symphysis in this projection along the balancing side. Twisting about a transverse axis occurs when the clockwise moment ($F_m \cdot x$) is either larger or smaller than the counter-clockwise moment ($F_c \cdot y$). Torsional stress is absent along the twisting axis of neutrality (the black dot) and it is maximal along the outer surface of the section closest to the twisting axis of neutrality (Fig. 3.1B).

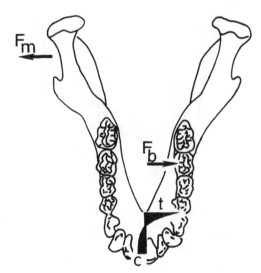

FIGURE 3.6. Bending of the mandibular symphysis in the transverse plane (wishboning). F_b, transverse component to the bite force; F_m, transverse component to the balancing-side jaw adductor muscle force; t, tensile bending stress; c, compressive bending stress. The darkly shaded areas indicate the distribution of bending stress if the symphysis was a symmetrical beam. Note that the distribution of bending stress is markedly nonlinear. Although bending stress increases as the distance from the neutral axis increases, the tensile stress along the outer surface of the lingual aspect greatly exceeds the compressive stress along the outer surface of the labial aspect of the symphysis.

of inertia for twisting about a transverse axis is to deposit additional bone along the periosteal surface in the area of maximal torsional stress, i.e., make the symphysis thicker in the labiolingual direction. In contrast, the most effective way to counter an increase in symphyseal bending due to twisting of the mandibular corpora about their long axes is to increase the vertical depth of the symphysis or deposit bone along the lower border of the symphysis so as to create a "simian shelf" (Hylander 1984).[2] Finally, the most effective way to counter wishboning of the symphysis is to deposit additional bone along the lingual periosteal surface in the region of the superior transverse torus, which also results in a symphysis that is thicker in the labiolingual direction (Hylander, 1984, 1985). In summary, a vertically deep symphysis or a symphysis with a "simian shelf" is an efficient solution to counter bending of the symphysis due to twisting of the mandibular corpora about their long axes, while a transversely thick symphysis is an efficient solution to countering both wishboning and symphyseal twisting about a transverse axis.

"Robust" Australopithecine Morphology

As noted earlier, the morphology of the teeth, jaws and face of "robust" australopithecines have led various workers to conclude that the chewing apparatus of these early hominids was especially adapted to generate and dissipate large forces during powerful postcanine biting and/or mastication. The purpose of this section is to analyze and discuss the functional significance of the mandibular corpus and symphyseal regions of "robust" australopithecines. Although there are two or more recognized species of "robust" australopithecines, only *Australopithecus boisei* will be analyzed because its masticatory apparatus is the most derived and highly specialized of this group. This section also includes an analysis of the mandibular corpus and symphysis of

---

[2]These two strategies are also efficient structural adaptations for countering symphyseal bending during incisal biting (Hylander, 1984), which is the dominant pattern of stress at this time.

Australopithecus afarensis, a species that is generally considered to be ancestral to all other australopithecines (Johanson and White, 1979).

Elsewhere I have argued that unlike limb–bone dimensions, there is no direct or predictable functional relationship between overall body size and variables that reflect the ability of the bony face to dissipate chewing or biting stress (Hylander, 1985). This is simply because primates do not transmit body weight or locomotor stress through their faces; thus, analyzing structural variables of the face relative to overall body size (cf., Smith, 1983) provides little insight into the biomechanics of facial design. A preferable strategy includes an evaluation of the presumed biomechanically related variables relative to conditions associated with masticatory or biting stress. For the mandibular corpus and symphysis this means that these variables should be analyzed relative to those conditions that result in maximal amounts of twisting and/or bending (cf., Hylander, 1979a, 1985; Bouvier, 1986a,b) because it is under these conditions that maximum levels of stress ordinarily occur.[3] The working hypothesis, then, is that symphyseal and corpus dimensions among closely related primates (e.g., catarrhines) scale according to the various bending and twisting moments acting about these regions, so that maximum levels of stress and strain in highly stressed regions are quite similar in both large and small forms (cf., Hylander, 1979a, 1985; Lanyon and Rubin, 1985; Biewener and Taylor, 1986). If this hypothesis is true, then we are faced with major problems when attempting to evaluate the functional significance of mandibular form because so little is known about the nature and morphological correlates of the various bending and twisting moments. These problems will become apparent in the following discussion.

The Mandibular Corpus

Corpus Depth. As noted above, the most efficient way to counter internal stress in the mandibular corpus due to an increase in balancing-side parasagittal bending is to deepen the corpus in the vertical direction. This is because the maximal amount of bending stress (σ) in an elliptical or rectangular section is directly proportional to the bending moment [$(F)(m)$] and inversely proportional to the relative section modulus (as approximated by a^2b), where F is the resultant vertical component of balancing-side muscle force, m is the moment arm associated with this force, a is the vertical depth of the corpus and b is its transverse thickness, i.e.,

$$\sigma \propto \frac{(F)(m)}{a^2 b}$$

Thus, an increase in the size of a results in a much greater decrease in the amount of bending stress than does the same increase in the size of b. One way to evaluate primate mandibular morphology is to plot the vertical depth of the corpus in the M_2 region relative to an estimate of the moment arm associated with the parasagittal bending moment acting about a vertical section in the M_2 region (Fig. 3.7A).[4] This moment arm is estimated by measuring the linear distance between the distal aspect of the M_2 and the tip of the most anterior incisor (Hylander, 1979a). Although this dimension is an accurate estimate of the moment arm acting about the M_2 region during incisor biting, it slightly overestimates the actual balancing-side moment arm during mastication (or molar biting).

Figure 3.8 is a plot of the mean values of the above described variables for 42 extant catarrhine species (males and females plotted separately), *A. boisei,* and *A. afarensis.* The composition of

---

[3] In certain regions of the craniofacial region (e.g. the upper jaw), direct shear stress, rather than twisting or bending, may be the predominant stress regime (Preuschoft *et al.,* 1986).

[4] Another approach is to plot the value of a^2b or, preferably, the actual second moment of inertia for parasagittal bending, relative to this bending moment arm. Doing the latter involves imaging procedures such as computed axial tomography or magnetic resonance techniques because it is impractical and undesirable to section the mandibles of a larger series of extant and extinct primates. Determining the actual second moment of inertia eliminates the need for making certain assumptions about the overall shape of a mandibular section (cf. Daegling, 1988).

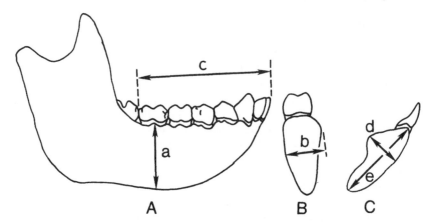

FIGURE 3.7. Mandibular dimensions. (A). Arch length (c) and corpus vertical depth (a). Arch length is the distance between the distal aspect of the M_2 and the labial aspect of the most anterior incisor projected on the midsagittal plane. Corpus depth is the minimum distance between the lower border of the mandible and the crest of the alveolar process opposite the middle of the M_2. (B). Transverse thickness of the corpus (b) is the maximum thickness of the corpus taken perpendicular to its vertical axis. (C). Symphyseal thickness (d) and depth (e). Symphyseal depth is the distance between the lower border of the symphysis and the crest of the alveolar process in the midline. Symphyseal thickness is the maximum distance between the labial and lingual aspects of the symphysis taken perpendicular to symphyseal depth.

the sample of extant catarrhines and fossil hominids for this and all succeeding figures is made up of only adults and is indicated in Appendices 3.1 and 3.2. The regression analysis for this and all succeeding figures is based only on extant species.

If the mandibular corpus is scaled according to parasagittal bending moments so that maximum levels of internal stress and strain remain the same in both large and small primates[5], then major

---

[5]The most convincing way to test this hypothesis is to compare maximum levels of *in vivo* mandibular bone strain during powerful mastication for a size-graded series of catarrhine primates. Unfortunately data such as these are currently unavailable. Another, although less persuasive way to test this hypothesis is to estimate relative levels of stress in the mandible within large and small closely related primates (cf., Hylander, 1985). This was done in the following way using the above formula. Mean mandibular corpus depth (a) and thickness (b) dimensions for *Miopithecus talapoin* and *Papio anubis* males were used to estimate the relative section modulus (a^2b) in the M_2 region for parasagittal bending. The moment arm acting about this region (m) was approximated by the M_2–I_1 dimension. A maximum muscle force (F) of 100 was arbitrarily assigned to *Miopithecus,* and the following procedure was used to estimate the maximum muscle force of *Papio.* Jaw muscle mass in anthropoids scales to the 1.0 power of body mass (Cachel, 1984). If muscle-fiber length also scales isometric to body mass (i.e., to the 0.33 power of body mass), then maximum muscle force scales to the 0.67 power of body mass. Since adult male *P. anubis,* which on average weigh about 28,000 grams, are about 24 times larger than male *Miopithecus* (Hill, 1970), then the maximum muscle force for *Papio* is about 8.4 times larger than it is for *Miopithecus*. Based on the above approximations of the relative section modulus, relative muscle force, and bending moment, the estimated levels of relative stress for these two species are rather similar with the baboon having about 13% less relative stress than the talapoin monkey (*Papio:* 4.2 units; *Miopithecus:* 4.8 units). Parenthetically it should be noted that if the dimensions of the mandibular corpus are indeed scaled according to bending moments, we would expect that the larger primates would have somewhat smaller estimates of relative stress based on the above method. This is because this method cannot account for allometric constraints on jaw muscle size and because larger primates include tougher and more fibrous foods in their diets. These two factors result in a relative increase in parasagittal bending of the balancing-side mandibular corpus because they are associated with a more frequent recruitment of maximal amounts of balancing-side jaw muscle force. Moreover, there is the distinct possibility that there is a significant error in the estimate of the relative section modulus between cercopithecine taxa (cf. Daegling, 1988).

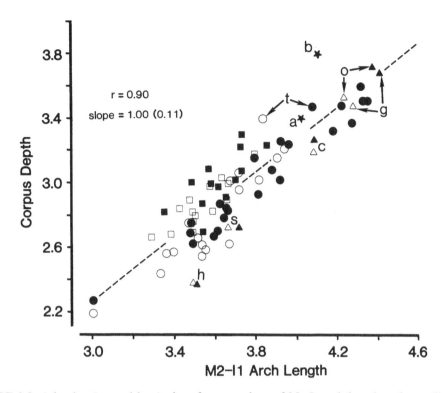

FIGURE 3.8. A log-log (natural logs) plot of mean values of M_2-I_1 arch length and mandibular corpus depth in the M_2 region for forty-two extant catarrhine species and two fossil hominid species. With the exception of the stars (fossil hominids), open and closed symbols are female and males, respectively. Squares, circles, and triangles are colobines, cercopithecines, and nonhuman hominoids, respectively. Stars labeled a and b are *Australopithecus afarensis* and *Australopithecus boisei*, respectively. Other abbreviations: t (gelada); g (gorilla), o (orangutan), c (chimpanzee), h (gibbon), and s (siamang). The dashed line indicates the least-squares regression line. Regression statistics are based solely on extant catarrhines. See Appendix 3.1 for details of complete sample.

deviations of data points above or below the regression line indicate a respective increase or decrease in muscle or reaction force because bending moments are a function of force multiplied by the moment arm (M_2-I_1 arch length). The plot clearly demonstrates that unlike *A. afarensis*, which falls slightly above the regression line, *A. boisei* has large vertical dimensions of the corpus (both relatively and absolutely), which suggests that the vertically directed muscle and/or reaction forces along the balancing side of *A. boisei* were unusually large. These large balancing-side forces were probably directly related to the mechanical properties of the foods eaten by *A. boisei*. That is, large masticatory muscle and reaction forces are associated with eating hard or tough foods. In addition to *A. boisei*, colobines also tend to have relatively deep jaws (Hylander, 1979a; Bouvier, 1986a), which suggests that they may have larger vertically directed muscle and reaction forces compared to other similar-sized catarrhine primates, whereas the unusually shallow jaws of gibbons suggest that their masticatory muscle and reaction forces are relatively small.

Table 3.1 presents the descriptive statistics of the regression analysis for extant males and females both combined and separated. The data in the table and Fig. 3.8 are relevant to hypotheses about the functional significance of mandibular corpus depth in the molar region. Following the observation that males have vertically deeper jaws than females (Kay *et al.*, 1981), it has been suggested that "perhaps it is the long canine roots of males that require great mandibular depth, simply so that they are enclosed within mandibular bone" (Smith, 1983: 327)(also cf., Wood, 1978; Chamberlain and Wood, 1985). There are at least two major problems with this interpretation. First, and most important, it is unclear why mandibular depth in the molar region of higher primates should bear any direct relationship to the housing of a long mandibular canine root, since the root does not extend back below the molar region. Second, because Kay *et al.* (1981) evaluated corpus depth relative to molar size, these data may simply indicate that males have relatively smaller teeth than females (cf., Goldstein *et al.*, 1978; Kay, 1978), rather than deeper jaws. Figure 3.8 demonstrates that relative to the parasagittal bending moment arm, there is no clear-cut division in jaw depth between males and females. Moreover, the regression analysis shown in Table 3.1 indicates that the scaling coefficients of males and females do not differ significantly from one another ($p > .05$). Thus, although males have deeper mandibles than females relative to molar size, they are very similar to one another when corpus depth is analyzed relative to the moment arm associated with parasagittal bending. Moreover, jaw depth among males and females scales in a very similar fashion, as would be expected if corpus depth scales proportional to parasagittal bending moments.

Corpus Thickness. The most efficient way to counter internal stress in the mandibular corpus associated with wishboning or twisting about its long axis is to increase the transverse thickness of the corpus. Attempting to evaluate transverse dimensions of the corpus in the molar region relative to twisting moments or wishboning bending moments is much more problematic than evaluating the corpus relative to parasagittal bending moments for a number of reasons. First, an estimate of twisting moment arms requires information about the resultant muscle-force position and direction and in some instances the bite-force point of application and direction (Hylander, 1979a; Demes *et al.*, 1984). Because so little is known about the nature of these forces, the moment arms for twisting about a section in the molar region cannot be estimated with any degree of confidence from the skeletal anatomy of extant primates. This problem is compounded when trying to approximate twisting moment arms from fragmentary fossil material. Second, an estimate of the wishboning bending moment arm about a section in the molar region requires information about the position of the resultant muscle force along the balancing side in the transverse or occlusal plane (Fig. 3.3). Although the linear distance between this section (e.g., in the M_2 region) and the posterior edge of the mandibular ramus may be a reasonable estimate of this moment arm for analyzing the effects of scale, this dimension is currently unavailable for the catarrhine series.[6] Thus, for a variety of reasons, it is not possible to perform a comparative analysis of transverse thickness of the corpus relative to transverse-bending and twisting moment arms so as to ascertain the nature of the muscle and reaction forces associated with these loading regimes. Instead, these transverse dimensions will be evaluated in the following manner.

Figure 3.9 is a plot of mean values of mandibular corpus vertical depth (Fig. 3.7A) versus transverse thickness in the M_2 region (Fig. 3.7B) for males of 28 extant catarrhine species, *A. afarensis*, and *A. boisei*. The cercopithecoid data are taken from Bouvier's (1986a) analysis of male cercopithecoids with the exception of *Procolobus verus*. (The data for *P. verus* and the hominoids can be found in Appendix 3.1.) The data in Figure 3.9 indicate that compared to male cercopithecoids, extant male hominoids have a relatively thick mandibular corpus. Smith's (1983) data indicate that this is also true for female hominoids. The data in Figure 3.9 substantiate the general impression that australopithecines have a relatively thick mandibular corpus compared to extant nonhuman catarrhines. Finally, this figure also demonstrates that both species of australopithecines have an absolutely thicker mandibular corpus than all other extant catarrhines.

The data presented here have no bearing on whether the unusually thick mandibular corpus of australopithecines represents a structural adaptation to counter bending stress during wishboning or torsional stress during twisting about its long axis. The morphology of their symphysis, however, indicates that it is more likely that their corpus morphology reflects a structural adaptation to powerful torsion and wishboning, rather than to wishboning alone. If so, then one important question is why was the australopithecine mandibular corpus, particularly the "robust" form,

---

[6]The distance between the M_2 and the posterior edge of the mandibular ramus slightly overestimates the moment arm associated with wishboning acting about the M_2 region of the mandibular corpus.

Table 3.1. Regression Analysis of Log-Transformed Mandibular Variables

Taxonomic group	Dependent variable	Independent variable	Least-squares slope (95% confidence interval)	Intercept	Reduced major-axis slope (estimate of 95% confidence interval)[a]	Correlation coefficient
Extant catarrhines (males and females)	Corpus depth	M_2–I_1 arch length	1.00(±0.11)	−0.76	1.11(±0.11)	0.90
Extant catarrhines (females only)	Corpus depth	M_2–I_1 arch length	1.00(±0.20)[b]	−0.92	1.17(±0.20)	0.88
Extant catarrhines (males only)	Corpus depth	M_2–I_1 arch length	0.96(±0.15)[b]	−0.60	1.07(±0.15)	0.90
Extant catarrhines (males only)	Corpus thickness	Corpus depth	0.92(±0.16)	−0.61	1.01(±0.16)	0.91
Extant hominoids (males only)	Corpus thickness	Corpus depth	0.86(±0.11)	−0.14	0.87(±0.11)	0.99
Extant catarrhines (males and females)	Symphyseal thickness	Symphyseal depth	1.04(±0.05)	−1.02	1.07(±0.05)	0.97
Extant catarrhines (females only)	Symphyseal thickness	Symphyseal depth	1.00(±0.09)[b]	−0.91	1.05(±0.09)	0.96
Extant catarrhines (males only)	Symphyseal thickness	Symphyseal depth	1.04(±0.07)[b]	−1.01	1.06(±0.07)	0.98
Cercopithecoids (males and females)	Symphyseal thickness	Symphyseal depth	1.13(±0.05)[c]	−1.29	1.15(±0.05)	0.98
Extant hominoids (males and females)	Symphyseal thickness	Symphyseal depth	0.92(±0.16)[c]	−0.69	0.94(±0.14)	0.98

[a] This estimate is the standard error of the slope multiplied by two.
[b] These paired slopes are not significantly different at the .05 confidence level.
[c] These paired slopes are significantly different at the .01 confidence level.

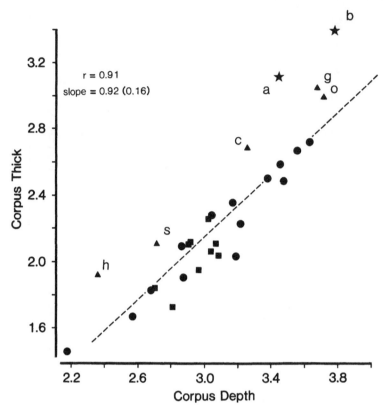

FIGURE 3.9. A log-log (natural logs) plot of mean values of mandibular corpus depth and thickness in the M_2 region for twenty-eight extant catarrhine species and two fossil hominid species. With the exception of the stars (fossil hominids), closed symbols are males. Squares, circles and triangles are colobines, cercopithecines and nonhuman hominoids, respectively. Stars labeled a and b are *Australopithecus afarensis* and *Australopithecus boisei*, respectively. Other abbreviations: g (gorilla), o (orangutan), c (chimpanzee), h (gibbon) and s (siamang). Regression statistics are based solely on extant male catarrhines. The dashed line indicates the least-squares regression line. See Appendix 3.1 for details of complete sample.

subjected to unusually large torsional or twisting moments during unilateral mastication? This question will be considered following an evaluation of the symphyseal region.

The catarrhine regression statistics in the table demonstrate that increased size is not necessarily associated with increased corpus "robusticity" (Smith, 1983). The data in the table indicate that the reduced major-axis slope for the regression analysis of corpus thickness vs. corpus depth for male catarrhines is not significantly different from 1.00. Moreover, for hominoids, the reduced major-axis slope is less than and significantly different from 1.00. This contrasts with Smith's analysis, which implies that corpus breadth scales positively allometric relative to corpus depth (cf., Chamberlain and Wood, 1985). Perhaps the inclusion of platyrrhines in Smith's (1983) study accounts for some of the differences in results simply because platyrrhines tend to have a less "robust" corpus than catarrhines (Bouvier, 1986b) and several platyrrhine species are much smaller than the smallest catarrhines. Moreover, some of the differences may be because Smith analyzed mandibular dimensions in the M_1 region of females whereas my measurements are taken in the M_2 region of males. A re-analysis of Smith's data on corpus thickness and corpus depth in the M_1 region indicates that for female catarrhines the reduced major-axis slope (1.06) is not significantly different from 1.00 (the estimate of the 95% confidence interval is ± 0.14). This is also true when only the hominoids are analyzed. (The reduced major-axis slope is 1.07 and the estimate of the 95% confidence interval is ± 0.13.) Therefore, since increased "robusticity" of the corpus is not necessarily associated with an increase in size, at least for catarrhines (and also among hominoids), there is no reason to accept the hypothesis that differences in the "robusticity" of the mandible between species of *Australopithecus* "are determined by the requirements of geometric scaling to body size rather than by differences in dietary adaptation" (Chamberlain and Wood, 1985: 404).

The Mandibular Symphysis

The most efficient way to counter symphyseal bending resulting from twisting of the mandibular corpora about their long axes is to increase the vertical depth of the symphysis or to add a simian shelf. Conversely, the most efficient way to counter twisting of the symphysis about a transverse axis or to counter wishboning is to increase the labiolingual thickness of the symphysis. In attempting to evaluate the morphology of the catarrhine mandibular symphysis relative to these bending and twisting moments, we find ourselves unable to do so for at least four reasons. First, an evaluation of bending of the symphysis because of twisting of the mandibular corpora about their long axes requires unavailable information about the nature of twisting moments acting about the corpus of extant and extinct primates (see discussion of transverse dimensions of the mandibular corpus). Second, symphyseal twisting about a transverse axis presents a somewhat different problem because it requires unavailable information about moment-arm lengths for two different twisting moments (Fig. 3.5) and muscle-force recruitment patterns between working and balancing sides (Hylander, 1984). Although a stress analysis of the symphysis in *M. fascicularis* suggests that these two twisting moments are close to being equal in magnitude and opposite in direction (Hylander, 1985), and therefore of little or no importance because they counter one another, this may not be true for all catarrhines because of different geometrical configurations of muscle and reaction forces. For example, compared to cercopithecines, colobines (and australopithecines) appear to have their jaw-closing muscles positioned further rostrally. This has the effect of altering the relative magnitude of the various twisting moments (see Fig. 3.5). These moments can also be altered relative to one another simply by changing the ratio of working to balancing-side muscle force. This is important because these ratios may vary between species if the mechanical properties of their diets vary considerably (cf., Hylander 1983, 1985). Third, evaluating symphyseal morphology relative to wishboning is a problem because the moment arm associated with this bending regime requires knowledge of the location of the resultant of the transverse component of the balancing-side muscle force (Fig. 3.6).[7] Even if its location was known, it cannot be reconstructed from fossilized bits and pieces of the mandibular symphysis or corpus alone. Finally, an additional complicating factor is that we do not have a reliable procedure for evaluating whether increased labiolingual thickness of the symphysis reflects a structural adaptation to counter wishboning or twisting about a transverse axis.[8] In summary, it is presently impossible to analyze the symphysis of catarrhine primates relative to bending and twisting moments. Instead, the symphysis will be analyzed in a fashion similar to the evaluation

---

[7]In an earlier study, total mandibular length was used as an estimate of the moment arm associated with this bending moment (Hylander, 1985). To analyze the effects of scale among a very closely related group of primates (e.g., female cercopithecines), this may be a reasonable thing to do, but using total mandibular length for this purpose with larger taxonomic groups (e.g., catarrhines) becomes increasingly problematic because this dimension may not reflect comparable functional attributes for all species.

[8]Perhaps this distinction can be made following an analysis of the distribution of secondary Haversian systems in symphyseal sections of adult catarrhines (cf., Bouvier and Hylander, 1981). If wishboning is the predominate stress regime, there should be a greater concentration of secondary osteons along the lingual aspect of the symphysis, as compared to its labial aspect. If transverse twisting is the predominant stress regime, secondary osteons will be scattered more uniformly along both lingual and labial aspects.

of transverse dimensions of the mandibular corpus, which is a procedure that has serious theoretical limitations.

Figure 3.10 is a plot of mean vertical depth vs. mean transverse thickness of the symphysis (Fig. 3.7C) for 46 extant catarrhine species (males and females), *A. afarensis,* and *A. boisei.* These data indicate no consistent separation of extant male and female catarrhines, suggesting that canine size dimorphism does not have a major effect on the relative size of these two variables. In addition, the data in Table 3.1 indicate that the scaling coefficients for extant catarrhines are not significantly different between males and females. These coefficients do differ, however, between extant cercopithecoids and hominoids. Among cercopithecoids, symphyseal thickness scales positively allometric to symphyseal depth, whereas among extant hominoids the relationship between these two variables is not significantly different from isometry (Table 3.1). Moreover, these dimensions are more intensely correlated among cercopithecoids than among hominoids (extant and extinct) (Fig. 3.10), and this, in turn, may reflect a greater amount of functional diversity among the hominoids. Note that the australopithecine species and the large-bodied cercopithecoids have relatively thick symphyses while the great apes [particularly chimpanzees (c) and orangutans (o)] have relatively thin symphyses. Note also that the symphysis of *A. boisei* is absolutely thicker than in any of the extant hominoids.

Evaluating the functional significance of these morphological differences is a problem because

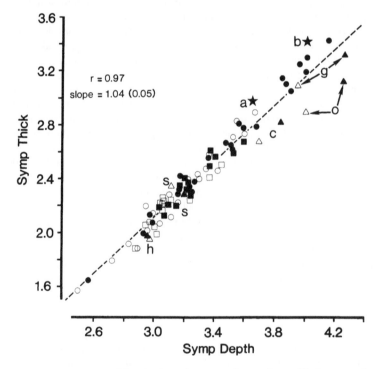

FIGURE 3.10. A log-log (natural logs) plot of mean values of mandibular symphysis depth and thickness for forty-six extant catarrhine species and two fossil hominid species. With the exception of the stars (fossil hominids), open and closed symbols are females and males, respectively. Squares, circles and triangles are colobines, cercopithecines and nonhuman hominoids, respectively. Stars labeled a and b are *Australopithecus afarensis* and *Australopithecus boisei,* respectively. Other abbreviations: g (gorilla), o (orangutan), c (chimpanzee), h (gibbon), and s (siamang). The dashed line indicates the least-squares regression line. Regression statistics are based solely on extant catarrhines. See Appendix 3.1 for details of complete sample.

there is no theoretical model for the interpretation of any sort of functional relationship between symphyseal thickness and symphyseal depth. However, if one assumes that the data in Fig. 3.10 indicate that the extinct hominids have an unusually thick symphysis while the great apes (particularly chimpanzees and orangutans) have an unusually deep symphysis, we might conclude that these apes have deep symphyses in order to counter symphyseal bending associated with twisting of the mandibular corpora about their long axes (Fig. 3.4) during powerful incisor biting, whereas the two hominid species have thick symphyses to counter bending stress associated with either wishboning or torsional stress associated with twisting of the symphysis about a transverse axis during powerful mastication (Figs. 3.5 and 3.6). The relatively large incisors of chimpanzees and orangutans, and the small incisors (and large molars) of *A. boisei* support this interpretation (cf., Hylander, 1975; Kay and Hylander, 1978; McHenry, 1984; Kay, 1985). Although the postcanine teeth of *A. afarensis* are also apparently large, the size of their incisors relative to body size is unclear and therefore the dental morphology of *A. afarenis* does not necessarily support this hypothesis (cf., Kimbel *et al.*, 1984; Kay, 1985).

The above explanation of the functional significance of hominoid symphyseal form would be more compelling if one could be confident as to the relative significance of each variable. For example, relative to bending or twisting moment arms, do orangutans have unusually thin symphyses, or do they have unusually deep symphyses, or some combination thereof? Moreover, because orangutans also eat very hard nuts (cf., Kay, 1981), their relatively deep symphysis may instead reflect an adaptation to counter symphyseal bending associated with twisting of the mandibular corpora about their long axes during powerful molar biting. (Or perhaps both powerful incision and powerful molar biting cause high levels of symphyseal bending stress in orangutans?) This illustrates one of the problems encountered when attempting to infer functional attributes of mandibular morphology: similar patterns of internal stress in the mandible can be associated with either incision or mastication. Moreover, this also illustrates that mandibular morphology alone cannot provide unambiguous clues about dietary adaptations in catarrhine primates (cf., Hylander, 1979a).

Interpretation of the cercopithecoid data also presents a problem that directly relates to our inability to analyze symphyseal dimensions relative to bending and twisting moments. Although the data in Fig. 3.10 indicate that large cercopithecoids have relatively thick symphyses compared to extant apes, these data do not necessarily indicate that cercopithecoids engage in more powerful mastication. Instead, because cercopithecoids have more sharply curved mandibles, i.e., longer mandibles relative to arch width (Fig. 3.11), if their symphyses were of a size similar to those of apes, this would result in excessively high levels of internal stress along the lingual aspect of their symphysis during comparable levels of mastication (cf., Hylander, 1985).

Although we are unable to analyze symphyseal dimensions relative to bending and twisting moments, additional information about symphyseal morphology among catarrhines can be gained by analyzing symphyseal dimensions relative to corpus dimensions. Figure 3.12 is a plot of mean symphyseal depth vs. mean corpus depth for 46 extant catarrhine species (males and females), *A. afarensis*, and *A. boisei*; Fig. 3.13 is a plot of mean symphyseal thickness vs. mean corpus depth for the same species. Note that in both instances colobines appear to have relatively small symphyseal dimensions. Presumably this is primarily a reflection of their deep mandibular corpus (Fig. 3.8), and should not necessarily be interpreted to mean that they have relatively small twisting and bending moments acting about their symphyses. Figure 3.12 also indicates that there are no major differences in the relative proportions of depth of the corpus and symphysis between the extant apes and large cercopithecoids, with the exception of female geladas, which plot very close to *A. afarensis* (open circle immediately below and to the left of star labeled a). The data in Fig. 3.12 indicate that compared to extant pongids the two extinct hominids appear to have vertically shallow symphyses, which is also supported by data in Fig. 3.10. In *A. boisei* this is at least partly a reflection of their unusually deep mandibular corpus. This may be true, at least in part, for colobines and for *A. afarensis*.

Most of the large male cercopithecines tend to have relatively thicker symphyses than do the large hominoids (Fig. 3.13). The two extinct hominid species have symphyseal thickness dimensions that fall close to the least-squares regression line. Because *A. boisei* has an unusually deep mandibular corpus, these data indicate that their symphyseal thickness dimensions may

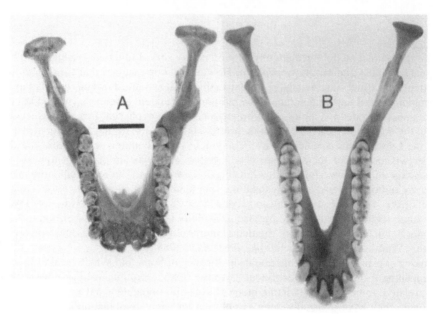

FIGURE 3.11. Photograph of adult female chimpanzee (A) and adult female baboon (B). Note that relative to jaw length the baboon mandible is much more sharply curved than is the chimpanzee mandible. Scale = 25 mm.

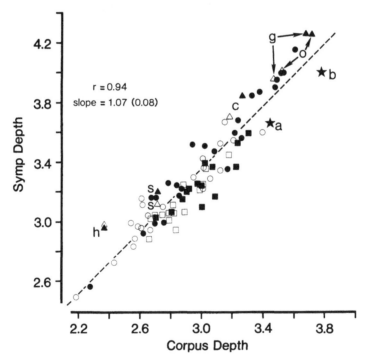

FIGURE 3.12. A log-log (natural logs) plot of mean values of mandibular symphysis depth and mandibular corpus depth for forty-six extant catarrhine species and two fossil hominid species. With the exception of the stars (fossil hominids), open and closed symbols are females and males, respectively. Squares, circles and triangles are colobines, cercopithecines and nonhuman hominoids, respectively. Stars labeled a and b are *Australopithecus afarensis* and *Australopithecus boisei,* respectively. Other abbreviations: g (gorilla), o (orangutan), c (chimpanzee), h (gibbon), and s (siamang). The dashed line indicates the least-squares regression line. Regression statistics are based solely on extant catarrhines. See Appendix 3.1 for details of complete sample.

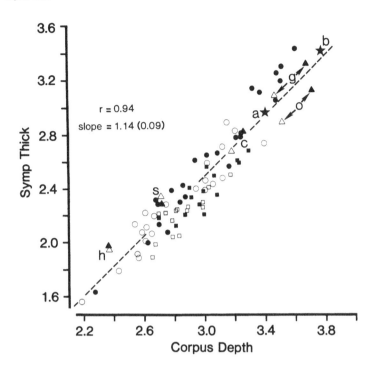

FIGURE 3.13. A log-log (natural logs) plot of mean values of mandibular symphysis thickness and mandibular corpus depth for forty-six extant catarrhine species and two fossil hominid species. With the exception of the stars (fossil hominids), open and closed symbols are females and males, respectively. Squares, circles and triangles are colobines, cercopithecines and non-human hominoids, respectively. Stars labeled a and b are *Australopithecus afarensis* and *Australopithecus boisei*, respectively. Other abbreviations: g (gorilla), o (orangutan), c (chimpanzee), h (gibbon), and s (siamang). The dashed line indicates the least-squares regression line. Regression statistics are based solely on extant catarrhines. See Appendex 3.1 for details of complete sample.

indeed be unusually large, as suggested by data in Fig. 3.10. Finally, Fig. 3.14 is a plot of mean corpus thickness vs. mean symphyseal thickness of males and females for six extant catarrhine species, *A. afarensis* and *A. boisei*. Species that fall above the solid black line have a corpus in the M_2 region that is thicker than the symphysis, while those that fall below this line have a symphysis that is thicker than the corpus. The dashed line indicates the location of the least-squares regression line based only on extant hominoids. Note the increased relative thickness of the symphysis in the male and female baboons (*Papio anubis*, open and closed circles labeled p), particularly the male. Note also that for the extinct hominids, corpus thickness slightly exceeds symphyseal thickness.

Summary and Interpretation of Morphology and Patterns of Stress

What, if anything, do the metrical data coupled with the stress analysis data tell us about the morphology of australopithecine jaws? Vertical dimensions of the mandibular corpus of *A. boisei* are large relative to parasagittal bending moment arms. In addition, the metrical data indicate that the thickness of their corpus is also unusually large. In contrast, only the thickness of the corpus of *A. afarensis* appears unusually large. For the symphysis, the metrical data suggest that the transverse thickness dimensions of both *A. afarensis* and *A. boisei* are unusually large (Figs. 3.10, 3.12 and 3.13). Unlike the interpretation of corpus depth, which suggests that jaw

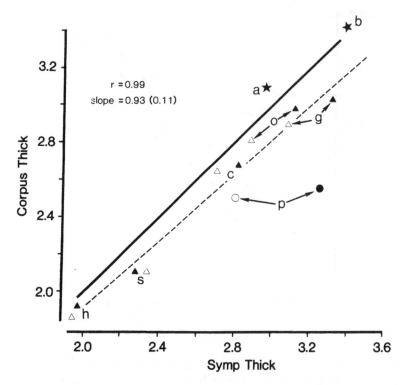

FIGURE 3.14. A log-log (natural logs) plot of mean values of mandibular symphysis thickness and mandibular corpus thickness for six extant catarrhine species and two fossil hominid species. With the exception of the stars (fossil hominids), open and closed symbols are females and males, respectively. Circles and triangles are cercopithecines and nonhuman hominoids, respectively. Stars labeled a and b are *Australopithecus afarensis* and *Australopithecus boisei*, respectively. Other abbreviations: p (baboon), g (gorilla), o (orangutan), c (chimpanzee), h (gibbon), and s (siamang). The dashed line indicates the least-squares regression line. Regression statistics are based solely on extant hominoids. See Appendix 3.1 for details of complete sample.

muscle and/or reaction forces were unusually large, it is not possible to determine whether the increased thickness of the corpus and symphysis of the australopithecines is due entirely to unusually large muscle and reaction forces or to some combination of large external forces and large twisting and bending moment arms.

The above overall pattern suggests that for *A. boisei*, the morphology of the corpus is best interpreted as a structural adaptation to counter internal stress associated with powerful parasagittal bending, twisting about its long axis, and wishboning. The symphysis of *A. boisei* is best interpreted as a structural adaptation to counter powerful bending stress associated with wishboning and perhaps torsional stress associated with symphyseal twisting about a transverse axis. A similar interpretation can be made for the morphology of the symphysis and corpus of *A. afarensis* except that its corpus is not as well designed to counter powerful parasagittal bending. As already noted, *in vivo* data on macaques indicate that with the exception of symphyseal twisting about a transverse axis, we can be reasonably confident that all of the above stress regimes occurred in australopithecines. Although twisting about a transverse axis does not appear to be very important for macaques, we cannot rule out the possibility of its existence in other catarrhines for reasons already given.

The data in Fig. 3.14 have interesting implications for mandibular biomechanics. This is most apparent when the mandibular morphology of cercopithecoids is compared to early hominids.

Morphologically these two groups are quite different from one another, with apes occupying somewhat intermediate positions. Unlike cercopithecoids, which have a symphysis that is much thicker than the corpus, the corpus and symphysis of hominids are of about equal thickness, or the hominid corpus may actually be thicker than its symphysis. If wishboning was the predominate pattern of stress in early hominids during mastication, then we would expect the thickness of their symphysis to be much greater than the thickness of their corpus because: (1) the moment arm (and therefore the moment) acting about the symphysis is much larger than the moment arm acting about the corpus; and (2) bending stress tends to be concentrated along the symphysis since the mandible functions as a sharply curved beam at this time. Both of these factors would result in much higher levels of stress along the symphysis than along the corpus unless the symphysis was much thicker than the corpus. Does the fact that symphyseal and corpus thickness dimensions are very similar among hominids indicate that wishboning is an unimportant stress regime for this group? The answer to that question is "no," although wishboning is surely less important among hominids than among cercopithecoids because hominid mandibles are relatively short and not nearly as sharply curved.

One way to account for a corpus and symphysis of about equal thickness in the presence of significant wishboning bending stress is to suggest that the hominid corpus also experienced large torsional moments because of twisting about its long axis. Increased twisting of the mandibular corpus about its long axis on the balancing side also stresses the symphysis by bending it (Fig. 3.4). If this type of symphyseal stress were to increase, the most efficient way to counter it would be to deepen the symphysis. The metrical data in Figs. 3.10 and 3.12, however, appear to indicate that these early hominids did not have deep symphyses. Thus, if there was a relatively large increase in torsion, it apparently was not associated with the balancing side. Increased twisting on the working side can have a similar effect on the symphysis, but it is more likely that a large portion of the twisting stress along the working-side mandibular corpus is countered, not at the symphysis, but instead between the working-side muscle force and the bite force (Hylander, 1979a, 1981; Demes et al., 1984). The symphysis, therefore, need not be subjected to additional bending stress because, although the resultant of the working-side muscle force tends to evert the lower border of the mandible and invert the postcanine alveolar process, the resultant bite force often has the opposite effect. It is hypothesized, therefore, that whereas the thick mandibular symphysis of "robust" australopithecines is a structural adaptation to counter wishboning due to powerful transverse jaw movements, their thick mandibular corpus is related primarily to countering both working-side twisting moments and bending stress during wishboning. This then brings us back to a question that was asked earlier. Why was the mandibular corpus of australopithecines, particularly the "robust" forms, subjected to such unusually large twisting moments?

Twisting Moments in "Robust" Australopithecines

There are a number of hypothetical reasons why, compared to living catarrhine primates, the "robust" australopithecine mandibular corpus might have been subjected to relatively large twisting moments (Hylander, 1979a). These reasons are related to (1) differential activity, position and/or enlargement of the muscles of mastication; (2) an increased emphasis on mastication or biting in the premolar region; and (3) a relatively large transverse component to the direction of the postcanine bite force. The last, and most obvious, reason (4) is that they were generating extremely large muscle and reaction forces within their masticatory apparatus because of the nature of their diet. All of these reasons will be discussed below.

Differential Activity, Enlargement and Position of Jaw Muscles

Whereas the masseter (and perhaps the anterior temporalis) muscle has a tendency to evert the lower border of the mandibular corpus and invert the alveolar process, the medial pterygoid muscle has the opposite effect; thus, it tends to counter twisting caused by the masseter. If "robust" australopithecines had a masseter muscle that was relatively more active than their medial pterygoid muscle during mastication (i.e., relative to the condition among other primates), twisting of the mandibular corpus would have been increased. This possibility is unlikely and untestable. Perhaps increased twisting resulted from an increase in the size of their masseter relative to their medial pterygoid muscles. It is difficult to estimate, however, solely on the basis of the skull, whether their masseter was relatively larger. Finally, the position of the masseter among "robust" australopithecines may have caused increased twisting of their mandibular corpus. This is because they apparently had an unusually large anterior temporalis muscle (Tobias, 1967) that caused the zygomatic arch to flare laterally, thereby displacing the upper portion of the masseter muscle laterally. The effect of this repositioning was a more transverse alignment of the masseter and, thus, possibly an increase in twisting moments about the mandibular corpus.

Premolar Mastication

The premolars of "robust" australopithecines are enlarged and heavily worn, indicating an increased functional emphasis on these teeth during mastication. The accentuation of the importance of the premolars during mastication probably resulted in increased twisting of the "robust" australopithecine mandibular corpus because, in contrast to the molars, the premolars of australopithecines (and most other anthropoids) are positioned more laterally relative to the long axis of the mandibular corpus. Thus, there is a larger twisting moment arm associated with the premolar bite force, even with a bite force that is primarily vertical.

Bite Force Direction

In addition to the magnitude of the bite force, the direction of the bite force can also influence twisting of the corpus. For example, if the direction of the postcanine bite force passes through the long axis of the mandibular corpus, this force will not cause the corpus to be twisted about its long axis. Under these conditions, twisting of the mandibular corpus owing to the position of the resultant muscle force will be countered in the symphysis, resulting in symphyseal bending (Fig. 3.4). On the other hand, if the mandibular bite force passes laterally to the long axis of the mandibular corpus, this force will have a tendency to twist the corpus in a direction opposite to that caused by the muscle force, i.e., it will twist the corpus by everting its alveolar process and inverting its lower border. Under these conditions, twisting stress will be concentrated along that portion of the corpus between the resultant bite force and the ipsilateral resultant muscle force (Hylander, 1979a; Demes et al., 1984). As seen in Fig. 3.15, a small increase in the transverse component of the bite force results in a much greater twisting moment about the mandibular corpus. That is, in this figure the twisting moment arm is doubled in size simply by changing the direction of the bite force from about 88° (F_1) to about 72° (F_2) relative to the occlusal plane. Based on current estimates of the frictional characteristics of teeth, these hypothetical bite-force directions are quite plausible (cf., Smith, 1983).

Among primates, the direction of the bite force during mastication has not been characterized, except for some preliminary data on humans (Graf, 1975) that indicate bite force directions similar to those seen in Fig. 3.15. Indirect evidence provides additional clues as to the direction of the bite force in primates. A consideration of primate jaw movement data and tooth morphology suggests that there must be a laterally directed component of force along the mandibular teeth during puncture–crushing because the working-side mandibular teeth are moving medially at this time. There must also be a laterally directed component of mandibular bite force for the period immediately prior to Phase I tooth contacts and also during Phase II movements. Although the working-side mandibular teeth are also moved medially during Phase I movements (Kay and Hiiemae, 1974; Luschei and Goodwin, 1974), it cannot be assumed that the mandibular bite force has a significant laterally directed component at this time because the morphology of the postcanine teeth of most primates will cause medial movement of the working-side teeth, even if the bite force is vertical. The only exception would be among those primates that have flat occlusal surfaces along their postcanine teeth. If it is assumed that the working-side mandibular teeth were moved medially during Phase I contacts among "robust" australopithecines (Grine, 1981, 1984), then there must have been a laterally directed component of force along their mandibular

FIGURE 3.15. Section through a catarrhine mandibular corpus in the M_1–M_2 region. The black dot represents the twisting axis of neutrality. F_1 and F_2 are two differently directed bite forces. As the bite force becomes more laterally directed (F_2), there is a greater tendency for twisting of the mandibular corpus because this twisting moment is increased. Note that if F_1 and F_2 are equal in magnitude the twisting moment associated with F_2 is about twice as large as the moment associated with F_1.

teeth at this time because the shape of the teeth did not automatically cause the working-side teeth to be moved medially during closure.

In addition to having a laterally directed component of bite force along mandibular teeth, there is some evidence that the direction of this bite force changes throughout the power stroke among primates. A combined bone strain, electromyographic and cinefluorgraphic analysis in macaques indicates that compared to the early portion of the power stroke, the later portion (late Phase I or early Phase II?) is characterized by a mandibular bite force that is decreasing in magnitude but has a relatively larger laterally directed component (Hylander et al., 1987). This change in bite force direction and magnitude is associated with an increase in activity of the working-side medial pterygoid, the cessation in activity of the balancing-side superficial masseter and medial pterygoid, and the persistent activity of the balancing-side deep masseter (Hannam and Wood, 1981; Wood, 1986; Hylander et al., 1987). Although this shift in bite force direction may not be associated with increased twisting of the mandibular corpus, it does cause a sharp increase in wishboning of the symphysis.

Among australopithecines, the flat biting surfaces of the postcanine teeth (and the reduced height of the canines?) indicate a large potential for medial movements of the working-side mandibular teeth during tooth–tooth contact. Compared to most other primates, relatively large lateral components of the mandibular bite force in "robust" australopithecines may have been associated with these movements, and if so, the mandibular corpus of "robust" australopithecines may have been twisted powerfully. This is in contrast to the condition of many other primates, such as colobines, where there is much less potential for such transverse movements during tooth–tooth contact because of trenchant steep-cusped cheek-teeth. For colobines, relatively small lateral components of the mandibular bite force and therefore relatively small twisting moments about the mandibular corpus are probably, although not necessarily, associated with these lesser transverse molar movements.

"Robust" Australopithecine Diet

Perhaps the most important cause of increased twisting of the corpus was the powerful masticatory forces routinely necessitated by the nature of the "robust" australopithecine diet. Their probable diet continues to be an important and much debated topic (cf., Walker, 1981; Grine, 1981; Lucas et al., 1985; Kay, 1985; Kay and Grine, this volume, chapter 26). There is, nevertheless, a general (although not uniform, cf., Walker, 1981) concensus that "robust" australopithecines ate unusually hard or tough objects. In addition, it has been generally assumed that the food objects they ate were small and therefore required little incisal preparation (Jolly, 1970; Hylander, 1979a; Grine, 1981; Sussman, 1987), although an equally plausible hypothesis is that they ate relatively large food objects (cf., Peters, 1987) that were reduced in size prior to ingestion. This does not mean that "robust" australopithecines were necessarily using tools for this purpose (cf., Izawa and Mizuno, 1977) although their breaking of large, hard food objects with tools is a distinct possibility and certainly cannot be ruled out at this time.

What additional clues, if any, does the preceding analysis and discussion provide as to the nature of their diet? Differences in the probable direction of the postcanine bite force between, for example, colobines and "robust" australopithecines, are presumably directly attributable to the physical properties of their diets. Leaves, which require little in the way of incisal preparation and make up a large portion of the colobine diet, are more easily reduced in size during chewing by shearing than by crushing and grinding primarily because of their two-dimensional nature (Walker and Murray, 1975). The small incisors of "robust" australopithecines are consistent with a primarily leaf-eating diet (cf., Hylander, 1975), but the morphology of their molars is not (Kay, 1985). Overall the morphology of "robust" australopithecine molars suggests that they were designed especially to reduce food particle size during mastication by crushing and/or grinding, rather than by vertical shear (cf., Kay and Hiiemae, 1974). For powerful crushing, the vertical component to the bite force would be emphasized. For powerful grinding, both the vertical and transverse components would be emphasized, and, as noted earlier these transverse components would be associated with increased torsional stress in the corpus and wishboning stress in the corpus and symphysis.

Although a significant portion of the diet of "robust" australopithecines may have consisted of hard or tough objects that required powerful crushing and grinding forces, one important question is whether their diet consisted largely of food objects that were rich or poor in dietary nutrients? That is, were they eating foods primarily rich in proteins, fats, or easy-to-digest storage carbohydrates, such as seeds or nuts (cf., Jolly, 1970; Kay, 1981, 1985; Lucas et al., 1985), or were these foods of a somewhat lower quality, perhaps containing a relatively large component of hard-to-digest structural carbohydrates (cf., Robinson, 1956, Grine, 1981; Lucas et al., 1985), such as highly fibrous roots? As already noted, the morphology of the molars of "robust" australopithecines suggests that their diet was not primarily composed of leaves or other fibrous materials. Nevertheless, "robust" australopithecine molar morphology does not preclude the possibility that a large portion of their diet consisted of structural carbohydrates simply because orangutans, who also have large flat molars with thick enamel, have a diet consisting of 25 to 30% leaves and about 5 to 10% bark (MacKinnon, 1974; Rodman, 1977; Rijksen, 1978).

The relative quality of their diet is an important question for a number of reasons although I will confine the following discussion to the importance of this question relative to the morphology of the craniofacial region. In an earlier paper I suggested that in addition to powerful twisting and bending regimes, another major factor that may influence structural adaptations in the craniofacial region is the number of masticatory or stress cycles (Hylander 1979a). This is because bone is remarkably susceptible to fatigue failure (Carter et al., 1977) and, therefore, the well-developed facial buttresses of "robust" australopithecines might be a structural adaptation to counter fatigue failure due to repeated cyclical loads rather than simply countering unusually large but less frequent loads. That is, if the diet of "robust" australopithecines consisted primarily of large amounts of low-quality foods that required extensive and prolonged chewing, then perhaps their jaws were primarily adapted to prevent fatigue failure due to an unusually large number of chewing cycles. Whether the australopithecine mandible was subjected to an unusually large number of stress cycles cannot be determined from the external morphology of their jaws, although perhaps an analysis of the frequency and occurrence of secondary Haversian systems in their jaws can provide clues to help resolve this problem (cf., Bouvier and Hylander, 1981).

In summary, the preceding analysis tells us little that we did not already suspect about the nature of the "robust" australopithecine diet. The morphological evidence suggests that these hominids were well adapted to chewing hard or tough foods that required powerful masticatory forces. This does not mean, however, that their entire diet was characterized as being unusually hard or tough. "Robust" australopithecines probably did not engage in extensive incisal preparation of food objects although this does not necessarily mean that all foods eaten were initially small. If some of their dietary items were large, perhaps these items were reduced in size with tools prior to ingestion. Finally, it is unclear as to whether their diet did or did not contain a large portion of low-quality food items. If it did, then perhaps the well-developed bony buttresses throughout the face of "robust" australopithecines may have also been an adaptation to counter fatigue failure due to highly repetitive masticatory stress cycles.

Conclusions

A review of *in vivo* experimental data on macaques and metrical analyses of the mandibular corpus and symphysis of catarrhine primates suggests the following: The vertical depth and transverse thickness of the mandibular corpus of "robust" australopithecines are unusually large. The increased vertical depth was primarily a structural adaptation to counter powerful parasagittal bending moments while the increased transverse thickness was primarily a structural adaptation to counter powerful twisting moments and wishboning. In contrast, only the transverse dimensions of their symphysis appear unusually large. Their unusually thick symphysis was primarily a structural adaptation to counter bending stress associated with wishboning. All of the above loading regimes are thought to have occurred during powerful mastication of unusually hard or tough food objects. Whether their diet consisted of foods that were initially large or small, or of high or low quality, is not known. The nature of this diet, however, is important for interpreting structural adaptations of the "robust" australopithecine face. If their diet consisted of very low-quality food items, this would have necessitated numerous highly repetitious chewing cycles each day, and therefore their facial buttresses may have been an adaptive response to prevent fatigue failure caused by these repeated loading regimes, rather than simply a response to unusually large masticatory forces.

Acknowledgments

I am grateful to Kirk Johnson for help on all aspects of this paper. Drs. Richard Kay and Pascal Picq kindly read and commented on this manuscript. I am also grateful to Kaye Brown and Matt Cartmill for measuring a number of monkey skulls housed in various European museums. I would like to thank the curators of the primate collections at the National Museums of Kenya, Field Museum of Natural History, the American Museum of Natural History, the United States Museum of Natural History, the Cleveland Museum of Natural History and the Museum of Comparative Zoology for their helpfulness. I am particularly grateful to Tim White for his generosity and helpfulness. He kindly provided me both the metrical data for the fossil hominids and information about the nature of the original specimens. Thanks go to John Graves and Dana Hall for typing numerous drafts of this manuscript. This study was supported by the Department of Anatomy, Duke University, and by research grants from NIH (DE04531) and NSF (BNS83-04708). Finally, I am grateful to Fred Grine for his helpful comments on this paper and for inviting me to participate in and providing support for such an interesting, informative and enjoyable workshop.

References

Arges, K. P. and Palmer, A. E. (1963). *Mechanics of materials*. McGraw-Hill, New York.
Biewener, A. A. and Taylor, C. R. (1986). Bone strain: A determinant of gait and speed? *J. Exp. Biol., 123*: 383–400.
Bouvier, M. (1986a). A biomechanical analysis of mandibular scaling in Old World Monkeys. *Amer. J. Phys. Anthropol., 69*: 473–482.

Bouvier, M. (1986b). Biomechanical scaling of mandibular dimensions in New World Monkeys. *Intern. J. Primatol.*, 7: 551–567.

Bouvier, M. and Hylander, W. L. (1981). Effect of bone strain on cortical bone structure in macaques *(Macaca mulata)*. *J. Morphol.*, 167: 1–12.

Cachel, S. (1984). Growth and allometry in primate masticatory muscles. *Arch. Oral Biol.*, 29: 287–293.

Carter, D. R., Spengler, D. M. and Frankel, V. H. (1977). Bone fatigue in uniaxial loading at physiologic strain rates. *IRCS Med. Sci.*, 5: 592.

Chamberlain, A. T. and Wood, B. A. (1985). A reappraisal of variation in hominid mandibular corpus dimensions. *Amer. J. Phys. Anthropol.*, 66: 399–405.

Crompton, A. W. and Hiiemae, K. M. (1969). How mammalian molar teeth work. *Discovery*, 5: 23–34.

Daegling, D. J. (1988). Mechanical modelling of hominoid mandibular cross-sections. *Amer. J. Phys. Anthropol.*, 75: 200–201.

Demes, B., Preuschoft, H., and Wolff, J. E. A. (1984). Stress-strength relationships in the mandibles of hominoids. In D. J. Chivers, B. A. Wood, and A. Bilsborough (eds.), *Food acquisition and processing in primates*, pp. 369–390. Plenum, New York.

DuBrul, E. L. (1977). Early hominid feeding mechanisms. *Amer. J. Phys. Anthropol.* 47: 305–320.

Goldstein, S., Post. D., and Melnick, D. (1978). An analysis of cercopithecoid odontometrics. I. The scaling of the maxillary dentition. *Amer. J. Phys. Anthropol.*, 49: 517–532.

Graf, H. (1975). Occlusal forces during function. In N. H. Rowe, (ed.), *Occlusion: Research in form and function*, pp. 90–109. University of Michigan School of Dentistry and the Dental Research Institute.

Grine, F. E. (1981). Trophic differences between "gracile" and "robust" australopithecines: a scanning electron microscope analysis of occlusal events. *S. Afr. J. Sci.*, 77: 203–230.

Grine, F. E. (1984). Deciduous molar microwear of South African australopithecines. In D. J. Chivers, B. A. Wood, and A. Bilsborough (eds.), *Food acquisition and processing in primates*, pp. 525–534. Plenum, New York.

Hannam, A. G. and Wood, W. W. (1981). Medial pterygoid muscle activity during the closing and compressive phases of human mastication. *Amer. J. Phys. Anthropol.* 55: 359–367.

Hiiemae, K. M. (1978). Mammalian mastication: A review of the activity of the jaw muscles and the movements they produce in chewing. In P. M. Butler and K. A. Joysey (eds.), *Development, function and evolution of teeth*, pp. 359–398. Academic Press, London.

Hill, W. C. Osman. (1970). *Primates: Comparative anatomy and taxonomy, vol. VIII: Cerecopithecinae*. Wiley-Interscience, New York.

Hylander, W. L. (1975). Incisor size and diet in anthropoids with special reference to Cercopithecidae. *Science*. 189: 1095–1098.

Hylander, W. L. (1979a). The functional significance of primate mandibular form. *J. Morphol.*, 160: 223–240.

Hylander, W. L. (1979b). Mandibular function in *Galago crassicaudatus* and *Macaca fascicularis*: An *in vivo* approach to stress analysis of the mandible. *J. Morphol.*, 159: 253–296.

Hylander, W. L. (1979c). An experimental analysis of tempore mandibular joint reaction force in macaques. *Amer. J. Phys. Anthropol.*, 51: 433–456.

Hylander, W. L. (1981). Patterns of stress and strain in the macaque mandible. In D. S. Carlson (ed.), *Craniofacial biology*. Monograph #10, Craniofacial Growth Series, Center for Human Growth and Development, pp. 1–37, Univ. of Michigan, Ann Arbor, MI.

Hylander, W. L. (1984). Stress and strain in the mandibular symphysis of primates: A test of competing hypotheses. *Amer. J. Phys. Anthropol.*, 64: 1–46.

Hylander, W. L. (1985). Mandibular function and biomechanical stress and scaling. *Amer. Zool.*, 25: 315–330.

Hylander, W. L. and Crompton, A. W. (1986). Jaw movements and patterns of mandibular bone strain during mastication in the monkey. *Arch. Oral Biol.*, 31: 841–848.

Hylander, W. L., Johnson, K. R. and Crompton, A. W. (1987). Loading patterns and jaw movements during mastication in *Macaca fascicularis:* A bone-strain, electromyographic and cineradiographic analysis. *Amer. J. Phys. Anthropol.,* 72: 287–314.

Izawa, K. and Mizuno, A. (1977). Palm-fruit cracking behavior of wild black-capped capuchin. *Primates,* 18: 773–792.

Johanson, D. C. and White, T. D. (1979). A systematic assessment of early African hominids. *Science,* 203: 321–330.

Jolly, C. J. (1970). The seed-eaters: A new model of hominid differentiation based on a baboon analogy. *Man,* 5: 5–26.

Kay, R. F. (1978). Molar structure and diet in extant Cercopithecidae. *In* P. M. Butler and K. A. Joysey (eds.), *Development, function and evolution of teeth,* pp. 309–339. Academic Press, London.

Kay, R. F. (1981). The nut-crackers: A new theory of the adaptations of the Ramapithecinae. *Amer. J. Phys. Anthropol.,* 55: 141–151.

Kay, R. F. (1985). Dental evidence for the diet of *Australopithecus. Ann. Rev. Anthropol.,* 14: 315–341.

Kay, R. F. and Hiiemae, K. M. (1974). Jaw movement and tooth use in recent and fossil primates. *Amer. J. Phys. Anthropol.,* 40: 227–256.

Kay, R. F. and Hylander, W. L. (1978). The dental structure of mammalian folivores with special reference to primates and phalangeroids (Marsupialia). *In* G. G. Montgomery (ed.), *The ecology of arboreal folivores,* pp. 173–191. Smithsonian Institution Press, Washington, D.C.

Kay, R. F., Fleagle, J. G., and Simons, E. L. (1981). A revision of the Oligocene apes of the Fayum Province, Egypt. *Amer. J. Phys. Anthropol.,* 55: 293–322.

Kelley, J. and Pilbeam, D. (1986). The dryopithecines: Taxonomy, comparative anatomy and phylogeny of Miocene large hominoids. *In* D. R. Swindler and J. Irwin (eds.), *Comparative primate biology, vol. 1: Systematics, evolution and anatomy,* pp. 361–411. Alan R. Liss, New York.

Kimbel, W. H., White, T. D. and Johanson, D. C. (1984). Cranial morphology of *Australopithecus afarensis:* A comparative study based on a composite reconstruction of the adult skull. *Amer. J. Phys. Anthropol.,* 64: 337–388.

Lanyon, L. E. and Rubin, C. T. (1985). Functional adaptation in skeletal structures. *In* M. Hildebrand, D. M. Bramble, K. F. Liem, and D. B. Wake (eds.), *Functional vertebrate morphology,* pp. 1–25. Harvard University Press, Cambridge, Mass.

Leakey, L. S. B. (1959). A new fossil skull from Olduvai. *Nature,* 184: 491–493.

Lucas, P. W., Corlett, R. T. and Luke, D. A. (1985). Plio-Pleistocene hominid diets: an approach combining masticatory and ecological analysis. *J. Hum. Evol.,* 14: 187–202.

Luschei, E. S. and Goodwin, G. M. (1974). Patterns of mandibular movement and jaw muscle activity during mastication in the monkey. *J. Neurophys.,* 37: 954–966.

McHenry, H. M. (1984). Relative cheek-tooth size in *Australopithecus. Amer. J. Phys. Anthropol.,* 64: 297–306.

MacKinnon, J. R. (1974). The behavior and ecology of wild orang-utans *(Pongo pygmaeus). Anim. Behav.* 22: 3–74.

Peters, C. R. (1987). Nut-like oil seeds: Food for monkeys, chimpanzees, humans and probably ape-men. *Amer. J. Phys. Anthropol.,* 73: 333–363.

Pilbeam, D. (1972). *The ascent of man: An introduction to human evolution.* Macmillan, New York.

Preuschoft, H., Demes, B., Meyer, M. and Bär, H. F. (1986). The biomechanical principles realised in the upper jaw of long-snouted primates. *In* J. G. Else and P. C. Lee (eds.), *Primate evolution,* pp. 249–264. Cambridge University Press, Cambridge.

Rak, Y. (1983). *The australopithecine face.* Academic Press, New York.

Rijksen, H. D. (1978). A field study on Sumatran orangutans *(Pongo pygmaeus abeli;* Lesson 1827). H Veenman and Zonen, Wageningen.

Robinson, J. T. (1956). The dentition of the Australopithecinae. *Mem. Tvl. Mus.,* 9: 1–179.

Rodman, P. S. (1977). Feeding behavior of orang-utans of the Kutai reserve. *In* T. H. Clutton-Brock (ed.), *Primate ecology: Studies of feeding and ranging behavior in lemurs, monkeys, and apes,* pp. 384–413. Academic Press, London.

Simons, E. L. (1972). *Primate evolution: An introduction to man's place in nature.* Macmillan, New York.

Smith, R. J. (1983). The mandibular corpus of female primates: Taxonomic, dietary and allometric correlates of interspecific variations in size and shape. *Amer. J. Phys. Anthropol., 61*: 315–330.

Sussman, R. W. (1987). Species-specific dietary patterns in primates and human dietary adaptations. *In* W. G. Kinzey (ed.), *The evolution of human behavior: Primate models,* pp. 151–179. State University of New York Press, Albany.

Tobias, P. V. (1967). *The cranium and maxillary dentition of Australopthecus (Zinjanthropus) boisei,* Vol. 2: Olduvai Gorge. Cambridge University Press, Cambridge.

Walker, A. C. (1981). Diet and teeth: Dietary hypotheses and human evolution. *Phil. Trans. Roy. Soc. (London), B292*: 57–64.

Walker, P. and Murray, P. (1975). An assessment of masticatory efficiency in a series of anthropoid primates with special reference to the Colobinae and Cercopithecinae. *In* R. H. Tuttle (ed.), *Primate functional morphology and evolution,* pp. 135–150. Mouton Publishers, The Hague.

Wolff, J. E. A. (1984). A theoretical approach to solve the chin problem. *In* D. J. Chivers, B. A. Wood and A. Bilsborough (eds.), *Food acquisition and processing in primates,* pp. 391–405. Plenum, New York.

Wolpoff, M. H. (1973). Posterior tooth size, body size, and diet in South African gracile australopithecines. *Amer. J. Phys. Anthropol., 39*: 375–394.

Wood, B. A. (1978). Allometry and hominid studies. *In* W. W. Bishop (ed.), *Geological background to fossil man,* pp. 125–128. Scottish Academic Press, Edinburgh.

Wood, W. W. (1986). Medial pterygoid muscle activity during chewing and clenching. *J. Prosthet. Dent., 55*: 615–622.

Appendix 3.1. Mean Values for Mandibular Dimensions (mm)"

Taxon	Sex	N	M_2-I_1 Length	Corpus Depth	Corpus Thickness	Symph Depth	Symph Thickness
Cercopithecines							
Miopithecus talopoin	M	6	20.3	9.7		13.0	5.2
Papio hamadryas	M	6	65.2	27.6		46.8	23.2
Papio ursinus	M	10	77.2	33.3		54.7	24.5
Theropithecus gelada	M	5	59.1	32.2		49.4	21.2
Mandrillus sphinx	M	6	74.9	36.5		63.3	30.9
Papio cynocephalus	M	1	71.7	29.0		47.9	22.4
Papio anubis	M	4	68.2	32.4	13.1	52.3	25.9
Cercopithecus cephus	M	5	32.7	15.6		19.9	8.0
Cercopithecus aethiops	M	6	36.4	14.5		23.7	10.2
Erythrocebus patas	M	7	45.4	18.8		33.8	13.6
Cercopithecus lhoesti	M	4	38.8	17.3		23.9	11.3
Allenopithecus nigroviridis	M	7	37.5	17.6		25.2	10.4
Cercopithecus hamlyni	M	5	37.2	14.9		23.6	9.9
Macaca nemestrina	M	5	52.5	25.4		39.5	16.3
Macaca fascicularis	M	5	38.3	16.1		26.2	10.9
Cercocebus torquatus	M	5	50.5	25.9		35.1	16.6
Cercopithecus mitis	M	5	38.9	17.0		25.7	10.0
Cercocebus albigena	M	7	44.7	23.6		28.8	13.0
Macaca hecki	M	4	50.5	20.4		33.5	14.2
Cercopithecus ascanius	M	5	32.8	13.7		18.7	7.4
Macaca niger	M	5	48.5	21.8		32.2	14.4
Cercopithecus denti	M	5	32.8	14.8		19.8	8.5
Macaca speciosa	M	4	—	24.8		36.4	16.1
Mandrillus leucophaeus	M	4	75.5	33.4		55.0	27.2
Miopithecus talopoin	F	5	20.3	8.9		12.1	4.8
Papio anubis	F	5	51.6	24.7	11.5	34.6	16.9
Papio cynocephalus	F	1	49.7	23.4		39.1	18.0
Theropithecus gelada	F	9	46.5	29.8		36.5	15.4
Macaca niger	F	5	41.3	19.1		27.0	11.0
Macaca fascicularis	F	5	32.0	15.6		22.2	9.8
Cercopithecus mitis	F	4	33.7	14.2		20.9	7.9
Cercopithecus aethiops	F	5	34.5	12.7		19.9	8.5
Erythrocebus patas	F	2	39.4	13.6		23.4	9.2
Macaca nemestrina	F	3	45.6	20.3		30.6	11.7
Cercocebus albigena	F	6	39.5	20.2		25.1	10.9
Cercopithecus lhoesti	F	7	35.0	13.3		19.4	8.0
Cercopithecus hamlyni	F	5	34.5	13.6		19.2	7.5
Cercopithecus neglectus	F	5	—	13.7		22.5	8.3
Cercopithecus ascanius	F	5	28.1	11.4		15.2	6.0
Cercopithecus cephus	F	5	29.9	13.0		17.9	6.6
Cercocebus torquatus	F	5	41.4	21.3		26.8	11.4
Macaca sylvana	F	4	—	22.7		28.3	11.9
Cercopithecus denti	F	5	29.0	12.9		17.0	6.8
Macaca speciosa	F	2	—	20.5		28.9	13.3
Allenopithecus nigroviridis	F	3	—	14.4		19.0	9.0
Mandrillus leucophaeus	F	4	—	22.7		33.9	15.0
Colobines							
Colobus badius	M	5	40.6	20.4		29.9	13.0
Procolobus verus	M	5	28.7	16.6	5.6	21.4	8.4
Rhinopithecus roxellanae	M	4	41.9	27.0		36.4	14.6

Continued

Appendix 3.1. *Continued*

Taxon	Sex	N	M_2–I_1 Length	Corpus Depth	Corpus Thickness	Symph Depth	Symph Thickness
Simias concolor	M	5	35.6	21.8		23.8	10.5
Nasalis larvatus	M	5	41.6	25.0		29.1	13.6
Presbytis obscurus	M	4	36.0	19.9		25.2	10.8
Pygathrix nigripes	M	5	—	18.1		24.5	11.1
Presbytis potenziani	M	5	34.6	17.6		23.3	9.1
Presbytis cristatus	M	5	32.7	20.0		22.1	9.1
Presbytis johni	M	5	37.2	19.5		25.8	9.8
Pygathrix nemaeus	M	5	38.7	18.4		25.0	10.3
Colobus guereza	M	5	41.8	21.5		28.9	12.1
Presbytis frontatus	M	4	34.5	14.8		20.8	8.9
Presbytis entellus	M	4	47.4	25.3		34.0	13.4
Pygathrix nigripes	F	4	—	16.2		21.2	8.7
Simias concolor	F	5	32.5	18.0		21.4	9.6
Presbytis potenziani	F	4	32.9	16.7		22.4	9.4
Presbytis cristatus	F	5	30.8	17.0		19.0	7.8
Presbytis entellus	F	6	44.5	24.0		31.4	12.2
Presbytis frontatus	F	5	33.0	14.9		21.1	9.1
Colobus badius	F	5	38.8	17.9		25.4	9.4
Colobus kirkii	F	2	29.6	14.6		20.5	7.3
Nasalis larvatus	F	5	36.3	16.8		21.4	9.4
Colobus guereza	F	5	39.1	19.9		24.8	9.6
Pygathrix namaeus	F	2	33.3	15.6		21.0	9.2
Rhinopithecus roxellanae	F	5	38.0	20.7		29.0	11.1
Presbytis johni	F	3	35.7	19.9		25.3	9.9
Presbytis obscurus	F	2	33.2	16.3		20.2	7.7
Procolobus verus	F	5	27.0	14.3	5.3	17.8	6.6
Hominoids							
Pan troglodytes	M	9	59.8	26.0	14.6	46.5	16.8
Gorilla gorilla	M	9	82.2	39.4	20.9	70.7	27.7
Pongo pygmaeus	M	9	79.5	40.9	19.8	70.4	22.7
Symphalangus syndactylus	M	6	41.4	15.1	8.2	24.5	9.8
Hylobates lar	M	10	33.5	10.6	6.9	19.3	7.2
Pan troglodytes	F	11	59.5	24.1	14.1	40.3	14.5
Gorilla gorilla	F	8	72.1	32.0	18.2	52.0	21.9
Pongo pygmaeus	F	12	69.0	33.7	16.7	54.9	18.0
Symphalangus syndactylus	F	6	39.1	15.1	8.2	22.5	10.4
Hylobates lar	F	5	33.0	10.7	6.4	19.6	7.0
Australopithecus boisei[b]			59.2(3)	43.8(11)	31.5(11)	54.9(3)	30.1(3)
Australopithecus afarensis[b]			56.0(4)	30.4(8)	22.5(8)	38.6(3)	19.3(5)

[a]Included are mean mandibular dimensions for all catarrhines in Figs. 3.8, 3.10, 3.12, 3.13 and 3.14. Also included are mean values for the hominoids and *Procolobus verus* in Fig. 3.9, but the remaining cercopithecoid values for Fig. 3.9 are from Bouvier (1986a).

[b]The sample size for each australopithecine mean value varies because not all dimensions are available for each specimen. The number in parentheses following each mean value indicates the sample size for that particular dimension. Appendix 3.2 indicates the source and the mandibular dimensions for the australopithecines.

Appendix 3.2. Australopithecine Mandibular Dimensions (mm)[a,b]

Specimen	M_2-I_1 length	Corpus		Symph	
		Depth	Thickness	Depth	Thickness
Australopithecus boisei					
WN-64 (Peninj)	59.1	38.2	28.0	50.1	26.0($\pm$2)
KNM-ER 729	60.9($\pm$2)	45.6	28.8	57.4($\pm$3)	30.0
L7A-125	57.7($\pm$2)	48.0(+2)[c]	33.4(+2)[c]	57.3($\pm$3)	34.4
KNM-ER 818		50.8($\pm$2)	35.3($\pm$2)		
KNM-ER 403		44.9(+2)[c]	30.8(+1)[c]		
KNM-ER 404		47.5($\pm$2)	35.5($\pm$2)		
KNM-ER 725		35.8($\pm$2)	30.9($\pm$1)		
KNM-ER 726		43.1(+2)[c]	29.6($\pm$1)		
KNM-ER 801A		39.8(+1)[c]	29.0($\pm$1)		
KNM-ER 805A		37.6($\pm$3)	31.0($\pm$1)		
KNM-ER 1468		43.8($\pm$2)	32.2($\pm$1)		
Australopithecus afarensis					
LH 4	56.7	29.4	22.4		19.1
AL 145-35			24.8		
AL 207-13		27.3(−2)[d]	20.4(−2)[d]		
AL 188-1		33.3(+1)[c]	22.3(+1)[c]		
AL 198-1		30.8	18.1		
AL 333w 1a+b		32.4	23.0		
AL 333w-32+60	62.8	35.4($\pm$1)	23.6($\pm$1)	47.0($\pm$2)	21.0($\pm$1)
AL 266-1		27.6	24.2(+2)[c]		20.3($\pm$2)
AL 400-1a	53.9			35.9	18.8
AL 288-1	50.6	27.6		32.8($\pm$1)	17.3($\pm$1)

[a] The numbers in parentheses indicate probable magnitude of error due to postmortem damage to the fossil. The mean values in Appendix 3.1 are based on the corrected values, not on the apparent values.

[b] All M_2-I_1 length measurements were taken on casts of the original specimens. The remaining measurements were taken on original specimens.

[c] Indicates that the number in parentheses should be added to the apparent value.

[d] Indicates that the number in parentheses should be subtracted from the apparent value.

Enlarged Occipital/Marginal Sinuses and Emissary Foramina: Their Significance in Hominid Evolution

4

DEAN FALK

To date, an enlarged occipital/marginal sinus system (O/M) occurs in 100% of scorable cranial remains of both "robust" australopithecines ($N = 7$) and the Hadar material attributed to *Australopithecus afarensis* ($N = 6$). This trait is not fixed in any other group of fossil hominids and is reduced to 6% in extant *Homo sapiens* (Falk, 1986a). A functional analysis of an enlarged O/M system in early hominids led to the conclusion that "*A. afarensis* was directly ancestral to, or shared a common ancestor with, 'robust' australopithecines" (Falk and Conroy, 1983), a view put forward earlier by Olson (1981, 1985) for entirely different reasons. The recent discovery of KNM-WT 17000, a "robust" australopithecine dated to 2.5 Myr (Walker *et al.*, 1986), is consistent with this view; and it now appears that, despite the earlier belief that "robust" australopithecines descended from *Australopithecus africanus* rather than *A. afarensis* (Johanson and White, 1979), many paleontologists believe that the population of "robust" australopithecines represented by KNM-WT 17000 descended from earlier Hadar hominids.

One purpose of this chapter is to assess the phylogenetic relationship of "robust" australopithecines vis-à-vis other hominids. A broader purpose is to clarify the interpretation of the functional significance of certain cranial features that are related to cranial blood flow and detectable in fossil hominids and to discuss the relative merits of utilizing functional morphological analyses versus recent cladistic approaches for assessing hominid systematics. Information about O/M systems and emissary foramina, brain size, and the concepts of paleospecies and race will be used to construct a "family plant" that summarizes one relatively parsimonious view of hominid evolution. In short, I hope to demonstrate that a functional analysis of cranial blood flow is worthy of consideration and that systematic conclusions based on such an analysis should not be viewed as doing cladistics based on one trait.

Functional Morphology versus Cladistic Analyses

Numerous workers who utilize a cladistic approach to hominid systematics expressed surprise at the discovery of a "robust" australopithecine with the morphological features of KNM-WT 17000 (Walker *et al.*, 1986). Specifically, a "robust" australopithecine of this antiquity contradicts the hypothesis that *A. africanus* was ancestral to "robust" australopithecines (Johanson and White, 1979). Since 1979, this hypothesis has repeatedly been affirmed by numerous workers (White *et al.*, 1981; Kimbel *et al.*, 1984; Skelton *et al.*, 1986; Kimbel, 1984), a situation that may be attributed to questionable applications of the cladistic method of analysis.

A cladogram is only as representative of reality as the trait list upon which it is based. Skelton *et al.* (1986) illustrate that one can construct trait lists to support almost any hypothetical tree for hominid evolution. However, proper cladistic analysis should be based on shared features that were independently derived. Therefore, for probabilistic reasons, highly correlated traits that may be part of one functional complex should *not* be weighed separately in constructing cladograms. Unfortunately, this rule is generally recognized but rarely practiced. For example, one paper recognizes that it is usually "not advisable to consider traits that appear to be evolving in tandem as separate, since this tends to inflate the importance of what may actually be only one phenomenon'' yet, in that same paper, five of seventeen traits on one list pertain to canines and three to P_3 (Skelton et al., 1986). Furthermore, no justification is offered for weighing these traits independently in the construction of a cladogram. Since these eight traits constitute 17% of the 45 features said to be consistent with the authors' preferred cladogram, and since that cladogram is favored over another that is supported by only one less trait, a discussion of whether or not each of these eight traits should really be counted independently is in order.

The practice of independently weighing traits that may be functionally correlated in cladistic analyses is common in the literature on hominid evolution. Although only one example has been provided above to illustrate the error of the practice, other problems are also apparent in current applications of cladistic analyses. Aside from the difficulty of accurately determining the polarity of features, many workers disagree on how character states should be scored and/or which character states should be ascribed to which hominids. For example, do *Homo, A. africanus,* and *A. robustus/boisei* share a P^3 occlusal profile that is symmetrical (Kimbel et al., 1984:375) or is the occlusal outline intermediate in *A. africanus* and *Homo habilis* and symmetrical (oval) in *A. robustus/boisei* (Skelton et al., 1986:25)? Is the mental foramen high on the mandibular corpus of *A. africanus* and *H. habilis* (Kimbel et al., 1984:375), or is its position variable in these two species (Skelton et al., 1986:25)? In short, given the above reservations about current applications of the cladistic method, it is not surprising that the field of hominid paleontology can be unsettled by the finding of one new skull.

Some workers have suggested that systematic conclusions determined from a study of cranial blood flow are suspect because they are based on cladistic analysis of one trait. This assessment is inaccurate for a number of reasons. Although earlier work focused on only a few traits including transverse and enlarged O/M sinuses (Falk and Conroy, 1983), the work on cranial blood flow has since been extended to include five other features that can be detected in osteological material: posterior condyloid, mastoid, occipital, and parietal foramina, and multiple hypoglossal canals (Falk, 1986a). One could also add other features to this list including the jugular foramen and the grooves for the superior sagittal, sigmoid, and superior petrosal sinuses. Although derived and primitive states for many of these features can be recognized, the traits are *not* proper subjects for cladistic analysis as separate entities because they are components of one functional complex, namely that involved in draining blood from the cranium.

This does not mean that the traits in question are without evolutionary significance. On the contrary, an effort can be made to understand the changes that occurred in the *system* of cranial blood flow during hominid evolution and to examine these changes from a functional point of view. Knowledge about evolutionary changes in the system can then be applied to the question of ancestral/descendant relationships among hominids. Although this knowledge incorporates recognition of derived and primitive character states, it differs from the cladistic method in that independently derived features are not enumerated and "weighted" in an effort to arrive at a most probable phylogeny.

Cranial Blood Flow: The Functional Approach

Tobias (1967, 1968) was the first to describe the enlarged O/M sinus system that characterizes "robust" australopithecines, and Holloway (1981) was the first to note that such a configuration is also present in the Hadar fossil material attributed to *A. afarensis.* As Fig. 4.1 shows, these early hominids possessed an enlarged O/M sinus system that either supplemented or replaced the transverse sinuses that connect with the sigmoid sinuses in most hominids and apes. Normally, living humans have O/M sinuses that are too small to leave marks on the insides of crania. Thus, more blood appears to have been channeled through the O/M sinus system in "robust" australopithecines and the Hadar hominids than is the case for living humans. Because of this, it is of interest to ask if blood that flows through the O/M sinus (enlarged or otherwise) is in any way distributed differently from that which flows through the transverse/sigmoid routes?

The answer to this question is "yes." Browning (1953) dissected the venous drainage systems in 100 human cadavers, of which nine possessed O/M sinus systems (termed "occipital sinus

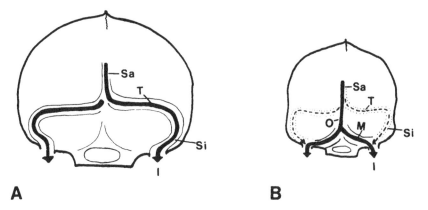

FIGURE 4.1. Venous drainage systems in humans (A) and Hadar fossil hominids and "robust" australopithecines (B), from Falk and Conroy (1983). Abbreviations: I, internal jugular vein; M, marginal sinus; O, occipital sinus; Sa, superior sagittal sinus; Si, sigmoid sinus; T, transverse sinus. See text for discusssion.

systems" by Browning) that he judged to be of significance in venous drainage from the cranium, i.e., "their combined capacity was similar to that of the straight sinus." Of these, the occipital sinuses in two cadavers connected the confluence with the jugular bulbs and therefore "served as collateral channels for the smaller of the transverse sinuses." In three cadavers, the occipital sinuses connected the confluence with the vertebral plexus and also delivered blood to the jugular bulbs. In the remaining four cadavers, the occipital sinuses connected the confluence with the vertebral plexus but did not deliver blood to the jugular bulbs. Browning therefore established that 9% of 100 human cadavers had occipital sinuses that participated significantly in draining blood from the cranium and that blood was drained into the vertebral plexus of veins in 78% of these specimens, or 7% of the total sample. Thus, in the majority of incidences where an even somewhat enlarged O/M sinus system occurs, blood is drained to the vertebral plexus of veins instead of, or in addition to, being delivered through the "normal" route, i.e., to the internal jugular system.

Other workers point out that an enlarged O/M sinus system is common in the human fetus and during the first few years of life (Woodhall, 1936) and note again that "connections of the occipital sinus with the internal vertebral plexus . . . are well known" (Das and Hasan, 1970). Padget (1957) illustrates connections between the O/M sinus and the vertebral plexus in humans (see Falk and Conroy, 1983 or Falk, 1986a for illustration) and Gray's Anatomy (1973) states that normally the occipital sinus connects the confluence with the posterior internal venous plexus. It should be noted that the O/M sinus system of early fossil hominids is dramatically enlarged compared to most of the human material presently under discussion. For instance, while Tobias (1967) noted an enlarged O/M sinus system in thirteen of two hundred and eleven skulls (6%) of black Africans, only 4 (2%) showed the dramatic enlargement that is typical of "robust" australopithecines. In sum, it seems reasonable to conclude that "robust" australopithecines and Hadar fossil hominids drained an appreciable amount of blood through their enlarged O/M sinuses and that it is probable that much of this blood was destined for the vertebral plexus of veins.

The Vertebral Plexus and Bipedalism

As detailed by Eckenhoff (1970), Gius and Grier (1950), and Falk and Conroy (1983), the vertebral plexus of veins is a major route of exit of cranial blood in individuals that are erect but such is not the case for individuals in the supine position, who instead deliver cranial blood mainly to their jugular systems. This finding is well documented and has important implications for patients undergoing surgery of the head and neck (Gius and Grier, 1950). Because pressure gradients differ across the vertebral venous plexus depending on gravity, posture and breathing, individuals in a vertical position drain cranial blood primarily to the vertebral plexus, especially during expiration and forced expiration (during exertion) (Eckenhoff, 1970).

A recent comparative angiographic study (Dilenge and Perey, 1973) of cranial blood flow in *Macaca mulatta* and humans shows the same results as the studies above on humans alone. Blood flows preferentially to the jugular system when monkeys are lying down (or, interestingly, in a vertical position but with the head down) and to the cervical vertebral plexus when they are standing. Dilenge and Perey (1973:336) note:

> Significantly, dye injected into the arteries of the brain in monkeys and man was found to favor the vertebral route in the vertical position while it returns exclusively through the jugular veins in the horizontal position. . . . *Indeed, since man is in the upright position most of the time, it is logical to think that it is the vertebral plexus rather than the internal jugular veins which is the major pathway of cerebral venous return in man.* (emphasis added)

Other experimental work (Epstein *et al.*, 1970) confirms that cerebral venous blood returns mainly to the vertebral plexus of veins in monkeys that are standing.

The above data concerning the anatomical connections of the O/M system and the physiology of cranial blood flow strongly support a functional explanation for the fixation of an enlarged O/M sinus system in two groups of early hominids, one of which includes some of the earliest known bipeds (Falk and Conroy, 1983). To wit, one must reject the notion that the "occipital/marginal and transverse/sigmoid sinus patterns are adaptively equivalent character states," as has been suggested by Kimbel (1984) without benefit of either the physiological data or information pertaining to the connections of O/M with the vertebral plexus. Rather, the transverse/sigmoid and O/M systems have different connections in living humans and, because of this, the situation should be interpreted similarly in fossil hominids. That is, an enlarged O/M in Hadar early hominids probably delivered a good deal of blood to the vertebral plexus of veins and, in light of the physiology of cranial blood flow in an upright position, it is logical to explain its fixation in those early bipeds as an epigenetic adaptation that accompanied selection for bipedalism. Conroy and I further interpret fixation of this feature in "robust" australopithecines as a retention in *one* lineage (see below) that descended from the Hadar fossil hominids.

Blood Flow in Other Hominid Groups

Since *H. sapiens* is a biped and since this species lacks fixation of an enlarged O/M sinus system, there must be another mechanism for delivering blood to the vertebral plexus of veins when humans are awake and moving about, i.e., in an upright position. According to Gray's Anatomy (1973:708), the internal vertebral plexus around the foramen magnum is connected above with the occipital sinus (small in most adults), the basilar plexus, the condyloid emissary vein, and the rete canalis hypoglossi. The vertebral plexus also receives blood from parietal, mastoid, and occipital emissary veins via their connections with the occipital vein (Gray, 1973:685,694). The contributions of these cranial emissary veins to the vertebral plexus are well known (Gius and Grier, 1950; Eckenhoff, 1970). Thus, instead of depending substantively on an enlarged O/M route as the Hadar bipeds apparently did, extant humans have a variety of small pathways (a network) for delivering blood to the vertebral plexus of veins. In fact, the total cross-sectional area of these contributions to the vertebral plexus as it leaves the cranium may actually exceed that of the two jugular veins (Batson, 1944).

The suggestion that a network of veins for delivering blood to the vertebral plexus may have been selected for during hominid evolution is supported by an earlier comparative study that showed that parietal, mastoid, and condyloid foramina exist in much higher frequencies in living humans than in gorillas or chimpanzees (Boyd, 1930). In 1984, I had the opportunity to confirm Boyd's findings and to extend the study of emissary foramina to the hominid fossil record. Table 4.1 and Fig. 4.2 summarize some of the results from the latter study (Falk, 1986a) and include data on the enlarged O/M sinus system.

Table 4.1. Features Present on One or Both Sides[a]

Groups	Enlarged O/M	Mastoid foramen	Parietal foramen
African apes	16 (43)	23 (88)	14 (94)
Hadar[b]	100 (6)		
"Robusts"	100 (7)	≥25 (8)	0 (3)
Laetoli[c]	0 (1)		
"Graciles"–*Homo erectus*	≥25 (12)[d]	≥50 (10)	≥0 (13)
Archaic–modern *Homo sapiens*	≥8 (225)	≥92 (76)	≥63 (60)

[a] Data in percentage (*N*). Most of the data from which this table is compiled are from Falk (1986a).

[b] From Kimbel *et al.* (1982); percentage cannot be calculated for emissary foramina since these have not been methodically scored for the Hadar specimens.

[c] LH-21 from Kimbel (1984).

[d] Sts 5 has been scored as missing O/M and added to the earlier sample (Falk, 1986a) for this table; Taung is included in "gracile" australopithecines (Tobias and Falk, 1988); ≥, frequency is greater than or equal to the entry since fossil crania are often represented by only one side.

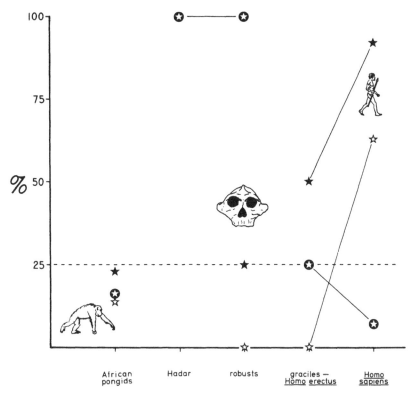

FIGURE 4.2. Graph of data from Table 4.2 (except LH-21), which are summarized from Falk (1986a). Solid stars represent mastoid foramina; open stars, parietal foramina; and stars within circles, enlarged O/M sinuses. The Hadar— "robust" australopithecine lineage and the "gracile" australopithecine—*Homo* lineage have different systems for draining blood from the cranium when hominids are upright. In the former group, an enlarged O/M sinus has been selected for and fixed whereas apelike frequencies are maintained for the emissary foramina. The reverse situation is true for the latter group. Apelike frequencies of enlarged O/M are maintained and high frequencies of emissary foramina are selected for. Besides draining blood to the vertebral plexus, the latter strategy has the added benefit of providing a means for cooling the brain under conditions of hyperthermia. See text for discussion.

The three structures shown in Fig. 4.2 are osteological indications of cranial blood routes to the vertebral plexus of veins, a structure that is known to be an important recipient of cranial blood in upright (but not supine) *H. sapiens*. The African ape frequencies for these structures are all below 25%. This is not surprising since African apes are not habitually bipedal and since they do not hold their heads upright even when their bodies appear to be in vertical positions (Goodall, 1986; Dilenge and Perey, 1973). "Robust" australopithecines retain apelike low frequencies for the two emissary foramina but are characterized by fixation of an enlarged O/M sinus system, which also characterizes the Hadar fossil hominids.

The situation for the lineage leading from "gracile" australopithecines to extant *H. sapiens* is different. The frequency of an enlarged O/M is within the ape range and appears to decline through time (although sample sizes are small for fossil material and Fig. 4.2 smooths data that are presented in more detail elsewhere (Falk, 1986a). The situation for the parietal and mastoid emissary foramina is the reverse, i.e., these two foramina appear to increase in the "gracile" australopithecine through *H. sapiens* line. Despite sample sizes, these latter trends are in keeping with expectations based on the comparative data from larger samples of *H. sapiens* and African apes (Boyd, 1930; Falk, 1986a).

It is important to note that we do not presently know the frequency of O/M in the ancestors of "gracile" australopithecines. Was it fixed, as is the case for the Hadar–"robust" australopithecine lineage, in which case fixation relaxed in "gracile" australopithecines, as was earlier suggested (Falk, 1986a)? Or did the ancestors of "gracile" australopithecines retain apelike frequencies of an enlarged O/M system as their circulatory systems were undergoing modifications in conjunction with refinement of bipedalism that differed from those of the "robust" lineage? (That is, were there two different solutions to the same problem in the two major groups of bipedal hominids?) As discussed below (p. 93), Kimbel's (1984) suggestion that the only Laetoli specimen scored for venous sinus system (LH-21) *lacked* the derived enlarged O/M system that characterizes 100% of scorable Hadar fossils and 100% of scorable "robust" australopithecines supports the latter hypothesis. It should be noted that this is a departure from my earlier (1986a) conclusion that fixation of O/M relaxed in "gracile" australopithecines. These questions bear on the issue of how many species are represented in so-called *A. afarensis* and on hominid systematics in general (discussed below).

The Simplest Phylogenetic Scheme

Although the acquisition of bipedalism may have been a relatively rapid event, selection for habitual bipedalism *had* to have been accompanied by vast changes in physiology as well as anatomy. Since blood drains preferentially to the vertebral plexus of veins when a standing position is assumed [even when quadrupedal monkeys are placed in standing positions (Dilenge and Perey, 1973)], this plexus and the routes leading to it are of great importance for a bipedal hominid. Presumably, they were subject to epigenetic selection and refinement as bipedalism was selected for, achieved, and modified.

Figure 4.3 presents an alternative "evolutionary plant" that is concordant with the functional/morphological model discussed in this paper, as well as with other information from the fossil record. Aptly, this plant is modeled after the "Old Man" cactus, *Opuntia vestita*, which has striking awl-shaped leaves on its new growth (Chidamian, 1984). In Fig. 4.3, each leaf represents a race, or subspecies. New branches of this cactus, on the other hand, are directed either straight ahead (anagenesis resulting in paleospecies) or off to one side (cladogenesis resulting in contemporary species).

The origin of bipedal hominids seems to have occurred around 5 Myr ago (Sarich, 1971). Bipedalism was accompanied by a dramatic event, as illustrated by the phylogenetic cactus, namely speciation of early hominids into two groups: more generalized ancestral "gracile" australopithecines, on one stem, and basal "robust" australopithecines, on the other. The Hadar specimens share fixation of a derived enlarged O/M sinus system with "robust" australopithecines from East and South Africa (Falk, 1986a). To date, fixation of this feature is not found in any

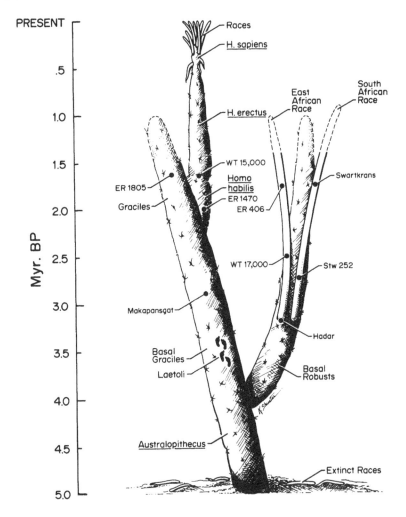

FIGURE 4.3. A phylogenetic plant that summarizes the view of hominid evolution discussed in the text. Modeled after the "Old Man" cactus (*Opuntia vestita*).

of the fossil hominids on the other stem. It is also important to note that, although brain size increases somewhat within the "robust" australopithecine lineage over time, there seems to be a constraint on its increase in this lineage as compared to the situation for later contemporary hominids on the other stem (Falk, 1987). So far, no "robust" australopithecine capacity (cm^3) has even reached the latter 500's (see below).

According to the view presented here, the oldest fossil evidence of "robust" australopithecines is from Hadar, and some if not all of these specimens are depicted in Fig. 4.3 as being members of the East African subspecies (race) of "robust" australopithecines. The KNM-WT 17000 specimen fits well within the East African race of "robust" australopithecines for two reasons. First, it is from the same geographical location as are the more recent specimens of this subspecies, as one might expect for the fossil record of a race (Wolpoff, 1985). Second, its cranial capacity of 410 cm^3 (Walker *et al.*, 1986) is between the mean of 404 cm^3 for the Hadar specimens ($N = 3$; Falk, 1987) and the mean of 512 cm^3 for East African "robust" australopithecines that lived about 1.76 Myr BP ($N = 4$; Falk, 1987). (I believe the mean for Hadar will decrease once the frontal part of the endocast of AL 333-45 is reconstructed on the basis of new information determined from the endocast of KNM-WT 17000. If AL 333-45 is eliminated from the Hadar sample, its mean falls to 359 cm^3.) As for South Africa, Clarke (this volume, chapter 18) believes that Stw 252 and Sts 71 from Sterkfontein exhibit features that are intermediate between *A. afarensis* and "robust" australopithecines. These data, together with Shipman and Harris's (this volume, chapter 24) findings that East and South African "robust" australopithecines may have lived in different habitats (wet vs. open, respectively), support the notion that geographical diversity may have resulted in racial variation in "robust" australopithecines, much as it does for extant hominids.

Why Subspecies and Not Species

In an effort to accommodate KNM-WT 17000, several workers suggest a polyphyletic phylogeny for "robust" australopithecines (Walker and Leakey, this volume, chapter 15; Wood, this volume, chapter 17). This suggestion requires a great deal of parallel evolution, which some workers find questionable (Eckhardt, 1986; Falk, 1986b). Interestingly, proponents of polyphyletic models seem to have rejected, or at least not to have considered, the possibility that the morphological differences and similarities of East and South African "robust" australopithecines might be explained as the result of adaptation to different geographic regions combined with gene flow within one species, i.e., as simply due to racial (subspecific) variation.

As is the case for "robust" australopithecines, Weidenreich observed regional continuity in geographically distinct hominid lineages, namely those in the Sunda subcontinent and those in north China (Wolpoff, 1985). Furthermore, as is the case with "robust" australopithecines, these fossil sequences spanned a good deal of time, more than half a million years in the case of Sundaland (Thorne and Wolpoff, 1981). Wolpoff (1985) points out that Weidenreich's observation of regional continuity should be distinguished from his orthogenetic explanation, and he further points out that the regional differences observed by Weidenreich were "maintained because of a balance between gene flow and selection that develops to form long-lasting morphological clines." I see no reason why this "multiregional evolution theory" (Wolpoff *et al.*, 1984) should not apply to South and East African "robust" australopithecines. It explains their similarities (one species), and it explains their differences (different races adapted to different regions). As was the case for the material examined by Weidenreich, the time depths for both groups are deep, and intermediary specimens *in both groups* show that evolutionary changes occurred within the species.

Hadar: One Species or Two?

As discussed above, I have placed Hadar on the "robust" stem in Fig. 4.3 on the basis of a functional analysis related to all six specimens that were scored for enlarged O/M sinuses (Falk, 1986a). However, other workers have suggested that there is too much variation in the Hadar material to be accounted for by sexual dimorphism and that the Hadar material therefore contains remains of two rather than one species (Olson, 1981; Zihlman, 1985; Senut and Tardieu, 1985). In particular, Zihlman has argued that humeral and femoral head diameters for the Hadar material suggest *greater* dimorphism than the sexual dimorphism that exists even for the most dimorphic of apes, the orangutans. Zihlman adds that since humans shared a more recent common ancestor with chimpanzees than with other pongids, one would *not* expect the hominid precursor to have been more sexually dimorphic than the pygmy or common chimpanzee, in both of whom mean average body weight is 80–84% of mean male body weight (Zihlman, 1985). However, according to McHenry (this volume, chapter 9), it is possible that sexual dimorphism has been reduced during the evolution of both chimpanzees and humans. Although most workers [such as McHenry (this volume, chapter 9) and Jungers (this volume, chapter 7) concerning the postcranial evidence, and White and Kimbel (this volume, chapter 11) on the craniodental evidence] suggest that the variation in the Hadar material may be due to sexual dimorphism, the question of how many species are represented at Hadar is in need of further investigation.

Laetoli and Hadar: Same or Different?

The question of how many species are represented at Hadar should be viewed separately from the question of whether or not the Laetoli and Hadar material represents one or more species. The latter is a difficult question to address because the Laetoli material consists mostly of jaws and teeth, whereas more cranial and postcranial specimens are included in the Hadar material. The overwhelming impression of the Laetoli remains is one of primitiveness (Kimbel et al., 1984), which generally does not provide much information for ascertaining specific phylogenetic relationships. In fact, according to Kimbel et al. (1984), "we have been unable to identify *any* shared-derived characters exclusive to *A. africanus* and *A. afarensis*" (but see below). It is perhaps noteworthy that, although postcrania are important for assessing whether or not variation in a sample is because of sexual dimorphism or multiple species, this was not considered in the construction of the *A. afarensis* hypodigm (White et al., 1981).

There are hints, however, that the Laetoli fossils may in fact differ from those at Hadar. For example, "some Laetoli dental metrics slightly exceed those of the Hadar sample . . ." (White et al., 1981). To be specific, four area measurements for premolars and first molars, out of sixteen (25%) tooth measurements, are significantly larger in the Laetoli sample (Blumenberg and Lloyd, 1983). Another example is provided by Tuttle (1985, 1987) who believes that the feet of the individuals that made the site G footprints at Laetoli were morphologically indistinguishable from those of extant humans and that it is unlikely that the curved phalanges from Hadar could have made such footprints. Although Tuttle's conclusions are controversial (Stern and Susman, 1983), if they prove to be correct, ancestors of "robust" australopithecines and those of other hominids would have comprised distinct groups by *at least* 3.5 Myr BP. [It should be noted that while Olson (1981) believed that the Laetoli and Hadar samples comprised two species, he placed the material from Laetoli with the Hadar remains from the 333 locality into a single taxon, *Paranthropus africanus*.]

Finally, the only cranial material from Laetoli that has been scored for the O/M sinus system is LH-21 and, according to Kimbel (1984: 253):

> The broad, low inferior sagittal limb of the cruciate eminence, although damaged, does not appear to bear an occipital sinus impression. Hence, there is no direct evidence for an enlarged occipital-marginal sinus system.

Although I have not had the opportunity to score this specimen, if Kimbel is right, the Laetoli specimen differs from all six of the scorable Hadar specimens that show a derived enlarged O/M sinus system. This, together with other evidence presented above, would strongly suggest that the Laetoli and the Hadar material should not be included in the same taxon.

Whence *A. africanus*?

Figure 4.3 shows "gracile" australopithecines on the same stem as Laetoli. Various authors have emphasized the primitive nature of the Laetoli dental and gnathic material and have listed numerous features of *A. africanus* that could have derived from such a primitive substrate (Johanson and White, 1979; White et al., 1981; Kimbel et al., 1984; Skelton et al., 1986). These data, together with the above "hints" regarding the possibility of humanlike footprints and of humanlike venous drainage systems at Laetoli as opposed to Hadar, are in keeping with Fig. 4.3.

The question of whither *A. africanus* is equally interesting. I have illustrated *H. habilis* branching from the "gracile" australopithecine stem (Fig. 4.3) because both groups share numerous derived features and the latter also has many specialized features that could have been derived from the former (Skelton et al., 1986). However, because of the angle at which the phylogenetic cactus is illustrated, we do not know how far down the *Homo* stem originates from the "gracile" australopithecine stem. It should be noted that KNM-ER 1805 and KNM-ER 1813 are located on the australopithecine stem instead of the *Homo* stem because of features reported for the endocast of KNM-ER 1805 (Falk, 1983) and because the two skulls resemble each other.

Discussion

For the physical, physiological, and anatomical reasons delineated above, greater volumes of cranial blood must have been channeled to the vertebral plexus of veins as bipedalism was selected for and modified in basal hominids. Basal "robust" australopithecines apparently delivered cranial blood to the vertebral plexus through an enlarged O/M sinus system, a trait that appears to have been fixed in their descendants. Basal "gracile" australopithecines and their *Homo* descendents, on the other hand, relied on a network of emissary (and other) veins to drain blood to the vertebral plexus. Besides delivering blood to the vertebral plexus, emissary veins serve to cool the brain by transporting cool surface blood into the skull under conditions of hyperthermia, but not under other thermal conditions (Cabanac and Brinnel, 1985). The latter is an important function since, as Baker (1979:136) notes:

> A rise of only four or five degrees C. above normal begins to disturb brain functions. For example, high fevers in children are sometimes accompanied by convulsions; these are manifestations of the abnormal functioning of the nerve cells of the overheated brain. Indeed, it may be that the temperature of the brain is the single most important factor limiting the survival of man and other animals in hot environments.

"Gracile" australopithecines eventually gave rise to the *Homo* lineage, in which brain size increased autocatalytically (Falk, 1987). Nothing like this increase in brain size has been seen in other mammals and, indeed, this phenomenon constitutes one of the greatest mysteries in paleontology. The work on cranial blood flow by Cabanac's team may provide a key for resolving this mystery. Cabanac (pers. comm.) suggests that "brain thermolytic needs have increased with its increasing size" and that it is possible that emissary veins "were developed for the defense of the brain temperature." Both comparative (Boyd, 1930) and direct fossil evidence (Falk, 1986a; fig. 2) suggest that the frequencies of emissary veins have increased dramatically during human evolution. (It should be noted, however, that elaboration of the full network of emissary veins did not occur until relatively late, since the parietal emissary vein has yet to be observed in populations prior to archaic *H. sapiens* [see Table 4.1 and Falk (1986a) for discussion].

Since brain size increases dramatically between "gracile" australopithecines and *H. sapiens*, it could be that selection for the web of veins that constributes blood to the vertebral plexus had the side benefit of cooling the brain during hyperthermia. If so, selection in this one lineage inadvertently removed a constraint that previously prevented brain enlargement. In other words, it may be that mammalian brains only grow to sizes that can be cooled by their circulatory systems under conditions of hyperthermia. This theory accounts for the "mosaic" nature of hominid evolution, i.e., it explains why bipedalism preceded brain enlargement, and it accounts for why brain size did not increase dramatically in "robust" australopithecines who lacked the cooling benefits provided by an emissary network of veins. A further implication of the present work is that basal "robust" and basal "gracile" australopithecines may have occupied distinct niches that were characterized by lifestyles associated with different thermoregulatory demands (e.g., Carrier, 1984).

Acknowledgments

I thank F. Grine for the invitation to participate in the International Workshop on "robust" Australopithecines, C. Helmkamp for preparing Fig. 4.2, and K. Shuster for Fig. 4.3. Comments by E. Delson and others at the Workshop caused me to reassess (and change my opinion about) whether or not O/M was likely to have been fixed in the ancestors of the "gracile"–*Homo* lineage.

This research is supported by National Science Foundation grant No. BNS-8796195 and Public Health Service award No. 1 RO1 NS24904.

References

Baker, M. A. (1979). A brain-cooling system in mammals. *Sci. Amer.*, 240: 130–139.
Baston, O. V. (1944). Anatomical problems concerned in the study of cerebral blood flow. *Fed. Proc.*, 3: 139–144.
Blumenberg, B. and Lloyd, A. T. (1983). *Australopithecus* and the origin of the genus *Homo*: Aspects of biometry and systematics with accompanying catalog of tooth metric data. *BioSystems*, 16: 127–167.
Boyd, G. I. (1930). The emissary foramina of the cranium in man and the anthropoids. *J. Anat.*, 65: 108–121.
Browning, H. (1953). The confluence of dural venous sinuses. *Amer. J. Anat.*, 93: 307–329.
Cabanac, M. and Brinnel, H. (1985). Blood flow in the emissary veins of the human head during hyperthermia. *Eur. J. Appl. Physiol.*, 54: 172–176.
Carrier, D. R. (1984). The energetic paradox of human running and hominid evolution. *Curr. Anthropol.*, 25: 483–495.
Chidamian, C. (1984). *The Book of cacti and other succulents.* Timber Press, Portland.
Das, A. C. and Hasan, M. (1970). The occipital sinus. *J. Neurosurg.*, 33: 307–311.
DiLenge, G. and Perey, B. (1973). An angiographic study of the meningorachidian venous system. *Radiology*, 108: 333–337.
Eckenhoff, J. E. (1970). The physiologic significance of the vertebral venous plexus. *Surg. Gynecol. Obstet.*, 131: 72–78.
Eckhardt, R. B. (1986). Hominid evolution. *Science*, 234: 11.
Epstein, H. M., Linde, H. W., Crampton, A. R., Ciric, I. S. and Eckenhof, J. E. (1970). The vertebral venous plexus as a major cerebral venous outflow tract. *Anesthesiology*, 32: 332–337.
Falk, D. (1983). Cerebral cortices of East African early hominids. *Science*, 221: 1072–1074.
Falk, D. (1986a). Evolution of cranial blood drainage in hominids: enlarged occipital/marginal sinuses and emissary foramina. *Amer. J. Phys. Anthropol.*, 70: 311–324.
Falk, D. (1986b). Hominid evolution. *Science*, 234: 11.
Falk, D. (1987). Hominid paleoneurology. *Annu. Rev. Anthropol.*, 16: 13–30.
Falk, D. and Conroy, G. C. (1983). The cranial venous sinus system in *Australopithecus afarensis*. *Nature*, 306: 779–781.
Gius, J. A. and Grier, D. H. (1950). Venous adaptation following bilateral radical neck dissection with excision of the jugular veins. *Surgery*, 28: 305–321.
Gray, H. (1973). *Anatomy of the human body*, 29th edition. Lea and Febiger, Philadelphia, PA.
Holloway, R. L. (1981). The endocast of the Omo L338y-6 juvenile hominid: gracile or robust *Australopithecus? Amer. J. Phys. Anthropol.*, 54: 109–118.
Johanson, D. C. and White, T. D. (1979). A systematic assessment of early African hominids. *Science*, 203: 321–330.
Kimbel, W. H. (1984). Variation in the pattern of cranial venous sinuses and hominid phylogeny. *Amer. J. Phys. Anthropol.*, 63: 243–263.
Kimbel, W. H., Johanson, D. C. and Coppens, Y. (1982). Pliocene hominid cranial remains from the Hadar Formation, Ethiopia. *Amer. J. Phys. Anthropol.*, 57: 453–500.
Kimbel, W. H., White, T. D. and Johanson, D. C. (1984). Cranial morphology of *Australopithecus afarensis:* A comparative study based on a composite reconstruction of the adult skull. *Amer. J. Phys. Anthropol.*, 64: 337–388.
Olson, T. R. (1981). Basicranial morphology of the extant hominoids and Pliocene hominids: The new material from the Hadar Formation, Ethiopia and its significance in early human evolution and taxonomy. *In* C. B. Stringer (ed.), *Aspects of human evolution*, pp. 99–128. Taylor and Francis, London.

Olson, T. R. (1985). Taxonomic affinities of the immature hominid from Hadar and Taung. *Nature, 316*: 539–540.

Padget, D. H. (1957). The development of the cranial venous system in man, from the viewpoint of comparative anatomy. *Contrib. Embryol., 36*: 79–140.

Sarich, V. (1971). A molecular approach to the question of human origins. *In* V. Sarich and P. Dolhinow (eds.), *Background for Man*, pp. 60–81. Little, Brown, Boston.

Senut, B. and Tardieu, C. Functional aspects of Plio-Pleutocere hominid limb bones: implication for taxonomy and phylogeny. (1985). *In* E. Delson (ed.), *Ancestors: the hard evidence*, pp. 193–201. Alan R. Liss, New York.

Skeleton, R. R., McHenry, H. M. and Drawhorn, G. M. (1986). Phylogenetic analysis of early hominids. *Curr. Anthropol., 27*: 21–43.

Stern, J. T. and Susman, R. L. (1983). The locomotor anatomy of *Australopithecus afarensis*. *Amer. J. Phys. Anthropol., 60*: 279–317.

Tobias, P. V. (1967). *The cranium and maxillary dentition of Australopithecus (Zinjanthropus) boisei*. vol. 2: Olduvai Gorge. Cambridge University Press, Cambridge.

Tobias, P. V. (1968). The pattern of venous sinus grooves in the robust australopithecines and other fossil and modern hominoids. *In* R. Peter, F. Schwartzfischer, G. Glowatzki, and G. Ziegelmayer (eds.), *Anthropologie und Humangenetik*, pp. 1–10. Gustav Fischer, Stuttgart.

Tobias, P. V. and Falk, D. (1988). Evidence for a dual pattern of cranial venous sinuses on the endocranial cast of Taung (*Australopithecus africanus*). *Amer. J. Phys. Anthropol., 76:* 309–312.

Thorne, A. G. and Wolpoff, M. H. (1981). Regional continuity in Australasian Pleistocene hominid evolution. *Amer. J. Phys. Anthropol., 55*: 337–349.

Tuttle, R. H. (1985). Ape footprints and Laetoli impressions: A response to the SUNY claims. *In* P. V. Tobias (ed.), *Hominid evolution: past, present and future*, pp. 129–133. Alan R. Liss, New York.

Tuttle, R. H. (1987). Kinesiological inferences and evolutionary implications from Laetoli bipedal trails G-1, G-2/3, and A. *In* M. D. Leakey and J. M. Harris (eds.), *Laetoli, a Pliocene site in northern Tanzania*, pp. 503–523. Clarendon Press, Oxford.

Walker, A., Leakey, R. E., Harris, J. M. and Brown, F. H. (1986). 2.5-Myr *Australopithecus boisei* from west of Lake Turkana, Kenya. *Nature, 322*: 517–522.

White, T. D., Johanson, D. C. and Kimbel, W. H. (1981). *Australopithecus africanus:* its phyletic position reconsidered. *S. Afr. J. Sci., 77*: 445–470.

Wolpoff, M. H. (1985). Human evolution at the peripheries: the pattern at the eastern edge. *In* P. V. Tobias (ed.), *Hominid evolution: past, present and future*, pp. 355–365. Alan R. Liss, New York.

Wolpoff, M. H., Wu Xinzhi and Thorne, A. G. (1984). Modern *Homo sapiens* origins: a general theory of hominid evolution involving the fossil evidence from east Asia. *In* F. H. Smith and F. Spencer (eds.), *The origins of modern humans: a world survey of the fossil evidence*, pp. 411–483. Alan R. Liss, New York.

Woodhall, B. (1936). Variations of the cranial venous sinuses in the region of the torcular herophili. *Arch. Surg., 33*: 297–314.

Zihlman, A. L. (1985). *Australopithecus afarensis:* two sexes or two species? *In* P. V. Tobias (ed.), *Hominid evolution: past, present and future*, pp. 213–220. Alan R. Liss, New York.

"Robust" Australopithecine Brain Endocasts: Some Preliminary Observations

5

RALPH L. HOLLOWAY

The purpose of this chapter is twofold: (1) to explore briefly the within-lineage evolution of the "robust" australopithecine brain, and (2) to provide newer information from comparative neuroanatomy that bears directly on problems of early hominid brain evolution in the broader sense.

It should be emphasized that this chapter deals exclusively with the brain endocasts of australopithecines, particularly "robust" ones, and also with brain endocasts as they relate to hominid evolution through neural structural and behavioral adaptations.

There are a number of events that make the topic of "robust" australopithecine brain evolution a possibility. First, the recent discovery of a 2.5 Myr-old australopithecine of undoubted "robust"-like morphology (Walker et al., 1986) allows for the new possibility of actually studying evolution within a lineage of hominid that was perhaps without evolutionary relevance to the origins of our own genus, *Homo*. That is to say, with OH 5 and SK 1585 dated at 1.7 to 1.8 Myr, there is almost one million years of evolutionary time between the new discovery and those previously described.

Second, the Hadar *Australopithecus afarensis* fossil discoveries, which clearly predate the new West Lake Turkana materials, have two very provocative endocast specimens, both of which suggest that a part of this group may have given rise to a "robust" line of australopithecine represented by the KNM-WT 17000 skull, the Koobi Fora material, such as KNM-ER 406, 407, 732, the Olduvai specimen, OH 5, and some of the specimens from Swartkrans, South Africa, particularly SK 1585.

Third, recent publications by Falk (1980, 1983, 1985a,b, 1986) have brought some well-deserved and timely controversy to the topics of australopithecine brain endocasts and their relevance to hominid evolution. Thanks to her persistence and courage, the topic is no longer moribund, moot or mute.

"Robust" Australopithecine Brains

The title of this section is somewhat presumptuous. It is worthwhile to be reminded that we have no hominid brains to study; rather, we study very imperfect and incomplete casts made from the internal table of cranial bone of those fragmented fossils that survive taphonomic and geologic processes until they are discovered. We all know that endocasts have their limitations and that their main value for most paleoanthropologists would appear to be their use in the estimate of cranial capacities. As the remaining morphology is commonly missing and/or damaged, and thus controversial, most workers not in the subdiscipline of brain studies tend to ignore the use of endocasts in determining the rest of the endocranial information.

Through the courtesy of Dr. Alan Walker, I have been able to briefly examine endocast materials of the following specimens: KNM-WT 17000 (i.e. the "black skull"), which is almost complete; KNM-WT 17400, a small frontal portion; and KNM-ER 13750, a somewhat distorted dorsal portion, which appears significantly larger than the other two specimens. [All of these materials have been published elsewhere. See in particular, Walker and Leakey (this volume, chapter 15; Walker et al., 1986).] Given the short time that I have had these specimens at my disposal, my comments must necessarily be cursory and preliminary to a more thorough study.

The KNM-WT 17000 Endocast Portion

According to Walker et al. (1986) this endocast measures exactly 410 ml in volume, thus making it some 120 ml less than either OH 5 (Tobias, 1967) or SK 1585 (Holloway, 1972). As all three specimens are most probably male, the differences between the KNM-WT 17000 and later endocasts is best explained as an evolutionary change, rather than a happenstance of statistical sampling. Whether this change of more than a 100 ml (roughly 25% increase) can be written off as a mere allometric increase related to increased body size remains to be seen. For reasons that will become apparent somewhat later, I would argue that this increase in cranial capacity is probably both allometric and organizational. Unfortunately, without whole skeletal materials, we will never know for certain how much of the increase was simply allometric and how much might be explained through selection pressures on behavioral attributes with some underlying neurological basic. I thus hypothesize that this increase is in part an example of evolution that is parallel and similar to brain evolution within early *Homo*.

There are four morphological aspects of this endocast that stand out in comparison with other "robust" endocasts. First, the frontal portion is quite narrow and pointed, with a frontal bec portion very similar to the OH 5 specimen (this portion is missing on the SK 1585 specimen). Second, the dorsal cerebral height above the cerebellar lobes appears proportionately less than in OH 5 and particularly SK 1585, a feature suggesting a more primitive condition in the KNM-WT 17000 endocast. Third, the cerebellar lobes show a lateral flare and posterior protrusion that is more reminiscent of the *Pan* condition (here considered more primitive) and not as well tucked under (and modern in appearance) as in OH 5 and SK 1585. This suggests some evolutionary change in the role the cerebellar lobes in later "robust" australopithecines. Fourth, I am struck by the presence, albeit small, of a combination of both left occipital petalias (in both posterior and lateral extent), and a right frontal petalia (again, both anterior and lateral) in KNM-WT 17000. This pattern, as explained earlier (Holloway and de LaCoste-Larymondie, 1982) and as demonstrated by LeMay (1976, 1985) and LeMay et al. (1982), is highly correlated with right-handedness and, thus, hemispheric dominance. It is this *combination* of petalias that is not found in the apes, even the highly asymmetrical endocasts of *Gorilla*.

Unfortunately, the degree of sulcal and gyral relief present on the KNM-WT 17000 endocast is so minimal that nothing can be said regarding the very important regions of the third inferior frontal convolution or the posterior parietal-anterior occipital regions. Thus, there is no morphology present that bears on the controversial problems of the lunate sulcus or Broca's and Wernicke's areas. The region of the third inferior frontal convolution shows a distinct bump, but neither a human pattern of a pars triangularis, pars orbitalis and pars orbicularis, nor the pongid fronto-orbital sulcus (as emphasized by Falk, 1980, 1983) can be seen on this or any of the other australopithecine endocasts.

Although the posterior region around the midline aspect of the cerebellar lobes is slightly damaged, I can find no evidence for the presence of an occipital/marginal (O/M) venous drainage pattern, which is clearly present on two of the Hadar *A. afarensis* specimens, the Olduvai and Koobi Fora *A. boisei* specimens, and the Swartkrans endocast fragments of *A. robustus*. The suggestion by Walker et al. (1986) of little contribution of a lateral sinus flow to the sigmoid as presumed evidence for a marginal accessory sinus in KNM-WT 17000 is very unconvincing. Insofar as the Hadar specimens share this drainage pattern with later "robust" forms, the absence of this feature from this intermediate group is highly intriguing. Either it is missing through sampling error or was not present in this line (which could therefore make it a possible candidate for later "gracile" australopithecines and thus perhaps early *Homo*), or the O/M sinus drainage pattern is a variably expressed epigenetic phenomenon that may have little value as a phylogenetic marker (Kimbel, 1984).

Finally, the KNM-WT 17000 endocast bears a strong resemblance to the Omo L338y-6 specimen described previously (Holloway, 1981b). In that description, it was concluded that Omo L338y-6 need not be a "robust" australopithecine, as was earlier declared by Rak and Howell (1978). That specimen also lacks an O/M drainage pattern and shares a low cerebral-to-cerebellar height with the KNM-WT 17000 specimen, as well as a low cranial capacity (e.g., 427 ml). I tend to believe that KNM-WT 17000 is ancestral to *A. boisei* and *A. robustus* and that while the Omo L338y-6 specimen is perhaps indeed a "robust" australopithecine, it is uncertain that Omo L338y-6 represents *A. boisei*.

It is not completely clear that this group must have derived from the Hadar hominids or that the Hadar group necessarily contained any "robust" australopithecines, despite the presence of the marginal sinus. As others have noted, it is not so much that we have an abundance of "robust" australopithecines as that we are now desperately in need of earlier fossils that provide some indication of the derivation of either *Homo habilis* or *A. africanus*. The recent announcement [Johanson *et al.*, 1987; see Wood (1987) for commentary] of a new *H. habilis* find from Olduvai with a small body, unusual limb proportions, and one assumes a larger brain volume than any australopithecine, underlines our remaining ignorance about the relationships of these hominid taxa. Unlike some (Walker *et al.*, 1986; Delson, 1986; Kimbel *et al.*, this volume, chapter 16), I am not completely ready to dismiss the population as represented in part by Walker and Leakey's discoveries as totally unrelated to *A. africanus* in both East and South Africa. In many ways the Sts 5 endocast is primitive, despite its 485 ml brain volume. For example, the cerebellar lobes are laterally flaring and posteriorly placed, the rostral bec is certainly pointed, but the endocast is of such poor quality regarding convolutional details and asymmetries that I cannot totally dismiss its derivation from an older KNM-WT 17000 kind of hominid. Other "gracile" australopithecine specimens, such as Sts 60, Sts 19, TM 1511 and Taung, while certainly preserving more convolutional detail than Sts 5, are not complete or undamaged in many critical regions such as the frontal and occipital (and cerebellar) regions.

The KNM-WT 17400 Endocast Portion

This specimen, composed of the frontal portion anterior to the coronal suture, is morphologically nearly identical to the KNM-WT 17000 specimen. The frontal lobe is pointed and slightly more narrow than the KNM-WT 17000 specimen, there is a slight right frontal pole petalia, and the lateral extent is just slightly more on the right side than the left. There is no clear-cut Broca's area in the third inferior frontal convolution, and there is no evidence for an apelike fronto-orbital sulcus. Thus, the question of a humanlike organization in this region remains both moot and mute. Given the smaller size of the frontal portion than KNM-WT 17000 specimen (both anteroposteriorly and laterally), I would estimate the cranial volume as roughly 390 to 400 ml.

The KNM-ER 13750 Endocast Portion

This specimen, although somewhat distorted (the right temporal is displaced superiorly over the lower parietal region) and missing the basal portion, strongly resembles the above two specimens from West Turkana. I estimate its capacity as minimally 450 ml and probably closer to 475 to 480 ml. (These estimates are based on the fact that the arc-chord measurements for this specimen are larger than those for the complete 410 ml KNM-WT 17000 specimen.) Interestingly, this specimen is clearly larger and later in time than either of the two above specimens.

As in the case of KNM-WT 17000, there is a clear right frontal petalia (both anterior and lateral) and the suggestion of a left occipital and parietal petalial pattern, strong evidence for cerebral asymmetry, left hemispheric dominance, and right-handedness.

As described above for the KNM-WT 17000 and KNM-WT 17400 specimens, there is no morphological detail present in the region of the third inferior frontal convolution that would suggest either a human or ape pattern.

The most interesting thing about this specimen, however, is the posterior region, where there is an extraordinarily well-marked groove on the endocast just anterior to the lambdoid suture on the posterior parietal. The morphology is not well preserved, but I sincerely doubt that this is the lunate sulcus. If this feature does represent the lunate, however, it is in a very posterior position, and thus it does not retain a primitive *Pan*-like morphology. In this respect, KNM-ER 13750 is very similar to the Hadar AL 162-28 specimen.

As the basal portion is missing, nothing can be said about cerebellar morphology. The dorsal surface, however, is relatively low; thus, KNM-ER 13750 does not share the more advanced parietal cerebral height present on the OH 5 and SK 1585 endocasts.

Preliminary Assessment of "Robust" Australopithecine Endocasts

Perhaps it is best to remember the very small sample with which these newer endocranial materials can be compared. From Hadar, Ethiopia, *A. afarensis* is represented by three endocranial fragments, each critically incomplete. Specimen AL 162-28, which preserves the best convolutional detail, has no base, no anterior portion starting posterior to bregma, and no lateral or basal portions of the cerebellar lobes. Thus there cannot be any evidence for an accessory O/M sinus, Broca's area or a fronto-orbital sulcus; nor can an accurate cranial capacity be provided for this specimen. The Hadar adult AL 333-45 specimen is also missing the anterior part, and the base is incomplete as well. There is an accessory O/M sinus, but no convolutional details that are truly unambiguous. Its size is rather large, being estimated at between 485 and 500 ml (Holloway, 1983). The infant AL 333-105 specimen lacks the entire dorsal surface; it is crushed and distorted, showing no convolutional patterns except in the inferior margin of the frontal lobe and the inferior aspect of the temporal lobe. It possesses an O/M sinus. Both AL 162-28 and AL 333-105 suggest small brain sizes, well within *Pan* limits (i.e., less than 400 ml). Therefore, the questions of a more primitive lateral flare of the cerebellar lobes, a Broca's area (rather than a pongid fronto-orbital sulcus), and a narrow pointed frontal lobe [although the frontal portion of AL 288-1 ("Lucy") suggests narrowness] cannot really be thoroughly demonstrated as having been identical to the geochronologically older of the "robust" australopithecines described above. At best, it is fair to say that, aside from the O/M sinus, there is nothing in the present morphology to lead one to believe that the Hadar materials could not be ancestral to the group represented by the KNM-WT 17000, KNM-WT 17400, and KNM-ER 13750 specimens.

The later "robust" group may be represented by two species, *A. boisei* and *A. robustus* (i.e., OH 5 and SK 1585, respectively). These two specimens are relatively complete, and in the latter case, preserve some fine convolutional detail. Their cranial capacities are identical at 530 ml, and both are presumed to be male. Each has an O/M sinus, and both show tendencies toward a left-occipital, right frontal petalial pattern, the left occipital petalia being particularly marked on the SK 1585 specimen. The frontal of SK 1585 is questionable but it is slightly present on the OH 5 specimen, where it is narrow and pointed, with a rostral bec portion very similar to KNM-WT 17000. The major differences between the earlier and later "robust" specimens are the very modern cerebellar morphology and the greater degree of parietal cerebral height of the recent endocasts. Thus, if these three groups are truly related in some time-lineal sequence, it would appear that the "robust" group was derived from a species whose brain size was within *Pan* limits but whose cortical components in the posterior parietal and anterior occipital regions had become reorganized toward a more human pattern. The latter groups (whether or not they are truly different species, which this writer sincerely doubts) clearly show refinements of cortical and cerebellar shape, as well as an important increment in size. Interestingly, in neither case can any selective pressures be recognized, particularly with regard to technological adaptations, such as tool use and making [although Brain's (this volume, chapter 20) and Susman's (this volume, chapter 10) analyses certainly suggest the former at Swartkrans]. In short, brain reorganization was an early hominid event, coming before rather than after brain enlargement. The brain enlargement within the "robust" group thus suggests that there was some parallel evolution with "gracile" forms that evolved into *Homo*. (In this regard, I am thinking specifically of an evolutionary trend from Sts 5, Sts 60 and Sts 70, here regarded as *A. africanus*, to early *H. habilis*, as represented by OH 7, OH 13, KNM-ER 1470 and perhaps OH 24, although it is too deformed to be certain of its taxonomic status. Specimen OH 16 is excluded from consideration given its incompleteness.) Specimens KNM-ER 1813 and KNM-ER 1805 remain problematic as to whether they are primitive habilines, or more advanced australopithecines, although I favor the former interpretation.

Australopithecine Brain Evolution in General

Reference has already been made to the possibility of cortical brain reorganization being present by some 3 million years ago, prior to cranial brain expansion, in the Hadar *A. afarensis* and *A. africanus* Taung specimens. Needless to say, this is a controversial area, but it nevertheless is an important consideration requiring some explication.

The Hadar AL 162-28 Endocast Shows Cortical Reorganization

There are at least three good reasons for discussing these controversial matters again. First, the KNM-ER 13750 specimen offers striking confirmation of the presence of a groove immediately anterior to the lambdoid suture that could be confused with the lunate sulcus. It is worth repeating that this sulcus forms the anterior boundary of the occipital primary visual striate cortex in all primates where it is present (Clark *et al.*, 1936), a point recently emphasized by Holloway (1985) and Holloway and Kimbel (1986) in response to Falk's (1986) insistence that such a groove is not present on the Hadar AL 162-28 specimen. In both of these specimens, the groove is placed in a position well posterior to any known ape lunate sulcus. If the groove is interpreted as the lunate sulcus (per Falk 1985, 1986), then the lunate is also well posterior to an ape position.

Second, it became clear during the Workshop upon which this volume is based that most, if not all, of the fossil hominid casts present (e.g., OH 5, SK 1585, MLD 37/38, Sts 5, KNM-ER 1813, KNM-ER 1805, and KNM-ER 406 to mention but a few) indicated that the squamous portion of the parietal bone was at a more superior level than the asterionic region. Falk's (1986) claim of a hominid described by Walker *et al.* (1983) showing a more or less parallel orientation of these two landmarks begs the question, as Walker *et al.* (1983) described a hominoid *(Proconsul)* specimen. Thus, the claim of Holloway and Kimbel (1986) regarding the orientation of AL 162-28 as being misoriented by Falk (1985a) by some 39° remains a valid criticism. Falk's (1986) suggestion of a small sulcus being the superior temporal sulcus is impossible to substantiate, given its incompleteness, and the complexity of that region in both humans and apes. (See Connolly, 1950 for numerous illustrations.) The so-called superior temporal sulcus of Falk could be several other sulci.

Third, as published previously (Holloway, 1985), the actual distance from the occipital pole to the lunate sulcus in *Pan* is about six standard deviations greater than the distance for the AL 162-28 specimen. The *Pan* sample included infants, juveniles and adults, thus alleviating any allometric effect.

Finally, as Table 5.1 demonstrates, the occipital pole to lunate sulcus distance, based on a sample of 32 *Pan* brain hemispheres is both absolutely and relatively larger than in either AL 162-28 or Taung.

The Taung Endocast and the Lunate Sulcus Problem

While it is beyond the scope of this paper to engage in yet more debate regarding the Taung specimen, Table 5.1 shows the results of measurements taken on 32 *Pan* casts of the distance from the occipital pole to the superior aspect of the lunate sulcus, i.e., where it would meet the posterior limb of the intraparietal sulcus. This distance is 42 mm on the Taung specimen rather than the 40 mm claimed by Falk (1985b). Even if the 40 mm measurement were correct, the ratio is still almost two standard deviations anterior to where it would be expected on *Pan*. The ratio of occipital pole–lunate sulcus to occipital pole–frontal pole distance is some 2.85 standard deviations *anterior* to the purported lunate sulcus claimed by Falk (1980, 1983, 1985). As the Taung specimen has sometimes been regarded as a "robust" australopithecine, even rumored to have an O/M sinus (Kimbel, 1984), it seems reasonable to bring up the matter once more of its endocast.

In sum, while neither the KNM-WT 17000 nor KNM-WT 17400 specimens show any evidence for a lunate sulcus, and the KNM-ER 13750 specimen shows only a groove that could be either a lunate sulcus or caused by the inferior posterior margin of the parietal bone, the latter is very posterior, and the antecedent Hadar AL 162-28 specimen and subsequent "robust" SK 1585 specimen clearly indicate a posteriorally located lunate sulcus. This strongly suggests that *reorganization* of the hominid brain had occurred early and was retained through the "robust" australopithecine lineage(s) (Holloway, 1984).

The Marginal O/M Sinus and Bipedalism

Arguments presented by Falk (this volume, chapter 4) expand upon the theme published earlier by Falk and Conroy (1984) regarding a possible relationship between this drainage system and

Table 5.1. Chord and Arc Measurements on Hominid Endocasts and Chimpanzee Brain Casts[a,b]

Specimen	Volume (ml)	Measurement OP to IP (mm)
AL 162-28[c]	ca. 375	16
Pan infant	140	25.5
Pan adults ($N=8$)	250–407	27–34
Pan adult mean ($N=8$)		30.39

Specimen	Volume (ml)	Measurement OP to LS (mm)[d]
Taung	404	42 (per Holloway); 40 (per Falk, 1986)
Pan brains ($N=14$)	200–407	23–32; mean = 28.3; SD = 2.57
Pan hemispheres ($N=32$)	—	23–35; mean = 28.8; SD = 3.36

[a]Some measurements on both the Hadar AL 162-28 and Taung endocasts, compared to a large sample of *Pan* brain endocasts are provided. The distance from the occipital pole to the intraparietal sulcus is exactly the same as the distance from occipital pole to the lunate sulcus in all pongids, but not hominids. (See Holloway, 1983, 1985 for diagrams, and Holloway, 1980, 1981a for further discussion.) Note that this chord distance on AL 162-28 is almost six standard deviations *posterior* to where it would be expected on a pongid.

The Taung specimen's occipital pole to lunate sulcus distance is an arc measurement, and is approximately the same as the occipital pole to intraparietal sulcus chord distance, the former being naturally somewhat larger. In this case, i.e., using my measurement of 42 mm from the occipital pole to Falk's lunate sulcus, Falk's (1980, 1983, 1985a) placement of the lunate sulcus is almost three standard deviations *anterior* to where it normally occurs on *Pan*. Obviously, the Hadar Al 162-28 LS would be in a very posterior position if ratio data were available as they are for Taung.

The point of these measurements is to show that some cortical *reorganization* took place early in hominid evolution, involving a reduction in the lateral extent of primary visual striate cortex. As this chapter shows, there are strong suggestions that this also occurred in "robust" australopithecines.

[b]OP, occipital pole; IP, intraparietal sulcus; LS, lunate sulcus; FP, frontal pole.

[c]Number of standard deviations of AL 162-28 OP to IP distance from above *Pan* average = 5.92. (See Holloway, 1985 for details.)

[d]Number of standard deviations of Falk's placement of the LS anterior to the *Pan* mean: $N=14$, 2.85; $N=32$, 2.25. These are based on the ratio of OP–LS to OP–FP, Taung per Falk being about .254 and the average *Pan* being about .217. The SD for the ratio $N=32$, .016; $N=14$, .013.

bipedalism. I regard this as perhaps a brave and interesting speculation but without convincing supporting evidence.

First, none of the new "robust" specimens show any marginal sinus, and one cannot simply accept the *lack* of a structure such as the transverse sinus as *evidence* for the presence of a marginal sinus. From my collection of well over 120 ape endocasts, there are cases where a transverse sinus is not visible, and where no O/M sinus can be found. In addition, the KNM-WT 17000 endocast is too eroded in the occipital region to be certain that a small elevation in the region of the transverse sinus was not indeed the transverse sinus. This does not help very much, as it is possible to have *both* a O/M and transverse sinus on the same endocast.

Second, I am not certain, how, physiologically, this drainage system could be either necessary or related to bipedal locomotion. *Which* neural structures in either the dura, the cerebellum, the spinal chord, or the remaining brain are both "serviced" by the O/M and relate to bipedal locomotion? Relationships with some kind of thermal efficiency seem similarly suspect without clear-cut evidence of the relationship between such variation and temperature variables within or between populations showing such morphological variation. Finally, I believe Kimbel's (1984) careful analysis of the variation of this pattern within large samples must be seriously considered and compared to Falk's (1986, this volume chapter 4) findings based on small samples.

Encephalization and Enhanced Sensorimotor Capabilities

I find it very remarkable that Susman (this volume, chapter 10) found so many modern (or advanced) postcranial attributes in both the hands and feet of purported "robust" australopithecines, for which McHenry (this volume, chapter 9) obtains relatively high encephalization coefficients (EQ's) that are greater than either *A. afarensis* or *A. africanus*, but less than *Homo*. Previous work by McHenry (1974) had suggested that the australopithecines were barely beyond ape levels in their EQ's, thereby concluding that the brain was the last organ to evolve during hominid evolution. Holloway (1981a), using McHenry's data on vertebral body size, and stepwise regression techniques, found smaller body weights for "robust" australopithecines, thus suggesting that "robust" australopithecines had relatively larger brains than apes, and EQ's values clearly advanced beyond apes, including *Pan* (see also Holloway, 1976.)

I am surprised, however, that McHenry (this volume, chapter 9) continues to uncritically use Jerison's (1973) formula for the encephalization coefficient, where EQ = .12 (body weight) raised to the .67 power. The equation was derived from drawing a line through a polygon with a preselected slope of .66 and is not an empirically generated formula for primates, mammals, or vertebrates, of which several such slopes exist (e.g, Martin 1983; Hofman, 1983; Holloway and Post, 1982). The empirically derived exponent is closer to .75, which suggests some interesting metabolic relationships between brain and body size. Furthermore, using different primate taxa formulae to derive EQ's makes a significant difference regarding how close a fossil form is to either *Pan* or to *H. sapiens*. It is well to remember that log–log regressions of limb–bone size against weight should provide all data for correlation coefficients and some idea as to the residuals that result when one chooses an equation for estimation purposes. From my own experiences with brain–body and brain-structure to brain-size allometry where correlation coefficients of greater than .97 can exist with residuals well over 150% in log–log plots [as in the case of *H. sapiens*, where the volume of primary visual cortex is reduced by that amount (Holloway, 1985)], one must be careful to look for possible outliers, and to provide data on residuals.

Summary and Conclusions

If one wishes to derive the new "robust" specimens from West Turkana from the Hadar Ethiopian specimens and have them ancestral to both the East and South African "robust" forms (e.g., OH 5 and SK 1585), then one has clear evidence for brain endocast evolutionary changes that may have paralleled those found in a "gracile" line such as *A. africanus* to *Homo*. These changes would include brain expansion, probable increase in EQ (i.e., relative brain size), increased dorsal cerebral height suggesting a relative expansion of the posterior parietal region, and thus perhaps some evolutionary refinement in Wernicke's region, and lateralization reflected in a consistent right-frontal–left-occipital petalial pattern of asymmetry. In addition, the cerebellar lobes become less flaring laterally and less projecting posteriorally. Both the Taung and AL 162-28 brain endocast portions strongly suggest some cortical reorganization toward a more *Homo* and less apelike pattern, in that the lateral extent of primary visual striate cortex is reduced, judging from the posterior position of the lunate sulcus. If this is so, then the earliest hominids for which we have endocranial evidence had already undergone a significant brain evolutionary change in the distribution of cortical tissues, and these changes were shared by subsequent hominids of both "robust" and "gracile" lines.

Acknowledgments

I am grateful to Fred Grine for his persistent encouragement to participate in the workshop, and for making my stay so pleasurable. I am grateful also for Bernard Wood's and Alan Walker's retention of medical as well as anthropological knowledge. Some of the work and newer analyses mentioned in this paper were in part supported by NSF Grant # BNS-8418921, for which support this author is thankful.

References

Clark, W. E. LeGros, Cooper, D. and Zuckerman, S. (1936). The endocranial cast of the chimpanzee. *J. Roy. Anthropol. Inst., 66*: 249–268.
Connolly, C. J. (1980). *External morphology of the primate brain.* Thomas, Springfield, IL.
Delson, E. (1986). Human phylogeny revised again. *Nature, 322*: 496–497.
Falk, D. (1980). A reanalysis of the South African australopithecine natural endocasts. *Amer. J. Phys. Anthropol. 60*: 525–539.
Falk, D. (1983). The Taung endocast. A reply to Holloway. *Amer. J. Phys. Anthropol. 60*: 479–490.
Falk, D. (1985a). Hadar AL 162-28 endocast as evidence that brain enlargement preceded cortical reorganization in hominid evolution. *Nature, 313*: 45–47.
Falk, D. (1985b). Apples, oranges, and the lunate sulcus. *Amer. J. Phys. Anthropol. 67*: 313–315.
Falk, D. (1986). Reply. *Nature, 321*: 536–537.
Falk, D. and Conroy, G. C. (1984). The cranial venous system in *Australopithecus afarensis. Nature, 306*: 779–781.
Hofman, I. (1983). Evolution of brain size in neonatal and adult placental mammals: a theoretical approach. *J. Theoret. Biol., 105*: 317–332.
Holloway, R. L. (1972). New australopithecine endocast, SK 1585, from Swartkrans, S. Africa. *Amer. J. Phys. Anthropol., 37*: 173–186.
Holloway, R. L. (1976). Some problems of hominid braincast reconstruction, allometry, and neural reorganization. Colloquium VI, IX Congress of U.I.S.P.P., Nice , pp. 69–119.
Holloway, R. L. (1981a). Revisiting the S. African australopithecine endocasts: results of stereoplotting the lunate sulcus. *Amer. J. Phys. Anthropol., 56*: 43–58.
Holloway, R. L. (1981b). The endocast of the Omo juvenile L338y-6 hominid specimen. *Amer. J. Phys. Anthropol., 54*: 109–118.
Holloway, R. L. (1983). Cerebral brain endocast pattern of *A. afarensis* hominid. *Nature, 303*: 420–422.
Holloway, R. L. (1984). The Taung endocast and the lunate sulcus: a rejection of the hypothesis of its anterior position. *Amer. J. Phys. Anthropol., 64*: 285–288.
Holloway, R. L. (1985). The past, present, and future significance of the lunate sulcus in early hominid evolution. In P. V. Tobias (ed.), *Hominid evolution: past, present, and future,* pp. 47–62. Alan R. Liss, New York.
Holloway, R. L. and de LaCoste-Larymondie, M. C. (1982). Brain endocast asymmetry in pongids and hominids: some preliminary findings on the paleontology of cerebral dominance. *Amer. J. Phys. Anthropol., 58*: 101–110.
Holloway, R. L. and Post, D. (1982). The relativity of relative brain measures and hominid evolution. In E. Armstrong, and D. Falk (eds.), *Primate brain evolution,* pp. 57–76, Plenum, New York.
Holloway, R. L. and Kimbel, W. H. (1986). Endocast morphology of Hadar hominid AL 162-28. *Nature, 321*: 536.
Jerison, H. (1973). *The evolution of brain and intelligence.* Academic Press, New York.
Johanson, D. C., Masao, F. T. Eck, G. G., White, T. D., Walter, R. C., Kimbel, W. H., Asfaw, B., Manega, P., Ndessokia, P. and Suwa, G. (1987). New partial skeleton of *Homo habilis* from Olduvai Gorge, Tanzania. *Nature, 327*: 205–209.

Kimbel, W. H. (1984). Variation in the pattern of cranial venous sinuses and hominid phylogeny. *Amer. J. Phys. Anthropol. 63*: 243–263.

LeMay, M. (1976). Morphological cerebral asymmetries of modern man, fossil man, and nonhuman primates. *Ann. N.Y. Acad. Sci., 280*: 349–366.

LeMay, M. (1985). Asymmetries of the brains and skulls of nonhuman primates. *In* S. D. Glick (ed.), *Cerebral lateralization in nonhuman species,* pp. 233–245. Academic Press, Orlando.

LeMay, M., Billig, M. S., and Geschwind, N. (1982) *In* E. Armstrong, and D. Falk (eds.), *Primate brain evolution,* pp. 263–278. Plenum, New York.

McHenry, H. M. (1974). Fossil hominid body weight and brain size. *Nature, 74*: 686–688.

Martin, R. D. (1983). Human evolution in an ecological context. James Arthur Lecture on the Evolution of the Human Brain (1982). Amer. Mus. Nat. Hist. New York.

Rak, Y. and Howell, F. C. (1978). Cranium of a juvenile *Australopithecus boisei* from the lower Omo basin, Ethiopia. *Amer. J. Phys. Anthropol. 48*: 345–366.

Tobias, P. V. (1967). *The cranium and maxillary dentition of Australopithecus (Zinjanthropus) boisei.* Olduvai Gorge, Vol.2. Cambridge University Press, Cambridge.

Walker, A., Leakey, R. E., Harris, J. M., and Brown, F. H. (1986). 2.5-Myr *Australopithecus boisei* from West Lake Turkana, Kenya. *Nature, 322*: 517–522.

Walker, A., Talk, D., Smith, R. and Pickford, M. (1983). The skull of *Proconsul africanus:* reconstruction and cranial capacity. *Nature 305*: 525–527.

Wood, B. (1987). Who is the 'real' *Homo habilis? Nature, 327*: 187–188.

Growth Processes in the Cranial Base of Hominoids and Their Bearing on Morphological Similarities that Exist in the Cranial Base of *Homo* and *Paranthropus*

6

M. CHRISTOPHER DEAN

The base of the cranium in modern humans has frequently been regarded as a "stable" part of the skull since Huxley (1863, 1867) focused attention upon it. The implication has been that, although situated between the face and the neurocranium, the basicranium is less variable than either, although this is not true in a comparative context: Major variations in the cranial base viewed in *norma basilaris* have been useful in phylogenetic analysis (Dean and Wood, 1981, 1982; Dean 1986). The endocranial aspects of the cranial base are profoundly influenced by the growth and development of the brain; indeed, the cranial fossae are a near perfect mold of the undersurface of the brain. Other extracranial aspects of the cranial base are greatly influenced by the masticatory system (Weidenreich, 1943, 1951) or the prevertebral and nuchal musculature (Dean, 1985a,b). It is increasingly important in studies of fossil hominids to try to distinguish between characters that are shared-derived (synapomorphic) and those that appear similar through convergent or parallel evolution. Resolving the factors that underlie or control morphological variation can assist in distinguishing these.

Evidence from the growth processes in the cranial base sheds some light on those factors that are likely to influence cranial-base morphology in hominoids so that it then becomes possible to comment about the degree of influence one functional unit has over another in this region. It may be impossible to assess precisely what effect other more distant influences such as posture or the degree of development of the masticatory system have upon the morphology of the skull base in early hominids, but careful comparisons between closely related species are likely to be helpful in this respect. For example, the consistent sagittal orientation of the petrous temporal bones in modern and fossil Old World monkeys, among which there is great variation in the size and disposition of the facial skeleton (as well as in the degree of endocranial base flexion), suggests some independence of the face from the cranial base (Dean, 1986). The apparently prognathic face and unflexed cranial base of KNM-WT 17000 (Walker *et al.*, 1986), which combines coronally oriented petrous bones and an anteriorly positioned foramen magnum, likewise suggest that the facial skeleton has no great influence over the petrous orientation in this specimen. In this way at least unlikely causal relationships can be dismissed more confidently and useful and important characters may be identified.

It transpires, from developmental studies of extant taxa, that some differences apparent between the adult cranial base of great apes and *Australopithecus,* on the one hand, and of *Paranthropus* and *Homo* (both modern and early), on the other, are already established at birth (Dean and Wood, 1984). For example, after birth the angle that the petrous temporal bone makes to the coronal plane shifts some 10° to the sagittal during the growth period in great apes and modern humans alike, but the orientation remains distinct in both groups, with ranges in angulation that do not overlap. Early minor changes to this angle in human infant skulls can perhaps be linked to changes in cranial base flexion that occur at the same time (George, 1978), but the disposition of homologous landmarks on ape and human cranial base is quite distinct, even at this early age. The hypothesis then, that the infant ape cranial base resembles the adult human cranial base [the latter being neotenous—see Gould (1977) for a good review] can be rejected as a generalization that applies only to the growth pattern of certain individual bones (Dean and Wood, 1984). These facts also clearly suggest that none of the differing postnatal growth changes that occur between apes and humans—for example, the lengthening basioccipital and backward migration of the foramen magnum in apes—can be linked to sagittally orientated petrous bones in adult apes. Indeed, this fact is borne out by the cranial base morphologies of *Pan paniscus* and *Gorilla*, which have differing positions of the foramen magnum and accompanying short and long basioccipitals, respectively, but near identical sagittally orientated petrous bones (Fenart and DeBlock, 1973; Cramer, 1977; Dean, 1986). Interestingly, differences in endocranial base flexion also exist between *P. paniscus* and other apes soon after birth (Fenart and DeBlock, 1973) and it would appear that this difference exists between these taxa before birth. In short, the fact that differences in petrous orientation exist between apes and humans before birth means that any "underlying cause" must be sought during fetal development.

It has been customary to account for coronally orientated petrous temporal bones in *Homo* and *Paranthroupus* and indeed, rats (Bateman, 1954), by presuming that skull base flexion together with an accompanying decrease in the length of the basioccipital has compressed them, so that they now come to lie across the cranial base rather than along it (Weidenriech, 1943; DuBrul, 1977). However, measuring cranial base flexion during growth is difficult and remodelling at sella (Latham, 1972) (one of the homologous landmarks often used endocranially) means that a small angle of *Nasion Sella Basion* may not accurately reflect the swinging down of the clivus and foramen magnum. Also, a long cranial base as reflected by the length of the basioccipital in, for example, adult great apes is often by mistake inextricably linked with the posterior placement of the foramen magnum. In fact expansion of the occipital region can occur *together* with an elongated basioccipital (as it does in KNM-WT 17000) so that the foramen magnum is still positioned relatively forward with respect to the bitympanic line and the petrous bones are still coronally orientated "despite" the long basioccipital. When there is occipital expansion together with a short basioccipital, as there is in the modern human cranial base, the brain encroaches on the petrous temporal bones so that the angle between the anterior and posterior endocranial surfaces becomes more acute and is then surmounted by a "sharp" petrous crest. *Homo erectus* crania, described by Weidenreich (1943) and Maier and Nkini (1984), and KNM-WT 17000 have long flattened basioccipitals but lack a vertical posterior surface to the petrous temporal bone in the posterior cranial fossa. This suggests there has been less compression of the petrous bones during postnatal growth due to a more marked elongation of the basioccipital.

Beigert (1963) has emphasized the importance of expansion of the neocortex during hominid evolution and the accompanying rotation of the supratentorial compartment. Moss *et al.* (1956) have shown that this sequence of events is recapitulated during human prenatal development as the tentorium cerebelli rotates backward through 90° "until the major portions of the human brain have assumed their definitive topological relations." During both hominid evolution and human development these supratentorial changes are accompanied by a separate neural expansion below the tentorium cerebelli in the posterior cranial fossa as the cerebellar hemispheres enlarge laterally. Moss (1958, 1963) has pointed out that there is some independence between the expansion of parts of the brain because of the dural tracts that divide it. Interestingly, Bull (1969) has defined another increasingly important function for the the tentorium cerebelli during evolution, that of a weight-bearing structure that transfers the weight of the cerebral hemispheres outward toward the lateral wall of the vault of the skull and away from an increasingly inferiorly orientated foramen magnum.

Both Holloway (1972) and Tobias (1967) have commented upon the expanded cerebellum in *Paranthopus* relative to *Australopithecus* and the great apes. In the light of the observations about petrous orientation in infant hominoids, a likely form of cranial base compression that might underlie coronally orientated petrous bones in *Homo* and *Parathropus* may take place in the posterior cranial fossa beneath the encroaching tentorium cerebelli. Rather than being associated with an increase in cranial base flexion however, petrous compression probably results from a prenatal flattening of the skull base as the cerebellum expands faster than the posterior part of the cranium is able to elongate (Ford, 1956).

Several authors have documented and commented upon the enormous rate of growth of the human cerebellum between the thirteenth week of gestation and the immediate postnatal period. Rakic and Sidman (1970) comment that the suface area of the cerebellum increases 2000 times during this period, and Dobbling and Sands (1973) noted that the human cerebellum grows more rapidly than the brain stem and forebrain in weight, cellularity and myelination. The cerebellum starts to grow later, grows more rapidly, and completes its growth earlier than other parts of the brain (see also Tanner, 1978). Norback and Moss (1956), in an analysis of brain growth, concluded that with two exceptions the ratios of the growth rates of specific regions of the brain were constantly proportional throughout growth. These two exceptions were cerebellar width, which showed a marked increase in growth rate at the end of the 4th fetal month, and the length of the corpus callosum. Moss *et al.* (1956) have also demonstrated that the growth in width of the occipital bone closely follows the growth pattern of the cerebellum. Moss (1958) has commented upon the fact that the occipital region (*i.e.*, the cerebellum and occipital bone) has the greatest postnatal neurocranial growth rate and that the downward and backward rotation of this region relative to the horizontal plane is rapidly completed during the first 2 years in humans (see also Cousin and Fenart, 1971).

The observations of Ford (1956) support these findings. Ford, in a study of the growth of the fetal skull, demonstrated that at 10 weeks the forehead is prominent, while the occipital region is underdeveloped and there is no clear demarcation from the neck. During growth, the frontal region becomes less prominent and the occipital region develops, becoming protuberant (Fig. 6.1). Ford explained this by noting that the anterior part of the cranial base from *nasion* to the pituitary fossa grows faster at this time than the posterior part of the cranial base from the pituitary fossa to *basion* so that there is a flattening of the cranial base to compensate for expansion of the brain. At the same time, there is a broadening of the posterior cranial fossa and a horizontal positioning of the occipital squama. These changes, Ford (1956) notes, give the fetal skull its increasing occipital fullness that is such a particular characteristic of modern humans. The fact that infant great ape crania appear to have relatively long basioccipitals and a foramen magnum already behind the bitympanic line suggests that the posterior part of the cranial base from the pituitary fossa to basion may grow at a faster rate than in modern humans. The illustration of a fetal chimpanzee by Stark (1963) seems to confirm this. It is likely that cerebellar growth patterns relative to the rest of the brain are similar in human and great ape fetuses. Differences in growth in length of the posterior part of the cranial base probably alter the forces generated within the cranial dural sheaths. The short posterior part of the cranial base of modern humans forces the

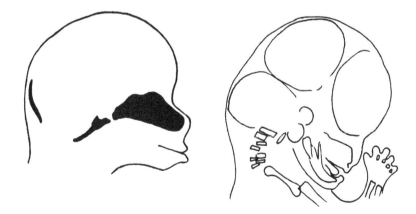

FIGURE 6.1. Diagram made from a cleared, alizarin red preparation of a 14-week human fetus, right, compared with a diagram of a 71 mm chimpanzee fetus redrawn from Stark (1976) left. The pronounced occipital protuberance of the human fetus is notable. (Drawings not to scale.)

cerebellum to expand laterally under the tentorium cerebelli, so orientating the petrous temporal bones coronally. A faster growing and so relatively longer posterior cranial base in great apes allows the cerebellum to expand posteriorly, and so reduces the lateral expansion, such that the petrous temporal bones are not reorientated coronally to the same extent they are in modern *Homo sapiens*.

A large part of the cerebellar hemispheres is concerned with the control of voluntary movements. Matano *et al.* (1985a,b) note that the size of the cerebellar hemispheres may be reflected in the size of the lateral cerebellar nuclei and that increases in this zone are correlated with the aquisition of planned skilled, diversified movements in primates, especially of finely tuned hand and finger movements, and opposability of the thumb. It may also be that the width of the spectrum of motor abilities, rather than narrowly specialized motor patterns, underlies lateral expansion of the cerebellum in primates. The ventral pons that lies against the clivus is a relay station between the cerebral neocortex and the cerebellar hemispheres so that expansion of the ventral pons relates directly to the increase in size of the neocortex and neocerebellum. This also has important implications for the degree of flexion of the clivus in hominids.

The evidence from new postcranial material at Swartkrans (Susman, this volume, chapter 10) raises the possibility that *Paranthropus* may have had a fully opposable thumb, precision grip, and by inference great diversity of hand and finger movements. It is tempting to speculate that were this so in *Paranthropus*, accompanying changes would have occurred in the cerebellum and that these changes might then underlie much of the morphology of the cranial base that appears so similar to *Homo*.

The role of the brain influencing the relative position of the foramen magnum has been stressed by Beigert (1957, 1963) who, along with others (Schultz, 1942; Clark, 1972:139; Adams and Moore, 1975), has cautioned against any close association between the orientation and position of the foramen magnum and habitual posture in primates. Previous explanations for morphological similarities between the cranial bases of *Paranthropus* and *Homo* that assume there are no differences in the degree of encephalization between great apes and early hominids and that attribute them to the proportions of the face and neurocranium to be balanced (DuBrul, 1979) should then now be questioned.

Conclusions

Studies of comparative growth and development can provide valuable information about the factors that underlie morphological variation in adult hominids. A clear theoretical link between manipulative skills and the morphology of the cerebellum exists, which might possibly explain some of the features that *Homo* and *Paranthropus* share in the cranial base. Evidence from future comparative morphological studies are likely to improve our understanding of other characters often used in phylogenetic analyses. Only the combined findings of such studies are likely to lead to a reliable distinction between synapomorphies and homoplasies in fossil hominids.

Acknowledgments

I am grateful to the Governments of Kenya and Tanzania and to the Governors and Directors of the National Museums of Kenya and the Transvaal Museum Pretoria. I thank the trustees of The British Museum (NH), The Powell Cotton Museum, Kent, UK, and The Royal College of Surgeons, London, for permission to work with valuable material in their care. I am particularly grateful to Fred Grine for inviting me to participate in the "robust" australopithecine workshop at Stony Brook, New York. This study was carried out while the author was recipient of grants from The Nuffield Foundation, The Wellcome Trust, The Royal Society, The Boise Fund, and the Leakey Trust.

References

Adams, L. M. and Moore, W. J. (1975). Biomechanical appraisal of some skeletal features associated with head balance and posture in the Hominoidea. *Acta. Anat.*, 92: 580–594.

Bateman, N. (1954). Bone growth: A study of the grey lethal and microphthalmic mutants of the mouse. *J. Anat.*, 88: 212–262.

Beigert, J. (1957). Der formuandel des primatenschädels und seine Beziehungen zur ontogenctischen Entwicklung und den phylogenetischen Specialisationen der Kopforgane. *Morph. Jb.*, 98: 77–199.

Beigert, J. (1963). The evaluation of characters of the skull, hands and feet for primate taxonomy. In S. L. Washburn (ed.), *Classification and human evolution*, pp.116–145. Aldine, Chicago, IL.

Bull, J. W. D. (1969). Tentorium cerebelli. *Proc. Roy. Soc. Med.*, 62: 1301–1310.

Clark, W. E. Le Gros. (1972). *The antecedents of man*, 3rd ed. Edinburgh University Press, Edinburgh.

Cousin, R. P. and Fenart, R. (1971). Etude ontogénetique des éléments sagittaux de le fosse cérébrale antérieure chez l'Homme orientation vestibullaire. *Arch. Anat. Pathol.*, 9: 383–395.

Cramer, D. L. (1977). Craniofacial morphology of *Pan paniscus:* a morphometric and evolutionary appraisal. *Contrib. primatol. 10.* S. Karger, Basel.

Dean, M. C. (1985a). The comparative myology of the hominoid cranial base. 1. The muscular relations and bony attachments of the digastric muscle. *Folia Primatol.*, 43: 157–180.

Dean, M. C. (1985b). The comparative myology of the cranial base in hominoids. 2. The muscles of the prevertebral and upper pharyngeal region. *Folia Primatol.*, 44: 40–51.

Dean, M. C. (1986). *Homo* and *Paranthropus*: similarities in the cranial base and the developing dentition. In B. A. Wood, L. B. Martin and P. Andrews (eds.),*Major topics in primate and human evolution*, pp.249–265. Cambridge University Press, Cambridge.

Dean, M. C. and Wood, B. A. (1981). Metrical analysis of the basicranium of extant hominoids and *Australopithecus. Amer. J. Phys. Anthropol.*, 54: 53–71.

Dean, M. C. and Wood, B. A. (1982). Basicranial anatomy of Plio-Pleistocene hominids from East and South Africa. *Amer. J. Phys. Anthropol.*, 59: 157–174.

Dean, M. C. and Wood, B. A. (1984). Phylogeny, neoteny and growth of the cranial base in hominoids. *Folia Primatol.*, 43: 157–180.

Dobbing, J. and Sands, J. (1973). Quantitative growth and development of human brain. *Arch. Dis. Children*, 48: 757–767.

DuBrul, E. L. (1977). Early hominid feeding mechanisms. *Amer. J. Phys. Anthropol.*, 47: 305–320.

DruBrul, E. L. (1979). Origin and adaptations of the hominid jaw joint. In B. G. Sarnat and D. M. Laskin (eds.), *The temporomandibular joint*, 3rd ed., pp. 85–113. Thomas, Springfield, IL.

Fenart, R. and DeBlock, R. (1973). *Pan paniscus* et *Pan troglodytes*-craniometrie:Etude comparative et ontogénetique selon les méthodes classiques et vestibulaire.Tome 1.*Musee Royale de l'Afrique Centrale.Tervuren,Belgique Annales-Serie IN-8 Sciences Zoologiques*, No.204.

Ford, E. H. R. (1956). The growth of the foetal skull. *J. Anat.*, 90: 63–72.

George, S. L. (1978). A longitudinal and cross-sectional analysis of the growth of the postnatal cranial base angle. *Amer. J. Phys. Anthropol.*, 49: 171–178.

Gould, S. J. (1977). *Ontogeny and Phylogeny*. Harvard University Press (Belknap Press), Cambridge, MA.

Holloway, R. L. (1972). New australopithecine endocast, SK 1585, from Swartkrans, South Africa. *Amer. Phys. Anthropol.*, 37: 73–186.

Huxley, T. H. (1863). *Evidence as to mans place in nature*. Williams and Norgate, London.

Huxley, T. H. (1867). On two widely contrasted forms of the human cranium. *J. Anat. Physiol.*, 1: 60–77.

Latham, R. A. (1972). The sella point and postnatal growth of the cranial base. *Amer. J. Ortho.*, *61*: 156–162.

Maier, N. and Nkini, A. (1984). Olduvai Hominid 9: new results of investigation. *Cour. Forsch. Inst. Senckenberg, 69*: 69–82.

Matano, S., Stephan, H. and Baron, G. (1985a). Volume comparisons in the cerebellar complex of primates. 1. Ventral pons. *Folia Primatol., 44*: 171–181.

Matano, S., Baron, G., Stephan, H. and Frahm, H. D. (1985b). Volume comparisons in the cerebellar complex of primates 11 cerebellar nuclei. *Folia Primatol., 44*: 182–203.

Moss, M. L. (1958). The pathogenesis of artificial cranial deformation. *Amer. J. Phys. Anthropol., 16*: 269–285.

Moss, M. L. (1963). Morphological variations of the crista galli and medial orbital margin. *Amer. J. Phys. Anthropol., 21*: 259–164.

Moss, M. L., Norback, C. B. and Robertson, G. G. (1956). Growth of certain human fetal cranial bones. *Amer. J. Anat., 97*: 155–176.

Norback, C. B. and Moss, M. L. (1956). Differential growth of the human brain. *J. Comp. Neurol., 105*: 539–552.

Rakic, P. and Sidman, R. L. (1970). Histogenesis of cortical layers in human cerebellum, particularly the lamina dissecans. *J. Comp. Neurol., 139*: 473–500.

Schultz, A. (1942). Conditions for balancing the head in primates. *Amer. J. Phys. Anthropol., 29*: 483–497.

Stark, D. (1973). The skull of the fetal chimpanzee. *In* G. H. Bourne (ed.), *The chimpanzee*, Vol. 6, pp 1–33. S. Karger, Basel.

Tanner, J. M. (1978). *Foetus into Man*. Fletcher and Son, Norwich, England.

Tobias, P. V. (1967). The cranium and maxillary dentition of *Australopithecus (Zinjanthropus) boisei*. Olduvai Gorge, vol. 2. Cambridge Univ. Press, Cambridge.

Walker, A., Leakey, R. E. F., Harris, J. M. and Brown, F. H. (1986). 2.5-Myr *Australopithecus boisei* from West of Lake Turkana, Kenya. *Nature, 322*: 517–522.

Weidenreich, F. (1943). The skull of *Sinanthropus pekinensis:* A comparative study on a primitive hominid skull. *Palaeontol. Sinica.* [NS 10] *127*, pp.1–134.

Weidenreich, F. (1951). Morphology of Solo man. *Anthropol. Pap. Amer. Mus. Nat. Hist., 43*: 205–290.

Studies of the Postcranial Evidence   II

II

Studies of the Fashion of Evidence

New Estimates of Body Size in Australopithecines

7

WILLIAM L. JUNGERS

> The importance of body size as a major aspect of an animal's adaptive strategy has long been recognized by paleontologists.
>
> J. G. Fleagle (1978)

> In actuality, size has until recently been one of the most neglected aspects of biology.
>
> W. A. Calder III (1984)

Calder's (1984) lament is curiously misplaced in the realm of paleontology (Fleagle, 1978), perhaps because "size" is one of those rare characteristics of an extinct organism that can be estimated even from fragmentary remains. The paleontologist's fascination with size in the fossil record also relates to the realization that body size plays a central role in an animal's overall adaptive strategy and can be described in many ways as an ecological variable (Peters, 1983; Calder, 1984; Smith, 1985). In other words, accurate estimates of body size in fossils can have important implications for a wide range of other biological factors: relative degree of encephalization (Jerison, 1973; Radinsky, 1974; Gingerich, 1977; Holloway and Post, 1982; McHenry, 1982; Hofman, 1983; Blumenberg, 1984); reconstruction of probable dietary preferences (Kay, 1984, 1985; Kay and Simons, 1980); habitat preference and probable positional repertoire (Kay and Simons, 1980; Smith and Pilbeam, 1980; Aiello, 1981; Andrews and Aiello, 1984; Jungers, 1984, 1988a); home range size and foraging radius (Foley, 1987); social and reproductive behavior (Foley, 1987); and sexual dimorphism (Leutenegger and Cheverud, 1985; Pickford and Chiarelli, 1986).

Although it is certainly true that estimating body size (e.g., stature or weight) of an extinct individual or species is potentially hazardous, few morphological questions in the fossil record appear as straightforward (Smith, 1985), and fewer still impact on so many other related issues of interest. In most of the contexts listed above—and for theoretical as well as pragmatic reasons (Schmidt-Nielsen, 1977; Jungers, 1984)—body mass or weight is the size variable of choice. Fewer biological relationships have been linked to stature, for example, and stature reconstruction in the hominid fossil record has its own host of complications and limitations (McHenry, 1974; Geissmann, 1986; Jungers, 1988b; Feldesman and Lundy, 1988).

Because teeth possess highly durable and resistant material properties, it is no surprise that they are better represented in the fossil record than any other skeletal part. It is no less surprising, therefore, that most systematic attempts to estimate body mass of fossil primates have relied on empirically determined relationships between tooth size and body mass in living species (Gingerich, 1977; Kay and Simons, 1980; Gingerich *et al.*, 1982; Fleagle and Kay, 1985; Conroy, 1987). Such dentally based avenues of body size estimation are almost certainly closed for early hominids inasmuch as all of the australopithecines appear to have possessed exceptionally large cheek-teeth for their body size (Wolpoff, 1973; McHenry, 1984; Kay, 1985). Regression predictions based on other craniofacial measurements (e.g., occipital condyle size (Martin, 1981); palate breadth, bizygomatic breadth and orbit width (Steudel, 1980) may also be inappropriate, but for another, more general reason. To paraphrase and extend the logical argument of Hylander (1985), because primates—early hominids included—do not transmit body weight (mass) through their skulls or teeth, there exists no biomechanical reason to expect a direct or especially predictable relationship between such variables and body size.

Weight-bearing portions of the locomotor skeleton are arguably better candidates for being reliable estimators of overall body size. Diaphyseal diameters or circumferences represent suitable alternatives to teeth (McHenry, 1976, 1988; Aiello, 1981; Rightmire, 1986)). Although they are more difficult to gather, cross-sectional geometrical data, such as cortical area or area moments of inertia, possess great logical appeal (Ruff, 1987). In the latter case, however, great caution needs to be exercised because virtually all fossil hominids seem to have been characterized by unusually thick long-bone cortices (Walker, 1973; Lovejoy and Trinkaus, 1980; Kennedy, 1983). Because the loads borne by articular elements of the locomotor skeleton are closely linked biomechanically to body mass (Alexander, 1980, 1981; Jungers, 1988c), measures of postcranial joint size should also be relatively reliable predictors of body size (McHenry, 1976; Scott, 1983; Rightmire, 1986; Jungers, 1988a).

The primary goal of the present study is to offer new estimates of body mass for extinct hominids from East and South Africa, including *Australopithecus afarensis*, *A. africanus*, *Paranthropus robustus* and *P. boisei*. These new estimates are based on the relationship between hindlimb (and lumbosacral) joint size and body mass in extant hominoids. Comparisons are made to previously published estimates, and several biological implications of these predictions are considered.

Materials and Methods

The Extant Samples

Seven species of living hominoids, including modern humans, were included in the present analyses. The nonhuman groups included representatives of all extant hominoid genera: *Gorilla gorilla, Pan troglodytes, Pan paniscus, Pongo pygmaeus, Hylobates (Symphalangus) syndactylus* and *Hylobates lar*. The nonhuman sample consists of wild-collected adult individuals with body mass recorded at death. The modern human skeletal sample is derived from a large cadaver population of European descent, and it also contains only specimens of known body mass (Jungers, 1982). Body mass and linear dimensions of the postcranium in this human sample are close to the mean values published for normal young adults of similar heritage (Reed and Stuart, 1959; Anderson *et al.*, 1964), and their ponderal indices are well within the ranges of most human groups (Bogin, 1988). The total sample size of all hominoids in this study is 87 individuals. Sex-specific means were computed and used throughout this study (Table 7.1). Average body mass ranges from 5.5 kg in the female lar gibbon to 164.3 kg in male gorillas.

Articular Dimensions

Nine linear dimensions of postcranial joints were measured (mm) on each extant specimen, and average, sex-specific values were computed (cf. Jungers, 1988c,d,):

1. Acetabulum height (ACET)
2. Femoral head diameter (FHD)
3. Posterior width of the medial femoral condyle (MCW)
4. Posterior width of the lateral femoral condyle (LCW)
5. Width of the patellar articular surface (PATW)
6. Anteroposterior length of the medial tibial condyle (MTB)
7. Anteroposterior length of the lateral tibial condyle (LTB)
8. Anteroposterior diameter of the distal tibial articulation (DTB)
9. Width of the lumbosacral articulation (LSW)

Table 7.1. Average Body Mass (by Sex) in Extant Hominoid Species[a]

Group	Body mass (kg)
Gorilla gorilla	
Males ($N=7$)	164.3
Females ($N=4$)	75.5
Pan troglodytes	
Males ($N=4$)	56.6
Females ($N=5$)	40.1
Pan paniscus	
Males ($N=6$)	46.3
Females ($N=6$)	33.2
Pongo pygmaeus	
Males ($N=8$)	81.2
Females ($N=11$)	37.2
Hylobates syndactylus	
Males ($N=3$)	11.4
Females ($N=3$)	11.3
Hylobates lar	
Males ($N=6$)	5.9
Females ($N=6$)	5.5
Homo sapiens	
Males ($N=9$)	68.2
Females ($N=9$)	50.0

[a]Total $N=87$; mean body mass range: 5.5–164.3 kg.

Regression Methods

Because of the fragmentary nature of most fossil hominid specimens, the source of information about joint size varied from individual to individual. In most instances, only one of the nine articular variables could be recorded with any degree of confidence. In such cases, simple least-squares regression was used to estimate mass (kg) from whichever joint dimension (mm) was preserved. However, some specimens presented the opportunity to combine information from more than one joint surface, and multiple regression was therefore employed to estimate body mass. Both raw and log-transformed variables were examined, but only the results based on the latter are reported here because this transformation tended to yield lower mean absolute percentage prediction errors and less serial correlation among residuals. Confidence limits on the individual fossil estimates are not provided because of reservations about underlying statistical assumptions that are critiqued elsewhere (Radinsky, 1982; Smith, 1985).

Two different extant samples were used. The first included all species (ALLHOM); the second included only nonhuman hominoids (NONHUM). The decision to omit *Homo sapiens* in the NONHUM sample is based on the observation that modern humans possess exceptionally large hindlimb joints for their body size (Jungers, 1988c,d), and also upon the prior demonstration that this highly derived condition does not characterize early hominids (Napier, 1964; McHenry and Corruccini, 1976; Corruccini and McHenry, 1978; Jungers , 1988d). Use of the NONHUM sample merely recognizes this distinction between australopithecines and modern humans and does *not* imply that these early hominids are viewed here as brachiators or quadrupeds. In some cases, the body size estimates differ substantially between ALLHOM and NONHUM samples. Intraspecific regression estimates based on the human sample were not attempted because many of the fossil specimens have joint dimensions far outside the extant sample limits; extrapolation down to diminutive individuals such as "Lucy" (AL 288-1) and Sts 14 seems inadvisable under such circumstances. Moreover, in this case it would presume incorrectly that all of the fossil hominids share relatively enlarged hindlimb joints with modern *H. sapiens*.

The Fossil Hominid Sample

The sample of fossil postcrania considered in this study could be subdivided in a variety of ways—taxonomically, geographically and temporally. The strategy used here is taxonomic; this is probably the most appropriate approach, although it is beset with its own problems. Thus, the sample of eight specimens from the Hadar Formation of Ethiopia is accepted as a monospecific group referred to as *A. afarensis*. The sample constituted by specimens from the South African sites of Sterkfontein ($N = 10$) and Makapansgat ($N = 1$) are presumed here to represent *A. africanus*. The sample of *P. robustus* specimens from Swartkrans is much smaller; only four individuals are represented by one of the variables discussed above. Most likely, all of these can be referred to the same taxon for reasons discussed elsewhere (Susman this volume, chapter 10). The East African australopithecine sample is the most problematic in terms of taxonomic attributions. KNM-ER 1500 is probably the least equivocal candidate for *P. boisei* (Grausz *et al.*, this volume, chapter 8), based on associated mandibular and postcranial remains. Two other specimens also make reasonable *P. boisei* candidates based on proximal femoral anatomy: KNM-ER 738 and KNM-ER 1503/1505. Assignment of specimens such as KNM-ER 741 is more equivocal, but this specimen still warrants consideration. Similarly, based on new conclusions by Grausz *et al.* (this volume, chapter 8), OH 35 needs to be entertained as a possible *P. boisei* individual.

Measurements for the Hadar sample were made by me on the reference casts at the Cleveland Museum of Natural History. J. T. Stern, Jr., kindly made available his measurements taken on the original fossils in the National Museums of Kenya. R. L. Susman and F. E. Grine provided the measurements on the South African fossils.

Results and Discussion

Body mass estimates for early hominids from lower limb joint size are presented in Table 7.2. ALLHOM and NONHUM predictions are listed separately for each specimen. The predictive regression equations used to calculate these values are provided in Table 7.3. The differences between ALLHOM and NONHUM estimates are most pronounced for the larger individuals. As noticeable outliers in most of the regressions, modern humans tend to depress the upper end of the quantitative relationships such that lower values are usually predicted by the ALLHOM sample (i.e., at a given joint size, humans tend to have smaller body masses than would be predicted from nonhuman regressions). This also leads to consistently higher standard errors of estimate and lower correlation coefficients for ALLHOM equations.

Estimated values for the Hadar sample range from 30.4 to 67.7 kg for ALLHOM (mean = 51.0 kg, midpoint = 49.1 kg) and from 30.4 to 80.8 kg for NONHUM (mean = 57.8 kg, midpoint = 55.6 kg). The *A. africanus* sample has values ranging from 33.0 to 57.6 kg for ALLHOM (mean = 46.0 kg, midpoint = 45.3 kg) and from 33.3 to 67.5 kg for NONHUM (mean = 52.9 kg, midpoint = 50.4 kg). The sample from Swartkrans ranges from 37.1 to 57.5 kg for ALLHOM (mean = 49.0 kg, midpoint = 47.3 kg) and from 42.2 to 88.6 kg for NONHUM (mean = 62.0 kg, midpoint = 65.4 kg); the upper ends of the ranges are due to a fragmentary patella of less than secure taxonomic affinities. The East African *P. boisei* hominid sample varies from 33.0 to 69.3 kg for ALLHOM (mean = 49.2 kg, midpoint = 51.2 kg) and from 36.9 to 88.6 kg for NONHUM (mean = 58.4 kg, midpoint = 62.8 kg); however, it should be noted that the two extremes of this group are both very problematic specimens in terms of taxonomic affinities. Regardless of extant sample, the relatively wide ranges suggest a considerable degree of sexual dimorphism *if* each group is indeed monospecific (see below). Despite the regrettably small sample sizes of the two putative *Paranthropus* groups, the averages and absolute ranges in all four groups are not terribly dissimilar. Thus, there is little evidence in these data for major differences in body size among australopithecine taxa. In other words, it would be premature to argue for the evolution of significantly increased body size from *A. afarensis* to either *A. africanus* or *Paranthropus*. Moreover, based on the ALLHOM results from this study and the predictions made recently for *Homo erectus* (Rightmire, 1986), average body size may have changed very little in

Table 7.2. Body Mass Estimates for Fossil Hominids[a,b]

Group	Specimen [dimension(s)]	Estimate (kg) ALLHOM	NONHUM
Australopithecus afarensis	AL 288-1 (FHD, ACET, MTB, LTB, DTB, LSW)	30.4	30.4
	AL 129-a,b (MCW, LCW, LTB)	35.2	36.6
	AL 333-3 (FHD)	67.7	80.8
	AL 333-4 (MCW)	48.3	50.3
	AL 333-6 (DTB)	47.8	56.3
	AL 333-7 (DTB)	64.7	79.5
	AL 333w-56 (LCW)	57.7	69.1
	AL 333x-26 (MTB, LTB)	56.3	59.2
Australopithecus africanus	Sts 14 (ACET, LSW)	34.8	33.3
	Sts 34 (MCW)	46.5	48.4
	Stw 300 (FHD)	41.6	47.0
	Stw 31 (FHD)	33.0	36.4
	Stw 99 (FHD)	49.9	57.6
	Stw 311 (FHD)	52.1	60.5
	Stw 392 (FHD)	37.2	41.5
	Stw 396 (LCW)	53.8	64.5
	Stw 389 (DTB)	55.5	66.8
	TM 1513 (MCW,LCW)	44.2	58.3
	MLD 17 (FHD)	57.6	67.5
Paranthropus robustus	SK 97 (FHD)	55.8	65.2
	SK 82 (FHD)	45.6	52.1
	SK 3155 (ACET)	37.1	42.2
	?SKX 1084 (PATW)	57.5	88.6
Paranthropus boisei	KNM-ER 738 (FHD)	47.7	54.8
	KNM-ER 1500 (DTB)	44.7	52.1
	KNM-ER 1503/1505	51.4	59.5

Continued

Table 7.2. *Continued*

Group	Specimen [dimension(s)]	Estimate (kg)	
		ALLHOM	NONHUM
	(FHD)		
	?KNM-ER 741	69.3	88.6
	(LTB)		
	?OH 35	33.0	36.9

[a](?) Denotes problematic assignment of fossil to the group.

[b]ACET, acetabulum height; FHD, femoral head diameter; MCW, posterior width of the medial femoral condyle; LCW, posterior width of the lateral femoral condyle; PATW, width of the patellar articular surface; MTB, anteroposterior length of the medial tibial condyle; LTB, anteroposterior length of the lateral tibial condyle; DTB, anteroposterior length of the distal tibial articulation; LSW, width of the lumbosacral articulation.

human evolution. If this proves to be the case based on other independent lines of evidence, then evolutionary scenarios that presuppose a trend of increasing size will require substantial rethinking (Pilbeam and Gould, 1974; Blumenberg, 1984; Foley, 1987). Perhaps it is the case that only females increase significantly in body size in later hominid evolution, resulting thereby in a trend toward reduced sexual dimorphism in size (and behavior?).

The extant data base was treated in a sex-specific fashion because it is not entirely clear what the biological meaning of an "average" or midpoint animal is in moderately to highly sexually dimorphic species. For the same reasons, one must question uncritical lumping of fossil size estimates into means and midpoints if there exists sufficient reason to suggest the presence of sexual size dimorphism. Given the number of assumptions and potential complications in making predictions in the fossil record, perhaps it is more advisable to concentrate instead on the ranges themselves. Regardless, the values reported here tend to be greater than most prior estimates, no matter whether ALLHOM or NONHUM is regarded as more appropriate.

Earlier reported values have relied at times on computation to make estimates of fossil hominid body size; sometimes, however, divination appears to have been the methodology of choice. The latter course is particularly popular in less technical volumes intended for the nonspecialist (e.g., Lambert, 1987). Regrettably, the analytical basis of some estimates is occasionally too vague to evaluate their reliability or to extend to other specimens (e.g., Pilbeam, 1972; Pilbeam and Gould, 1974; Latimer *et al.*, 1987). Even when made explicit, the methods employed can sometimes border on the bizarre (Krantz, 1977). Noteworthy exceptions include the predictions made by Wolpoff (1973), McHenry (1976), Steudel (1980) and Suzman (1980). I have serious reservations as to the number of intervening variables employed and the logical basis of some of the estimates provided by Wolpoff (1973) and Steudel (1980, 1985). There is a modest degree of reassuring convergence between my estimates and the new ones generated independently by McHenry (this volume, chapter 9) using different critera. My results and those of Suzman (1980), based on nonhuman analogies, are at least in the same ballpark for the South African material. As a final point, and to use a quote from McHenry (1982) out of context but in similar spirit, exclusive use of models based on modern *H. sapiens* to estimate quantitative relationships in the australopithecines is unwarranted: "It is a false bias to treat them as humans."

Returning to the issue of sexual dimorphism vs. multiple taxa in the hominid fossil record, some common sense seems in order. For example, I fail to see the logic of an analysis that compares the ratio of smallest over largest observed values for hominid fossils (e.g., *A. afarensis*) only to the ratio of female mean over male mean for extant species of hominoids (Zihlman, 1985). Given the probable inflation of variability in any terrestrial fossil assemblage due to averaging over considerable amounts of time (see Bell *et al.*, 1988), the *maximum* ranges observed within living species provide a more realistic and conservative analogy than do extant means (Martin and Andrews, 1984). Table 7.4 summarizes a comparison of max/min values in the fossil groups

Table 7.3. Regression Equations Used to Predict Mass in this Study[a]

Dimension(s)	All hominoids (ALLHOM)	Nonhuman hominoids (NONHUM)
FHD	2.6142 lnFHD − 5.4282 ($r = 0.974$, SEE $= 0.243$)	2.9047 lnFHD − 6.3233 ($r = 0.997$, SEE $= 0.081$)
ACET	2.8025 lnACET − 6.6459 ($r = 0.967$, SEE $= 0.274$)	3.1824 lnACET − 7.9090 ($r = 0.997$, SEE $= 0.085$)
MCW	2.1224 lnMCW − 2.6824 ($r = 0.978$, SEE $= 0.225$)	2.1743 lnMCW − 2.8023 ($r = 0.979$, SEE $= 0.231$)
LCW	1.9335 lnLCW − 1.7269 ($r = 0.950$, SEE $= 0.335$)	2.1865 lnLCW − 2.3033 ($r = 0.977$, SEE $= 0.242$)
DTB	2.5037 lnDTB − 3.9397 ($r = 0.957$, SEE $= 0.309$)	2.8561 lnDTB − 4.8747 ($r = 0.988$, SEE $= 0.177$)
ACET and LSW	3.3223 lnACET − 0.5268 lnLSW − 6.6753 ($r = 0.967$, SEE $= 0.285$)	2.9752 lnACET + 0.2115 lnLSW − 7.9025 ($r = 0.997$, SEE $= 0.089$)
MCW and LCW	2.0780 lnMCW + 0.0429 lnLCW − 2.6691 ($r = 0.978$, SEE $= 0.235$)	1.2217 lnMCW + 0.9783 lnLCW − 2.6370 ($r = 0.983$, SEE $= 0.222$)
MTB and LTB	3.0942 lnMTB − 0.6079 lnLTB − 5.1965 ($r = 0.968$, SEE $= 0.278$)	1.2564 lnMTB + 1.8900 lnLTB − 7.0237 ($r = 0.996$, SEE $= 0.108$)
MCW, LCW, and LTB	1.9314 lnMCW − 0.7601 lnLCW + 1.4068 lnLTB − 4.7511 ($r = 0.981$, SEE $= 0.224$)	0.3657 lnMCW + 0.0266 lnLCW + 2.7630 lnLTB − 6.7175 ($r = 0.995$, SEE $= 0.128$)
FHD, ACET, MTB, LTB, DTB, and LSW	4.8888 lnFHD − 1.0150 lnACET + 3.3811 lnMTB − 1.0643 lnLTB − 2.3456 lnDTB − 1.5531 lnLSW − 5.2628 ($r = 0.982$, SEE $= 0.267$)	1.5920 lnFHD + 0.7905 lnACET + 0.4687 lnMTB + 0.8010 lnLTB − 0.3257 lnDTB − 0.2698 lnLSW − 7.0649 ($r = 0.998$, SEE $= 0.0923$)

[a] r, correlation coefficient; SEE, standard error of estimate. See footnote b of Table 7.2 for other abbreviations.

Table 7.4. Ratio of Max/Min Body Mass in the Fossil and Extant Samples

Group	Observed values	Estimated values	
		NONHUM	ALLHOM
Australopithecus afarensis	—	2.66	2.23
Australopithecus africanus	—	2.03	1.75
Paranthropus robustus	—	2.10	1.55
Paranthropus boisei	—	2.40	2.10
Gorilla gorilla	3.02	—	—
Pan troglodytes	2.23	—	—
Pongo pygmaeus	2.78	—	—
Homo sapiens	1.65	—	—

(ALLHOM and NONHUM again treated separately) to max/min values of body mass in the living ape and human samples used in this study. The ranges of the extant samples would be even greater if reliable values from the literature were included. It also seems certain that the ranges would expand somewhat in the "robust" fossil groups if the sample sizes were larger. The ratios observed in gorillas and orangutans are greater than those observed in any of the fossil hominid groups, regardless of whether ALLHOM or NONHUM values are used in the contrasts. If ALLHOM estimates are preferred, then even chimpanzees are as variable as any of the fossil groups. Admittedly, this set of simple comparisons does not resolve the complicated issue of species number in fossil samples such as those from either the Hadar Formation or the Sterkfontein Formation, but it does eliminate arguments for more than one species based largely or solely on excessive variability in size. As such, these results are more in agreement with the conclusions about *A. afarensis* reached by McHenry (1986) than with those of Zihlman (1985) or Senut (1986).

Although it is now part of the conventional wisdom that all australopithecines were "megadont" in terms of their postcanine dentition, a precise consideration of relative tooth size is severely hampered by a paucity of associated dental and postcranial remains (McHenry 1984). Specimen AL 288-1 represents a welcome exception to such limitations, and it is instructive to compare body size estimates derived from tooth-size regressions to those presented above. Using "Lucy's" first mandibular molar and the relevant equation from Gingerich *et al.* (1982), one predicts a body mass of 52.8 kg. This value is almost 75% greater than the 30.4 kg estimate based on six articular dimensions of the hindlimb; megadontia in molars appears to remain a reasonable inference. On average, both South and East African "robust" australopithecines have greatly enlarged cheek-teeth compared to either *A. afarensis* or to *A. africanus* (Jungers and Grine, 1986). If, as the tentative results presented here suggest, there is little absolute difference in body size among all these groups, the trend toward even more pronounced megadontia in *Parathropus* obviously cannot be related simply or causally to major body size differences (Pilbeam and Gould, 1974; Gould, 1975). Any quantitative study of the evolution of relative brain size from the early hominid fossil record appears to be subject to a similar caveat.

Acknowledgments

I wish to express my sincere appreciation to Fred Grine for the invitation to participate in The International Workshop on The Evolutionary History of the "Robust" Australopithecines. Many thanks are also due to Jack T. Stern, J., Randall L. Susman and Fred E. Grine for providing measurements from original fossil material. I thank Joan Kelly for typing the tables and bibliography. This research was supported by NSF Grants BNS 8519747 and BNS 8606781.

References

Aiello, L. C. (1981). Locomotion in the Miocene Hominoidea. *In* C. B. Stringer (ed.), *Aspects of human evolution*, pp. 63–97. Taylor and Francis, London.

Alexander, R. McN. (1980). Forces in animal joints. *Eng. Med.*, 9: 93–97.

Alexander, R. McN. (1981). Analysis of force platform data to obtain joint forces. *In* D. Dowson and V. Wright (eds.), *An introduction to the biomechanics of joints and joint replacements*, pp. 30–35. Mechanical Engineering Publishers, London.

Anderson, M., Messener, M. B. and Green, W. T. (1964). Distribution of lengths of the normal femur and tibia in children from one to eighteen years of age. *J. Bone Joint. Surg.*, 46A: 1197–1202.

Andrews, P. and Aiello, L. C. (1984). An evolutionary model for feeding and positional behavior. *In* D. J. Chivers, B. A. Wood and A. Bilsborough, (eds.), *Food acquisition and processing in primates*, pp. 429–466. Plenum, New York.

Bell, M. A., Sadagursky, M. S. and Baumgartner, J. V. (1988). Utility of lacustrine deposits for the study of variation within fossil samples. *Palaios*, in press.

Blumenberg, B. (1984). Allometry and evolution of Tertiary hominoids. *J. Hum. Evol.*, 13: 613–676.

Brogin, B. (1988). *Patterns of human growth*. Cambridge University Press, Cambridge.

Calder, W. A., III (1984). *Size, function, and life history*. Harvard University Press, Cambridge, MA.

Conroy, G. C. (1987). Problems of body-weight estimation in fossil primates. *Intern. J. Primatol*, 8: 115–137.

Corruccini, R. S. and McHenry, H. M. (1978). Relative femoral head size in early hominids. *Amer. J. Phys. Anthropol.*, 49: 145–148.

Feldesman, M. R. and Lundy, J. K. (1988). Stature estimates for some African Plio-Pleistocene fossil hominids. *J. Hum. Evol.*, in press.

Fleagle, J. G. (1978). Size distributions of living and fossil primate faunas. *Paleobiology*, 4: 67–76.

Fleagle, J. G. and Kay, R. F. (1985). The paleobiology of catarrhines. *In* E. Delson (ed.), *Ancestors: the hard evidence*, pp. 23–36. Alan R. Liss, New York.

Foley, R. (1987). *Another unique species*. Wiley, New York.

Geissman, T. (1986). Estimation of australopithecine stature from long bones: A. L. 288-1 as a test case. *Folia Primatol.*, 47: 119–127.

Gingerich, P. D. (1977). Correlation of tooth size and body size in living hominoid primates, with a note on relative brain size in *Aegyptopithecus* and *Proconsul*. *Amer. J. Phys. Anthropol.*, 47: 395–398.

Gingerich, P. D., Smith B. H. and Rosenberg, K. (1982). Allometric scaling in the dentition of primates and prediction of body weight from tooth size in fossils. *Amer. J. Phys. Anthropol.*, 58: 81–100.

Gould, S. J. (1975). On the scaling of tooth size in mammals. *Amer. Zool.*, 15: 351–362.

Hofman, M. A. (1983). Encephalization in hominids: evidence for the model of punctuationalism. *Brain Behav. Evol.*, 22: 102–117.

Holloway, R. L. and Post, D. G. (1982). The relativity of relative brain measures and hominid mosaic evolution. *In* E. Armstrong and D. Falk (eds.), *Primate brain evolution*, pp. 57–76. Plenum, New York.

Hylander, W. L. (1985). Mandibular function and biomechanical stress and scaling. *Amer. Zool.*, 25: 315–330.

Jerison, H. J. (1973). *Evolution of the brain and intelligence*. Academic Press, New York.

Jungers, W. L. (1982). Lucy's limbs: Skeletal allometry and locomotion in *Australopithecus afarensis*. *Nature*, 297: 676–678.

Jungers, W. L. (1984). Aspects of size and scaling in primate biology with special reference to the locomotor skeleton. *Yrbk. Phys. Anthropol.*, 27: 73–97.

Jungers, W. L. (1988a). Body size and morphometric affinities of the appendicular skeleton in *Oreopithecus bambolii* (IGF 11778). *J. Hum. Evol.*, 16: 445–456.

Jungers, W. L. (1988b). Lucy's length: status reconstruction in *Australopithecus afarensis* (A. L. 288-1) with implications for other small-bodied hominids. *Amer. J. Phys. Anthropol., 76:* 227–231.

Jungers, W. L. (1988c). Scaling of postcranial joint size in hominoid primates. *Hum. Evol.,* in press.

Jungers, W. L. (1988d). Relative joint size and hominoid locomotor adaptations with implications for the evolution of hominid bipedalism. *J. Hum. Evol., 17:* 247–265.

Jungers, W. L. and Grine, F. E. (1986). Dental trends in the australopithecines: The allometry of mandibular molar dimensions. *In* B. Wood, L. Martin and P. Andrews (eds.), *Major topics in primate and human evolution,* pp. 203–219. Cambridge University Press, Cambridge.

Kay, R. F. (1984). On the use of anatomical features to infer foraging behavior in extinct primates. *In* P. S. Rodman and J. G. H. Cant (eds.), *Adaptations for foraging in nonhuman primates,* pp. 21–53. Columbia University Press, New York.

Kay, R. F. (1985). Dental evidence for the diet of *Australopithecus. Ann. Rev. Anthropol., 14:* 315–341.

Kay, R. F. and Simons, E. L. (1980). The ecology of Oligocene African Anthropoidea. *Intern. J. Primatol., 1:* 21–37.

Kennedy, G. E. (1983). Some aspects of femoral morphology in *Homo erectus. J. Hum. Evol., 12:* 587–616.

Krantz, G. S. (1977). A revision of australopithecine body size. *Evol. Theory, 2:* 65–94.

Lambert, D. (1987). *The Cambridge guide to prehistoric man.* Cambridge Univ. Press, Cambridge.

Latimer, B., Ohman, J. C. and Lovejoy, C. O. (1987). Talocrural joint in African hominoids: implications for *Australopithecus afarensis. Amer. J. Phys. Anthropol., 74:* 155–175.

Leutenegger, W. and Cheverud, J. M. (1985). Sexual dimorphism in primates. The effects of size. *In* W. L. Jungers (ed.), *Size and scaling in primate biology,* pp. 33–50. Plenum, New York.

Lovejoy, C. O. and Trinkaus, E. (1980). Strength and robusticity of the Neandertal tibia. *Amer. J. Phys. Anthropol., 53:* 465–470.

Martin, L. and Andrews, P. (1984). The phyletic position of *Graecopithecus freybergi* Koenigswald. *Cour. Forsch. Inst. Senckenberg, 69:* 25–40.

Martin, R. A. (1981). On extinct hominid population densities. *J. Hum. Evol., 10:* 427–428.

McHenry, H. M. (1974). How large were the australopithecines? *Amer. J. Phys. Anthropol., 40:* 329–340.

McHenry, H. M. (1976). Early hominid body weight and encephalization. *Amer. J. Phys. Anthropol., 45:* 77–84.

McHenry, H. M. (1982). The pattern of human evolution: studies on bipedalism, mastication and encephalization. *Ann. Rev. Anthropol., 11:* 151–173.

McHenry, H. M. (1984). Relative cheek-tooth size in *Australopithecus. Amer. J. Phys. Anthropol., 64:* 297–306.

McHenry, H. M. (1986). Size variation in the postcranium of *Australopithecus afarensis* and extant species of Hominoidea. *In* M. Pickford and A. B. Chiarelli (eds.), *Sexual dimorphism in living and fossil primates,* pp. 183–189. Il Sedicesimo, Firenze.

McHenry, H. M. and Corruccini, R. S. (1976). Fossil hominid femora and the evolution of walking. *Nature, 259:* 657–658.

Napier, J. R. (1964). The evolution of bipedal walking in the hominids. *Arch. Biol. (Liege), 75:* 673–708.

Peters, R. H. (1983). *The ecological implications of body size.* Cambridge University Press, Cambridge.

Pickford, M. and Chiarelli, B. (1986). *Sexual dimorphism in living and fossil primates.* Il Sedicesimo, Firenze.

Pilbeam, D. (1972). *The ascent of Man. An introduction to human evolution.* MacMillan, New York.

Pilbeam, D. and Gould, S. J. (1974). Size and scaling in human evolution. *Science, 186:* 892–901.

Radinsky, L. (1974). The fossil evidence of anthropoid brain evolution. *Amer. J. Phys. Anthropol., 41*: 15–27.

Radinsky, L. (1982). Some cautionary notes on making inferences about relative brain size. *In* E. Armstrong and D. Falk (eds.), *Primate brain evolution: methods and concepts*, pp. 29–37. Plenum, New York.

Reed, R. B. and Stuart, H. L. (1959). Patterns of growth in height and weight from birth to eighteen years of age. *Pediatrics, 24*: 904–921.

Rightmire, G. P. (1986). Body size and encephalization in *Homo erectus*. *Anthropos, 23*: 139–150.

Ruff, C. (1987). Structural allometry of the femur and tibia in Hominoidea and *Macaca*. *Folia Primatol., 48*: 9–49.

Schmidt-Nielsen, K. (1977). Problems of scaling: locomotion and physiological correlates. *In* T. J. Redley (ed.), *Scale effects in animal locomotion*, pp. 1–21. Academic Press, New York.

Scott, K. (1983). Body weight prediction in fossil Artiodactyla. *Zool. J. Linn. Soc., 77*: 197–228.

Senut, B. (1986). Long bones of the primate upper limb: monomorphic or dimorphic? *In* M. Pickford and B. Chiarelli (eds.), *Sexual dimorphism in living and fossil primates*, pp. 7–22. Il Sedicesimo, Firenze.

Smith, R. J. and Pilbeam, D. R. (1980). Evolution of the orang-utan. *Nature, 284*: 447–448.

Smith, R. J. (1985). The present as a key to the past: body weight of Miocene hominoids as a test of allometric methods for paleontological inference. *In* W. L. Jungers (ed.), *Size and scaling in primate biology*, pp. 437–447. Plenum, New York.

Steudel, K. (1980). New estimates of early hominid body size. *Amer. J. Phys. Anthropol., 52*: 63–70.

Steudel, K. (1985). Allometric perspectives on fossil catarrhine morphology. *In* W. L. Jungers (ed.), *Size and scaling in primate biology*, pp. 449–475. Plenum, New York.

Suzman, I. M. (1980). A new estimate of body weight in South African australopithecines. *Proc. Pan-Afr. Cong. Prehist. Quat. Stud. (Nairobi)*, pp. 175–179.

Walker, A. (1973) New *Australopithecus* femora from East Rudolf, Kenya. *J. Hum. Evol., 2*: 545–555.

Wolpoff, M. H. (1973). Posterior tooth size, body size, and diet in South African gracile australopithecines. *Amer. J. Phys. Anthropol., 39*: 375–394.

Zihlman, A. L. (1985). *Australopithecus afarensis*: two sexes or two species? *In* P. V. Tobias (ed.), *Hominid evolution: past, present and future*, pp. 213–220. Alan R. Liss, New York.

Associated Cranial and Postcranial Bones of *Australopithecus boisei*

8

HANNAH M. GRAUSZ, RICHARD E. LEAKEY,
ALAN C. WALKER AND CAROL V. WARD

Bones of *Australopithecus boisei* are the most common hominid fossils from the Lake Turkana Plio-Pleistocene deposits. Many isolated postcranial bones have been attributed to this species, possibly correctly, but there are only two sets of associated cranial and postcranial bones found so far in Kenya. The first is a set of four bones associated with mandibles and teeth of at least four individuals from the KBS Member in Area 6A at Ileret (Leakey, 1972; Leakey and Walker, 1973, 1985; Day et al., 1976; Leakey and Leakey, 1978). The second is a fragmentary skeleton, KNM-ER 1500, from the Upper Burgi Member of Area 130 at Koobi Fora (Leakey, 1973; Day et al., 1976; Leakey and Leakey, 1978).

The four bones from Ileret are a complete right talus, KNM-ER 1464, a proximal part of a right third metatarsal, KNM-ER 1823, a fragment of distal humerus, KNM-ER 1824, and much of the left half of an atlas, KNM-ER 1825. These are probably all from adults. However, they are certainly not from only one individual since the talus and metatarsal are not of the right sizes to have belonged to one foot. KNM-ER 1464 is extremely similar in shape to the talus from the OH 8 foot (Day and Napier, 1964), but is considerably larger. The talus is distinguished by a very strongly developed lateral flange, an extreme degree of trochlear wedging and a short, dorsiflexed neck. The first of these features indicates that the distal tibiofibular articulation was different from that usually seen in *Homo sapiens*. The third metatarsal is only a proximal fragment, but it resembles that from Olduvai foot closely in shape and size, except that the shaft is more compressed mediolaterally and is more concave plantarward. The humeral fragment preserves part of the distal articular surface and the medial epicondyle with a markedly narrow trochlea. The lateral mass of the atlas is relatively very deep and its occipital condyle facets are very concave to the point of being V-shaped in horizontal section.

The partial skeleton KNM-ER 1500 has not been formally assigned before to *A. boisei*, although Leakey (1973) did assign it to *Australopithecus*. The only associated cranial fragment is a small piece of mandible, recognized as such only recently. The relative thickness of the mandibular body under the mental foramen and the presence of a blunt marginal crest in *A. boisei* specimens, including this one, are unusual for hominid species and make *A. boisei* easily identifiable (Fig. 8.1). Mandibular measurements in this species are very variable because there is probably a good deal of sexual dimorphism (Walker and Leakey, this volume, chapter 15). The distance from the inferior margin to the mental foramen in a sample of 27 *A. boisei* mandibles has a mean value of 27.18 mm with a standard deviation of 4.74. At 19.0 mm, the KNM-ER 1500 mandible has the smallest value, being 1 mm less than the value for Omo 1967-18. If the mandible and skeleton were roughly proportional in this species, and we expect that they were, then the KNM-ER 1500 individual was a small female.

Apart from the mandibular fragment, the following pieces of the skeleton have been identified: fragment of clavicle, left scapula glenoid, fragment of left distal humerus shaft, proximal right radius, right radius shaft fragment, left radius shaft fragment, proximal right ulna, right ulna shaft fragment, proximal left femur, distal left femur, nearly complete left tibia, fragment of distal right tibia shaft, distal right tibia, distal right fibula, proximal right metatarsal III. Figure 8.2 shows the position of the fragments on a fanciful outline. For the most part, surface detail of the fossil

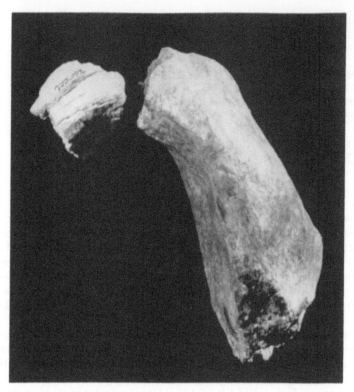

FIGURE 8.1. Basal view of KNM-ER 1500 mandibular fragment (left) compared to presumptive female *Australopithecus boisei* corpus KNM-ER 15930 (right) showing thickness of corpus and blunt marginal crest.

FIGURE 8.2. Drawing showing the pieces of KNM-ER 1500 arranged in a schematic body outline to illustrate skeletal representation.

Table 8.1. Postcranial Measurements (mm)

	Measurement	KMN-ER 1500	AL 288-1	Pan troglodytes	Homo sapiens
1.	Mid-sigmoid width of ulna	17.5	12.5	17.5	21.0
2.	Maximum olecranon length	22.0	16.5	22.0	24.0
3.	Proximodistal length at sigmoid notch of ulna	17.5	16.0	24.0	25.0
4.	Maximum diameter of radial head	20.2	15.1	22.5	21.0
5.	Minimum diameter of radial head	19.5	14.0	21.5	20.0
6.	Distance from center of radial head to center of bicipital tuberosity	32.6	31.3	36.0	30.0
7.	Maximum diameter of radial neck	11.3	9.6	12.7	13.0
8.	Minimum diameter of radial neck	9.1	8.6	11.5	12.0
9.	Maximum scapular glenoid width	21.0	18.2	26.5	27.0
10.	Anteroposterior diameter of femoral shaft at lesser trochanter	22.4	18.3	24.0	25.0
11.	Mediolateral diameter of femoral shaft at lesser trochanter	26.3	27.0	27.0	29.0
12.	Estimated length of tibia	250	245	243	370
13.	Maximum anteroposterior width at distal epiphysis of tibia	28.0	23.9	26.0	47.0
14.	Maximum mediolateral width at distal epiphysis of tibia	34.0	32.2	35.0	58.0
15.	Anteroposterior breadth of malleolus of tibia	17.0	14.5	18.6	28.0
16.	Mediolateral breadth of malleolus of tibia	11.0	9.0	13.5	14.0
17.	Projection of malleolus below articular surface of tibia	13.7	11.3	12.0	17.0
18.	Maximum mediolateral width at proximal epiphysis of tibia	53.0	50.7	52.0	76.0
19.	Maximum anteroposterior width at proximal epiphysis of tibia	(39.0)	34.0	40.0	55.0
20.	Dorsoplantar width of base of third metatarsal	17.3	—	18.0	21.0
21.	Dorsal width at base of third metatarsal	12.9	—	14.0	15.0
22.	Anteroposterior midshaft diameter of tibia	26.6	(22.0)	22.4	31.0
23.	Mediolateral midshaft diameter of tibia	15.8	(14.5)	15.7	22.0

is poor. The bones were weathered before sedimentary burial, and the specimen had been exposed on the surface for a long time before discovery.

It is difficult to make comparisons with such fragmentary material. We have attempted to assess the relative proportions of the bones of this skeleton by making direct comparisons with one individual male chimpanzee, one relatively small individual human, and the *Australopithecus afarensis* partial skeleton, AL 288-1 (Johanson *et al.*, 1982). The measurements are listed in Table 8.1. Measurements on AL 288-1 were taken from the literature and, where not available, from casts. We did not attempt to use either a statistically average chimpanzee, an average human, or a complex size-adjusted standard such as that used by McHenry (1984). These comparisons must be taken, therefore, as being only illustrative.

Figure 8.3 shows the relative sizes of the measurements in KNM-ER 1500 and *H. sapiens* compared with *Pan troglodytes*, which is used as the standard (zero on the horizontal scale). On the whole, forelimb parts of KNM-ER 1500 are smaller than *Pan* and are nearly always smaller and have different relative proportions from those of *Homo*. Hindlimb parts are roughly the size of *Pan*, but are very different in relative proportions. It is striking that the pattern with which KNM-ER 1500 deviates from the *Pan* standard resembles that of the human, although the human dimensions are much larger.

Using the same *Pan* standard, Figure 8.4 shows the same values for KNM-ER 1500 compared to AL 288-1. Although the values for the *A. afarensis* specimen are nearly all slightly smaller than either of the others, the pattern for the two *Australopithecus* individuals is nearly identical. Note that most of the forelimb measurements are smaller than those of *Pan*. The hindlimb measurements are about the same size, but of different proportions than those of *Pan*.

Figure 8.5 shows measurements of the same two fossils with *Homo* as the standard. Note that with the exception of measurement number 6, which reflects the length of the *biceps brachii* lever arm, both fossils have values less than our human. Further, the pattern of deviation from our human standard is very similar for both. Both fossils have relatively larger values for forelimb than for hindlimb measurements. This complements Jungers' (1982) finding that the hindlimbs in AL 288-1 were relatively short compared with the forelimbs. It should also be noted that the postcranial skeleton of *A. africanus* is very similar indeed to that of AL 288-1 (McHenry, 1986).

In summary, these comparisons suggest that the postcranial skeleton of female *A. boisei* is

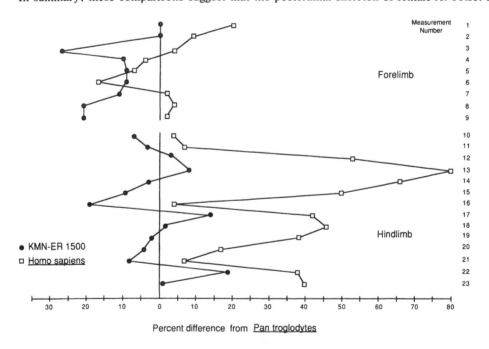

FIGURE 8.3. Measurements of KNM-ER 1500 and reference *Homo sapiens* expressed as a percentage of the *Pan troglodytes* standard.

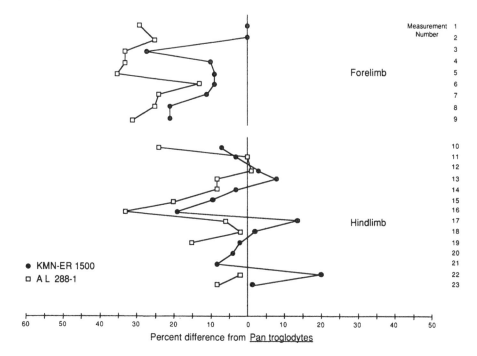

FIGURE 8.4. Measurements of KNM-ER 1500 and AL 288-1 expressed as a percentage of the *Pan troglodytes* standard.

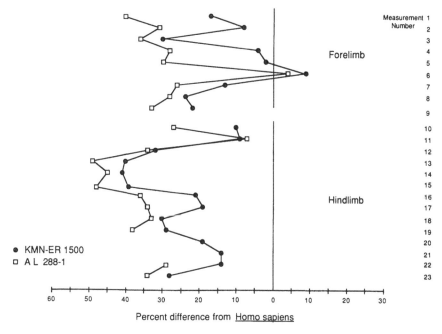

FIGURE 8.5. Measurements of KNM-ER 1500 and AL 288-1 expressed as a percentage of the *Homo sapiens* standard.

very similar to that of female *A. afarensis* and *A. africanus*. Further, they show that the proportions of these species were not like those of *H. sapiens* or *P. troglodytes*. Johanson *et al.* (1987) estimate the humerofemoral index of OH 62, a partial skeleton of *Homo habilis*, at 95%, which is almost identical to the value given by Schultz (1937) for *Pan* and higher than that of AL 288-1 (Jungers, 1982). Since the proportions of the limbs of KNM-ER 1500 closely resemble those of AL 288-1, we can predict that *A. boisei* probably had less chimpanzee-like limb proportions than did *H. habilis*. Finally, our comparisons between these *A. boisei* specimens and OH 8 and OH 35 suggest one of two possibilities. Either OH 8 and OH 35 should be attributed to *A. boisei*, as suggested by Wood (1974) following his work on the talus, or the lower limb skeletons of *H. habilis* and *A. boisei* cannot be easily distinguished.

Acknowledgments

We thank the Government of Kenya and the Governors of the National Museums of Kenya. The field work which led to the finding of most of the fossils discussed here was funded by the National Geographic Society, Washington, D.C., The William Donner Foundation, The Garland Foundation and the National Museums of Kenya. A. W. was the recipient of a John Simon Guggenheim Memorial Fellowship. Many colleagues have helped in many different ways, but we thank especially Bw. Kamoya Kimeu and his field team and Bw. Joseph Mutaba, Mrs. E. Mbua and M. G. Leakey for technical and curatorial help in Nairobi.

References

Day, M. H. and Napier, J. R. (1964). Hominid fossils from Bed I, Olduvai Gorge, Tanganyika. Fossil foot bones. *Nature*, 201: 967–970.
Day, M. H., Leakey, R. E. F., Walker, A. C. and Wood, B. A. (1976). New hominids from East Turkana, Kenya. *Amer. J. Phys. Anthropol.*, 45: 369–436.
Johanson, D. C., Lovejoy, C. O., Kimbel, W. H., White, T. D., Ward, S. C., Bush M. E., Latimer, B. M. and Coppens, Y. (1982). Morphology of the Pliocene partial hominid skeleton (A. L. 288-1) from the Hadar Formation, Ethiopia. *Amer. J. Phys.Anthropol.*, 57: 403–451.
Johanson, D. C., Masao, F. T., Eck, G. G., White, T. D., Walter, R. C., Kimbel, W. H., Asfaw, B., Manega, P., Ndessokia, P. and Suwa, G. (1987). New partial skeleton of *Homo habilis* from Olduvai Gorge, Tanzania. *Nature*, 327: 205–209.
Jungers, W. L. (1982). Lucy's limbs: skeletal allometry and locomotion in *Australopithecus afarensis*. *Nature*, 297: 676–678.
Leakey, M. G. and Leakey, R. E. (1978). *Koobi Fora research project* vol. 1, p. 91. Clarendon Press, Oxford.
Leakey, R. E. F. (1972). Further evidence of Lower Pleistocene hominids from East Rudolf, North Kenya, 1971. *Nature*, 237: 264–269.
Leakey, R. E. F. (1973). Further evidence of Lower Pleistocene hominids from East Rudolf, North Kenya, 1972. *Nature*, 242: 170–173.
Leakey, R. E. F. and Walker, A. C. (1973). New australopithecines from East Rudolf, Kenya (III). *Amer. J. Phys. Anthropol.*, 39: 205–222.
Leakey, R. E. F. and Walker, A. C. (1985). Further hominids from the Plio-Pleistocene of Koobi Fora, Kenya. *Amer. J. Phys. Anthropol.*, 67: 135–163.
Leakey, R. E. F. and Walker, A. C. (1988). New *Australopithecus boisei* specimens from East and West Lake Turkana, Kenya. *Amer. J. Phys. Anthropol.*, 76: 1–24.
McHenry, H. M. (1984). The common ancestor. *In* R. L. Susman (ed.), *The pygmy chimpanzee*, pp. 201–230. Plenum, New York.
McHenry, H. M. (1986). The first bipeds: a comparison of the *A. afarensis* and *A. africanus* postcranium and implications for the evolution of bipedalism. *J. Hum. Evol.*, 15: 177–191.
Schultz, A. H. (1937). Proportions, variability and asymmetries of the long bones of the limbs and the clavicles in man and the apes. *Hum. Biol.*, 9: 281–328.
Wood, B. A. (1974). Olduvai Bed I post-cranial fossils: a reassessment. *J. Hum. Evol.*, 3: 373–378.

New Estimates of Body Weight in Early Hominids and Their Significance to Encephalization and Megadontia in "Robust" Australopithecines

9

HENRY M. MCHENRY

Most studies of body weight in species of early hominids report that "gracile" australopithecines were smaller in body size than "robust" australopithecines (e.g., Genet-Varcin, 1967; Lovejoy and Heiple, 1970; Robinson, 1972; Pilbeam and Gould, 1974; McHenry, 1974, 1975a,b, 1976, 1982, 1984; McHenry and Temerin, 1979; Boaz, 1977, 1985; Steudel, 1980; Suzman, 1980). This assertion is often taken to account for the larger cheek-teeth and masticatory complex of the latter (e.g., Pilbeam and Gould, 1974). It is also common to state that body size increased through time in human evolution as would be expected from Cope's Law.

All of these studies, however, have been limited by the rarity of early hominid postcrania. Body weight estimates are usually based on only a very few specimens. This is a special problem because by chance the two most complete specimens representing the two earliest hominid species (AL 288-1 for *Australopithecus afarensis* and Sts 14 for *A. africanus*) are among the smallest hominid individuals ever found. Much larger but more fragmentary hominid postcranial specimens are known from the same sites as these partial skeletons.

This study attempts to partially remedy the problems created by small sample sizes by using the most commonly preserved portion of the postcranial skeleton in the fossil hominid collections. There are over two dozen proximal portions of the femoral shaft from hominids dating between 3.4 and 1.4 Myr (Table 9.1). Fortunately, the size of the femoral shaft has a high correlation with body weight among species of Hominioidea and within *Homo sapiens*. Confining the study to a single element avoids some of the problems brought on by the fact that body proportions within the early hominids differed from those in modern humans. These problems include, for example, the fact that estimates of body weight extrapolated from a sample modern *H. sapiens* based on early hominid forelimbs will tend to give larger results than those based on hindlimbs. Estimates derived from joint sizes in the hindlimb tend to give smaller weights than those based on shaft widths (compare this study to McHenry, 1976, for example). Basing estimates on a single element for all species of early hominid should result in more reliable comparability among them.

The approach here is twofold: (1) to document the variability in hominid femoral shaft size from 3.4 to 1.4 Myr, and (2) to derive an estimation of body weight from femoral shaft size. From the former, one can learn with some confidence the pattern of size variation through time and compare it with the variation in extant hominoids, particularly with respect to sexual dimorphism. The latter approach (estimating body weight) requires more assumptions but is of great importance to many questions in human evolution.

Material and Methods

This study makes use of two comparative samples. The first (Table 9.2) is a generous collection of all extant large bodied hominoids. This sample provides the background with which to compare the variability in fossil hominid femora. The human sample, housed at the Peabody Museum,

Table 9.1. Fossil Femora Used in Estimates of Early Hominid Body Weight

Specimen	Portion of shaft measured	Age of specimen (Myr)	Group
Hadar			
AL 128-1[a]	L. proximal	3.1–3.4[b]	*Australopithecus afarensis*[c]
AL 129-1c[a]	R. proximal	3.1–3.4[b]	*Australopithecus afarensis*[c]
AL 211-1	R. proximal	3.1[d]	*Australopithecus afarensis*[c]
AL 288-1ap	L. complete	3.1[d]	*Australopithecus afarensis*[c]
AL 333-3	R. proximal	3.1[d]	*Australopithecus afarensis*[c]
AL 333-95	R. proximal	3.1[d]	*Australopithecus afarensis*[c]
AL 333w-40	L. proximal	3.1[d]	*Australopithecus afarensis*[c]
East Turkana			
KNM-ER 736	L. shaft	1.6–1.9[e]	*Homo* or *Australopithecus*[f]
KNM-ER 737	L. shaft	1.4–1.6[g]	*Homo* sp.[f]
KNM-ER 738	L. proximal	1.6–1.9[g]	*Australopithecus* sp.[f]
KNM-ER 803	L. shaft	1.4–1.6[g]	*Homo* sp.[f]
KNM-ER 815	L. proximal	1.6–1.9[e]	*Australopithecus* sp.[f]
KNM-ER 993	R. shaft	1.4–1.6[g]	*Australopithecus* sp.[f]
KNM-ER 1463[h]	R. shaft	1.4–1.6[g]	*Australopithecus* sp.[f]
KNM-ER 1465[h]	L. proximal	1.4–1.6[g]	*Australopithecus* sp.[f]
KNM-ER 1472	R. complete	1.9–2.0[i]	*Homo* sp.[f]
KNM-ER 1475	R. proximal	1.9–2.0[i]	*Homo* sp.[f]
KNM-ER 1481a	L. complete	1.9–2.0	*Homo* sp.[f]
KNM-ER 1500d	L. proximal	1.9–2.0[i]	*Australopithecus*[f]
KNM-ER 1503	R. proximal	1.6–1.9[e]	*Australopithecus*[f]
KNM-ER 1809[h]	R. shaft	1.6–1.9[e]	
KNM-ER 3728	R. shaft	1.9–2.0[i]	
Olduvai			
OH 20	L. proximal	1.6–1.7[j]	*Australopithecus*[k]
OH 62	L. shaft	1.7–1.8[l]	*Homo habilis*[l]
Swartkrans			
SK 82	R. proximal	1.6–1.8[m]	*Australopithecus robustus*[n]
SK 97	R. proximal	1.6–1.8[m]	*Australopithecus robustus*[n]
Sterkfontein			
Sts 14	L. shaft	2.4–2.8[m]	*Australopithecus africanus*[o]

[a]AL 128-1 and 129-1c are probably the same individual (Johanson and Coppens, 1976).

[b]Between the Sidi Hakoma Tuff (which Brown, 1982, correlates with the Tulu Bor Tuff, the age of which is 3.35 Myr according to Brown *et al.*, 1985) and the Kadada Moumou Basalt (estimated age at 3.12 Myr by Brown, 1982).

[c]Johanson *et al.* (1982).

[d]Between the Kadada Moumou Basalt (see note b above) and BKT-2 which is dated at 3.14 Myr (Hall *et al.*, 1985).

[e]Above the KBS Tuff dated to 1.88 Myr (McDougall, 1985) and below the Koobi Fora/Okote/Ileret Tuff Complexes dated to 1.64 Myr (McDougall *et al.*, 1985).

[f]Day (1976b).

[g]Above the Koobi Fora/Okote/Ileret Tuff Complexes (see note e above) and below Chari Tuff dated to 1.39 Myr (McDougall, 1985).

[h]Maturity unknown.

[i]Not more than 36 m below KBS Tuff which is dated to 1.88 Myr (McDougall, 1985).

[j]From Lower Bed II (Hay, 1976).

[k]Day (1969).

[l]Johanson *et al.* (1987).

[m]Vrba (1985).

[n]McHenry and Corruccini (1978) and references therein.

[o]Howell (1978) and references therein.

Table 9.2. Femoral Shaft Width in Hominoidea

Species	N	Femoral shaft width (mm)				Sexual dimorphism (male product/ female product) × 100
		Transverse		Anteroposterior		
		$\bar{X}$	SD	$\bar{X}$	SD	Product (mm²)	
Homo sapiens							
Male	20	30.4	(2.6)	24.6	(1.5)	746.8	130.3
Female	23	26.8	(2.0)	21.4	(1.3)	573.0	
Pan troglodytes							
Male	21	27.5	(1.3)	22.7	(1.2)	623.8	111.2
Female	21	26.1	(1.6)	21.5	(1.5)	561.1	
Pan paniscus							
Male	7	23.5	(2.1)	21.1	(1.4)	495.9	106.6
Female	10	22.8	(1.8)	20.4	(1.4)	465.1	
Gorilla gorilla							
Male	25	41.3	(2.5)	33.6	(2.8)	1386.7	153.8
Female	41	32.9	(2.3)	27.4	(1.9)	901.5	
Pongo pygmaeus							
Male	13	26.2	(2.4)	19.6	(1.0)	513.1	154.0
Female	21	20.7	(1.6)	16.1	(1.6)	333.3	

Harvard University, consists of Amerindians. The common chimpanzee sample is *Pan troglodytes troglodytes* and the gorilla sample is the lowland subspecies housed at the Powell-Cotton Museum, Birchington. *Pongo pygmaeus* derives from the Smithsonian collection from both Sumatra and Borneo. Table 9.3 presents a smaller comparative sample of hominoid femora with associated body weights. In this sample the numbers are much smaller due to the rarity of body weights associated with skeletons. In this study the disadvantages of the small samples are outweighed by the advantages of using these body weights instead of those derived from the literature: the correlation between shaft widths and body weight is much higher in this sample than in one using the femoral data from Table 9.2 and body weights from McHenry (1984) or from Jungers (1985). The human sample (Table 9.3) is of American Blacks from the Terry Collection, now housed at

Table 9.3. Associated Body Weights and Femoral Cross-Sectional Areas in Comparative Sample

Species	N	Body weight (kg)		Femoral cross-sectional area (mm²)	
		$\bar{X}$	SD	$\bar{X}$	SD
Homo sapiens					
Male	15	59.3	5.98	846.1	74.2
Female	15	53.3	5.86	724.9	79.2
Pan paniscus					
Male	3	43.3	5.03	521.6	53.3
Female	5	33.7	4.69	515.9	39.3
Pan troglodytes					
Male	6	44.9	7.80	558.2	63.9
Female	10	35.6	6.74	520.5	48.0
Gorilla gorilla					
Male	3	168.0	19.64	1485.6	81.0
Female	1	68.0		1126.3	
Pongo pygmaeus					
Male	9	77.5	9.31	500.6	70.1
Female	8	40.2	10.80	362.2	60.7

the Smithsonian Institution, which derives from dissecting room cadavers. The records are very complete for each specimen including body weight, stature, cause of death, age at death, and a photograph. Some selection had to be done to eliminate excessively emaciated or obese individuals. Body weight ranges between 42.2 and 71.0 kg and age ranges between 22 and 50 years. The sample of *Pan paniscus* contains three males and five females all of whom were wild-shot adults. The *P. troglodytes* sample has twelve captive and four wild-shot specimens. In this sample the captive specimens do not differ significantly from the wild-shot specimens. The gorilla sample is regretably small, but at least all are wild-shot body weights. The orangutan sample is the most complete of the apes with nine males and eight females, all of which are wild-shot adults.

The femoral measurements are the transverse and anteroposterior diameters of the shaft just below the lesser trochanter (Martin measurements 9 and 10, Martin and Saller, 1957). The product of these two dimensions gives an estimate of the cross-sectional area of the shaft (actually closer to 4/3 the area).

The fossils are listed in Table 9.1. Twenty-six femora preserve the portion of the proximal shaft well enough to measure (or in some cases carefully estimate) diameters. The author took all measurements on the original specimens except those reported for OH 62 (see footnote c in Table 9.1). The fossils span the time between 3.4 and 1.4 Myr. The taxonomic designations for the East Turkana sample are primarily from Day (1969, 1976b). The sources of the dating are given in the footnotes to the table.

The method of relating femoral shaft size to body weight is by the least-squares analysis of logarithmically transformed variables. This method is called for since the primary purpose is to predict one variable (body weight) from another (femoral shaft diameters).

As with any study of rare fossils, this study must confine itself to limited sample sizes. Obviously there are sampling biases. For example, the million years between 3.1 and 1.9 Myr has only one proximal femoral shaft (Sts 14) and it is very badly preserved. Therefore, conclusions must be regarded as tentative.

A further problem is the fact that femoral shaft widths are related to body weight only within broad error limits. Even within a known species, some specimens lie far from the regression line (see the scatter plot of Fig. 9.1). Furthermore, in extinct species, one must assume that body proportions can be predicted from known species. At least some species of early hominids had quite different body proportions than those seen in modern humans (McHenry, 1978; Jungers, 1982; Johanson et al., 1987). The effort here must be seen as an attempt to narrow the outer limits of what might be possible. At least variability in skeletal size in the available sample is documented reliably even if the extension of this evidence to predictions of body weight may seem too speculative to be useful.

Results and Discussion

Femoral Size and Body Size

Table 9.2 presents the average femoral shaft widths for the larger sample of hominoid species. The right hand column gives the ratio of male to female within each species for the product of the two shaft diameters. The greatest sexual dimorphism is in *Gorilla gorilla* and *P. pygmaeus*, and the least in the two species of *Pan*. The sample of *H. sapiens* (Amerindians) displays a level of sexual dimorphism which is approximately halfway between the chimpanzee and the other great apes.

Table 9.3 presents the average body weights and femoral shaft size in the comparative sample in which the body weights at death are known for each specimen.

Figure 9.1 displays the relationship between femoral shaft size and body weight on a double logarithmic scale within the human sample summarized in Table 9.3. The correlation between the log of body weight and log of the product of the two femoral diameters is $r = 0.67$. The formula, log body wt = 0.624 log femur $-$ 0.0562, describes the least-squares relationship between the variables. Since body weight is proportional to the cube of length and the product of the two shaft femoral diameters is proportional to the square of length, then one expects a slope of 1.5

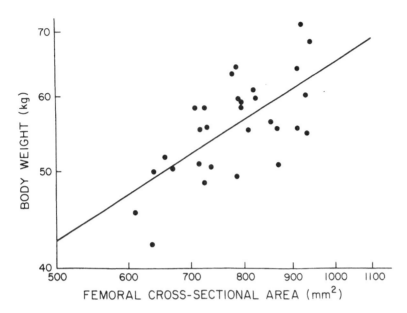

FIGURE 9.1. Bivariate log-log plot of body weights and femoral cross-sectional areas in a sample of *Homo sapiens*. The least-squares regression is log(body weight) = 0.624 log (femur area) − 0.0562. The correlation coefficient is 0.67.

in this equation if the two variables are isometric. The low slope observed here (0.624) results in part from the low level of sexual dimorphism in body weight in *H. sapiens*. In this sample the ratio of male to female body weight ($\times$ 100) is 111.3. By contrast, *P. pygmaeus* is 192.8. The slope of the log-log regression within *Pongo* is 1.53. As Fleagle (1985) notes, intragroup allometry can reflect the degree of sexual dimorphism.

There is considerable evidence that sexual dimorphism was much greater in early hominids than in modern ones (Frayer and Wolpoff, 1985; McHenry, 1986). The effect of using the intragroup regression formula based on modern *H. sapiens* with its low slope is that it overestimates the weight from the small sized femora and underestimates the weight from the large femora. Figure 9.2 presents one solution to this problem by plotting an intergroup regression analysis. The axes are the same as in Fig. 9.1 but only species' means are plotted. Using all large-bodied species of extent Hominoidea, the correlation between log body weight and log femoral shaft area is $r = 0.829$. From the plot it is clear that *P. pygmaeus* is an outlier: body weight and femoral size have a different relationship to one another in this highly specialized quadrumanual climber than they have in the other large-bodied hominoids. Ruff (1987) reaches the same conclusion about *Pongo* in a study of body weight, bone length, and cross-sectional dimensions in the lower limb in the large-bodied hominoids and *Macaca*. In my study the least-squares regression for male and female means of the species of African apes and humans has a very high correlation ($r = 0.94$). The formula for this equation is log body weight = 1.189 log femur − 1.663.

The fossils appear in Table 9.4 with their femoral dimensions and predicted body weight derived from both the intrahuman regression and the intergroup regression *(P. paniscus, P. troglodytes, H. sapiens, G. gorilla)*. Twenty-six fossil individuals span between 3.4 and 1.4 Myr. The extremes in estimated body weight range between 37.2 kg (Sts 14) and 70.7 kg (KNM-ER 736) for the intrahuman regression and between 27.4 and 93.3 kg (for the same specimens) for the interspecific regression. The two formulae give similar predictions in the middle range of variation but diverge at the extremes due to the fact that they have different slopes. The intergroup regression appears to give more reasonable predictions. For example, AL 288-1 ("Lucy") is 38.9 kg by the intrahuman regression but 29.9 kg by the interhominoid formula. The latter is closer to what other investigators have estimated by various methods (Jungers, 1982; McHenry, 1984). In a previous analysis

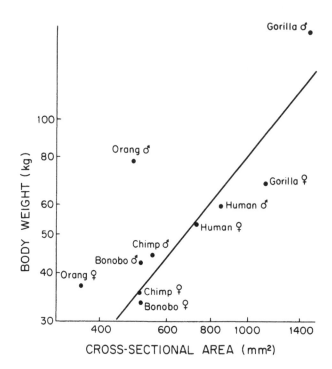

FIGURE 9.2. Bivariate log-log plot of the species average from Table 9.3 of body weights and femoral cross-sectional areas in associated skeletons. The least-squares regression line is calculated on the basis of the African Hominoidea and human samples only and the formula is log(body weight) = 1.189 log (femur area) − 1.663. The correlation coefficient is 0.94.

(McHenry, 1975a), the vertebrae of Sts 14 were used to predict a weight of 27.6 kg, which is closer to the intergroup prediction in this study (27.4 kg) than to the intrahuman prediction (37.2 kg). The intergroup predictions accord better with the expectation that body weight should vary at least as much as femoral shaft area (i.e., have an allometric coefficient greater than one). The ratio of largest to smallest femoral shaft is 2.8. This ratio for body weights predicted by the intrahuman regression is 1.9 and for the intergroup formula, 3.7.

Body Weight and Geological Time

Table 9.5 presents femoral shaft area and predicted body weight divided into five time periods. Within each time period the averages appear followed by the range, the average for those specimens clustering at the small size range, and those closer to the large pole. The last entry in each time category is the mid-value between the average large and small morph. In all five time zones the fossils cluster at either the large or the small pole, which appears to make this midpoint value the best estimate of the true average. At three (and possibly more) time categories, more than one species is probably being sampled. This problem is addressed below.

When divided into time categories as in Table 9.5, the sample sizes become distressingly low but some overall conclusions seem to be suggested. Most striking is the apparent fact that the average hominid has remained a relatively large animal throughout most of its known evolutionary record. The midpoint between the large and small morphs between 3.1 to 3.4 Myr was 50.4 and 50.6 kg by the two prediction methods. These figures dip to slightly above 44 kg in the next two time periods, which have very small samples, but rise back to about 50 kg by 1.6 to 1.9 Myr. By 1.4 to 1.6 Myr the midpoint increases to between 53.7 and 55.6 kg. In the better sampled time zones, the range remains very large. Even at sites thought to sample only one species (i.e.

Table 9.4. Femoral Shaft Size and Predicted Body Weight

	Femoral shaft width			Estimated body wt (kg)	
Specimen	Transverse (mm)	Anteroposterior (mm)	Product (mm²)	Intrahuman regression	Interhominoid regression
AL 128-1	24.4	19.1	466.0	40.6	32.3
AL 211-1	33.0	24.1	795.3	56.6	61.1
AL 288-1ap	24.5	17.8	436.1	38.9	29.9
AL 333-3	35.0[a]	27.9[a]	976.5	64.4	77.9
AL 333-95	31.8	26.7	849.1	59.0	66.0
AL 333W-40	35.2	27.0	950.4	63.3	75.5
KNM-ER 736	38.0	29.9	1136.2	70.7	93.3
KNM-ER 737	38.0	26.0	988.0	64.8	77.0
KNM-ER 738	26.9	21.7	583.7	46.7	42.3
KNM-ER 803	34.5	26.7	921.2	62.1	72.7
KNM-ER 815	26.7	18.9	504.6	42.6	35.5
KNM-ER 993	32.6[a]	25.2[a]	821.5	57.8	63.5
KNM-ER 1463[b]	27.3	21.9	597.9	47.4	43.5
KNM-ER 1465[b]	28.9[a]	26.2	757.2	54.9	57.6
KNM-ER 1472	31.4	21.8	684.5	51.6	51.1
KNM-ER 1475	28.9	24.0[a]	693.6	52.0	51.9
KNM-ER 1481a	31.3	21.0	657.3	50.3	48.7
KNM-ER 1500d	25.7	20.0	514.0	43.1	36.3
KNM-ER 1503	30.7[a]	22.3	684.6	51.6	51.1
KNM-ER 1809[b]	26.5	20.7	548.6	44.9	39.3
KNM-ER 3728	30.4	18.4	559.4	45.5	40.2
OH 20	29.6	22.8	674.9	51.1	50.2
OH 62	21.0[c]	21.0[c]	441.0	39.3	30.3
SK 82	30.4	25.0	760.0	55.1	57.8
SK 97	32.6	24.3	792.2	56.5	60.8
Sts 14	22.0[a]	28.4[a]	404.8	37.2	27.4

[a] Estimated.
[b] May not be fully mature.
[c] Given in Johanson *et al.* (1987:208) as the "shaft diameters at the base of the OH 62 lesser torchanter." These may have been taken slightly more proximally than those taken by the author on other femora as judged by the values for AL 288-1 reported by Johanson *et al.* (1987).

Hadar), the difference between the large and small morphs exceeds the average difference between males and females in the present samples of *Gorilla* and *Pongo*.

Taxonomy of Postcrania

It is difficult to classify isolated hominid postcranial specimens into species when the fossils derive from mixed sites. The postcrania of all early hominid species were probably quite similar. All species were bipedal, all were roughly the same size, and all apparently had strong sexual dimorphism. Variability within species was probably very large, which further confounds efforts to classify isolated specimens. The extent of this variability can be seen at sites that apparently sample only one species such as Hadar where elbows, knees, and ankles all vary from being rather *Homo*-like to quite primitive (Senut and Tardieu, 1985; Stern and Susman, 1983).

The femora listed in Table 9.1 from East Turkana have been assigned to either *Australopithecus* or *Homo* by Day (1976b), but even complete femora are hard to classify. Day (1976b, 1985) has built on the work of Napier (1964) in distinguishing the two genera on the basis of morphology, but every character is problematic. So, for example, *Australopithecus* is distinguished from *Homo* by a relatively long femoral neck, but this is true if only *H. sapiens* is taken to represent our

Table 9.5. Geological Time and Body Size

Time period	N	Femoral shaft area (mm²)	Estimated body wt (kg) Intrahuman regression	Interhominoid regression
2.9–3.4 Myr				
Average	6	745.6	53.8	57.1
Range		436.1–976.5	38.9–64.4	29.9–77.9
Small morph	2	451.1	40.0	31.1
Large morph	4	892.8	60.8	70.1
Midpoint		672.0	50.4	50.6
2.4–2.8 Myr				
Small morph	1	404.8	37.2	27.4
Large morph[a]	1	821.5	57.8	63.5
Midpoint		613.2	47.5	45.5
1.9–2.0 Myr				
Average	5	621.8	48.5	45.7
Range		514.0–693.6	43.1–52.0	36.3–51.9
Small morph	2	536.7	44.3	38.3
Large morph	3	678.5	51.3	50.6
Midpoint		607.6	47.8	44.4
1.6–1.9 Myr				
Average	9	620.0	46.0	46.8
Range		441.0–1136.2	39.3–70.7	30.3–93.3
Small morph	3	510.8	42.9	36.1
Large morph	5	809.6	57.0	62.6
Midpoint		660.2	50.0	49.4
1.4–1.6 Myr				
Average	5	817.2	57.4	62.9
Range		597.9–988.0	47.4–64.8	43.5–77.0
Small morph[b]		597.9	47.4	43.5
Large morph	1	872.0	59.9	67.7
Midpoint	4	734.9	53.7	55.6

[a]The size of the large morph at Sterkfontein is estimated from Sts 34, which is approximately the same size as KNM-ER 993.

[b]The small morph between 1.4 and 1.6 my is represented by KNM-ER 1463, which may be immature.

genus. All the early *Homo* femora, including KNM-ER 1472, KNM-ER 1481a, and KNM-WT 15000, have exceptionally long necks. Anteroposteriorly flattened femoral necks have been proposed as an *Australopithecus* character, but it is also true of *H. erectus* (KNM-WT 15000) and is not particularly true of some *Australopithecus* femora (e.g., AL 288-1 AL 333-3, KNM-ER 738). *Australopithecus* femora are supposed to be distinguishable from *Homo* in having no lateral trochanteric flare, but at least one femur classified as *Homo* (KNM-ER 1472) also lacks such flare. Day (1969) and Napier (1964) report that the intertrochanteric line and femoral tubercle are absent in "robust" australopithecine femora, but McHenry (1972), Lovejoy and Heiple (1972), Robinson (1972), Walker (1973), and Day (1976a) show that these features can be seen on the original specimens. Day (1969) and Napier (1964) state that the posterior position of the lesser trochanter is unlike modern humans, but Lovejoy and Heiple (1972) show that his is not a diagnostic trait. A relatively small femoral head is often claimed to be a characteristic of *Australopithecus* (e.g., Napier, 1964; Day, 1976a,b, 1985; Wood, 1976; McHenry and Corruccini, 1976, 1978; Corruccini and McHenry, 1978; Kennedy, 1983), but Walker (1973), Lovejoy (1975) and Wolpoff (1976) report that the head size relative to femoral length is the same in the two genera of hominids, although this point has been refuted (Corruccini and McHenry, 1978). Relative to other dimensions of the proximal end, femoral head size is quite small in all known *Australopithecus* femora and large in early *Homo* (with the possible exception of KNM-ER 1465, which is thought to be *Australopithecus*) but the small fragment of head has a radius of curvature of 20 mm, which is quite large (Day et al., 1976).

There are a few specimens in which taxonomically diagnostic craniodental material is associated with postcrania of the same individual in the time period between 2.3 and 1.3 Myr. The genus *Homo* is represented by KNM-ER 3735, 1808, 803, KNM-WT 15000, and OH 62. Unfortunately, the "robust" australopithecines are represented by only two individuals, and one of them is a questionable association: Omo 323 preserves craniodental material with a talus and calcaneus and TM 1517 is questionably an associated skeleton with craniodental, talar, and humeral material together in addition to a left hallucial terminal phalanx (according to Day and Thorton, 1986). There are other "clusters" of *A. boisei* specimens found in close proximity such as above Middle Tuff in Area 1, just below Lower Tuff in Area 6A, and in the channel above the KBS Tuff in Area 105 at East Turkana (Howell, 1982), but all of these deposits also contain *Homo* (i.e., KNM-ER 820, 731, and 1480).

More convincing is the material from Swartkrans. Although *Homo* is known in Members 1 and 2, it is quite rare and there is as yet no evidence of it in Member 3 (Grine, this volume, chapter 14). This rarity favors the probability that isolated hominid postcrania found at that site belongs to the "robust" australopithecine (Susman, this volume, chapter 10). The conspicuous morphologic difference between the Swartkrans femora (SK 82 and 97) and those attributed to *Homo* at about the same geologic age in the Turkana Basin (KNM-ER 1472, 1481a, and KNM-WT 15000) reinforce the inference that the hominid postcrania sampled at Swartkrans belong to the "robust" australopithecines. Likewise, the pelvic remains from Swartkrans (SK 50 and 3155b) are distinctly different from a *Homo* pelvic specimen of approximately the same geological age (KNM-ER 3228). If these Swartkrans specimens are taken as representing the "robust" australopithecine, then one feature clearly distinguishes them from *Homo*: The size of the "robust" australopithecine hip joint is small relative to either the size of the iliac blade (McHenry, 1975b) or the size of the proximal end of the femur (McHenry and Corruccini, 1978).

The relative size of the femoral head does vary considerably in the collections from East Turkana. If it is true that the "robust" australopithecines can be distinguished from *Homo* by their relatively small femoral heads, then at least some of the femora can be classified. In this case, KNM-ER 1472 and 1481a are *Homo* and presumably so is KNM-ER 1475 on the basis of its similarity to these specimens. KNM-ER 1503 is clearly *A. boisei*. KNM-ER 738 is intermediate in its head size relative to other proximal dimensions, and its classification is ambivalent. Unfortunately the rest of the specimens lack heads, so one must rely on comparisons of the neck and shaft to KNM-ER 1472, 1475, 1481a, and 1503. This leads to a classification similar to that given by Day (1976b) and is presented in the footnotes to Table 9.6. One important difference is that KNM-ER 738 is used to represent the small morph of *H. erectus*.

Body Weight, Brain Size and Megadontia

The first column of Table 9.6 presents the estimates for the average weight of each species. These estimates are actually the midpoints between the average small and large morphs for each species. There is, of course, a major margin of uncertainty around each value. Some of the estimates rely on a single individual specimen for the estimate of one or both of the morphs. Only the two earliest species *(A. afarensis* and *A. africanus)* come from sites thought to contain only one species of hominid, so that the taxonomic allocation of the later postcrania further confounds the estimates.

Assuming that these body weight estimates for early hominid species are reasonable, some interesting features emerge. Most obvious is the fact that the terms "gracile" and "robust" refer to craniodental features only. The bodies of the males of all species were large, making the averages for both "gracile" *(A. afarensis* and *A. africanus)* and "robust" *(A. robustus* and *A. boisei)* australopithecines between 46 and 51 kg. The estimate of *H. habilis* is smaller (40.5 kg) partly because the small morph is represented only by the tiny OH 62, but also because the femora assigned to the large morph (KNM-ER 1472, 1475, and 1481a) are smaller than the large

Table 9.6. Body Weights, Brain Sizes, and Tooth Areas of Plio-Pleistocene Hominid Species

Species	Estimated body weight (kg)	Cranial capacity (cc)	Encephalization quotient (EQ)[a]	Postcanine tooth area (mm^2)	Megadontia quotient (MQ)[b]
Australopithecus afarensis	50.6[c]	415[d]	2.44	460[e]	1.30
Australopithecus africanus	45.5[f]	442[g]	2.79	516[e]	1.59
Australopithecus robustus	47.7[h]	530[i]	3.24	588[e]	1.74
Australopithecus boisei	46.1[j]	515[k]	3.22	799[i]	2.44
Homo habilis	40.5[m]	631[n]	4.30	479[o]	1.63
Homo erectus	58.6[p]	826[q]	4.40	372[r]	0.92

[a]Encephalization quotient = observed cranial capacity/predicted cranial capacity = endocranial volume/.12 (body wt, g)$^{.67}$ after Jerison (1973). Other estimates of relative brain size give similar results. Martin (1981) reports that an exponent of 0.76 describes the relationship between brain and body size in 309 species of placental mammals. One can derive a modified encephalization quotient from the regression equation where actual brain weight (mg) is divided by 58.9 times body weight (g) raised to the 0.76 power. The resulting values for the fossil species listed above are 1.87, 2.16, 2.50, 3.37, and 3.33, respectively. However, the high exponent may not be appropriate because the interest here is in comparing closely related species with similar body weights. An index such as Hemmer's (1971) Constant of Cephalization might be more appropriate. In this case the exponent is 0.23 and the values for the species listed above are 34.4, 37.5, 44.5, 43.6, 55.0, and 66.1, respectively.

[b]Megadontia quotient = observed tooth area P_4 to M_2/predicted tooth area = P_4 to M_2 area/12.15 (body wt)$^{.86}$ derived from the regression formula log (predicted tooth area) = 0.8566 log (body wt) + 1.0846 given in McHenry (1984, p. 299) using extant large-bodied hominoids.

[c]The midpoint between the large morph (AL 211-1, 333-3, 333-95, 333w-40) and small morph (AL 128-1, 288-1ap).

[d]From Johanson and Edey (1981).

[e]From McHenry (1984).

[f]The midpoint between the large morph (Sts 34 which is approximately the same size as KNM-ER 993) and small morph (Sts 14).

^aFrom Holloway (1970).
^bThe midpoint between the large morph (SK 82, 97) and small morph (SK 3981) vertebrae estimates in McHenry (1975).
^cFrom Holloway (1972).
^dThe midpoint between the large morph (KNM-ER 993, 1503; OH 20) and the small femora (KNM-ER 815, 1500, 3728).
^eBased on KNM-ER 406, 732; OH 5.
^fBased on KNM-ER 727, 802, 3230.
^gThe midpoint between the large morph (KNM-ER 1472, 1475, 1481a) and small morph (OH 62).
^hBased on KNM-ER 1470, 1813; OH 13, 24. The midpoint between what is thought to be a male (1470) and females (1813, 13, 24) is 678.8 kg which makes the EQ estimate 4.63.
ⁱBased on KNM-ER 1802, 3734; OH 7, 13, 16.
^jThe midpoint between the large morph (KNM-ER 737, 803) and the small morph (KNM-ER 738).
^kBased on KNM-ER 3733 and 3883.
^lBased on KNM-ER 992, and KNM-WT 15000.

morphs of other species. Clearly, early *H. erectus* is larger (58.6 kg) than earlier species since the largest femora in the sample are attributed to it. Quite possibly early *H. erectus* was considerably larger: I have excluded the enigmatic and huge KNM-ER 736 femur and included the ambiguous and small KNM-ER 738.

With the femora sorted into species the trend in body weight through time can be clarified. Stasis characterizes the australopithecine species: From 3.4 to 1.4 Myr the average body weight remains between 46 and 51 kg and the difference between the average small and large morph in each species stays very high. The two species of *Homo* go from smaller than *Australopithecus* to considerably larger than this genus. These estimates show, therefore, that there is not a gradual increase in body size through time in human evolution, but body size appears to have increased rapidly between 2.0 and 1.6 Myr in the *Homo* lineage. Body size sexual dimorphism apparently remained very strong from 3.4 to 1.6 Myr, although the evidence for a small *H. erectus* morph is ambiguous and further discoveries may show that the female of this species was much larger than the female of *H. habilis*.

The second and third columns of Table 9.6 contain estimates of relative and absolute endocranial volumes for the early hominid species. It is obvious that there has been significant absolute brain size increase through this series from *A. afarensis* with an endocranial volume of 415 cc, up to 515 cc for the East African "robust," and 631 cc and 745 cc for the earliest species of *Homo*. The estimates of body weight show that this increase in brain size is not simply a result of increasing body size: relative to estimated body weight, the cranial capacity increases dramatically through the series. The third column presents values for the Encephalization Quotient (EQ), which is one measure of relative brain size (Jerison, 1973). This index is simply the ratio of observed endocranial volume divided by endocranial volume predicted from body weight. The prediction formula is $0.12 \text{ (body wt in g)}^{0.67}$. Other measures of relative brain size give similar results as noted in footnote *a* of Table 9.6.

From column 3 of Table 9.6 it appears that increase in relative endocranial volume is true of both the "robust" australopithecine and the *Homo* lineages. The two earliest species of *Australopithecus* have EQ values of 2.44 and 2.79. The values for the two "robust" species (*A. robustus* and *A. boisei*) are 3.24 and 3.22. The two species of *Homo* are 4.30 and 4.40.

The last line of Table 9.6 gives an estimate of relative cheek-tooth size. I have devised a Megadontia Quotient (MQ) that is analogous to the encephalization quotient. It is the observed area of the mandibular last premolar and first two molars divided by the area predicted from the body weight of the animal. For this prediction I take the regression formula given in Table 1 of McHenry (1984): log tooth area = 0.8566log(body wt) + 1.0846. By analogy with the EQ, the MQ is defined as: MQ = observed tooth area/$12.15 \text{(body wt)}^{.86}$. By this definition the average large-bodied extant hominoid has an MQ of 1.0, *H. sapiens* has 0.89, *P. troglodytes* has 1.14, *G. gorilla* has 0.98, and *P. pygmaeus* has 1.06 using the data in McHenry (1984, table 1).

The megadontia quotients for the fossil species show that relative cheek-tooth size increased dramatically from *A. afarensis* (1.30) to *A. africanus* (1.59) to *A. robustus* (1.74) to *A. boisei* (2.24). The earliest species of *Homo* has very large cheek-teeth relative to estimated body weight (MQ = 1.63), but megadontia falls off dramatically in early *H. erectus* (0.92). The latter value may be artificially low because only two mandibular specimens are included (KNM-ER 992 and KNM-WT 15000) which happen to be very small. If the three Ternifine mandibles are included, then the MQ for *H. erectus* is 1.07.

Conclusions and Summary

This study explores the relationship between the size of the most commonly preserved hominid postcranial element (femoral shaft) and body weight in order to estimate the body size of the Plio-Pleistocene hominids. Within a sample of modern *H. sapiens* a measure of femoral shaft size is significantly correlated with body weight ($r = 0.67$), but the resulting regression equation is of limited value for predicting body weight of the very small early hominids. The two variables are highly correlated ($r = 0.94$) in a sample of femora derived from specimens of known body weight from species of African apes and modern *H. sapiens*. The resulting interspecific regression predicts body weights for the early hominids that are reasonable on independent lines of evidence.

The predicted body weights from the interspecific regression equation show that from 3.4 to 1.4 Myr there were two distinct size morphs of hominids with the average large morph being twice or more the size of the average small morph. The average large morph remains quite large throughout this time period (between 50 and 70 kg) and the average small morph remains quite small (between 30 and 44 kg). There is no obvious size trend through time: The earliest hominids (*A. afarensis*) range in size from 30 to 78 kg and the later hominids (those between 1.4 and 1.6 Myr) range between 44 and 77 kg. When separated into species the body weight estimates show stasis through time in australopithecine species but a rapid increase in the *Homo* lineage of almost 20 kg in the period between 2.0 and 1.6 Myr. Body weight sexual dimorphism is apparently very strong in all species with the possible exception of *H. erectus*.

Species attribution of unassociated fossil femora from sites yielding more than one hominid species is uncertain, but some aspects of the proximal morphology appear to distinguish "robust" australopithecines from *Homo*. Taking the midpoint between the large and small morph within each species to represent the average, the body weights (kg) are as follows: *A. afarensis*, 50.6; *A. africanus*, 45.5; *A. robustus*, 47.7; *A. boisei*, 46.1; *H. habilis*, 40.5; *H. erectus*, 58.6.

Relative to these estimates of body weight, relative endocranial volume increases in both the "robust" australopithecine and *Homo* lineages. The earliest hominid species (*A. afarensis*) is about 2.5 times more encephalized than the average mammal by one measure of relative brain size. This figure rises to 3.25 for the two species of "robust" australopithecine (*A. boisei* and *A. robustus*). The earliest species of *Homo* reach between 4.0 and 4.3 times larger than predicted from their body size.

The enormous absolute size of the posterior dentition seen in the "robust" australopithecines is far more than would be expected from their estimated body weight. The size of these teeth is 2.2 times larger in *A. boisei* than would be expected for a modern hominoid of its body size. Relative cheek-tooth size increases steadily in the series from *A. afarensis* to *A. africanus* to *A. robustus* to *A. boisei*. The earliest species of *Homo* retains moderately large cheek-teeth relative to estimated body weight, but the relative size of these teeth decreases dramatically in early *H. erectus*.

Acknowledgments

I thank C. K. Brain, P. V. Tobias, R. E. Leakey, M. D. Leakey, the late L. S. B. Leakey, D. C. Johanson, Tadesse Terfa, and Mammo Tessema for permission to study the fossils in their charge; C. Powell-Cotton, M. Rutzmoser, R. Thorington, D. J. Ortner, W. W. Howells, D. R. Pilbeam, D. F. E. T. van den Audenaerde, J. Biegert, and the late A. H. Schultz for the use of the comparative material in their charge; F. E. Grine for inviting me to the "Robust" Australopithecine Workshop; C. Ruff, E. Trinkaus, F. Brown, and F. E. Grine for valuable comments; L. J. McHenry and K. Phillips for assistance; and S. Williams for typing. Partial funding was provided by the Committee on Research, University of California, Davis.

References

Boaz, N. T. (1977). Paleoecology of early Hominidae in Africa. *Kroeber Anthropol. Soc. Pap.*, **50**: 37–62.

Boaz, N. T. (1985). Early hominid paleoecology in the Omo Basin, Ethiopia. *In L'environnement des hominides au Plio-Pleistocene*, pp. 279–308. Masson, Paris.

Brown, F. H. (1982). Tulu Bor Tuff at Koobi Fora correlated with the Sidi Hakoma Tuff at Hadar. *Nature*, **300**: 631–633.

Brown, F. H., McDougall, I., Davies, T. and Maier, R. (1985). An integrated Plio-Pleistocene chronology for the Turkana Basin. *In* E. Delson (ed.), *Ancestors: the hard evidence*, pp. 82–90. Alan R. Liss, New York.

Corruccini, S. and McHenry, M. (1978). Relative femoral head size in early hominids. *Amer. J. Phys. Anthropol.*, 49: 145–148.

Day, M. H. (1969). Femoral fragment of a robust australopithecine from Olduvai Gorge, Tanzania. *Nature*, 221: 230–233.

Day, M. (1976a). Hominid postcranial material from Bed I, Olduvai Gorge. *In* G. L. Isaac and E. R. McCown (eds.), *Human origins: Louis Leakey and the East African evidence*, pp. 363–374. Benjamin, Menlo Park, CA.

Day, M. H. (1976b). Hominid postcranial from the East Rudolf succession. *In* Y. Coppens, F. C. Howell, G. L. Issac, and R. F. Leakey (eds.), *Earliest man and environments in the Lake Rudolf Basin*, pp. 507–521. Univ. Chicago Press, Chicago, IL.

Day, M. H. (1985). Hominid locomotion—from Taung to the Laetoli footprints. *In* P. V. Tobias (ed.), *Hominid evolution: past, present, and future*, pp. 115–128. Alan R. Liss, New York.

Day, M. H. and Thorton, C. M. D. (1968). Extremity bones of *Paranthropus robustus* from Kromdraai B, East Formation Member 3, Republic of South Africa—a reappraisal. *Anthropos*, 23: 91–100.

Day, M. H., Leakey, R. E. F., Walker, A. C. and Wood B. A. (1976). New hominids from East Turkana, Kenya. *Amer. J. Phys. Anthropol.*, 45: 369–436.

Fleagle, J. G. (1985). Size and adaptation in primates. *In* W. L. Jungers (ed.), *Size and scaling in primate biology*, pp. 1–20. Plenum, New York.

Frayer, D. W. and Wolpoff, M. H. (1985). Sexual dimorphism. *Ann. Rev. Anthropol.*, 14: 429–474.

Genet-Varcin, E. (1967). De quelques problèmes posēs pan les Australopitheques. In: *Problèmes actuels de paléontologie (Evolution des vertēbrēs)*, pp. 649–653. Paris, 1966. C.N.R.S.

Hall, C. M., Walter, R. C. and York, D. (1985). Tuff above "Lucy" is 3 Ma old. *Eos*, 66: 257.

Hay, R. L. (1976). *Geology of the Olduvai Gorge*. University of California Press, Berkeley.

Hemmer, H. (1971). Beitrag zur Erfassung der Progressiven Cephalisation bei Primaten. *Proc. 3rd Congr. Primatol., Zürich 1970*, Vol. 1, pp. 99–107. Karger, Basel.

Holloway, R. L. (1970). New endocranial values for the australopithecines. *Nature*, 227: 199–200.

Holloway, R. L. (1972). New australopithecine endocast, SK 1585, from Swartkrans, South Africa. *Amer. J. Phys. Anthropol.*, 37: 173–186.

Howell, F. C. (1978). Hominidae. *In* V. J. Maglio and H. B. S. Cooke (eds.), *The evolution of African mammals*, pp. 154–248. Harvard Univ. Press, Cambridge, MA.

Howell, F. C. (1982). Origins and evolution of African Hominidae. *In* Clark, J. D. (ed.), *The Cambridge history of Africa*, Vol. 1, pp. 70–156. Cambridge, University Press, Cambridge.

Jerison, H. (1973). *Evolution of the brain and intelligence*. Academic Press, New York.

Johanson, D. C. and Coppens, Y. (1976). A preliminary anatomical diagnosis of the first Plio/Pleistocene hominid discoveries in the central Afar, Ethiopia. *Amer. J. Phys. Anthropol.*, 45: 217–234.

Johanson, D. C. and Edey, M. (1981). *Lucy: the beginnings of humankind*. Simon and Schuster, New York.

Johanson, D. C., Masao, F. T., Eck, G. G., White, T. D., Walter, R. C., Kimbel, W. H., Asfaw, B., Manega, P., Ndessokia, R. and Suwa, G. (1987). New partial skeleton of *Homo habilis* from Olduvai Gorge, Tanzania. *Nature*, 327: 205–209.

Johanson, D. C., Taieb, M. and Coppens, Y. (1982). Pliocene hominids from the Hadar formation, Ethiopia (1973–1977). Stratigraphic chronologic, and paleoenvironmental contexts, with notes on hominid morphology and systematics. *Amer. J. Phys. Anthropol.*, 57: 373–402.

Jungers, W. L. (1982). Lucy's limbs: skeletal allometry and locomotion in *Australopithecus afarensis*. *Nature*, 297: 676–678.

Jungers, W. L. (1985). Body size and scaling of limb proportions in primates. *In* W. L. Jungers (ed.), *Size and scaling in primate biology*, pp. 345–382. Plenum, New York.

Kennedy, G. E. (1983). A morphometric and taxonomic assessment of a hominine femur from the Lower Member, Koobi Fora, Lake Turkana. *Amer. J. Phys. Anthropol.*, 61:429–436.

Lovejoy, C. O. (1975). Biomechanical perspectives on the lower limb of early hominids. *In* R. H. Tuttle (ed.), *Primate functional morphology and evolution*, pp. 291–326. Mouton, The Hague.

Lovejoy, C. O. and Heiple, K. G. (1970). A reconstruction of the femur of *A. africanus*. *Amer. J. Phys. Anthropol.*, 32: 33–40.

Lovejoy, C. O. and Heiple, K. G., (1972). The proximal femoral anatomy of *Australopithecus*. *Nature*, 235: 175–176.

Martin, R. and Saller, K. (1957). *Lehrbuch der Anthropologie*. Gustav Fisher Verlag, Stuttgart.

Martin, R. D. (1981). Relative brain size and basal metabolic rate in terrestrial vertebrates. *Nature*, 293: 57–60.

McDougall, I. (1985). K-Ar and $^{40}Ar/^{39}Ar$ dating of the hominid-bearing Pliocene-Pleistocene sequence at Koobi Fora, Lake Turkana, northern Kenya. *Geol. Soc. Amer. Bull.*, 96: 159–175.

McDougall, I., Davies, T., Maier, R. and Rudowski, R. (1985). Age of the Okote Tuff complex at Koobi Fora, Kenya. *Nature*, 316: 792–794.

McHenry, H. M. (1972). The postcranial anatomy of early Pleistocene Hominids. Ph.D. Thesis, Harvard University, Cambridge, MA.

McHenry, H. M. (1974). How large were the australopithecines? *Amer. J. Phys. Anthropol.*, 40: 329–340.

McHenry, H. M. (1975a). Fossil hominid body weight and brain size. *Nature*, 254: 686–688.

McHenry, H. M. (1975b). A new pelvic fragment from Swartkrans and the relationship between the robust and gracile australopithecine. *Amer. J. Phys. Anthropol.* 43: 245–262.

McHenry, H. M. (1976). Early hominid body weight and encephalization. *Amer. J. Phys. Anthropol.*, 45: 77–84.

McHenry, H. M. (1978). Fore- and hindlimb proportions in Plio-Pleistocene hominids. *Amer. J. Phys. Anthropol.*, 49: 15–22.

McHenry, H. M. (1982). The pattern of human evolution: studies on bipedialism, mastication and encephalization. *Ann. Rev. Anthropol.*, 11: 151–173.

McHenry, H. M. (1984). Relative cheek-tooth size in *Australopithecus*. *Amer. J. Phys. Anthropol.*, 64: 297–306.

McHenry, H. M. (1986). Size variation in the postcranium of *Australopithecus afarensis* and extant species of Hominoidea. *In* M. Pickford and B. Chiarelli (eds.), *Sexual dimorphism in living and fossil primates*, pp. 183–189. Giugno, Il Sedicesimo.

McHenry, H. M. and Corruccini, R. S. (1976). Fossil hominid femora and the evolution of walking. *Nature*, 259: 657–658.

McHenry, H. M. and Corruccini, R. S. (1978). The femur in early human evolution. *Amer. J. Phys. Anthropol.*, 49: 473–488.

McHenry, H. M. and Temerin, A. (1979). The evolution of hominid bipedalism: evidence from the fossil record. *Yrbk. Phys. Anthropol.*, 22: 105–131.

Napier, J. R. (1964). The evolution of bipedal walking in the hominids. *Arch. Biol. (Suppl) (Liege)*, 75: 673–708.

Olson, T. R. (1985). Cranial morphology and systematics of the Hadar formation hominids and *Australopithecus africanus*. *In* E. Delson (ed.), *Ancestors: the hard evidence*, pp. 102–119. Alan R. Liss, New York.

Pilbeam, D. and Gould, S. J. (1974). Size and scaling in human evolution. *Science*, 186: 892–901.

Robinson, J. T. (1972). *Early hominid posture and locomotion*. Univ. Chicago Press, Chicago, IL.

Ruff, C. (1987). Structural allometry of the femur and tibia in Hominoidea and *Macaca*. *Folia Primatol.*, 48: 9–49.

Senut, B. and Tardieu, C. (1985). Functional aspects of Plio-Pleistocene hominid limb bones: implications for taxonomy and phylogeny. *In* E. Delson (ed.), *Ancestors: the hard evidence*, pp. 193–201. Alan R. Liss, New York.

Stern, J. T., Jr. and Susman, R. L. (1983). The locomotor anatomy of *Australopithecus afarensis*. *Amer. J. Phys. Anthropol.*, 60: 279–318.

Steudel, K. (1980). New estimates of early hominid body size. *Amer. J. Phys. Antropol., 52*(1): 63–70.

Suzman, I. M. (1980). A new estimate of body weight in South African australopithecines. *Proc. 8th Pan. Afr. Cong, Nairobi, Sept. 1977*, pp. 175–179.

Vrba, E. S. (1985). Early hominids in southern Africa: Up-dated observations on chronological and ecological background. *In* P. V. Tobias (ed.), *Hominid evolution: past, present and future*, pp. 195–200. Alan R. Liss, New York.

Walker, A. (1973). New *Australopithecus* femora from East Rudolf, Kenya. *J. Hum. Evol, 2*: 545–555.

Walter, R. C. and Aronson, J. L. (1982). Revisions of K/Ar ages for the Hadar hominid site, Ethiopia. *Nature, 296*: 122–127.

Wolpoff, M. H. (1976). Fossil hominid femora. *Nature, 264*: 812–813.

Wood, B. A. (1975). Remains attributable to *Homo* in the East Rudolf succession. *In* Y. Coppens, F. C. Howell, G. L. Issac and R. E. F. Leakey (eds.), *Earliest man and environments in the Lake Rudolf Basin*, pp. 490–506. Univ. Chicago Press, Chicago, IL.

New Postcranial Remains from Swartkrans and Their Bearing on the Functional Morphology and Behavior of *Paranthropus robustus*

10

RANDALL L. SUSMAN

> . . . there is no good evidence in support of the thesis that australopithecines were stone toolmakers . . . this group consisted essentially of tool-users.
>
> J. T. Robinson (1961)

It has been generally assumed that large-toothed, small-brained hominids comprised the genus *Paranthropus*. These early hominids subsisted principally on a vegetarian diet. As the above quote indicates, early workers believed that neither *Paranthropus* nor *Australopithecus* were toolmakers, and some inferred that neither were they tool-users to any appreciable degree (e.g., Mason, 1962; Tobias, 1965). The tools that were recovered from Swartkrans, Sterkfontein and sites in East Africa were attributed to early *Homo* (Robinson, 1953, 1961, 1962a,b, 1972). It was thought that with a small brain and a vegetarian diet, *Paranthropus* had neither the intellectual capacity nor the need to engage in tool behavior (Robinson, 1962a,b). Robinson suggested that increasingly dry conditions through the Pliocene produced *Australopithecus* and from this genus *Homo*. With the advent of *Homo*, culture became the "vital factor." According to Robinson:

> The need for tool-using in successfully adapting to the different way of life would, as indicated, place a high premium on intelligence. As this improved, presumably by an increase in size of the cerebral cortex so as to provide increased correlation and association areas, cultural facility also improved. When the modification of the brain had proceeded to the point where hominine levels of intellectual ability began to appear— apparently when the brain volume reached the order of 750–1,000 cm^3—facility with tools reached a point where a characteristically hominine phenomenon appeared: the deliberate manufacture of tools for particular purposes (1962a:137).

Paranthropus was not covered by Robinson's statement. Instead, *Paranthropus* was thought to be adapted to wetter conditions with a more apelike skull (Le Gros Clark, 1947; 1949), a more specialized megadont dentition, and a vegetarian diet. Robinson (1962a) suggested that overall, *Paranthropus* might represent most closely the primitive australopithecine condition. The cultural status of *Paranthropus* was relatively poor according to Robinson (1962a).

Most of our knowledge of *Paranthropus* derives from the work of Broom and Robinson (1952), Tobias (1967) and Rak (1983), which has been based principally on studies of craniodental remains. This emphasis is natural since the vast majority of the fossil record of *Paranthropus* consists of skulls and teeth. Until now only seven postcranial fossils attributable to *Paranthropus* were known from Swartkrans. Only five postcranials of *Paranthropus* are known from Kromdraai. Speculations on the functional morphology and cultural status of *Paranthropus* were based on limited and indirect evidence. As Tobias (1965) noted, without the remains of hands and other postcranial fossils, statements that *Paranthropus* was or was not a toolmaker could not be confirmed.

Fortunately, we now have a significant new collection of *Paranthropus* fossils that bear directly on interpretations of the habitus of *Paranthropus* in southern Africa, and which provide the first direct evidence that *Paranthropus* not only engaged in tool behavior, but that *Paranthropus* was as capable a biped as other contemporary Plio-Pleistocene hominids such as *Australopithecus* and early *Homo*.

Thirty-seven fossils representing the postcranial skeleton of *Paranthropus robustus* and *Homo cf. erectus* were recovered at Swartkrans from 1979 through 1983 (Table 10.1; Fig. 10.1). These remains come from the earliest three of the five Members at Swartkrans (Brain, this volume, chapter 20). The earliest, Member 1, is considered to date to approximately 1.8 Myr BP (Vrba, 1982). Brain (1981) has provided an analysis and historical account of Member 1, formerly known as the "pink" or "orange" breccia. Member 1 consists of a "Hanging Remnant" that has yielded the remains of over 80 *Paranthropus* individuals (Brain, 1985), and a "Lower Bank" moiety that has yielded hominids, stone artifacts and bone tools (Brain, 1982, 1985). At latest count the remains of approximately 125 *Paranthropus* individuals have been recovered from Member 1 out of a total of one hundred and thirty hominids. Thus, in Member 1 over 95% of hominid specimens represent *Paranthropus*. (Of the most recent craniodental material from Member 1, all but one individual are *Paranthropus* according to Grine, this volume, chapter 14.)

Also from Member 1 are twenty-five to thirty bone tools and several stone artifacts. These tools have been described by Leakey (1970) and Brain (1982). Leakey considered only specimens "unquestionably" derived from Member 1 in her assessment of the stone artifacts from Swartkrans. The assemblage at that time consisted of twenty-seven artifacts excavated by Brain. Leakey (1970:16) noted that: "One of the most striking features in the assemblages from Swartkrans [and Sterkfontein] is the relatively large size of the tools when compared with similar specimens from Olduvai . . .", where the mean diameter of side choppers from Swartkrans was 79 mm compared

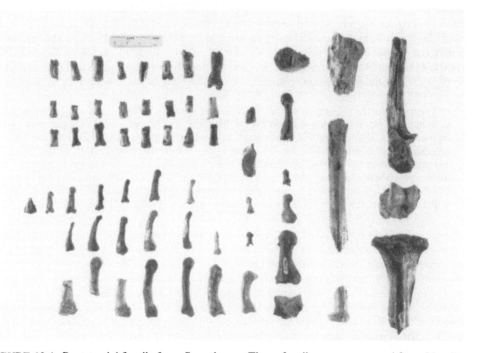

FIGURE 10.1. Postcranial fossils from Swartkrans. These fossils were recovered from Members 1 to 3. Most of these fossils represent *Paranthropus*. Three specimens from Member 2 and possibly one from Member 3 represent *Homo* cf. *erectus* (see Table 10.1). One of these specimens (4th from right, top row) is a leopard phalanx; some of the manual middle phalanges (upper left) are those of baboons.

Table 10.1. New Hominid Postcranial Fossils from Swartkrans

Specimen		Description[c]	Attribution
Member 1[a]			
SKX	5016	Pollical distal phalanx	*Paranthropus*
	5017	Hallucal metatarsal (I)	*Paranthropus*
	5018	Manual proximal phalanx (IV?)	*Paranthropus*
	5019	Manual middle phalanx	?
	5020	Pollical metacarpal (I)	*Paranthropus*
	5021	Manual middle phalanx	*Paranthropus*
	5022	Manual middle phalanx	*Paranthropus*
	3602	Distal radius	*Paranthropus*
	8761	Proximal ulna	*Paranthropus*
	8963	Manual distal phalanx	*Paranthropus*
	9449	Manual middle phalanx	*Paranthropus*
	12814	Radius (shaft)	*Paranthropus*
	13476	Manual Middle phalanx	*Paranthropus*
	3774	Distal humerus	*Paranthropus*
SKW	34805	Distal Humerus	*Paranthropus*
SK	45690	Hallucal proximal phalanx	*Paranthropus*
Member 2[b]			
SKX	247	Metatarsal III	?
	344	Pedal middle phalanx	?
	1084	Patella	*Paranthropus*
	3062	Middle phalanx	?
	3342	Thoracic vertebra	?
	3498	Triquetral	?
	3699	Radius (proximal)	*Paranthropus*
SKW	1261	Pedal middle phalanx	?
	2954	Metacarpal IV	?
	3646	Metacarpal III	?
Member 3			
SKX	19576	Manual proximal phalanx	?
	22511	Manual proximal phalanx	?
	22741	Manual proximal phalanx	*Paranthropus*
	27431	Manual proximal phalanx III	*Homo*
	27504	Manual distal phalanx	?
	31117	Manual medial cuneiform	*Paranthropus*
	33355	Manual Middle phalanx	?
	33380	Manual Metatarsal V	*Paranthropus*
	35439	Manual middle phalanx	?
	35822	Manual proximal phalanx	?
	36712	Manual middle phalanx	?

[a]Specimens SK 50, 82, 84, 85, 97, 853, 854, 14147, and 3155b, recovered earlier, are from Member 1; SK 84, 85, 853, and 3155b are attributed to *Paranthropus* (Robinson, 1972; Brain et al., 1974).

[b]SK 18b, was also recovered earlier from Member 2. This specimen was attributed to *Homo* (Robinson, 1953; Brain, 1978).

[c]I to V = individual manual and pedal digits.

to a mean of 65 mm for those from Bed I at Olduvai (DKI). Overall, however, Leakey (1970) observed a similarity in the stone tools from East and South Africa. The present collection of stone artifacts from Swartkrans is being analyzed by J. D. Clark.

The bone tools that were found in association with hominid remains in the "Lower Bank" of Member 1 have been analyzed experimentally by Brain (1982, 1985). The bone artifacts take the form of worn points, and Brain has experimentally duplicated these "by taking a piece of long bone and digging out edible bulbs on a rocky hillside." He notes that "it takes 20 minutes of digging to get out a bulb, and after several hours or days of use the bones start to acquire . . . remarkable wear on the ends." These bone "digging" implements persist into the two succeeding units (Members 2 and 3) at Swartkrans (Brain, 1985).

Postcranial Fossils from Members 1, 2 and 3

Member 1 Remains of Paranthropus

Metatarsal I (SKX 5017). This is a complete and undistorted left first metatarsal; it is 43.5 mm long, and both articular surfaces are intact (Fig. 10.2). The basal surface has a prominent tubercle for the insertion of m. peroneus longus and has a well-developed lateral margin, both of which contribute to the mildly concave appearance of the articular surface. The shaft is thick on its basoplantar surface, indicating the presence of a well-developed plantar aponeurosis. In profile, the distal articular surface (head) of SKX 5017 reveals a broad extension onto the dorsum of the bone, while in distal (end on) view it is narrow dorsally and widens markedly on its plantar extent (Fig. 10.2c). Morphologically, SKX 5017 is essentially like OH 8-H in size and shape, and both are essentially like modern humans. The only exception is seen in the morphology of the head of SKX 5017 which resembles, in distal view, that of African apes and AL 333-115, rather than later hominids (Fig. 10.2c).

Hallucal Proximal Phalanx (SK 45690). This is a complete, undistorted left proximal phalanx that measures 23.2 mm in length. The basal articular surface opens dorsally to accommodate extension. There is a prominent dorsal tubercle for the insertion of m. extensor digitorum communis. The plantar tubercles for the attachment of the collateral ligaments are well-developed. Morphologically, SK 45690 is humanlike in both size and shape. When combined with SKX 5017, it yields a length ratio (45690/5017 $\times$ 100) of 53, similar to that of modern humans (mean = 52; N = 18).

Pollical Distal Phalanx (SKX 5016). An undistorted phalanx, missing only a small portion of the dorsobasal margin of the proximal articular surface (Fig. 10.3). SKX 5016 has an extensive fossa for the insertion of m. flexor pollicis longus and a massive distal apical tuft. The closest morphological counterparts of SKX 5016 are OH 7 (FLK NN-A) and to a lesser extent, modern humans (Fig. 10.3, left).

Manual Distal Phalanx (SKX 8963). The distal two-thirds of a terminal phalanx (Fig. 10.4, right). The base is missing, and the well-developed apical tuft displays slight erosion on its margins. Morphologically, what little there is of SKX 8963 is clearly humanlike. Although the apical tuft is not "spade-like" as in most modern humans, its development relative to radioulnar midbody diameter is clearly nonpongid and recalls the condition in humans (Susman and Creel, 1979; Stern and Susman, 1983) (Fig. 10.4).

Metacarpal I (SKX 5020). A right metacarpal, 42.4 mm in length, missing the dorsomedial part of the basal articular surface and body and the distal articular surface; there are a number of longitudinal cracks in the cortex of the body, which extend into the medullary cavity (Fig. 10.5 left, a, b, d). SKX 5020 possesses a broad, concavoconvex basal articular surface (Fig. 10.5c) and a prominent crest, which marks the insertion of the m. opponens pollicis on the lateral margin of the body (Fig. 10.5a). SKX 5020 is missing much of the distal articular surface (Fig. 10.5d) but it would appear that the head was broader than that of SK 84 from Member 1 (Napier, 1959 and Fig. 10.5d right). The base is relatively broad and saddle-shaped, the body is stout with

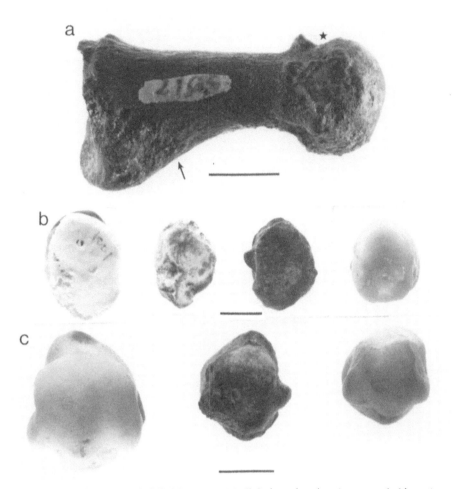

FIGURE 10.2. Hallucal metatarsal (SKX 5017). (a) Medial view showing the expanded base (arrow) and dorsal extension of the distal articular surface (star) (both characteristic of modern humans). The former trait indicates the presence of a plantar aponeurosis; the latter humanlike toe extension at toe-off. (b) Basal articular view of modern human, OH 8, SKX 5017, and pygmy chimpanzee (left to right). (c) Distal view of modern human, SKX 5017, and pygmy chimpanzee (left to right). Note the apelike shape of the metatarsal head in SKX 5017. The narrow dorsal aspect of the articular surface in SKX 5017, which resembles the pygmy chimpanzee, indicates the lack of close-packing at extension of the first metatarsophalangeal joint at toe-off (see Susman et al. (1984) for a full explanation). Scale bars = 1 cm.

a broad, flat dorsal surface. The distal articular surface is expanded anteroposteriorly. SKX 5020 resembles modern human homologues.

Middle Phalanges II to V (SKX 5019, SKX 5021, SKX 5022, SKX 9449). These bones are mostly complete, undistorted manual phalanges, representing at least two individuals (Fig. 10.6). SKX 5019 and SKX 5021 measure 17.5 and 26.4 mm in length, respectively. All have well-developed markings for the insertion of m. flexor digitorum superficialis and pronounced basal collars for the attachment of the interphalangeal joint capsules (Fig. 10.6b). SKX 5022 represents the proximal half of a manual middle phalanx. These bones do not appear "bottle-shaped" as do the middle phalanges of OH 7, although it is important to note that the OH 7

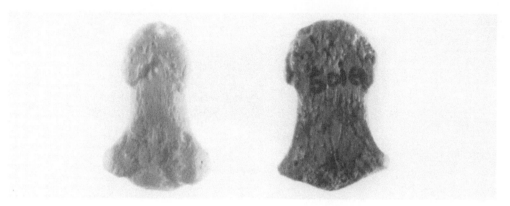

FIGURE 10.3. Pollical distal phalanx (SKX 5016, right) compared with that of a modern human (left). Note the broad apical tuft on SKX 5016 and the extensive concavity for insertion of the tendon of m. flexor pollicis longus, a uniquely human thumb muscle.

FIGURE 10.4. Manual distal phalanges (SKX 27504, left; SKX 8963, right) compared with pygmy chimpanzee (above) and modern human (below).

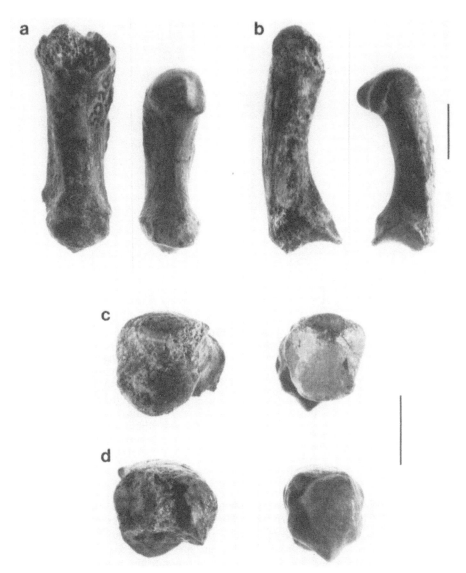

FIGURE 10.5. Pollical metacarpal (SKX 5020, left, SK 84, right). (a) palmar view; (b) medial view; (c) basal view; (d) distal view. Note the size difference in these two specimens and the "beak" on the head of SK 84.

individual was a subadult. The closest morphological counterparts to the Member 1 middle phalanges are seen in modern human homologues (Fig. 10.6c, right).

Proximal Phalanx (SKX 5018). A complete, undistorted proximal phalanx (II or IV) with only minor erosion of the dorsal rim of the basal articular surface and of the dorsal margins of the trochlea (Fig. 10.7, left). The length of SKX 5018 is 33.4 mm. The basal tubercles for the attachment of the interossei and the collateral ligaments are well developed, while the scars for the attachment of the flexor retinaculum are only moderately developed. The included angle (= longitudinal curvature) of SKX 5018 is 27° compared to the modern human mean of 26° and the mean of 42° for chimpanzees (Fig. 10.8). The cortex is extremely thick anteroposteriorly in proportion to overall anteroposterior midshaft diameter. SKX 5018 is unique among extant hominoids in its mosaic of primitive and derived features. In terms of size, curvature, and muscle markings this phalanx is humanlike, whereas the thickened cortex is more apelike than humanlike (Susman and Creel, 1979).

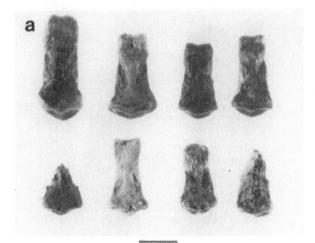
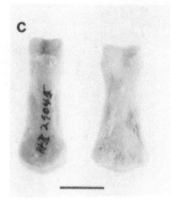
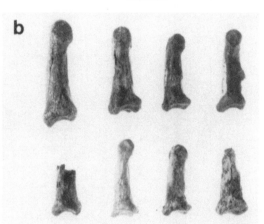

FIGURE 10.6. Manual middle phalanges (a) Palmar view of SKX 5021, 36712, 9449, 35439 (left to right above); SKX 33355, 13476, 5019, 5022 (left to right, below). (b) Medio-lateral view—same as above. (c) Middle phalanx III—pygmy chimpanzee (left), modern human (right).

Radius (SKX 3602, 12814). These specimens represent two-thirds of a right radius (Fig. 10.9). The shaft (SKX 12814) has a sharp, well-defined interosseus border and the faint insertion of m. pronator teres is visible on its dorsolateral aspect. The distal end (SKX 3602) displays a prominent tubercle of Lister and a prominent (but smaller) tubercle for the partial attachment of the extensor retinaculum and which also serves to separate the tendons of mm. extensor indicis proprius and extensor digitorum communis on one side from m. extensor pollicis longus on the other. SKX 3602 has a bony crest for the insertion of m. brachioradialis on its lateral side which extends proximally from the anterolateral aspect of the well-developed radial styloid process for a distance of 19.5 mm. The "set" of the radiocarpal joint is neutral (rather than flexed). Morphologically, the shaft and distal end are more human- than apelike. The shaft has a sharp interosseous border (unlike that of apes) and the distal end has a neutral set to the radiocarpal joint (unlike the flexed set of the radiocarpal joint seen in the knuckle-walking African apes). The distal radius bears a close morphological similarity to Stw 46, a distal radius from Member 4, Sterkfontein (Fig. 10.9d). In both SKX 3602 and Stw 46 the joint morphology (articulations with the distal ulna, scaphoid and lunate) is essentially humanlike.

Ulna (SKX 8761). The proximal third of a left ulna that has suffered numerous small, longitudinal weathering cracks on its surface (Fig. 10.10). The trochlear notch measures 14.1 mm anteroposteriorly. The olecranon process measures 10.4 mm. There is a very deep fossa for the insertion of the m. brachialis. Morphologically, SKX 8761 is the same size as the TM 1517 ulnar fragment; the olecranon process is similar in both. Both are about the size of a small modern human. Among extant hominoids the closest morphologic counterpart of SKX 8761 is found in modern humans.

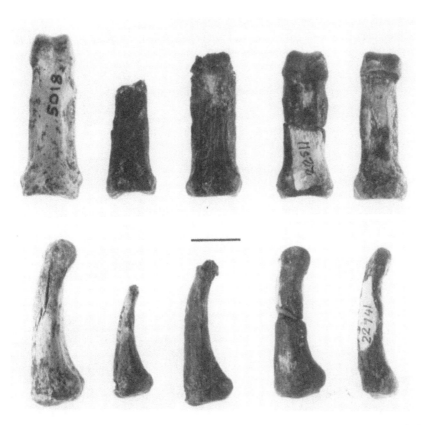

FIGURE 10.7. Manual proximal phalanges (above) palmar view; (below) mediolateral view, SKX 5018, 19576, 27431, 22511, 22741 (left to right).

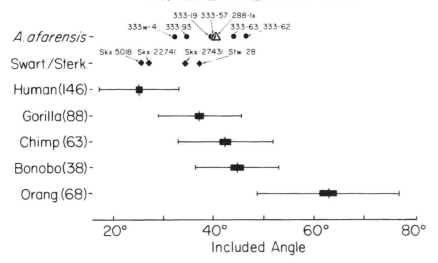

FIGURE 10.8. Included angle of Swartkrans proximal phalanges SKX 5018, 22741, and 27431 compared to *A. afarensis* and extant hominoids. Mean for extant species is indicated by vertical bar; horizontal bar = 95% fiducial limits of population; black bar = 95% confidence limits of the mean. Sample sizes are in parenthesis. The included angles for the extant hominoids are indicative of the relative degree of arboreality in these species. Note that SKX 5018 and SKX 22741 have essentially humanlike curvatures while SKX 27431 (and one phalanx from Sterkfontein, Stw 28) are more apelike.

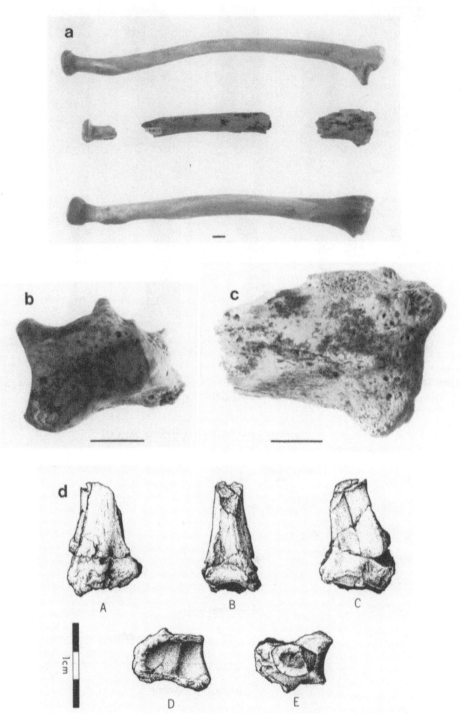

FIGURE 10.9. Radius. (a) Dorsal view of the radius of pygmy chimpanzee (top), Swartkrans (composite) radius (center), and modern human (below). The shaft and distal end on the Swartkrans radius are humanlike; the proximal end is closer to that of pygmy chimpanzees (see text). (b) End on view of radius (SKX 3602). (c) Dorsal view of SKX 3602. (d) Sterkfontein distal radius, Stw 46 (from Member 4) (A) posterior; (B) medial; (C) anterior; (D) distal; (E) proximal views. SKX 3602 and Stw 46 are similar and both most closely resemble modern humans.

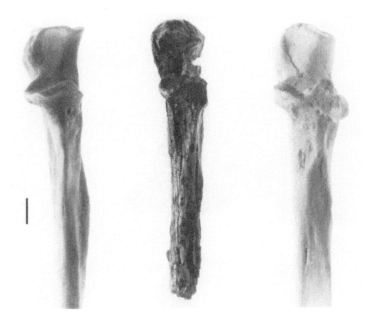

FIGURE 10.10 Ulna (anterior view). Pygmy chimpanzee, SKX 8761, and modern human (left to right).

Humerus (SKW 34805). Distal part of a right humerus, lacking part of each condyle, the trochlea, and the capitulum (Fig. 10.11). SKW 34805 has a biepicondylar diameter of 47.2 mm and a shaft diameter of 18.2 mm (mediolaterally) × 17.4 mm anteroposteriorly. There is a well-defined supinator ridge which extends for 39.7 mm along the lateral side of the shaft to a point just above the lateral epicondyle. This specimen is a reasonably good match for TM 1517 in both size and shape (Fig. 10.12), and on this basis it is attributed to *Paranthropus*. Overall, this distal humerus, is like that of small modern humans.

Humerus (SKW 3774). Small fragment of a right humerus preserving the trochlea, olecranon fossa, and part of the medial epicondyle (Fig. 10.12). The trochlea measures 17.4 mm from the medial edge to the center of the keel while the anteroposterior diameter of the medial trochlea is 21.3 mm. The inferior margin of the medial epicondyle forms a right angle with the medial edge of the trochlea. Although this specimen is very fragmentary, the trochlea, and especially inferior margin of the medial epicondyle are decidedly humanlike (rather than chimpanzee-like) in appearance. The trochleae of SKX 3774 and TM 1517 are similar in size.

Member 2 Remains

The taxonomic attribution of Member 2 remains is more problematic than the assignment of Member 1 fossils for two reasons. First, there is a greater relative proportion of craniodental fossils attributed to *Homo* in Member 2. Grine (this volume, chapter 14) notes that of the 17 to 22 new hominid individuals from Member 2, 4 or 5 are attributable to *Homo*. This represents a minimum of 18% to a maximum of 29% of the new hominid sample (compared to, at most, 5% of the hominids in Member 1). As Grine (this volume, chapter 14) points out, if the new Swartkrans remains are combined with the older material described by Broom and Robinson (1949, 1952) then the percentages of *Homo* in Member 2 rise to 26–31% of the total. Moreover, three of the postcranial fossils from Member 2 (SKX 3342, SKW 2954, and SKW 3646) might be assigned to *Homo* on morphological criteria (none of the Member 1 remains can be so assigned). Thus, when one considers the increased proportion of *Homo* in Member 2 (as judged by teeth and skulls) and the postcranial remains themselves, there may be a greater occurrence of *Homo* in proportion to *Paranthropus* in Member 2 than in Member 1.

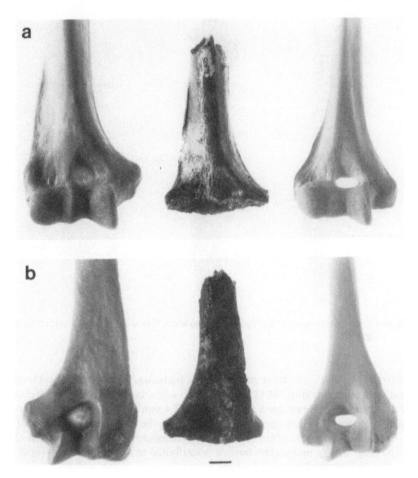

FIGURE 10.11. Humerus. (a) Anterior view. Pygmy chimpanzee, SKW 34805, modern human (left to right). (b) Posterior view of same.

Member 2 Fossils Attributed to Paranthropus

Radial Head (SKX 3699). A right radial head which retains a virtually intact articular surface measuring 19.5 mm in diameter (Fig. 10.9a). The neck is long; the radial tuberosity is missing. SKX 3699 is essentially chimpanzee-like with a thin rim for articulation with the radial notch on the ulna and a long neck. It is interesting to note that although SKX 3699 was recovered from Member 2, it was found in the same grid square, and at roughly the same 10 cm level, as SKX 3602. The ratio of SKX 3602 and SKX 3699 (SKX 3699/SKX 3602) is 0.64 and falls well within the 95% confidence limits of the mean of modern humans (0.67, $N = 15$) and outside the range of pygmy chimpanzees (0.77, $N = 15$). It thus appears that SKX 3602, SKX 3699, and SKX 12814 may represent three-fourths of a single hominid right radius. If so, this radius displays a unique combination of primitive and derived features. The proximal radius is very chimpanzee-like, with a small head and long neck, while the shaft and distal end are essentially humanlike.

Patella (SKX 1084). Upper half of a left patella (Fig. 10.13). Most of the superior surface is intact. The mediolateral breadth of SKX 1084 is 30.2 mm. The posterior surface is well represented, particularly the articular portion, which contacts the lateral trochlea of the femur. There is considerable erosion of what remains of the anterior surface. Clearly, the closest morphological counterpart of SKX 1084 is found in modern humans.

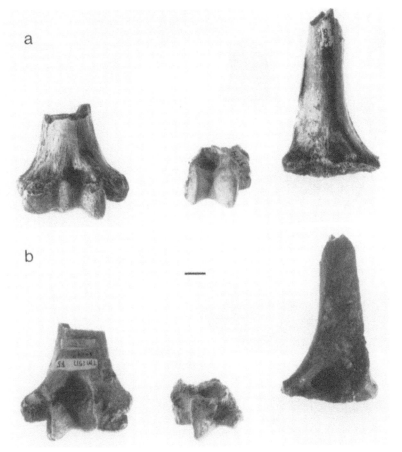

FIGURE 10.12. Humerus. (a) Anterior view of *Paranthropus robustus* TM 1517 from Kromdraai (cast), SKW 3774, and SKW 34805 (left to right). (b) Posterior view of same.

Member 2 Fossils of Uncertain Attribution

Vertebra (SKX 3342). Right half of the body of a thoracic vertebra (T6–9), with only a remnant of the right pedicle; vestiges of the epiphyseal plates are seen on the superior and inferior surfaces (Fig. 10.13). Vestiges of the epiphyseal plates are visible on the superior and inferior surfaces. The height of the body measures 13.7 mm at the anterior margin. There is a large demifacet above and a faint remnant of one below. SKX 3342 is similar in size to its counterpart in Sts 14, *Australopithecus africanus*. Sts 14-m measures 13.2 mm in height at the anterior margin of the body. SKX 3342 is considerably smaller than its counterparts in modern humans.

Metacarpal IV (SKW 2954). A complete right fourth metacarpal with erosion of the head on the medial and lateral sides. Four large weathering cracks are present on the proximal part of the shaft (Fig. 10.14). There appears to be a crook in the specimen which imparts an abnormally strong curvature to the shaft, and although radiographs do not reveal evidence of a healed fracture, this seems the most likely cause of distortion. (Flexion deformities of the metacarpals are not uncommon in humans due to the bowstring effect of the interosseus muscles acting on a fractured shaft.) The interosseous markings on the dorsolateral aspects of the shaft are marked. The basal and distal morphology are unremarkable. SKW 2954 is smaller than any fourth metacarpal in the present human sample ($N = 50$). It is also smaller than Stw 65, a right fourth metacarpal from Member 4 at Sterkfontein. Otherwise, SKW 2954 is essentially humanlike in morphology. SKW 2954 and SKW 3646 are similar in size and shape and they articulate well.

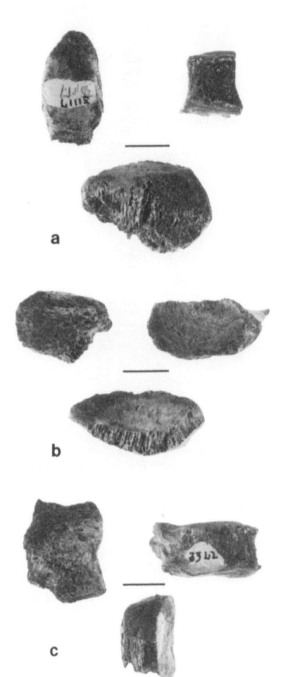

FIGURE 10.13. (a) Anterior view of: medial cuneiform SKX 31117 (left); thoracic vertebra SKX 3342 (right); patella SKX 1084 (center below). (b) Superior view of same. (c) Mediolateral view of same.

Metacarpal III (SKW 3646). Proximal half of a right third metacarpal (Fig. 10.14). The shaft is curved with a mediolateral midshaft diameter of 8.1 mm and an anteroposterior midshaft diameter of 8.9 mm. The base has a faint styloid process, a single facet for contact with metacarpal IV, and a bipartite facet for articulation with the capitate. SKW 3646 falls within the size range of small modern human females ($N = 50$). The lines on the dorsal surface of the shaft marking the origins of the second and third dorsal interossei resemble their counterparts in modern humans.

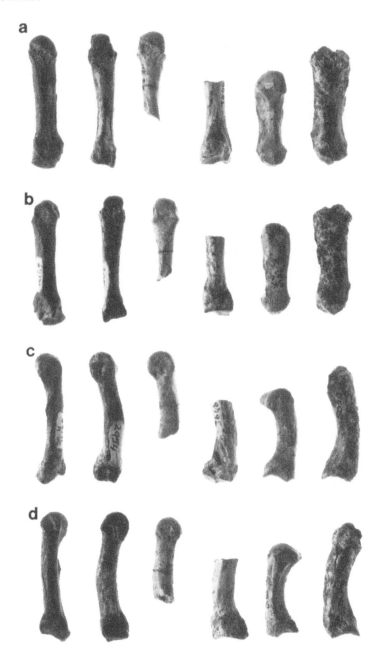

FIGURE 10.14. Metacarpals SKW 14147 (V), SKW 2954 (IV), SK 85 (IV), SKX 3646 (III), SK 84 (I), SKX 5020 (I) (left to right, a–d.) (a) palmar (anterior) view (b) dorsal (posterior) view. (c, d) mediolateral views.

The shaft curvature exceeds that of modern humans and recalls chimpanzees. SKW 3646 is very similar in morphology to Stw 68, a complete right third metacarpal from Member 4 at Sterkfontein.

Metatarsal III (SKX 247). Left base and proximal third of a left third metatarsal. One demifacet on the medial aspect of the base serves as the articulation for metatarsal II and a single demifacet for metatarsal IV is found on the lateral side. SKX 247 is similar in morphology to modern humans. Its size is also within the range of modern humans.

Manual Middle Phalanx (SKX 3062). A complete, undistorted manual middle phalanx (V?) measuring 15.3 mm in length. The specimen is short and stout, with a thick basal collar for attachment of the joint capsule. The m. flexor digitorum superficialis insertions are only faintly represented. SKX 3062 is stout, as are the middle phalanges of modern humans, although it is considerably smaller (shorter) than even the shortest (fifth) middle phalanges in the present sample of modern humans ($N = 50$)

Triquetral (SKX 3498). A complete, undistorted right triquetral that measures 13.3 mm (proximodistally), by 8.0 mm mediolaterally at its base (Fig. 10.15). The pisiform facet is relatively small compared to the size of this facet in other hominoids, and it is elliptical (long axis = 6.9 mm; short axis = 4.3 mm) and convex (Fig. 10.15b). The lunate surface is slightly concave and the hamate surface is faintly saddle-shaped. There is a large posterosuperior surface for contact with the articular disc.

SKX 3498 is essentially humanlike with regard to its overall shape; the lunate and hamate surfaces match those of modern humans. The size of SKX 3498 places it in the range of small (~5'0") modern humans. The small pisiform facet, however, is oval in shape and is unlike that of either humans or chimpanzees ($N = 15$).

Pedal Middle Phalanges (SKX 344; SKW 1261). SKX 344 is a complete, undistorted pedal middle phalanx that measures 13.2 mm. The specimen is short and stout with a well-defined ridge for the insertion of the extensor assembly. The basal articular surface slopes distalward at its dorsal edge, reflecting the capacity for a broad range of dorsiflexion at the proximal interphalangeal joint in this individual. SKW 1261 measures 12.3 mm in length, and is less robust than SKX 344. The base is thick (dorsoplantarly) and the basal articular surface, like that of SKX 344, opens dorsally to accommodate full extension at the proximal interphalangeal joint. Both SKX 344 and SKW 1261 are unique morphologically, as well as in their sizes and proportions. SKX 344 most closely approximates pedal middle phalanx II in modern humans, while SKX 1261 is more attenuated than its counterparts in humans. Neither specimen resembles pedal middle phalanges of any of the extant great apes.

Member 3 Remains

Craniodental remains from Member 3 consist entirely of *Paranthropus* (Grine, this volume, chapter 14). Of the eleven new postcranial remains from Member 3 (Table 10.1), three represent *Paranthropus*, and nine are unattributed.

Member 3 Remains Attributed to Paranthropus

Manual Proximal Phalanx (SKX 22741). This phalanx is a essentially complete and undistorted (Fig. 10.7). It lacks only the dorsalmost cortex on its base and some the dorsal surface of the trochlea. The bone is 30.1 mm in length and has a curvature (included angle) of 27° (Fig. 10.8). SKX 22741 (and the other Member 3 phalanges) are smaller than those of modern humans. The curvature of SKX 22741 is similar to that of SKX 5018 from Member 1 (see above); the value of 27° places SKX 22741 close to the mean for modern humans (26°, $N = 146$). The very species specific nature of phalangeal curvature suggests that SKX 22741 and SKX 5018 are from the same taxon.

Medial Cuneiform (SKX 31117). A fragmentary left medial cuneiform, which lacks the inferior-most portion of the metatarsal articulation; the proximal and dorsal-most parts of the specimen are eroded (Fig. 10.13). The bone corresponds in size to SKX 5017. The distal articular surface is flat and measures 12.7 mm (mediolaterally) by approximately 23 mm (dorsoproximally). Although SKX 31117 is badly damaged, it is unlike extant hominoids (or cercopithecoids). Nevertheless, the distal articular surface, the flatness of which indicates an adducted, rather than an opposable hallux, is distinctly humanlike.

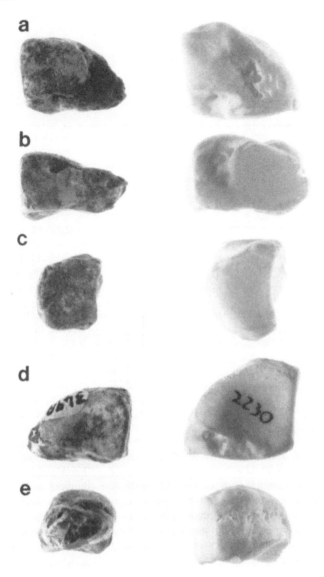

FIGURE 10.15. Triquetral SKX 3498 (left), modern human (right). (a) medial, (b) palmar (anterior), (c) proximal, (d) lateral, (e) distal views.

Metatarsal (SKX 33380). Distal three-fourths of a left fifth metatarsal (Fig. 10.16). The shaft is broad (mediolaterally) and flat dorsally, and there is a pronounced curvature (concave laterally) and a crest on the plantar surface that separated the third plantar from the fourth dorsal interosseus muscle. On the dorsal surface there is a faint ridge for the insertion of m. peroneus tertius. SKX 33380 is essentially humanlike in morphology. The peroneus tertius insertion is a humanlike attribute as is the dorsally extended articular surface of the metatarsal head.

Member 3 Remains of Uncertain Attribution

Manual Proximal Phalanx (SKX 27431). This specimen has suffered some loss of cortex on its dorsal surface; it also lacks the distalmost portion of the trochlea (Fig. 10.7). SKX 27431

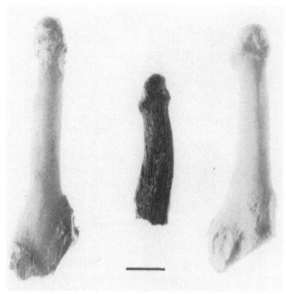

FIGURE 10.16. Metatarsal V. Dorsal view of chimpanzee, SKX 33380, and modern human (left to right).

measures approximately 30.1 mm, similar to SKX 22741, but it has an included angle of 34° (Fig. 10.8). The included angle of 34° places it outside the 95% confidence limits of the human mean and squarely within the 95% confidence limits of the means for gorillas and chimpanzees (but outside the range of pygmy chimpanzees). Thus the curvature of SKX 27431 differs from that of SKX 22741: the former resembles African apes and Stw 28 from Member 4 at Sterkfontein. Given the species specific nature of phalangeal curvature, it is most likely that SKX 27431 and 22741 represent different taxa in Member 3, viz. *Homo* and *Paranthropus*, respectively. Notwithstanding the apelike curvature of SKX 27431, all of the Member 3 phalanges are morphologically more human- than apelike.

Manual Proximal Phalanges (SKX 19576, SKX 22511, SKX 35822). SKX 19576 is an eroded proximal fragment, similar in size and shape to the other Member 3 proximal phalanges (Fig. 10.7). SKX 22511 together with SKX 30220 constitute a complete, albeit distorted, proximal phalanx that measures 30.3 mm in length (Fig. 10.7). This specimen is similar to the others from Member 3. SKX 35822 is a distal fragment (not pictured) that is smaller than the others. Computation of phalangeal curvature is not possible for these specimens because of their fragmentary and distorted condition. Although clearly human- rather than apelike, these proximal phalanges are smaller than those of modern humans.

Manual Middle Phalanges (SKX 36712, SKX 35439, SKX 33355). SKX 36712 and SKX 35439 are complete and undistorted middle phalanges; SKX 33355 represents the proximal half of a middle phalanx (Fig. 10.6). All three specimens possess very prominent bases, with thick basal collars for attachment of the proximal interphalangeal joint capsules, and all have prominent dorsobasal beaks for insertion of the central tendon of the extensor assembly. SKX 36712 and SKX 35439 have very pronounced impressions that mark the insertion of the m. flexor digitorum superficialis. These specimens lack any hint of longitudinal curvature. These manual middle phalanges from Member 3 are unique morphologically. While they are more human- than apelike, they have stouter bases (and basal collars) and better developed flexor impressions than do the middle phalanges of modern humans.

Manual Distal Phalanx (SKX 27504). This specimen lacks only its lateral basal tubercles (Fig. 10.4). It measures 14.0 mm in length, 3.8 mm at midbody (mediolaterally), and has an apical tuft that is 5.5 mm (mediolaterally). The apical tuft is well developed. This bone is decidedly humanlike in size and shape. The pronounced apical tuft and its development relative to the body place it well within the modern human range.

Discussion

Cultural Status and Functional Morphology of Paranthropus

As noted above, Robinson (1961) suggested that early *Homo* rather than australopithecines made the artifacts found at Swartkrans and Sterkfontein. The presence of both *Homo* and *Paranthropus* at sites in South (and East) Africa led Robinson to the belief that *Paranthropus* was not a toolmaker:

> ... clear proof of the presence of a relatively large-brained hominid at (the lower levels in Bed I at Olduvai and at South African sites) would make it very difficult indeed to regard the australopithecines as tool-makers. To my mind the Sterkfontein evidence shows that *Australopithecus* was not the toolmaker there. If *it* was not, then it is difficult to acccept *Paranthropus* as a toolmaker since it was a vegetarian with small need for tools under normal conditions. An omnivore which is an active meateater (viz. *Australopithecus africanus*) has far more need for tools than has a vegetarian (1962:106).

Mason (1961, 1962) echoed Robinson's (1961) opinion on the toolmaking potential of *Australopithecus*, and suggested that stone tools at Sterkfontein were made by a creature such as early *Homo*, then known in South Africa only at Swartkrans. Mason (1962:123) thought that the tools at Sterkfontein reflected a brain and hand that were "closer to our own hands and brain than those of the australopithecines. For these reasons we believe that the Sterkfontein toolmaker was a hominid of the pithecanthropine grade in conflict with *Australopithecus* whom he eventually forced into extinction." Mason thought that "pithecanthropines" such as those from Swartkrans might have made the artifacts and killed *Australopithecus* in competition for meat and plant food on the open plains of southern Africa, ". . . we may now see [australopithecines] as the unsuccessful competitors of progressive pithecanthropines who were making tools . . ." (Mason, 1962:124). (It is interesting to note that at the time Mason offered the above, *Homo* fossils were not yet known from Sterkfontein!)

Tobias, who adopted the view of Robinson (1961) and Mason (1961, 1962) noted that:

> The evidence of (South African) cultural-hominid associations, coupled with the *a priori* considerations advanced by Robinson and Mason and those which we have inferred from morphological and comparative evidence, all conspire to leave little doubt that the earliest and most primitive stone implements found in Africa belonging to the Oldowan culture, were made by the hominid we have called *Homo habilis* (1965:188).

The assertion by Robinson and Mason that australopithecines did not make tools was not based on anatomical evidence. Tobias (1965), on the other hand, claimed to have based his inference on "morphological and comparative evidence." This "evidence" consisted mainly of assumptions about the cultural capacities of *Australopithecus* based on brain size and intellectual capacity, toolmaking in living primates, and perhaps most important, a discussion of the anatomy of the hand. The fatal flaw in the discussion of the hand was the fact that while the hand of early *Homo* (i. e., OH 7) was known, there were no significant hand fossils of either *Australopithecus* or *Paranthropus* with which to compare the OH 7 bones. Tobias himself recognized this shortcoming when he stated, "Just how hominized the hand of *Australopithecus* was we cannot yet say. . . . Only more specimens can answer the problem of what sort of hand *Australopithecus* possessed" (1965:184).

The specimens that Tobias alluded to are now known, and indicate that the hand of *Paranthropus* was indeed "hominized." The morphology of the hand bones from Member 1 (and those from Members 2 and 3 although we cannot assume with the same confidence that these are attributed to *Paranthropus* and not *Homo*) now reveals the capacity for refined manipulation commensurate with that of *Homo habilis* (e.g., OH 7) and later hominids. The humanlike first metacarpal (SKX 5020), the humanlike pollical distal phalanx (SKX 5016), the straight, short fingers (SKX 5018), and the broad fingertips reveal a distinctively humanlike hand of *Paranthropus* in Member 1 times (Susman, 1988).

Taxonomy of the Swartkrans Hominids

Taxonomic attribution of individual specimens is made difficult by the presence of two hominid taxa in Members 1 and 2. The difficulty is lessened, however, with regard to Member 1 by the fact that over 95% of the individual hominids from that member represent *Paranthropus* and only some 5% are attributable to *Homo* on the basis of craniodental remains (Brain 1981; this volume, chapter 20; Grine, this volume, chapter 14). In addition to the simple statistical probability that Member 1 remains come from *Paranthropus*, there are additional morphological criteria which aid in taxonomic assessment. One such example regards assignment of first metacarpals SKX 5020 and SK 84. In this case there are two morphs: one in the form of the larger SKX 5020 and the other in the form of the smaller SK 84. SK 84 has a peculiar "beak," which separated what must have been prominent sesamoid bones of the metacarpophalangeal joint (Broom and Robinson, 1949; Napier, 1959). The possession of this beak in the first metacarpal of the *Homo erectus* skeleton KNM-WT 15000 from the west side of Lake Turkana (A. Walker, pers. comm.) suggests that SK 84 shares this unique trait with KNM-WT 15000 and likewise represents *Homo cf. erectus*. Size and morphological differences between SK 84 and SKX 5020 suggest that the latter is that of *Paranthropus* (or perhaps, at minimum, they represent a very dimorphic *Paranthropus robustus*).

Taxonomic assessment of Member 2 hominids is much more problematic. As of 1981, there were but a few hominids recovered from Member 2, and while Brain (1982) noted the relative dearth of *Paranthropus* remains from Member 2, he suggested the possibility that the *Paranthropus* remains in Member 2 could have been recycled from Member 1. The hominids described earlier from Member 2 include the type of *Telanthropus*, SK 15, parts of two lower premolars, SK 43 and SK 18a, and a proximal radius, SK 18b (Brain, 1981, 1982). On the basis of recently discovered craniodental remains (Grine, this volume, chapter 14), there are now roughly twelve to seventeen individuals representing *Paranthropus* and four or five individual *Homo* identified in Member 2. Thus the statistical grounds for allocating Member 2 fossils to either taxon are weaker than they are with respect to Member 1. On the basis of comparisons with specimens that are likely to be those of *Paranthropus* (see above), I attribute the patella (SKX 1084), and the radial head (SKX 3699) to *Paranthropus*.

All of the new postcranial fossils from Member 3 could likely belong to *Paranthropus*. Grine (this volume, chapter 14) attributes all of the new craniodental remains from Member 3 to *Paranthropus*. There is direct morphological justification to so assign the medial cuneiform (SKX 31117), since it fits well the first metatarsal from Member 1 (although the author fully realizes the circularity of such a rationale). On the other hand, metatarsal V (SKX 33380) and the manual distal phalanx (SKX 27504) are essentially modern humanlike, nothwithstanding their small size. An apelike phalangeal curvature in manual proximal phalanx SKX 27431 indicates a primitive retention that is perhaps more likely to have been manifested in *Homo* than in the somewhat larger *Paranthropus* (Susman et al., 1984; Johanson et al., 1987). There is a hint (based on its difference with SKX 27431 and its similarity to SKX 5018 from Member 1) that one proximal phalanx from Member 3 (SKX 22741) might be attributable to *Paranthropus*.

Morphology and Behavior of Paranthropus robustus in Member 1 Times

Three factors were responsible historically for the idea that *Paranthropus* was not a toolmaker. The first was the small cranial capacity of both the East and South African "robusts." They clearly had not reached the "human" cerebral Rubicon of 750 cc (Keith, 1948). Most earlier workers simply believed that the "robusts" lacked the intelligence (as judged by small brain size) for stone toolmaking. Second was the fact that Broom (1942) had described earlier what was thought to be the hand of *Paranthropus* (TM 1517) from Kromdraai. This hand, which is in fact that of an extinct baboon (Day and Thornton, 1986), was recovered in close association with the type skull and other remains of *Paranthropus* and thus presented a very primitive, to say nothing of incorrect, picture of the hand of *Paranthropus*. Finally, there was the *H. habilis* hand (OH 7), which revealed a very humanlike capacity for fine manipulation and seemed to confirm the hypotheses of Robinson (1961) and Mason (1961, 1962) that *Homo*-like hominids rather than australopithecines, had the anatomical potential for tool behavior.

It has been suggested that the locomotion of *Paranthropus* differed from that of early *Homo*, and that the lower limb of the former was manifestly less "hominized" than the latter (Tobias, 1965). Once again, as was noted with regard to the attribution of tool behavior, only half of the comparative story was known, viz. that of *Homo* as seen in the OH 8 foot from Bed I at Olduvai (Leakey, 1960; Susman and Stern, 1982). Until now, direct comparisons of the limb bones of *Paranthropus* and *Homo* have been limited to relatively few features of locomotor anatomy (see Robinson, 1972). Robinson, in a comprehensive study of the skeleton of Plio-Pleistocene hominids concluded that there were important differences in the locomotion and posture of *Paranthropus* and *Homo*. He suggested that, although (*Paranthropus*) clearly must have had appreciable capability as an erect biped, it was not as effectively specialized in this direction as was *Homo africanus*, (Robinson, 1972:251). Unlike most who relied on generalized assertions, Robinson advanced morphological grounds for his functional inferences. He surmised that *Paranthropus* had a long ischium and "therefore still used its propulsive mechanism at least partly in a power-specialized manner," and that:

> This suggests a compromise adaptation—perhaps not completely efficient bipedality on the ground, coupled with spending some time in the trees. The small amount of evidence of the hand suggests that this was powerful but elongate with a short thumb. The evidence of the foot suggests mobility and flexibility and perhaps an incompletely adducted great toe—also a compromise between ape and human characteristics. Such hands and feet would be consistent with some climbing activity (Robinson, 1972:251).

It is important to remember that there was a virtual absence of fossil material most appropriate to test the hypothesis of Robinson. With only two hand bones potentially attributable to *Paranthropus* (Robinson, 1972), and only one foot bone, it was difficult at best to confirm hypotheses as to the positional behavior of *Paranthropus*. Moreover, the improperly assigned TM 1517 (baboon) hand obfuscated our understanding of *Paranthropus* hand anatomy.

The prevailing notion of the habitus of *Paranthropus* is that they coexisted with *Homo* and that they occupied a vegetarian niche and did not possess toolmaking (or perhaps even tool-using) capabilities (Robinson, 1972). [A similar conception of the east African "robust" australopithecines dates to the early 1960's (Leakey, 1961).] The large teeth and jaws, the small relative brain size of *Paranthropus* (Tobias, 1971), coupled with the presence of a more advanced hominid such as *Homo* at sites in both South and East Africa are the principal reasons that tool behavior is not imputed to *Paranthropus* (Toth, 1987).

In Member 1 at Swartkrans, where remains of more than one hundred individual hominids have been recovered, the bone and stone tools have been attributed to *Homo*, despite the fact that *Paranthropus* outnumbers *Homo* twenty to one. Until now, and the recovery of fossil hand bones attributable to *Paranthropus*, it was not possible to address the issue of which hominid at Swartkrans made the tools. With the recovery of fossil hand bones such as SKX 5016 and SKX 5020 (thumb), SKX 8963 (distal phalanx of the finger), and others which represent the hand, we can now comment on the morphological evidence for tool behavior in *Paranthropus*. As the preceding descriptions and comments on morphologic affinities suggest, the fossil hand bones of *Paranthropus* from Member 1 reveal a precision grip as refined as that in later hominids such as *H. habilis* and modern humans. The thumb of *Paranthropus* was stout with a broad saddle-shaped carpometacarpal joint, a well-developed m. opponens pollicis, an extensive area of insertion for m. flexor pollicis longus on the distal phalanx, and a prominent apical tuft on the tip of the thumb. The fingers were straight (not curved as in apes), and had broad apices, to match that of the thumb. The metacarpus was also like that of later hominids, including modern humans.

The above traits are the very ones upon which Napier (1962a,b) and others based their belief that OH 7 was a toolmaker. These represent many of the features on which Leakey *et al.* (1964) based their suggestion that the OH 7 juvenile was a member of our own genus. Napier noted, with regard to the OH 7 hand, that it is in the functional morphology of the hand itself that one must look for evidence of toolmaking. Furthermore, he noted that:

> On anatomical grounds there is no doubt that the Olduvai hand was sufficiently advanced in terms of the basic power and precision grips to have used naturally occurring

objects as tools to good advantage . . . the construction of the crude, rather small pebble-tools of the type found on the (FLK NN level 3) living floor, is well within the physical capacity of the Olduvai hand (Napier, 1962a:411).

All of the salient features that are indicative of refined precision grip (Napier 1959, 1962a,b), and that are in evidence in *H. habilis* at 1.76 Myr in Bed I at Olduvai Gorge (and in later hominids), are in evidence in the fossil hand bones from a similar time period in Member 1 at Swartkrans. There is every morphological reason to impute tool behavior to *Paranthropus* that there is to assign tool behavior to *H. habilis* or any other early hominid. It is plausible, given the fact that *Paranthropus* is regarded largely as a vegetarian, that tool behavior in the "robust" australopithecines was an adaptation for the procurement and perhaps processing of plant foods. Indeed, the microwear on the bone tools suggests that they were used for digging (Brain, 1985). The stone tools as well, could have been used to exploit plant resources rather than for scavenging or meat eating; the lack of cut marks on the faunal remains at Swartkrans is suggestive in this regard. A study of the edge wear and the function of the Swartkrans stone tool assemblage should help clarify this issue. On the other hand, there is no reason to reject *Paranthropus* as toolmakers simply because they were vegetarians (c.f. Robinson, 1962a, 1972).

Finally, it should be noted that there are sites in both South and East Africa where stone artifacts have been recovered and where *Paranthropus* is the only hominid known. These include Kromdraai (Vrba, 1981:18) and FxJj 20 East at Koobi Fora (Harris and Isaac, 1976; Leakey and Leakey, 1978:71). A handful of *P. robustus* individuals are the only hominids known from Kromdraai, which also yielded an "unquestionable" artifact. A "robust" hominid mandible (KNM-ER 3230) was excavated at FxJj 20 East from a level that also yielded some fifty artifacts.

Conclusions

In sum, the postcranial morphology from *Paranthropus* in Member 1, together with diet reconstruction based on the microscopic anatomy of the tooth surfaces and evidence from bone artifacts, all point to the liklihood that *P. robustus* was a tool-using (and likely tool–making), terrestrial biped with a relatively small component of climbing in its locomotor repertoire compared to other Plio-Pleistocene hominids such as *Australopithecus afarensis* and *A. africanus*. *Paranthropus* could well have been a toolmaking/-using vegetarian, while its less megadont hominid counterparts may have enjoyed a more omnivorous lifestyle and employed their tool behavior as much, or more, for scavenging and meat eating[1] as for plant procurement. If *Paranthropus* at Swartkrans and other "robust" australopithecines in South and East Africa employed tool behavior in conjunction with their craniodental specializations for a diet composed in large part of hard, fibrous plant foods (Grine, 1981, 1986), then tool behavior may have constituted an important adaptation of both the "gracile" and "robust" hominid lineages (Fig. 10.17). The notion that tool behavior provided the adaptive wedge that separated the "robust" australopithecines from the *Homo* lineage (Leakey, 1961; Tobias, 1965; Robinson, 1972) may no longer be tenable.

Acknowledgments

I thank Bob Brain for the opportunity to study the new Swartkrans hominids and for sharing his insights into the geology ecology and taphonomy of Swartkrans. I also thank Alan Turner, Elisabeth Voigt and members of the staff of the Transvaal Museum for their warm hospitality and help during our stay. Fred Grine not only conceived this project to study new Swartkrans fossils, but he also offered valuable insights into the systematics and craniodental morphology of the South African hominids. I am grateful to him for his colleagueship and for organizing the workshop on "robust" australopithecines and this volume. This research was made possible by a BRSG (NIH) grant from the School of Medicine at Stony Brook and NSF grant BNS 8311206.

---

[1]Not to be confused with hunting and killing (large) prey.

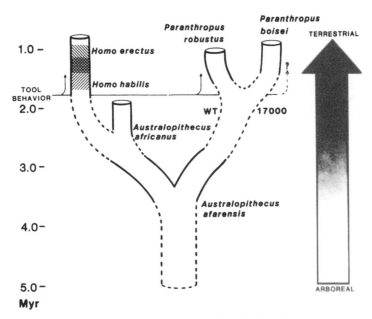

FIGURE 10.17. Phylogeny of Plio-Pleistocene Hominidae indicating approximate range in time for which fossils are known (solid lines), inferred relationships (dotted lines), occurrence of artifacts and inferred tool behavior and increasing trend toward terrestriality (arrow, right). Tool behavior is well-documented at 1.9 Myr, and tools are dated as early as 2.4 Myr at sites in East Africa.

References

Brain, C. K. (1978). Some aspects of the South African australopithecine sites and their bone accumulations. *In* C. J. Jolly (ed.), *Early hominids of Africa*, pp. 131–161. Duckworth, London.

Brain, C. K. (1981). *The Hunters or the hunted? An introduction to African cave taphonomy.* University of Chicago Press, Chicago, IL.

Brain, C. K. (1982). The Swartkrans site: stratigraphy of the fossil hominids and a reconstruction of the environment of early *Homo. 1st Congr. Intl. Paleont. Hum.*, pp. 676–706. Nice.

Brain, C. K. (1985). Cultural and taphonomic comparisons of hominids from Swartkrans and Sterkfontein. *In* E. Delson (ed.), *Ancestors: the hard evidence*, pp. 72–75. Alan R. Liss, New York.

Brain, C. K., Vrba, E. S. and Robinson, J. T. (1974). A new hominid innominate bone from Swartkrans. *Ann. Tvl. Mus.*, 29: 55–66.

Broom, R. (1942). Tha hand of the ape-man, *Paranthropus robustus. Nature*, 149: 513–514.

Broom, R. and Robinson, J. T. (1949). Thumb of the Swartkrans ape-man. *Nature*, 164: 841–842.

Broom, R. and T. Robinson, J. T. (1952). Swartkrans ape-man *Paranthropus crassidens. Mem. Tvl. Mus.*, 6: 1–123.

Clark, W. E. Le Gros (1947). Observations on the anatomy of the fossil Australopithecinae. *J. Anat.*, 81: 300–335.

Clark, W. E. Le Gros (1949). *History of the primates.* British Museum of Natural History, London.

Day, M. H. and Thornton, C. M. B. (1986). The extremity bones of *Paranthropus robustus* from Kromdraai B, East Formation Member 3, Republic of South Africa—A reappraisal. *Anthropos*, 23: 91–99.

Grine, F. E. (1981). Trophic differences between ''gracile'' and ''robust'' australopithecines: a scanning electron microscope analysis of occlusal events. *S. Afr. J. Sci.*, 77: 203–230.

Grine, F. E. (1986). Dental evidence for dietary differences in *Australopithecus* and *Paranthropus*: a quantitative analysis of permanent molar microwear. *J. Hum. Evol.*, 15: 783–822.

Harris, J. W. K. and Isaac, G. (1976). The Karari industry: early Pleistocene archaeological evidence from the terrain east of Lake Turkana, Kenya. *Nature, 262*: 102–107.

Johanson, D. C., Masao, F. T., Eck, G. G., White, T. D., Walter, R. C., Kimbel, W. H., Asfaw, B., Manega, P., Ndessokia, P. and Suwa, G. (1987). New partial skeleton of *Homo habilis* from Olduvai Gorge, Tanzania. *Nature, 327*: 205–209.

Keith, A. (1948). *A new theory of human evolution*. Watts & Co., London.

Leakey, L. S. B. (1960) Recent discoveries at Olduvai Gorge. *Nature, 188*: 1050–1052.

Leakey, L. S. B. (1961). The juvenile mandible from Olduvai. *Nature, 191*: 417–418.

Leakey, L. S. B., Tobias, P. V., and Napier, J. R. (1964). A new species of the genus *Homo* from Olduvai Gorge. *Nature, 202*: 7–9.

Leakey, M. D. (1970). Stone artefacts from Swartkrans. *Nature, 225*: 13–16.

Leakey, M. G. and Leakey, R. E. F. (1978). *Koobi Fora Research Project, Vol 1: The fossil hominids and an introduction to their context, 1968–1974*. Clarendon Press, Oxford.

Mason, R. J. (1961). *Australopithecus* and the beginning of the Stone Age in South Africa. *S. Afr. Archaeol. Bull., 17*: 8–14.

Mason, R. J. (1962). *Australopithecus* and artefacts at Sterkfontein. Part II. The Sterkfontein stone artefacts and their makers. *S. Afr. Arch. Bull., 17*: 109–125.

Napier, J. R. (1959). Fossil metacarpals from Swartkrans. *Fossil Mamm. Africa No. 17. British Mus. Nat. Hist.,* pp. 1–18.

Napier, J. R. (1962a). Fossil hand bones from Olduvai Gorge. *Nature 196*: 409–411.

Napier, J. R. (1962b). The evolution of the hand. *Sci. Amer. 205*: 2–8.

Rak, Y. (1983). *The australopithecine face*. Academic Press, New York.

Robinson, J. T. (1953). *Telanthropus* and its phylogenetic significance. *Amer. J. Phys. Anthropol, 11*: 445–501.

Robinson, J. T. (1961). The Australopithecinae and their bearing on the origin of man and of stone tool-making. *S. Afr. J. Sci., 57*: 3–13.

Robinson, J. T. (1962a). The origin and adaptive radiation of the australopithecines. *In* G. Kurth (ed.), *Evolution und Hominisation,* pp. 120–140. Gustav Fischer, Stuttgart.

Robinson, J. T. (1962b). *Australopithecus* and artefacts at Sterkfontein. Part I. Sterkfontein stratigraphy and the significance of the Extension Site. *S. Afr. Archaeol. Bull., 17*: 85–108.

Robinson, J. T. (1972). *Early hominid posture and locomotion*. University of Chicago Press, Chicago, IL.

Stern, J. T., Jr., and Susman, R. L. (1983). The locomotor anatomy of *Australopithecus afarensis*. *Amer. J. Phys. Anthropol. 60*: 279–317.

Susman, R. L. (1988). Hand of *Paranthropus robustus* from Member 1, Swartkrans: fossil evidence for the behavior. *Science, 240*: 781–784.

Susman, R. L. and Creel, N. (1979). Functional and morphological affinities of the subadult hand (O.H. 7) from Olduvai Gorge. *Amer. J. Phys. Anthropol., 51*: 311–332.

Susman, R. L. and Stern, J. T., Jr. (1982). Functional morphology of *Homo habilis*. *Science, 217*: 931–934.

Susman, R. L., Stern, J. T. and Jungers, W. L. (1984). Arboreality and bipedality in the Hadar hominids. *Folia Primatol., 43*: 113–156.

Tobias, P. V. (1965). *Australopithecus, Homo habilis*, tool-using and tool-making. *S. Afr. Archaeol. Bull., 20*: 167–192.

Tobias, P. V. (1967). *The cranium and maxillary dentition of Australopithecus (Zinjanthropus) boisei*. Olduvai Gorge Volume 2. Cambridge University Press, Cambridge.

Tobias, P. V. (1971). *The brain in hominid evolution*. Columbia University Press, New York.

Toth, N. (1987). The first technology. *Sci. Amer., 256*: 112–121.

Vrba, E. S. (1981). The Kromdraai *Australopithecus* site revisited in 1980: recent investigations and results. *Ann. Tvl. Mus., 33*: 17–60.

Vrba, E. S. (1982). Biostratigraphy and chronology, based particularly on bovidae of southern African hominid-associated assemblages Makapanssat, Sterkfontein, Taung, Kromdraai, Swartkrans; also Elandsfontein (Saldanhn), Broken Hill (now Kabwe) and Cave of Hearths. *Congr. Intl. de Paleontol. Hum.,* pp. 707–752, CNRS, Nice.

Studies of Variation and Taxonomy   III

III

Variation, Sexual Dimorphism and the Taxonomy of *Australopithecus*

11

WILLIAM H. KIMBEL AND TIM D. WHITE

After more than a century of research, the extent of sexual dimorphism in the early Hominidae continues to concern anthropologists. In paleoanthropology, the phenomenon of intraspecific variation presents a difficult and pivotal challenge. As Walker and Leakey (1978) recognized, the first and most fundamental issue in the analysis of fossil hominid samples is the determination of how many species they contain. Discerning sexual dimorphism from interspecific variation, for example, currently plagues attempts to circumscribe taxonomically the recently enlarged samples of Miocene Hominoidea. Analogous problems have been prominent in debates concerning the status of the Hadar collection of *Australopithecus afarensis* and the Koobi Fora collection of *Homo habilis*. This chapter presents quantitative and qualitative data bearing on variation in the skull and dentition of *Australopithecus*. Our purpose is to examine the contribution of sex to this variation in the light of ongoing discussions about the taxonomy of early hominids.

Methods

It is general paleoanthropological practice to employ estimates of intraspecific skeletal variation in extant great ape species as a test of hypotheses regarding the taxonomic composition of fossil hominid samples. The working assumption is that if homologous features vary within the fossil sample significantly more than in closely related extant taxa, then this constitutes *prima facie* evidence for the presence of more than one species in the fossil sample. There are, however, problems with this approach. Fossil hominid samples are not biologically equivalent to the geographically restricted populations from which comparative specimens of extant great ape taxa are usually drawn. Rather, they represent evolutionary lineages, often with considerable time depth in addition to geographic range, and thus potentially harbor a level of variation not documented in living great ape species. These observations, while hardly arguable, are rarely acknowledged in practice. Circumscribing intraspecific variation is vital to an understanding of the evolutionary dynamics of a species in space and ecological time, including the speciation process itself. Such variation can be considerable, but closely related species usually differ in only a few characters, the majority showing substantial overlap (Tattersall, 1986). Certainly this is true of the African apes, the most frequently employed outgroup in systematic studies of fossil hominids.

It is known that primate sexual dimorphism has a multifactorial basis, including (minimally) body size, ontogenetic constraints, habitat, socioeconomic structure and reproductive strategy (e.g., Clutton-Brock and Harvey, 1977; Leutenegger and Chevrud, 1982; Gaulin and Sailer, 1984; Harvey and Bennett, 1985; Willner and Martin, 1985; Pickford, 1986). Notwithstanding that none of these parameters is (or could be) adequately known for most of the hominid fossil record, we may rely on the proposition that no extinct hominid species corresponded in the totality of these attributes to any living hominoid. Thus, attempts to ascertain the expected level of sexual dimorphism in extinct hominid species based on biochemical evidence of human–great ape relationships (e.g., Zihlman, 1985) are clearly ill-founded.

There are a number of quantitative techniques available for estimating sexual dimorphism of skeletal attributes (e.g., Wood, 1976, 1985a). A common method is the comparison of the ratio of female to male means, but this is difficult to apply to the paleontological record because the sex of fossil specimens is not known (Kay, 1982a). Attempts to estimate dimorphism in early hominid species by comparing the ratio of the *extreme* dimensions to the ratios of female to male *means* of extant species (Zihlman, 1985; McHenry, 1986) rest on the explicit, but unwarranted, assumption that the difference between the size extremes in the fossil sample mirrors the mean difference between the sexes. This method also implicitly assumes no overlap between the sexes. Simpson *et al.* (1960:94) note that, "for . . . small samples, such a measure bears little rational relationship to the real variability."

The coefficient of variation *(CV)* is a superior method of describing the relative dispersion of sample data (Simpson *et al.,* 1960). From the perspective of the fossil record, this method has the advantage of not requiring that the sex of specimens be known a priori (Kay, 1982a). Gingerich (1974) and Gingerich and Schoeninger (1979) found that among primate postcanine teeth the size of M1 and M2 is generally least variable (lowest *CV*) and thus is most suitable for discriminating between extinct sympatric species that differ only in size. Furthermore, intraspecific frequency distributions of primate M1 and M2 size are apparently not bimodal (Pilbeam and Zwell, 1973; Gingerich and Schoeninger, 1979; Kay, 1982a,b).

These guidelines provide a useful framework in which to test propositions about the taxonomic composition of fossil hominid samples. It should be noted, however, that a low *CV* and/or a unimodal distribution do not by themselves confirm a hypothesis of a sample's taxonomic unity (e.g., Pilbeam and Zwell, 1973: fig. 2). Rather, they fail to refute the hypothesis under the assumption that size is a valid discriminator among the samples considered. Given the history of emphasis on postcanine tooth size as a criterion of distinction among early hominid taxa, and repeated assertions that the degree of size variation in *A. afarensis* is too great to be accommodated in a single species' range, we believe that this approach has intrinsic merit.

In this chapter we employ the *CV* in a study of metric variation in *Australopithecus* canine and postcanine teeth. We then examine variation in mandible breadth, height and robusticity. Finally, we discuss qualitative variation in several cranial regions, and conclude with an assessment of recently published taxonomies of *Australopithecus* in light of our results.

Table 11.1 lists the taxa and hypodigms discussed here. Comparative metric data for extant primates were extracted from the literature as referenced, except for mandibular measurements, which were taken by the authors on specimens in the Hamann-Todd Osteological Collection, Cleveland Museum of Natural History.

Results

Postcanine Tooth Breadths[1]

Summary statistics for *Australopithecus* postcanine tooth breadths (i.e., buccolingual diameter) are presented in Table 11.2. Table 11.3 provides a comparison of *CV*s for a number of extant

Table 11.1. Hominid Taxa and Hypodigms

Australopithecus afarensis: Hadar Formation, Ethiopia; Laetolil Beds, Tanzania
Australopithecus africanus: Taung; Sterkfontein Type Site (Member 4); Makapansgat, grey and pink breccias (Members 3 and 4)
Australopithecus robustus: Kromdraai, B; Swartkrans, pink breccia (Member 1) (*not* including SK 27, SK 45, SK 847, SK 2635, or any SKW specimens)
Australopithecus boisei: non-*Homo* specimens from: Koobi Fora Formation, Kenya (upper Burgi, KBS and Okote Members); Shungura Formation, Ethiopia (Members G–K); Humbu Formation (Peninj), Tanzania; Chemoigut Formation (Chesowanja), Kenya; Olduvai Gorge, Tanzania (Beds I and II)

---

[1]Variation in crown length (mesiodistal diameter) is not discussed here because interproximal wear is not consistently controlled for in published studies of catarrhine dental metrics.

Table 11.2. Summary Statistics for *Australopithecus* Postcanine Tooth Breadths (Buccolingual Diameters)[a]

	N	$\bar{X}$ ± SE	SD	CV ± SE[b]
P₃				
A. afarensis	18	10.6 ± .21	0.89	8.5 ± 1.44
A. africanus	8	11.8 ± .28	0.78	6.8 ± 1.70
A. robustus	18	11.7 ± .26	1.10	9.5 ± 1.59
A. boisei	6	13.2 ± .38	0.92	7.3 ± 2.10
P₄				
A. afarensis	13	11.0 ± .25	0.89	8.3 ± 1.62
A. africanus	10	11.6 ± .22	0.71	6.3 ± 1.39
A. robustus	17	12.9 ± .24	0.98	7.7 ± 1.32
A. boisei	10	14.5 ± .46	1.41	9.9 ± 2.22
M₁				
A. afarensis	15	12.7 ± .22	0.86	6.9 ± 1.26
A. africanus	17	12.7 ± .23	0.94	7.5 ± 1.29
A. robustus	19	13.6 ± .20	0.85	6.4 ± 1.03
A. boisei	7	15.5 ± .40	1.07	7.1 ± 1.91
M₂				
A. afarensis	18	13.5 ± .24	1.00	7.5 ± 1.25
A. africanus	29	14.2 ± .12	0.90	6.4 ± 0.84
A. robustus	20	14.8 ± .18	0.80	5.5 ± 0.86
A. boisei	7	17.1 ± .44	1.17	7.1 ± 1.91
M₃				
A. afarensis	11	13.4 ± .28	0.93	7.1 ± 1.50
A. africanus	21	14.4 ± .24	1.10	7.7 ± 1.18
A. robustus	21	14.5 ± .26	1.20	8.2 ± 1.27
A. boisei	10	16.5 ± .47	1.48	9.2 ± 2.06
P³				
A. afarensis	6	12.4 ± .30	0.74	6.2 ± 1.80
A. africanus	12	12.5 ± .15	0.53	4.3 ± 0.88
A. robustus	19	13.8 ± .20	0.89	6.5 ± 1.05
A. boisei	5	15.2 ± .60	1.35	9.3 ± 2.96
P⁴				
A. afarensis	6	12.1 ± .24	0.58	5.0 ± 1.45
A. africanus	16	13.0 ± .22	0.87	6.8 ± 1.20
A. robustus	23	14.9 ± .21	0.99	6.7 ± 0.98
A. boisei	5	16.2 ± .78	1.56	10.1 ± 3.19
M¹				
A. afarensis	8	13.2 ± .31	0.87	6.8 ± 1.70
A. africanus	23	13.9 ± .17	0.80	5.9 ± 0.86
A. robustus	19	14.6 ± .20	0.85	5.9 ± 0.95
A. boisei	5	16.2 ± .46	1.03	6.7 ± 2.13
M²				
A. afarensis	6	14.7 ± .26	0.64	4.6 ± 1.32
A. africanus	30	15.8 ± .23	1.28	8.2 ± 1.06
A. robustus	21	15.8 ± .21	0.95	6.1 ± 0.94
A. boisei	3	18.8	—	—
M³				
A. afarensis	7	13.9 ± .37	0.99	7.4 ± 1.97
A. africanus	21	15.5 ± .31	1.44	9.4 ± 1.46
A. robustus	22	16.8 ± .15	0.71	4.3 ± 0.64
A. boisei	3	18.6	—	—

[a] N = Number of individuals; measurements are those described by White (1977).
[b] Coefficients of variation (CV) are adjusted for bias according to equation 4.10 in Sokal and Rolf (1981).

Table 11.3. Coefficients of Variation for Postcanine Tooth Breadths[a]

Taxon	P_3	P^3	P_4	P^4	M_1	M^1	M_2	M^2	M_3	M^3
Australopithecus afarensis	8.5	6.2	8.3	5.0	6.9	6.8	7.5	4.6	7.1	7.4
Australopithecus africanus	6.8	4.3	6.3	6.8	7.5	5.9	6.4	8.2	7.7	9.4
Australopithecus robustus	9.5	6.5	7.7	6.7	6.4	5.9	5.5	6.1	8.2	4.3
Australopithecus boisei	7.3	9.3	9.9	10.1	7.1	6.7	7.1	—	9.2	—
Pan troglodytes[b]	10.9	7.0	7.2	6.2	6.4	6.4	7.3	9.1	9.5	10.7
Pan paniscus[c]	12.7	5.4	7.9	5.6	6.8	5.0	6.6	5.9	5.9	5.9
Gorilla gorilla[b]	10.3	7.4	8.4	6.7	6.4	6.1	7.1	6.8	8.1	7.8
Pongo pygmaeus[b]	9.1	9.0	8.3	7.3	4.8	6.1	5.9	6.7	6.6	7.7
Homo sapiens[d]	6.2	5.5	6.8	7.0	5.1	5.5	5.9	7.0	7.6	8.1
Hylobates agilis[b]	6.9	7.8	2.6	7.2	3.7	5.9	3.2	6.1	5.3	3.0
Cercopithecus mitis[b]	13.2	10.0	7.2	7.0	5.9	6.6	7.3	6.9	8.4	7.6
Papio anubis[b]	17.1	7.8	6.9	6.6	6.9	6.4	8.2	8.2	8.2	9.9
Macaca fascicularis[b]	13.7	9.0	10.1	9.1	8.2	8.5	9.0	9.2	11.8	9.7
Colobus badius[b]	17.6	8.5	10.2	7.4	6.6	5.8	7.8	6.3	6.7	8.9
$\bar{X}$ for (N) primate species[e]	11.0	8.6	8.7	7.9	7.1	7.1	7.1	7.0	7.2	7.8
	(40)	(43)	(42)	(43)	(44)	(45)	(45)	(45)	(37)	(37)

[a] Adjusted for hominids according to equation 4.10 in Sokal and Rolf (1981).

[b] Data from Kay (1982b). Kay does not provide sample sizes, but his data are derived from the large samples published by Mahler (1973) and Swindler (1976) and are adjusted to balance sex compositions.

[c] Data from Johanson (1974); minimum N for any position is 22, but samples are generally weighted in favor of females.

[d] Data from Wolpoff (1971) for the Dickson Mound sample; minimum N for any position is 89.

[e] Data from Gingerich and Schoeninger (1979).

catarrhine taxa and *Australopithecus*. On the basis of these data, no *Australopithecus* species exhibits an unusually high CV for any postcanine tooth.[2] Examination of the CVs for first and second molar breadths in *Australopithecus* fails to reveal obvious taxonomic heterogeneity within each species sample, as we have stressed elsewhere (Kimbel et al., 1985; White, 1985) with respect to claims that *A. afarensis* craniodental remains represent more than one species (Olson, 1981, 1985).

In four of six molar positions (three of four when only M1 and M2 are considered) *A. africanus* is more variable than *A. robustus*, confirming Robinson's (1956:151) impression based on more limited samples. At 8.2, the CV for *A. africanus* M^2 breadth ($N = 30$) is the largest for any first or second molar position among the *Australopithecus* samples.[3] The extremes of the *A. africanus* ranges for these four molar breadths are represented by teeth from Sterkfontein Type Site (Member 4) and for M^2, the Sterkfontein range (13.7–18.3) completely engulfs that for the combined (Swartkrans + Kromdraai) *A. robustus* sample. Figure 11.1A is a frequency histogram of M^2 breadth for the *A. africanus* sample. This is the only clearly bimodal distribution among *Australopithecus* first and second molar breadth histograms (compare to Figs. 11.1B and 11.1C).[4] Indeed, if the Makapansgat specimens are removed from the sample, the class in which the mean value (15.8) falls disappears, resulting in nonoverlapping Sterkfontein distributions. It is unlikely

---

[2] High CVs for *A. boisei* should be viewed cautiously because of the very small samples at most tooth positions.

[3] The CV for mesiodistal length of M^2 (8.0) is also highest among the *Australopithecus* first and second molar samples.

[4] We are aware of the problems inherent in discerning bimodality in a histogram of this nature. However, varying interval definition fails to remove the bimodal structure of the *A. africanus* M^2 distribution.

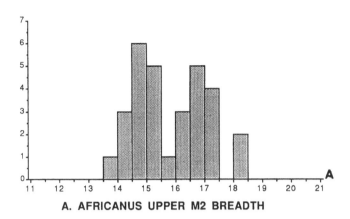
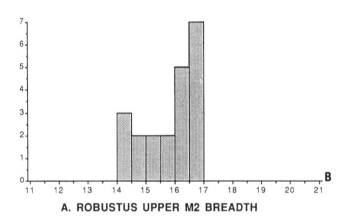
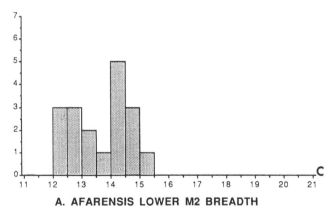

FIGURE 11.1. Frequency distributions for M2 breadth in *Australopithecus*. A, *A. africanus* M^2; B, *A. robustus* M^2; C, *A. afarensis* M_2. Abscissa scaled in 0.5mm intervals. See text for discussion.

that a nonoverlapping distribution of molar breadth can be attributed to sexual dimorphism (Pilbeam and Zwell, 1973).

Canine Breadth

Data on canine breadth variation in *Australopithecus* and the extant great apes are presented in Table 11.4. As a group, *Australopithecus* species exhibit less variation in canine breadth than *Pan, Gorilla* and *Pongo*. Because the degree of canine size variation in hominoids chiefly reflects sexual dimorphism, it is apparent that canine sexual dimorphism in *Australopithecus* is relatively reduced (see also Leutenegger and Shell, 1987).

Australopithecus afarensis has the most dimorphic canines ($CV = 12.1$ for mandibular teeth) of the hominid taxa considered here (disregarding *A. boisei*, the sample sizes for which are too small to produce meaningful CV values), although not markedly so.[5] *Australopithecus africanus* has the least dimorphic canines, especially in the mandible ($CV = 9.0$). Although the contrast between these species may reflect a real biologic difference, examination of the data in Table 11.4 and Fig. 11.2 suggests that the disparity is, at least in part, artificially inflated by unequal sample sizes. In the mandible, *A. afarensis* and *A. africanus* canine breadth mean and range are very similar (10.4, 8.8 to 12.4; 10.0, 8.7 to 12.0, respectively) but *A. africanus* is represented by twice as many teeth ($N = 22$ vs. $N = 11$). Figure 11.2A shows that the collection of *A. afarensis* mandibular canines does not sample specimens in the 10.5–11.4 range, resulting in the three

Table 11.4. Summary Statistics for *Australopithecus* and Great Ape Canine Breadths[a,b]

Taxon	N	$\bar{X} \pm SE$	SD	$CV \pm SE$
$\overline{C}$				
A. afarensis	11	10.4 ± .37	1.23	12.1 ± 2.60
A. africanus	22	10.0 ± .19	0.89	9.0 ± 1.35
A. robustus	9	8.5 ± .32	0.96	11.6 ± 2.74
A. boisei	3	8.9	—	—
Pan paniscus[c]	27	—	—	11.8
Pan troglodytes[d]	—	—	—	16.7
Gorilla gorilla[d]	—	—	—	19.1
Pongo pygmaeus[d]	40	—	—	18.6
$\underline{C}$				
A. afarensis	10	10.9 ± .35	1.11	10.5 ± 2.34
A. africanus	12	10.2 ± .29	1.01	10.1 ± 2.06
A. robustus	14	9.1 ± .25	0.92	10.3 ± 1.94
A. boisei	4	8.9 ± .56	1.11	13.3 ± 4.69
Pan paniscus[c]	26	—	—	13.6
Pan troglodytes[d]	—	—	—	15.5
Gorilla gorilla[d]	—	—	—	19.9
Pongo pygmaeus[d]	40	—	—	14.6

[a] N = number of individuals; measurements for *Australopithecus* are those described by White (1977).
[b] Adjusted for hominids according to equation 4.10 in Sokal and Rolf (1981).
[c] Data from Almquist (1974); mandibular $N = 15$ females and 12 males; maxillary $N = 16$ females and 10 males.
[d] Data from Kay (1982b); see Table 11.3, Footnote a.

---

[5] The CV for *A. afarensis* lower canine length is much higher than for breadth. As noted by Kay (1982a), this discrepancy is due to the very small number of teeth ($N = 5$) for which length can be measured.

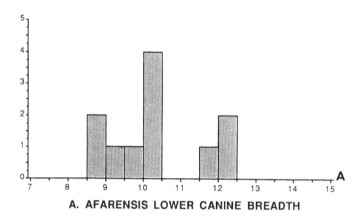

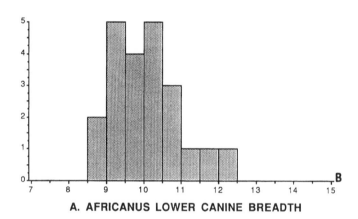

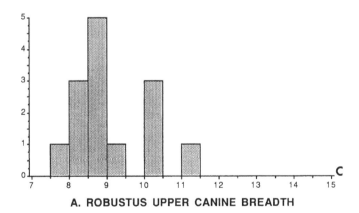

FIGURE 11.2. Frequency distributions for canine breadth in *Australopithecus*. A, *A. afarensis* C̄; B, *A. africanus* C̄; C, *A. robustus* C̱. Abscissa scaled in 0.5mm intervals. See text for discussion.

largest teeth having a disproportionate effect on the standard deviation.[6] A similar phenomenon affects the *A. robustus* mandibular canine data. In the maxillary canine comparisons (Table 11.4), the sample sizes for the three species are more nearly equivalent, and the *CV* values do not differ appreciably.

Pilbeam and Zwell (1973) have emphasized that in sexually dimorphic primate species the distribution of dental variates is either sexually overlapping (nondiscriminating) or sexually nonoverlapping (discriminating). The distribution of canine size in *Gorilla* is sexually nonoverlapping (Pilbeam and Zwell, 1973; Wolpoff, 1975; Kay, 1982a), in *P. troglodytes* it is sexually overlapping, but bimodal (Wolpoff, 1975; Oxnard *et al.*, 1985), and in modern humans it is sexually overlapping and unimodal (Wolpoff, 1975). Where samples are adequate, frequency distributions generated from the early hominid data indicate sexual overlap. The *A. africanus* mandibular canine distribution (Fig. 11.2) closely resembles that presented by Wolpoff (1975: fig. 3) for modern humans (Libben Amerindians), where distinct, but overlapping, male and female distributions produce a unimodal combined distribution. Bimodality cannot be ruled out in other examples (e.g., Figs. 11.2A and 11.2C), but a doubling or trebling of sample sizes is critical to a more precise characterization of the degree of canine dimorphism in *A. afarensis*, *A. robustus* and *A. boisei*.

Mandible Corpus Breadth, Height and Robusticity[7]

Data on mandible corpus size and robusticity (we use the term "robust" specifically in reference to corpus proportions, not size) in *Australopithecus*, *P. troglodytes* and *Gorilla* are presented in Table 11.5. In all *Australopithecus* species the *CV*s for breadth, height, and robusticity fall either below or within the range for the two extant African apes.

The *CV* for corpus breadth slightly exceeds that for height in *Australopithecus*, whereas the opposite relationship obtains in *P. troglodytes* and *Gorilla*. Following Wood (1978) and Kay *et al.* (1981), we interpret the relatively higher *CV* for corpus height in the apes as a function of large, sexually dimorphic canines, with longer canine roots effecting a generally deeper corpus in males (Smith, 1983). This is particularly true in *Gorilla*, where the large degree of corpus height variation ($CV = 12.3$) parallels especially strong canine size and body weight dimorphism (Leutenegger, 1982).

Among the *Australopithecus* species, only *A. afarensis* and *A. boisei* are represented by sufficiently large samples to provide a basis for comparison. These taxa are equally variable in corpus breadth (*A. afarensis* $CV = 11.5$, $N = 12$; *A. boisei* $CV = 11.7$, $N = 19$), but *A. boisei* exhibits greater corpus height variation (*A. afarensis* $CV = 10.5$, $N = 10$; *A. boisei* $CV = 11.5$, $N = 18$).[8] As measured by the ratio of extreme values (max./min. × 100), *A. boisei* emerges as the most variable of the hominoid taxa under consideration (see Table 11.5). Chamberlain and Wood (1985) also note the extreme degree of corpus size variation in *A. boisei*, and this variation will increase when the diminutive KNM-WT 16005 specimen is included (Walker and Leakey this volume, Chapter 15).

Data on mandible robusticity (breadth/height × 100, Table 11.5) indicate that in the African apes the corpus at M1 is twice as deep as it is broad. Within the ape samples there is virtually no mean sex difference in corpus robusticity, and frequency distributions are unimodal (see also Wood, 1976). In these species the male canine root strongly influences corpus proportions. As a consequence, the largest male mandibles tend to be significantly less robust than smaller female mandibles whose height is relatively small.

Compared to the apes, *Australopithecus* mandibles are more robust, with robusticity increasing

---

[6]These three teeth are all from Hadar; as shown by White (1985), Laetoli specimens define the high end of the *A. afarensis* range in mandibular canine length and maxillary canine breadth and length.

[7]Our measure of corpus breadth is *minimum* breadth, as defined by White (in Leakey and Leakey, 1978:172). Chamberlain and Wood (1985) do not define their technique, but discrepancies between the data sets suggest a different method was used by these authors.

[8]Our measurements correct for expansion cracking affecting some mandibles of *A. boisei* (see Leakey and Leakey, 1978:172–173).

Table 11.5. Mandible Breadth, Height, and Robusticity in *Australopithecus* and Extant Hominoids[a]

Taxon	N	$\bar{X} \pm$ SE	SD	$CV \pm$ SE	max/min $\times$ 100
A. afarensis					
Breadth (br)	12	19.0 ± 0.63	2.2	11.7 ± 2.40	149.4
Height (ht)	10	32.9 ± 1.10	3.4	10.5 ± 2.34	135.2
br/ht × 100	10	57.8 ± 2.00	6.3	11.2 ± 2.50	142.4
A. africanus					
Breadth	6	21.5 ± .90	2.2	10.5 ± 3.04	125.0
Height	4	36.1 ± .85	1.7	5.0 ± 1.76	110.6
br/ht × 100	4	59.8 ± .30	5.9	10.4 ± 3.68	121.8
A. robustus					
Breadth	7	24.5 ± 0.94	2.5	10.3 ± 2.77	133.6
Height	5	39.0 ± 1.30	3.0	8.0 ± 2.56	121.9
br/ht × 100	5	62.5 ± 2.10	4.6	7.6 ± 2.43	121.5
A. boisei					
Breadth	19	27.6 ± 0.73	3.2	11.7 ± 1.89	168.0
Height	18	43.1 ± 1.10	4.9	11.5 ± 1.91	164.1
br/ht × 100	17	64.6 ± 1.20	4.9	7.7 ± 1.32	132.3
Pan troglodytes[b]					
Breadth	20	14.2 ± 0.29	1.3	9.1 ± 1.44	132.2
Height	20	28.8 ± 0.60	2.7	9.4 ± 1.49	143.8
br/ht × 100	20	49.5 ± 1.20	5.5	11.2 ± 1.78	158.1
Gorilla gorilla[b]					
Breadth	20	19.3 ± 0.50	2.2	11.7 ± 1.85	155.6
Height	20	38.2 ± 1.00	4.6	12.3 ± 1.93	153.7
br/ht × 100	20	50.9 ± 1.00	4.7	9.3 ± 1.47	146.2

[a] Measured at M_1; measurement technique of White (in Leakey and Leakey, 1978).
[b] N = 10 males and 10 females.

in the series *A. afarensis–A. africanus–A. robustus–A. boisei*, as quantified by Chamberlain and Wood (1985). According to our data there is a fairly close relationship between corpus height and breadth in *A. boisei*, and, thus, the largest mandibles (e.g., probable males such as KNM-ER 818 and Omo L7a-125) do not exhibit the ape-like tendency to be among the least robust specimens (see also Chamberlain and Wood, 1985). In *A. afarensis* the relationship between corpus dimensions is not as strong (reflected in the higher robusticity index *CV*). Indeed, one of the deepest *A. afarensis* mandibles (AL 277-1) is the *least* robust specimen in the entire sample. Thus, compared to *A. boisei*, *A. afarensis* appears to be more ape-like in the relation of corpus dimensions to robusticity. This difference is possibly related to the much larger canines in *A. afarensis* (see Table 11.4).

Variation in Facial Morphology

In his 1956 monograph, John Robinson contrasted variation in facial prognathism in *A. africanus* (Sts 5) and *A. robustus* (SK 46) (his *Paranthropus*). Although Sts 5 was noted to be much more prognathic than SK 46 (as determined by the angle between the prosthion–nasion line and the Frankfurt Horizontal), Robinson (1956:14) stressed that:

> This should not be taken as a clear indication of a difference between the two genera, for, while it is true that none of the *Paranthropus* specimens now known has anything like as high a degree of prognathism as Sts 5, it is also true that Sts 5 is the most prognathous specimen of *Australopithecus* at present known. The least prognathous specimens of the latter form have much the same degree of prognathism as the general run of *Paranthropus* specimens . . . Visual comparison of some half-dozen specimens

of *Australopithecus* with about twice that number of *Paranthropus* indicates that the range of variation in the former is greater than in the latter and that the most prognathous *Paranthropus* specimen is appreciably less prognathous than the *Australopithecus* equivalent . . . Sts 5 is the most prognathous *Australopithecus* at present known, while SK 46 is approximately average for *Paranthropus*, but the most prognathous specimen does not differ greatly from it.

Among specimens of *A. africanus*, the relatively orthognathic Sts 71 and Sts 52a provide the most dramatic contrasts with Sts 5 (Fig. 11.3). The angle between the sellion–prosthion line and the Frankfurt Horizontal is 53° in Sts 5 and 65° in Sts 71 (Kimbel *et al.*, 1984). Although the Frankfurt Horizontal cannot be established for Sts 52a, the angle between the sellion–prosthion line and the line tangent to the external postcanine alveolar rim is 69° in this specimen.[9] The 15° range for these three specimens is large, but can be matched in pooled-sex samples of great apes (Kimbel *et al.*, 1984). Only two (patently male) *A. boisei* crania can be compared metrically: OH 5 and KNM-ER 406. The Olduvai specimen is the most orthognathic *Australopithecus* cra-

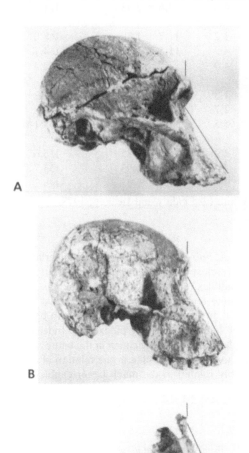

FIGURE 11.3. Variation in facial prognathism in *A. africanus*. A, Sts 5; B, Sts 71; C, Sts 52a. Diagonal line connects sellion to prosthion; vertical line passes through sellion. 1/3×. See text for discussion.

[9]We use the external alveolar rim as a reference for Sts 52a since this line essentially parallels the Frankfurt Horizontal in Sts 5 and Sts 71. Judging by extant hominoids, the incompletely erupted M3s in Sts 52 should not unduly bias measures of facial prognathism, especially since the permanent canines are in full occlusion.

nium (75°), but KNM-ER 406 (65°) is at least as prognathic as two of three *A. africanus* specimens.

We estimate the prognathism in KNM-WT 17000 (Walker *et al.*, 1986) at 45° based on the craniogram made available by Alan Walker (see Walker and Leakey, this volume, chapter 15: fig. 15.5). This probable male specimen is the most prognathic *Australopithecus* cranium presently known. If it is considered to belong to *A. boisei*, the resulting 30° range of variation is approximately double that observed among the extant great apes (Kimbel *et al.*, 1984).

Rak (1983: table 3) determined the proportion of the palate that projects anterior to sellion in *Australopithecus*. His results show the difference in prognathism between OH 5 (35.4%) and KNM-ER 406 (53.4%). Out estimate for KNM-WT 17000 is 65%, nearly identical to the great ape mean (Rak, 1983). Rak's index also demonstrates the great disparity between Sts 5 (67.7%) and Sts 71 (42.6%), and our estimate for Sts 52a of 40% confirms that this specimen is the least prognathic *A. africanus* cranium. It is noteworthy that when the projection of the palate is judged relative to the position of the anterior masseter origin (Rak, 1983), the difference between OH 5 (27.8%) and KNM-ER 406 (26%) vanishes because the masseter origin has a relatively more anterior position in the more prognathic KNM-ER 406. In this comparison *A. africanus* specimens remain highly variable [Sts 5 = 64.5%; Sts 71 = 47.1%; Sts 52a = 50% (our estimate)] owing to marked variation in the position of the zygomatic process of the maxilla (Kimbel *et al.*, 1984: table 4), independent of prognathism. Our estimate for KNM-WT 17000 is 40%, which is much more similar to the *A. robustus* mean of 37.7% than to *A. boisei*.[10]

Following the great ape model, under the hypothesis that the differences in facial prognathism in *A. africanus* are attributable to sexual dimorphism, we should expect females to be less prognathic than males (Schultz, 1962; Wood, 1976; Kimbel *et al.*, 1984). Broom *et al.* (1950) viewed both Sts 5 and Sts 71 as females. Based on the size of the canine (or its alveolus), Wolpoff (1975) listed Sts 5 as a female and both Sts 71 and Sts 52 as males. Wallace (1972) identified Sts 71 as the occlusal partner of the large mandible Sts 36, reinforcing its male attribution. He also claimed that Sts 52 is a male, based on the less-than-secure criterion of canine eruption timing.

In view of the relatively reduced canine dimorphism in *A. africanus* (see above), it is unlikely that canine size will prove to be a reliable guide to sexing many members of this species. On the basis of facial robusticity and postcanine tooth size, we view Sts 71 as a male. But if Sts 5 is a female, as is commonly thought, then the difference in facial prognathism between these specimens is opposite that which characterizes the sexes in the great apes. If Sts 52 is a male, then this conflicting pattern becomes even more dramatic.

Rak (1983, 1985) has addressed the problem of sexing *A. africanus* crania from a different perspective. He identified in *A. africanus* derived facial characters (anterior pillars and a flat nasoalveolar clivus combining to form the "nasoalveolar triangular frame") and stated that "those specimens in [the *A. africanus*] hypodigm which exhibit fewer structures interpreted as derived are regarded . . . as females" (Rak, 1985: 233). On this basis Rak identified TM 1512, Sts 52 and Sts 53 as females; by implication, the more strongly derived Sts 5, Sts 17, Sts 71, Stw 13 and TM 1511 are regarded as males. The females, in Rak's view, retain more primitive aspects of facial morphology and, thus, in the absence of the nasoalveolar triangular frame, tend to resemble *A. afarensis* and the great apes.

Rak is probably correct in labeling TM 1512 and Sts 53 as females (they are, in palate and tooth size, among the smallest *A. africanus* individuals), but using the presence or absence of discrete, derived structures to sex larger specimens such as Sts 5, Sts 17 and Sts 52a is problematic. Within Rak's male group, variation in facial morphology ("dishing" and maxillary prognathism) is considerable, as the comparison between Sts 5 and Sts 71 illustrates. Moreover, in the great apes sexual dimorphism in facial size is not accompanied by such distinct structural differences as distinguish Rak's male and female morphs of *A. africanus*. The massiveness of the male canine root, a strong influence on facial dimorphism in the apes, is not a major factor in *A. africanus* because of reduced canine dimorphism and the apparent independence of the anterior pillar from the canine jugum (Rak, 1983). It should also be noted that if the bimodal distribution of maxillary

---

[10]Rak's (1983) estimates of these indices for KNM-ER 732 depend on a complete reconstruction of the subnasal region and anterior dental arch and, thus, should be used cautiously.

M2 breadth shown in Fig. 11.1A is taken as a function of sexual dimorphism, then both males (TM 1511, Sts 17, Stw 13) *and* females (Sts 52a, Sts 53) cluster around the "female" mode.

In its markedly uneven distribution of derived facial morphology, to a large extent independent of size, *A. africanus* stands in strong contrast to the other species of *Australopithecus*. In *A. boisei* and *A. afarensis,* as in other anthropoids (Schultz, 1962; Wood, 1976), size is the principal basis of sexual variation in facial form (Rak, 1983: tables 1 and 2; Kimbel *et al.,* 1982; Wood, 1985a). Thus, in our view it is unlikely that the total variation in facial morphology in the Sterkfontein Type Site collection is attributable to sexual dimorphism.

Variation in the Cranial Base

A number of investigators have recently turned to the cranial base of *Australopithecus* as a source of phylogenetic information (Du Brul, 1977; Olson, 1981, 1985; Dean and Wood, 1982; Falk and Conroy, 1983; Picq, 1985; Kimbel *et al.,* 1984; Kimbel, 1986). Olson's phylogenetic analysis of the mastoid region and Falk and Conroy's conclusions based on cranial venous flow patterns have been critically examined in other publications (Kimbel, 1984; Kimbel *et al.,* 1985) and will not be addressed here.

In its narrow base, sagittally oriented petrous bones and transversely aligned tympanic plates, *A. africanus* specimen Sts 5 resembles the extant great apes (Dean and Wood, 1982). Dean and Wood note variation within the *A. africanus* sample, contrasting the longer base and more sagittally inclined petrous in Sts 5 with the opposite, more derived conditions in Sts 19. The "robust" *Australopithecus* species show a mixed pattern of derived and primitive states. The base is generally wider and shorter than in the apes. The tympanic and petrous axes are more coincident, and the latter more coronally aligned, than in the apes. In all of these features, putative females of *A. boisei* (KNM-ER 407 and KNM-ER 732) are firmly united with "robust" specimens such as TM 1517, OH 5, and KNM-ER 406.

In a qualitative assessment of glenoid morphology, we described *A. afarensis* as primitive relative to other hominids (White *et al.,* 1981; Kimbel *et al.,* 1984; Kimbel, 1986). A summary of the major features of this primitive pattern includes a shallow mandibular fossa with a low, weakly inclined articular eminence; a massive, inferiorly projecting entoglenoid process; a large postglenoid process, which lies completely anterior to a shallow, tubular tympanic plate; and a sagittally oriented petrous temporal. In *A. robustus, A. boisei* and *Homo* accentuated mandibular fossa topography is produced by a more pronounced, steeply inclined articular eminence and a deeper, concave, and more vertically oriented tympanic plate, the superior margin of which partially or entirely merges (in basal view) with a variably reduced postglenoid process. Several of these observations have been verified by Picq (1985).

The *A. boisei* glenoid region possesses additional derived conditions which distinguish it from that of other *Australopithecus* species, including *A. robustus* (Du Brul, 1977; Kimbel, 1986). Thus, in OH 5 and KNM-ER 406 the preglenoid plane is very restricted due to a combination of tight postorbital constriction and a mandibular fossa that projects nearly entirely lateral to the brain case. The articular eminence, which is extremely steep, is truncated abruptly by the temporal foramen, and it "twists" about its transverse axis to face posteriorly. The entoglenoid process itself projects backward to overlap the medial section of the tympanic plate, producing what Du Brul (1977) called the "medial glenoid plane." This morphological pattern is clearly evident in KNM-ER 407 and the less well preserved KNM-ER 732 (Fig. 11.4), substantiating the attribution of these specimens to *A. boisei*. Interestingly, KNM-WT 17000 exhibits none of these specializations and, indeed, resembles *A. afarensis* in several aspects (see also Kimbel *et al.,* this volume, chapter 16).

The glenoid region of *A. africanus* is highly variable. Tympanic plate morphology ranges from the primitive (*A. afarensis*-like) shallow and tubular form in Sts 25 to the strongly derived deep and completely vertical condition in Sts 19. The vertical tympanic and the postglenoid process are coplanar in Sts 19; but MLD 37/38 and Sts 5 are intermediate, with anteroposteriorly compressed plates, which, while vertically inclined, recede behind the postglenoid processes (Fig. 11.4). In Sts 19 the reduced sphenoid body length, the coronally aligned petrous axis (Dean and

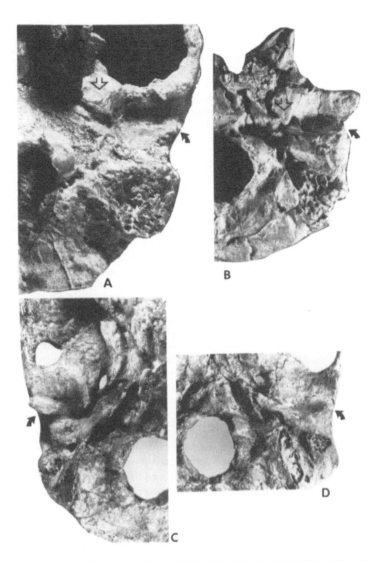

FIGURE 11.4. Basal views of *A. boisei* (A, KNM-ER 406; B, KNM-ER 407) and *A. africanus* (C, Sts 5; D, Sts 19) aligned on the spheno-occipital synchondrosis. Solid arrows indicate position of postglenoid process and tympanic plate; open arrows on (A) and (B) mark "medial glenoid plane." Despite sex-related size differences, KNM-ER 406 and KNM-ER 407 are very similar morphologically. Sts 5 and Sts 19 are equivalent in size, but differ considerably in morphology. Note difference in cranial base length and relationship of tympanic to postglenoid process. 2/3×. See text for additional details.

Wood, 1982), and the large flange-like vaginal process of the styloid (Clarke, 1977) combine with tympanic plate form and orientation to produce a remarkably humanlike glenoid morphology (as initially noticed by Broom *et al.*, 1950) that is not encountered in any other *A. africanus* specimen. The resulting range of variation in the *A. africanus* cranial base is greater than that encountered in other hominoids. Relative conservatism in this part of the cranial base makes it difficult to attribute such variation to sexual dimorphism.

Discussion and Conclusions

Variation and Taxonomy of Australopithecus afarensis

Our results do not suggest species-level taxonomic heterogeneity in the Hadar and Laetoli collections of *A. afarensis*. In postcanine tooth size, mandible corpus size and proportions, and facial morphology, *A. afarensis* is no more variable than extant hominoids that exhibit moderate to large amounts of sexual dimorphism and is, in many respects, less variable than *A. africanus*. Detailed study of the *A. afarensis* dental samples reveals no significant metric or morphological criteria that distinguish the Hadar and Laetoli hominids taxonomically; directional change is thus not discernible over the species' ca. 0.5-Myr temporal span (White, 1985). Olson's (1981, 1985) repeated challenges to the single species hypothesis have now been thoroughly rebutted (see also Kimbel *et al.*, 1985).

Our research shows that *A. afarensis* retains the apelike tendency for some large (male) mandibles to be relatively "gracile." This is not the case in the large *A. boisei* sample, as already discussed by Chamberlain and Wood (1985). While canine size dimorphism in *Australopithecus* is less than in the extant great apes, the absolutely much larger canine crowns and roots in *A. afarensis* males (as compared to *A. boisei* males) may be an important influence on mandible corpus proportions.

Although postcranial variation is not addressed here, we emphasize that some investigators who have assessed variation in the *A. afarensis* postcranium (Zihlman, 1985; McHenry, 1986) employed methods that inflate estimates of sexual dimorphism (see Jungers, this volume, chapter 7 concerning estimates of dimorphism from postcranial measurements). More precise techniques for estimating dimorphism in fossil hominid postcranial samples are currently under investigation by Lovejoy and colleagues. Their results will be of great interest in view of the finding that reduced canine size dimorphism characterizes the earliest known definitive hominids.

Variation and Taxonomy of Australopithecus robustus

On the basis of postcanine dental size and facial morphology, we find no clear evidence of taxonomic variation within the combined Swartkrans/Kromdraai sample of *A. robustus*. When the Swartkrans dental sample (including new SKW and SKX specimens) is analyzed separately, *CV*s for postcanine tooth breadths are only slightly less than for the pooled sample (Grine, pers. comm.), but Jungers and Grine (1986) and Grine (this volume, chapter 14) argue that subtle metrical differences separate the Kromdraai and Swartkrans samples. Moreover, Grine (1985) has described morphological details in the deciduous dentition to support separate specific designations for the Kromdraai and Swartkrans hominid assemblages (Howell, 1978). Although the Kromdraai sample is small, it is significant that the deciduous canines and molars lack specializations that unite the Swartkrans homologues with *A. boisei*. If these distinctions withstand the test of further study and additional discoveries, then overlapping metrical variation may mask important underlying biologic differences between the Swartkrans and Kromdraai site samples (see also Grine and Martin, this volume, chapter 1). Tattersall's (1986) comment regarding the broad morphological overlap between closely related species is relevant here.

Variation and Taxonomy of Australopithecus boisei

Despite apparent size and robusticity differences, males and females of *A. boisei* are united by highly derived facial morphology (Rak, 1983) and by a set of specializations in the glenoid region first identified by Du Brul (1977) on OH 5. The relatively low degree of morphological variation in the very small available sample of *A. boisei* crania contrasts with the large amount of size variation in the mandibles of this species, a pattern that would be expected in a highly dimorphic hominoid such as *Gorilla* or *Pongo*. The major departure from this common pattern (aside from reduced canine size dimorphism) is the trend toward increased robusticity in large (male) mandibles, as discussed above.

These observations are relevant to discussions regarding KNM-WT 17000, a 2.5-Myr-old *Australopithecus* cranium from west of Lake Turkana, Kenya. Walker *et al.* (1986) and Walker and Leakey (this volume, chapter 15) regard this specimen as an early *A. boisei*. However, it is clear that KNM-WT 17000 lacks most of the specializations in the face and cranial base that characterize the *A. boisei* hypodigm (Kimbel *et al.*, this volume, chapter 16). To subsume KNM-WT 17000 in *A. boisei* would so elevate the level of morphological variation in this species that it could no longer be defined by the uniquely derived character states that distinguish it from *A. robustus*. Primarily for this reason, we regard KNM-WT 17000 as a distinct species, *A. aethiopicus* (see Kimbel *et al.*, this volume, chapter 16 for additional details). Although the phylogenetic position of KNM-WT 17000 is still a topic of debate, the assumption that this specimen is an early member of the *A. boisei lineage* poses no obstacle to recognizing it as representing a separate *species* (Rose and Bown, 1986).

Variation and Taxonomy of Australopithecus africanus

Among the taxa discussed here, *A. africanus* presents the most unusual pattern of variation, nearly all of which is contained in the Sterkfontein sample. Canine size dimorphism is low, but maxillary M2 breadth values divide into two essentially nonoverlapping clusters. Structural variation in the face is greater than is the rule even in highly size-dimorphic hominoids. Morphological variation in the cranial base, which is notably nondimorphic in primates, is excessive. Most important, structural variation among *A. africanus* crania is of the order used to distinguish *between* species of early hominids. To us it appears inescapable that such variation cannot be attributed to sexual dimorphism.

We suggest that the nature and extent of variation in the skull of *A. africanus* is responsible for the observation that this species possesses "few, if any autapomorphies" (Wood, 1985b: 228). Moreover, it has been argued that *A. africanus* is a species whose phylogenetic position is obscure (Kimbel *et al.*, this volume, chapter 16). We further suggest that the basis of the wide range of variation currently attributed to *A. africanus* is one of two, not necessarily mutually exclusive, hypotheses: (1) the Sterkfontein Type Site (Member 4) assemblage samples representatives of temporally mixed populations of an evolving lineage; (2) this assemblage contains more than one hominid species.

Clarke (1985) has alluded to the considerable time span over which the Member 4 talus cone may have accumulated. This is supported by the presence in Member 4 of *Equus* together with archaic bovids, suggesting that the *A. africanus*-bearing Type Site deposit may contain a temporally mixed faunal assemblage (Vrba, 1982). Variation in the hominid sample may reflect the same phenomenon, thus adding a chronological dimension to Clarke's (this volume, chapter 18) morphologically-based proposal that the Sterkfontein Member 4 hominid sample contains more than one species.

While evaluation of the details of Clarke's (this volume, chapter 18) arguments are beyond the scope of this paper, our results indicate that they merit serious consideration. It is evident that if there are two hominid species in the Sterkfontein Member 4 deposit, they are very similar morphologically and may be difficult to define unequivocally, at least at present. But the inability to adequately diagnose closely related species in practice in no way diminishes their reality in nature. As Tattersall (1986:168) has stated succinctly: "To brush morphological diversity under the rug of an all-encompassing species is simply to blind oneself to the complex realities of phylogeny."

Acknowledgments

Thanks go to Fred Grine for helpful discussion and Carl Swisher for valuable assistance on the Mac. Thanks to Fred Grine for inviting us to participate in the workshop at which this paper was initially read; and to the governments, institutions, and individuals providing access to the original fossils incorporated into this study.

References

Almquist, A. J. (1974). Sexual differences in the anterior dentition in African primates. *Amer. J. Phys. Anthropol., 30*:359-359-368.

Broom, R., Robinson, J. T. and Schepers, G. (1950). Sterkfontein ape-man *Plesianthropus. Tvl. Mus. Mem., 4*:1-117.

Chamberlain, A. T. and Wood, B. A. (1985). A reappraisal of variation in hominid mandibular corpus dimensions. *Amer. J. Phys. Anthropol., 66*:399-405.

Clarke, R. J. (1977). The cranium of the Swartkrans hominid SK 847 and its relevance to human origins. Ph.D. Thesis, University of the Witwatersrand, Johannesburg.

Clarke, R. J. (1985). Early Acheulean with *Homo habilis* at Sterkfontein. In P. V. Tobias (ed.), *Hominid evolution: past, present and future*, pp. 287-298. Alan R. Liss, New York.

Clutton-Brock, T. H. and Harvey, P. H. (1977). Primate ecology and social organization. *J. Zool. (London), 183*:1-39.

Dean, M. C. and Wood, B. A. (1982). Basicranial anatomy of Plio-Pleistocene hominids from East and South Africa. *Amer. J. Phys. Anthropol., 59*:157-174.

Du Brul, E. L. (1977). Early hominid feeding mechanisms. *Amer. J. Phys. Anthropol., 47*:305-320.

Falk, D. and Conroy, C. G. (1983). The cranial venous sinus system in *Australopithecus afarensis. Nature, 306*:779-781.

Gaulin, S. J. C. and Sailer, L. D. (1984). Sexual dimorphism in weight among primates: the relative impact of allometry and sexual selection. *Int. J. Primatol., 5*:515-535.

Gingerich, P. D. (1974). Size variability of the teeth in living mammals and the diagnosis of closely related sympatric fossil species. *J. Paleontol., 48*:895-903.

Gingerich, P. D. and Schoeninger, M. J. (1979). Patterns of tooth size variability in the dentition of primates. *Amer. J. Phys. Anthropol., 51*:457-466.

Grine, F. E. (1985). Australopithecine evolution: the deciduous dental evidence. In E. Delson (ed.), *Ancestors: the hard evidence*, pp. 153-167. Alan R. Liss, New York.

Harvey, P. H. and Bennett, P. M. (1985). Sexual dimorphism and reproductive strategies. In J. Ghesquiere, R. D. Martin, and F. Newcombe (eds.), *Human sexual dimorphism*, pp. 43-59. Taylor and Francis, London.

Howell, F. C. (1978). Hominidae. In V. J. Maglio, and H. B. S. Cooke (eds.), *Evolution of African mammals*, pp. 154-248. Harvard University Press, Cambridge, MA.

Johanson, D. C. (1974). An odontological study of the chimpanzee with some implications for hominoid evolution. Ph.D. Dissertation, University of Chicago, Chicago, IL.

Jungers, W. L. and Grine, F. E. (1986). Dental trends in the australopithecines: the allometry of mandibular molar dimensions. In B. A. Wood, L. B. Martin and P. Andrews (eds.), *Major topics in primate and human evolution*, pp. 203-219. Cambridge University Press, Cambridge.

Kay, R. F. (1982a). Sexual dimorphism in Ramapithecinae. *Proc. Natl. Acad. Sci. U.S., 79*:209-212.

Kay, R. F. (1982b). *Sivapithecus simonsi*, a new species of Miocene hominoid, with comments on the phylogenetic status of the Ramapithecinae. *Int. J. Primatol., 3*:113-173.

Kay, R. F., Fleagle, J. G. and Simons, E. L. (1981). A revision of the Oligocene apes of the Fayum Province, Egypt. *Amer. J. Phys. Anthropol. 55*:293-322.

Kimbel, W. H. (1984). Variation in the pattern of cranial venous sinuses and hominid phylogeny. *Amer. J. Phys. Anthropol., 63*:243-263.

Kimbel, W. H. (1986). Calvarial morphology of *Australopithecus afarensis:* a comparative phylogenetic study. Ph.D. Dissertation, Kent State University, Kent, OH.

Kimbel, W. H., Johanson, D. C. and Coppens, Y. (1982). Pliocene hominid cranial remains from the Hadar Formation, Ethiopia. *Amer. J. Phys. Anthropol., 57*:453-499.

Kimbel, W. H., White, T. D. and Johanson, D. C. (1984). Cranial morphology of *Australopithecus afarensis:* a comparative study based on a composite reconstruction of the adult skull. *Amer. J. Phys. Anthropol., 65*:337-388.

Kimbel, W. H., White, T. D., and Johanson, D. C. (1985). Craniodental morphology of the hominids from Hadar and Laetoli: evidence of *"Paranthropus"* and *Homo* in the mid-Pliocene of eastern Africa? *In* E. Delson (ed.), *Ancestors: the hard evidence*, pp. 120–137. Alan R. Liss, New York.

Leakey, R. E. and Leakey, M. G. (eds.) (1978). *The fossil hominids and an introduction to their context, 1968–1974*. Koobi Fora Research Project, Vol. 1. Clarendon Press, Oxford.

Leutenegger, W. (1982). Scaling of sexual dimorphism in body weight and canine size in primates. *Folia primatol.*, 37:163–176.

Leutenegger, W. and Chevrud, J. (1982). Correlates of sexual dimorphism in primates: ecological and size variables. *Int. J. Primatol.*, 3:387–402.

Leutenegger, W. and Shell, B. (1987). Variability and sexual dimorphism in canine size of *Australopithecus* and extant hominoids. *J. Hum. Evol.*, 16: 359–367.

McHenry, H. M. (1986). Size variation in the postcranium of *Australopithecus afarensis* and extant species of Hominoidea. *Hum. Evol.*, 1:149–156.

Mahler, P., (1973). Metric variation in the pongid dentition. Ph.D. Dissertation, University of Michigan, Ann Arbor.

Olson, T. R. (1981). Basicranial morphology of the extant hominoids and Pliocene hominids: the new material from the Hadar Formation, Ethiopia and its significance in early human evolution and taxonomy. *In* C. B. Stringer (ed.), *Aspects of human evolution*, pp. 99–128. Taylor and Francis, London.

Olson, T. R. (1985). Cranial morphology and systematics of the Hadar Formation hominids and *"Australopithecus" africanus*. *In* E. Delson (ed.), *Ancestors: the hard evidence*, pp. 102–119. Alan R. Liss, New York.

Oxnard, C. E., Lieberman, S. S. and Gelvin, B. R. (1985). Sexual dimorphisms in dental dimensions of higher primates. *Amer. J. Primatol.*, 8:127–152.

Pickford, M. (1986). On the origins of body size dimorphism in primates. *Hum. Evol.*, 1:77–90.

Picq, P. (1985). L'articulation temporo-mandibulaire d'*Australopithecus afarensis*. *C. R. Acad. Sci. (Paris)*, 300:469–474.

Pilbeam, D. R. and Zwell, M. (1973). The single species hypothesis, sexual dimorphism, and variability in early hominids. *Yrbk. Phys. Anthropol.*, 16:69–79.

Rak, Y. (1983). *The australopithecine face*. Academic Press, New York.

Rak, Y. (1985). Sexual dimorphism, ontogeny and the beginning of differentiation of the robust australopithecine clade. *In* P. V. Tobias (ed.), *Hominid evolution: past, present and future*, pp. 233–237. Alan R. Liss, New York.

Robinson, J. T. (1956). The dentition of the Australopithecinae. *Tvl. Mus. Mem.*, 9:1–179.

Rose, K. D. and Bown, T. M. (1986). Gradual evolution and species discrimination in the fossil record. Contrib. Geol., University of Wyoming, Spec. Paper 3, pp. 119–130.

Schultz, A. J. (1962). Metric age changes and sex differences in primate skulls. *Z. Morphol. Anthropol.*, 52:239–255.

Simpson, G. G., Roe, A. and Lewontin, R. C. (1960). *Quantitative zoology*. Harcourt, Brace and World, New York.

Smith, R. J. (1983). The mandibular corpus of female primates: taxonomic, dietary, and allometric correlates of interspecific variation in size and shape. *Amer. J. Phys. Anthropol.*, 61:315–330.

Sokal, R. R. and Rolf, F. J. (1981). *Biometry*. Freeman, New York.

Swindler, D. P. (1976). *Dentition of living primates*. Academic Press, New York.

Tattersall, I. (1986). Species recognition in human paleontology. *J. Hum. Evol.*, 15:165–175.

Vrba, E. S. (1982). Biostratigraphy and chronology, based particularly on Bovidae, of southern hominid-associated assemblages: Makapansgat, Sterkfontein, Taung, Kromdraai, Swartkrans; also Elandsfontein (Saldanha), Broken Hill (now Kabwe) and Cave of Hearths. *Prétirage, 1er Congr. Inter. Paléont. Humaine*, pp. 707–752. C.N.R.S., Nice.

Walker, A. C. and Leakey, R. E. (1978). The hominids of East Turkana. *Sci. Amer.*, 239:54–66.

Walker, A. C., Leakey, R. E. Harris, J. M. and Brown, F. H. (1986). 2.5-Myr *Australopithecus boisei* from west of Lake Turkana, Kenya. *Nature*, 322:517–522.

Wallace, J. A. (1972). The dentition of the South African early hominids: a study of form and function. Ph.D. Thesis, University of the Witwatersrand, Johannesburg.

White, T. D. (1977). New fossil hominids from Laetolil, Tanzania. *Amer. J. Phys. Anthropol.*, 46: 197–230.

White, T. D. (1985). The hominids of Hadar and Laetoli: an element-by-element comparison of the dental samples. *In* E. Delson (ed.), *Ancestors: the hard evidence*, pp. 138–152. Alan R. Liss, New York.

White, T. D., Johanson, D. C. and Kimbel, W. H. (1981). *Australopithecus africanus:* its phylogenetic position reconsidered. *S. Afr. J. Sci.*, 77:445–470.

Willner, L. A. and Martin, R. D. (1985). Some basic principles of mammalian sexual dimorphism. *In* J. Ghesquiere, R. D. Martin, and F. Newcombe (eds.), *Human sexual dimorphism*, pp. 1–42. Taylor and Francis, London.

Wolpoff, M. H. (1971). Metric trends in hominid dental evolution. *Case Western Reserve University Studies in Anthropology*, no. 2.

Wolpoff, M. H. (1975). Sexual dimorphism in the australopithecines. *In* R. Tuttle (ed.), *Paleoanthropology, morphology and paleoecology*, pp. 245–284. Mouton, The Hague.

Wood, B. A. (1976). The nature and basis of sexual dimorphism in the primate skeleton. *J. Zool. (London)*, 180:15–34.

Wood, B. A. (1978). Allometry and hominid studies. *In* W. W. Bishop (ed.), *Geological background to fossil man*, pp. 125–138. Scottish Academic Press, Edinburgh.

Wood, B. A. (1985a). Sexual dimorphism in the hominid fossil record. *In* J. Ghesquiere, R. D. Martin and F. Newcombe (eds.), *Human sexual dimorphism*, pp. 105–123. Taylor and Francis, London.

Wood, B. A. (1985b). A review of the definition, distribution and relationships of *Australopithecus africanus*. *In* P. V. Tobias (ed.), *Hominid evolution: past, present and future*, pp. 227–232. Alan R. Liss: New York.

Zihlman, A. L. (1985). *Australopithecus afarensis:* two sexes or two species? *In* P. V. Tobias (ed.), *Hominid evolution: past, present and future*, pp. 213–220. Alan R. Liss, New York.

On Variation in the Masticatory System of *Australopithecus boisei*

12

YOEL RAK

Several elements join to make the masticatory system of *Australopithecus boisei* a most specialized apparatus, in which the jaws are immensely powerful, the occlusal load is evenly distributed along longer segments of the dental arcade, and the occlusal contact between the postcanine teeth is simultaneous. These properties are achieved by the occlusal plane being positioned far beneath the transverse plane of the fulcrum of the lower jaw (i.e., the articular eminence), by the extension of the masticatory muscles away from that point, and by the retraction of the dental arcade closer to the coronal level of the eminence. The biomechanical advantage of these changes in the mandibular lever is simple, easily comprehensible from a mechanical point of view, and has been demonstrated experimentally (Ward and Molnar, 1980). The modifications are realized anatomically through features such as an extremely high, very wide mandibular ramus, a very long upper face, and a concave facial topography in which the central portion lies deeper than the periphery. In addition, the M3 is situated (hidden) behind the anterior margin of the ramus, close to the coronal plane of the articular eminence.

Australopithecus boisei is the most advanced of the australopithecines in attaining this level of masticatory specialization, and its topographic manifestation in the *A. boisei* face is, indeed, the most dramatic (Rak, 1983). Surprisingly, however, close examination of the admittedly small *A. boisei* sample reveals tremendous variation in the manner in which this goal is achieved; corresponding variation in facial topography is also observed, as expected from the close relationship between topography and the biomechanical properties of the masticatory system. The different biomechanical strategies that are employed by individuals of a species to achieve the same end are in themselves interesting as a phenomenon and shed more light on the nature of these fascinating early hominids. Furthermore, the strategies convey a significant message concerning the way we relate to so-called taxonomic characters and character states, especially in light of the increasing use of cladistic analysis, whether formal, such as in Skelton *et al.* (1986) regarding the phylogenetic and taxonomic status of *A. africanus,* or indirect, such as the presentation of a list of characters and character states and their comparison with those of other taxa, as exemplified by Walker *et al.* (1986) in reference to KNM-WT 17000. Recognition and appreciation of these rather extreme variations in the masticatory system of *A. boisei* are vital to any attempt to evaluate the unexpected mixture of features in KNM-WT 17000 or, for that matter, to clarify the taxonomy and phylogeny of the "robust" australopithecines.

Two specimens whose designation as *A. boisei* seems to be unquestionable are OH 5 from Olduvai and KNM-ER 406 from East Turkana. Their sex is universally agreed upon: both are mature males that died after the fusion of spheno-occipital synchondrosis, though OH 5 appears to have died younger than KNM-ER 406. Chronologically, both are about 1.8 Myr old, although KNM-ER 406 may be as much as 100,000 years younger. Examination of their masticatory system, however, reveals that the protrusion of the anterior dental arcade relative to the coronal plane of sellion (the deepest point in the mid-sagittal facial contour) is very modest in OH 5, with only 35.4% of the length of the dental arcade protruding beyond the coronal plane of sellion (Table 12.1, Fig. 12.1). KNM-ER 406, on the other hand, is more prognathous; at least 53.4% of the dental arcade length protrudes beyond the level of sellion. (The percentage value in KNM-ER 406 is based on a measurement of the length of the palate as far as the missing anterior piece.

Table 12.1. Comparison of Indices Expressing Relationships within the Masticatory System of the Gorilla, *Australopithecus robustus* and *Australopithecus boisei*[a]

	1 Index of palate position	2 Index of M3 position	3 Index of masseter position	4 Index of overlap
Gorilla (N=12)				
$\bar{X}$	66.8	59.7	90.7	16.7
SD	5.3	8.3	8.5	4.8
Range	57.8–78.0	48.0–75.6	78.7–105.8	8.0–24.1
Gorilla ♂ (N=6)				
$\bar{X}$	70.4	15.2	93.8	15.2
Range	64.3–78.0	56.9–75.6	82.1–105.8	8.0–17.3
Gorilla ♀ (N=6)				
$\bar{X}$	63.3	53.7	87.6	18.2
Range	57.8–66.3	48.0–60.6	78.7–96.9	11.1–24.1
A. robustus				
$\bar{X}$	45.6	40.4	119.3	44.1
Range	—	—	—	37.4–50.0
	(SK 48)	(SK 48)	(SK 48)	(SK 48, TM 1517, SK 52)
A. boisei (N=3)				
$\bar{X}$	43.3	38.3	113.3	49.9
Range	35.4–41.2	30.5–48.5	100.0–130.3	43.3–57.0
A. boisei ♂ (N=2)				
$\bar{X}$	44.4	39.5	120.0	53.0
Range	35.4–53.4	30.5–48.5	109.7–130.3	49.5–57.0
	(OH 5, KNM-ER 406)	(OH 5, KNM-ER 406)	(OH 5, KNM-ER 406)	(KNM-ER 406, OH 5)
A. boisei ♀ (N=1)				
$\bar{X}$	41.2	36.0	100.0	43.4
	(KNM-ER 732)	(KNM-ER 732)	(KNM-ER 732)	(KNM-ER 732)

[a] Index numbers 1–4 correspond to Figure numbers 12.1–12.4.

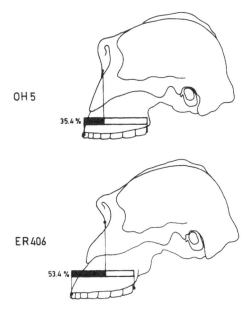

FIGURE 12.1. Index of prognathism; the segment of the palate protruding beyond the sellion is expressed as a percentage of palate length (prosthion–distal end of M3).

Reconstruction of the upper jaw will increase the value of the index, i.e., will make it slightly more prognathous.) This extreme protrusion (prognathism) exceeds that of the less specialized species *Australopithecus robustus*.

The discrepancy between the values of prognathism for the two *A. boisei* males (that of OH 5 is 34% less than that of KNM-ER 406) is found to be greater than between the mean values for male vs. female gorillas (the mean for females is 10% less than for males) and even greater than between the extreme values for the males and the females (the least prognathous female value is 26% lower than the most prognathous male value). Gorillas were used for comparison precisely because of their pronounced sexual dimorphism.

Because it is the whole dental arcade that is retracted, one should not be surprised to find corresponding differences between the two *A. boisei* specimens in the distance of M3 from the coronal plane of the articular eminence: short in OH 5 and very long in KNM-ER 406. The index value that expresses these distances is 37% less in the former than in the latter (Table 12.1, Fig. 12.2), whereas the average index value of female gorillas is 18.2% less than that of male gorillas.

The degree of the masseter's advancement varies as well in regard to sellion in the two specimens (Table 12.1, Fig. 12.3). It extends forward slightly in OH 5 and considerably in KNM-ER 406; the index value expressing masseter position is 16% less for OH 5 than for KNM-ER 406. The corresponding index for female gorillas is 6.6% less than for males.

However, what is really significant biomechanically is the position of the masseter relative to that of the palate (Table 12.1; Fig. 12.4). In spite of the enormous discrepancies in each of the components discussed above, the two *A. boisei* specimens reach almost the same level of biomechanical efficiency regarding the mandibular lever. The values expressing this index of overlap are 57 in OH 5 and 49.5 in KNM-ER 406; the latter is only 13% less than the former. The mean index for male gorillas stands at 15.2, where the lowest is 50% less than the highest.[1] In light of the very similar values of what is here called the index of overlap, it is clear that none of the other components constituting this overlap is of any taxonomic significance when considered at face value detached from the whole functional complex.

Certain modifications have far-reaching effects on other elements of the skull, some of which are well removed from the cause of the change itself. The affected features may be mistakenly added as independent taxonomic characters in the formal list of features. For example, the degree

---

[1] To demonstrate the similarity between the two *A. boisei* males, they are compared here to the extreme values of male gorillas only.

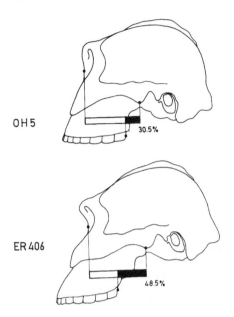

FIGURE 12.2. Index of M3 position whereby the distance of M3 from the coronal plane of the articular eminence is expressed as a percentage of the distance between the coronal plane of the articular eminence and sellion.

to which the masseter-bearing bone extends anteriorly influences the shape of the infraorbital bone plate. In KNM-ER 406, the extreme anterior position of the masseter necessitates the transformation of the bone plate into a most dramatic visorlike structure (Rak, 1983), whereas the masseter of OH 5 is less extended because of the posterior position of the dental arcade, which results in a much more modestly developed visor. Furthermore, the manner in which the infraorbital region merges with the body of the maxilla (the zygomatic process) is greatly affected by the degree to which the visor is developed (Rak, 1983; fig. 25). Hence, it is not strange to find that the extreme specialization of the masticatory system of OH 5 manifests itself in a relatively generalized infraorbital region and in a primitive-looking zygomatic process of the maxilla.

The advancement of the infraorbital region and the retraction of the dental arcade also influence other facets of facial topography. The conspicuous anterior pillars of *Australopithecus africanus* and *A. robustus* fade away topographically at least in the males of *A. boisei;* and the maxillary

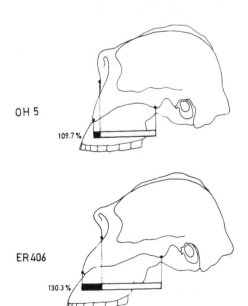

FIGURE 12.3. Index of masseteric position relative to the upper face, whereby the distance between the coronal plane of the articular eminence and the most anterior edge of masseter origin is expressed as a percentage of the distance between the coronal plane of the articular eminence and sellion.

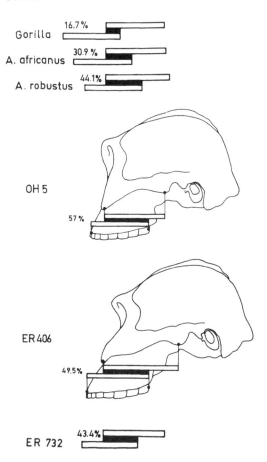

FIGURE 12.4. Index of overlap, expressing the degree of overlap between the palate length and the distance from articular eminence to zygomatic tubercle (most anterior edge of the masseteric origin).

fossula in *A. robustus*, which in itself constitutes a modification of the maxillary furrow typical of *A. africanus*, completely disappears in *A. boisei*.[2]

We may conclude, therefore, that it is solely in the index of overlap (because of its biomechanical implications) that the similarity between the two *A. boisei* specimens is apparent. This index legitimately permits us to assign them a place together at the extreme end of the morphocline as the most advanced stage in a sequence of masticatory specialization (Table 12.1, Fig. 12.4). The masticatory system of KNM-ER 406 is highly specialized biomechanically (as seen in the index of overlap) in comparison to that of *A. robustus*, although KNM-ER 406 is more prognathous than any available specimen of *A. robustus* (Rak, 1983). It is in the index of overlap that the female *A. boisei* specimen, KNM-ER 732, emerges as generalized when compared to the two males. In the other indices, female and male vary in their position in the morphocline. For example, according to the index of palate protrusion, KNM-ER 406 appears more prognathous than the female KNM-ER 732, while OH 5 is more orthognathous; the index of masseter position yields the sequence KNM-ER 406, OH 5, KNM-ER 732, revealing the former as considerably more specialized than the two others in this respect.

It is thus in the index of overlap that we identify the female, KNM-ER 732, as the most biomechanically generalized of the small *A. boisei* sample. The value for the female is actually

---

[2]From the way in which the zygomatic process flares in specimen KNM-ER 405 (which consists solely of a badly eroded *A. boisei* palate), it is apparent that *A. boisei* populations included some individuals with an even more dramatic infraorbital visor than is found in KNM-ER 406. This in itself serves as evidence that individuals with greater prognathism than KNM-ER 406 existed within *A. boisei*.

very similar to the mean value for the *A. robustus* sample (Fig. 12.4). It is not surprising, therefore, that the female exhibits the most generalized topography. KNM-ER 732 still maintains anterior pillars similar to those found in the face of *A. robustus* (Rak, 1983), whereas in the male *A. boisei* specimens they disappear with the increased specialization of the masticatory system, i.e., the extreme advancement of the infraorbital region relative to the palate.

It is not the purpose of this paper to discuss the phylogenetic or taxonomic status of specimen KNM-WT 17000, or for that matter, any other particular specimen or species of the genus *Australopithecus*. My intention in presenting these observations is merely to advocate the use of extreme caution—as many have done before—in taking anatomical features at face value and employing them as taxonomic characters detached from their functional complex. The fact that *Australopithecus afarensis* and KNM-WT 17000 are extremely prognathous, with the M3 far away from the articular eminence [two anatomical landmarks that are counted as distinct features by Walker *et al.* (1986), though they represent one trait], is still not sufficient to place the former taxonomically closer to the latter than to the orthognathous *A. robustus*. Similarly, there is no justification for using the simple fact that no pillar exists in the face of *A. boisei* males to declare their similarity in this feature to *A. afarensis* (Skelton *et al.*, 1986; Walker *et al.*, 1986). The word "no" ("no pillar") has a profoundly different meaning for *A. afarensis* than for *A. boisei*. The lack of pillars, therefore, cannot serve as a character state, since it is simply not homologous. Similarly, the "absence" of the maxillary fossula in the face of both KNM-WT 17000 and *A. afarensis* is of no significance; the determining issue is whether the *nature* of the absence in KNM-WT 17000 is the same as in *A. boisei* or as in *A. afarensis*. In *A. boisei* the advancement of the infraorbital plate is so extreme that there is no longer any trace of the fossula, whereas in *A. afarensis* the face is so generalized that the fossula has not yet formed. Similar examples abound in paleoanthropological literature. Through the analysis of the masticatory systems of *A. boisei* and the gorilla, I have attempted to demonstrate that an understanding of the complexity of anatomical features and the awareness of the intricate relationships between them are prerequisite to any taxonomic and phylogenetic analysis, particularly when the formal number of characters is of utmost importance and the principle of maximum parsimony is employed in determining our preference for one cladogram over another.

References

Rak, Y. (1983). *The australopithecine face*. Academic Press, New York.
Skelton, R. R., McHenry, H. and Drawhorn, G. (1986). Phylogenetic analysis of early hominids. *Curr. Anthropol.*, 27:21–43.
Walker, A., Leakey, R. E., Harris, J. M. and Brown, F. H. (1986). 2.5-Myr *Australopithecus boisei* from west of Lake Turkana, Kenya. *Nature, 322*:517–522.
Ward, S. C. and Molnar, S. (1980). Experimental stress analysis of topographic diversity in early hominid gnathic morphology. *Amer. J. Phys. Anthropol.*, 53:383–395.

Evolution of the "Robust" Australopithecines in the Omo Succession: Evidence from Mandibular Premolar Morphology

13

Gen Suwa

The Shungura and Usno Formations in the Omo Valley, Ethiopia, have yielded abundant but fragmentary hominid remains. The majority of these specimens occupy the time range 3 to 2 Myr BP and provide us with important glimpses of a little-known phase of hominid evolution in eastern Africa. In light of recent geochronological and stratigraphic work in the Turkana basin (Brown and Feible, 1986), over 97% of the hominid remains from the Omo succession can be interpreted to predate the hominid sample from the Upper Burgi Member (= "sub-KBS" levels) and higher stratigraphic levels at East Turkana. With the exception of the L 894-1 fragmentary cranium from upper Member G, all hominid specimens from Member G derive from levels below the lacustrine sequences. The chronological age of these specimens is greater than 2 Myr BP (Brown et al., 1985; Brown and Feibel, this volume, chapter 22).

The current hominid dental sample from the Omo succession consists of approximately 235 teeth (excluding antimeres) in approximately 210 specimens of which over 190 are isolated teeth. The majority of the specimens have been only briefly described or discussed (Howell, 1969a; Coppens, 1970, 1971, 1973a,b), listed in catalogs (Howell and Coppens, 1974; Boaz, 1977), or mentioned in summary articles (Howell and Coppens, 1976; Coppens, 1980). Well described dental specimens are confined to the deciduous teeth (Howell and Coppens, 1973; Grine, 1984), the majority of the specimens from the Usno Formation (Howell, 1969b), and those of the L 894-1 fragmentary cranium (Boaz and Howell, 1977). A handful of specimens have not been referred to in the literature. I am currently engaged in a systematic description and analysis of the Omo dental sample, and the present chapter focuses on mandibular premolar morphology.

Morphological differences between "robust" and "non-robust" australopithecines are more conspicuous in some dental elements than in others (e.g., Robinson, 1956). For example, the dm_1 is known to show a relatively large morphological difference among the australopithecine taxa (Robinson, 1956; von Koenigswald, 1967; Grine, 1984). The P_3 is another element that exhibits a considerable morphological "gap" between "robust" and "non-robust" australopithecines. The present paper outlines the suite of features that characterizes the "robust" australopithecine condition. An attempt is made to assign some mandibular specimens and isolated P_3s from the Omo succession to "robust" australopithecines. In the P_4s, contrary to the condition seen in P_3s, the most conspicuous morphological distinction occurs between *Australopithecus boisei* and all other specimens. *Australopithecus robustus* P_4s from Swartkrans and Kromdraai are relatively conservative compared to those of *A. boisei* and exhibit a greater amount of morphological overlap with "non-robust" specimens.

Materials and Methods

Mandibular premolars of the Omo collection are represented by 15 P_3s and 22 P_4s, not counting antimeres. These are summarized in Table 13.1 by stratigraphic level. As will be shown below, 15 of these specimens from the Shungura Formation are considered to represent "robust" australopithecine remains. Of these specimens, seven were included in the "non-robust" austral-

Table 13.1. Mandibular Premolars of the Omo Hominid Dental Collection[a]

Stratigraphic level	Isolated dentition		In situ in mandibles		Mandibular specimen numbers
	P_3	P_4	P_3	P_4	
Member H	0	1	0	0	
Member G	2	4	3	4	Omo 75-14, L 7a-125, L 74a-21, L 427-7
Member F	5	4	1	1	L 860-2
Member E	1	2	0	1	Omo 57.4-41
Member D	0	0	0	0	
Member C	2	2	0	1	L 55-53
Member B	0	0	0	0	
Usno	1	2	0	0	
TOTAL	11	15	4	7	

[a] Numbers do not include antimeres.

opithecine sample of Hunt and Vitzthum (1986). The comparative sample is summarized in Table 13.2 and includes all East and South African Pliocene and early Pleistocene hominid dental specimens available for study during spring/summer of 1986.

The present study is confined to the macroscopic morphology of the crowns. Various morphological features of the mandibular premolars have been proposed to differentiate a/some "ro-

Table 13.2. Summary of the Comparative Sample[a]

Taxon	Site	P_3	P_4
A. afarensis	Hadar	13	12
	Laetoli	5	3
Total		18	15
A. africanus	Sterkfontein	8	7
	Makapansgat	3	4
Total		11	11
A. robustus	Kromdraai	3	2
	Swartkrans	19	19
Total		22	21
A. boisei	East Turkana	6	10
	Olduvai	1	0
	Peninj	1	1
Total		8	11
Homo[b]	East Turkana	5	6
	Olduvai	4	4
	Sterkfontein	0	2
	Swartkrans	1	0
Total		10	12
Others		2	3
Grand Total		71	73

[a] Numbers do not include antimeres.
[b] The Homo specimens from Olduvai are restricted to those from Bed I and Bed II.
[c] Specimens listed as "others" include KNM-ER 1482, KNM-ER 1801 and KNM-ER 5431.

bust" australopithecine sample(s) from a/some "non-robust" australopithecine sample(s) (Robinson, 1956; Wallace, 1972; Sperber, 1974; Howell, 1978; White et al., 1981). These studies were based on smaller subsets of the present comparative sample due to the ever-increasing number of Pliocene and early Pleistocene hominid dental remains. Therefore all such features were re-evaluated for the entire comparative sample. Features that characterize both A. robustus and A. boisei but none of the other samples, or those that characterize only A. robustus or A. boisei, can be considered as being potentially diagnostic of "robust" australopithecines in general or of a "robust" australopithecine species.

Table 13.3 lists features, derived from personal observations and from the literature, that exhibit significant intersample trends to qualify as potentially useful features in discriminating "robust" from "non-robust" australopithecine specimens. For the P_3s, intra- and intersample variation is evaluated by defining character states for each morphological feature and scoring specimens according to that scale. For the P_4s, metric variables were devised to quantify aspects of the following features; metaconid position, lingual notch position, relative metaconid size and relative talonid size. Other features were evaluated by the scoring of character states as for the P_3.

Mandibular P3: Nonmetric Analysis

1. Metaconid position. The metaconid position was evaluated in relation to the axis of the mesial and distal protoconid crests and the position of the protoconid. It is scored as (1) transverse to protoconid, (2) weak mesial placement, and (3) significant mesial placement. Worn specimens were evaluated whenever possibly by reference to either remaining cuspal and occlusal ridge structure or to dentine exposure.

2. Lingual segment of the mesial marginal ridge. Regardless of the height and development of the mesial marginal ridge as a whole, the lingual segment (i.e., the continuation of the mesial metaconid crest) may be variably developed. Its development is scored as (1) lacking, (2) incipiently developed thin crest, (3) thick but ill-defined ridge, and (4) well-developed crest/ridge.

3. Mesial marginal ridge height. The height of the lingual end of the buccal segment of the mesial marginal ridge was scored relative to the total crown height as (1) low, (2) intermediate, and (3) high. Worn specimens preserving the lingual end of the buccal segment were scored by gauging crown height in reference to unworn specimens of the same sample.

4. Transverse crest. The transverse crest may be a high prominent crest connecting the protoconid and metaconid apices, minimally incised by the longitudinal groove, and forming a clear demarcation between the anterior and posterior foveae. Alternatively, it may be an ill-defined crest incised deeply by the longitudinal groove, hardly forming a separation between the foveae. Specimens were scored as (1) prominent, (2) moderately prominent, (3) weak prominence, and

Table 13.3. Features of the Mandibular Premolars Considered in the Present Study

Mesial marginal ridge development	P_3
Transverse crest prominence	$P_3 + P_4$
Metaconid position	$P_3 + P_4$
Lingual notch position	P_4
Relative talonid development	P_4
Relative protoconid/metaconid size	P_4
Distolingual cusplet	P_3
Buccal basal contour	P_3
Mesiobuccal groove expression	P_3
Buccal wall slope	P_4
Crown wall convexity	P_4
Wear pattern	$P_3 + P_4$

(4) virtual lack of prominence. Estimations were made on worn specimens whenever the slopes of the transverse crest were sufficiently preserved.

5. Distolingual cusplet. The development of the distolingual cusplet was scored as (1) absent/minimally developed, (2) moderately developed, and (3) well developed. Separate categories, absent and minimally developed, were not established because such a distinction becomes impossible even with minimal wear. Relative development was evaluated in reference to size, prominence, and individualization. Weak to moderately worn specimens were scored based on fissure pattern. In such cases, specimens were scored as well developed if a substantial area of the lingual talonid was delineated both mesially and distally by fissures or constrictions.

6. Buccal basal contour. The buccal basal contour of P_3s often exhibits an asymmetric profile relative to the axis of the mesial and distal protoconid crests. Varying degrees of asymmetry were observed with the mesial basal crown projecting more buccally than the distal portion. The degree of asymmetry was scored as (1) strong, (2) moderately strong, (3) weak, and (4) minimal/lacking.

7. Mesiobuccal groove. The development of the mesiobuccal groove was scored as (1) strong, (2) moderate, (3) weak, and (4) minimal/lacking. Specimens with ill-defined depressions were scored as weak.

Mandibular P4: Nonmetric Analysis

1. Buccal wall slope. The slope of the buccal crown wall was evaluated by viewing each specimen in mesial view. The buccal slope was scored as (1) strong buccal slope starting immediately above basal enamel protuberance, (2) strong slope but less defined because of convexity, or overall slope moderate to weak, and (3) convex buccal wall with overall vertical disposition.

2. Crown wall convexity. Convexity of the mesial, distal, or lingual crown walls was examined. Specimens were scored as (1) minimal convexity, (2) moderate convexity, and (3) marked convexity.

3. Transverse crest prominence/mesial marginal ridge height. The transverse crest may be formed by appressed protoconid and metaconid portions minimally incised by the longitudinal groove. Alternatively, the protoconid and metaconid portions of the crest may curve deeply into the longitudinal groove submerging to below the level of the mesial marginal ridge. Specimens were scored as (1) appressed transverse crest above mesial marginal ridge level, (2) appressed transverse crest at level of mesial marginal ridge, (3) deeply incised crest at level of mesial marginal ridge, and (4) deeply incised crest significantly below mesial marginal ridge level.

Mandibular P4: Metric Analysis

A total of nine variables were analyzed. These are summarized in Fig. 13.1 and include (A) projected crown area in occlusal view, (B) metaconid position as measured by the angle between the protoconid crests and the line connecting protoconid and metaconid apices, (C) lingual notch position as measured by the angle between the protoconid crests and the line connecting the lingual and buccal notches, (D) relative "functional" talonid area, i.e., the ratio of crown area distal and mesial to the transverse crest, (E) relative posterior crown area, i.e., the ratio of crown area distal and mesial to the transverse groove, (F) relative metaconid area, i.e., ratio of metaconid to protoconid areas, (G) length/breadth index, i.e., ratio between mesiodistal diameter at mid-crown and the larger of trigonid and talonid breadths, (H) buccal length index, i.e., the ratio between the mesiodistal dimensions taken at mid-crown and at the position of the protoconid crests, and (I) trigonid/talonid breadth index, i.e., the ratio between the trigonid and talonid breadths. For variables that required the location of cusp apices or the position of occlusal ridges, only unworn to minimally worn specimens in which these structures could be identified, or moderately worn specimens in which reasonable estimations could be made from dentinal perforations were measured.

Quantification was performed by obtaining linear, areal, and angular measurements from occlusal view diagrams of the specimens. When specimens exhibit little to moderate interproximal wear, a conservative correction was made on the occlusal diagrams as in Wood and Abbott

(1983). The ratios and indices listed above were calculated from the raw linear and crown component areal measurements. The occlusal view diagrams were obtained by use of a stereoscopic microscope with drawing tube attachment. Specimens were oriented by controlling the tilt of the specimen relative to the optical axis in both mesiodistal and buccolingual directions. The mesiodistal tilt was determined in reference to the vertical axis of the buccal protoconid wall, i.e., a line bisecting the buccal face of the protoconid from the midbasal buccal point toward the protoconid apex. The buccolingual tilt was determined by the mesial cervical line but by disregarding its buccalmost portion, which dips inferiorly. Specimens that do not allow observation of the mesial cervical line were oriented by a "best fit" of the buccal and lingual crown contours in mesial view in reference to those specimens that preserve the cervical line. The occlusal view diagrams were made at a magnification of approximately 6 to 6.5×. The magnification factor was recorded for each specimen. The linear, angular, and areal measurements were made from these diagrams using a Neumonics digitizing tablet (0.025 mm resolution) run by commercially available software (Sigmascan, Jandel Scientific). The individual measurements were then reduced to natural size.

The degree of measurement error involved in the above method (which includes error in orientation, location of landmarks, drawing, calibration and digitizing) was evaluated by estimating error variances from duplicate measurements of 16 specimens. In the linear measurements, the magnitude of measurement error was broadly comparable to that involved in conventional crown diameter metrics taken by calipers (e.g., Mahler, 1973). Measurement error involved in crown component areal measurements was of slightly larger magnitude relative to total intraspecies variance than that in the linear measurements. It was, however, within the relative magnitude reported in craniometric studies (e.g., Utermohle and Zegura, 1982; Kouchi and Koizumi, 1985).

Statistical analyses include univariate statistics and principal component analysis of the cor-

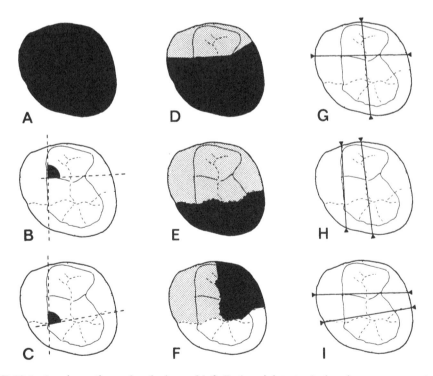

FIGURE 13.1. A schematic occlusal view of left P_4 (mesial to top) showing representations of the metric variables. A, crown area; B, metaconid position angle; C, lingual notch position angle; D, relative "functional" talonid area; E, relative posterior crown area; F, relative metaconid area; G, length/breadth index; H, buccal length index; and I, trigonid/talonid breadth index.

relation matrix. Measurements from antimeres, when available, were averaged in the statistical analyses. Principal component analysis was performed on the combined sample to summarize the metric variation represented by the above variables without the effect of *a priori* grouping.

Results

Mandibular P3: Analysis of the Comparative Sample

The distribution of character states for the seven variables are summarized in Table 13.4. In the first five features, minimal overlap occurs between *A. boisei* and *A. robustus*, on the one hand, and *A. afarensis*, *A. africanus*, and early *Homo*, on the other. A moderate degree of overlap between "robust" and "non-robust" samples occurs for basal buccal asymmetry, due mostly to a considerable percentage of *Homo* specimens showing weak asymmetry, and for mesiobuccal groove expression because some *A. afarensis* and *Homo* specimens exhibit morphologies scored as weak or minimal/lacking. *Australopithecus boisei* and *A. robustus* P_3s are characterized by a mesially positioned metaconid, a high mesial marginal ridge, a nonprominent transverse crest, a large distolingual cusplet, reduced mesiobuccal asymmetry, and a weak expression of the mesiobuccal groove. Most of these features can be interpreted as relating to the buccolingual expansion of the crown with increased talonid size and to the reduction of occlusal surface relief. They almost certainly represent the derived condition relative to other hominids.

For specimens on which all seven or the first five variables could be scored, the degree of expression of "robust" australopithecine features was evaluated by the "summary score" derived from the original scores (see Table 13.5). The summary score is produced by adding the individual scores for each variable. Since scoring for each feature was defined so that the "robust" australopithecine condition has a high score, the summary scores indicate the degree to which a specimen exhibits "robust" australopithecine trends.

In Table 13.5 a clear separation is seen between "robust" and "non-robust" samples, indicating that when multiple features are considered, what little overlap in range of variation is seen between "robust" and "non-robust" samples in the individual features disappears. The two specimens from Kromdraai lie toward the lower end of the range of the Swartkrans sample.

Many specimens for which the summary scores are not available are also expected to conform to this pattern of "robust" versus "non-robust" dichotomy. Over half of the Swartkrans specimens are too worn to be scored for mesial marginal ridge development, transverse crest prominence or distolingual cusplet size. In many of these specimens, however, the derived "robust" australopithecine condition can be inferred. For example, the degree of wear on the transverse crest can be gauged from the extent of longitudinal groove preserved adjacent to the crest. From such considerations, a relatively low transverse crest and high mesial marginal ridge can be inferred for some of the unscored specimens from Swartkrans. Similar inferences can be drawn regarding the distolingual cusplet since "robust" australopithecine P_3s with a large cusplet exhibit a modified transverse groove configuration with a mesially deflected lingualmost portion that forms the mesial boundary of the distolingual cusplet. Worn specimens of *A. robustus* that have not been scored for this feature do exhibit a mesially deflected lingual segment of the transverse groove and are likely to have had a sizable distolingual cusplet. From the above considerations, approximately ten additional specimens can be said to conform to the "robust" australopithecine pattern. On the other hand, none of the *A. robustus* or *A. boisei* specimens for which summary scores are not available show any suggestion of exhibiting "non-robust" australopithecine overall morphology.

Mandibular P3: Omo Specimens

Of the fifteen P_3s known in the Omo collection, at least five specimens exhibit the "robust" australopithecine condition. The P_3s of the L 7a-125 mandible are too worn for a morphological evaluation, but the specimen is clearly a "robust" australopithecine from mandibular morphology and size (Howell, 1969a; White, 1977; Chamberlain and Wood, 1985). The stratigraphic position, character states and summary scores for these specimens are given in Table 13.6.

Table 13.4. Nonmetric Analysis of Mandibular P3s[a,b]

Taxon	Metaconid position			Mesial marginal ridge lingual segment				Mesial marginal ridge height			Transverse crest prominence				Distolingual cusplet			Buccal basal asymmetry				Mesiobuccal groove			
	1	2	3	1	2	3	4	1	2	3	1	2	3	4	1	2	3	1	2	3	4	1	2	3	4
A. afarensis	11	6	0	5	8	1	1	14	2	0	9	1	0	0	10	2	0	7	7	2	1	9	5	2	2
A. africanus	5	1	0	0	5	1	0	5	1	0	2	3	0	0	4	1	0	2	4	1	1	6	1	0	0
Homo	4	6	2	1	1	2	1	2	4	0	1	4	0	0	6	3	0	2	3	2	4	2	4	1	1
A. robustus	1	0	15	0	0	0	7	0	1	6	0	3	4	1	1	2	4	0	1	11	6	0	2	5	12
A. boisei	0	0	8	0	0	0	6	0	2	4	0	0	3	2	0	3	5	0	1	3	1	0	1	2	1
"Non-robust" combined	23	10	2	6	14	5	2	21	7	0	12	8	0	0	20	6	0	11	14	5	6	17	10	3	3
"Robust" combined	1	0	23	0	0	0	13	0	3	10	0	3	7	3	1	5	9	0	2	14	7	0	3	7	13

[a] Numbers within body of table represent the number of specimens for each category, antimeres not included.
[b] Numbers 1, 2, 3, and 4 beneath categories represent scores for each category, see text pp. 201–202 for key to scores.

Table 13.5. Summary Scores of the Nonmetric Analysis[a] of P_3s

	Five variables			Seven variables		
Taxon	N	$\bar{X}$	Range	N	$\bar{X}$	Range
A. afarensis	12	6.6	5–10	11	10.4	7–12.5
A. africanus	6	7.3	7–8	5	10.6	9–12
Homo	8	9.1	5.5–12	8	14.4	10–17.5
A. robustus	7	14.5	12–17	7	21.3	19–24
A. boisei	6	15.6	14–17	3	22.3	20.5–24
"Non-robust" combined	26	7.5	5–12	24	11.8	7–17.5
"Robust" combined	13	15.2	12–17	10	21.7	19–24

[a]See the text for the definition of the summary score.

Omo 18-31, L 398-120, L 465-111 and the P_3 of the L 427-7 mandible are unworn or little worn specimens. Their occlusal morphology clearly indicates "robust" australopithecine affinities. L 860-2 is a strongly weathered/abraded mandible with crowns of P_3, P_4 and M_2 of moderately advanced wear. Although surface enamel has been abraded away, it must have had minimal relief judging from its completely flattened preserved occlusal surface. Small dentine exposures are present for both protoconid and metaconid. Similar sized dentine exposures occur in comparably worn "robust" australopithecine P_3s with flattened protoconids and metaconids. "Non-robust" australopithecine P_3s exhibit comparable protoconid dentine exposures at an earlier wear stage before the cusp is worn flat. Occlusal morphology of the L 860-2 P_3 can be evaluated for metaconid position and distolingual cusplet development. Judging from the shape of the protoconid dentine exposure as well as the remaining buccal wall, the metaconid is placed distinctly mesial relative to the protoconid. Remnants of the transverse groove show a mesially deflected lingual segment delineating a large distolingual portion of the talonid. This configuration is comparable to the majority of similarly worn "robust" australopithecine P_3s and most likely is indicative of a large distolingual cusplet in its unworn stage.

Mandibular P4: Comparative Analysis

The distribution of character states for the three nonmetric variables is summarized in Table 13.7. Separation of "robust" and "non-robust" samples is not as clear as in the nonmetric analysis of the P_3. Clear trends, however, do exist with both *A. robustus* and *A. boisei* exhibiting higher frequencies of specimens with relatively vertical buccal wall slope (Fig. 13.2), convex crown walls, and weak transverse crests.

Univariate statistics for the metric variables are presented in Table 13.8. The *A. boisei* sample exhibits little to moderate overlap with the *A. robustus* sample in crown area, metaconid position, relative "functional" talonid area, relative posterior crown area, relative metaconid area, and buccal length index. In these same features, the "non-robust" samples overlap considerably with *A. robustus* but not with *A. boisei*. In the remaining three variables, *A. boisei* again exhibits the extreme condition although overlap in the range of variation between the *A. boisei* and other samples is greater. In relative metaconid area, trigonid/talonid breadth index, buccal length index, and lingual notch position angle, the mean values of the current small samples of *A. africanus* and/or *Homo* fall closer to *A. boisei* than do the means of the *A. robustus* sample.

Principal component analyses of these variables produce a distinct clustering of *A. boisei* specimens. Analyses were performed on a number of variable combinations. All analyses produced broadly comparable results, including those that excluded absolute crown area. Results of a six-variable analysis are presented in Fig. 13.3 and Table 13.9. The component loadings indicate that the first principal component that separates *A. boisei* from all other samples has equally strong correlation with relative area of the metaconid, relative area of the posterior crown, buccal length index, length/breadth index, trigonid/talonid breadth index, and absolute crown area. *A.*

Table 13.6. Nonmetric Features of Omo Mandibular P3s Attributed to "Robust" Australopithecines[a]

Specimen	Stratographic level	Metaconid position	Mesial marginal ridge lingual segment	Mesial marginal ridge height	Transverse crest prominence	Distolingual cusplet	Buccal basal asymmetry	Mesial buccal groove	Five variables	Seven variables
Omo 18-31	C8	3	4	3	3	2	(3/4)	3	15	21.5
L 398-120	F0	3	4	2	3	1	(3/4)	3	13	19.5
L 465-111	F1	3	4	(3)	(3)	—	(3/4)	3	13+	19.5+
L 860-2	F1	3	—	—	—	—	(3/4)	(2/3)	—	—
L 427-7	G4	3	4	3	3	3	4	1	16	21
L 7a-125	G5	—	—	—	—	—	4	(3/4)	—	—

[a] Numbers represent scores for each feature. See text pp. 201–202 for the key to the scores.

Table 13.7. Nonmetric Analysis of Mandibular P4s[a,b]

Taxon	Buccal wall slope			Crown wall convexity			Transverse crest prominence				Summary score		
	1	2	3	1	2	3	1	2	3	4	N	$\bar{X}$	Range
A. afarensis	11	0	0	2	3	0	3	1	0	0	6	3.8	3–4
A. africanus	4	1	0	0	2	0	0	4	0	0	3	5.0	4–6
Homo	5	3	0	2	5	0	1	4	1	0	4	4.8	4–5
A. robustus	5	16	0	0	7	2	0	4	6	3	11	7.0	5–9
A. boisei	0	7	1	0	1	4	0	0	2	2	4	8.8	8–9
"Non-robust" combined	20	4	0	4	10	0	4	9	1	0	13	4.4	3–6
"Robust" combined	5	23	1	0	8	6	0	4	8	5	15	7.5	5–9

[a] Numbers within body of table represent the number of specimens for each category, antimeres not included.
[b] Numbers 1, 2, 3, and 4 beneath categories represent scores for each category. See the text for keys to the scores (pp. 201–202) and summary score (p. 204).

robustus specimens overlap extensively with the "non-robust" sample in the first principal component scores. The two specimens from Kromdraai fall within the observed Swartkrans sample range. Average values of the first principal component scores of the *A. robustus* and *A. africanus* samples are close to each other while those of *A. afarensis* and *Homo* samples are smaller. Thus, it can be said that *A. robustus* shows only incipiently the talonid expansion so conspicuous in *A. boisei* P$_4$s. The second and third components show a wide range of component score distribution

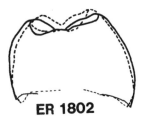

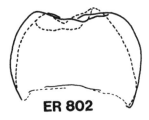

FIGURE 13.2. Mesial view diagrams of P$_4$s. Comparison of *H. habilis* (KNM-ER 1802), *A. robustus* (SK 826), and *A. boisei* (KNM-ER 802) with *A. afarensis* (LH 3, interrupted lines).

Table 13.8. Summary Statistics for Mandibular P4s

		Crown area				Metaconid position angle				Lingual notch position angle		
	N	$\bar{X}$	SD	Range	N	$\bar{X}$	SD	Range	N	$\bar{X}$	SD	Range
A. boisei	10	162.7	24.91	127.1–195.0	4	60.6	11.07	44.2–67.8	4	71.4	7.58	63.0–79.5
A. robustus	18	116.0	12.92	97.0–137.4	7	74.0	8.53	61.5–83.3	7	77.7	9.39	62.5–91.6
A. africanus	4	99.8	9.12	92.3–113.1	4	79.9	4.97	74.2–84.1	4	83.3	5.63	75.2–87.4
A. afarensis	11	87.1	14.18	62.8–105.2	6	87.8	2.47	84.8–92.0	6	84.5	6.01	76.9–91.7
Homo	8	86.5	12.87	72.5–115.3	5	78.6	3.22	74.8–82.7	5	75.7	3.19	72.4–79.4
"Non-robust" combined	23	89.1	13.45	62.8–115.3	15	82.6	5.47	74.2–92.0	15	81.2	6.27	72.4–91.7

		Relative "functional" talonid area				Relative posterior crown area				Relative metaconid area		
	N	$\bar{X}$	SD	Range	N	$\bar{X}$	SD	Range	N	$\bar{X}$	SD	Range
A. boisei	5	3.308	.301	3.030–3.788	8	.729	.102	.579–.857	7	.950	.082	.811–1.006
A. robustus	11	2.255	.350	1.799–3.006	18	.531	.062	.407–.618	14	.737	.082	.622–.880
A. africanus	4	2.022	.448	1.431–2.516	4	.485	.118	.323–.602	4	.774	.083	.702–.879
A. afarensis	9	2.180	.236	1.906–2.603	10	.469	.054	.392–.557	6	.680	.060	.608–.766
Homo	5	1.934	.200	1.673–2.176	7	.427	.086	.293–.536	7	.656	.070	.564–.765
"Non-robust" combined	18	2.077	.289	1.431–2.603	21	.458	.079	.293–.602	17	.693	.081	.564–.879

		Length/breadth index				Buccal length index				Trigonid/talonid breadth index		
	N	$\bar{X}$	SD	Range	N	$\bar{X}$	SD	Range	N	$\bar{X}$	SD	Range
A. boisei	8	.984	.043	.937–1.057	8	1.206	.040	1.168–1.287	8	.934	.032	.897–.976
A. robustus	18	.937	.042	.870–1.019	18	1.108	.036	1.044–1.202	18	1.013	.036	.962–1.087
A. africanus	4	.910	.027	.872–.931	4	1.128	.028	1.100–1.156	4	1.009	.062	.947–1.093
A. afarensis	11	.898	.062	.790–.977	11	1.098	.036	1.045–1.175	11	1.037	.050	.938–1.089
Homo	8	.910	.062	.818–.981	8	1.110	.050	1.040–1.183	7	1.097	.071	1.000–1.202
"Non-robust" combined	23	.904	.056	.790–.981	23	1.107	.040	1.040–1.183	22	1.051	.066	.938–1.202

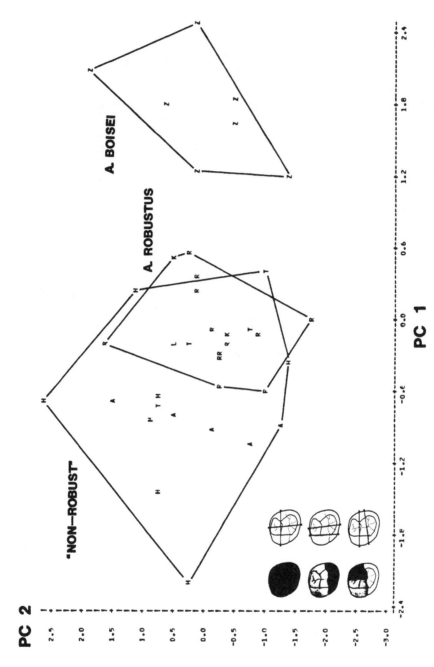

FIGURE 13.3. Principal component analysis of P_4s, six-variable analysis. Plot of first and second principal components. Z, *A. boisei*; R, *A. robustus* (Swartkrans); K, *A. robustus* (Kromdraai); T, *A. africanus*; A, *A. afarensis* (Hadar); L, *A. afarensis* (Laetoli); H, early *Homo*. The six variables are schematically shown at the left lower corner of the plot and include absolute crown area, relative posterior crown area, relative metaconid area, length/breadth index, buccal length index, and trigonid/talonid breadth index.

Table 13.9. Principal Component Analysis, Eigenvalues, and Component Loadings

	Principal component			
	1	2	3	4
Eigenvalues	4.015	.730	.522	.395
Proportion (%)	66.900	12.200	8.700	6.600
Cumulative proportion (%)	66.900	79.100	88.800	94.400
Component loadings				
Absolute crown area	.861	−.022	.090	.432
Relative posterior crown area	.861	−.357	−.230	−.101
Relative metaconid area	.849	.085	.412	.122
Length/breadth index	.706	.499	−.480	.071
Buccal length index	.790	.363	.231	−.395
Trigonid/talonid breadth index	−.830	.463	.088	.152

within all samples. This indicates that these components represent intrasample variation present in all samples and are not particularly useful for distinguishing taxa.

The above metric results confirm the visual observation of the highly distinctive occlusal morphology of *A. boisei* P_4s. This is schematically illustrated in Fig. 13.4. The *A. boisei* morphology may be related to an extremely expanded talonid both buccolingually and distally and consists of the following features: (1) the position of the metaconid relative to the orientation of the protoconid crests is shifted to an extreme mesiolingual position; (2) the lingual notch is also positioned mesially creating a mesially deflected lingual segment of the transverse groove; (3) the absolute and relative area of the talonid is increased with extreme development of the distolingual cusplet as well as a multitude of accessory cuspules on the distal marginal ridge; (4) this talonid expansion is reflected metrically in the crown component area measurements, the large talonid breadth relative to that of the trigonid, and the trend for mesiodistal elongation relative to buccolingual breadth and especially to buccal crown length; (5) the transverse crest

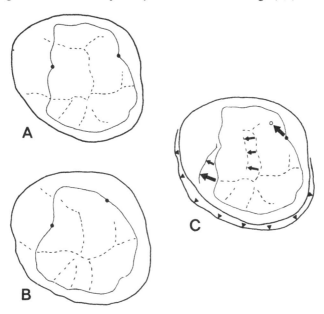

FIGURE 13.4. A schematic comparison of P_4s of A: *H. habilis*, KNM-ER 1802, B: *A. boisei*, KNM-ER 1171, and C: superimposition of A and B by an approximate matching of the anterior crown contour. The KNM-ER 1171 P_4 is reduced so that the combined protoconid/metaconid area is equivalent to that of KNM-ER 1802. Left crowns, mesial to top.

is virtually lacking or reduced to weak structures hardly recognizable as occlusal ridges; (6) in contrast to this, the accessory occlusal ridges on the posterior slope of the protoconid and metaconid are well developed so as to "fill" the occlusal topography; and (7) the size of the metaconid relative to that of the protoconid is increased.

These features are only incipiently expressed in *A. robustus* specimens, in which a much more conservative overall morphology is observed, although in isolated features some specimens may approximate the *A. boisei* condition.

A total of seven specimens from the KBS and Okote Members of the Koobi Fora Formation clearly exhibit the above-outlined morphological pattern. Two heavily worn specimens (KNM-ER 1509 and KNM-ER 5429) from the KBS Member can also be considered to exhibit the *A. boisei* pattern, judging from the relative positions of the metaconid and distolingual cusplet enamel perforations or from the dentine cores exposed by crown damage. The above occlusal morphology can also be inferred for the premolars of the Peninj mandible judging from dentinal exposure pattern and remnants of the transverse groove. The P_4 of KNM-ER 818 from the Okote Member is too worn and weathered to evaluate its occlusal morphology. However, its occlusal view crown contour is similar to that of the KNM-ER 1171 P_4 and does not contradict the *A. boisei* morphological pattern.

The only post-2-Myr East African specimen commonly considered to represent a "robust" australopithecine (White, 1977; Howell, 1978; Chamberlain and Wood, 1985) that does not exhibit the above-outlined P_4 morphological pattern is KNM-ER 1482. The overall occlusal morphology of this specimen is comparable to that seen in *A. robustus*. Since the absolute size of the KNM-ER 1482 P_4 is much smaller than the *A. boisei* homologues, the question arises whether or not the extreme talonid expansion and associated morphologies seen in *A. boisei* specimens are actually an allometric phenomenon related to increase in absolute size. The validity of such a suggestion can be evaluated by examining the intrasample allometric trend in the "robust" australopithecine species. The fact that, in both *A. boisei* (KNM-ER 3229) and *A. robustus* (SK 1587, TM 1601), some small specimens have large factor scores for the first principal component renders an allometric explanation highly unlikely. It should also be noted that large specimens of *A. robustus* (SK 88, SK 34, SK 826) that overlap in size with small specimens of *A. boisei* do not approximate the *A. boisei* P_4 morphology. Wood and Uytterschaut (1987) reached similar conclusions in reporting no significant allometric relationship between crown area and posterior crown area within their combined *A. africanus/Homo* sample.

The size of the P_4 relative to M_1 measured in associated specimens is summarized in Fig. 13.5.

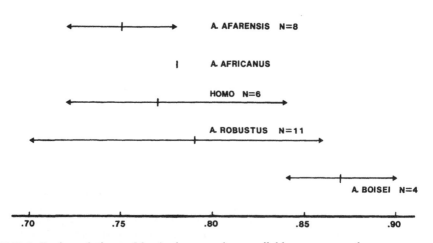

FIGURE 13.5. P_4 size relative to M_1. Antimeres when available are averaged.

Australopithecus boisei exhibits a noticeable tendency for a relatively large P_4. Much of this increase in relative P_4 size would appear to be produced by the extreme talonid expansion and related morphological features outlined above.

Mandibular P4: Omo Specimens

Of the 22 P_4s of the Omo collection, ten specimens are sufficiently preserved and exhibit morphologic features that suggest "robust" australopithecine affinities. Stratigraphic, metric, and nonmetric data of these specimens are given in Tables 13.10 and 13.11. Results of principal component analyses are shown in Fig. 13.6. In addition to these specimens, "robust" australopithecine mandibles L 7a-125 and Omo 57.4-41 also preserve P_4s but are heavily worn and damaged, making morphological comparisons difficult.

Specimen L 55-33 from Member C6, a small fragment of the left mandibular corpus with most of the P_4 crown, was briefly discussed but not attributed to taxon by White (1977). The P_4 crown is moderately worn with fused dentine exposures of the protoconid and distobuccal cusplet. The metaconid is worn to minimal saliency but lacks enamel perforation. Flattening of the metaconid prior to dentine exposure occurs only in *A. boisei* and *A. robustus* specimens and is not known in the other samples. *Australopithecus afarensis, A. africanus* and early *Homo* specimens that exhibit comparable or greater metaconid saliency possess enamel perforations. The lingual extremity of the transverse groove is placed in an extreme mesial position, impinging on the mesially positioned metaconid. Although quantification is not possible because of wear, such a configuration is not known in any of the "non-robust" samples of the present study, while it is rare in *A. robustus* and frequent in *A. boisei*. Estimated values of relative posterior crown area of the L 55-33 P_4 are greater than the maximum value of the *A. robustus* sample, but the estimated crown area lies toward the lower end of the range of *A. robustus*.

The Omo 57.4-41 mandible is a massive specimen commonly attributed to *A. boisei* (White, 1977; Howell, 1978; Coppens, 1980). The lingual half of a worn P_4 is preserved. It is worn flat with minimal surface relief and is therefore consistent with a "robust" australopithecine attribution. Judging from the remaining fissure pattern, however, it is clear that the extreme talonid expansion of later *A. boisei* specimens was lacking.

Isolated P_4s, L 338x-40, Omo 33-508, and L 628-4 exhibit similar morphologies. These specimens all lack the talonid expansion seen in *A. boisei* specimens but most likely represent "robust" australopithecines based on the following observations. All three exhibit relatively flat wear on which Phase I and Phase II facets are not clearly delineated. Grine (1981) reported a similar

Table 13.10. Nonmetric Features of Omo Mandibular P4s Attributed to "Robust" Australopithecines[a]

Specimen	Stratigraphic level	Buccal wall slope	Crown wall convexity	Transverse crest prominence	Summary score
L 55-33	C6	(2)	—	—	—
L 338x-40	E3	3	(2)	(2)	7
Omo 57.4-41	E4	—	—	—	—
Omo 33-507	F0	—	3	3	—
Omo 33-508	F0	2	(1/2)	(2/3)	6
L 398-1223	F0	—	3	3	—
L 420-15	F3	—	3	2	—
L 860-2	F1	(2/3)	—	—	—
L 7a-125	G5	(2/3)	—	—	—
L 74a-21	G3-5	3	(3)	—	—
L 427-7	G4	(2/3)	(3)	2	7.5
L 628-4	G3	2	(2)	(2/3)	6.5

[a]Numbers represent scores for each feature. See text, pp. 202 for key to the scores.

Table 13.11. Metric Data for the Omo "Robust" Australopithecine Mandibular P4s

Specimen	Absolute crown area	Relative "functional" talonid area	Relative posterior crown area	Relative metaconid area	Length/breadth index	Buccal length index	Trigonid/talonid breadth index	Metaconid position angle	Lingual notch position angle
L 55-33	105.8	—	.64–.66[b]	—	.88–.90	—	—	—	—
L 338x-40	119.0[a]	2.71[a]	.561	.616	.922	1.16	1.03	73.1[a]	72.2[a]
Omo 33-508	123.4[a]	2.33	.649[b]	.779	1.025[b]	1.15	1.03	77.0	81.9
L 860-2	125.3[a]	2.1–2.4	.60–.62[a]	c.78	.94–.96	1.12–1.14	c.1.00	—	—
L 74a-21	163.1[b]	2.9–3.2[a]	.646[b]	.756	.982[a]	1.13–1.18	.95–.96	—	—
L 427-7	138.8[b]	3.38[b]	.937[b]	.794	.926	1.19[a]	.95	69.3[a]	70.9[a]
L628-4	99.3	2.31	.668[b]	.831	.920	1.20–1.25[a]	.99	—	—

[a]Outside the range of the "non-robust" sample.
[b]Outside the range of the A. robustus and the "non-robust" samples.

relationship between Phase I and II facets in "robust" australopithecine deciduous molars. Of the three specimens, Omo 33-508 is most advanced in wear stage. In this specimen, both the protoconid and metaconid are reduced to minimal saliency, a small dentine exposure is present on the protoconid, and enamel perforation has just commenced on the metaconid. As discussed above, in "non-robust" australopithecine specimens, dentine exposure commences when the cusps are much more salient. The buccal wall of all three specimens is steep, lacking a significant slope. In the metric analysis, each specimen fell outside the range of variation of the current "non-robust" sample for at least one variable. These include the metaconid position angle and relative "functional" talonid area of L 338x-40 and the relative posterior crown area of Omo 33-508 and L 628-4. In the various principal component analyses, the first principal component scores of L 338x-40 lie at the mid-range of the *A. robustus* sample, while the corresponding values of Omo 33-508 and L 628-4 lie at or slightly above the upper range of the current *A. robustus* sample (Fig. 13.6a).

Omo 33-507, L 398-1223, and L 420-15 are similarly preserved unworn P_4s lacking the buccal third of the crown. These specimens also lack the talonid expansion of *A. boisei* homologues. "Robust" australopithecine affinities are suggested by the fact that all three specimens exhibit extremely convex crown walls matched commonly in *A. boisei* and rarely in *A. robustus*, but not in the other samples. Moreover, L 420-15 and Omo 33-507 possess mesially positioned metaconids almost certainly outside the known range of variation of the "non-robust" samples (although quantification is difficult due to buccal crown loss), and the transverse crest of L 398-1223 is minimally prominent.

The P_4 of the L 860-2 mandible is abraded with no original surface enamel preserved except for minute patches close to the cervix of the distal face. In occlusal view, this portion of the crown protrudes distally suggesting dimensional reduction in other portions of the crown because of abrasion. Metric analysis was undertaken by making a reasonable compensation for the supposed dimensional reduction of the distal crown but not for the remainder of the crown. According to analysis based on this estimation, the L 860-2 P_4 lies within the range of variation of *A. robustus*. Visual observation confirms this metric assessment. Metaconid area is small relative to protoconid area, the lingual termination of the transverse groove is only moderately anteriorly placed, and the shape of the metaconid dentinal exposure suggests a metaconid placement slightly distal to the mesiolingual corner of the crown.

The P_4 of the L 427-7 mandible is an unerupted germ. Most of the crown is exposed, although in occlusal view the lingual margin cannot be observed. This premolar clearly exhibits an expanded talonid comparable to the condition seen in *A. boisei* homologues. The metaconid and lingual termination of the transverse groove is placed extremely mesially in conjunction with a very large crown area posterior to the transverse groove. Metric analyses were performed on several reconstructions of the lingual margin, and all produced comparable results aligning this specimen with the *A. boisei* P_4s. In some nonmetric features, such as the expression of the transverse crest, the L 427-7 P_4 exhibits a more conservative morphology than seen in the current small *A. boisei* sample.

L 74a-21 is a large right mandibular corpus with crowns of $\overline{C}$ and P_4. The P_4 crown is worn flat but lacks enamel perforation on either the protoconid or metaconid. Because of wear, metric analysis of variables excepting crown area, relative metaconid area, and relative posterior crown area were based on approximate estimates. Results place L 74a-21 in a position intermediate between *A. robustus* and *A. boisei*. In crown area and relative posterior crown area its affinities are with *A. boisei*, while relative metaconid size is more comparable with the other samples. In the principal component analysis of five variables without absolute crown area (Fig. 13.6a), L 74a-21 falls close to the *A. robustus* range. When absolute crown area or estimated values of relative "functional" talonid area are included in the analysis, then the specimen falls much closer to the *A. boisei* sample (Fig. 13.6b). These results reflect the combination of a modest degree of *A. boisei*-like talonid expansion and a conservative relative metaconid size. Other features align this specimen with *A. boisei*. The buccal wall shows an extremely vertical disposition matched only in some *A. boisei* specimens, and to judge from the preserved occlusal morphology and buccal crown wall, the position of the metaconid most likely was at the mesiolingual corner of the crown.

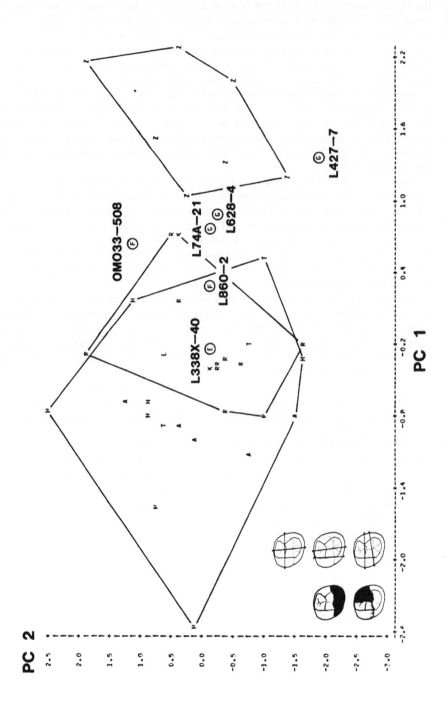

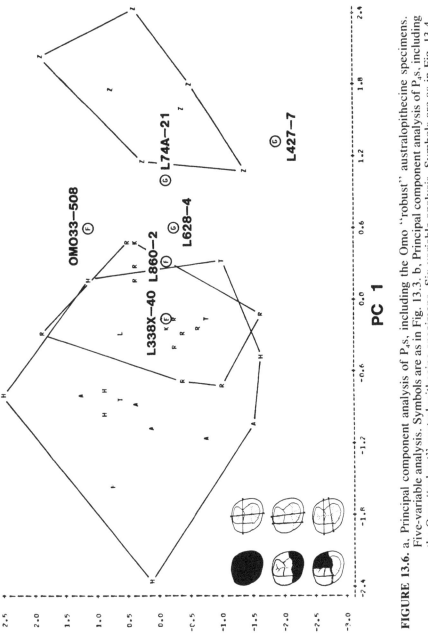

FIGURE 13.6. a, Principal component analysis of P_4s, including the Omo "robust" australopithecine specimens. Five-variable analysis. Symbols are as in Fig. 13.3. b, Principal component analysis of P_4s, including the Omo "robust" australopithecine specimens. Six-variable analysis. Symbols are as in Fig. 13.4.

The L 7a-125 mandible has been attributed to *A. boisei* (White, 1977; Howell, 1978; Chamberlain and Wood, 1985). Both right and left P_4s are preserved but wear and damage preclude a definitive morphologic analysis. The remnants of the occlusal fissures do not suggest an expanded talonid to the degree seen in later *A. boisei* specimens, although such a statement must remain tentative.

Discussion and Conclusions

The morphological conclusions outlined above for both P_3 and P_4 may prove to be of utility in phylogenetic considerations. This is because they are based on a modest but reasonably large sample that enables some appreciation of variability within taxa, and because morphocline polarities seem to be unequivocal. A preliminary outgroup comparison with Miocene hominoids and modern apes reinforces the suggestion made above that the "robust" australopithecine P_3 morphology represents a uniquely derived condition and is clearly apomorphic relative to all other hominids. The same can be said of the advanced P_4 morphology seen in the *A. boisei* specimens. Within hominoids, talonid size varies significantly among taxa. However, no other taxon manifests the extreme talonid expansion of *A. boisei* involving marked change in cuspal arrangement.

The major overviews of the Omo hominid collection placed the unequivocal appearance of *A. boisei* at Member E (Howell and Coppens, 1976; Coppens, 1980). Grine (1985) attributed two deciduous molars from Member D to *A. boisei*. The Omo 18-18 mandible from upper Member C deposits (Arambourg and Coppens, 1968) was considered to represent a "robust" australopithecine by White (1977), Coppens (1980), and Chamberlain and Wood (1985). The present study further documents the presence of "robust" australopithecines in the lower Omo valley at upper Member C times based on the Omo 18-31 P_3, and most likely at middle Member C times based on the L 55-33 fragmentary mandible.

Grine (1984) outlined detailed similarities in the deciduous canine and molars of *A. boisei* and *A. robustus/crassidens* and interpreted them to be synapomorphies. Wood *et al.* (1983) documented relative cusp proportions of the mandibular molars shared by East and South African "robust" australopithecines. The derived P_3 morphology discussed above can be added to these lists of features shared by *A. robustus* and *A. boisei*. The presence of these detailed morphological similarities in clearly apomorphic character complexes strongly suggests that these features were present in the last common ancestor of *A. robustus* and *A. boisei*. If *A. africanus* is excluded from the ancestry of *A. boisei* (Walker *et al.*, 1986; Kimbel *et al.*, this volume, chapter 16), then the evolution of *A. robustus* from *A. africanus* would require a substantial amount of homoplasy in detailed dental morphology.

The timing of the emergence of these apomorphic features is of prime importance in discussing phylogenetic relationships. If the Omo 18-31 P_3 is taken to correspond to the actual first appearance of its derived morphology, then the cladistic event between East and South African "robust" australopithecines must postdate upper Member C times. Such a circumstance would not be compatible with a phylogenetic hypothesis placing the divergence between South and East African "robust" australopithecine lineages substantially prior to the KNM-WT 17000 cranium, which derives from levels just below or equivalent to tuff D (Walker *et al.*, 1986; Brown and Feibel, this volume, chapter 22). Alternatively, if Omo 18-31 merely reflects the known first appearance of its derived P_3 morphology and the actual first appearance were to be appreciably earlier, then it would be compatible with phylogenies placing the East and South African "robust" australopithecine divergence either before or after c. 2.5 to 2.6 Myr. The P_3s known from Hadar and Laetoli as well as the single specimen from the Usno Formation (W 978) do not exhibit the "robust" australopithecine morphology. However, the lack of P_3s in the Shungura Formation from levels below upper Member C precludes favoring either of the above alternatives. Thus, accumulation of larger samples, even of isolated teeth, will contribute to refine our phylogenetic hypotheses.

It has been shown above that *A. boisei* possesses a unique morphology of the P_4 involving extreme expansion of the talonid. It was also shown that *A. boisei* P_4s are larger than those of *A. robustus* relative to M_1 size. Thus, it is most natural to interpret the *A. boisei* P_4 morphology as part of an increased specialization involving absolute and relative premolar expansion. This is in accord with suggestions by Tobias (1967), Rak (1983) and Grine (1984) for a more intensified masticatory specialization in *A. boisei* compared to the condition seen in *A. robustus*. The P_4 morphology of *A. robustus* is more conservative than that of *A. boisei*, and in many features expressing talonid expansion, *A. robustus* P_4s overlap in range of variation extensively with the "non-robust" samples. Of the "non-robust" samples, the current *A. africanus* sample overlaps most with that of *A. robustus*. It can be argued that the degree of talonid expansion seen in *A. robustus* most likely represents the primitive condition for "robust" australopithecines in general. Wood and Uytterschaut (1987) report metric analysis of the P_4 broadly comparable to the above results but with less overlap between *A. africanus* and *A. robustus*.

The present study identified twelve P_4 specimens from the Shungura Formation as "robust" australopithecines. Of these twelve, only two, L 427-7 and L 74a-21, approximate the *A. boisei* condition seen in the Koobi Fora and Peninj P_4s. Both specimens derive from lower Member G. The P_4 of the L 7a-125 mandible, also from lower Member G, is of indeterminate morphology. The other specimens, deriving from Members C, E, F, and G, do not exhibit extreme talonid expansion. If identification of these specimens as "robust" australopithecines is correct, they document the continuous presence of "robust" australopithecines in East Africa spanning the time range of approximately 2.6 to 2.2 Myr. These forms seem to lack the extreme specialization of the post-2-Myr *A. boisei* specimens.[1]

The temporal distribution of the "advanced" *A. boisei* P_4 and the more "conservative" P_4 is summarized in Fig. 13.7. Below Member G, i.e., prior to c. 2.3 Myr, only P_4s with the "conservative" morphology are known. From the KBS Member and higher strata, only P_4s with the "advanced" morphology are found. Member G shows a mixture of the two. The presence of some conservative features in the overall advanced P_4 morphology exhibited by L 427-7 and L 74a-21 makes lower Member G a prime candidate for a time period when this morphological specialization actually appeared in the East African "robust" australopithecine lineage. The precise nature of transition to the more specialized condition, however, may be clarified only with an increase in sample size including more complete specimens. The derived P_4 morphology unique to post-2-Myr *A. boisei* and the two specimens from Member G suggests a biologically significant difference between earlier and later East African "robust" australopithecines. It is of interest to document what other anatomical modifications occurred synchronously with premolar specialization. There are indications, however, that other anatomical specializations evolved at different times. For example, Grine and Martin (this volume, chapter 1) report "hyperthick" enamel and a deciduous pattern of amelogenesis in mandibular molars from the Shungura Formation Members E and F. These features are otherwise known in later *A. boisei* specimens (Beynon and Wood, 1986, 1987). Thus, results of the present study support suggestions of specific distinction between earlier and later "robust" australopithecines of East Africa (Arambourg and Coppens, 1968; Walker *et al.*, 1986), a suggestion which has been strongly advocated by Kimbel *et al.* (this

---

[1]Such a conclusion is warranted despite the largely fragmentary fossil evidence and the reasonably large degree of morphological overlap between "conservative" "robust" australopithecine P_4s, on one hand, and *A. africanus* or early *Homo*, on the other. First, two of the above specimens, Omo 57.4-41 and L 860-2, are mandibular specimens identifiable as "robust" australopithecine on grounds other than P_4 morphology, i.e., mandibular or P_3 morphology. Identification of the isolated P_4s may be less secure. As demonstrated above, however, all of these specimens possess significant numbers of features that fall at the limits or outside the range of variation of the present "non-robust" hominid samples. Thus, if less fragmentary fossils are found in the future that necessitate an overturn of the present identifications, that would at the same time indicate that a "non-robust" australopithecine lineage existed in East Africa that shows stronger convergence toward the "robust" australopithecine condition than in any of the known "non-robust" samples. With lack of such evidence and the presence of Omo 57.4-41 and L 860-2, it is best to consider the isolated P_4s discussed in the present paper to represent a "robust" australopithecine with a conservative P_4 morphology broadly comparable to *A. robustus* in degree of talonid expansion. Finally, the KNM-WT 16005 "robust" australopithecine mandible derived from Member D equivalent strata at West Turkana (Walker *et al.*, 1986) also exhibits a P_4 morphology lacking the extreme talonid expansion of later *A. boisei* homologues.

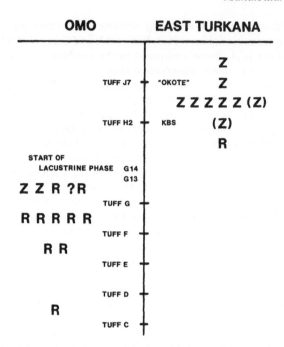

FIGURE 13.7. Temporal distribution of the derived *A. boisei* P_4s (Z) and the more conservative "robust" australopithecine P_4s (R) that lack extreme talonid expansion.

volume, chapter 16). However, the above results do not provide an unequivocal solution as to where to divide the evolving lineage. Finally, the *A. boisei* P_4 morphological complex may be used to delimit the possible divergence date between East and South African "robust" australopithecines. If it is considered that the transition in P4 morphology occurred anagenetically during lower Member G times, then the cladistic event between East and South African "robust" australopithecines must have taken place prior to 2.2 or 2.3 Myr. Undoubtedly, further discussion on the phylogenetic relationship and the timing of cladistic events of Pliocene hominids must be performed by taking into account additional specimens and a wider range of character complexes.

Acknowledgments

I thank the governments of Kenya and Tanzania for permission to conduct research in their countries. For generous access to the original fossils and/or research facilities, I thank the National Museums of Kenya, the National Museums of Tanzania, the Transvaal Museum, the Department of Anatomy, University of the Witwatersrand, and the Museum of Vertebrate Paleontology, University of California, Berkeley. Appreciation is expressed to F. Grine for inviting me to participate in the workshop "The Evolutionary History of the "Robust" Australopithecines." I thank F. C. Howell and T. White for encouragement and support throughout the present study. This research was supported by NSF grant BNS-85-13835 and the University of California Fellowships.

References

Arambourg, C. and Coppens, Y. (1968). Decouverte d'un australopithecien nouveau dans les gisements de l'omo (Ethiopie). *S. Afr. J. Sci.*, *64*:58–59.
Benyon, A. D. and Wood, B. A. (1986). Variations in enamel thickness and structure in East African hominids. *Amer. J. Phys. Anthropol.*, *70*:177–193.

Beynon, A. D. and Wood, B. A. (1987). Patterns and rates of enamel growth in the molar teeth of early hominids. *Nature, 326*:493–496.
Boaz, N. T. (1977). Paleoecology of Plio-Pleistocene Hominidae in the Lower Omo basin, Ethiopia. Ph.D. Dissertation, University of California, Berkeley.
Boaz, N. T. and Howell, F. C. (1977). A gracile hominid cranium from Upper Member G of the Shungura Formation, Ethiopia. *Amer. J. Phys. Anthropol., 46*:93–108.
Brown, F. H. and Feible, C. S. (1986). Revision of lithostratigraphic nomenclature in the Koobi Fora region, Kenya. *J. Geol. Soc. London, 143*:297–310.
Brown, F. H., McDougall, I., Davies, T. and Maier, R. (1985). An integrated Plio-Pleistocene chronology for the Turkana Basin. *In* E. Delson (ed.), *Ancestors: the hard evidence*, pp. 82–90. Alan R. Liss, New York.
Chamberlain, A. T. and Wood, B. A. (1985). A reappraisal of variation in hominid mandibular corpus dimensions. *Amer. J. Phys. Anthropol., 66*:399–405.
Coppens, Y. (1970). Les restes d'hominides des series inferieures et moyennes des formations plio-villafranchiennes de l'Omo en Ethiopie. *C. R. Acad. Sci. (Paris), 271*:2286–2289.
Coppens, Y. (1971). Les restes d'hominides des series superieures des formations plio-villafranchiennes de l'Omo en Ethiopie. *C. R. Acad. Sci. (Paris), 272*:36–39.
Coppens, Y. (1973a). Les restes d'hominides des series inferieures et moyennes des formations plio-villafranchiennes de l'Omo en Ethiopie (recoltes 1970, 1971 et 1972). *C. R. Acad. Sci. (Paris), 276*:1823–1826.
Coppens, Y. (1973b). Les restes d'hominides des series superieures des formations plio-villafranchiennes de l'Omo en Ethiopie (recoltes 1970, 1971 et 1972). *C. R. Acad. Sci. (Paris), 276*:1981–1984.
Coppens, Y. (1980). The difference between *Australopithecus* and *Homo;* preliminary conclusions from the Omo research expedition's studies. *In* L. Konigsson (ed.), *Current argument on early Man*, pp. 207–225. Pergamon Press, New York.
Grine, F. E. (1981). Trophic differences between "gracile" and "robust" australopithecines: a scanning electron microscope analysis of occlusal events. *S. Afr. J. Sci., 77*:203–230.
Grine, F. E. (1984). The deciduous dentition of the Kalahari San, the South African Negro and the South African Plio-Pleistocene hominids. Ph.D. Thesis, University of the Witwatersrand, Johannesburg.
Grine, F. E. (1985). Australopithecine evolution: the deciduous dental evidence. *In* E. Delson (ed.), *Ancestors: the hard evidence*, pp. 153–167. Alan R. Liss, New York.
Howell, F. C. (1969a). Remains of Hominidae from Pliocene/Pleistocene formations in the Lower Omo Basin, Ethiopia. *Nature, 223*:1234–1239.
Howell, F. C. (1969b). Hominid teeth from White Sands and Brown Sands localities, Lower Omo Basin (Ethiopia). *Quaternaria, 11*:47–64.
Howell, F. C. (1978). Hominidae. *In* V. J. Maglio and H. B. S. Cooke (eds.), *Evolution of African mammals*, pp. 154–248. Harvard University Press, Cambridge, MA.
Howell, F. C. and Coppens, Y. (1973). Deciduous teeth of Hominidae from the Pliocene/Pleistocene of the Lower Omo Basin, Ethiopia. *J. Hum. Evol., 2*:461–472.
Howell, F. C. and Coppens, Y. (1974). Inventory of remains of hominidae from Pliocene/Pleistocene formations of the Lower Omo Basin, Ethiopia (1967–1972). *Amer. J. Phys. Anthropol., 40*:1–16.
Howell, F. C. and Coppens, Y. (1976). An overview of Hominidae from the Omo succession, Ethiopia. *In* Y. Coppens *et al.* (eds.), *Earliest Man and environments in the Lake Rudolf Basin*, pp. 522–532. University of Chicago Press, Chicago, IL.
Hunt, K. and Vitzthum, V. J. (1986). Dental metric assessment of the Omo fossils: implications for the phylogenetic position of *Australopithecus africanus. Amer. J. Phys. Anthropol., 71*:141–155.
Kouchi, M. and Koizumi, K. (1985). An analysis of error in craniometry. *J. Anthropol. Soc. Nippon, 93*:409–424.
Mahler, P. (1973). Metric variation in the pongid dentition. Ph.D. Dissertation, University of Michigan, University Microfilms, MI.
Rak, Y. (1983). *The australopithecine face.* Academic Press, New York.

Robinson, J. T. (1956). The dentition of the Australopithecinae. *Mem. Tvl. Mus.*, 9:1–179.
Sperber, G. H. (1974). Morphology of the cheek teeth of early South African hominids. Ph.D. Thesis, University of the Witwatersrand, Johannesburg.
Tobias, P. V. (1967). *The cranium and maxillary dentition of* Australopithecus (Zinjanthropus) boisei. vol. 2, Olduvai Gorge. Cambridge University Press, Cambridge.
Utermohle, C. J. and Zegura, S. L. (1982). Intra- and interobserver error in craniometry: a cautionary tale. *Amer. J. Phys. Anthropol.*, 57:303–310.
von Koenigswald, G. H. R. (1967). Evolutionary trends in the deciduous molars of the Hominidae. *J. Dent. Res., Suppl.* 46:779–786.
Walker, A., Leakey, R. E., Harris, J. M. and Brown, F. H. (1986). 2.5-Myr *Australopithecus boisei* from west of Lake Turkana, Kenya. *Nature*, 322:517–522.
Wallace, J. A. (1972). The dentition of the South African early hominids: a study of form and function. Ph.D. Thesis, University of the Witwatersrand, Johannesburg.
White, T. D. (1977). The anterior mandibular corpus of early African Hominidae: functional significance of shape and size. Ph.D. Dissertation, University of Michigan, Ann Arbor, MI.
White, T. D., Johanson, D. C., and Kimbel, W. H. (1981). *Australopithecus africanus:* its phyletic position reconsidered. *S. Afr. J. Sci.*, 77:445–470.
Wood, B. A. and Abbott, S. A. (1983). Analysis of the dental morphology of Plio-Peistocene hominids. I. Mandibular molars: crown area measurements and morphological traits. *J. Anat.*, 136:197–219.
Wood, B. A. and Uytterschaut, H. (1987). Analysis of the dental morphology of Plio-Pleistocene hominids. III. Mandibular premolar crowns. *J. Anat.* 16:625–641.
Wood, B. A., Abbott, S. A. and Graham, S. H. (1983). Analysis of the dental morphology of Plio-Pleistocene hominids. II. Mandibular molars: study of cusp areas, fissure pattern, and cross-sectional shape of the crown. *J. Anat.*, 137:287–314.

New Craniodental Fossils of *Paranthropus* from the Swartkrans Formation and Their Significance in "Robust" Australopithecine Evolution

14

FREDERICK E. GRINE

The site of Swartkrans has long been recognized as the richest single depository of early hominid fossils in southern Africa. Work by R. Broom and J. T. Robinson between 1948 and 1953 and by C. K. Brain between 1965 and 1979 has resulted in the recovery of well over 200 craniodental fossils of *Paranthropus* from the Member 1 "Hanging Remnant" breccia. Mann (1975) has estimated that at least 113 individuals are represented by this sample, while more recent estimates by Brain (1981) place this number between 85 and 87.

Work by Brain from 1979 to 1986 concentrated on the excavation of *in situ* decalcified breccia and/or unconsolidated sediments within the cave, and these efforts have served not only to further clarify the complex stratigraphy of the site, but to produce over 100 new hominid fossils from this already rich locality (Brain, 1982, this volume, chapter 20).

Prior to these most recent excavations, the Swartkrans Formation was considered to comprise two Members (Brain, 1976; Butzer, 1976); all *Paranthropus* fossils were recognized as having derived from the Member 1 breccia, and only a handful of fossils, probably representing a single individual of *Homo* (Broom and Robinson, 1949; Robinson, 1953), were known from Member 2. Brain's work, however, has revealed the presence of five distinct Members within the cave (Brain, this volume, chapter 20), and he has demonstrated that Member 1 consists of two separate masses: the "Hanging Remnant," which clings unsupported to the northern dolomite wall of the cave and from which all of the Member 1 fossils procured earlier derive, and the "Lower Bank," which remains in place on the cave floor (Brain, 1982, 1985, this volume, chapter 20). Brain's excavations also have recovered *Paranthropus* fossils from the Member 1 "Lower Bank" and Member 2 deposits (Brain, 1982, 1985) as well as from the previously unknown Member 3 (Brain, 1985, this volume, chapter 20). Over 30 postcranial bones (Susman, this volume, chapter 10) and some 71 individually numbered jaws and teeth have been recovered *in situ* from these sediments.

With regard to the craniodental fossils, some 20 specimens come from Member 1 "Lower Bank," 32 derive from Member 2, 11 come from Member 3, and nine isolated teeth derive from co-ordinates along the Member 1–2 Interface where the separation between these units is not readily apparent. These specimens are described in detail elsewhere (Grine, 1988).

Comparisons between the fossils comprising the Member 1 "Hanging Remnant" sample and those from the Member 1 "Lower Bank" revealed no instance of definite or even probable individual association. Similarly, none of the jaws and teeth from the Member 1–2 Interface, Member 2, or Member 3 appears to be associated with any of the fossils from the Member 1 "Hanging Remnant." There are, however, associations between the fossils comprising each of the recently excavated lithostratigraphic samples and between specimens from the Member 1–2 Interface and Member 2 samples.

Thus, the 20 individually numbered fossils that constitute the Member 1 "Lower Bank" sample are considered to represent between 13 and 16 individuals. One fragmentary dental specimen from this sample cannot be identified at the generic level, and one specimen is attributable to *Homo;* the remaining 11 to 14 individuals are attributable to *Paranthropus*. Of the nine teeth from the Member 1–2 Interface, all but three are likely associated with fossils from Member 2, and of the remaining three, one cannot be identified at the generic level while the other two teeth belong to *Paranthropus*. The 32 numbered jaws and teeth from Member 2 are considered to represent between 19 and 24 individuals; two fragmentary dental specimens cannot be identified at the genus level, five specimens are attributable to *Homo*, while the remaining 12 to 17 represent *Paranthropus*. Of the 11 numbered fossils from Member 3, 10 are identifiable at the generic level, and these appear to represent eight or nine individuals of *Paranthropus*. Thus, between 32 and 41 individuals of *Paranthropus* are likely represented by the new fossil collections from Member 1 "Lower Bank," Member 2, and Member 3. While the Member 1 "Lower Bank" sample does not in itself add significantly to the size of the Member 1 "Hanging Remnant" sample, it is of potential interest in view of the possibility that the "Lower Bank" specimens may have accumulated at a somewhat earlier time and under different environmental conditions than the "Hanging Remnant" sample (Brain, this volume, chapter 20). In addition, the new specimens provide evidence of *Paranthropus* in the time-successive Member 2 and 3 sediments, and this affords the unique opportunity to analyze the relationships between chronologically successive samples of a single hominid genus at a single southern African locality.

Comparisons of Swartkrans *Paranthropus* Samples

Only four of the new *Paranthropus* specimens preserve gnathic parts, and all are from Member 2. Specimen SKX 162 (Fig. 14.1), which consists of an incomplete right maxilla with the $\overline{C-P^4}$, and which is likely associated with SKX 163 (an incomplete Rdm^2), SKX 308 (a RI^2) and $\overline{SKX\ 1788}$ (a LI^2), represents a young juvenile individual in which the lateral incisors were erupting at the time of death. Morphologically, this maxillary fragment is similar to Member 1 "Hanging Remnant" specimens in the slope of the naso-alveolar clivus, the slight convexity present over the I^1 and concavity over the I^2 roots, and in the anteriorly flat nature of the palate.

Specimen SKX 265, an edentulous left maxilla of a subadult individual (Fig. 14.2), is similar to Member 1 "Hanging Remnant" specimens except in the relatively anterior position of the anterior nasal spine, which is situated in the same coronal plane as the root of the lateral nasal margin. The floor of the nose of SXK 265 is demarcated from the naso-alveolar clivus by a low ridge that runs from the anterior nasal spine to the lateral nasal margin, and this degree of demarcation tends to be somewhat more marked than in other Swartkrans maxillae. In addition, the naso-alveolar clivus is virtually flat from the midline to a sharp, anterolateral projecting bony ridge over the $\underline{C}$ root.

The right and left mandibular corpora, SKX 4446 and SKX 5013 (Fig. 14.3), from Member 2

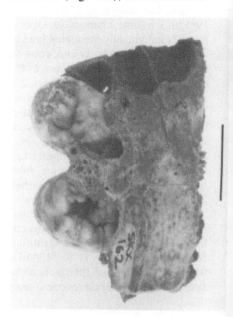

FIGURE 14.1. Palatal view of SKX 162 right maxilla with P^3 and P^4 in place. Scale bar = 1 cm.

FIGURE 14.2. Vertical view of SKX 265 left maxilla. Scale bar = 1 cm.

are similar to *Paranthropus* specimens from Member 1 "Hanging Remnant" in both size and morphology.

The vast bulk of the new collection is made up of isolated and/or associated teeth. These specimens find morphologic equivalents among the Member 1 "Hanging Remnant" specimens, and while an exhaustive element-by-element morphological analysis is beyond the scope of this paper, it is evident that the morphological variability within the Member 1 "Hanging Remnant" sample is as great as, if not greater than, the variability evinced by the samples from Member 1 "Lower Bank," Member 2, and Member 3.

The mesiodistal (MD) and buccolingual (BL) diameters of the new Swartkrans *Paranthropus* specimens are recorded in Table 14.1. With regard to these specimens, SKX 240 and SKX 242 are I^1 antimeres, SKX 308 and SKX 1788 are probable I^2 antimeres (associated with SKX 162 and SKX 163), and SKX 310 and SKX 312 are $\underline{C}$ antimeres. The diameters of the specimens listed in Table 14.1 are compared to the corresponding values for the Member 1 "Hanging Remnant" sample in Figs. 14.4 through 14.8. It is apparent that several of the recently acquired teeth

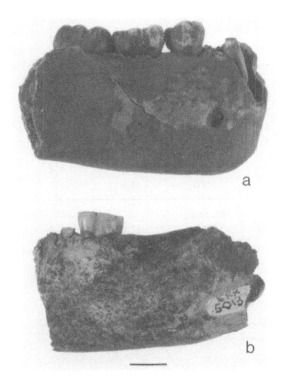

FIGURE 14.3. (a) SKX 4446; and (b) SKX 5013, mandibular corpora in lateral view. Scale bar = 1 cm.

Table 14.1. Tooth Diameters of New *Paranthropus* Specimens from Swartkrans

Tooth	Member	Specimen	MD measurement	MD estimate	BL measurement	BL estimate
Maxillary permanent						
I¹	1	SKX 3300	8.1	8.3	7.0	7.0
	1	SKX 19031	9.6	9.8	7.4	7.4
	2	SKX 240	8.6	8.8	7.0	7.0
	2	SKX 242	8.5	8.8	7.2	7.2
	2	SKX 271	9.5	9.8	8.1	8.1
	2	SKX 313	8.5	—	7.4	7.4
	3	SKX 27524	8.0	8.2	6.3	6.3
I²	1–2ª	SKX 1788	6.3	6.3	5.7	5.8
	2	SKX 308	6.5	6.5	6.4	6.4
C	2	SKX 162	8.8	8.8	9.4	9.4
	2	SKX 310	7.8	—	9.1	9.1
	2	SKX 312	7.4	—	9.4	9.4
	3	SKX 25296	9.7	9.7	9.1	9.1
	3	SKX 28724	8.9	8.9	9.8	8.8
P³	1	SKX 7781	9.6	9.6	—	—
	2	SKX 162	10.0	10.0	14.5	14.5
P⁴	1	SKX 3354	9.9	10.5	14.2	14.2
	2	SKX 162	10.8	10.8	15.8	15.8
	3	SKX 26625	12.1	12.1	16.1	16.1
M¹	1	SKX 3601	11.5	(12.4)	14.4	14.4
M²	1	SKX 3355	12.3	(12.8)	15.0	15.0
M³	3	SKX 21841	15.5	15.5	16.6	16.6
Maxillary deciduous						
di¹	1	SKX 16060	6.6	6.8	4.3	4.3
	1–2	SKX 2003	5.4	—	4.2	4.2
dm²	2	SKX 163	—	—	12.7	12.9
	3	SKX 32832	10.1	10.7	11.8	11.8
Mandibular permanent						
I₁	1	SKX 5004b	5.2	5.4	6.2	6.2
	2	SKX 3559	5.2	—	6.7	6.7
	3	SKX 26979	5.7	(5.9)	6.1	6.1
	3	SKX 35416	—	—	6.1	6.1
I₂	1–2	SKX 7992	5.8	—	6.5	6.5
	2	SKX 1017	5.7	—	6.5	6.5
	2	SKX 1313	5.8	6.5	7.2	7.2
C	1	SKX 5007	7.3	7.5	8.2	8.2
	1	SKX 6013	6.8	6.9	7.7	7.7
	2	SKX 241	6.7	7.1min.	7.4	7.4
P₃	2	SKX 311	8.7	9.0	11.9	11.9
P₄	2	SKX 4446	11.7	11.9	12.5	12.5
	3	SKX 32162	10.8	11.3min.	—	—
M₁	1	SKX 5023	13.2	(14.1)	12.8	12.8
	2	SKX 4446	15.1	15.5	14.3	14.3
	2	SKX 5013	13.2	(13.6)	12.0	12.0
M₂	2	SKX 4446	17.1	17.1	15.8	15.8
	3	SKX 19892	—	—	14.8	14.8
M₃	1	SKX 5002	17.6	17.8	13.9	13.9
	1	SKX 5014	17.2	17.2	15.0	15.0

ªMember by individual association.

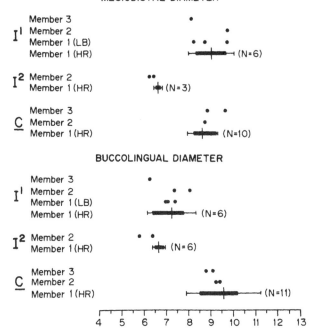

FIGURE 14.4. Comparison of maxillary anterior permanent tooth diameters recorded for new Swartkrans *Paranthropus* specimens and the Member 1 "Hanging Remnant" (HR) sample. Vertical line = sample mean; horizontal line = mean ± 2SD; horizontal bar = observed sample range. Scale in mm.

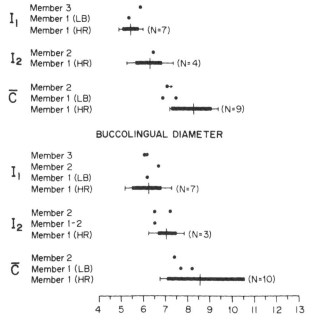

FIGURE 14.5. Comparison of mandibular anterior permanent tooth diameters recorded for new Swartkrans *Paranthropus* specimens and the Member 1 "Hanging Remnant" (HR) sample. Key same as for Fig. 14.4.

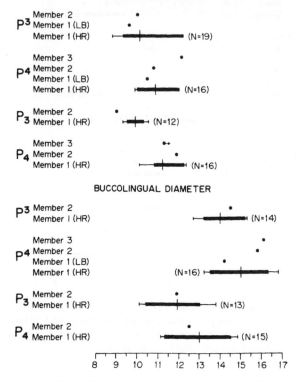

FIGURE 14.6. Comparison of maxillary and mandibular premolar diameters recorded for new Swartkrans *Paranthropus* specimens and the Member 1 "Hanging Remnant" (HR) sample. Key same as for Fig. 14.4.

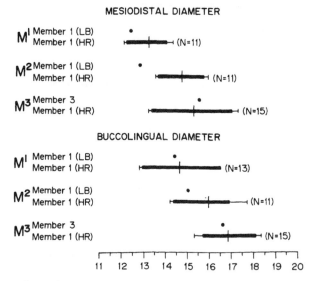

FIGURE 14.7. Comparison of maxillary permanent molar diameters recorded for new Swartkrans *Paranthropus* specimens and the Member 1 "Hanging Remnant" (HR) sample. Key same as for Fig. 14.4.

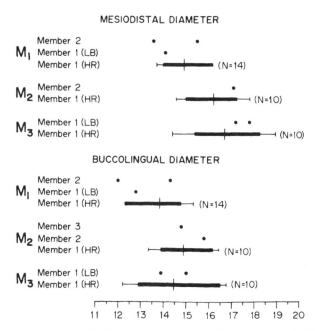

FIGURE 14.8. Comparison of mandibular permanent molar diameters recorded for new Swartkrans *Paranthropus* specimens and the Member 1 "Hanging Remnant" (HR) sample. Key same as for Fig. 14.4.

increase slightly the observed ranges for some of the tooth diameters from the Member 1 "Hanging Remnant" sample (most notably in the MD and BL diameters of the I^2, the MD diameters of the C, P_3, P^4 and M^2, and the MD and BL diameters of the M_1). In most of the instances in which diameters recorded for the new teeth fall outside the observed ranges for the Member 1 "Hanging Remnant" samples, the latter consist of fewer than nine specimens. Notwithstanding the rather small sample of homologous elements from Members 2 and 3, there is no discernible tendency for tooth size to either increase or decrease through time from Member 1 through Member 3.

In view of their morphological and metrical similarities, the recently excavated *Paranthropus* fossils undoubtedly represent the same species as the Member 1 "Hanging Remnant" specimens. The permanent and deciduous tooth diameters for the combined Swartkrans sample are recorded in Tables 14.2–14.4. It is evident that when sample sizes are taken into consideration, the variation in tooth diameters, as expressed by the coefficients of variation (*CV*), displayed by the combined Swartkrans *Paranthropus* sample is of comparable magnitude to that shown by other early hominid samples. Indeed, the Swartkrans sample *CV*s for permanent cheek-tooth diameters tend to be lower than those recorded for the *Australopithecus afarensis* sample from the Laetolil Beds and Hadar Formation, and for the Sterkfontein Member 4/Makapansgat Member 3 *Australopithecus africanus* sample (cf. Tables 14.6–14.9 below) (see also Kimbel and White, this volume, chapter 11). Since the *A. afarensis* and *A. africanus* samples each appear to span some 0.5 to 0.7 Myr, the amount of time that is spanned by the Swartkrans *Paranthropus* sample from Member 1 through Member 3 is of some significance in comparisons of variability within species samples.

Ages of the Swartkrans *Paranthropus* Samples

At present, it is unclear as to precisely how much time elapsed during and between the accumulations of the Member 1, 2 and 3 sediments. Vrba (1975, 1982) found the Swartkrans Member 1 "Hanging Remnant" bovids to be most similar to those from Olduvai Beds I and lower II and

Table 14.2. MD Diameters of Swartkrans *Paranthropus* Permanent Teeth

Tooth	N	$\bar{X}$	SD	SE	CV	Observed range
I^1	11	9.04	0.61	0.18	6.75	8.2– 9.8
I^2	5	6.58	0.19	0.09	2.89	6.3– 6.8
$\underline{C}$	13	8.80	0.40	0.11	4.55	8.3– 9.7
P^3	21	10.10	0.65	0.14	6.44	9.3–12.2
P^4	19	10.91	0.54	0.12	4.95	10.2–12.1
M^1	12	13.12	0.57	0.17	4.34	12.2–14.0
M^2	12	14.55	0.79	0.23	5.43	12.8–15.7
M^3	16	15.28	0.99	0.25	6.48	13.4–17.0
I_1	9	5.52	0.28	0.09	5.07	5.1– 5.9
I_2	5	6.37	0.47	0.21	7.38	5.7– 6.9
$\underline{C}$	11	8.12	0.69	0.21	8.50	6.9– 9.1
P_3	13	9.85	0.39	0.11	3.96	9.0–10.5
P_4	17	11.27	0.57	0.14	5.06	10.5–12.5
M_1	17	14.85	0.69	0.17	4.65	13.6–16.2
M_2	11	16.30	0.82	0.25	5.03	15.0–17.3
M_3	12	16.83	1.09	0.31	6.48	15.4–18.3

Table 14.3. BL Diameters of Swartkrans *Paranthropus* Permanent Teeth

Tooth	N	$\bar{X}$	SD	SE	CV	Observed range
I^1	12	7.24	0.55	0.16	7.60	6.3– 8.1
I^2	8	6.53	0.33	0.12	5.05	5.8– 6.9
C	15	9.45	0.75	0.19	7.94	8.5–11.1
P^3	15	14.04	0.64	0.16	4.56	13.2–15.2
P^4	19	15.08	0.90	0.21	5.97	13.5–16.3
M^1	14	14.61	0.89	0.24	6.09	12.9–16.5
M^2	12	15.88	0.88	0.25	5.54	14.5–16.9
M^3	16	16.82	0.72	0.19	4.28	15.7–18.1
I_1	11	6.25	0.45	0.13	7.20	5.5– 6.8
I_2	6	6.89	0.40	0.16	5.80	6.5– 7.5
C	13	8.38	0.89	0.25	10.62	7.1–10.5
P_3	14	11.94	0.89	0.24	7.25	10.7–13.0
P_4	16	12.96	0.91	0.23	7.02	11.3–14.5
M_1	17	13.69	0.86	0.21	6.28	12.0–14.8
M_2	12	14.96	0.76	0.22	5.08	13.9–16.2
M_3	12	14.46	1.06	0.31	7.33	12.9–16.5

Table 14.4. MD and BL Diameters of Swartkrans *Paranthropus* Deciduous Teeth

Tooth	N	$\bar{X}$	SD	SE	CV	Observed range
MD diameters						
di^1	2^a	6.78				6.7–6.8
di^2	0					
d^c	0					
dm^1	1	9.10				
dm^2	6^a	10.90	0.43	0.18	3.94	10.4–11.3
di_1	1	4.00				
di_2	2	4.90				4.8–5.0
d_c	4	6.03	0.13	0.06	2.16	5.9–6.2
dm_1	6	10.38	0.43	0.18	4.14	10.0–11.1
dm_2	10	12.77	0.45	0.14	3.52	12.0–13.5
BL diameters						
di^1	3^a	4.60	0.53	0.31	11.52	
di^2	0					
d^c	0					
dm^1	2	10.05				10.0–10.1
dm^2	6^a	12.10	0.41	0.17	3.39	11.5–12.7
di_1	1	3.90				
di_2	2	4.25				3.8–4.7
d_c	4	5.13	0.50	0.25	9.75	4.6–5.8
dm_1	6	8.43	0.65	0.27	7.71	7.9–9.6
dm_2	9	10.83	0.70	0.23	6.46	9.7–12.0

[a] Includes new Swartkrans specimens.

to those from Shungura Member G, and Harris and White (1979) correlated the Swartkrans Member 1 suid with forms of sub-Okote age in the Koobi Fora Formation, suggestive of an age of between about 1.64 and 1.8 Myr (Brown et al., 1985; Brown and Feibel, this volume, chapter 22). Delson (1984) has noted that while the Swartkrans Member 1 monkeys show strong ties to his African cercopithecid (AC) zone 4, this fauna is best correlated with AC zone 5 because of the absence of *Parapapio* and *Cercopithecoides williamsi*; this suggests a correlative age of between about 1.8 and 2.0 Myr. Thus, on faunal evidence, Swartkrans Member 1 "Hanging Remnant" would appear to date between about 1.64 and 2.0 Myr, with a suggested midpoint of some 1.8 Myr.

The Swartkrans Member 2 bovids were divided into earlier and later "SKB" groups by Vrba (1975), who suggested that they spanned the last 0.5 Myr, while Delson (1984) has posited a cercopithecid correlation with the upper part of his AC zone 6 or perhaps the lower part of AC zone 7, with an approximate age of 1.5 Myr. Most recently, however, Delson (this volume, chapter 21) has argued that there are no observable differences between the cercopithecid fauna from Member 1 "Hanging Remnant" and that excavated by Brain from Member 2.

Attempts to obtain a paleomagnetic sequence for the Swartkrans deposits have been unsuccessful (Brock et al., 1977), while Vogel (1985) has provided preliminary relative thermoluminescence (TL) signal values for quartz sand grains from Swartkrans Members 1, 2, and 3 of 12.80, 12.10 and 6.21, respectively. Assuming a constant rate of change in the thermoluminescence signal with time, and assuming that TL can be used for a range of more than a few hundred thousand years, if the TL signal of 12.80 obtained by Vogel (1985) for the Member 1 breccia is related to an age of between 1.8 and 1.7 Myr, the respective ages for Member 2 would be on the order of 1.7–1.6 Myr, and those for Member 3 would be on the order of 0.88–0.82 Myr. While these suggestions are indeed tantalizing, the assumptions underlying their validity have yet to be substantiated with reference to a system such as that at Swartkrans.

Brain (this volume, chapter 20) has suggested that the cyclic nature of the depositional and erosional episodes that characterize the Swartkrans Formation may reflect wider climatic events such as the glacial and interglacial episodes. Glacial and interglacial fluctuations have been shown to have followed cycles of 40 kyr from the early Pliocene to about 0.9 Myr ago, and cycles of 100 kyr from 0.9 Myr ago to the present (Prentice and Denton, this volume, chapter 24). At least 17 cycles are known to have occurred since the Olduvai Event (Fink and Kukla, 1977), which would appear to correlate with the age of Swartkrans Member 1. As noted by Brain (this volume, chapter 20), however, not all of these cycles can be discerned within the Swartkrans deposits, and thus it is not possible to utilize them in any attempt to date the sediments. It would even appear that the suggestion relating deposition to interglacial events and erosion to cold and damp glacial times (Brain, this volume, chapter 20) posits an unusual relationship for African fossil accumulations (Klein, this volume, chapter 29; pers. comm.).

It would appear, then, that as with all South African sites, the best estimates for the ages of the Swartkrans deposits are those based upon faunal correlations with radiometrically dated sediments in eastern Africa. While such estimates are available for Swartkrans Member 1 (c. 1.8 Myr ± 0.2) and Member 2 (c. 1.7–1.5 Myr), they have not yet been determined for Member 3. Nevertheless, the TL estimate obtained by Vogel (1985) for Member 2 corresponds reasonably well with the faunal age and thus an age of approximately 1.0 Myr for Member 3 might be considered reasonable. It would certainly seem that at least several hundred thousand years separated Member 1 from Member 3.

Comparison of the Swartkrans and Kromdraai *Paranthropus* Samples

As was noted above, the morphological equivalence and metrical similarities among the Swartkrans Member 1 "Hanging Remnant," Member 1 "Lower Bank," Member 2 and Member 3 *Paranthropus* samples indicate that all represent a single species. This is recognized as *P. robustus* (or *Australopithecus robustus*) by most workers, including those contributing to this volume, although Howell (1978) and Grine (1982, 1985) have argued that the Swartkrans sample should be accorded separate status as *P. crassidens*, the specific designation applied originally by Broom (1949), and that the name *P. robustus* Broom, 1938 should be utilized in reference to the Kromdraai hominid sample.

Given the expanded sample sizes now available for the Swartkrans *Paranthropus* teeth, it is well to compare these with the specimens comprising the other "australopithecine" samples. For present purposes, such comparisons are restricted to the sizes of teeth for which specimens from Kromdraai are available. Maxillary permanent incisors and canines and mandibular permanent canines are not known for Kromdraai *Paranthropus* specimens; similarly, maxillary deciduous teeth and mandibular deciduous central incisors are unknown from Kromdraai.

The MD diameters of the mandibular permanent incisors from Swartkrans (*P. crassidens*) are compared to those for other "australopithecine" samples in Table 14.5. The I_1s of *P. crassidens*

Table 14.5. MD Diameters of Mandibular Permanent Incisors[a]

Tooth	Sample	N	$\overline{X}$	SD	SE	CV	Observed range
I_1	A. afarensis	3	6.63	1.23	0.71	18.55	5.6–8.0
	A. africanus	3	6.00	0.35	0.20	5.83	5.6–6.2
	P. robustus	1	4.85				
	P. crassidens	9	5.52	0.28	0.09	5.07	5.1–5.9
	P. boisei	5	5.36	0.68	0.29	12.69	4.2–5.9
I_2	A. afarensis	5	6.46	0.66	0.30	10.22	5.7–7.2
	A. africanus	3	7.27	0.06	0.03	0.83	7.2–7.3
	P. robustus	1	5.65				
	P. crassidens	5	6.37	0.47	0.21	7.38	5.7–6.9
	P. boisei	4	6.30	0.26	0.13	4.13	6.0–6.6

[a]Data in this and succeeding tables for *A. afarensis* and *P. boisei* after White *et al.* (1981); data for *A. africanus* includes Stw specimens of certain Sterkfontein Member 4 derivation.

and *P. boisei* tend to be smaller than those of *A. africanus* and *A. afarensis*, while the *P. crassidens*, *P. boisei* and *A. afarensis* means for the I_2s are similar to one another and smaller than the corresponding *A. africanus* sample mean. In both instances, the incisors from the single *P. robustus* individual (KB 5223) are slightly smaller than the smallest Swartkrans homologues.

The MD and BL diameters of the maxillary and mandibular permanent cheek-teeth are compared in Tables 14.6–14.9 and Figs. 14.9–14.12. In all instances the *A. afarensis* means are the smallest and those for *P. boisei* the largest, and the *A. africanus* sample means are larger than those for *A. afarensis* except for the MD diameter of the P_3 and the BL diameter of the M_1, in which instances the sample averages are nearly identical. At the same time, the *P. crassidens* sample means fall between the larger *P. boisei* and smaller *Australopithecus* values except for the MD and BL diameters of the P_3 and the BL diameter of the M_3, in which instances the *A. afarensis* and/or *A. africanus* means are nearly the same as those for the *P. crassidens* sample.

Metric values for the Kromdraai *P. robustus* teeth are most similar to the *P. crassidens* sample means with regard to the MD and BL diameters of the upper and lower premolars, the BL diameter of the M^2, and the BL diameter of the M_3, although it should be noted that the *P. robustus*, *P. crassidens* and *A. africanus* values are all similar to one another in the MD and BL diameters of the P_3 and the BL diameter of the M_3. On the other hand, the *P. robustus* values tend to be smaller than those of *P. crassidens* in permanent molar diameters (save for the BL diameters of the M^2 and M_3, in which cases the sample means are nearly identical). The *P. robustus* permanent molar values are intermediate between the smaller *A. africanus* and larger *P. crassidens* means (MD and BL M^3), most similar to the *A. africanus* means (MD and BL M^1 and MD M_3), or even smaller than the *A. africanus* means (MD and BL diameters of M_1 and M_2). Only with regard to the MD and BL diameters of the M_2 and the MD diameter of the M_1 do all Kromdraai specimens fall below the observed *P. crassidens* sample ranges. Nevertheless, it is clear that in a number of instances the *A. africanus* and *P. crassidens* sample ranges overlap

Table 14.6. MD Diameters of Maxillary Permanent Cheek-Teeth

Tooth	Sample	N	$\bar{X}$	SD	SE	CV	Observed range
P^3	*A. afarensis*	7	8.73	0.57	0.22	6.53	7.5–9.3
	A. africanus	15	9.26	0.32	0.08	3.46	8.9–9.8
	P. robustus	1	10.50				
	P. crassidens	21	10.10	0.65	0.14	6.44	9.3–12.2
	P. boisei	6	10.62	0.88	0.36	8.29	9.5–11.8
P^4	*A. afarensis*	10	8.98	0.61	0.19	6.79	7.6–9.7
	A. africanus	11	9.61	0.48	0.15	5.00	9.0–10.6
	P. robustus	1	10.75				
	P. crassidens	19	10.91	0.54	0.12	4.95	10.2–12.1
	P. boisei	4	12.08	0.30	0.15	2.48	11.7–12.4
M^1	*A. afarensis*	8	12.21	1.06	0.38	8.68	10.8–13.8
	A. africanus	18	12.54	0.73	0.17	5.82	11.2–13.8
	P. robustus	4	12.55	0.98	0.49	7.81	11.1–13.2
	P. crassidens	12	13.12	0.57	0.17	4.34	12.2–14.0
	P. boisei	5	14.96	0.89	0.40	5.95	13.4–15.6
M^2	*A. afarensis*	5	12.78	0.50	0.22	3.91	12.1–13.5
	A. africanus	14	13.89	0.77	0.21	5.54	12.6–15.2
	P. robustus	1	13.80				
	P. crassidens	12	14.55	0.79	0.23	5.43	12.8–15.7
	P. boisei	3	16.50	0.82	0.47	4.79	15.6–17.2
M^3	*A. afarensis*	7	12.04	1.35	0.51	11.21	10.9–14.3
	A. africanus	16	13.60	1.61	0.40	11.84	10.8–16.8
	P. robustus	3	14.53	0.15	0.09	1.03	14.4–14.7
	P. crassidens	16	15.28	0.99	0.25	6.48	13.4–17.0
	P. boisei	4	16.93	0.44	0.22	2.60	16.3–17.3

Table 14.7. BL Diameters of Maxillary Permanent Cheek-Teeth

Tooth	Sample	N	$\bar{X}$	SD	SE	CV	Observed range
P^3	A. afarensis	6	12.40	0.74	0.30	5.97	11.3–13.4
	A. africanus	13	12.71	0.95	0.26	7.47	10.7–14.4
	P. robustus	1	13.80				
	P. crassidens	15	14.04	0.64	0.16	4.56	13.2–15.2
	P. boisei	5	15.18	1.35	0.60	8.89	13.8–17.0
P^4	A. afarensis	6	12.08	0.58	0.24	4.80	11.1–12.6
	A. africanus	12	13.05	0.59	0.17	4.52	12.4–14.2
	P. robustus	1	15.30				
	P. crassidens	19	15.08	0.90	0.21	5.97	13.5–16.3
	P. boisei	4	16.23	1.56	0.78	9.61	14.2–17.6
M^1	A. afarensis	8	13.20	0.87	0.31	6.59	12.0–14.6
	A. africanus	15	13.79	0.69	0.18	5.00	12.8–15.2
	P. robustus	4	13.78	0.84	0.42	6.10	12.8–14.8
	P. crassidens	14	14.61	0.89	0.24	6.09	12.9–16.5
	P. boisei	5	16.18	1.03	0.20	6.37	14.9–17.7
M^2	A. afarensis	6	14.65	0.64	0.26	4.37	13.4–15.1
	A. africanus	16	15.38	0.85	0.21	5.53	13.8–17.1
	P. robustus	1	15.95				
	P. crassidens	12	15.88	0.88	0.25	5.54	14.5–16.9
	P. boisei	3	18.77	2.10	1.16	10.71	17.1–21.0
M^3	A. afarensis	7	13.93	1.00	0.38	7.15	13.0–15.5
	A. africanus	17	15.34	1.49	0.36	9.71	13.2–18.1
	P. robustus	3	16.33	0.42	0.24	2.57	16.0–16.8
	P. crassidens	16	16.82	0.72	0.19	4.28	15.7–18.1
	P. boisei	3	18.57	1.69	0.97	9.10	17.4–20.5

Table 14.8. MD Diameters of Mandibular Permanent Cheek-Teeth

Tooth	Sample	N	$\bar{X}$	SD	SE	CV	Observed range
P$_3$	A. afarensis	18	9.61	1.00	0.24	10.41	8.2–12.6
	A. africanus	7	9.74	0.71	0.27	7.29	9.0–11.3
	P. robustus	3	10.25	0.76	0.44	7.42	9.4–10.9
	P. crassidens	13	9.85	0.39	0.11	3.96	9.0–10.5
	P. boisei	5	11.10	1.66	0.74	14.96	8.9–13.0
P$_4$	A. afarensis	14	9.68	1.00	0.27	10.31	7.7–11.1
	A. africanus	8	10.06	0.60	0.21	5.96	9.8–11.3
	P. robustus	1	11.90				
	P. crassidens	17	11.27	0.57	0.14	5.06	10.5–12.5
	P. boisei	9	13.52	1.76	0.59	13.02	10.1–15.6
M$_1$	A. afarensis	18	12.90	0.99	0.23	7.67	10.1–14.6
	A. africanus	12	14.14	1.02	0.29	7.21	12.4–15.2
	P. robustus	3	13.93	1.07	0.62	7.68	12.7–14.6
	P. crassidens	17	14.85	0.69	0.17	4.65	13.6–16.2
	P. boisei	10	16.53	0.97	0.31	5.87	15.4–18.6
M$_2$	A. afarensis	19	14.06	1.05	0.24	7.47	12.1–15.4
	A. africanus	13	15.70	1.00	0.28	6.37	14.4–17.0
	P. robustus	1	14.70				
	P. crassidens	11	16.30	0.82	0.25	5.03	15.0–17.3
	P. boisei	8	18.16	1.33	0.47	7.32	16.4–20.0
M$_3$	A. afarensis	11	14.62	0.76	0.23	5.20	14.0–16.3
	A. africanus	14	15.90	1.28	0.34	8.05	13.7–17.4
	P. robustus	2	15.95				15.5–16.4
	P. crassidens	12	16.83	1.09	0.31	6.48	15.4–18.3
	P. boisei	12	19.98	1.71	0.49	8.56	17.6–22.4

Table 14.9. BL Diameters of Mandibular Permanent Cheek-Teeth

Tooth	Sample	N	$\bar{X}$	SD	SE	CV	Observed range
P_3	A. afarensis	18	10.62	0.89	0.21	8.38	9.5–12.6
	A. africanus	10	11.82	1.00	0.32	8.46	10.2–13.6
	P. robustus	3	11.87	0.75	0.43	6.32	11.1–12.6
	P. crassidens	14	11.94	0.89	0.24	7.45	10.7–13.0
	P. boisei	5	13.04	0.93	0.42	7.13	11.4–13.7
P_4	A. afarensis	13	10.95	0.89	0.25	8.13	9.8–12.8
	A. africanus	10	11.78	0.50	0.16	4.24	10.8–12.4
	P. robustus	1	13.15				
	P. crassidens	16	12.96	0.91	0.23	7.02	11.3–14.5
	P. boisei	9	14.28	1.28	0.43	8.96	12.3–16.5
M_1	A. afarensis	15	12.69	0.86	0.22	6.78	11.0–13.9
	A. africanus	11	12.65	0.92	0.28	7.27	11.3–14.2
	P. robustus	3	12.43	0.65	0.38	5.23	11.8–13.1
	P. crassidens	17	13.69	0.86	0.21	6.28	12.0–14.8
	P. boisei	6	15.48	1.17	0.48	7.56	14.4–17.6
M_2	A. afarensis	18	13.51	1.01	0.24	7.48	12.1–15.2
	A. africanus	15	14.52	0.69	0.18	4.75	13.2–15.5
	P. robustus	2	14.25				13.8–16.2
	P. crassidens	12	14.96	0.76	0.22	5.08	13.9–16.2
	P. boisei	6	16.92	1.11	0.45	6.56	15.8–18.6
M_3	A. afarensis	11	13.38	0.93	0.28	6.95	12.1–14.9
	A. africanus	12	14.40	0.91	0.26	6.32	12.9–15.9
	P. robustus	2	14.35				14.0–14.7
	P. crassidens	12	14.46	1.06	0.31	7.33	12.9–16.5
	P. boisei	9	16.43	1.56	0.52	9.49	14.7–19.2

to a considerable extent (e.g., MD and BL diameters of M^1–M^3, MD and BL diameters of P_3, MD diameter of M_1, and BL diameter of M_3), such that the *P. robustus* values fall beyond the limits of the *P. crassidens* sample ranges about as often as they fall beyond the limits of the *A. africanus* sample ranges.

The MD and BL diameters of mandibular deciduous teeth for which Kromdraai specimens are available are recorded in Tables 14.10 and 14.11 and depicted in Figs. 14.13 and 14.14. As was noted for the mandibular permanent incisors, the Swartkrans and Kromdraai di_2s tend to be smaller than *A. africanus* and/or *A. afarensis* homologues, and the single known Kromdraai tooth is slightly smaller than Swartkrans crowns. *P. crassidens* and *P. boisei* d_cs tend to be smaller than those of *A. africanus* and *A. afarensis* in their MD diameters, and the single *P. robustus* crown is notably the smallest in this diameter. The observed Swartkrans sample range for the BL diameter of the d_c encompasses all other known "australopithecine" values. The comparative size patterns shown by the deciduous molars of these early hominids are similar to those displayed by the permanent cheek-teeth, where *P. boisei* values tend to be larger than those of *P. crassidens;* but whereas *A. afarensis* means are almost uniformly smaller than those of *A. africanus* in the permanent dentition, the deciduous mandibular molars of *A. africanus* tend to be smaller than those of *A. afarensis* (Figs. 14.13 and 14.14), although this is not the case for the deciduous maxillary molars (Grine, 1984). The mandibular deciduous molars of *P. robustus* are on average smaller than those comprising the *P. crassidens* sample. The Kromdraai means for the MD and BL diameters of the dm_1 are intermediate between the corresponding *A. afarensis* and *P. crassidens* sample averages, while the Kromdraai means for the dm_2 are more similar to the *A. afarensis* than to the *P. crassidens* values. Only with regard to the MD diameter of the dm_1 do all *P. robustus* specimens fall outside the limits of the observed Swartkrans sample range; in the other instances the Kromdraai values fall within the lower limits of the Swartkrans sample ranges as well as within the *A. afarensis* sample ranges.

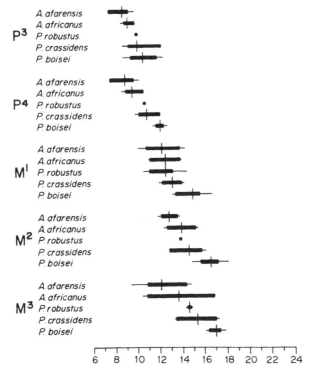

FIGURE 14.9. Comparisons of MD diameters of maxillary permanent cheek-teeth. Vertical line = sample mean, horizontal line = mean ± 2SD, horizontal bar = observed sample range. Data for *A. afarensis* and *P. boisei* from White et al. (1981).

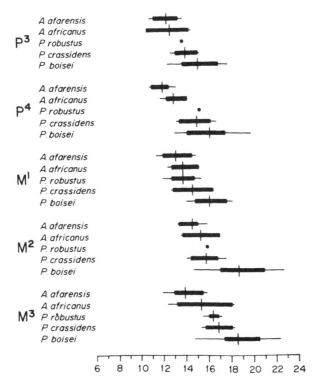

FIGURE 14.10. Comparisons of BL diameters of maxillary permanent cheek-teeth. Key same as for Fig. 14.9.

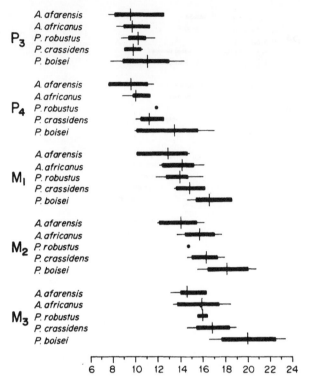

FIGURE 14.11. Comparison of MD diameters of mandibular permanent cheek-teeth. Key same as for Fig. 14.9.

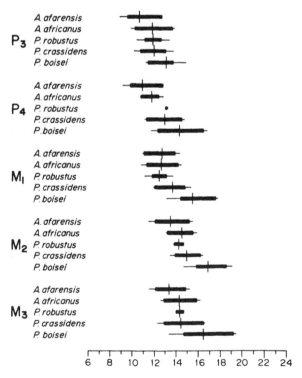

FIGURE 14.12. Comparison of BL diameters of mandibular permanent cheek-teeth. Key same as for Fig. 14.9.

Table 14.10. MD Diameters of Mandibular Deciduous Teeth

Tooth	Sample	N	$\bar{X}$	SD	SE	CV	Observed range
di_2	A. afarensis	2	5.25				4.8–5.7
	A. africanus	1	5.50				
	P. robustus	1	4.70				
	P. crassidens	2	4.90				4.8–5.0
d_c	A. afarensis	3	6.43	0.21	0.12	3.27	6.2–6.6
	A. africanus	2	6.50				6.4–6.6
	P. robustus	1	5.00				
	P. crassidens	4	6.03	0.13	0.06	2.16	5.9–6.2
	P. boisei	1	6.00				
dm_1	A. afarensis	4	9.18	0.49	0.24	5.34	8.5–9.6
	A. africanus	3	8.83	0.38	0.22	4.30	8.5–9.2
	P. robustus	3	9.70	0.30	0.17	3.09	9.4–10.0
	P. crassidens	6	10.38	0.43	0.18	4.14	10.0–11.1
	P. boisei	2	11.65				11.0–12.3
dm_2	A. afarensis	2	12.08				11.7–12.5
	A. africanus	7	11.88	0.73	0.28	6.14	10.9–12.7
	P. robustus	2	12.20				11.9–12.5
	P. crassidens	10	12.77	0.45	0.14	3.52	12.0–13.5
	P. boisei	3	13.72	0.10	0.06	0.73	13.6–13.8

While the *P. robustus* premolars tend to be large by comparison to the Swartkrans sample (i.e., the Kromdraai values are equivalent to or exceed the corresponding Swartkrans sample means), in contrast to the situation regarding the permanent molars, the deciduous molars from Kromdraai tend to be smaller than Swartkrans homologues. Thus, compared to the size relationships of the permanent molars of *P. robustus* and *P. crassidens* the premolars of the former tend to be large, but the comparatively large sizes of the *P. robustus* premolars are not reflected by their deciduous precursors.

The differences in deciduous and permanent cheek-tooth sizes between the Swartkrans and Kromdraai *Paranthropus* samples would seem to be especially noteworthy in light of the ob-

Table 14.11. BL Diameters of Mandibular Deciduous Teeth

Tooth	Sample	N	$\bar{X}$	SD	SE	CV	Observed range
di_2	A. afarensis	3	4.63	0.40	0.23	8.64	4.2–5.0
	P. robustus	1	3.90				
	P. crassidens	2	4.25				3.8–4.7
d_c	A. afarensis	2	5.80				5.8–5.8
	A. africanus	2	5.48				5.3–5.7
	P. robustus	1	5.00				
	P. crassidens	4	5.13	0.50	0.25	9.75	4.6–5.8
	P. boisei	1	5.80				
dm_1	A. afarensis	4	7.95	0.34	0.17	4.28	7.6–8.4
	A. africanus	3	7.62	0.38	0.22	4.99	7.3–8.0
	P. robustus	3	8.15	0.31	0.18	3.80	7.9–8.5
	P. crassidens	6	8.43	0.65	0.27	7.71	7.9–9.6
	P. boisei	2	9.28				9.1–9.5
dm_2	A. afarensis	2	10.08				9.6–10.6
	A. africanus	7	10.17	0.64	0.24	6.29	9.0–10.9
	P. robustus	3	10.03	0.31	0.18	3.09	9.7–10.3
	P. crassidens	9	10.83	0.70	0.23	6.46	9.7–12.0
	P. boisei	3	12.13	0.23	0.13	1.90	12.0–12.4

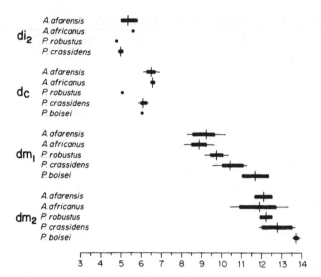

FIGURE 14.13. Comparison of MD diameters of mandibular deciduous teeth. Key same as for Fig. 14.9.

servation that there is no discernible tendency for tooth size to change (either decrease or increase) through time within the Swartkrans Formation. If cheek-teeth displayed a general tendency to increase in size in the *Paranthropus* lineage from sizes such as those displayed by *A. afarensis* or *A. africanus*, then the notable tendency for Kromdraai teeth to be smaller than those of *P. crassidens* and *P. boisei* may be of some evolutionary significance.

In addition to the absolute size differences discussed here, the Kromdraai and Swartkrans mandibular molars display proportional differences that do not appear to be allometrically related (Jungers and Grine, 1986). Thus, it has been noted that with regard to the BL diameters of the trigonid and talonid of the M_1, the specimens from Kromdraai fall at the inflection between the major axis slopes (for MD vs. BL diameters) defined by the *A. africanus* and *P. crassidens* samples. It has been noted also that while the Swartkrans crowns display positive allometric scaling of the BL diameters, the *P. boisei* and smaller *P. robustus* homologues appear to evince

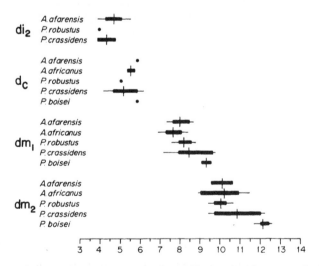

FIGURE 14.14. Comparison of BL diameters of mandibular deciduous teeth. Key same as for Fig. 14.9.

negative allometric increase in breadth. Moreover, the *A. afarensis* molars display a substantial shape difference from the Kromdraai specimens, being relatively broader for a given length. While the observed Kromdraai molar BL diameters are within about 3% of the values that would be predicted for their MD diameters using either the *A. africanus* or *P. crassidens* sample slopes (Jungers and Grine, 1986), the data recorded in that paper indicate that the *A. afarensis* slopes would substantially overpredict BL values for the Kromdraai specimens.

The Kromdraai and Swartkrans teeth also differ in a number of morphological features (Howell, 1978; Grine, 1982, 1984, 1985). In some instances, where discrete traits display intrasample variability (e.g., the presence and development of the tuberculum sextum and tuberculum intermedium on lower molar crowns), the Kromdraai and Swartkrans samples differ in terms of "central tendency," and in most of these cases the Kromdraai teeth resemble more closely those of *A. africanus* and/or *A. afarensis* than those from Swartkrans (Grine, 1982, 1984). In a number of morphological characters the states possessed by the Kromdraai specimens are either shared with *A. africanus* and *A. afarensis* specimens, in contrast to the presumably derived conditions displayed by the Swartkrans and *P. boisei* crowns, or the states evinced by the Kromdraai teeth appear to be intermediate along a morphocline between the states shown by the *A. afarensis* and *A. africanus* specimens, on the one hand, and by Swartkrans and *P. boisei* specimens, on the other.

The morphological and metrical differences between the Kromdraai and Swartkrans dental samples, taken in conjunction with the morphometric similarities among the samples from each of the three time successive Swartkrans Members, suggest very strongly that distinction at the species level should be retained for *P. robustus* and *P. crassidens*. To subsume *P. crassidens* into *P. robustus* serves only to mask the possible evolutionary significance of the Kromdraai hominid sample.

Relationships of *Paranthropus robustus* and *Paranthropus crassidens*

Elsewhere (Grine, 1985) it has been argued that *P. robustus* represents the sister taxon of the *P. crassidens*–*P. boisei* pair, and that *P. crassidens* likely evolved from *P. robustus* or a *P. robustus*-like ancestor (Grine, 1981, 1982, 1985). It has also been proposed that *P. robustus* was morphologically and phylogenetically intermediate between *A. africanus* and *P. crassidens* (Grine, 1982, 1985), in view of the hypothesized sister relationship of *A. africanus* to the "robust" australopithecines (Johanson and White, 1979; White *et al.*, 1981; Rak, 1983; Grine, 1985).

The hypothesis of a sister relationship of *A. africanus* to the "robust" australopithecines, and which posited *P. robustus* as a phylogenetic intermediate between *A. africanus* and *P. crassidens*, however, was based on the strength of only four apparent synapomorphies linking *A. africanus* and the "robust" australopithecines (Grine, 1985): a shorter and more obliquely disposed distal apical d_c edge, an expanded dm^1 distal marginal ridge, a decrease in the degree of lingual protoconal bevelling, and a relative mesial shift in the position of the dm_2 protoconid. The latter two were held to be recognizable along morphoclines (Grine, 1985). By contrast, *P. robustus*, *P. crassidens* and *P. boisei* were linked by some 15 hypothesized synapomorphies of the deciduous dentition (Grine, 1985).

The hypothesis put forward by Johanson and White (1979) and supported by White *et al.* (1981), Rak (1983), and Grine (1985) linking *A. africanus* as a sister of the "robust" australopithecines would seem to be effectively falsified by the unique combination of primitive (i.e., shared with *A. afarensis*) and derived *Paranthropus* features exhibited by KNM-WT 17000 (Walker *et al.*, 1986; Walker and Leakey, this volume, chapter 15; Kimbel *et al.*, this volume, chapter 16; Wood, this volume, chapter 17). Based upon the cranial similarities shown by *P. robustus*, *P. crassidens* and KNM-WT 17000, which almost certainly represents a species distinct from *P. boisei*, *P. robustus* and *P. crassidens* (Kimbel *et al.*, this volume, chapter 16), it could be argued that *P. robustus* and the species represented by KNM-WT 17000 shared a common ancestor that would have been essentially *A. afarensis*-like with the derived features shared by *P. robustus* and KNM-WT 17000, or that *P. robustus* evolved from the species represented by KNM-WT 17000.

The question of the possible evolutionary relationships of *P. robustus* and the species represented by KNM-WT 17000, and the question of whether *P. boisei* is the sister of *P. crassidens* or the species represented by KNM-WT 17000, must await detailed evaluation (or re-evaluation) of the cranial features of these "paranthropine" taxa and the recovery of additional specimens (hopefully including those with teeth) of the species represented by KNM-WT 17000. For the present, however, it is urged that due consideration be given to the possible specific distinctiveness of *P. robustus* and *P. crassidens* in phylogenetic analyses.

Summary and Conclusions

The gnathic and dental remains of between 32 and 41 individuals of *Paranthropus* have been recovered by C. K. Brain from Member 1 "Lower Bank," Member 2 and Member 3 sediments of the Swartkrans Formation. These specimens, which probably span a period of at least several hundred thousand years, are morphologically and metrically very similar to the *Paranthropus* specimens comprising the Member 1 "Hanging Remnant" assemblage, and all are considered to represent a single species. These newly discovered, diachronous Swartkrans samples, in which there are no discernible tendencies for morphological or metrical change, serve to highlight the differences between the Swartkrans and Kromdraai specimens of *Paranthropus*. Taken in conjunction with the morphometric similarities among the time-successive Swartkrans specimens, the differences between the Swartkrans and Kromdraai teeth suggest that species-level distinction be retained for *P. robustus* and *P. crassidens*.

The unique combination of primitive characters, together with derived *Paranthropus* features displayed by the KNM-WT 17000 cranium effectively falsifies the hypothesis that *P. robustus* was an evolutionary intermediate between *A. africanus* and *P. crassidens* (Grine, 1982, 1985). It suggests instead that the species represented by KNM-WT 17000 and *P. robustus* shared a common ancestor that would have been essentially *A. afarensis*-like but with the derived features shared by KNM-WT 17000 and *P. robustus*, or that *P. robustus* evolved from the species represented by KNM-WT 17000. The phylogenetic relationships of *P. robustus*, and the question of whether *P. boisei* represents the sister of *P. crassidens* or the species represented by KNM-WT 17000 must await detailed evaluation (or re-evaluation) of the cranial characters possessed by these fossils. Due consideration should be given to the question of the specific distinctiveness of *P. robustus* and *P. crassidens* in any phylogenetic analysis.

Acknowledgments

I am grateful to C. K. Brain for his invitation to study the new hominid fossils that have been brought to light through his painstaking work at the site of Swartkrans and for his generous hospitality during my numerous visits to the Transvaal Museum. I thank R. L. Susman for many hours of fruitful discussion and debate over the new Swartkrans fossils, F. C. Howell and W. L. Jungers for their invaluable comments on this paper, L. Jungers for the artwork and M. Stewart for photographic assistance. This work was supported by grants from SUNY, Stony Brook and the L. S. B. Leakey Foundation.

References

Brain, C. K. (1976). A re-interpretation of the Swartkrans site and its remains. *S. Afr. J. Sci.,* 72:141–146.
Brain, C. K. (1981). *The hunters or the hunted? An introduction to African cave taphonomy.* University of Chicago Press, Chicago.
Brain, C. K. (1982). The Swartkrans site: stratigraphy of the fossil hominids and a reconstruction of the environment of early *Homo. Pretirage ler Cong. Internat. Paleont. Humaine,* Nice, vol. 2, pp. 676–706.

Brain, C. K. (1985). Cultural and taphonomic comparisons of hominids from Swartkrans and Sterkfontein. *In* E. Delson (ed.), *Ancestors: the hard evidence*, pp. 72–75. Alan R. Liss, New York.

Brock, A., McFadden, P. L. and Partridge, T. C. (1977). Preliminary palaeomagnetic results from Makapansgat and Swartkrans. *Nature, 266*:249–250.

Broom, R. (1949). Another type of fossil ape-man *(Paranthropus crassidens). Nature, 163*:57.

Broom, R. and Robinson, J. T. (1949). A new type of fossil man. *Nature, 164*:322–323.

Brown, F. H., McDougall, I., Davies, T. and Maier, R. (1985). An integrated Plio-Pleistocene chronology for the Turkana Basin. *In* E. Delson (ed.), *Ancestors: the hard evidence*, pp. 82–90. Alan R. Liss, New York.

Butzer, K. W. (1976). Lithostratigraphy of the Swartkrans Formation. *S. Afr. J. Sci., 72*:136–141.

Delson, E. (1984). Cercopithecid biochronology of the African Plio-Pleistocene: correlation among eastern and southern hominid-bearing localities. *Cour. Forsch. Inst. Senckenberg, 69*:199–218.

Fink, J. and Kukla, G. J. (1977). Pleistocene climates in central Europe: at least 17 interglacials after the Olduvai Event. *Quat. Res., 7*:363–371.

Grine, F. E. (1981). Trophic differences between "gracile" and "robust" australopithecines: a scanning electron microscope analysis of occlusal events. *S. Afr. J. Sci., 77*:203–230.

Grine, F. E. (1982). A new juvenile hominid (Mammalia; Primates) from Member 3, Kromdraai Formation, Transvaal, South Africa. *Ann. Tvl. Mus., 33*:165–239.

Grine, F. E. (1984). The deciduous dentition of the Kalahari San, the South African Negro and the South African Plio-Pleistocene hominids. Ph.D. Thesis, University of the Witwatersrand, Johannesburg.

Grine, F. E. (1985). Australopithecine evolution: the deciduous dental evidence. *In* E. Delson (ed.), *Ancestors: the hard evidence*, pp. 153–167. Alan R. Liss, New York.

Grine, F. E. (1989). New hominid fossils from the Swartkrans Formation (1979–1986 excavations): craniodental specimens. *Amer. J. Phys. Anthropol.*, in press.

Harris, J. M. and White, T. D. (1979). Evolution of the Plio-Pleistocene African Suidae. *Trans. Am. Phil. Soc., 69*:1–128.

Howell, F. C. (1978). Hominidae. *In* V. J. Maglio and H. B. S. Cooke (eds.), *Evolution of African mammals*, pp. 154–248. Harvard University Press, Cambridge, MA.

Johanson, D. C. and White, T. D. (1979). A systematic assessment of early African hominids. *Science, 203*:321–330.

Jungers, W. L. and Grine, F. E. (1986). Dental trends in the australopithecines: the allometry of mandibular molar dimensions. *In* B. A. Wood, L. B. Martin and P. Andrews (eds.), *Major topics in primate and human evolution*, pp. 203–219. Cambridge University Press, Cambridge.

Mann, A. E. (1975). *Paleodemographic aspects of the South African australopithecines*. University of Pennsylvania Pubs. Anthropol., no. 1. University of Pennsylvania, Philadelphia, PA.

Rak, Y. (1983). *The australopithecine face*. Academic Press, New York.

Robinson, J. T. (1953). *Telanthropus* and its phylogenetic significance. *Amer. J. Phys. Anthropol., 11*:445–501.

Vrba, E. S. (1975). Some evidence of chronology and palaeoecology of Sterkfontein, Swartkrans and Kromdraai from the fossil Bovidae. *Nature, 254*:301–304.

Vrba, E. S. (1982). Biostratigraphy and chronology, based particularly on Bovidae, of southern hominid-associated assemblages: Makapansgat, Sterkfontein, Kromdraai, Swartkrans; also Elansdfontein (Saldanha), Broken Hill (now Kabwe) and Cave of Hearths. *Pretirage ler Cong. Internat. Paleont. Humaine* Nice, vol. 1, pp. 707–752.

Vogel, J. C. (1985). Further attempts at dating the Taung tufas. *In* P. V. Tobias (ed.), *Hominid evolution: past, present and future*, pp. 189–194. Alan R. Liss, New York.

Walker, A. C., Leakey, R. E. F., Harris, J. M. and Brown, F. H. (1986). 2.5-Myr *Australopithecus boisei* from west of Lake Turkana, Kenya. *Nature, 322*:517–522.

White, T. D., Johanson, D. C. and Kimbel, W. H. (1981). *Australopithecus africanus:* its phyletic position reconsidered. *S. Afr. J. Sci., 77*:445–470.

Studies of Evolutionary Relationships IV

The Evolution of *Australopithecus boisei*

ALAN C. WALKER AND RICHARD E. LEAKEY

The type of *Australopithecus boisei* was discovered at Olduvai Gorge in 1959 by M. D. Leakey. L. S. B. Leakey (1959) originally named it *Zinjanthropus boisei* and Robinson (1960) immediately proposed that it should be *Paranthropus boisei*. OH 5 became the subject of an exhaustive monograph by Tobias (1967) as *Australopithecus boisei*. Until recently, most authorities except Robinson have accepted that the genus *Australopithecus* could serve to hold both East and South African "robust" species as well as *Australopithecus africanus*, but there is developing a fashion to use *Paranthropus* again for the "robust" group (e.g., Grine, 1985; Dean, 1985, 1986; Clarke, 1977, 1985; Olson, 1985). Despite nomenclatural differences, practically everyone has viewed *A. boisei* as being just a northern species that was perhaps a little more "robust" than *A. robustus*. They have often been considered together in phylogenetic reconstruction as *A. robustus/boisei* or *A. robustus + boisei* (e.g., White *et al.*, 1981; Kimbel *et al.*, 1984; Skelton *et al.*, 1986) as though the relationship between them is well-known and that they are a clade. In Rak's (1983) study of the australopithecine face he developed a scheme whereby the *A. boisei* condition was derived from the *A. robustus* one. Recent discoveries of hominid fossils in the Turkana Basin, Kenya (Walker *et al.*, 1986; Leakey and Walker, 1988) have focused our attention on the relationships between the East and South African species and have also helped us to understand some enigmatic fossils from earlier in the stratigraphic record. It is a measure of the significance of these new fossils that some members of the Workshop that led to this volume now consider that *A. boisei* and *A. robustus* might have evolved many of their common features in parallel.

Australopithecus boisei specimens have now been found at Olduvai and Peninj in Tanzania, at Chesowanja, on both sides of Lake Turkana in Kenya and in the Lower Omo Valley of Ethiopia. The Turkana and Lower Omo systems are part of the same sedimentary deposits in the Turkana Basin, for which an integrated stratigraphy has been determined (Brown *et al.*, 1985). Altogether the cranial sample is impressive, with parts (not counting isolated teeth) representing about 60 individuals, although many of these specimens are only broken bits of edentulous mandibular body. Any particular cranial feature is usually represented in only a few individuals. By far the largest sample comes from the Turkana Basin where the species is known from over 2.5 to 1.4 Myr.

Of the more important specimens out of the eight new individuals that have been found in the Turkana Basin in the last few years (Leakey and Walker, 1988) are two partial crania and one mandible. KNM-WT 17400 is much of the facial skeleton and part of the anterior calvaria of a juvenile male individual (Fig. 15.1). It comes from strata 12 m above the Malbe Tuff (= H4) and so is close to 1.8 Myr in age. The palate contains all the relatively unworn teeth except the I¹s. The M³s are still in their crypts and skiagrams reveal no root formation. The supraorbital tori and the zygomatic arches were weathered away before exposure, but the breaks show extensive development of the maxillary and frontal sinuses. Breaks through the squamous temporal show very extensive pneumatization. The anterior part of the endocranial surface is in reasonable condition; on the left it runs forward from the posterior surface of the petrous pyramid, and on the right forward from foramen ovale. The cranial base is strongly flexed. This can be seen externally by the angulation of the sphenoid and internally by the almost vertical course of the grooves for the cavernous part of the internal carotid arteries.

KNM-ER 13750 is the top of an adult male calvaria from below the KBS Tuff in Area 105, Koobi Fora so it is about 2.0 Myr in age (Fig. 15.2). It is almost identical in size, crest development,

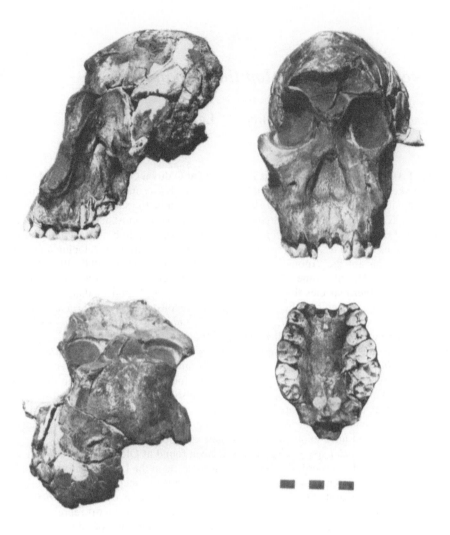

FIGURE 15.1. Lateral, facial, and vertical views of cranium and occlusal view of palate of KNM-WT 17400. Scale in cm.

and overall shape to KNM-ER 406 (Leakey, 1970; Leakey et al., 1971). The cranium has allowed us to make an endocast from above the petrous crest on the right. KNM-ER 15930 is an adult left mandibular body with the three relatively unworn molars from the KBS Member of Area 104. It is important because, for whatever taphonomic reasons, mandibles at the small end of the size range [i.e., those that could reasonably articulate with crania such as KNM-ER 732 (Leakey, 1971; Leakey et al., 1972)] are not common in the collections. It is about the same size as the Omo 18-1967-18 mandible from Member C of the Shungura Formation (Arambourg and Coppens, 1967).

There are three specimens (not counting isolated teeth) that relate to the earlier part of this species' record. They are mandible Omo 18-1967-18 from the Shungura Formation (Submember C8) (Arambourg and Coppens, 1967), mandible KNM-WT 16005 (= middle Member D), and cranium KNM-WT 17000 (Fig. 15.3) from the equivalent of Submember C9 (Walker et al., 1986). The first has been placed in its own genus and species, *Paraustralopithecus aethiopicus*, by its describers, in *A. africanus* by some (e.g., Howell, 1978; Johanson and White, 1979: figs. 7 and

FIGURE 15.2. Lateral, vertical, facial, and occipital views of cranium KNM-ER 13750. Scale in cm.

10b), and in *A. boisei* by others (e.g., Coppens, 1979, 1980; Chamberlain and Wood, 1985). Judging by the broken roots, this edentulous mandible had very small canines and incisors and relatively large cheek-teeth. With the discovery of KNM-ER 15930, a presumed female mandible from the period 1.9 to 1.6 Myr, KNM-WT 16005, a presumed male mandible from about 2.45 Myr, and knowing that KNM-WT 17000 (another presumed male) would have had an extremely large mandible, the most likely attribution of Omo 18-1967-18 is that it is a female from the early part of the *A. boisei* lineage. It is close in form and size to later presumed females. It must be remembered that the large Peninj specimen was the only *A. boisei* mandible available for comparison when the Omo 18-1967-18 mandible was first found. The presumed male mandible KNM-WT 16005 is about the same size as the Peninj mandible (Leakey and Leakey, 1964), and three of the four teeth are outside the size range of *A. robustus* in BL diameter (White *et al.*, 1981). The mandible of KNM-WT 17000 would have been as big as the largest *A. boisei* ever found (e.g., KNM-ER 3230). As judged from the broken roots, its palate is as big as that of OH 5. The alveolar processes are broken away obliquely, so that more is missing posteriorly than anteriorly, and this makes the palate appear much flatter posteriorly than it was when complete. The anterior teeth are represented by alveoli and the root of the left I^2. Mesiodistal diameters can be taken of the alveoli, however, and these can be compared with the MD diameters of the anterior tooth roots of OH 5 (Table 15.1). The canine roots in KNM-WT 17000 are marginally bigger than those of OH 5, those of the lateral incisors are about the same size, and those of the central incisors are smaller than in OH 5. A maximum bicanine breadth is 55.0 mm for KNM-WT 17000 and 47.0 mm for OH 5, so despite the similarity in root size, the anterior tooth crowns were larger in the former.

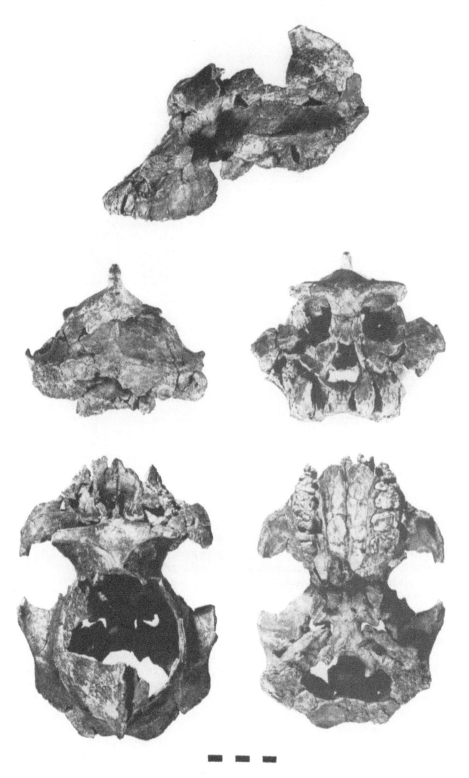

FIGURE 15.3. Facial, occipital, lateral, vertical, and basal views of cranium KNM-WT 17000. Scale in cm.

Table 15.1. Mesiodistal Diameters of Anterior Tooth Roots or Alveoli of KNM-WT 17000 and OH 5

Specimen	Left			Right		
	C	I^2	I^1	C	I^2	I^1
KNM-WT 17000	7.2	5.5	6.4			
OH 5	7.0	5.0	8.0	6.6	5.0	8.1

As was pointed out in a previous brief discussion (Walker et al., 1986), future finds should show whether KNM-WT 16005, KNM-WT 17000 and Omo 18-1967-18 are within the range of variation of *A. boisei*, or whether they should be called *Australopithecus aethiopicus*. There are dangers in giving different names to parts of short-lived evolving lineages, especially when samples are small. Rose and Bown (1986) have recently reviewed the problem of species discrimination when the fossil record is fairly continuous, and in a situation very much like the one we have for the *A. boisei* lineage, they chose to use informal, numbered stages of evolutionary development. On the question of sample size it should be considered, for instance, that only about six decent facial skeletons and another six braincases of *A. boisei* serve to document over a million years of evolution. Whatever names we call them, it seems most likely that the early "hyper-robust" fossils represent early populations of the later ones. With the present sample and the variation seen in it, and given that some evolution probably took place in over a million years, it seems unwise at present to cut up the lineage into successive species.

Apart from demonstrating that the Omo 18-1967-18 mandible is representative of the early part of the *A. boisei* lineage, the new fossils also clear up the minor disagreement about another Omo fossil, L 338y-6. This partial juvenile cranium from Member E of the Shungura Formation was originally described by Rak and Howell (1978) as *A. boisei*. Holloway (1981) described the endocast and came to the conclusion that it more likely represented either *A. afarensis* or *A. africanus*, because compared with known later *A. boisei* individuals, Omo L 338y-6 had (1) a smaller endocranial volume, (2) no occipital/marginal venous sinus system, (3) a cerebellum not tucked under the occipital lobes of the cerebrum, and (4) a meningeal vessel pattern like that of *A. africanus*. There is no doubt that had the individual represented by Omo L 338y-6 lived to maturity it would have developed massive sagittal and nuchal crests, as it had already developed the very overlapping and striated squamosal suture typical of *A. boisei*. It also possesses a heart-shaped foramen magnum, a feature unique to *A. boisei*. With the discovery of KNM-WT 17000 we can now see that, as might be expected, individuals from the early part of the lineage retained more primitive traits. With the exception of the condition of the venous sinuses, which cannot be confirmed on KNM-WT 17000, these two crania are alike.

Saban (1983) has shown that *A. robustus* and *A. boisei* have a middle meningeal vessel pattern that is derived relative to that seen in *A. africanus*. There are just two main branches in the latter, an anterior and a posterior one. In the "robust" forms there is a strong middle branch developed from the posterior branch. However, since the main function of these vessels is to supply and drain the bones of the skull, and since the vault bones of *A. africanus* and early *A. boisei* specimens are about the same size, it follows that the derived pattern will only be expected in the individuals with larger calvariae, i.e., those with larger cranial capacities. If our present small sample is an accurate guide, then we would expect to find these more often in the later members of the lineage. Fig. 15.4 shows the meningeal vessel pattern in KNM-WT 17000. The pattern is clearly very like that described by Saban (1983) for *A. africanus*, with an anterior branch supplying most of the frontal and parietal and a posterior branch supplying the temporal, posterior parietal and superior occipital areas. There is no middle branch arising from the posterior branch.

There seem to have been small females and large males throughout the time that we have sampled the lineage. Chamberlain and Wood (1985) analyzed mandibular body shape of *A. boisei* specimens and concluded that female mandibles were underrepresented in the fossil record. We believe that had they first carried out their analysis on a dimorphic sample where the sexes were known, they would have come to different conclusions. We present the *A. boisei* data (brought

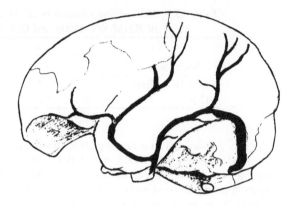

FIGURE 15.4. Drawing of left lateral view of the KNM-WT 17000 endocast to show the pattern of meningeal vessel distribution.

up-to-date and replotted) together with similar data on a set of male and female gorilla mandibles in Fig. 15.5 and 15.6. The measurements for the gorillas are from Pilbeam (1969). As can be seen, when the male and female distributions for the gorilla data are plotted together, there is no easy way of deciding (without some other knowledge) whether mandibles in the middle of the distribution are males or females. We believe that the level of sexual dimorphism in *A. boisei* was about the same as that seen today in gorillas. A presumed male and a presumed female mandible are shown in Fig. 15.7.

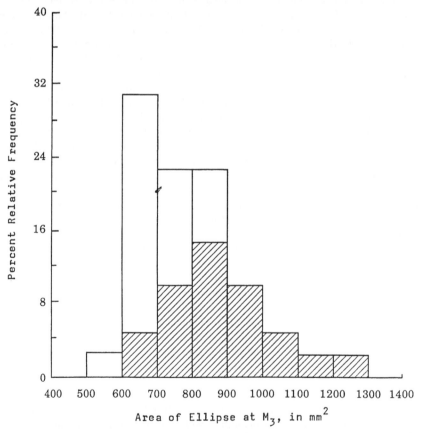

FIGURE 15.5. Distribution of mandibular body dimensions in male (hatched) and female (open) gorillas. Data from Pilbeam (1969); ellipse calculated after Chamberlain and Wood (1985).

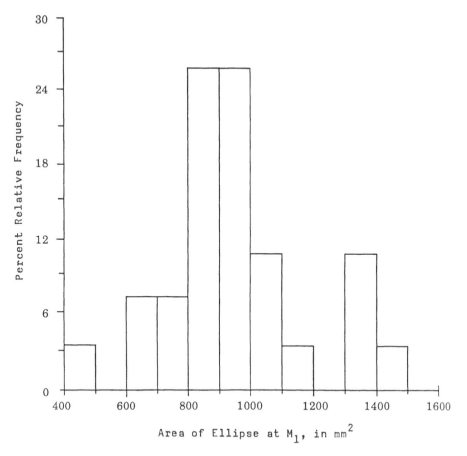

FIGURE 15.6. Distribution of mandibular body dimensions in *A. boisei*. Ellipse calculated after Chamberlain and Wood (1985). Far left column, KNM-ER 15930. Far right column, KNM-ER 3230.

Detailed analyses of tooth proportions have not been completed, but the earliest skull has, as far as can be judged from the roots and alveoli, anterior tooth to cheek-tooth proportions very similar to those from the middle and end phases of the species record.

Cranial capacity can be measured or estimated in only a few specimens, but it appears that it increased considerably in 500,000 years. The endocranial volume of KNM-WT 17000 is 410 ml, and the estimate made by Holloway (1981) for the Omo L 338y-6 cranium 427 ml. These values are roughly 100 ml less than those from the later half of the lineage. However, the expected range of capacities for a hominoid with an average of 500 ml could include the only known capacity of an early male at 410 (e.g., for *Gorilla* see Tobias, 1971). This fully adult individual has a small cranial capacity and an extremely large palate. The temporal muscles were extremely well-developed and they, in turn, stimulated the development of a huge sagittal and strong nuchal crests in KNM-WT 17000. There is no bare area on KNM-WT 17000, but neither is there in KNM-CH 304, a late member of the species from Chesowanja. In general, *A. boisei* cresting patterns are extremely variable.

Pneumatization of the face and cranial base ranges from slight, as in KNM-ER 733 (Leakey and Walker, 1973), to very extensive, as in KNM-WT 17400 (Leakey and Walker, 1987). It has been claimed (Kimbel *et al.*, 1984) that pneumatization in the temporal bone of *A. boisei* tends to be restricted to the mastoid region. KNM-WT 17400 shows that some late *A. boisei* specimens have extensive pneumatization here and so the relevance of the same pneumatization in KNM-WT 17000 is only that it is primitive. The configuration of the infraorbital facial skeleton of KNM-WT 17000 is that described by Rak (1983) as representing *A. boisei* (Fig. 15.8). There are no

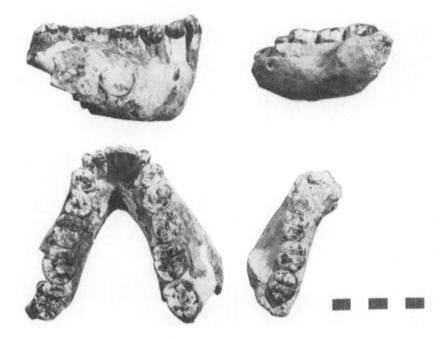

FIGURE 15.7. *A. boisei* mandibles illustrating postulated sexual dimorphism. Left: KNM-ER 3230 (presumed male) in lateral and occlusal views. Right: KNM-ER 15930 (presumed female) in lateral and occlusal views.

FIGURE 15.8. Left three-quarters view of KNM-WT 17000 with missing parts of vault and zygomatic filled in. Note similarity of facial features to those in *A. boisei* specimens.

anterior pillars, and the edges of the pyriform aperture are sharp and slightly everted superiorly with a nasomaxillary basin on either side. The nasals are extremely expanded superiorly, reaching above glabella and very narrow inferiorly, as in *A. boisei*. Whatever this says about the relationships between *A. boisei* and *A. robustus*, it does appear to indicate that the evolutionary progression of functionally related features is not as Rak has presented them. In KNM-WT 17000 we have a typical *A. boisei* face on a very prognathic palate, such that the whole face in this early specimen is set in a different way from later ones (Fig. 15.9). If the face of KNM-WT 17000 is rotated about the supraorbital tori by 15°, the resulting profile is very like those of later *A. boisei* such as OH 5 and KNM-ER 406. The biomechanical scheme given by Rak (1983: fig. 27) does seem to hold up in the *A. boisei* lineage since anterior pillars are not needed if the zygomatic take-off (and the masseter origin) are situated very far forward. But in the earlier *A. boisei* individual the whole face is placed far from the temporomandibular joints, rather than being very retracted towards them as in later individuals. It does not follow, as Rak (1983) suggests, that *A. robustus*, or something with facial skeleton like it, had to precede *A. boisei*.

Finally, with regard to basicranial flexion, the condition in KNM-WT 17000 is remarkably unflexed. This can be seen externally on the sphenoid and vomer and internally by the oblique grooves for the cavernous part of the internal carotid artery as well as the angle of the pontine clivus. In later *A. boisei* specimens where the condition can be seen (and this is only in OH 5, KNM-ER 406, KNM-WT 17400 and possibly KNM-ER 407 and KNM-CH 1) there is considerable basicranial flexion. Midsagittal craniograms of several specimens (Fig. 15.9) show that the main difference between, say, OH 5 or KNM-ER 406 and KNM-WT 17000 is that the whole facial skeleton of the latter is unflexed by about 15° relative to the others.

The endocasts of KNM-WT 17000 (Leakey and Walker, 1987) and later *A. boisei* (Tobias, 1967) are very informative. The brain is low in lateral profile and relatively unflexed in the early specimen. In the later ones the cerebellum comes to lie more under the occipital poles of the cerebrum (and the posterior faces of the petrous bones become more vertical to accommodate this), and the rostral portion of the frontal lobes tucks under toward the temporal lobes. This gives the endocast a higher lateral profile and makes it wider across the temporal lobes. Is the change in facial angle related to changes in the flexion of the brain, or vice-versa, or are they both part of some greater remodeling process? As far as masticatory function is concerned, the

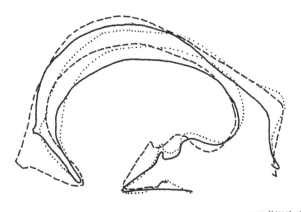

FIGURE 15.9. Mid-sagittal craniograms of KNM-ER 406 (dotted line), OH 5 (dashed line), and KNM-WT 17000 (solid line). Craniogram for OH 5 from Tobias (1967).

A. boisei condition (i.e., having the masseter situated as far forward as possible and the premolars molarized to the extent that the P^3s mark the corner of the palate) was already developed by 2.5 Myr. This makes it unlikely that selection acting on the masticatory system was the force behind the change in the set of the face on the braincase.

Because of the *A. boisei* condition of the teeth, palate, face, and foramen magnum, we think that KNM-WT 17000 was from a population after *A. robustus* had already diverged. Several features that are common to both "robust" species must have evolved in parallel; this would have to be true even if KNM-WT 17000 represents the ancestral "robust" condition (unless these features developed after 2.5 Myr in the *A. boisei* lineage before *A. robustus* split off). These parallelisms include: (1) increase in cranial capacity; (2) changes in endocranial shape; (3) development of middle branch of meningeal vessels; (4) possible changes in the venous sinuses; (5) change in orientation of the posterior surface of the petrous pyramid; (6) change in the set of the facial skeleton on the braincase; (7) development of articular eminence; (8) change in orientation of tympanic plate to become more vertical; and (9) changes to make the cranial base more flexed.

These changes are not trivial morphological ones. They affect several functional and developmental systems. They also must have been taking place at about the same time in South Africa. If this amount of parallel development is shown to have occurred, then it becomes possible that the development of the "robust" features in both species is a parallel development and that they need not have shared a "robust" common ancestor. The claims for *A. africanus* being the ancestor of the South African "robust" species (Johanson and White, 1979) might still be warranted, but this can no longer be true for the ancestry of the East African "robust" species, as *A. africanus* demonstrates many derived features relative to the condition seen in KNM-WT 17000.

These new fossils suggest that homoplasy might be common in the "robust" lineages; indeed, it might have characterized all of later hominid evolution. Dean (1986) has argued that the genus *Homo* and the two "robust" species share apparently derived incisor and first molar eruption patterns, and he has pointed out that they share sagittal orientation of the petrous temporal bones and loss of pronounced muscle markings in the upper pharyngeal region. The list of features that must have evolved in parallel in the two "robust" species, is not much different from the list of features that also evolved in parallel between them and *Homo,* as follows: (1) brain size increases; (2) cranial flexion increases; (3) orthognathism decreases; (4) there is reflexion of the rostral part of the frontal lobes; (5) the cerebellum retreats under the occipital poles; (6) the petrous temporal crest becomes sharper; (7) middle meningeal vessel pattern develops a middle branch; (8) the articular eminence develops; (9) the axes of petrous temporals become more coronally oriented; (10) longus capitis markings get weaker; (11) eruption patterns become identical [note that this feature is under review following the work of Smith (1986) and Grine (1987)].

There are two ways to explain this. Either these features evolved in parallel and at the same time in the "robust" lineages and the *Homo* one, or *Homo* and the "robust" species shared a last common ancestor to the exclusion of *Australopithecus africanus*. In the unlikely event that the second alternative is true, it follows that KNM-WT 17000, which has not developed most of these features, must be close to that ancestral condition.

Acknowledgments

We thank the Government of Kenya and the Governors of the National Museums of Kenya. Most of the work that led to this article was funded by the National Geographic Society, Washington, D.C. and The William Donner Foundation; and A. W. is the recipient of a John Simon Guggenheim Memorial Fellowship. Many colleagues have given invaluable help, but we thank especially Kamoya Kimeu and his field team and Joseph Mutaba, Emma Mbua, and M. G. Leakey for technical and curatorial help in Nairobi. We also thank Alan Mann, Janet Monge and Nancy Minugh for taking the measurements on the OH 5 Master cast. Lastly we thank F. E. Grine for his kind invitation to participate in the Workshop and the participants who made many valuable suggestions.

References

Arambourg, C. and Coppens, Y. (1967). Sur la découverte dans le Pléistocène inférieur de la vallée de l'Omo (Éthiopie) d'une mandibule d'australopithecien. *C. R. Acad. Sci. (Paris)*, 265: 1981–1986.
Brown, F. H., McDougall, I., Davies, T. and Maier, R. (1985). An integrated Plio-Pleistocene chronology for the Turkana Basin. *In* E. Delson (ed.), *Ancestors: the hard evidence*, pp. 82–90. Alan R. Liss, New York.
Chamberlain, A. T. and Wood, B. A. (1985). A reappraisal of variation in hominid mandibular corpus dimensions. *Amer. J. Phys. Anthropol.*, 66: 399–405.
Clarke, R. J. (1977). The cranium of Swartkrans Hominid SK 847 and its relevance to human origins. Unpublished Ph.D. Thesis, University of the Witwatersrand, Johannesburg.
Clarke, R. J. (1985). *Australopithecus* and early *Homo* in southern Africa. *In* E. Delson (ed.), *Ancestors: the hard evidence*, pp. 171–177. Alan R. Liss, New York.
Coppens, Y. (1979). Les hominidés du Pliocène et de Pléistocène de la Rift Valley. *Bull. Soc. Geol. Fr.*, 21: 313–320.
Coppens, Y. (1980). The differences between *Australopithecus* and *Homo*: preliminary conclusions from the Omo Research Expeditions studies. *In* L. -K. Königsson (ed.), *Current argument on early man*, pp. 207–225. Pergamon, Oxford.
Dean, M. C. (1985). The eruption pattern of the permanent incisors and first permanent molars in *Paranthropus robustus*. *Amer. J. Phys. Anthropol.*, 54: 63–71.
Dean, M. C. (1986). *Homo* and *Paranthropus*: similarities in the cranial base and developing dentition. *In* B. A. Wood, L. B. Martin and P. Andrews (eds.), *Major topics in primate and human evolution*, pp. 248–265. Cambridge University Press, Cambridge.
Grine, F. E. (1985). Australopithecine evolution: the deciduous dental evidence. *In* E. Delson (ed.), *Ancestors: the hard evidence*, pp. 153–167. Alan R. Liss, New York.
Grine, F. E. (1987). On the eruption pattern of the permanent incisors and first permanent molars in *Paranthropus*. *Amer. J. Phys. Anthropol.*, 72: 353–360.
Holloway, R. L. (1981). The endocast of the Omo L 338y-6 juvenile hominid: gracile or robust *Australopithecus? Amer. J. Phys. Anthropol.*, 54: 109–118.
Howell, F. C. (1978). Hominidae. *In* V. J. Maglio and H. B. S. Cooke (eds.), *Evolution of African mammals*, pp. 154–258. Harvard University Press, Cambridge, MA.
Johanson, D. C. and White, T. D. (1979). A systematic assessment of early African hominids. *Science*, 203: 321–330.
Kimbel, W. H., White, T. D. and Johanson, D. C. (1984). Cranial morphology of *Australopithecus afarensis*: a comparative study based on a composite reconstruction of the adult skull. *Amer. J. Phys. Anthropol.*, 64: 337–388.
Leakey, L. S. B. (1959). A new fossil skull from Olduvai. *Nature*, 184: 491–493.
Leakey, L. S. B. and Leakey, M. D. (1964). Recent discoveries of fossil hominids in Tanganyika: at Olduvai Gorge and near Lake Natron. *Nature*, 202: 5–7.
Leakey, R. E. F. (1970). Fauna and artefacts from a new Plio-Pleistocene locality near Lake Rudolf in Kenya. *Nature*, 226: 223–234.
Leakey, R. E. F. (1971). Further evidence of Lower Pleistocene hominids from East Rudolf, North Kenya. *Nature*, 231: 241–245.
Leakey, R. E. F. and Walker, A. C. (1973). New australopithecines from East Rudolf, Kenya (III). *Amer. J. Phys. Anthropol.*, 39: 205–222.
Leakey, R. E. F. and Walker, A. C. (1988). New *Australopithecus boisei* specimens from East and West Lake Turkana, Kenya. *Amer. J. Phys. Anthropol.* 76: 1–24.
Leakey, R. E. F., Mungai, J. M., and Walker, A. C. (1971). New australopithecines from East Rudolf, Kenya. *Amer. J. Phys. Anthropol.*, 35: 175–186.
Leakey, R. E. F., Mungai, J. M. and Walker, A. C. (1972). New australopithecines from East Rudolf, Kenya (II). *Amer. J. Phys. Anthropol.*, 36: 235–252.
Olson, T. R. (1985). Cranial morphology and systematics of the Hadar Formation hominids and "*Australopithecus*" *africanus*. *In* E. Delson (ed.), *Ancestors: the hard evidence*, pp. 102–119. Alan R. Liss, New York.

Pilbeam, D. R. (1969). Tertiary Pongidae of East Africa: evolutionary relationships and taxonomy. *Peabody Mus. Bull., 31*: 1–185.

Rak, Y. (1983). *The australopithecine face*. Academic Press, London.

Rak, Y. and Howell, F. C. (1978). Cranium of a juvenile *Australopithecus boisei* from the lower Omo Basin, Ethiopia. *Amer. J. Phys. Anthropol., 48*: 345–366.

Robinson, J. T. (1960). The affinities of the new Olduvai australopithecine. *Nature, 186*: 456–458.

Rose, K. D. and Bown, T. M. (1986). Gradual evolution and species discrimination in the fossil record. *Contrib. Geol. Wyoming., 3*: 119–130.

Saban, R. (1983). Les veines méningées moyennes des australopitheques. *Bull. Mém. Soc. d'Anthrop. (Paris), 10*: 313–324.

Skelton, R. R., McHenry, H. M. and Drawhorn, G. M. (1986). Phylogenetic analysis of early hominids. *Curr. Anthropol., 27*: 21–43.

Smith, B. H. (1986). Dental development in *Australopithecus* and early *Homo*. *Nature, 323*: 327–330.

Tobias, P. V. (1967). *The cranium and maxillary dentition of Australopithecus (Zinjanthropus) boisei*. Olduvai Gorge, vol. 2. Cambridge University Press, Cambridge.

Tobias, P. V. (1971). *The brain in hominid evolution*. Columbia University Press, New York.

Walker, A., Leakey, R. E., Harris, J. M. and Brown, F. H. (1986). 2.5- Myr *Australopithecus boisei* from west of Lake Turkana, Kenya. *Nature, 322*: 517–522.

White, T. D., Johanson, D. C. and Kimbel, W. H. (1981). *Australopithecus africanus:* its phyletic position reconsidered. *S. Afr. J. Sci., 77*: 445–470.

Implications of KNM-WT 17000 for the Evolution of "Robust" *Australopithecus*

16

WILLIAM H. KIMBEL, TIM D. WHITE AND DONALD C. JOHANSON

In previous contributions we have stressed the need for new East African hominid fossils in the 2.0–3.0 Myr time range to test competing phylogenetic hypotheses (Johanson and White, 1979; White *et al.*, 1981; Kimbel *et al.*, 1984, 1986). The recent recovery of an *Australopithecus* cranium from 2.5 ± 0.07 Myr sediments at Lomekwi, west of Lake Turkana, Kenya, provides a basis for this test (Walker *et al.*, 1986). The cranium, KNM-WT 17000, already has stimulated considerable discussion (Bower, 1986, 1987; Delson, 1986; Falk, 1986; Johanson and White, 1986; Lewin, 1986; Shipman, 1986). The occasion of a workshop devoted to the evolution of "robust"[1] early hominids affords an appropriate opportunity to assess the taxonomic and phylogenetic implications of KNM-WT 17000.

Background

The ancestry of "robust" *Australopithecus* has been a subject of vigorous debate during the past several years. Most investigators recognize a "robust" *Australopithecus* clade (including *A. robustus* and *A. boisei*)[2] derived from a species similar or identical to *A. africanus*, as known from Taung, Makapansgat and Sterkfontein. There are two variations on this view.

The hypothesis proposed initially by Johanson and White (1979) placed *A. africanus* in a single "robust" *Australopithecus* lineage by virtue of a series of shared-derived craniodental characters for which *A. afarensis*, *Homo* and the extant great apes exhibit the primitive condition (White *et al.*, 1981). Rak (1983, 1985) and Kimbel *et al.* (1984) modified this tree by recognizing *A. boisei* as a distinct East African species partly contemporary with South African *A. robustus*.

Emphasizing other shared-derived characters in the skull and dentition that unite all non-*A. afarensis* hominids, some investigators have chosen to retain *A. africanus* (or a hypothetical population or species similar to it) as the last common ancestor of "robust" *Australopithecus* and *Homo* lineages (Tobias, 1980; Wolpoff, 1983; Yaroch and Vitzthum, 1984; Skelton *et al.*, 1986; Kimbel, 1986).

The major challenge to these phylogenetic hypotheses contends that *A. afarensis* (in part or in whole) is linked exclusively with "robust" *Australopithecus* species. This linkage is postulated by Olson (1981, 1985) on the basis of mastoid, nasal, and dental morphology [*contra* Kimbel *et al.*, 1985] and by Falk and Conroy (1983) on the basis of cranial venous drainage patterns.

Evidence to test these alternative hypotheses has, to date, been limited. Grine (1981, 1985) has attributed two mandibular deciduous molars from upper Member D of the Shungura Formation (Howell and Coppens, 1973) to *A. boisei*. Recent redating of Tuff D, which defines the base of Member D, has yielded an age of 2.52 ± 0.05 Myr (Brown *et al.*, 1985). A small sample of isolated

---

[1] One of us (TDW) objects to the use of quotation marks around the adjective robust.
[2] Some investigators (Howell, 1978; Grine, 1985) distinguish the two South African "robust" site samples at the species level, i.e., Kromdraai = *A. robustus* (Broom, 1938); Swartkrans = *A. crassidens* (Broom, 1949). For present purposes, we follow more traditional taxonomic practice and combine the two in *A. robustus*.

teeth from Members C and D of the Shungura Formation may also be attributable to a "robust" *Australopithecus* species (Howell and Coppens, 1976; Coppens, 1977). Suwa (this volume, chapter 13) has confirmed the "robust" *Australopithecus* status of several of these dental elements.

An edentulous mandible, Omo 18-1967-18, from Member C (Unit C-8) of the Shungura Formation was made the holotype of *Paraustralopithecus aethiopicus* by Arambourg and Coppens in 1968.[3] Based on its stratigraphic position, this specimen is slightly older than Tuff D (c. 2.6 Myr) (Brown *et al.,* 1985: fig. 2). Coppens (1970) noted similarities of the Omo 18-1967-18 mandible to *A. robustus* and *A. boisei,* but Howell and Coppens (1976) saw a closer affinity to *A. africanus* [to which species Howell (1978) has attributed the specimen]. White (1977) placed the jaw in his mandible "Set C" (= "robust" *Australopithecus*), and Johanson and White (1979: fig. 7) explicitly referred it to the early part of the "robust" lineage. Wood and Chamberlain (1986) list it as *A. boisei.*

The 1968 diagnosis of *P. aethiopicus* highlighted several features of the mandible (e.g., oblique inclination of the *planum alveolare;* strong inferior transverse torus; low, squat horizontal rami; convex, receding anterior symphyseal profile) that indicated to Arambourg and Coppens a more primitive morphology than was observed in the mandibles of *Australopithecus* (1968: 59) (see also Coppens, 1970: 2287; Howell and Coppens, 1976: 524–525).

In sum, discoveries made since the 1960's demonstrate the presence of a "robust" hominid taxon in the East African late Pliocene. With the chronological refinements introduced during the past few years (Cerling and Brown, 1982; Brown *et al.,* 1985; McDougall, 1985; Hillhouse *et al.,* 1986), it has become clear that the record of "robust" early hominids extends back to at least 2.6 Myr in the Turkana basin. Prior to the discovery of KNM-WT 17000, this record was too incomplete to permit detailed assessment of its taxonomic and phylogenetic implications.

Taxonomy of KNM-WT 17000

The KNM-WT 17000 specimen consists of most of an undistorted adult cranium lacking tooth crowns except for one premolar and a partial molar. Walker *et al.* assigned the cranium to *A. boisei,* emphasizing "the massive size, extremely large palate and teeth, the build of the infraorbital and nasal areas and the anterior position and low take-off of the zygomatic root" (1986: 521). In a table they listed the states of forty characters, culled from the analysis of Skelton *et al.* (1986)[4], for four *Australopithecus* species and KNM-WT 17000. Walker *et al.* concluded: "for most features the specimen resembles *A. boisei*" (1986: 521). This statement, while not inaccurate, is misleading since sixteen of the forty listed character states exhibited by KNM-WT 17000 are primitive traits shared with *A. afarensis* (thirteen of these are shared *exclusively* with *A. afarensis* among hominids). In addition, thirteen of sixteen *derived* features that link KNM-WT 17000 to *A. boisei* are also present in at least one other *Australopithecus* species included in the comparisons. Based on the data supplied by Walker *et al.* (1986), only three character states listed actually link KNM-WT 17000 exclusively with *A. boisei.* Delson (1986) has questioned the validity of some of these alleged similarities to *A. boisei,* and elsewhere (Kimbel *et al.,* 1986) we have criticized aspects of the Skelton *et al.* (1986) character analysis, which Walker *et al.* (1986) incorporated into their tabulated comparisons.

We have re-analyzed the data of Walker *et al.* (1986), eliminating characters that we believe

---

[3]Arambourg and Coppens first published their diagnosis of *Paraustralopithecus aethiopicus* in 1967. In this publication, however, the authors erected the species on a provisional basis. Their subsequent paper, which includes a nearly verbatim diagnosis, renders the species name available as of 1968.

[4]Most of the characters used by Skelton *et al.* (1986) come from the analyses by White *et al.* (1981) and Kimbel *et al.* (1984).

to be redundant or misleading,[5] and adding our own observations. Figure 16.1 summarizes data important in any assessment of the affinities of KNM-WT 17000. Of the thirty-two traits listed in Fig. 16.1, twelve are primitive resemblances to *A. afarensis*, six are derived characters shared with *A. africanus*, *A. robustus*, and *A. boisei* (and to some extent *Homo*), twelve are derived features shared with *A. robustus* and *A. boisei*, and only two are derived characters shared exclusively with *A. boisei*.

The facial skeleton of KNM-WT 17000 differs from that of *A. boisei* in two important respects. First, Rak (1983: 48) has shown that the infraorbital region of the *A. boisei* face can be likened to a "visor" in that the facial and lateral surfaces of the cheek merge indistinguishably in a smooth, continuous curve. In our judgment, KNM-WT 17000 more closely resembles other hominids in retaining the primitive condition of distinct facial and lateral surfaces in this region. In this regard, KNM-WT 17000 compares most favorably to *A. robustus*, with marked a "zygomatic prominence" delimiting these surfaces from one another (Rak, 1983: 34). Second, KNM-WT 17000 shares with *A. robustus* a distinct "zygomaticomaxillary step" which, corresponding to the zygomaticomaxillary suture, denotes a sharp break in elevation between the infraorbital surfaces of the maxilla and zygomatic (Rak, 1983: 31–32). Given these fundamental departures from the *A. boisei* facial morphology, the absence in the KNM-WT 17000 face of an anterior pillar and a maxillary fossula, which Walker *et al.* (1986) argue link this cranium to *A. boisei*, are not valid shared-derived characters unless one accepts Rak's (1983) functional interpretation of the *A. africanus* to *A. robustus* character transformation (see also Rak, this volume, chapter 12). Comparison with extant and fossil outgroups would, on the contrary, establish the absence of these features as primitive.

Beyond the pattern documented in Fig. 16.1, which argues against attribution of KNM-WT 17000 to *A. boisei*, there are additional implications to consider. *Australopithecus robustus* and *A. boisei* share many derived characters not present in KNM-WT 17000. If the new cranium were attributed to *A. boisei*, it would, on the same morphological basis, be justifiable to synonymize this species in favor of a broadly defined *A. robustus*. Yet, as Walker *et al.* (1986) note, growing evidence from diverse studies supports the species-level distinction between *A. robustus* and *A. boisei* (Howell, 1978; Rak, 1983; Kimbel *et al.*, 1984; Grine, 1985; Kimbel, 1986; Skelton *et al.*, 1986; Wood and Chamberlain, 1986). We therefore conclude that there are few grounds for attributing KNM-WT 17000 to *A. boisei*. This conclusion has already been reached by Johanson and White (1986), who emphasized the numerous primitive characters shared by KNM-WT 17000 and other *Australopithecus* species (especially *A. afarensis*) to the exclusion of *A. boisei*.

If KNM-WT 17000 is not *A. boisei*, then to what species does it belong? Walker *et al.* (1986) recognize that the extraordinary amalgam of characters linking KNM-WT 17000 to "robust" *Australopithecus*, on the one hand, and *A. afarensis*, on the other, might require due consideration in the evaluation of its taxonomic status:

> Although future finds may show that KNM-WT 17000 is well within the range of variation of *A. boisei*, it is also possible that the differences will prove sufficient to warrant specific distinction. If the latter proves to be the case we suggest that some specimens *from the same time period and from the same sedimentary basin* (for example, Omo 1967-18

---

[5]*Redundant* characters in Table 2 of Walker *et al.* (1986) include: depth of mandibular fossa = articular eminence; divergence of temporal lines relative to lambda = relative sizes of posterior to anterior temporalis; relative size of canine = relative size of postcanine teeth.

Misleading characters include: "lateral concavity of nuchal plane" which reflects depth of the superior oblique muscle insertion, and thus is of questionable phylogenetic value. In the analysis of a single specimen, "length of nuchal plane relative to occipital" [a mean estimate which obscures species-specific sex distribution patterns (see Kimbel *et al.*, 1984)] and "common origin of zygomatic arch" [a mean estimate which obscures the overlap of species' ranges (see Kimbel *et al.*, 1984)] are difficult to evaluate.

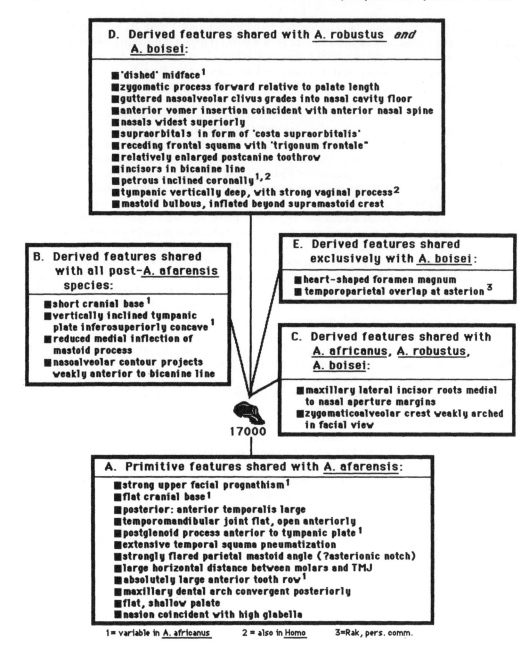

FIGURE 16.1. Affinities of KNM-WT 17000. See text for discussion.

[sic] from the Shungura Formation) will be included in the same species . . . In our view, the appropriate name then would be *Australopithecus aethiopicus*. (1986: 521, emphasis added).

We concur with this judgment, but believe that species distinction is justified on the basis of evidence at hand, and that KNM-WT 17000 validates the taxonomic status of *A. aethiopicus*.

Some colleagues have objected to the assignment of the KNM-WT 17000 cranium to *A. aethiopicus* because the holotype is a mandible (e.g., see Bower, 1987). We would note, however, that Arambourg and Coppens' (1968) diagnosis of *P. aethiopicus* includes a number of primitive characters that distinguished Omo 18-1967-18 from the mandibles of other "robust" hominids then known. Subsequent discoveries of *A. boisei* mandibles have not diminished the validity of this diagnosis. We further contend that if KNM-WT 17000 is not *A. aethiopicus*, then a new species would have to be erected for it. This would imply the existence of two essentially sympatric and synchronic, primitive "robust" hominid lineages during the time period represented by Members C and (lower) D of the Shungura Formation. While this is possible, we predict that future discoveries will show that mandibles very similar to Omo 18-1967-18, taking into account sex-related size differences, are associated with the cranial morphology of KNM-WT 17000. For these reasons we refer KNM-WT 17000 to *A. aethiopicus*.

KNM-WT 17000 and Hominid Phylogeny

The mosaic of primitive and derived characters exhibited by KNM-WT 17000 indicates that *A. aethiopicus* is phylogenetically intermediate between *A. afarensis* and "robust" *Australopithecus* species. We contend that in morphology and known temporal range, *A. aethiopicus* could have been ancestral to *either* or *both A. robustus* and *A. boisei*.

The species usually nominated for this ancestral role is *A. africanus* (either as an exclusive ancestor or as the last common ancestor of the *Homo* and "robust" clades). In light of the KNM-WT 17000 discovery, the phylogenetic position of *A. africanus* continues to be enigmatic. Walker et al. (1986) suggest that *A. africanus* is best considered the exclusive ancestor of *A. robustus*, implying the existence of independent South and East African "robust" lineages. Delson (1986), in contrast, attributes *A. africanus* to the *Homo* clade and views KNM-WT 17000 as representing the ancestor of *A. robustus* and *A. boisei*.

With the discovery of KNM-WT 17000, and the consequent legitimization of *A. aethiopicus*, hypotheses that describe *A. africanus* as either the exclusive ancestor of "robust" *Australopithecus* species (Johanson and White, 1979; White et al., 1981; Rak, 1983; Kimbel et al., 1984) or the last common ancestor of the *Homo* and the "robust" clades (Tobias, 1980; Wolpoff, 1983; Yaroch and Vitzthum, 1984; Skelton et al., 1986; Kimbel, 1986) are, in our judgment, no longer tenable, and should be considered effectively refuted. The basis for this refutation is the large number of primitive characters (listed in Fig. 16.1) shared by *A. afarensis* and KNM-WT 17000, implying extensive character reversal in a transition from *A. afarensis* through *A. africanus* to *A. aethiopicus*. A further implication is that the numerous derived craniodental characters that unite *A. africanus* and all other post-*A. afarensis* hominids evolved in parallel in at least two lineages (Wolpoff, 1983; Kimbel et al., 1984; Skelton et al., 1986).

Four alternative phylogenetic hypotheses to account for the newly available evidence are depicted in Fig. 16.2. It should be noted that each of these phylogenetic hypotheses assumes that *A. afarensis* was the last common ancestor of subsequent hominids. New information gleaned from the KNM-WT 17000 cranium does not alter this conclusion.

In Fig. 16.2A, *A. africanus* is the exclusive ancestor of *Homo*. Derived characters that support this hypothesis include: an increase in frontal steepness, a mean increase in the relative length of the upper occipital scale, an overall reduction in pneumatization of the *pars mastoidea* of the temporal bone, and a distinct (but variably developed) inferior nasal sill (Olson, 1981; Rak, 1983; Wood, 1985; Kimbel, 1986). However, in addition to the parallel evolution of features that link all post-*A. afarensis* hominids (*group 1* parallelisms), this hypothesis requires a second group of parallelisms (*group 2*) comprised of the derived dentognathic characters that White et al. (1981) and Rak (1983) argue unite *A. africanus* and the "robust" *Australopithecus* species. This hypothesis also requires that the primitive resemblances between *A. afarensis* and *Homo* would have reversed through *A. africanus* (White et al., 1981; Rak, 1983; Kimbel et al., 1984).

In Fig. 16.2B *A. africanus* is the exclusive ancestor of *A. robustus*. This hypothesis is implied by Walker et al.'s (1986) analysis (see Lewin, 1986). It features *group 1* parallelisms and introduces three other sets of parallelisms: *group 3*, comprised of all those derived characters (listed in Fig. 16.1) that unite KNM-WT 17000 with *A. robustus* and *A. boisei*; *group 4*, comprised of all those

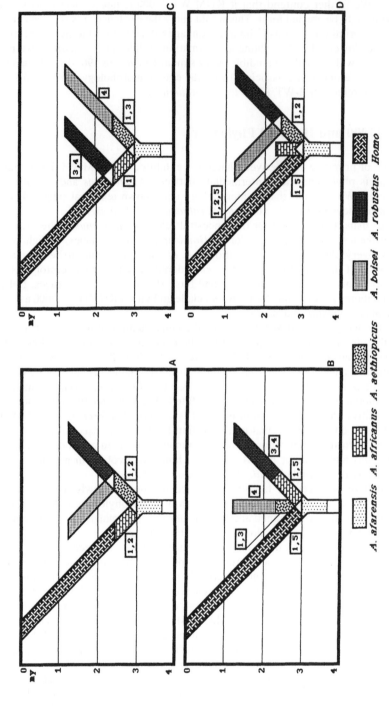

FIGURE 16.2. Alternative phylogenies of the Hominidae, incorporating *Australopithecus aethiopicus* (KNM-WT 17000). Numbers refer to groups of parallelisms discussed in the text.

derived characters that unite *A. robustus* and *A. boisei* (e.g., strongly flexed cranial base; "tucking in" of the face), but for which KNM-WT 17000 retains the primitive state; and *group 5*, consisting of those derived characters that link *A. africanus* to *Homo* (and a consequent character reversal in the transition from *A. afarensis* through *A. africanus* to *A. robustus*).

In Fig. 16.2C *A. africanus* is ancestral to both *A. robustus* and *Homo*, a relationship that is supported by the diversity of cranial and dental morphologies in *A. africanus* from Sterkfontein and Makapansgat (Kimbel and White, this volume, chapter 11; Clarke this volume, chapter 18). In addition to *group 1* parallelisms, this phylogeny requires *group 2* and *group 4* parallelisms due to the independent derivation of the *A. aethiopicus–A. boisei* clade. The reversal of primitive characters in the transition from *A. afarensis* to *Homo* via *A. africanus* is also a feature of this phylogeny.

In Fig. 16.2D *A. africanus* is an independent clade. This three lineage phylogeny has not been proposed previously. It requires *group 1, 2,* and *5* parallelisms, all of which characterize the *A. africanus* clade.

Prior to the interpretation of the Hadar/Laetoli hominids by Johanson and White (1979), *A. africanus* was commonly viewed as the last common ancestor of later taxa, because of its prevailingly primitive craniodental anatomy (e.g., Eldredge and Tattersall, 1975; Delson *et al.*, 1977; Tattersall and Eldredge, 1977). At present, many investigators would place the more primitive *A. afarensis* in this ancestral role. Under this condition, one of the most important implications of KNM-WT 17000 is that parallel evolution characterized hominid evolution to a much greater degree than previously recognized. We do not find this conclusion surprising, since hominid systematics has traditionally emphasized features of adaptive importance. For example, known "robust" *Australopithecus* species are differentiated from other hominid taxa predominantly by features related to maximizing and spreading vertical occlusal forces along the postcanine tooth row (White *et al.*, 1981). From Darwin on, it has been acknowledged that such adaptive morphology in a closely related group of organisms is particularly susceptible to parallelism. The KNM-WT 17000 cranium shows that the Hominidae is no exception to this pattern.

Although future discoveries may more conclusively demonstrate an *A. aethiopicus–A. boisei* lineage, we urge caution in interpreting KNM-WT 17000 as evidence for this link. The currently available evidence does not, in our judgment, unambiguously favor any one of the phylogenies shown in Fig. 16.2 over the alternatives (indeed, the three of us have not reached a consensus as to which phylogeny best represents this evidence). Central to the resolution of this issue will be further discoveries of specimens contemporary with KNM-WT 17000 in eastern Africa and renewed evaluation of South African remains attributed to *A. africanus*.

KNM-WT 17000 and the Status of *Australopithecus afarensis*

Since the naming of *Australopithecus afarensis* in 1978, some investigators have sought to demonstrate the presence of more than one hominid species at the Ethiopian site of Hadar, the source of the majority of *A. afarensis* remains (R. E. Leakey, 1981; Olson, 1981, 1985; Senut and Tardieu, 1985; Zihlman, 1985—see Kimbel *et al.*, 1985 and Kimbel and White, this volume, chapter 11, for responses). This issue has been raised once again, this time in the context of the new West Turkana cranium (Walker *et al.*, 1986; Shipman, 1986).

Walker *et al.* (1986) allude to the contention, made explicit by Shipman (1986), that the composite reconstruction of a male *A. afarensis* skull (Kimbel *et al.*, 1984) may combine remains of two taxa, one of which was the ancestor of East African "robust" *Australopithecus*. Walker *et al.* (1986: 522) note that many of the primitive features shared by KNM-WT 17000 and *A. afarensis* are confined to the calvaria, and they assert that despite the existence at Hadar of several adult facial specimens, not one of them is directly associated with a calvaria. The implication is that the *A. afarensis* reconstruction incorrectly joins the facial skeleton of one (unidentified) taxon with the calvaria of a second, "robust" taxon. Shipman (1986: 93) appeals to taphonomic bias as a possible explanation for the differential loss of "two or more *boisei*-type faces and at least six braincases of more gracile fossils" that could have led to the mistaken combination of two taxa in the reconstruction.

The appeal to selective preservation fails to account for important information derived from the Hadar hominid sample:

1. There is, in fact, an adult facial specimen attached to a partial calvaria in the Hadar collection. This specimen, AL 58-22, exhibits primitive calvarial characters that are shared by *A. afarensis* and KNM-WT 17000 plus primitive facial characters (deep canine fossa, posteriorly positioned zygomatic root, multiple infraorbital foramina) that distinguish *A. afarensis* from any "robust" *Australopithecus* specimen (Kimbel *et al.*, 1982, 1984; Rak, 1983).

2. Among the five adult facial specimens, 27 adult mandibles, and more than 200 teeth in the Hadar hominid collection, no derived feature shared exclusively with "robust" *Australopithecus* species has been identified (White *et al.*, 1981; Kimbel *et al.*, 1982, 1984, 1985; White and Johanson, 1982; Johanson *et al.*, 1982; Rak, 1983; Grine, 1985; Skelton *et al.*, 1986; Wood and Chamberlain, 1986; *contra* Olson, 1981). In fact, Wood and Chamberlain's (1986: 237) cladistic analysis showed that "the cladograms for the three regions most involved with mastication—that is, those for the face, palate, and mandible—are the ones congruent with the cladogram of White *et al.* (1981)" which portrays *A. afarensis* as primitive relative to other *Australopithecus* species. Indeed, Walker *et al.*'s (1986) analysis fails to detect derived facial features that link *A. afarensis* and "robust" *Australopithecus* species. In this light, it strains credulity to claim that the "robust" dentognathic elements from the Hadar site have been lost through taphonomic bias.

3. Although calvarial remains are relatively uncommon in the known Hadar hominid sample, fragmentary specimens such as AL 288-1 and 162-28 are clearly smaller and less heavily built than AL 333-45 (which forms the rear of the composite reconstruction). Insofar as comparisons can be made, there is no derived feature that divides the Hadar calvarial sample into two subsets (Kimbel *et al.*, 1984; Kimbel and Rak, 1985; Kimbel, 1986). Indeed, Olson (1981, 1985) never used calvarial (or nasal) characters to diagnose two taxa at Hadar [Zihlman's (1985: 217) assertion to the contrary notwithstanding], and Olson's claims, which were based on dental, mandibular and palatal characters, have been refuted (Kimbel *et al.*, 1985).

Thus, the logic employed in this argument is not compelling. It is clearly inappropriate to use a single 2.5-Myr-old cranium to judge the taxonomic unity of a fossil sample comprised of more than 250 specimens that is some 0.5–1.0 Myr older. If the Hadar *A. afarensis* sample really does contain more than one taxon, then obviously this must be demonstrated on the basis of the Hadar fossils themselves. In fact, while Walker *et al.* (1986) note that only the largest known *A. boisei* mandibles would fit the KNM-WT 17000 cranium, they go on to imply that the Omo 18-1967-18 mandible is conspecific with KNM-WT 17000. We agree with both observations, and point out the obvious consequences—mandible size and shape variation in *A. aethiopicus* was probably at least as great as that documented for *A. afarensis* (e.g., AL 288-li or 333w-12 vs. AL 333w-32/60) and other species of *Australopithecus* (Kimbel and White, this volume, chapter 11).

In sum, the discovery of KNM-WT 17000 does not require a re-evaluation of the taxonomic unity of *A. afarensis*. We further maintain that such a re-evaluation can only be based on the Hadar remains themselves; to employ KNM-WT 17000 as a test of the taxonomic composition of the Hadar sample constitutes an inappropriate use of this otherwise important hominid fossil.

Conclusions

In conclusion, our study indicates that KNM-WT 17000: (1) confirms *Australopithecus aethiopicus* as a valid taxon; (2) represents a link between *A. afarensis* and *A. robustus* and/or *A. boisei;* (3) eliminates *A. africanus* from consideration as the last common ancestor of *Homo* and all "robust" *Australopithecus* species; (4) eliminates *A. africanus* from consideration as the exclusive ancestor of "robust" *Australopithecus* species; (5) intensifies problems regarding the place of *A. africanus* in hominid phylogeny; (6) demonstrates considerable parallelism in early hominid evolution.

On the other hand, KNM-WT 17000 does not: (1) belong to *A. boisei;* (2) necessitate the addition of a third lineage at the mid-Pliocene branching of the hominid clade; (3) imply that the *A. afarensis* cranial reconstruction combines two taxa; (4) provide evidence of more than one hominid species at Hadar.

Acknowledgments

We thank Fred Grine, Eric Delson, Alan Walker, Yoel Rak, Lawrence Martin, and Georges Merenfeld for helpful discussion. Alan Walker kindly allowed us to examine a cast of KNM-WT 17000. Thanks to Fred Grine for inviting us to participate in the workshop at which this paper was initially read.

References

Arambourg, C. and Coppens, Y. (1967). Sur la découverte, dans le pléistocène inférieur de la vallée de l'Omo (Ethiopie), d'une mandibule d'Australopithécien. *C. R. Acad. Sci. (Paris)*, 265: 589–590.
Arambourg, C. and Coppens, Y. (1968). Découverte d'un Australopithécien nouveau dans les gisements de l'Omo (Ethiopie). *S. Afr. J. Sci.*, 64: 58–59.
Bower, B. (1986). Skull gives hominid evolution new face. *Sci. News, 130*: 100.
Bower, B. (1987). Family feud: enter the "Black Skull." *Sci. News, 131*: 58–59.
Brown, F. H., McDougall, I., Davies, T. and Maier, R. (1985). An integrated Plio-Pleistocene chronology for the Turkana Basin. *In* E. Delson (ed.), *Ancestors: the hard evidence*, pp. 82–90. Alan R. Liss, New York.
Cerling, T. E. and Brown, F. H. (1982). Tuffaceous marker horizons in the Koobi Fora region and the Lower Omo Valley. *Nature, 299*: 216–221.
Coppens, Y. (1970). Les restes d'hominidés des séries inférieurs et moyennes des formations plio-villafranchiennes de l'Omo en Ethiopie. *C.R. Acad. Sci. (Paris), 271*: 2286–2287.
Coppens, Y. (1977). Hominid remains from the Plio/Pleistocene formations of the Omo Basin, Ethiopia. *J. Hum. Evol.*, 6: 169–173.
Delson, E. (1986). Human phylogeny revised again. *Nature, 322*: 496–497.
Delson, E., Eldredge, N. and Tattersall, I. (1977). Reconstruction of hominid phylogeny: a testable framework based on cladistic analysis. *J. Hum. Evol., 6*: 263–278.
Eldredge, N. and Tattersall, I. (1975). Evolutionary models, phylogenetic reconstruction and another look at hominid phylogeny. *Contrib. Primatol., 5*: 218–242.
Falk, D. (1986). Hominid evolution. *Science. 234*: 11.
Falk, D. and Conroy, G. (1983). The cranial venous sinus system in *Australopithecus afarensis*. *Nature, 306*: 779–781.
Grine, F. E. (1981). Trophic differences between "gracile" and "robust" australopithecines: a scanning electron microscope analysis of occlusal events. *S. Afr. J. Sci., 77*: 203–230.
Grine, F. E. (1985). Australopithecine evolution: the deciduous dental evidence. *In* E. Delson (ed.), *Ancestors: the hard evidence*, pp. 153–167. Alan R. Liss, New York.
Hillhouse, J. W., Cerling, T. E. and Brown, F. H. (1986). Magnetostratigraphy of the Koobi Fora Formation, Lake Turkana, Kenya. *J. Geophys. Res., 91*: 11,581–11,595.
Howell, F. C. (1978). Hominidae. *In* V. Maglio and H. B. S. Cook (eds.), *Evolution of African mammals*, pp. 154–248. Harvard University Press, Cambridge, MA.
Howell, F. C. and Coppens, Y. (1973). Deciduous teeth of Hominidae from the Pliocene/Pleistocene of the Lower Omo Basin, Ethiopia. *J. Hum. Evol., 2*: 461–472.
Howell, F. C. and Coppens, Y. (1976). An overview of Hominidae form the Omo succession, Ethiopia. *In* Y. Coppens, F. C. Howell, G. Isaac and R. E. Leakey (eds.), *Earliest Man and environments in the Lake Rudolf Basin*, pp. 522–531. University of Chicago Press, Chicago.
Johanson, D. C. and White, T. D. (1979). A systematic assessment of early African hominids. *Science, 203*: 321–330.
Johanson, D. C. and White, T. D. (1986). Fossil debate. *Discover, 7*: 116.
Johanson, D. C., White, T. D. and Coppens, Y. (1982). Dental remains from the Hadar Formation: Ethiopia: 1974–1977 collections. *Amer. J. Phys. Anthropol., 57*: 545–603.
Kimbel, W. H. (1986). The calvarial remains of *Australopithecus afarensis:* a comparative phylogenetic study. Ph.D. Dissertation, Kent State University, Kent, Ohio.

Kimbel, W. H. and Rak, Y. (1985). Functional morphology of the asterionic region in extant hominoids and fossil hominids. *Amer. J. Phys. Anthropol.*, 66: 31–54.

Kimbel, W. H., Johanson, D. C. and Coppens, Y. (1982). Pliocene hominid cranial remains from the Hadar Formation, Ethiopia. *Amer. J. Phys. Anthropol.*, 57: 453–499.

Kimbel, W. H., White, T. D. and Johanson, D. C. (1984). Cranial morphology of *Australopithecus afarensis:* a comparative study based on a composite reconstruction of the adult skull. *Amer. J. Phys. Anthropol.*, 64: 337–388.

Kimbel, W. H., White, T. D. and Johanson, D. C. (1985). Craniodental morphology of the hominids from Hadar and Laetoli: evidence of *"Paranthropus"* and *Homo* in the mid-Pliocene of eastern Africa? *In* E. Delson (ed.), *Ancestors: the hard evidence*, pp. 120–137. Alan R. Liss, New York.

Kimbel, W. H., White, T. D. and Johanson, D. C. (1986). On the phylogenetic analysis of early hominids. *Curr. Anthropol.*, 27: 361–362.

Leakey, R. E. (1981). *The making of mankind.* Dutton, New York.

Lewin, R. (1986). New fossil upsets human family. *Science, 233:* 720–721.

McDougall, I. (1985). K-Ar and ^{40}Ar/^{39}Ar dating of the hominid-bearing Pliocene-Pleistocene sequence at Koobi Fora, Lake Turkana, northern Kenya. *Geol. Soc. Amer. Bull.*, 96: 159–175.

Olson, T. R. (1981). Basicranial morphology of the extant hominoids and Pliocene hominids: the new material from the Hadar Formation, Ethiopia, and its significance in early human evolution and taxonomy. *In* C. B. Stringer (ed.), *Aspects of human evolution*, pp. 99–128. Taylor and Francis, London.

Olson, T. R. (1985). Cranial morphology and systematics of the Hadar Formation hominids and *"Australopithecus" africanus. In* E. Delson (ed.), *Ancestors: The hard evidence*, pp. 102–119. Alan R. Liss, New York.

Rak, Y. (1983). *The australopithecine face.* Academic Press: New York.

Rak, Y. (1985). Australopithecine taxonomy and phylogeny in light of facial morphology. *Amer. J. Phys. Anthropol.*, 66: 281–287.

Senut, B. and Tardieu, C. (1985). Functional aspects of Plio-Pleistocene hominid limb bones: implications for taxonomy and phylogeny. *In* E. Delson (ed.), *Ancestors: the hard evidence*, pp. 193–201. Alan R. Liss, New York.

Shipman, P. (1986). Baffling limb on the family tree. *Discover, 7:* 87–93.

Skelton, R. R., McHenry, H. M. and Drawhorn, G. M. (1986). Phylogenetic analysis of early hominids. *Curr. Anthropol.*, 27: 21–43.

Tattersall, I. and Eldredge, N. (1977). Fact, theory, and fantasy in human paleontology. *Amer. Sci.*, 65: 204–211.

Tobias, P. V. (1980). *"Australopithecus afarensis"* and *A. africanus:* critique and an alternative hypothesis. *Palaeontol. Afr.*, 23: 1–17.

Walker, A. C., Leakey, R. E., Harris, J. M. and Brown, F. H. (1986). 2.5-Myr *Australopithecus boisei* from west of Lake Turkana, Kenya. *Nature, 322:* 517–522.

White, T. D. (1977). The anterior mandibular corpus of early African Hominidae: functional significance of shape and size. Ph.D. Thesis, University of Michigan, University Microfilms, Ann Arbor, MI.

White, T. D., Johanson, D. C. and Kimbel, W. H. (1981). *Australopithecus africanus:* its phylogenetic position reconsidered. *S. Afr. J. Sci.*, 77: 445–470.

Wolpoff, M. H. (1983). Australopithecines: the unwanted ancestors. *In* K. Reichs (ed.), *Hominid origins*, pp. 109–126. University Press of America, Washington, D.C.

Wood, B. A. and Chamberlain, A. T. (1986). *Australopithecus:* grade or clade? *In* B. Wood, L. Martin and P. Andrews (eds.), *Major topics in primate and human evolution*, pp. 220–248. Cambridge University Press, Cambridge.

Yaroch, L. A. and Vitzthum, V. J. (1984). Was *Australopithecus africanus* ancestral to the genus *Homo? Amer. J. Phys. Anthropol.*, 63: 237.

Zihlman, A. (1985). *Australopithecus afarensis:* two sexes or two species? *In* P. V. Tobias (ed.), *Hominid evolution: past, present and future*, pp. 213–220. Alan R. Liss, New York.

Are "Robust" Australopithecines a Monophyletic Group?

17

BERNARD A. WOOD

It is presently conventional to subsume three formal taxa into the informal category known as "robust" australopithecines. These are *Australopithecus (Paranthropus) robustus, A. (P.) crassidens,* and *A. (P.)boisei*. The constituent hypodigms of the three taxa were comprehensively reviewed in Howell (1978) and summarized in Wood and Chamberlain (1987). More recent additions are reported in this volume (Grine, this volume, chapter 14; Susman, this volume, chapter 10). In his detailed review, Howell (1978) chose to retain a specific distinction between *A. (P.) robustus* and *A. (P.) crassidens;* otherwise, with the exceptions discussed below, it has become conventional to subsume the latter taxon into *A. (P.) robustus*. The discovery of *A. (P.) robustus* was first recorded in 1938, whereas it was not until 11 years later that Broom reported the holotype of *A. (P.) crassidens* (Broom, 1938, 1949). Two species of "robust" australopithecines from the Transvaal cave deposits have more recently been recognized by Grine (1981, 1982, 1984, 1985), but most other studies pool the two taxa for the purposes of analysis and discussion. A cranium and a mandible from the localities of Lomekwi and Kangatukuseo on the west side of Lake Turkana are the most recent discoveries germane to the "robust" australopithecines. They have been provisionally placed in *A. (P.) boisei* (Walker *et al.*, 1986), but a more detailed assessment of what, hopefully, will soon be an enlarged sample from these sites may result in the erection of a fourth taxon to accommodate this material.

"Robust" Australopithecine Phylogenetic Affinities

Since their discovery half a century ago, opinion about the place of "robust" australopithecines within the overall context of hominid phylogeny has undergone several major changes. Some of these shifts of opinion are the result of genuinely new insights, based either on reassessments of existing material, or on new fossil evidence. Other changes in interpretation owe more to differences in the way hominids as a whole are perceived. An example of the importance of noting this wider context is the discussion of the affinities of *Paranthropus crassidens,* which is presented in Broom and Robinson (1952). Whereas today we would assume that such a discussion would deal with the affinities *between* hominid taxa, Broom and Robinson devote most of this section of the monograph to rebutting suggestions that *P. crassidens* is, at least dentally, indistinguishable from extant anthropoid apes. Surprisingly, there is virtually no explicit discussion of the relationships between what were then termed *Australopithecus, Plesianthropus* and *Paranthropus,* except to note the larger brain size of *Paranthropus*. It was only when the hominid status of the *Paranthropus* remains was more widely accepted that we find any extended discussion of the relationships between the two major categories of australopithecines.

Robinson (1954) saw sufficient similarities between the remains from Swartkrans and Kromdraai to link them in the same lineage, or clade. He was virtually alone in regarding them as congeneric with *Meganthropus palaeojavanicus;* but it is clear from the evolutionary scheme that he illustrates, that he regarded the two "robust" taxa known at that time, *Paranthropus robustus* and *P. crassidens,* as time-successive components of a lineage that was quite separate from that containing the "gracile" australopithecine remains (Robinson, 1954: fig. 7). Robinson's views on the extent of the difference between *Paranthropus* and *Australopithecus* were subsequently amplified (Robinson, 1961, 1962, 1963), and his proposition that the morphologic differences are best interpreted as adaptations to different climates and habitats (Robinson, 1968) is well known. However, it is of interest that Robinson (1968:167) associated *Australopithecus* with "the more arid periods" and *Paranthropus* with "wetter periods." As early as 1965, Robinson had advocated the retention of *Paranthropus* (to include the, by now, three "robust" australopithecine species) and the assimilation of *Australopithecus* into *Homo* (Robinson, 1965). Details of the subdivisions within *Homo,* and the nomenclatural devices used to describe them, have varied from time-to-time, but the essence of his proposal remained the same (Robinson 1972a). However, what is abundantly clear is that Robinson was to come to modify the 1954 scheme to the extent of proposing that "a population of *Paranthropus* was directly ancestral to *Homo africanus*" (Robinson 1972b: 248) (Fig. 17.1A). The majority of Robinson's (1965) hypodigm of *H. africanus* comprised *A. africanus* remains from southern African sites, the balance being the specimens from Bed I at Olduvai Gorge, which he could not accept as *Homo habilis.* It must be understood that Robinson referred specifically to a "population" of *Paranthropus* being ancestral to *H. africanus,* and elsewhere he makes it clear how that population may have differed from the majority of the *Paranthropus* hypodigm, when he suggests that *Australopithecus* "arose from the *Paranthropus* line well before the reduction of the anterior teeth in the latter had reached the stage found in the known forms" (Robinson 1968:167). Thus, it is clear that Robinson regarded this ancestral population as a hypothetical one at the time of writing.

Tobias (1967) also saw the "robust" australopithecines as evolving from a relatively unspecialized ancestral form. He regarded both *A. africanus* and *A. boisei* as being derived from such an australopithecine ancestor; the former representing "a more conservative residual line" and the latter a "megadontic" form with "specialized dentition." The latter was believed to have undergone "moderate reduction in cheek-tooth size . . . shortening of the face and reduction of the jaws" to become *A. robustus* (Tobias, 1967:243) (Fig. 17.1B). In subsequent papers, Tobias was more specific about the identity of the unspecialized australopithecine ancestor, for he came to regard *A. africanus* as fulfilling the criteria of a common ancestral form (Tobias, 1973, 1978, 1980). The sequence *A. boisei* to *A. robustus* is also given less emphasis in subsequent phylogenetic trees (Tobias, 1978) (Fig. 17.1C).

There is yet a fourth interpretation of the affinities of the "robust" australopithecines. This sees *A. africanus* not as a generalized common australopithecine ancestor but as a form that is already derived in the direction of the "robust" australopithecines. The specializations are considered to be such that they would preclude it being the ancestor of the *Homo* lineage. Instead, that role was accorded to *Australopithecus afarensis,* with *A. africanus* being displaced to the base of a "robust" australopithecine clade (Johanson and White, 1979) (Fig. 17.1D), with much of the decisive evidence for this assessment coming from the work of Rak (1981, 1983).

Relationships between "Robust" Australopithecine Taxa

One of the few common threads which runs through the various phylogenetic interpretations of the "robust" australopithecines presented above is the belief that the two main taxa, *A. (P.) robustus* and *A. (P.) boisei,* share a common ancestor which is exclusive to them. That is, in cladistic parlance, they are regarded as "sister taxa."

The technique of phylogenetic, or cladistic, analysis explores the affinities between taxa by concentrating on the distribution of those features that have become established since the emergence of a common ancestor, in this case the common hominid ancestor. It has been used for more than a decade to attempt to resolve debates about the affinities of fossil hominids (Eldredge and Tattersall, 1975; Bonde, 1976, 1977; Delson *et al.,* 1977; Tattersall and Eldredge, 1977, Olson, 1978, 1985; Johanson and White, 1979; Corruccini and McHenry, 1980; White *et al.,* 1981, Grine, 1981, 1982, 1984, 1985; Kimbel *et al.,* 1984; Wood and Chamberlain, 1986; Skelton *et al.,* 1986; Chamberlain and Wood, 1987). Of these eighteen studies, eleven make a distinction between *A. (P.) robustus* and *A. (P.) boisei,* and the studies of Grine (1981, 1982, 1984, 1985) go further by distinguishing between *A. (P.) robustus* and *A. (P.) crassidens.* However, in all the analyses that

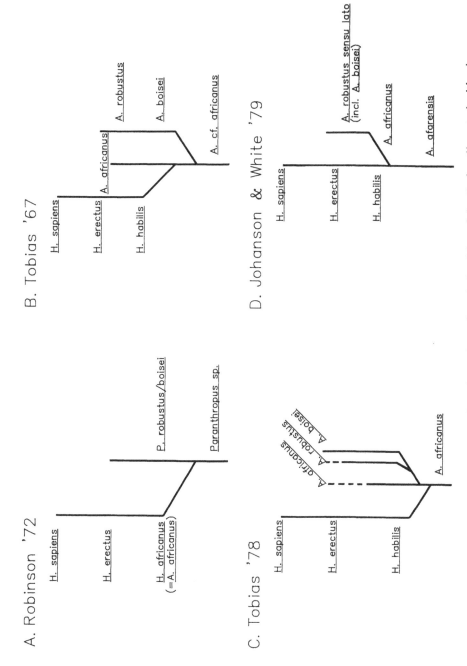

FIGURE 17.1. Diagrams of the changing views about the relationships of the "robust" australopithecines.

make these distinctions, the two major "robust" australopithecine taxa [i.e., *A. (P.) robustus* and *A. (P.) boisei*] consistently emerge as sister taxa. This evidence is not as conclusive about the affinities of the "robust" australopithecines as it may at first seem, because several of the assessments use the same data. Nonetheless, there seems to be overwhelming evidence to suggest that the "robust" australopithecines constitute a distinct clade or monophyletic group.

Why, then, given the apparent strength of the evidence, should the hypothesis of "robust" australopithecine monophyly be challenged? First, because it is becoming increasingly clear that a significant number, perhaps up to one third, of the characters that are apparently shared-derived among hominid taxa may, in fact, have been independently acquired (Corruccini and McHenry, 1980; Wood and Chamberlain, 1986; Chamberlain and Wood, 1987). If that is so, and if there is no *a priori* reason to believe that the association between the two "robust" australopithecine taxa is immune from the confounding influence of independently acquired characters, or homoplasy, then logically we must entertain the possibility that homoplasies may play a part in linking "robust" taxa.

The second major reason to question the hypothesis of monophyly is that most of the characters that link the "robust" australopithecines so strongly (and that are also responsible for associating *A. africanus* with the "robust" clade) are related to a single functional complex, the masticatory system (see Kimbel *et al.*, 1984:381). Wood and Chamberlain's (1986) more formal cladistic analysis was based on a different set of characters than that of White *et al.* (1981). The results were presented as a series of anatomical "regional" cladograms, as well as in the form of a summary cladogram, and while all regional cladograms (i.e., vault, face, palate, base, and mandible) supported the association of the two "robust" taxa as sister groups, it was the three regions most strongly linked with mastication—i.e., the face, palate and mandible—that contributed most in the summary cladogram to both the hypothesis of "robust" monophyly and to the incorporation of *A. africanus* in the "robust" australopithecine clade. In other vertebrate mammalian groups there are several examples of convergence in the gross morphology of masticatory features. In the Bovidae, for example, Vrba (1984) reports that a reduction in the relative contribution of the premolars occurs within the three lineages, *Parmularius, Megalotragus,* and *Connochaetes*. It is clear from a cladistic scheme of the Bovidae, which was generated using a total of 58 characters (Vrba, 1979), that this and other dental and facial features associated with mastication are likely to be homoplasies. Gentry (pers. comm.) also points to a suite of mastication-related cranial and dental characters that can be identified in five lineages of African large-mammal herbivores. Thus, it is reasonable to conclude that analogous, if not homologous, mastication-linked similarities in hominids should at least be considered candidates for convergence or parallelism.

Detection of Homoplasies

If it is accepted that up to one third of the characters in a cladistic analysis of hominids that are apparently shared-derived characters may in fact be homoplasies, then the next task must be the identification of those characters. Patterson (1982) suggested that there were three categories of tests that could be applied to potential homologies: (a) conjunction, (b) congruence, and (c) similarity. The first category, *conjunction*, rests on the simple principle that if more than one expression of the proposed homologous series occurs in the same organism, then the hypothesis of homology fails. The second category, *congruence*, is somewhat circular, but Patterson (1982:66) claims it is the strongest of the three tests. In brief, it proposes that the hypothesis of homology is supported when the pattern of association between taxa indicated by accepting the proposition is supported by other proposed homology schemes. Its weakness must be its circularity, in that it requires a known homology to establish the likelihood of an unknown homology. The third category of tests, *similarity*, involves the detailed investigation of the apparently homologous features to establish whether their superficial phenetic similarity is paralleled by detailed similarities of, for example, ontogeny and detailed morphology. In the next two sections of this paper, Patterson's *similarity* criteria will be used to examine different aspects of the ontogenetic basis of two features that are apparently shared by *A. (P.) robustus* and *A. (P.) boisei*. The first feature is the large size and peculiar shape of the crowns of the mandibular molar and premolar teeth

Table 17.1. Results of Ten Tests of Allometric Scaling Comparing Differences between the Tooth Crowns of *A.(P.) boisei* (EAFROB) and *A.(P.) robustus* (SAFROB)[a,b]

	P$_3$				P$_4$				M$_1$				M$_2$			
	SAFROB		EAFROB		SAFROB		EAFROB		SAFROB		EAFROB		SAFROB		EAFROB	
	r	α	r	α	r	α	r	α	r	α	r	α	r	α	r	α
MCBA vs. TALA	0.6 N=13	1.0	NS	—	0.9 N=9	1.5	0.9	0.6	—	—	—	—	—	—	—	—
MCBA vs. ENTA	—	—	—	—	—	—	—	—	0.7 N=20	0.8	NS	—	0.8 N=17	1.5	NS	—
MCBA vs. HYLDA	—	—	—	—	—	—	—	—	0.7 N=20	1.2	NS	—	0.6 N=17	1.3	NS	—
MD vs. TRIB	—	—	—	—	—	—	—	—	0.9 N=18	0.9	NS	—	0.8 N=17	0.7	0.9 N=7	0.8[c]
MD vs. TALB	—	—	—	—	—	—	—	—	0.9 N=18	1.0	NS	—	0.8 N=17	0.8	0.9 N=7	0.9[c]

[a] r, Pearson Correlation Coefficient. Value given if significant at <0.05. α, Least-squares regression slope of log-transformed data. N, numbers of specimens in each test pair.

[b] MCBA, Measured Crown Base Area; MD, mesiodistal length corrected for interproximal wear; TALA, talonid area; ENTA, entoconid area; HYLDA, hypoconulid area; TRIB, buccolingual breadth of trigonid; TALB, buccolingual breadth of talonid.

[c] Cases in which EAFROB slope values do not differ at 5% level from SAFROB slopes, i.e., EAFROB conforms to the criterion of a scaled variant.

(Wood and Abbott, 1983; Wood and Uytterschaut, 1987). The second concerns the peculiar root form of the mandibular premolars of the two taxa (Abbott, 1984; Abbott and Wood, 1985; Wood et al., 1988).

For the investigation of tooth crown morphology, the premise or "null hypothesis" is the following. If the two "robust" australopithecines are similarly adapted sister taxa, with *A. (P.) boisei* being the larger toothed (Wood and Abbott, 1983) and more dentally derived (Grine, 1985; Suwa, this volume, chapter 13) of the pair, then it is reasonable to suppose that the larger taxon should be what Pilbeam and Gould (1974) called a scaled variant of the smaller. In this case, the proposition would be that the shape differences between the tooth crowns of *A. (P.) boisei* and *A. (P.) robustus* could have been predicted from any relative size, or allometric, relationships present in the latter. In other words, for the larger teeth of *A. (P.) boisei* to be homologous, it is suggested that they should be the result of an extrapolation of the allometric trends in *A. (P.) robustus*. In order to test this, we have adapted the approach used by Jungers and Grine (1986). In the present study ten tests were conducted using the crown dimensions of the P_3, P_4, M_1, and M_2 of *A. (P.) robustus*—designated SAFROB; and *A. (P.) boisei*—designated EAFROB (Table 17.1). Four of the tests of the postcanine crown data involved dimensions used by Jungers and Grine (1986); the remainder employed more precise area estimates of the size of the tooth crown and its components. Regression analysis was the method chosen for estimating the slope of the line because the problem is essentially a predictive one, e.g., for a given size of tooth, what is the value of a component breadth or area (Hills and Wood, 1984). Thus, least-squares slopes were fitted to logarithmically transformed data, and the results are presented in Table 17.1.

It has first to be established that the two variables being investigated are significantly correlated

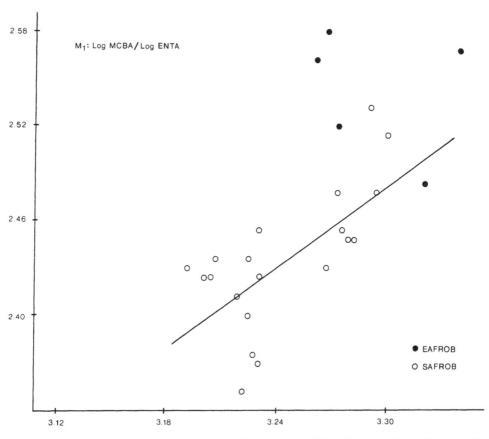

FIGURE 17.2. Least-squares plot of logarithmically transformed data for the scaling of entoconid area (ENTA) with measured M_1 crown base area (MCBA) in "robust" australopithecines. *A. (P.) robustus*: SAFROB; *A. (P.) boisei*: EAFROB.

in the SAFROB sample. Without this one cannot talk of an allometric relationship. The results show that the pairs of variables are significantly related in all ten tests. Then it has to be demonstrated that the two variables are also correlated in EAFROB. This, however, is only the case in the three tests in which the correlation coefficient is significant at the 5% level. An example of poor correlation between the test variables in EAFROB is given in Fig. 17.2. Thus, for seven of the pairs of relationships studied, despite there being a predictive relationship between the reference variable and the character under investigation in SAFROB, there is no such relationship in the EAFROB sample. Therefore, in these seven cases, EAFROB cannot be regarded as a scaled variant of SAFROB. Of the three remaining tests in which the variables are significantly correlated, the distributions of the EAFROB data are such that they are compatible with the null (i.e., scaled variant) hypothesis in two of them [M_2: mesiodistal length (MD) vs. buccolingual trigonid breadth (TRIB), and MD vs. buccolingual talonid breadth (TALB)] (Fig. 17.3), but not in the third [P_4: measured crown base area (MCBA) vs. talonid area (TALA)] (Fig. 17.4). Thus, this limited investigation of relative size relationships in the tooth crowns of mandibular postcanine teeth suggests that the morphology of the tooth crowns of *A. (P.) boisei* corresponds to that expected of a larger-sized scaled variant in only two of the ten examples considered, the other eight being rejected because there was a lack of correlation between the variables, or because the slopes of the correlated variables were significantly different in the two taxa.

The second investigation for possible homoplasies concerns the ontogenetic basis for, and the detailed morphology of, mandibular premolar root form. It has long been known that the root form of the mandibular premolars of *A. (P.) robustus* is more complex than that of modern humans and extant apes. A recent study has surveyed the root morphology of *A. (P.) boisei* and

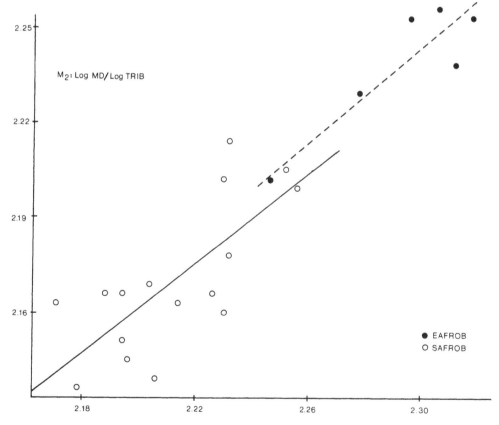

FIGURE 17.3. Least-squares plot of logarithmically transformed data for the scaling of buccolingual trigonid breadth (TRIB) with M_2 crown mesiodistal length (MD) in "robust" australopithecines. *A. (P.) robustus:* SAFROB; *A. (P.) boisei:* EAFROB.

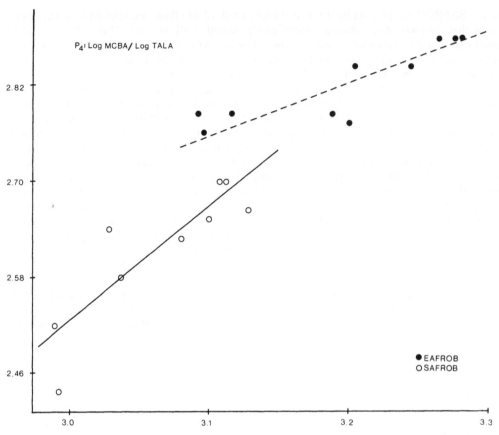

FIGURE 17.4. Least-squares plot of logarithmically transformed data for the scaling of talonid area (TALA) with measured P_4 crown base area (MCBA) in "robust" australopithecines. *A. (P.) robustus:* SAFROB; *A. (P.) boisei:* EAFROB.

attempted to deduce the probable mandibular premolar root form of a common hominid ancestor (Wood *et al.*, 1988). This evidence, combined with a knowledge of root development, allowed the authors to propose two morphoclines. One is a trend toward root molarization, which links the presumed ancestral condition with the apparently derived root form seen in *A. (P.) boisei* (Fig. 17.5); the second trend is towards root reduction. The necessary morphological changes probably resulted from relatively minor adjustments to the extent of cellular proliferation at the margins of the primary apical foramen of the developing teeth. It is against this background that the premise for this second investigation was formulated. The proposition is that if *A. (P.) robustus* is the less-derived sister taxon of *A. (P.) boisei*, then the prediction would be that the mandibular premolar root form of *A. (P.) robustus* would be located along the most ontogenetically parsimonious morphocline between the ancestral condition and the derived molarized character state.

The crucial evidence to test this proposal comes from the root form of the P_3 of *A. (P.) robustus.* Wood *et al.* (1988) showed that whereas four *A. (P.) robustus* P_3s have separate mesial and distal roots, the majority of P_3s of that taxon (nine out of fourteen) have a Tomes' root form in which mesiobuccal and distolingual root moieties are separated by a cleft; one of the nine is a tooth from Kromdraai. However, reference to Fig. 17.5 shows that such a root form is part of a second trend toward root reduction and not part of the most parsimonious pathway to root molarization. Thus, we must conclude that the twin-rooted P_3s of *A. (P.) robustus* and *A. (P.) boisei* are not homologous character states when subjected to the rigorous testing of the sort proposed by Patterson (1982).

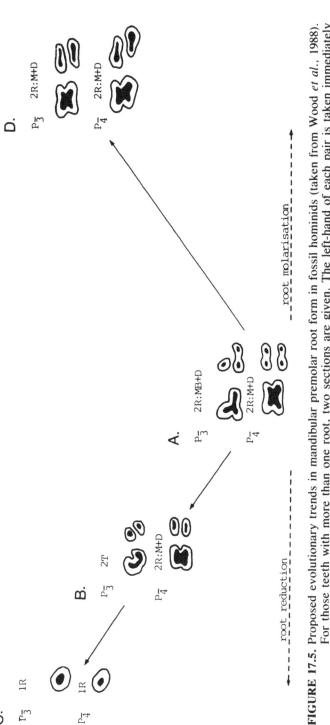

FIGURE 17.5. Proposed evolutionary trends in mandibular premolar root form in fossil hominids (taken from Wood et al., 1988). For those teeth with more than one root, two sections are given. The left-hand of each pair is taken immediately above the bifurcation; the right-hand section below it. A, The inferred primitive condition of premolars for a hominid ancestor. This is the arrangement for the majority of the Hadar specimens. B, Ontogenetic evidence suggests that Tomes' root form for P_3, and single root canals for the P_4 roots, are a stage between the inferred primitive condition and the derived condition of single-rooted premolars. The majority of the small sample of A. africanus teeth have this root form. C, Derived condition of root reduction in both P_3 and P_4 seen in Homo sapiens. D, Derived condition of root elaboration seen in A. (P.) boisei.

Parsimony and Monophyly

The second way in which the hypothesis of "robust" australopithecine monophyly has been investigated in this study was the use of the potential of a computer program for cladogram construction—PAUP (Swofford, 1985)—to compare cladograms. The PAUP program includes the facility to use the same distribution of character states to test a series of specified cladograms. The efficiency of cladograms can be compared using two criteria, the Consistency Index (CI) and the Minimum Tree Length (TL); the higher the CI, the fewer the homoplasies. In this way, we can measure the impact of imposing paraphyly on the "robust" australopithecines by comparing the CI and TL values of such "constrained" trees with those that are compatible with a hypothesis of monophyly. If the latter hypothesis is significantly better at explaining the data, then we would predict that the CI values of such trees would be greater than those determined from a series of imposed paraphyletic solutions.

Three data sets were used for this part of the investigation. The first comprises the distribution of the states of the 40 cranial characters that Walker et al. (1986) extracted from a larger study by Skelton et al. (1986) of four hominid taxa [*A. afarensis, A. africanus, A. (P.) robustus, A. (P.) boisei*] and KNM-WT 17000. The second set comprised the distribution among the same taxa of the states of 39 cranial characters selected relatively evenly across five anatomical areas of the skull (Wood and Chamberlain, 1986). The third data set is made up of the distributions of the states of 90 cranial characters (Chamberlain and Wood, 1987). The second and third data sets sample similar taxonomic hypodigms, but the data are independent. Each data set was submitted to the PAUP analysis using the branch-and-bound algorithm (Swofford, 1985).

The results are presented in diagrammatic form in Figs. 17.6 through 17.8. In each pair of diagrams two unrooted trees are presented. The diagram on the left-hand side of each figure is the most parsimonious arrangement, i.e., the one with the fewest evolutionary steps. The right-hand pair of each figure portrays the shortest unrooted cladogram that results from the imposition of "robust" australopithecine paraphyly. Clearly, with five taxa in two data sets, and four in the third, there are many combinations of taxa that would result in paraphyly. However, when these various combinations were tested, the shortest paraphyletic trees were associated with the cladograms that constrain *A. afarensis* to be the sister group of *A. (P.) boisei*, and *A. africanus* the sister group of *A. (P.) robustus*.

The results show that for all three data sets the shortest trees are associated with the hypothesis of monophyly. However, for two of the data sets, that is, the 40 characters of Walker et al. (1986), and the 90 characters of Chamberlain and Wood (1987), the paraphyletically constrained trees are only marginally less parsimonious and have similar levels of homoplasy. Thus, for these data sets, paraphyly of the "robust" australopithecines is only marginally less compatible with the distribution of the character states than is a monophyletic arrangement. For the third data set the differences in CI and TL values are greater. It is worth noting, *en passant*, that when it is included in the analysis (Fig. 17.6) KNM-WT 17000 is regarded as the sister group of *A. afarensis* in both the most parsimonious tree and the shortest constrained tree that can be derived from the data set.

Conclusions

The results of this investigation are salutary for several reasons. First, they show that the case for "robust" australopithecine paraphyly is stronger than many may have believed. Nonetheless, it must not be forgotten that the two main "robust" australopithecine taxa share a formidable suite of detailed cranial and dental characters (e.g., Rak, 1983; Grine, 1985). It is clearly not impossible that they all arose by parallel or convergent evolution, but for this to have been the case any evidence for paraphyly must be both plentiful and compelling. Second, the results emphasize that the phylogenetic relationships of the "robust" australopithecines are no clearer than their cladistic relationships. For example, if further studies confirm the apparently close relationship between *A. afarensis* and the "robust" taxa (Wood and Chamberlain, 1986), as well as between *A. afarensis* and the taxon to which KNM-WT 17000 belongs, then they would put

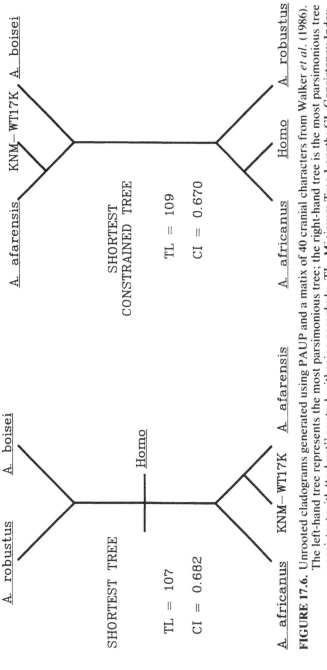

FIGURE 17.6. Unrooted cladograms generated using PAUP and a matix of 40 cranial characters from Walker *et al.* (1986). The left-hand tree represents the most parsimonious tree; the right-hand tree is the most parsimonious tree consistent with "robust" australopithecine paraphyly. TL, Minimum Tree Length; CI, Consistency Index.

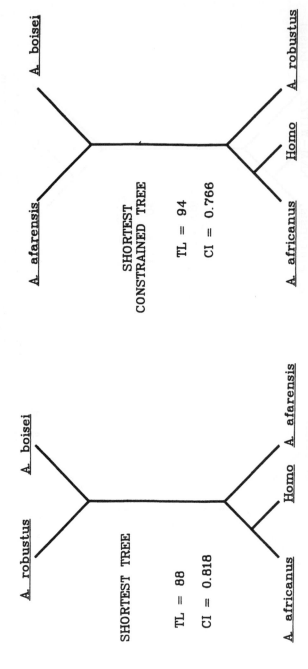

FIGURE 17.7. Unrooted cladograms generated using PAUP and a matrix of 39 cranial characters from Wood and Chamberlain (1986). See caption to Fig. 17.6 for explanation.

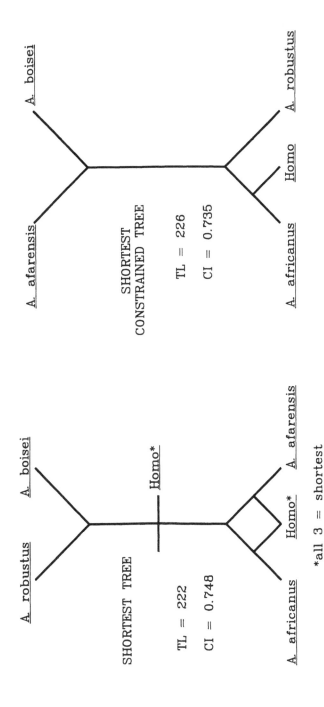

FIGURE 17.8. Unrooted cladograms generated using PAUP and a matrix of 90 cranial characters from Chamberlain and Wood (1987). See caption to Fig. 17.6 for explanation.

in question the polarity of morphoclines that presently see *A. (P.) robustus* as less derived than *A. (P.) boisei*. Those who continue to favour "robust" australopithecine monophyly must then explain the nature of the relationship between *A. afarensis* and *A. africanus*, and between *A. africanus* and *A. (P.) robustus*. Which of the former taxa is primitive with respect to the obviously more derived hominids? Is either taxon a suitable common hominid ancestor? Is *A. (P.) robustus* the sister taxon of *A. africanus* or of *A. (P.) boisei?*

The third possible lesson to be learned from these results is that those undertaking cladistic studies should avoid making prior assumptions about either sister group relationships or the degree of primitiveness of taxa, as opposed to characters. It is necessary to avoid the pitfalls consequent on such assumptions in order to prevent the subsequent analyses from being biased and the results inevitably flawed. Skelton *et al.* (1986) effectively did this when they assumed that *A. afarensis* epitomized the primitive hominid. Likewise, the present results demonstrate that it would be wise not to assume "robust" australopithecine monophyly, even though it may continue to prove to be the most likely cladistic hypothesis.

The evidence of this, and other studies in this volume, demonstrates that the "robust" australopithecines are far from being a "backwater" of hominid evolutionary studies. Indeed, their evolutionary history is proving to be as exciting and intriguing a research problem as any other in hominid palaeontology.

Acknowledgments

I am grateful to Fred Grine for his invitation to what proved to be an excellent and constructive workshop. My sincere thanks are due to Richard Leakey for his invitation to study the Koobi Fora hominids and for his encouragement and hospitality during the period of the study. The Trustees of the National Museums of Kenya, the NERC, and The Leverhulme Trust have all facilitated or supported research incorporated in this paper.

References

Abbott, S. A. (1984). A comparative study of tooth root morphology in the great apes, modern man and early hominids. Ph.D. Thesis, University of London.
Abbott, S. A. and Wood, B. A. (1985). Mandibular premolar root form and evolution in the Hominoidea. *J. Anat., 140*: 536–537.
Bonde, N. (1976). Nyt om menneskets udviklingshistorie. *Dansk. Geol. Foren. Arsskr., f.*: 19–34.
Bonde, N. (1977). Cladistic classification as applied to vertebrates. *In* M. K. Hecht, P. C. Goody and B. M. Hecht (eds.), *Major patterns in vertebrate evolution*, pp. 741–804. Plenum, New York.
Broom, R. (1938). The Pleistocene anthropoid apes of South Africa. *Nature, 142*: 377–379.
Broom, R. (1949). Another new type of fossil ape-man *(Paranthropus crassidens)*. *Nature, 163*: 57.
Broom, R. .and Robinson, J. T. (1952). Swartkrans ape-man, *Paranthropus crassidens. Tvl. Mus. Mem., 6*: 1–123.
Chamberlain, A. T. and Wood, B. A. (1986). Phylogenetic analysis of early hominids: comments. *Curr. Anthropol., 27*(1): 36–37.
Chamberlain, A. T. and Wood, B. A. (1987). Early hominid phylogeny. *J. Hum. Evol., 16*: 119–133.
Corruccini, R. S. and McHenry, H. M. (1980). Cladometric analysis of Pliocene hominids. *J. Hum. Evol., 9*: 209–221.
Delson, E., Eldredge, N. and Tattersall, I. (1977). Reconstruction of hominid phylogeny: a testable framework based on cladistic analysis. *J. Hum. Evol., 6*: 263–278.
Eldredge, N. and Tattersall, I. (1975). Evolutionary models, phylogenetic reconstruction and another look at hominid phylogeny. *In* F. S. Szalay (ed.), *Approaches to primate paleobiology*, pp. 218–242. Karger, Basel.

Grine, F. E. (1981). Trophic differences between "gracile" and "robust" australopithecines: a scanning electron microscope analysis of occlusal events. *S. Afr. J. Sci.*, 77: 203–230.

Grine, F. E. (1982). A new juvenile hominid (Mammalia: Primates) from Member 3, Kromdraai Formation, Transvaal, South Africa. *Ann. Tvl. Mus.*, 3: 165–239.

Grine, F. E. (1984). Deciduous molar microwear of South African australopithecines. In D. J. Chivers, B. A. Wood and A. Bilsborough (eds.), *Food acquisition and processing in primates*, pp. 525–534. Plenum, London.

Grine, F. E. (1985). Australopithecine evolution: the deciduous dental evidence. In E. Delson (ed.), *Ancestors: the hard evidence*, pp. 153–167. Alan R. Liss, New York.

Hills, M. and Wood, B. A. (1984). Regression lines, size and allometry. In D. J. Chivers, B. A. Wood and A. Bilsborough (eds.), *Food acquisition and processing in primates*, pp. 557–567. Plenum, London.

Howell, F. C. (1978). Hominidae. In V. J. Maglio and H. B. S. Cooke (eds.), *Evolution of African mammals*, pp. 154–248. Harvard University Press, Cambridge, MA.

Johanson, D. C. and White, T. D. (1979). A systematic assessment of early African hominids. *Science*, 202: 321–330.

Jungers, W. L. and Grine, F. E. (1986). Dental trends in the australopithecines: the allometry of mandibular molar dimensions. In B. Wood, L. Martin and P. Andrews (eds.), *Major topics in primate and human evolution*, pp. 203–219. Cambridge University Press, Cambridge.

Kimbel, W. H., White, T. D. and Johanson, D. C. (1984). Cranial morphology of *Australopithecus afarensis*: a comparative study based on a composite reconstruction of the adult skull. *Amer. J. Phys. Anthropol.*, 64: 337–388.

Olson, T. R. (1978). Hominid phylogenetics and the existence of *Homo* in Member 1 of the Swartkrans Formation, South Africa. *J. Hum. Evol.*, 7: 159–178.

Olson, T. R. (1985). Cranial morphology and systematics of the Hadar Formation hominids and "*Australopithecus*" *africanus*. In E. Delson (ed.), *Ancestors: the hard evidence*, pp. 102–119. Alan R. Liss, New York.

Patterson, C. (1982). Morphological characters and homology. In K. A. Joysey and A. Friday (eds.), *Problems of phylogenetic reconstruction*, pp. 21–74. Academic Press, London.

Pilbeam, D. and Gould, S. J. (1974). Size and scaling in human evolution. *Science*, 186: 892–901.

Rak, Y. (1981). The morphology and architecture of the australopithecine face. Ph.D. Thesis, University of California, Berkeley.

Rak, Y. (1983). *The australopithecine face*. Academic Press, New York.

Robinson, J. T. (1954). The genera and species of the Australopithecinae. *Amer. J. Phys. Anthropol.*, 12: 181–200.

Robinson, J. T. (1961). The australopithecines and their bearing on the origin of man and of stone tool-making. *S. Afr. J. Sci.*, 57: 3–13.

Robinson, J. T. (1962). The origin and adaptive radiation of the australopithecines. In G. Kurth (ed.), *Evolution und Hominisation*, pp. 120–140. Fischer Verlag, Stuttgart.

Robinson, J. T. (1963). Adaptive radiation in the australopithecines and the origin of man. In F. C. Howell and F. Bourliere (eds.), *African ecology and human evolution*, pp. 385–416. Aldine, Chicago.

Robinson, J. T. (1965). *Homo 'habilis'* and the australopithecines. *Nature*, 205: 121–124.

Robinson, J. T. (1968). The origin and adaptive radiation of the australopithecines. In G. Kurth (ed.), *Evolution und Hominisation*. 2nd Ed., pp. 150–175. Fischer Verlag, Stuttgart.

Robinson, J. T. (1972a). The bearing of East Rudolf fossils on early hominid systematics. *Nature*, 240: 239–240.

Robinson, J. T. (1972b). *Early hominid posture and locomotion*. University of Chicago Press, Chicago.

Skelton, R. R., McHenry, H. M. and Drawhorn, G. M. (1986). Phylogenetic analysis of early hominids. *Curr. Anthropol.*, 27(1): 21–43.

Swofford, D. L. (1985). *PAUP. Phylogenetic analysis using parsimony. Version 2.4.* Illinois Natural History Survey.

Tattersall, I. and Eldredge, N. (1977). Fact, theory and fantasy in human paleontology. *Amer. Scient.*, 65: 204–211.

Tobias, P. V. (1967). *The cranium and maxillary dentition of Australopithecus (Zinjanthropus) boisei. Olduvai Gorge, Volume II.* Cambridge University Press, Cambridge.

Tobias, P. V. (1973). Implications of the new age estimates of the early South African hominids. *Nature, 246*: 79–83.

Tobias, P. V. (1978). The South African australopithecines in time and hominid phylogeny, with special reference to the dating and affinities of the Taung skull. *In* C. J. Jolly (ed.), *Early hominids of Africa,* pp. 45–84. Duckworth, London.

Tobias, P. V. (1980). "*Australopithecus afarensis*" and *A. africanus:* critique and an alternative hypothesis. *Palaeont. Afr., 23*: 1–17.

Vrba, E. S. (1979). Phylogenetic analysis and classification of fossil and recent Alcelaphini (Family Bovidae, Mammalia). *Biol. J. Linn. Soc., 11*(3): 207–228.

Vrba, E. S. (1984). Evolutionary pattern and process in the sister-group Alcelaphini - Aepycerotini (Mammalia: Bovidae). *In* N. Eldredge and S. M. Stanley (eds.), *Living fossils,* pp. 62–79. Springer-Verlag, New York.

Walker, A., Leakey, R. E., Harris, J. M. and Brown, F. H. (1986). 2.5-Myr *Australopithecus boisei* from west of Lake Turkana, Kenya. *Nature, 322*: 517–522.

White, T. D., Johanson, D. C., and Kimbel, W. H. (1981). *Australopithecus africanus:* its phyletic position reconsidered. *S. Afr. J. Sci., 77*: 445–470.

Wood, B. A. and Abbott, S. A. (1983). Analysis of the dental morphology of Plio-Pleistocene hominids. I. Mandibular molars—crown area measurements and morphological traits. *J. Anat., 136*: 197–219.

Wood, B. A. and Chamberlain, A. T. (1986). *Australopithecus:* grade or clade? *In* B. Wood, L. Martin and P. Andrews (eds.), *Major topics in primate and human evolution,* pp. 220–248. Cambridge University Press, Cambridge.

Wood, B. A. and Chamberlain, A. T. (1987). The nature and affinities of the "robust" australopithecines: a review. *J. Hum. Evol., 16*: 625–641.

Wood, B. A. and Uytterschaut, H. (1987). Analysis of the dental morphology of Plio-Pleistocene hominids. III. Mandibular premolar crowns. *J. Anat., 154:* 121–156.

Wood, B. A., Abbott, S. A., and Uytterschaut, H. (1988). Analysis of the dental morphology of Plio-Pleistocene hominids. IV. Mandibular postcanine tooth morphology. *J. Anat., 156:* 107–139.

A New *Australopithecus* Cranium from Sterkfontein and Its Bearing on the Ancestry of *Paranthropus*

18

Ronald J. Clarke

The assemblage of hominid fossils from the Member 4 breccia of the Sterkfontein Formation, South Africa, has traditionally been accepted as representing one variable hominid species, *Australopithecus africanus*. This species has been taken to include fossils with diverse cranial morphologies and tooth sizes, with some individuals possessing cheek-teeth of a size and morphology that could place them within the genus *Homo*, and others in possession of cheek-teeth that are comparable in size, but not morphology, to those of *Paranthropus* (Table 18.1). Differing views on the relative significance of traits within this sample have inspired three main interpretations of the phylogenetic status of *Australopithecus africanus*. In the first instance, those who attach greatest weight to the *Homo*-like traits consider that *A. africanus* was solely a *Homo* ancestor and should be classified as *Homo* (Robinson, 1967; Olson, 1985), while in the second instance those who interpret *A. africanus* as a highly variable species believe that it was ancestral to both *Homo* and *Paranthropus* (Tobias, 1980; Skelton *et al.*, 1986). Those who attach greatest significance to the *Paranthropus*-like features of *A. africanus* envisage it as solely a *Paranthropus* ancestor (Johanson and White, 1979; White *et al.*, 1981; Rak, 1983).

Another possible explanation for the great variability in the Sterkfontein Member 4 sample has been advanced by Clarke (1985, 1986), who suggested that this assemblage might contain two species that had been grouped as one. Such an error would not be without precedent. For example, specimens of early *Homo erectus* from Swartkrans were previously classified as *Paranthropus robustus* and interpreted as being indicative of the range of variation of that taxon (Clarke *et al.*, 1970; Clarke, 1977). A fragmented partial cranium (Stw 252) excavated on June 21, 1984 by A. R. Hughes and his staff, from deep within the Sterkfontein Member 4 talus cone has given further support to Clarke's suggestion.

The fragments of Stw 252 consist of a virtually complete upper dentition, anterior palate, medial portion of the face and frontal bone, left parietal and the superomedial portion of the occipital. These have been reconstructed by the author (Fig. 18.1). It was possible to make a good restoration of the palate as the anterior portions preserved the left incisor sockets into which these teeth were replaced, and both canines and all four premolars were still in their sockets. The RI^2 and the distal half of the RI^1 crown were isolated, but could be positioned symmetrically opposite the left incisors. Portions of the maxillary suture were preserved so that two halves of the palate could be aligned correctly, and the M^1's and M^2s could be positioned relative to the premolars by their interproximal wear facets. The M^3s, which were unerupted, were placed in an estimated position.

A portion of the facial surface extends upward from the superior quarter of the $\underline{LC}$ socket; it preserves the anterior portion of the left nasal floor and the lower lateral nasal margin, and it almost contacts on the left nasal margin with the frontal process of the maxilla. That fragment has been placed symmetrically relative to the right fragment of the frontal process of the maxilla, which is keyed into the frontal bone by the frontomaxillary suture. Thus, the palate, mid-face and frontal were all reconstructed with confidence. The posterior portion of the frontal, which comprises the sagittal region between and including the temporal lines, does not quite contact

Table 18.1. Dental Measurements of Stw 252 Compared with Some *Paranthropus* and Smaller-Toothed Sterkfontein Specimens

	Sterkfontein large-toothed hominid		*Paranthropus*								Sterkfontein small-toothed hominids			
	Stw 252		SK 13/14		SK 46		SK 831		TM 1517		Sts 52		Sts 17	
Tooth	BL	MD	BL	MD	BL	MD	BL	MD	BL	MD	BL	MD	BL	MD
Left														
I¹	8.0	11.0									8.3	9.5		
I²	7.7	7.9									6.8	7.4		
C	11.4	10.9									10.0	9.8		
P³	13.9	9.6	13.2	9.6							12.9	8.6	12.9	8.7
P⁴	15.2	10.9	15.2	10.5	14.6	9.0			13.8	10.3	13.0	9.0		
M¹	15.2	14.6	15.1	14.0	14.5	11.3			15.3	10.3	14.1	12.4	13.5	
M²	17.0	15.9	16.6	16.0	16.3	13.0	>17.0	>16.0	14.7	12.5	15.3	13.5		
M³	18.2	15.9	17.6	14.5	16.3	14.8	17.8	17.3	15.9	13.5	14.5	12.6		
Right														
I¹	7.6	8.0									8.4	9.3		
I²	11.3	11.2									6.1	6.6		
C	14.2	9.9	13.2	9.9	13.2	8.4					10.0	10.0		
P³	15.4	10.9	15.1	10.7	14.6	9.2					12.8	8.6		
P⁴			15.0	13.1	15.2	11.8					13.5	9.7	12.5	8.2
M¹	17.0	15.8	16.3	15.1	16.1	13.0					14.4	12.3	13.2 est	12.0 est
M²	17.7	15.9	17.6	14.7	16.8	14.7					15.3	13.0	15.4	13.0
M³											14.6	12.5		

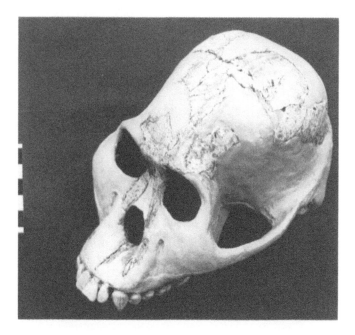

FIGURE 18.1. Superior oblique view of reconstruction of Stw 252. Scale in cm.

the anterior portion of the frontal, but it is firmly connected to the left parietal by means of the coronal suture. The right parietal fragment, which preserves a portion of the coronal suture and a fragment of temporal line, was placed symmetrically opposite the left parietal (Fig. 18.2). The occipital fragment retains part of the lambdoid suture, and although it does not contact the left parietal, it was possible to place it in an anatomically correct position relative to that bone.

The reconstructed Stw 252 cranium exhibits a complex of features that trend toward the *Paranthropus* condition. Thus, the frontal fragment displays a slight concavity of the frontal squame behind glabella, a thin and flattened supra-orbital margin, and marked encroachment of the temporal lines just posterior to the supra-orbital margin. Glabella is not prominent, and nasion is situated above the frontomaxillary suture in close proximity to glabella. The frontal processes of the maxilla present a concave facial surface, and the premolars and molars are very large and *Paranthropus*-like in their proportions (Fig. 18.3 and Table 18.1). Finally, the zygomatic process of the maxilla begins to curve laterally at a point anterior to the mesiobuccal root of the P^4.

The anterior dentition of Stw 252 is, however, very unlike that of *Paranthropus*. The canines and incisors are large, the incisors are procumbent, there is marked alveolar prognathism, and a wide diastema between the I^2 and $\underline{C}$ (Figs 18.3 and 18.4). Thus, the anterior dentition is more reminiscent of the apes.

The relative positions of the frontal and maxillary fragments indicate that the malar region probably would have been anteriorly prominent such that in side view the nasal bones could not have been seen. In this respect, and in the concave frontal process of the maxilla, Stw 252 is similar to another Sterkfontein cranium, Sts 71 (Fig. 18.5). These two specimens are similar also in their frontal morphology, their high, gently curved occipital profiles, large cheek-teeth, and in the anterior position of the zygomatic process of the maxilla. They contrast in these features with three other Sterkfontein Member 4 cranial specimens—Sts 5, Sts 17 and Sts 52—which, taken as a group, exhibit a complex of features trending toward the *Homo habilis* morphology (i.e., relatively small cheek-teeth, nasal skeleton that displays slight anterior prominence, zygomatic process that curves laterally over M^1, thickened supra-orbital margin, convex frontal squame, reduced sagittal encroachment of temporal lines postorbitally, and sharply angled occipital profile).

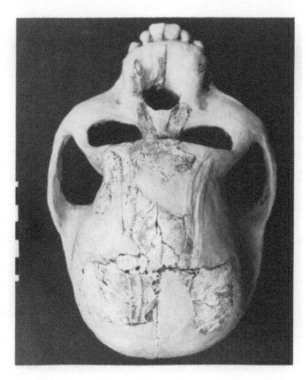

FIGURE 18.2. Superior view of reconstruction of Stw 252. Scale in cm.

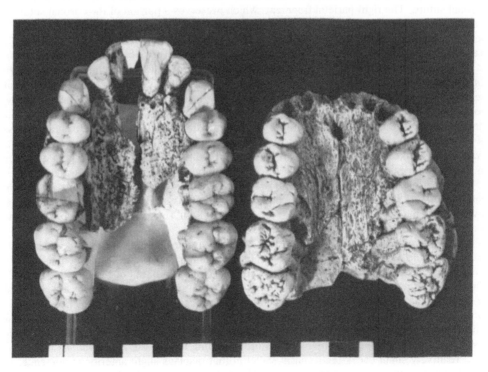

FIGURE 18.3. Palate of Stw 252 (left) compared to palate of *Paranthropus* specimen SK 13/14 (cast). Scale in cm.

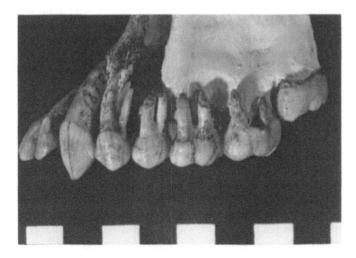

FIGURE 18.4. Lateral view of left dentition of Stw 252. Scale in cm.

Thus, there are two distinct morphological complexes in the Sterkfontein Member 4 assemblage: (A) a small-toothed hominid with thick supraorbital margin and prominent nasal skeleton, and (B) a large-toothed hominid with thin supraorbital margin and flat or concave nasal skeleton. There are four possible ways in which one can explain such variation within this assemblage: sexual dimorphism, individual variation, change through time, and different species. Sexual dimorphism seems highly improbable because in the Sterkfontein sample thin supraorbital margins are associated with large teeth and thick margins with small teeth. This is precisely the reverse of the situation found in males and females of extant primates. Individual variation also seems unlikely in view of the degree of dental size variation, which is coupled with major structural differences of the cranium. With regard to change through time, the small- and large-toothed morphotypes are found in close proximity both high and low in the talus cone, which suggests that one was not ancestral to the other. Although it could be argued that a talus stratigraphy is not simple and that older and younger fossils could theoretically be found at one level, the anat-

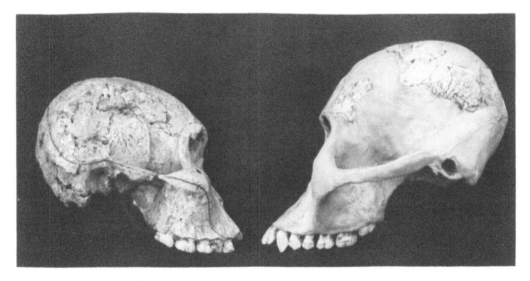

FIGURE 18.5. Lateral view of reconstruction of Stw 252 (right) compared to Sts 71 (cast).

omies of the morphotypes argue against the one having been ancestral to the other. The concept of the more specialized and large-tooth *Paranthropus*-like form being ancestral to the less specialized and small-toothed *Homo*-like form is contrary to what one would anticipate in an evolutionary sequence from less to more specialized. The alternative concept of the smaller-toothed morphotype having been ancestral to the larger-toothed is untenable in view of the more apelike anterior dentition of the larger-toothed morphotype.

Different species seems to be the most plausible explanation for the variation seen in the Sterkfontein Member 4 assemblage. It is my contention that hominid (A), as represented by fossils such as Sts 5, Sts 17 and Sts 52, together with other small-toothed specimens from Sterkfontein Member 4, Makapansgat (e.g., MLD 6) and Taung, constitutes *A. africanus* and presents characters trending toward *Homo*. Hominid (B), as represented by Stw 252, Sts 71, and other large-toothed fossils from Sterkfontein Member 4 (e.g., Sts 36) and Makapansgat (e.g., MLD 2), is, I believe, a distinct species of *Australopithecus* with characters trending toward *Paranthropus* yet with the ape-like anterior dentition retained from an ancestral species of *Australopithecus*. Reservations about the sympatry of two *Australopithecus* species at Sterkfontein can be countered by the argument that we do not know exactly where and when they would have diverged from a common ancestor, and the presence of their remains in one talus cone by no means implies that the two species occupied the Sterkfonteein area in precisely the same season, year, decade, or even century. Such temporal separation would not be detected in the Sterkfontein stratigraphy, and the home ranges of two contemporary *Australopithecus* species could have overlapped at different seasons or years, making it possible for their remains to have been deposited in the same site through carnivore or scavenger activity.

Sterkfontein Member 4 is believed to be between 3.0 and 2.5 Myr old (Vrba, 1985). Phylogenetically, the proposed new species from that Member would be situated between an ancestral *Australopithecus*, possibly *A. afarensis*, and the earliest *Paranthropus*. It is worthy of note that the earliest *Paranthropus* cranium so far discovered is that of a *Paranthropus boisei* or *Paranthropus aethiopicus* (Walker *et al.*, 1986; Walker and Leakey, this volume, chapter 15; Kimbel *et al.*, this volume, chapter 16) dated to some 2.5 Myr. The cranium, KNM-WT 17000, has the

FIGURE 18.6. Palatal view of *Paranthropus* specimen SKX 265 to show proximity of the premaxillary suture to the incisor sockets. Scale in cm. Photo courtesy of F. E. Grine.

typical *Paranthropus* characters of massive, enlarged cheek-teeth, mid-facial hollowing and depressed supraglabellar region, yet it is remarkable for its prognathism and wide intercanine region. *Paranthropus* of 2.0 to 1.5 Myr is typified by its orthognathism and relatively small canine and incisor region. Two subadult *Paranthropus* specimens (SKX 162 and SKX 265) recovered recently by C.K. Brain at Swartkrans preserve a very important feature not previously observed. Both specimens clearly display the premaxillary suture on the palate running immediately behind the incisor sockets (Grine, 1988) (Fig. 18.6). This is very different from the pattern in all other primates, including *A. africanus* (e.g., Taung), where the premaxillary suture forms the posterior boundary of a much larger area of premaxilla behind the incisors. These two new Swartkrans specimens demonstrate that not only did the premolars and molars of *Paranthropus* become enlarged in the course of evolution, but that the premaxilla reduced dramatically in anteroposterior dimension. The KNM-WT 17000 cranium shows that in East African *Paranthropus* at 2.5 Myr, cheek-tooth enlargement was well advanced but premaxillary retraction was not. It therefore follows that in an earlier ancestor of *Paranthropus* one could expect to find trends toward enlargement of cheek-teeth and mid-facial and supraglabellar hollowing, together with a retention of the primitive characters of prognathism and large incisors and canines. Stw 252 exhibits just such a character complex and is thus ideally placed morphologically and temporally to be a member of a species that was ancestral to and directly on the lineage of *Paranthropus*. That a mandible with cheek-teeth of *Paranthropus* proportions would articulate well with Stw 252 is illustrated in Fig. 18.7; a reduction in the sizes of the incisors and canines of Stw 252 would bring them into line with those of SK 23 without altering the size or position of the cheek-teeth.

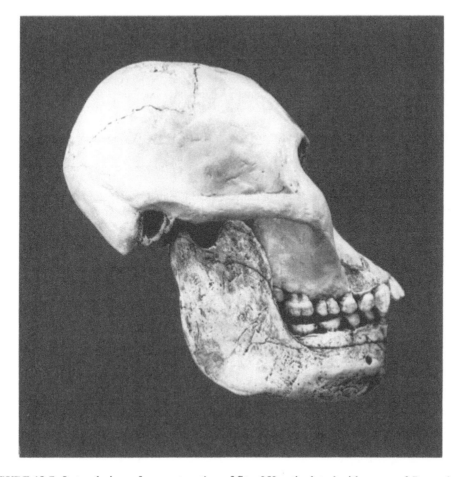

FIGURE 18.7. Lateral view of reconstruction of Stw 252 articulated with a cast of *Paranthropus* mandible SK 23.

Acknowledgments

I thank F.E. Grine for permission to publish here my observation on SKX 162 and SKX 265, which he is describing, and for providing Figure 18.6. The reconstruction and description of STW 252 were completed while I was Senior Research Officer in the Palaeoanthropology Research Unit of the University of the Witwatersrand. I thank P.V. Tobias and A.R. Hughes for providing this opportunity and C.K. Brain for access to Transvaal Museum specimens.

References

Clarke, R. J. (1977). A juvenile cranium and some adult teeth of early *Homo* from Swartkrans, Transvaal. *S. Afr. J. Sci., 73*: 46–49.
Clarke, R. J. (1985). *Australopithecus* and early *Homo* in southern Africa. *In* E. Delson (ed.), *Ancestors: the hard evidence*, pp. 171–177. Alan R. Liss, New York.
Clarke, R. J. (1986). Early Acheulean with *Homo habilis* at Sterkfontein. *In* P. V. Tobias (ed.), *Hominid evolution: past, present, and future*, pp. 287–298. Alan R. Liss, New York.
Clarke, R. J., Howell, F. C. and Brain, C. K. (1970). More evidence of an advanced hominid at Swartkrans. *Nature, 225*: 1219–1222.
Grine, F. E. (1988). New hominid specimens from Swartkrans (1979–1986 excavations): craniodental remains. *Amer. J. Phys. Anthropol.* (in press).
Johanson, D. C. and White, T. D. (1979). A systematic assessment of early African hominids. *Science, 202*: 321–330.
Olson, T. R. (1985). Cranial morphology and systematics of the Hadar Formation hominids and "*Australopithecus*" *africanus*. *In* E. Delson (ed.), *Ancestors: the hard evidence*, pp. 102–119. Alan R. Liss, New York.
Rak, Y. (1983). *The australopithecine face*. Academic Press, New York.
Robinson, J. T. (1967). Variation and taxonomy of the early hominids. *In* T. Dobzhansky, M. K. Hecht and W. C. Steere (eds.), *Evolutionary biology*, vol. 1, pp. 69–100. Appleton-Century-Crofts, New York.
Skelton, R. R., McHenry, H. M. and Drawhorn, G. M. (1986). Phylogenetic analysis of early hominids. *Curr. Anthropol., 27*: 21–43.
Tobias, P. V. (1980). "*Australopithecus afarensis*" and *A. africanus*: critique and an alternative hypothesis. *Palaeont. Afr., 23*: 1–17.
Vrba, E. S. (1985). Early hominids in southern Africa: updated observations on chronological and ecological background. *In* P. V. Tobias (ed.), *Hominid evolution: past, present and future*, pp. 195–200. Alan R. Liss, New York.
Walker, A. C., Leakey, R. E. F., Harris, J. M. and Brown, F. H. (1986). 2.5-Myr *Australopithecus boisei* from west of Lake Turkana, Kenya. *Nature, 322*: 517–522.
White, T. D., Johanson, D. C., and Kimbel, W. H. (1981). *Australopithecus africanus*: its phyletic position reconsidered. *S. Afr. J. Sci., 77*: 445–470.

Numerous Apparently Synapomorphic Features in *Australopithecus robustus, Australopithecus boisei* and *Homo habilis:* Support for the Skelton-McHenry-Drawhorn Hypothesis

19

PHILLIP V. TOBIAS

In the comparative analysis of the Olduvai and other specimens of *Homo habilis*, cranial, encephalic and dental features of this species and of other early African and Asian hominids have been tabulated (Tobias, 1988). Such lists of traits have been compiled rather frequently in the last decade, most commonly in cladistic analyses of hominid phylogenetic relationships (e.g., Delson *et al.*, 1977; Ciochon, 1983; Wood, 1984; Stringer, 1984; Kimbel *et al.*, 1984; Skelton *et al.*, 1986; Walker *et al.*, 1986).

As a step toward such an analysis, the author prepared a list of 148 cranial, mandibular, dental, and endocast characters that he found useful in distinguishing among *H. habilis, Australopithecus* spp. and *H. erectus* (Tobias, 1980c). This list was comprehensive though not exhaustive. It was stressed that not all of the 148 characters were necessarily independent variables, from a genetic or epigenetic viewpoint. At that stage, only the list of traits was given, but no attempt was made to show *how* each trait sorted *H. habilis* from other hominids. Later, the author selected seven characters of the 148 and tried to show how each expressed itself in *H. habilis* and other hominid taxa (Tobias, 1985a). In his latest analytical and comparative summary, the author has endeavored to give the average or modal expression of each of 350 traits in *H. habilis* and in *Australopithecus africanus, A. robustus, A. boisei* and *H. erectus* (Tobias, 1988).

Caveat on the Comparison of Data Tabulated by Different Investigators

As several different investigators have presented analyses of hominid phylogenetic relationships, it becomes necessary to compare their analyses and results. However, in making such comparisons, one must be aware of a number of possible pitfalls.

1. The entries for different taxa may vary greatly in value, within any analysis or between two or more analyses. Some entries are less meaningful than others because they are based on very few specimens or even on a single fossil (e.g., Sts 5 for *A. africanus*), which may be unrepresentative. Sometimes, one is forced to leave blank a cell in the table, as the condition of that trait in a particular taxon is not known, either because the anatomical part is lacking from the fossil assemblages, or because this detail has not been recorded or observed and the specimens in question are not readily accessible.

2. If a table lists only australopithecine species (e.g., Walker *et al.*, 1986), the standards of comparison within that hominid genus may lead to a different categorization of a trait from that arrived at in a comparison including both early *Homo* and the australopithecines. For example, Walker *et al.* (1986) list the "depth of the mandibular fossa" in *A. africanus* as "deep," the same category they apply to *A. robustus* and *A. boisei*. From the author's measurements of all available early hominid fossae, however, that of *A. africanus* is *absolutely* rather shallow ($\bar{X}$ = 9.6 mm), whereas those of *A. robustus* and *A. boisei* are moderately deep ($\bar{X}$ = 11.0 and 11.3

mm, respectively); while in *relative* depth (depth as compared with either length or breadth of the fossa), all are shallow as compared with the deep fossae of *H. erectus!*

3. Traits listed separately may be biologically interrelated, i.e., they may constitute items in a single structural, structural–functional, or regional–functional complex. A tally of traits that does not take this point into consideration might inadvertently give undue weight to the number of resemblances or differences between taxa. Wood (1981, 1984) and Skelton *et al.* (1986) have been at pains to avoid this problem.

4. Two traits may be stochastically one, e.g., where the axial orientation of the petrous pyramid is expressed either by the petromedian angle or by the petrotympanic axial angle. Yet, it may be justified to include both items in a table of traits, as in this article, if only since published data have been expressed in either manner.

5. Comparisons between different tabular analyses may be vitiated because two investigators may use a different technique to determine a feature (e.g., flexion of the cranial base).

6. A classificatory difficulty may inhibit comparisons among the results of different analyses, i.e., the hypodigm of a taxon used by different investigators may vary. For example, Blumenberg and Lloyd (1983) include Swartkrans specimens SK 15 and SK 847 in their hypodigm of *H. habilis,* whereas Howell (1978) places the former in *H. erectus* and the latter in *H. habilis,* while the author considers both as *H. erectus.*

7. Comparisons may betray greater intraspecific, synchronic, or allochronic variability and, perhaps, mosaic effects, than have been thought to exist. In the extreme, the trait complex may vary so highly and unsystematically in a number of hominid taxa that its phyletic valency is low and it should not be included in the list.

The Neglect of Plesiomorphic Traits

In the determination of character polarity, the variates are usually considered as discrete entities. We should do well to remember, however, the interrelatedness of the components of biological systems. It may happen, for instance, that a trait or trait-complex A has been found to be a primitive or plesiomorphic character of the Hominidae or even of the Hominoidea. In cladistic phylogenetic studies there might follow a tendency to discount its phylogenetic value or its taxonomic relevance (Le Gros Clark, 1964).

A weakness in this reasoning may occasionally be encountered: now and again, a primitive character may be shown to have assumed a new significance, by virtue of its persistence in a taxon alongside and interrelated with a derived feature B. The complex (A + B) may be such as to permit the emergence of a new functional pattern, a new behavioral modality, even a new survival strategy. It may be of such importance to the organism that (A + B) may not validly be excluded from the data base upon which phylogenetic inferences are constructed. Examples of this come readily to mind in the development of various parts of the brain, with its faculty for permitting evolutionary novelty to spring from odd blends of the primitive and the derived. The evolution of the vertebrate urogenital system may provide another illustration.

For this reason, no trait should be discarded or discounted at this stage of our ignorance on the interrelatedness of structural minutiae and functional nuances. It would be wrong to infer from this belief that the author "has suggested that the emphasis in human paleontological analysis should be shifted away from this principle" ("of taxonomic relevance") (*pace* Olson, 1985: 103). However, to keep a wary eye on the "less relevant" and primitive traits is to keep one's mind open to the possibility that, now and again in phylogeny, one or another of the plesiomorphic traits might have gained promotion. Through a structural, developmental, or functional link with an autapomorphic acquisition, the trait might have been elevated to the status of an (A + B) novelty, which may be the key to new evolutionary success and perhaps even to the attainment of a new level of organization. In such an event it would clearly be of little value to continue discussing the "primitive" trait A in isolation, as a discrete entity. The author believes that the attainment of the capacity for spoken language probably arose in this way, though this was more likely to have been an (A + B + C + D) combination, involving a blend of several plesiomorphic with autapomorphic traits.

It is suggested, then, that the principle of biologic interrelatedness may sometimes invalidate the handling of primitive traits in isolation, and the assumption that such traits, considered thus, are irrelevant in the search for affinities and phylogenies. To paraphrase a famous *cri de coeur*, one is tempted to exclaim, Cherish your primitive traits!—they may yet have their day.

The Exclusion of *Australopithecus afarensis* from the Author's Tables

The author has been hesitant at this stage to include a column on *A. afarensis*, which Johanson et al. (1978) proposed for the accommodation of hominid specimens from Hadar and Laetoli. This reluctance is *not* due to the published differences in odontometric features between the Laetoli and Hadar samples (Tobias, 1980a, 1988; Blumenberg and Lloyd, 1983; White, 1985). Blumenberg and Lloyd (1983) have shown, in comparisons of the mean MD and BL diameters for sixteen permanent teeth, that the Hadar and Laetoli series show fewer significant differences *inter se* (7/32) than the figure of 16/32 significant differences between means that "appears to represent the degree of tooth metric differentiation frequently characteristic of taxa that are considered separate species by many taxonomists" (Blumenberg and Lloyd, 1983: 159–160). Their study does not, however, take into consideration that some of the most striking differences among early hominid taxa are in step indices, crown shape indices, and "tooth material" values (Tobias, 1967, 1980c): many of these are missed by a comparison of one crown diameter of one tooth at a time.

However, odontometric contrasts form only a small part of the unresolved problem of the relationships of the Hadar and Laetoli series (Tobias, 1980b; cf. Blumenberg and Lloyd, 1983: 160). Studies from France, based largely on postcranial bones, have suggested that not one but at least two hominid species may be represented at Hadar (Coppens, 1981, 1982, 1983a,b; Senut, 1978, 1981a,b; 1982; Senut and Tardieu, 1985; Tardieu, 1979, 1982, 1983). Olson (1981, 1985), also has claimed from his studies on basicranial morphology that two hominid species are included in the Hadar assemblage. Moreover, the earlier publications on the Hadar hominids were inclined to recognize two or even three hominids in the Hadar assemblages (Johanson and Taieb, 1976; Johanson et al., 1976).

It is not the author's intention here to assess the whole *A. afarensis* problem with its taxonomic, procedural, morphological, and phylogenetic difficulties (cf. Day et al., 1980; Leakey and Walker, 1980). The question is briefly ventilated only to indicate why the author has not at this stage included a separate column for *A. afarensis* in his comparative table. In the *H. habilis* volumes, however, he has inserted cross-references, where possible, to the Laetoli and the Hadar assemblages (Tobias, 1988), while Skelton et al. (1986), like a number of others, have continued to regard *A. afarensis* as an entity and to include it in their data lists.

Resemblances among *Australopithecus robustus*, *Australopithecus boisei* and *Homo habilis*

Of the 350 items tabulated (Tobias, 1988), in 73 items *H. habilis* differs from *A. africanus* and resembles either *A. boisei* or *A. robustus* or both. In Table 19.1, data for only these 73 traits are offered. The presentation of the data has here been simplified by use of a (+) and (−) system. The sign (+) connotes the presence and the sign (−) the absence of the trait named; (+ +) and (+ + +) represent stronger degrees of expression of the trait; (±) denotes minimal expression of a trait. For most of the traits, the practice has been adopted of naming the trait as the version appearing in *H. habilis* and/or *A. robustus* and/or *A. boisei:* in such instances, (−) represents the expression of the trait in *A. africanus*, that is, the presumed primitive state, while (+), (+ +),

Table 19.1. Traits in which *Homo habilis* and *Australopithecus robustus* and/or *Australopithecus boisei* Agree with One Another and Depart from the Condition in *Australopithecus africanus*[a]

Character	H. habilis	A. africanus	A. robustus	A. boisei
1. Subangularity of parietooccipital contour	+	−		+
2. Cranial breadth enlargement	+ +	−	+	+
3. Tranverse expansion of frontal part of calvaria	+	−	+	±
4. Tendency for parietal sagittal arc to predominate over occipital	− to +	+ +		+
5. Nearly horizontal nuchal plane	+ to ±	−	+	+ to −
6. Inion low on occipital bone	+ +	−	+	+ +
7. Foramen magnum anteriorly placed	+ +	−	+	+
8. Very wide nasion–basion–opisthion angle	+	−		+
9. Anterior position of porion	+	−		+
10. Thin temporal squama	+	−	+	+
11. Long occipital condyle (absolute)	+	−	+	+
12. Slender occipital condyle (relative)	+	−	−	+
13. Strongly recurved medial end of anterior wall of mandibular fossa	+	−	+	+
14. Tendency to development of infratemporal crest	±	+ +		+
15. Tendency to doubling of external acoustic porus and meatus	+	−	−	+
16. Tendency to anterior rotation of lower part of external acoustic porus and meatus	+	−	+	−
17. Total basicranial shortening	+ + +	−		+ +
18. Shortening of posterior segment of basis cranii (basion to sphenobasion)	+ + +	−		+ +
19. Shortening of anterior segment of basis cranii (hormion to sphenobasion)	+ +	−		+ + +
20. Coronal lie of petrous temporal				
(by petrotympanic angle)	+	−	+	+
(by petromedian angle)	+	−	+	+

Table 19.1. Continued

Character	H. habilis	A. africanus	A. robustus	A. boisei
21. Trend toward expanded frontal sagittal arc and chord (nasion to bregma)	+	−		+
22. Trend toward expanded superior facial breadth (frontomaxillare temporale-frontomaxillare temporale)	+ +	−	+ +	± to + +
23. Trend toward expanded inner biorbital breadth (frontomaxillare orbitale-frontomaxillare orbitale)	+	−	+	+
24. Degree of curvature of frontal margin of parietal bone	+	+ +		±
25. Elevated breadth–length of parietal bone	+	−		+
26. Elevated biasterionic breadth of calvaria	+	−		+
27. Expanded cerebellar impressions on occipital	± to +	±		+ +
28. Primary bifurcation of middle meningeal vessels tending to lie inferior to temporal lobe	+ +	±	±	+ +
29. Tendency to more complex middle meningeal vascular pattern	+ +	−	+	+
30. Endocranial capacity higher than in pongids	+ +	−	+	+
31. Raised constant of cephalization (Hemmer)	+ +	−	+	
32. Increase in cerebral length	+	−		+
33. Increase in cerebral breadth	+ +	−		+
34. Frontal lobe transverse expansion	+ +	±		+
35. Parietal lobe transverse expansion	+ + +	±		+ +
36. Development of inferior parietal lobule	+ +	−	±	±
37. Increased zygomatic height	+ +	−	+ +	+ to + + +
38. Increased facial breadth	+	−	+	+ to + +
39. More vertical nasal component of upper face	+	−		+ +
40. Reduced prognathism of total upper face (facial angle)	+	−	+ +	+ +
41. More obtuse nasal profile angle	+	−		+ +

continued

Table 19.1. Continued

Character	H. habilis	A. africanus	A. robustus	A. boisei
42. Low infraorbital foramen	+	−	+	+ +
43. Very high nasion	+	−	+ +	+ +
44. Malar notch (incisura malaris) present	+	−	−	+
45. Low curve of anterior teeth	+	− to ±	+ to + +	+ to +,+
46. Presence of torus palatinus medianus	+	−	+ +	+ + +
47. Presence of torus maxillaris medianus	− to +	−	+	− to +
48. Sharp medial convergence of anterior part of inner contour of mandibular arch	+ +	−	+	+
49. Small angle between endocoronoid and endocondyloid crests	+	−	+	+
50. High relatively narrow mandibular ramus	+	−	+ +	+ +
51. Mesiodistal expansion of I^1	+ +	±	−	+ +
52. Labiolingual diminution of I^1	+	−	+	+
53. Mesiodistal expansion of I^2	+	−	−	+
54. High ratio of root module to crown module in $\underline{C}$	+	−	+	
55. Increase in crown dimensions P^3 to P^4	+	−	+ +	+ + +
56. Buccal cusp predominance of P^3	+ +	−	+	+ + +
57. Reduced crown size (absolute) of $\overline{C}$	+	−	+	+ +
58. Increased MD diameter of P_3	+	−	+	+ +
59. Increased MD diameter of P_4	+	−	+ +	+ + +
60. Reduction in crown height of P_3	+	−	+	
61. Marked MD increase from M_1 to M_2	±	+ +	±	±
62. High MD/LL index of I^1	+ +	−	+	+ +
63. High MD/LL index of I^2	+	−	−	+
64. Low MD/LL index of $\underline{C}$	±	−	+ +	+
65. High MD/BL index of $\overline{M}^1$	+ +	−	±	+
66. Reduced MD/LL index of I_2	+	−	+ +	+ + +
67. High MD/LL index of $\overline{C}$	+ +	−	+ +	−
68. Elevation of MD/BL index of M_1	+ +	−	± to +	+
69. Raised MD/BL index of M_3	+	−	− to +	+
70. Tendency to 3-rooted P^4	+	−	+ +	+ + +

Table 19.1. *Continued*

Character	*H. habilis*	*A. africanus*	*A. robustus*	*A. boisei*
71. Degree of development of central lingual ridge of $\overline{C}$	± to +	+ +	±	+
72. Tendency to closed mesial fovea of P_3	+	±	+ +	+ +(?)
73. Some reduction of protoconidal cingulum on $\overline{M}$	+	− to ±	±	+

"Data for the respective taxa are based as follows:
H. habilis: mainly Olduvai specimens and where applicable additional specimens, notably L 894-1 from Omo, KNM-ER 1470 and KNM-ER 1813 from East Turkana and Stw 53 from Sterkfontein.
A. africanus: Sterkfontein and Makapansgat australopithecines, supplemented where possible by Taung.
A. robustus: Kromdraai and Swartkrans australopithecines.
A. boisei: Australopithecines from Olduvai, Peninj, Chesowanja, Koobi Fora, and Omo.

and (+ + +) represent degrees of expression of the inferred derived state. In several instances it has been more convenient to name the trait after the version in *A. africanus:* in these cases, (+) represents the primitive and (−) the derived state. [In the full version of the table (Tobias, 1988), more detailed data including metrical details are given for each quantifiable trait and a column is included for *H. erectus*.]

Of the 73 items listed in the table, data are available for all four taxa for 51 traits; for 19 traits we have data for *A. africanus*, *A. boisei* and *H. habilis;* while for three characters there are data for *A. africanus*, *A. robustus*, and *H. habilis*. For all 73 traits, *H. habilis* departs from *A. africanus* and, in the same direction, so do *A. robustus* and/or *A. boisei*. (There are scores, if not hundreds, of other respects in which *H. habilis* differs from *A. africanus* and also from the "robust" or "hyper-robust" australopithecines; these presumed autapomorphies of *H. habilis* or, more broadly, of *Homo* are not discussed here.)

Considered in relation to the character state in *A. africanus*, the table provides 42 items for which *H. habilis*, *A. robustus*, and *A. boisei* show departures that are similar in direction (though not necessarily in degree). These may be regarded as synapomorphies of the three taxa, in relation to the apparently plesiomorphic condition in *A. africanus*.

In an additional 22 items, data are tabulated for *A. africanus*, *H. habilis*, and *one or the other* of the two "robust" species. Of these 22 traits, 19 are presumptive synapomorphies of *H. habilis* and *A. boisei* (with respect to *A. africanus*) and three are synapomorphies of *H. habilis* and *A. robustus*.

Thus, we have 61 traits shared by *H. habilis* and *A. boisei;* 45 by *H. habilis* and *A. robustus*. It is possible that, when further data become available, most, if not all, of the additional 22 characters will prove to be shared by all three taxa, *A. robustus*, *A. boisei*, and *H. habilis*. Hence, without our correcting at this stage for biological or stochastic interrelatedness between some traits listed, we have a minimum of 42 and a maximum of 64 apparently synapomorphic character states or trends shared by *H. habilis*, *A. robustus* and *A. boisei*, as contrasted with the condition of those traits in *A. africanus*.

Only nine traits for which data are available for all four taxa show *H. habilis* sharing an apparently derived state with *one* of the two "robust" species, *but not with the other*. Shared between *H. habilis* and *A. robustus*, but not with *A. africanus* or *A. boisei* are characters 16 and 64. Shared between *H. habilis* and *A. boisei*, but not with *A. africanus* or *A. robustus*, are characters 12, 15, 28, 44, 51, 53 and 63.

These 42–64 resemblances between *H. habilis*, *A. robustus*, and *A. boisei*, derived with respect to *A. africanus*, form an astonishingly long list and demand an explanation.

The Evidence for a Cladogenetic Emergence of *Homo habilis*, *Australopithecus robustus* and *Australopithecus boisei*

In 1965 the author claimed that

> The major implication of the recognition of *Homo habilis* is that, already in the Lower Pleistocene, we have evidence for the coexistence of at least 2 lines of hominids. One is the australopithecine line, itself highly diversified, and the other is a hominine line. Where previously the *Homo erectus* remains of the Djetis Beds, agreed by most to belong to the earliest part of the Middle Pleistocene, represented the earliest known hominine, it now seems clear that hominines were already present in Africa, and perhaps in Asia, during at least the Upper Villafranchian, i.e., the second half of the Lower Pleistocene . . . It was previously not strictly essential to postulate that the hominine line arose earlier than the end of the Lower Pleistocene. The recognition of *H. habilis* makes it now seem inevitable that the separation of the hominine line from a common hominid ancestral stock must have occurred appreciably earlier, not later than the Upper Pliocene or the first part of the Lower Pleistocene. . . . (Tobias 1965: 397)

Since that was written, it has become increasingly clear that the evolving *Homo* lineage was synchronic with the later phases of the australopithecines. In South Africa, we have evidence of *H. habilis* (mainly or perhaps exclusively from Sterkfontein Member 5) and *H. erectus* (from Members 1 and 2 of the Swartkrans Formation): at Swartkrans, certainly, and at Sterkfontein, probably, *Homo* was synchronic with *A. robustus*.

There is still doubt about the dating of the Taung type specimen of *A. africanus*. Among the currently competing estimates of the age of the Taung fossil deposit are < 1.0 Myr (Partridge, 1973); no older than the Transvaal "robust" australopithecine sites (Butzer, 1974); between Sterkfontein Member 4 (2.4–2.8 Myr) and Swartkrans Member 1 (1.6–1.8 Myr) but decidedly closer to the latter (Vrba, 1982, 1985); about 1.0 Myr (Vogel and Partridge, 1984; Vogel, 1984, 1985); questionably 1.0–1.2 Myr (Butzer, 1980); questionably contemporaneous with the later Swartkrans and Kromdraai deposits (Butzer, 1974); and somewhat older than the 1.9-Myr KBS tuff unit (Delson, 1984). In round figures we have estimates of about 1.0 Myr, about 2.0 Myr, and one or two dates in between. Hence, at this stage we are not able to attest with which other hominids the Taung population was synchronic. If the earliest faunal datings estimated for Taung are correct, its population might have existed at a time which was only several hundred thousand years later than the inferred time of the hominid branching. If the later dating of Taung is correct, it would be reasonable to infer that the Taung hominid was synchronic with one or more other, long-established hominid lineages.

In East Africa, we have evidence of at least two lineages from Omo, East Turkana, and Olduvai: at these sites, *A. boisei* and either *H. habilis* (prior to about 2.0 Myr) or *H. erectus* (after about 1.6 Myr) were synchronic and sympatric. Their synchrony and sympatry seem to have lasted from some time earlier than 2.0 to about 1.0 Myr.

This statement is predicated upon the author's belief that the "robust-looking" elements among the hominids of Makapansgat (Tobias, 1967, 1969, 1973, 1980b; Aguirre, 1970; Wallace, 1978) and of Hadar (Johanson and Taieb, 1976; Johanson et al., 1976) do *not* reflect the occurrence of a "robust" alongside a "gracile" lineage at 3.0 Myr BP, but rather bespeak an expression of polymorphism. However, the claims that two different populations have been sampled in the Hadar assemblages may, if corroborated, require that the duration of the synchrony of two or more hominid lineages be reconsidered. If there do turn out to be two hominid populations[1] at

---

[1] In view of the great number of pongid-like features manifest in the Hadar assemblage, Ferguson (1983) has suggested that there are two populations at Hadar and that one population comprises pongids and the other hominids. If the latter hypothesis were confirmed, the duration of synchrony of the two (or more) hominid lineages would remain at just over 1 Myr.

Hadar, we may have to infer that they had arisen from a latest common ancestral population by a splitting of the hominid lineage at some time earlier than 3.0 Myr BP.

If for the moment we accept that the case for two populations at Hadar is not proved, then we may infer that about 2.5 Myr BP a common ancestral population of hominids underwent a major cladogenetic splitting into two (or more) derivative lineages. The recent discovery of KNM-WT 17000, an apparent variant of *A. boisei* from Lomekwi I, West Turkana (Walker *et al.*, 1986), seems to argue for a slightly earlier date than 2.5 Myr for that splitting event—if the date of 2.5 Myr claimed for this specimen proves to be correct.

The Nature of the Last Common Ancestral Population

What was the nature of the population whose lineage underwent splitting? Over the past 30 years a substantial consensus has been attained, that all australopithecines are hominids and, for most though not all palaeoanthropologists, all may be accommodated in a single genus of micrencephalic hominids, *Australopithecus*. Another major aspect of this paradigm is that some populations of one australopithecine species were ancestral to the genus *Homo*. The most likely claimant for this distinction was *A. africanus* (Mayr, 1970; Campbell, 1974).[2]

It is the view of a number of workers that *A. africanus* or a form close to *A. africanus* is not only the most likely ancestor of *Homo* (more specifically of *H. habilis*), but is similarly a very probable ancestor of *A. robustus* and *A. boisei*. In other words, *A. africanus* is the most likely claimant for last common ancestral status (Tobias, 1968, 1973, 1975, 1980c, 1985b; Campbell, 1972; Wallace, 1975, 1978; Eldredge and Tattersall, 1975; Tattersall and Eldredge, 1977; Delson *et al.*, 1977; Delson, 1978; Yaroch and Vitzthum, 1984).

An important variation on the latter theme has recently emerged (Skelton *et al.*, 1986). It flows from the growing body of evidence that there are many close resemblances between *A. robustus* and *A. boisei*, on the one hand, and *H. habilis*, on the other hand, which are shared-derived characters relative to *A. africanus*. In the study of the *A. boisei* type specimen (Tobias, 1967), attention was drawn to several *Homo*-like features, such as the more vertical face [emphasized again by Wood (1981)], and the nearly coronal orientation of the axis of the petrous part of the temporal bone [confirmed by Dean and Wood (1981)]. Over the years, the list of such features possessed both by early *Homo (H. habilis)*, and by *A. robustus* and *A. boisei*, has grown. Skelton *et al.* (1986) have listed no fewer than 16 such items. Moreover, the present author's tabular summaries enumerate up to 64 traits in which *H. habilis*, *A. robustus* and/or *A. boisei* share presumptively derived features as compared with *A. africanus*. In 22 of these, the trait is shared by *H. habilis* with one or other of the two "robust" taxa. The other 42 are more compelling, and it is inferred that they are shared-derived characters of all three taxa. It should be stressed again that not all of the 64 traits are necessarily biologically or stochastically independent characters.

Both Skelton's and the author's lists reflect a staggering number of derived resemblances. As long as there were only three or four such features, the invoking of parallel evolution to explain these seems reasonable. On the other hand, to explain 16 or 44 or even 64 derived resemblances other than by inheritance from a common ancestor would be to stretch the bounds of credulity inordinately.

For each of the 16 traits listed by Skelton *et al.* (1986) as comprising their Complex 2, a

---

[2]The ultimate in the expression of the close ancestor-descendant relationship of *A. africanus* and *Homo* is the submission of Robinson (1972) that the fossils assigned to *A. africanus* should be reclassified in the genus *Homo* as *Homo africanus*. For him there is no question of *A. africanus* being in the position of a common ancestor to "robust" australopithecines and hominines. On the contrary, he holds that the ancestral form was similar to the known later "robust" australopithecines—in other words that *A. robustus* and *A. boisei* (or *Paranthropus*) traits are primitive, not derived. The morphocline Robinson sees extends from *A. robustus*-like forms through "gracile australopithecines" (*"Homo africanus"*) to later hominines.

morphocline extends from its most primitive manifestation in *A. afarensis*, through an intermediate expression in *A. africanus*, to a more derived state shared by *A. robustus*, *A. boisei*, and *H. habilis*. Skelton *et al.* (1986) have invoked a number of principles to derive a phylogeny from their most parsimonious cladogram. By the judicious application of these principles, skelton *et al.* (1986) are led to state that "the last common ancestor of *A. robustus/A. boisei* and *H. habilis* was derived with reference to *A. africanus*." In this respect, Skelton *et al.* (1986) have made a most important contribution to our thinking about the relationships between *A. africanus* and the *A. robustus*, *A. boisei* and *Homo* lineages. On the question of the nature of the derived *A. africanus* envisaged as the last common ancestor, Skelton *et al.* (1986: 30) have this to say:

> We feel, however, that a population of hominids resembling *A. africanus* in all respects except for having the *robustus/boisei-habilis* condition of the 16 traits in Complex 2 would still fit comfortably within the definition of *A. africanus*. The existing specimens of *A. africanus* were all recovered from a small geographic area and perhaps a relatively small time span as well. Therefore, it is probable that a population of *A. africanus* existed at some time and place not yet sampled which had the derived condition for the traits represented in Complex 2.

Four major proposals have been made regarding the relationship of *A. africanus* to the *Homo* lineage and to the *A. robustus* and *A. boisei* lineage(s): Robinson (1972) placed *A. africanus* on the *Homo* branch, but not on the implied "paranthropine" branch; Johanson and White (1979) confined *A. africanus* to the *A. robustus–A. boisei* branch; Tobias (1968, 1973, 1975, 1980c) placed *A. africanus* as the antecedent of both derived branches; while Skelton *et al.* (1986) accept the latter interpretation, but point up a number of derived states in the postulated last common ancestral population, as compared with the *available* populations of *A. africanus*. Of the four proposals, the author now feels that that advanced by Skelton *et al.* (1986) meets more of the known facts than any other interpretation.

Is Taung a Representative of the "Derived *Australopithecus africanus*" Population?

Skelton *et al.* (1986) speak of the supposed "derived *A. africanus*" population having existed "at some time and place not yet sampled." The major assemblage of fossils assigned to *A. africanus* (*sensu stricto*) is that of Sterkfontein Member 4, now dated faunally to 2.4–2.8 Myr BP (Vrba, 1982, 1985). It seems that it is some 0.5 Myr (0.2–0.6 Myr) more recent than the smaller assemblage from Makapansgat, dated at c 3.0 Myr BP (Brock *et al.*, 1977; McFadden *et al.*, 1979; Partridge, 1982; Vrba, 1982, 1985). It may be noted in passing that the Makapansgat *A. africanus*, in the author's opinion, bears the closer resemblance to some fossils of Hadar, for example in the morphology of the maxilla and palate, and in the presence of "robust-like" elements (Tobias, 1967; Aguirre, 1970). We may suppose that, if the phylogenetic schema proposed by Skelton *et al.* (1986) is correct, it would be a more recent population of *A. africanus* than those of Makapansgat or Sterkfontein, to which we should look for the "derived *A. africanus*" claimed as the last common ancestor.

Inevitably, the Taung skull suggests itself as a possible representative of such a population. Its range of estimated datings is wide, but none of them extends as far back in time as 2.4 Myr BP, the youngest dating proposed by Vrba (1982, 1985) for Sterkfontein Member 4. The oldest of the dates proposed for Taung are those based on palaeontological data. Delson (1984), as we have seen, has set its age as "somewhat older than 1.9 Myr" on the evidence of the Taung Cercopithecoidea. Vrba (1982, 1985), admittedly on "very slender evidence," suggests tentatively that the Taung hominid may be close in time to Kromdraai B East, Member 3, a little earlier than Swartkrans Member 1 but not as early as Sterkfontein Member 4. That is, her provisional faunal dating sets the Taung hominid at somewhat older than the 1.6–1.8 Myr which she assigns to Swartkrans Member 1. Thus, on Delson's and Vrba's provisional dates, the age of the Taung hominid could be fairly close to the supposed time of branching of the "derived *A. africanus*."

(It must be recalled that the results obtained by K. W. Butzer, J. C. Vogel, and T. C. Partridge by other geochronological methods suggest that Taung is not older than Swartkrans or Kromdraai and possibly substantially more recent. The paradox of the contradictory results remains unresolved.)

On morphological grounds, a careful re-examination of the Taung child skull will have to include a special search for those shared-derived features, that would not be expected to be significantly and relevantly affected by ontogenetic changes. In a first assay, the author finds that one of the shared-derived traits, the near-coronal orientation of the petrous pyramids, seems to be present in the Taung cranium as in *A. robustus, A. boisei* and *H. habilis*, and in contrast with the pongidlike nearly sagittal orientation in the Sterkfontein *A. africanus* (Fig. 19.1). If the characters cited by Skelton *et al.* (1986) for the lower permanent canines apply also to the lower deciduous canines, in one of these traits, the reduced lingual ridge (my character 71), and perhaps in the other, a reduced length of the distal occlusal edge (cf. Grine, 1985: fig. 12.1), the position in Taung approximates that of the apomorphic state of *A. robustus/A. boisei* and *H. habilis*. In short, a cursory survey suggests that at least several of the traits of Skelton *et al.* (1986) and of this author may, in Taung, manifest the derived state, as in *A. robustus, A. boisei* and *H. habilis*.

It is suggested that the Taung child may represent a population of *A. africanus* somewhat closer in morphology to the postulated last common ancestor of the *A. robustus, A. boisei* and *H. habilis* lineages than are the populations represented by *A. africanus* of Makapansgat and Sterkfontein. The suggestion that Taung may represent a population of *A. africanus* that has moved further in the direction of derived morphology may gain support from a fourth morphological trait, namely the presence of anterior pillars in the young child's face of Taung, as recognized by Rak (1985) who infers therefrom that Taung could perhaps have attained "a more advanced position in the robust sequence."

In sum, the author's analysis of the characters that are apparently synapomorphic between the *A. robustus, A. boisei* and *H. habilis* lineages has supported and strengthened the hypothesis of Skelton *et al.* (1986), that the last common ancestral population was basically *A. africanus*, modified by its further evolutionary acquisition of a number of derived features. Moreover, if the dating of the Sterkfontein Member 4 *A. africanus* is set at the earlier to middle date in Vrba's 2.8–2.4 Myr BP, namely 2.8–c 2.6 Myr BP, and if the critical branching of the lineage is dated at about 2.5 Myr BP, then the evolutionary advancement of *A. africanus* to the "derived state" postulated by Skelton *et al.* (1986) must have taken place between c 2.8 and c 2.3 Myr BP.

The author suggests that the Taung child skull may represent a population that was somewhat closer in morphology and possibly in time to the posited "derived *A. africanus*," the last common ancestral population. If the older dating that the cercopithecoids of Taung have suggested to Delson, and other Taung fauna to Vrba, proves correct, it may well be that Taung represents a population that was fairly close in time and in morphology to the actual last common ancestor's. If, on the other hand, the younger dating of the Taung hominid that several investigators (Butzer, Vogel, Partridge) have inferred proves correct, then Taung itself might be seen as a late survivor of the last common ancestral population of "derived *A. africanus*."

It would be intriguing and important, also, to re-examine the position of *A. robustus robustus* of Kromdraai Member 3, which Vrba (1982, 1985), using Grine's (1982) work on the hominids, considers may be close in time to Taung. In morphology, as Grine (1982, 1985) pointed out, Kromdraai may lie very close to the root of the *A. robustus* lineage that reached its full South African flowering in the Swartkrans *A. robustus crassidens*. If phylogenetically it does lie at or close to that root, the fossil population of Kromdraai may help to throw light on the nature of the cladogenetic transpecific change from the "derived *A. africanus*" population postulated by Skelton *et al.* (1986) to the earliest member of the "robust" lineage *sensu stricto*.

Summary and Conclusions

In the course of an exhaustive comparative analysis of the skulls, endocasts, and teeth of *H. habilis*, the author tabulated some 350 traits, for each of which were given the conditions in three

FIGURE 19.1. Norma basilaris of the Taung cranium, showing the inferior face of the pars petrosa of the temporal bone (x), and other parts of the basis cranii externa. The lie of the petrous part may be determined by measuring the petromedian angle (of Tobias, 1967) or the CC to PA/CC to CC angle (of Dean and Wood, 1981, called by the author of the petrocoronal angle, Tobias, 1988).

The petromedian angle of Taung is 42°, identical to the angle in *H. habilis* OH 24 of Olduvai (42° left, 41° right), and similar to that of *H. erectus pekinensis* (e.g., 40° in Zhoukoudian III). The values in *A. boisei* are similar (e.g., 47.5° in OH 5; 40°–45° in KNM-ER 407). All of these values contrast with the low values in Sterkfontein and Makapansgat *A. africanus* (32° in Sts 5; 32° in Sts 19; 33.5° in MLD 37/38) and still lower values in chimpanzee (16.5°–20°), gorilla (10°–23°) and orangutan (17.5°–30°) (Tobias 1988).

The petrocoronal angle of Taung is 47°. Again, this is closer to values given by Dean and Wood (1981) for *A. robustus, A. boisei* and *H. habilis* than to the higher values for *A. africanus* (59°–72°) and for the pongids (60°–81°).

In sum, the orientation of the Taung petrous temporal is closer to the coronal plane, as in *H. habilis* and the "robust" australopithecines, and unlike the more sagittal lie of the petrous temporal in Sterkfontein and Makapansgat *A. africanus* and great apes. Dean and Wood (1984) have shown that in great apes, the petrocoronal angle increases from birth to adulthood by about 10°, whereas in modern *H. sapiens* the mean value drops after birth by about 8° and rises again by 8° to attain, in adulthood, the same mean value as at birth. However, as they point out, we lack growth curves for the fossil hominids; thus we cannot speculate whether, nor to what degree, the petrous angulation of a Taung child might have changed by the time it reached adulthood.

australopithecine and two early *Homo* taxa. Among these were 73 characters in which *H. habilis* shared a presumptively derived state with either or both of *A. robustus* and *A. boisei*, relative to *A. africanus*. For 42 traits, data were available for all four taxa and *H. habilis*, *A. robustus*, and *A. boisei* shared a common pattern of departure from the *A. africanus* condition. For 19 more characters *H. habilis* and *A. boisei* shared a derived state, while data for *A. robustus* were not available; similarly, for three traits *H. habilis* and *A. robustus* shared a derived condition, data for *A. boisei* being unavailable. Thus for 42 and possibly as many as 64 traits, *H. habilis* and the two "robust" taxa were characterized by presumptive synapomorphies with respect to *A. africanus*. Even though this list may be somewhat reduced because of biological or stochastic interrelatedness between some items, it remains, like the list of 16 such traits offered by Skelton *et al.* (1986), a multiplicity of synapomorphies considered to be too great to be ascribed to homoplasy: it would seem to demand an explanation in terms of common inheritance. The author's results therefore lend strength to the most parsimonious hypothesis of Skelton *et al.* (1986), namely that the most likely common ancestor of *H. habilis*, *A. robustus* and *A. boisei* is *A. africanus*, but a variant of *A. africanus* that is derived with respect to the assemblages known from Makapansgat and Sterkfontein. Skelton *et al.* (1986) opined that representatives of the postulated "derived *A. africanus*" had yet to be discovered. The present author proposes that the Taung child, differing as it does in a number of respects from the Sterkfontein and Makapansgat samples, may be close in morphology to, or may even represent, the "derived *A. africanus*" variant. Uncertainties surrounding the dating of the Taung child (though all estimates agree it is younger than Sterkfontein Member 4) would not disqualify its morphology from approximating to that of the "derived *A. africanus*" variant. If we accept Grine's (1982, 1985) view that *A. robustus robustus* of Kromdraai was at or close to the root of the "robust" australopithecine radiation, the population represented by the Kromdraai hominid may throw light on the nature of the cladogenetic trans-specific change from the postulated "derived *A. africanus*" to the earliest "robust" australopithecine *sensu stricto*.

Acknowledgments

My grateful appreciation is extended to my former research student, the indefatigable organizer of the Workshop on the "robust" australopithecines, Frederick E. Grine, for inviting me to participate. Although university and family problems compelled me to withdraw from the Stony Brook meeting at the last moment, he none the less invited me to submit a chapter for the volume. I extend my thanks to J. K. McKee and to my incomparable assistants, Heather White and Val Strong.

References

Aguirre, E. (1970). Identification de "Paranthropus" en Makapansgat. *Cronica de XI Congr. Natl. de Arqueologie, Merido 1969*, pp. 98–124.
Blumenberg, B. and Lloyd, A. T. (1983). *Australopithecus* and the origin of the genus *Homo*: aspects of biometry and systematics with accompanying catalogue of tooth metric data. *BioSystems, 16*: 127–167.
Brock, A., McFadden, P. L. and Partridge, T. C. (1977). Preliminary palaeomagnetic results from Makapansgat and Swartkrans. *Nature, 266*: 249–250.
Butzer, K. W. (1974). Paleoecology of South African australopithecines: Taung revisited. *Curr. Anthropol, 15*: 367–382.
Butzer, K. W. (1980). The Taung australopithecine: contextual evidence. *Palaeontol. Afr., 23*: 59.
Campbell, B. G. (1972). Conceptual progress in physical anthropology: fossil man. *Ann. Rev. Anthropol, 1*: 27–54.

Ciochon, R. (1983). Hominoid cladistics and the ancestry of modern apes and humans: a summary statement. *In* R. L. Ciochon and R. S. Corruccini (eds.), *New interpretations of ape and human ancestry*, pp. 781–844. Plenum. New York.
Coppens, Y. (1981). Le cerveau des hommes fossiles. *C. R. Acad. Sci. Suppl. (Paris)* 292: 3–24.
Coppens, Y. (1982). Les origines de l'homme. *Histoire et Archeologie,* 60: 8–17.
Coppens, Y. (1983a). Les plus anciens fossiles d'hominidés. *Pontif. Acad. Scient. Scripta Varia,* 50: 1–9.
Coppens, Y. (1983b). *Le Singe, l'Afrique et l'homme*, pp. 1–148. Fayard, Paris.
Day, M. H., Leakey, M. D., and Olson, T. R. (1980). On the status of *Australopithecus afarensis*. *Science,* 207: 1102–1103.
Dean, M. C. and Wood, B. A. (1981). Metrical analysis of the basicranium of extant hominids and *Australopithecus*. *Amer. J. Phys. Anthropol,* 54: 63–71.
Dean, M. C. and Wood, B. A. (1984). Phylogeny, neoteny and growth of the cranial base in hominoids. *Folia Primatol.,* 43: 157–180.
Delson, E. (1978). New fossil cercopithecoids from East Africa. *Amer. J. Phys. Anthropol.,* 48: 389.
Delson, E. (1984). Cercopithecid biochronology of the African Plio-Pleistocene: correlation between eastern and southern hominid-bearing localities. *Cour. Forsch. Inst. Senckenberg,* 69: 199–218.
Delson, E., Eldredge, N. and Tattersall, I. (1977). Reconstruction of hominid phylogeny: a testable framework based on cladistic analysis. *J. hum. Evol.,* 6: 263–278.
Eldredge, N. and Tattersall, I. (1975). Evolutionary models, phylogenetic reconstruction and another look at hominid phylogeny. In: F. S. Szalay (ed.), *Approaches to primate paleobiology*, pp. 218–242. Karger, Basel.
Ferguson, W. W. (1983). An alternative interpretation of *Australopithecus afarensis* fossil material. *Primates,* 24: 397–409.
Grine, F. E. (1982). A new juvenile hominid (Mammalia, Primates) from Member 3, Kromdraai Formation, Transvaal, South Africa. *Ann. Tvl. Mus. 33*: 165–239.
Grine, F. E. (1985). The Deciduous Dentition of the Kalahari San, the South African Negro and the South African Plio-Pleistocene Hominids. Ph.D. Thesis, University of the Witwatersrand, Johannesburg.
Howell, F. C. (1978). Hominidae. *In* V. J. Maglio and H. B. S. Cooke (eds.), *Evolution of African mammals*, pp. 154–248. Harvard University Press, Cambridge, MA.
Johanson, D. C. and Taieb, M. (1976). Plio-Pleistocene hominid discoveries in Hadar, Ethiopia. *Nature,* 260: 293–297.
Johanson, D. C. and White, T. D. (1979). A systematic assessment of early African hominids. *Science,* 202: 321–330.
Johanson, D. C., Coppens, Y. and Taieb, M. (1976). Pliocene hominid remains from Hadar, Central Afar, Ethiopia. *In* P. V. Tobias and Y. Coppens (eds.), *Les Plus Anciens Hominidés*, pp. 120–137. *Cong. Internat. des Sci. Préhist. et Protohist.,* Nice, Septembre 1976.
Johanson, D. C., White, T. D., and Coppens, Y. (1978). A new species of the genus *Australopithecus* (Primates: Hominidae) from the Pliocene of eastern Africa. *Kirtlandia,* 28: 1–14.
Kimbel, W. H., White, T. D., and Johanson, D. C. (1984). Cranial morphology of *Australopithecus afarensis*: a comparative study based on a composite reconstruction of the adult skull. *Amer. J. Phys. Anthropol.,* 64: 337–388.
Leakey, R. E. F. and Walker, A. C. (1980). On the status of *Australopithecus afarensis*. *Science,* 207: 1102–1103.
LeGros Clark, W. E. (1964). *The fossil evidence for human evolution: an introduction to the study of paleoanthropology,* 2nd edit. University of Chicago Press, Chicago.
McFadden, P. L., Brock, A., and Partridge, T. C. (1979). Palaeomagnetism and the age of the Makapansgat hominid site. *Earth Planet. Sci. Lett.,* 44: 373–382.
Mayr, E. (1970). *Population, species and evolution*, pp. 1–453. Harvard University Press, Cambridge, MA.

Olson, R. T. (1981). Basicranial morphology of the extant hominids and Pliocene hominids: the new material from the Hadar Formation, Ethiopia, and its significance in early human evolution and taxonomy. In C. B. Stringer (ed.), *Aspects of human evolution*, pp. 99–128. Taylor and Francis, New York.

Olson, T. R. (1985). Cranial morphology of the Hadar Formation hominids and *Australopithecus africanus*. In E. Delson (ed.), *Ancestors: the hard evidence*, pp. 102–119. Alan R. Liss, New York.

Partridge, T. C. (1973). Geomorphological dating of cave opening at Makapansgat, Sterkfontein, Swartkrans and Taung. *Nature*, 246: 75–79.

Partridge, T. C. (1982). The chronological positions of the fossil hominids of southern Africa. *Proc. 1st Inter. Congr. Hum. Palaeont.*, 2: 617–675.

Rak, Y. (1985). Sexual dimorphism, ontogeny and the beginning of differentiation of the robust australopithecine clade. In P. V. Tobias (ed.), *Hominid evolution: past, present and future*, pp. 233–237. Alan R. Liss, New York.

Robinson, J. T. (1972). The bearing of East Rudolf fossils on early hominid systematics. *Nature*, 240: 239–240.

Senut, B. (1978). Contribution a l'étude de l'humérus et de ses articulations chez les Hominidés de Plio-Pléistocène. Thése Doctorat 3éme cycle, Paris: Université Pierre et Marie Curie.

Senut, B. (1981a). Humeral outlines in some hominoid primates and in Plio-Pleistocene hominids. *Amer. J. Phys. Anthropol.* 56: 275–282.

Senut, B. (1981b). L'humérus et ses articulations chez les Hominidés Plio-Pleistocènes. *Cah. Paleoanthropol. Paris: C.N.R.S.*

Senut, B. (1982). Réflexions sur la brachiation et l'origine des Hominidés a la lumiére des Hominoïdes Miocénes et des Hominidés Plio-Plèistocénes. *Gébios. Mém. Spéc.*, 6: 335–344.

Senut, B. and Tardieu, C. (1985). Functional aspects of Plio-Pleistocene hominid limb bones: implications for taxonomy and phylogeny. In E. Delson (ed.), *Ancestors: the hard evidence*, pp. 193–201. Alan R. Liss, New York.

Skelton, R. R., McHenry, H. M. and Drawhorn, G. M. (1986). Phylogenetic analysis of early hominids. *Curr. Anthropol.*, 27: 21–43.

Stringer, C. B. (1984). Human evolution and biological adaptation in the Pleistocene. In R. Foley (ed.), *Human evolution and community ecology*, pp. 55–83. Academic, London.

Tardieu, C. H. R. (1979). Aspects biomécaniques de l'articulation du genou chez les primates. *Bull. Soc. Anat. Paris*, 4: 66–86.

Tardieu, C. H. R. (1982). Caractéres plésiomorphes et apomorphes de l' articulation du genou chez les primates hominoïdes. *Géobios, Mém. Spéc.*, 6: 321–334.

Tardieu, C. H. R. (1983). Analyse morpho-fonctionnelle de l'articulation du genou chez les primates et les hominidés fossiles. *Cah. Paleoanthropol., Paris, C.N.R.S.*

Tattersall, I. and Eldredge, N. (1977). Fact, theory and fantasy in human paleontology. *Amer. Scient.*, 65: 204–211.

Tobias, P. V. (1965). New discoveries in Tanganyika, their bearing on hominid evolution. *Curr. Anthropol.*, 6: 391–399; and reply to comments, 406–411.

Tobias, P. V. (1967). *The Cranium and Maxillary Dentition of* Australopithecus (Zinjanthropus) boisei. Olduvai Gorge vol. 2. Cambridge University Press, Cambridge.

Tobias, P. V. (1968). The taxonomy and phylogeny of the australopithecines. In B. Chiarelli (ed.), *Taxonomy and phylogeny of Old World primates with references to the origin of Man*, pp. 277–315. (Supplement to 1967 volume of *Rivista di Antropologia*). Rosenberg and Sellier, Turin.

Tobias, P. V. (1969). Commentary on new discoveries and interpretations of early African fossil hominids. *Yrbk. Phys. Anthropol.*, 15: 24–30.

Tobias, P. V. (1973). New developments in hominid palaeontology in South and East Africa. *Ann. Rev. Anthropol.*, 2: 311–324.

Tobias, P. V. (1975). New African evidence on the dating and the phylogeny of the Plio-Pleistocene Hominidae. In R. P. Suggate and M. N. Cresswell (eds.), *Quaternary studies*, pp. 289–296. Bulletin No. 13 of Royal Society of New Zealand.

Tobias, P. V. (1980a). *"Australopithecus afarensis"* and *A. africanus:* critique and an alternative hypothesis. *Palaeont. Afr., 23*: 1–17.

Tobias, P. V. (1980b). The natural history of the helicoidal occlusal plane and its evolution in early *Homo. Amer. J. Phys. Anthropol., 53*: 173–187.

Tobias, P. V. (1980c). A survey and synthesis of the African hominids of the late Tertiary and early Quaternary periods. *In* L.-K. Konigsson (ed.), *Current argument on early Man*, pp. 86–113. Pergamon, Oxford.

Tobias, P. V. (1985a). Single characters and the total morphological pattern redefined: the sorting effected by a selection of morphological features of the early hominids. *In* E. Delson (ed.), *Ancestors: the hard evidence*, pp. 94–101. Alan R. Liss, New York.

Tobias, P. V. (1985b). Ten climacteric events in hominid evolution. *S. Afr. J. Sci., 81*: 271–272.

Tobias, P. V. (1988). *Olduvai Gorge*, Vol. 4A and 4B: *Homo habilis: Skulls, Endocasts and Teeth.* Cambridge University Press, Cambridge (in press).

Vogel, J. C. (1984). Did *Australopithecus* and *Homo* co-exist? *Nuclear Active, 31*: 19–21.

Vogel, J. C. (1985). Further attempts at dating the Taung tufas. *In* P. V. Tobias (ed.), *Hominid evolution: past, present and future*, pp. 189–194. Alan R. Liss, New York.

Vogel, J. C. and Partridge, T. C. (1984). Preliminary radiometric ages for the Taung tufas. *In* J. C. Vogel (ed.), *Late Cainozoic palaeoclimates of the southern hemisphere*, pp. 507–514. A. A. Balkema, Rotterdam.

Vrba, E. S. (1982). Biostratigraphy and chronology, based particularly on Bovidae, of southern hominid-associated assemblages: Makapansgat, Sterkfontein, Taung, Kromdraai, Swartkrans; also Elandsfontein (Saldanha), Broken Hill (now Kabwe) and Cave of Hearths. *Proc. 1st Inter. Cong. Hum. Palaeont. Prétirage 2*: 707–752.

Vrba, E. S. (1985). Ecological and adaptive changes associated with early hominid evolution. *In* E. Delson (ed.), *Ancestors: the hard evidence*, pp. 63–71. Alan R. Liss, New York.

Walker, A., Leakey, R. E. F., Harris, J. M. and Brown, F. H. (1986). 2.5-Myr *Australopithècus boisei* from west of Lake Turkana, Kenya. *Nature, 322*: 517–522.

Wallace, J. A. (1975). Dietary adaptations of *Australopithecus* and early *Homo. In* R. H. Tuttle (ed.), *Palaeoanthropology, morphology and palaeoecology*, pp. 203–233. Mouton, The Hague.

Wallace, J. A. (1978). Evolutionary trends in the early hominid dentition. *In* C. J. Jolly (ed.), *Early hominids of Africa*, pp. 285–310. Duckworth, London.

White, T. D. (1985). The hominids of Hadar and Laetoli: an element-by-element comparison of dental samples. *In* E. Delson (ed.), *Ancestors: the hard evidence*, pp. 138–152. Alan R. Liss, New York.

Wood, B. A. (1981). Human origins: fossil evidence and current problems of analysis and interpretation. *In* R. J. Harrison and R. L. Holmes (eds.), *Progress in anatomy*, Vol. 1, pp. 229–245. Cambridge University Press, Cambridge.

Wood, B. A. (1984). The origin of *Homo erectus. Cour. Forsch. Inst. Senckenberg, 69*: 99–111.

Yaroch, L. A. and Vitzthum, V. J. (1984). Was *Australopithecus africanus* ancestral to the genus *Homo? Amer. J Phys. Anthropol., 63*:237.

Studies of Paleogeography, Paleoecology and Natural History

V

New Information from the Swartkrans Cave of Relevance to "Robust" Australopithecines

20

CHARLES K. BRAIN

What can the Swartkrans cave tell us about "robust" australopithecines? Potentially a good deal, since this site has produced remains of more of these individuals than any other single locality yet known. Discoveries there go back to 1948 when the site was first worked by Robert Broom and John Robinson who, over a 5-year period, recovered a spectacular haul of *Australopithecus robustus* fossils and demonstrated, for the first time, the coexistence of this hominid with early *Homo* (Broom and Robinson, 1950).

The current site investigation, for which I have been responsible, goes back to 1965 and has continued unbroken since then. Actual excavation was terminated at the end of 1986, and the preceding 21-year-long field investigation fell naturally into three 7-year episodes. The first 7 years were spent in removing and sorting miners' rubble from the cave and its vicinity, in fact restoring order after traumatic exploitation; the second 7-year period was taken up with the systematic removal of natural overburden, which obscured the extent of the cave deposit, while the last 7 years were devoted to meticulous excavation of the fossiliferous deposits still remaining on the cave floor. Conclusions drawn from the results of the first two periods of fieldwork have been reported in detail (Brain, 1981), but those from the last phase of excavation are now being evaluated. Some of these new conclusions are reported here.

Cave Form and Stratigraphy

The form of the cave, which occurs in Pre-Cambrian dolomitic limestone, has been described elsewhere (e.g., Brain, 1981) but a simplified plan is given in Fig. 20.1. Almost all of the fossiliferous deposits occur in the so-called Outer Cave, which has lost most of its roof through erosion.

The excavation has shown that the stratigraphy of the cave is exceptionally complex and that this complexity is due to the fact that, following the first filling of the cave with surface-derived sediment, the site was subjected to successive cycles of erosion and deposition affecting different parts of the fossiliferous matrix (Brain, 1982). The resulting stratigraphic situation may be expressed in scheme 20.1. Apparently, at the time of the start of the first sediment accumulation, the floor of the cave was composed of a jumble of massive dolomite blocks—a boulder choke—that blocked the inclined passageway down to lower caverns. Sediment resting on this boulder choke may be said to have been in a state of metastable equilibrium, so that in times of increased water flow through the cave, sediment was likely to have been carried through the porous floor, leaving cavities in the overlying breccia mass. At other times, such an erosional episode was replaced by a depositional one, when sediment washed into the cave and remained there, for some time at least.

So it appears that after the first infilling of the cave with Member 1 debris, and after its partial calcification, the middle levels of this mass were carried away to the lower cavern system leaving Member 1 divided into two separate masses: one, the "Hanging Remnant" clinging unsupported to the north wall (see Fig. 20.1) and the other, the "Lower Bank," remaining in place on the floor. Between them was an irregular gap up to 4-meters wide into which the Member 2 sediment was deposited during the next depositional cycle. Much later in the cave's history the process

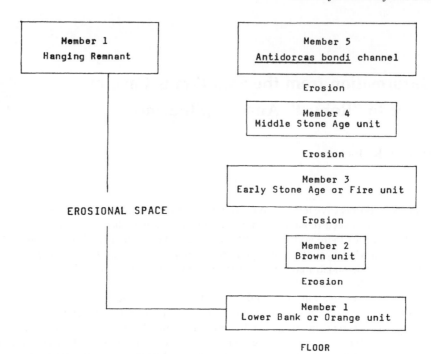

SCHEME 20.1. Representation of Swartkrans strata and the relationship between the Member 1 units.

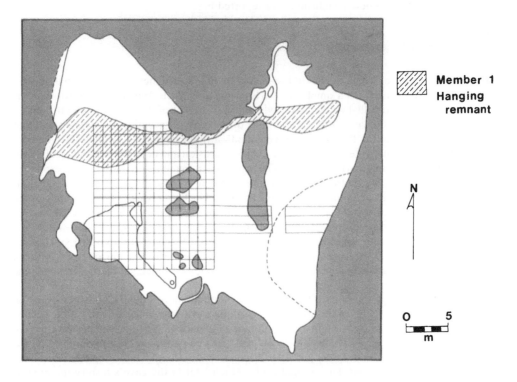

FIGURE 20.1. A plan of the Swartkrans cave showing the excavation grid and the position of the "Hanging Remnant" of Member 1.

was repeated, the erosion this time cutting a steep-sided gulley along the cave's west wall extending down about 9 meters to the interlocking boulders of the floor. This was filled with Member 3 debris. Such a process occurred on at least two further occasions with a large space created for the Member 4 infilling in the northeast corner of the cave and a smaller one, the so-called *Antidorcas bondi* channel, disappearing under the north wall in the central area, filled with Member 5 sediment as shown in Fig. 20.2.

Judging from the composition of the dumps that resulted from the first lime-mining operations at Swartkrans during the 1930's, it is highly likely that the miners removed calcified channel fillings other than those currently recognized. For instance, some excellent Early Stone Age handaxes and cleavers have been found on the dumps but they have not been found in any of excavated deposits and presumably came from a sediment body removed entirely by the miners.

Fossil Provenance, Age and Environment

It is now clear that the whole of the very important fossil sample obtained during the Broom/Robinson operations between 1948 and 1953 came from the western end of the "Hanging Remnant" of Member 1, with the exception of the *Homo ("Telanthropus")* mandible (SK 15) and associated fragments, which came from a later sediment pocket intrusive into the "Hanging Remnant." It has been assumed that this pocket was the same age as the main Member 2 sediment mass, but this is by no means certain in view of the complexity of the cave filling as a whole. Unfortunately, now that this part of the deposit has been removed, there is no certainty as to the stratigraphic relationships there.

The composition of the fossil assemblage from the "Hanging Remnant," both in terms of species and skeletal parts represented, has been fully described (Brain, 1981). On faunal grounds the most reasonable age estimate for this assemblage appears to be between 1.8 and 1.5 Myr

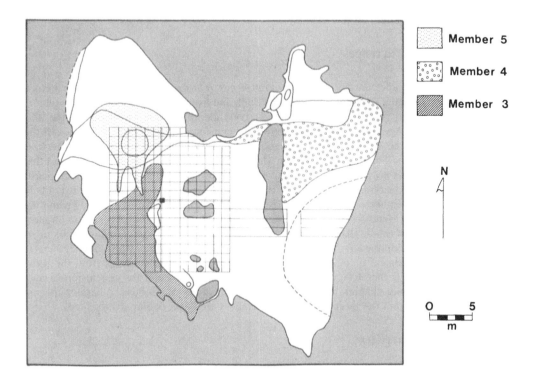

FIGURE 20.2. A plan of the Swartkrans cave showing the positions of the infillings attributed to Members 3, 4, and 5. Members 1 and 2 jointly occupy virtually all the remaining area in the cave.

(Vrba, 1982). Recently, Vogel (1985) has attempted relative thermoluminescence dating of quartz sand grains from Members 1 to 4 and has deduced the following approximate relative values: Member 1 = 10, Member 2 = 9, Member 3 = 5, and Member 4 = 1. It is not known whether the ratio of thermoluminescent signal strength to age is a linear one, but if it is, the indicated ages of the members are Member 1: 1.7 Myr (based on a faunal estimate); Member 2: 1.5 Myr; Member 3: 0.85 Myr; and Member 4: 0.17 Myr. The absence of significant differences in the faunal and cultural assemblages among these members, however, suggests that the ages of Members 1–3 are closer than suggested by the thermoluminescence figures.

The cyclic nature of the depositional and erosional episodes reflected in the Swartkrans deposit strongly suggests that these may have been synchronous with wider climatic events such as glacial/interglacial cycles of which 17 are known to have occurred during the last 1.7 Myr (Fink and Kukla, 1977). Clearly not all of these are reflected in the Swartkrans deposit or, if they are, the record is too subtle to read. But if there was a link between Swartkrans cycles and certain glacials and interglacials, how was the link aligned? The age of the Swartkrans Member 5 infilling provides a clue. The fossils in this member come almost exclusively from the small extinct springbuck *Antidorcas bondi*, and recently a radiocarbon age has been determined for these specimens by J. C. Vogel (pers. comm.). A 495 g sample of *A. bondi* bone was treated for 24 hours with 10% acetic acid to remove mineral carbonate before a ^{14}C age was determined on the bone apatite, giving a result of 11,100 ± 100 yrs BP. This suggests that the Member 5 infilling occurred rapidly around the peak of the current interglacial, or the so-called climatic optimum. It seems likely that the Swartkrans depositional events were brief interglacial ones, while erosion occurred during the long cold and damp glacial times.

A study of the complete faunal assemblages from each Swartkrans member is being made at present, and when this study is completed, some new paleoenvironmental conclusions may be reached. The environmental indications of the "Hanging Remnant" sample are of an open grassland habitat with some patches of wooded cover (Vrba, 1985; Brain, 1985). Of interest is the fact that the Lower Bank of Member 1 contains numerous waterworn pebbles, together with the remains of a hippo and an otter. Thus the first opening of the cave must have been close to a Bloubank River much larger than it is today.

Hominid Fossil Occurrences

Hominid remains, representing both *Australopithecus robustus* and *Homo* are now known from Member 1 "Hanging Remnant" and Lower Bank and from Member 2 (Grine, this volume, chapter 14). So far only *A. robustus* has been positively identified from Member 3 (Grine, this volume, chapter 14). No hominid remains have yet been found in Members 4 and 5. All the hominid fossils from the 1979 to 1986 excavations have recently been studied by Grine (this volume, chapter 14) and Susman (this volume, chapter 10). Of particular interest is the fact that the Swartkrans deposits provide three successive glimpses of hominid and faunal populations at the same geographic location but spanning perhaps more than half-a-million years. It is likely that the Member 3 assemblage includes the most recent *A. robustus* remains yet known and the three collections from Members 1, 2, and 3 can potentially provide useful information on morphological changes and trends that may have occurred during the last half-a-million years or so of the "robust" australopithecine lineage. A conclusion reached by Grine (this volume, chapter 14) that little morphological or metrical change is detectable in the dentitions from the three Swartkrans members, is therefore of special interest. Likewise some of the conclusions drawn by Susman (this volume, chapter 10) on the functional anatomy of "robust" australopithecine hand and foot bones from the three Swartkrans members are of particular significance.

Archeological Occurrences

Stone Artifacts

Worked stone pieces occur in fair abundance in Members 1, 2 and 3, and these have recently been studied in detail by J. D. Clark, whose full description and interpretation will be published shortly. Middle Stone Age artifacts are known to occur in Member 4, but none of these has been described from an excavation context.

Bone Artifacts

Pieces of bone that show characteristic wear and scratches on their tips, and which I interpret as having been used as digging tools, occur in Members 1, 2 and 3. To date 60 have been found, of which 44 are pieces of long bone shaft, 13 are horncores, and 2 are other skeletal parts. The numbers of bone digging tools recovered from the Members are shown in the following tabulation:

	Members			
Tools	1	2	3	Total
Long-bone flakes	11	8	25	44
Horncores	3	4	7	14
Other bones	0	0	2	2
Totals	14	12	34	60

The persistent presence of these tools, associated with *A. robustus* and *Homo* remains in Swartkrans Members 1, 2 and 3, suggest that the digging of vegetable food from the ground was a long-lived tradition of particular importance in the economy of these early hominids. The bone tools are currently being studied in detail by P. Shipman and myself, and descriptions will be published elsewhere.

A Case for Early Hominid Carrying Bags

A number of bone tools from each of the Swartkrans Members show a degree of wear that could only have resulted from many successive days of digging use. This implies that each tool in question was carried by a hominid during successive days or weeks of foraging and, if this is so, it is highly probable that the hominids had some sort of carrying bag. Of suggestive significance is the presence in Member 3 of two very delicate, awl-like bone tools that were probably used for making holes in leather. They are currently being subjected to a SEM study, and one is tempted to surmise that it is not only modern humans that have made use of leather handbags.

Evidence of Fire in the Cave

Although we have examined over 100,000 pieces of bone from Members 1 and 2, none of these shows signs of having been in a fire. Yet the position in Member 3 is very different. Here the excavation has produced 270 pieces of bone that, on the basis of gross appearance and structural changes visible in thin section, have almost certainly been burned. In addition to these, a larger number of bone pieces show probable signs of heating. The presumed burned bones are scattered through a vertical profile of 5 meters, but they suggest that fire was only very intermittently made in the cave although hominid occupation, on the basis of flaked stone, was far more regular. It is probable that hominids at this stage did not have the capability of starting a fire at will but that they occasionally brought back to the cave fire that they had gathered after lightning had ignited the grass, as it occasionally does on the highveld today. Of interest is the possibility that "robust" australopithecines had the opportunity of witnessing the controlled use of fire at Swartkrans, though they may not have participated in this technology.

It may be surmised that the management of fire by early human populations could have contributed to the extinction of the "robust" australopithecines if these two kinds of hominid had been in competition. Therefore, it may be more than a coincidence that the last glimpse one has of *A. robustus* in Member 3 at Swartkrans coincides with the first appearance of evidence for controlled fire in the cave's stratigraphic record.

Acknowledgments

F. E. Grine is to be commended for his initiative in organizing the highly successful international workshop on the evolutionary history of the "robust" australopithecines. I would like to thank him for the invitation to participate in this meeting and also to express my gratitude to him and R. L. Susman for their detailed studies of the Swartkrans hominids.

References

Brain, C. K. (1981). *The hunters or the hunted? An introduction to African cave taphonomy.* University of Chicago Press, Chicago.
Brain, C. K. (1982). Cycles of deposition and erosion in the Swartkrans cave deposit. *Palaeoecol. Afr., 15*: 27–29.
Brain, C. K. (1985). New insights into early hominid environments from the Swartkrans cave. *In* Y. Coppens (ed.), *L'Environment des Hominidés au Plio-Pléistocene*, pp. 325–343. Masson, Paris.
Broom, R. and Robinson, J. T. (1950). Man contemporaneous with the Swartkrans ape-man. *Amer. J. Phys. Anthropol., 8*: 151–156.
Fink, J. and Kukla, G. J. (1977). Pleistocene climates in Central Europe: at least 17 interglacials after the Olduvai Event. *Quart. Res., 7*: 363–371.
Vogel, J. C. (1985). Further attempts at dating the Taung tufas. *In* P. V. Tobias (ed.), *Hominid evolution: past, present and future*, pp. 189–194. Alan R. Liss, New York.
Vrba, E. S. (1982). Biostratigraphy and chronology, based particularly on Bovidae, of southern African hominid-associated assemblages. *Congres Int. Paléontol. Hum. 1er Congres*, pp. 707–752. Nice.
Vrba, E. S., (1985). Palaeoecology of early Hominidae, with special reference to Sterkfontein, Swartkrans and Kromdraai. *In* Y. Coppens (ed.), *L'Environment des Hominidés au Plio-Pléistocene*, pp. 345–369. Masson, Paris.

Chronology of South African Australopith Site Units

21

Eric Delson

Important refinements of the Turkana Basin chronology have recently been presented by Frank Brown and colleagues (Brown *et al.*, 1985; Brown and Feibel, 1986; this volume, chapter 22; Feibel *et al.*, 1988; Harris *et al.*, 1988). These have led to several minor changes in the ages of South African site units I correlated to that sequence, based on the distribution patterns of cercopithecid monkeys (Delson, 1984). Moreover, through the courtesy of C. K. Brain, I have been able to examine the new sample of cercopithecid fossils from Members 1 through 3 of the Swartkrans deposits (see Brain, this volume, chapter 20), which permits comments on the relative ages of these Members. In addition, as the result of discussions at the Stony Brook workshop on which this volume is based, some comments can be offered on the relative ages of Swartkrans and Kromdraai and on environmental indications suggested by the monkeys.

Cercopithecid Fossils from South and East African Site Units

In 1984, I presented a table of taxa known from 17 site units (individual localities or Members or horizons of longer sequences). The data are unchanged for the majority of these units. I have not seen any of the newer material from Sterkfontein or Makapansgat—unfortunately, the sample from undoubted Sterkfontein Member 5 (ST 5) levels remains too minimal to identify, but specimens from ST 1 through 3 will be of great interest. New data provide additional information for several sets of units, while for others only minor clarification is necessary.

Taung

My study of the entire Taung sample is still incomplete, but I continue to infer that the small *Papio* from Taung and ST 4 is a distinct species *Papio izodi*, while the comparably sized population (*P. angusticeps*) known best from Kromdraai and the nearby Cooper's A (COA) site is morphologically and probably taxonomically closer to the living *P. hamadryas kindae* (see Fig. 21.1). About 120 Taung cercopithecids are now known from the following collections: University of the Witwatersrand Anatomy Department (20), Bernard Price Institute (2), South African Museum (18), Transvaal Museum (29) and the University of California Museum of Paleontology (50, but some matrix is unprepared). Of these, three partial maxillae (all from the Transvaal Museum) are referred to a small cercopithecine, cf. *Cercocebus* or *Parapapio;* 25 crania and mandibles are allocated with reasonable certainty to *Parapapio antiquus;* 20 crania and jaws are identified as *Papio izodi* and the remainder (partial jaws, juvenile crania, neurocrania, endocasts, and postcranial elements) are as yet unallocated to taxon. The presence of the two common papionins in roughly equal numbers and in both types of matrix discerned in the Berkeley collection suggests that they were contemporaries throughout the span of time represented at Taung.

Swartkrans

Four taxa are well represented in the "Hanging Remnant" of Swartkrans Member 1 (SK 1): *Papio hamadryas robinsoni* (over 100 specimens), *Papio (Dinopithecus) ingens* (48 specimens),

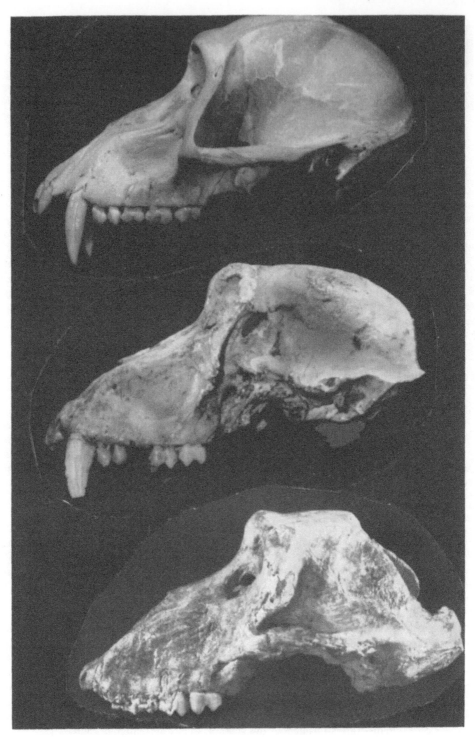

FIGURE 21.1. Left lateral view of three male *Papio* crania, oriented in the occlusal plane and to scale (about two-thirds natural size). Top: modern *P. hamadryas kindae* from Angola; center: *P. angusticeps* from Cooper's A, latest Pliocene?; bottom: *P. izodi* from Taung, late Pliocene. The relatively shorter (and broader) snout and larger teeth and orbits of *P. izodi* suggest it to be distinct from *P. angusticeps*, which may, in fact, best be included within *P. hamadryas*.

Theropithecus oswaldi (25 specimens), and cf. *Cercocebus* sp. (17 specimens). The number of specimens represents identifications of measurable fossils in 1981 and is meant to serve only as a guide to approximate frequency; further analysis is needed to estimate the more meaningful minimum number of individuals. These numbers differ from those of Brain (1981) due to his inclusion of more fragmentary and questionably identifiable fossils, separation of some numbered pieces that fit together, and minor differences in allocation. A "large variant" of *Cercopithecoides williamsi* (4 or 5 specimens, previously termed *C. molletti*) occurs in the Swartkrans deposits, and Brain (pers. comm.) has suggested they may have come from what he then termed Member 2 (but see below). A few specimens comparable to modern *Papio hamadryas ursinus* may be of similar provenance. One crushed palate was identified by Clarke (pers. comm.) as cf. *Gorgopithecus major* on the basis of an apparently large maxillary fossa, and although this specimen is probably correctly allocated, it could have been mislabeled as to origin.

In a sample of several hundred specimens Brain has collected from the Lower Bank of Member 1, as well as Members 2 through 3, only two taxa appear to be present. The great majority of identifiable remains from all three units may be allocated to *P. hamadryas robinsoni*. Several specimens from each of Members 1, 2 and 3 seem to be identical to the *T. oswaldi* material known from the "Hanging Remnant"; they are mostly quite fragmentary, but one partial maxilla is known from Member 2. The lack of any colobine specimens even potentially identifiable as *C. williamsi* and the identity of known taxa from the new and old samples suggest to me that Swartkrans Members 1 through 3 may be more similar in age than Brain (this volume, chapter 20) has inferred. The apparent lack of *P. (Dinopithecus) ingens* in the new sample is surprising, as it is somewhat more common than *Theropithecus* in the "Hanging Remnant"; the smaller papionins *P. angusticeps* and cf. *Cercocebus* sp. also seem absent, but metrical analysis of the dental remains might indicate their rare presence. Perhaps the "Hanging Remnant" is slightly older than the Lower Bank or at least samples a time interval when neither *Papio (Dinopithecus)* nor cf. *Cercocebus* inhabited the Swartkrans area (or died there). On the other hand, if any of Members 1 through 3 were younger than the "Hanging Remnant," I would have expected them to yield the large *C. williamsi*, *P. angusticeps* and/or *Gorgopithecus*, as found at the Kromdraai and Cooper's site units. This may have further implications for the relative age of these units (see below).

West Turkana

Harris *et al.* (1988) have recently presented a summary of the mammalian faunas from the Nachukui Formation on the west side of Lake Turkana. They divided the collection into nine assemblages spanning approximately the 4- to 1-Myr interval. Their tabulated distribution of taxa reveals that no primates at all have yet been collected from the Lokalalei Member, equivalent to Omo Shungura D and E. The Kalochoro Member, equivalent to Shungura F, G and lowermost H (up to the KBS = H2 Tuff level), has yielded only a fragment attributed to *Homo habilis* and a few *Theropithecus oswaldi* specimens (J. M. Harris, pers. comm.). This interval (D through G) is among the most productive of monkeys at Omo, representing my African Cercopithecid (AC) Zones 3 (upper) through 5. The rarity of both hominid and cercopithecid material from this interval in the Nachukui Formation is disappointing. It is a time about which more needs to be known. I look forward to further remains from at least the more fossiliferous Kalochoro Member and also to more details on the *Parapapio* material reported from levels in the Lomekwi Member.

Relative Dating and Chronometric Calibration

Taung

The studies by Brown and colleagues have provided a recalibration of the middle part of the Shungura Formation, which results in some differences in the estimated ages of my Zones AC 3 through AC 5. Figure 21.2 indicates the probable placement of those zones and several South African site units in the 2.5 to 2.0 Myr range. Perhaps the most interesting change from my

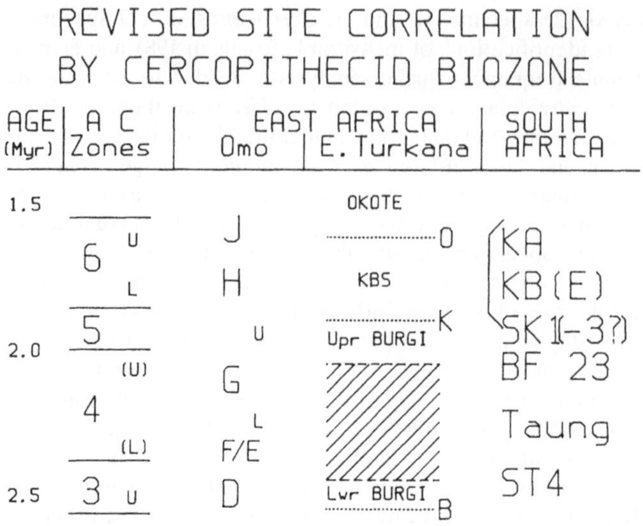

REVISED SITE CORRELATION BY CERCOPITHECID BIOZONE

AGE (Myr)	A C Zones	EAST AFRICA Omo	EAST AFRICA E. Turkana	SOUTH AFRICA
1.5			OKOTE	
	6 U	J	············O	KA
	6 L	H	KBS	KB (E)
2.0	5	U	············K Upr BURGI	SK 1(–3?)
	(U)	G	//////	BF 23
	4	L	//////	Taung
	(L)	F/E	//////	
2.5	3 U	D	Lwr BURGI ············B	ST 4

FIGURE 21.2. A revised correlation of later Pliocene South African site units based on the cercopithecid biozonation of Delson (1984), with East African levels recalibrated after the work of Brown and colleagues (see text). Note that while Taung can more confidently be placed around 2.3 Myr, there is less certainty about the relative ages of Swartkrans (SK) Members 1 through 3 and the several Kromdraai sites (KA, KBE). ST 4 is Member 4 ("Type Site" and correlated dumps) of Sterkfontein; BF 23 is Pit 23 at Bolt's Farm.

previous chart, is the somewhat greater age accorded to the Taung faunal assemblage. This is especially important in light of the continued suggestion by Partridge (1982, 1985, 1986) and others (e.g., Vogel, 1985) that the Taung hominid lived around 1 Myr ago. In terms of morphology, the Taung specimen appears to be indistinguishable from *Australopithecus africanus* as known at Sterkfontein and Makapansgat (Grine, 1985). I have argued (1984) that there is little geological evidence to support the separation of the Taung fauna and hominid. Vrba (1982) was unable to go farther than to say that the few Taung bovids agree with an age between ST 4 and SK 1 or Kromdraai B East (KBE), as Wells (1967, 1969) had previously concluded. In his review of micromammals, Pocock (1987: 69) wrote that "suggestions that Taung is significantly younger than other australopithecine sites are not supported," although this subject is not treated in detail by him.

How then does Partridge support this view? Without arguing against my contention that the total Taung fauna is of one age, Partridge (1986: 82) opined that the assemblage may have been part of a relict local fauna which persisted in this supposedly well-watered region long after the surrounding province had become aridified late in the Pliocene. In turn, the age of 1 Myr was obtained from work by Vogel (1985), who estimated uranium-series and thermoluminescence dates for the levels penecontemporaneous with the Taung fauna by extrapolation back from younger dates. Vogel (1985) has also applied this approach to Swartkrans, in part based on estimates for the age of the layers derived from paleontology. The approach would appear to be both circular and fragile; it especially reminds me of the attempts over a decade ago to explain away the younger estimates of the age of the circum-KBS horizons at Koobi Fora compared to the more "accurate" radiometric dates, which, as is well known, were erroneous. Much more detailed analyses must be brought to bear on the Taung deposits if dates younger than 2 Myr, much less close to the 1 Myr championed by Partridge and colleagues, are to be taken seriously.

Relative to Taung, the Makapansgat and ST 4 assemblages are clearly older. Vrba (this volume, chapter 25) and Kimbel and White (this volume, chapter 11) have recently suggested that the ST 4 fauna may be mixed, with some younger elements alongside a generally older set of taxa. If that is the case, then the (apparently) unique juvenile *Papio izodi* and perhaps the small sample of *P. humadryas robinsoni* might be intrusive. These specimens were first identified by Eisenhart (1974), and although they are based on fragments from dumps, they appear secure. They are the oldest South African members of the genus, however, and among the oldest in all of Africa (in part depending on the age and generic allocation of *Papio baringensis*).

Kromdraai and Swartkrans

The two sets of site units most relevant to this volume, and about which dating is perhaps least certain, are Swartkrans (Members 1 through 3) and Kromdraai (especially B East, Member 3, but also Kromdraai A and Cooper's A). There is almost no taxonomic overlap among the cercopithecids between the Swartkrans and Kromdraai/Cooper's groups (see Table 21.1). As indicated above, four taxa are known from the SK 1 "Hanging Remnant" (SK 1"HR"): *P. h. robinsoni, T. oswaldi, P. (D.) ingens* and cf. *Cercocebus* sp., and a fifth, *G. major*, may also be present but rare. Only the first two are known also in the samples recently excavated by Brain from SK 1 through 3 (not included in the table), and *Theropithecus* is very rare. The Kromdraai A (KA) ("faunal site") assemblage includes about 29 *P. angusticeps;* 2-plus *P. h. robinsoni;* another 13 specimens for which allocation is to either *P. angusticeps* or perhaps *P. h. robinsoni,* 21 *G. major,* and 1 jaw with an M_3 of *Cercocebus* sp. The Kromdraai B, Member 3 East (KBE) ("hominid site") excavations by Vrba have produced an assemblage only slightly different from Brain's previously recovered (KB) sample which crosscut the layers distinguished by Vrba. *Papio angusticeps* and *P. h. robinsoni* are probably both present, with the former predominant; *G. major* is rare; and *C. williamsi* ("large variant") is common. The small sample from nearby Cooper's A (COA) has a few specimens of each of the last two taxa and perhaps a dozen each of the two *Papio* species, but as at KA, no hominid fossils have yet been found.

Of these site units, SK 1"HR" and KA are the most distinctive; each has unique taxa in some quantity, and they share only rare specimens of each other's most common elements. The KB (including KBE) and COA units are linked by the large *C. williamsi,* but they share some features of each of the two previous units as well. All are clearly younger than AC 4 (meaning Taung and Bolt's Farm in South Africa) due to their lack of *Parapapio*. In 1984, I suggested that this pattern led to the sequence SK 1("HR"); KB(E); ?COA; KA. Swartkrans was linked to the older ST 4 assemblage (AC 3 upper) by *P. h. robinsoni,* and to Shungura C through G by *Papio (Dinopithecus)* (albeit a different species). But it was placed in the younger AC 5 Zone due to the presence of *T. oswaldi* (without any earlier *Theropithecus*) and *Cercocebus* (rather than *Parapapio*). The large *Cercopithecoides* and unique *Gorgopithecus* were said to imply a still younger

Table 21.1. Estimated Number of Cercopithecid Specimens by Taxon for Swartkrans and Kromdraai Site Units

Site unit[a]	Taxon[b]							
	Pa	Phr	Pa/Phr	P(D)i	Gm	To	Csp	Cw(L)
KA	29	2+	13		21		1	
KB	7		14		2			15
KBE			33		2			30
COA	16	10	?		3			2
SK1 "HR"		100+		48	1?	25	17	

[a]KA = Broom's original sample from the Kromdraai "faunal site" assemblage; KB = Brain's excavated sample, which crosscut layers discerned in Vrba's later work; KBE = sample from Kromdraai B East of Vrba (Member 3); COA = Cooper's A site; SK 1 "HR" = Swartkrans Member 1 "Hanging Remnant" collection, not including newer collections.

[b]Pa = *Papio angusticeps*; Phr = *Papio hamadryas robinsoni*; Pa/Phr, allocation to preceding taxa uncertain; P(D)i = *Papio (Dinopithecus) ingens*; Gm = *Gorgopithecus major*; To = *Theropithecus oswaldi*; Csp = cf. *Cercocebus* sp.; Cw(L) = *Cercopithecoides williamsi* ("large variant").

age for KB and especially KA, in part because the former taxon also occurred in what was then called SK Member 2.

These interpretations are less secure today, however. If, as noted above, *P. h. robinsoni* is potentially "intrusive" in ST 4, there is little to determine which of the three sites (SK 1 "HR", KA or KB) is the oldest. If KB were older, as suggested by Grine (1982) and followed by Vrba (1982; Vrba and Panagos, 1983) on the basis of the stage of evolution of the australopiths, this would appear to imply discontinuities in the distribution of the Swartkrans taxa *Papio (Dinopithecus)* (known at Omo through AC 4 only), *Cercopithecoides* ("typical" form known in AC 3 and in AC 4 at Bolt's), and perhaps *P. h. robinsoni* (known in AC 4 at Bolt's Farm). The first taxon is known only from SK 1"HR" and does suggest greater age. The provenance of the Swartkrans specimens of *Cercopithecoides* is now unclear, so that its implication for relative age is cast into question. *Papio h. robinsoni* is widespread, but its relative frequency compared to *P. angusticeps* may be important. If this decreases over time, it suggests that Swartkrans is early, but if variation is random, there is obviously no biochronological value in the relative frequency of *P. h. robinsoni*.

Vrba and I now agree that there is no clear indication as to whether KBE is older or younger than SK 1"HR". However, Vrba (e.g., 1982) has argued that KA is younger than KB from bovid evidence. Although the monkeys do not confirm or reject that hypothesis, I do think that KB and COA are likely to be intermediate in age between KA and SK 1"HR". Combining these two rather weak views leads me to support my 1984 sequence, but with far less certainty (Fig. 21.2). Overall, I agree with Vrba (1982) and others that these site units probably are between 1.9 and 1.65 Myr old, that is, in the latest Pliocene.

It further appears to me that the consistent nature of the cercopithecid and hominid sample from Brain's new SK 1 and SK 2 levels contradicts the view that a long time may have elapsed between them. If Brain (this volume, chapter 20) is correct that each level represents a warm interval, perhaps three successive warm phases are being sampled. Prentice and Denton (this volume, chapter 24) indicate that at this time perhaps only 50 kyr separated warm peaks, so that all of the lower Swartkrans deposits might have been laid down within as little as a 100 kyr span, with the SK 1"HR" (and the unknown *Cercopithecoides* level) and even the Kromdraai units broadly contemporaneous as well. I wonder if we can yet discern differences on this order of fine structure either in the environment or in evolutionary sequence when dealing with events 2 Myr old. I plan to continue analysis of patterns of species distribution, relative frequency, and phyletic relationship in hopes of clarifying this question.

In other paleontological situations, there is always the possibility that differing environments may have been sampled, but because these site units are so close geographically, the environmental differences must be under temporal control; that is, the site units differ in paleoenvironment because of their age differences, rather than being contemporaneous on a scale of 10 kyr with regionally different climates. Unfortunately, little can be added about these local paleoenvironments. Vrba and Panagos (1983) suggested that increasing aridity was indicated by the decreased presence of *C. williamsi* from ST 4 through KBE to SK 1"HR" and KA. I responded (1984) that this species was highly terrestrial and may not have required as much environmental water as other colobines. I tentatively postulated an *increase* of surface water from SK 1"HR" through KB to KA, but that also is probably too simplistic. All the *Papio* species were probably quite terrestrial, as was *Theropithecus* (and presumably *Gorgopithecus*, which is unknown postcranially), so that there is no clear distinction among these site units.

Summary and Conclusions

New cercopithecid specimens recovered by Brain from Members 1 through 3 of the Swartkrans deposit are almost all identifiable as *P. hamadryas robinsoni*, with rare elements attributed to *T. oswaldi* in Members 1 and 2. This suggests some differences from the earlier collected SK 1"HR" sample, which includes many *P. (D.) ingens*, some cf. *Cercocebus* sp. and one probable *G. major*. Other Swartkrans specimens identified as a "large variant" of *C. williamsi* were originally attributed to Member 2, but no similar material has been identified in the new sample.

This suggests that at least Members 1 to 3 of the new sample are similar in age to each other but possibly slightly different in age from the SK 1"HR" sample.

Based on the recalibration of Omo Group members by Brown and colleagues in a series of papers since 1985, my 1984 cercopithecid zonation must be slightly modified. Taung, in lower AC 4, is probably about 2.3 Myr old. Attempts to chronometrically date the Taung tufas about 1 Myr and explain the cercopithecid correlation in terms of a relict fauna are tenuous at best. If some of the ST 4 assemblage is mixed, then cercopithecid candidates for such "intrusion" include the unique *P. izodi* and perhaps the *P. hamadryas robinsoni* specimens, which are now among the oldest members of this genus in Africa.

It is not yet feasible to unequivocally determine the age sequence of the group of site units from Swartkrans and Kromdraai. Few species occur in common, and several of the known taxa are rare elsewhere. A more complete analysis of the relative frequency of taxa may prove helpful in the future, but as yet, neither the bovids nor the cercopithecids (or hominids) indicate a clear pattern. If Vrba's (this volume, chapter 25) suggestion is accepted that KA is younger than KB, and if KB (and the similar COA) is intermediate between SK 1"HR" and KA, then my 1984 suggestion of a SK 1"HR"–KB–KA sequence is weakly supported. These site units may be quite close in age, probably between 1.9 and 1.65 Myr. Given my suggestion of age similarity among Members 1 through 3 at Swartkrans and Brain's idea that each was formed during a warm interval, perhaps several successive warm phases sampled over only 100 kyr are involved in the combined Swartkrans (and Kromdraai?) complex.

The major implication of this work for the evolution of "robust" *Australopithecus* is to confirm the young age of the South African forms by comparison to those of East Africa. It is almost certain that the "robust" clade is monophyletic (Grine, this volume, chapter 30), and if the 2.6 Myr old *A. aethiopicus* is morphologically close to the common ancestor of *A. boisei* and *A. robustus*, it is surprising that no "robust" hominid is yet known in the south older than 2 Myr. The only known site units at which such an earlier "robust" form might be expected are ST 4, Taung and the Bolt's Farm complex; the latter two have together yielded only one hominid specimen, of course, and despite suggestion in this volume, it is unclear that anything other than *A. africanus* occurs in ST 4. Perhaps the time is right to search for new sites in the Sterkfontein valley in hopes of tying in the Swartkrans and Kromdraai forms to the better-documented line of "robust" *Australopithecus* evolution in the east.

Acknowledgments

I thank Fred Grine for so ably organizing the "robust" australopithecine workshop, for inviting me to attend and present this work, for much patience during its production, and for editorial suggestions. I also thank Frank Brown, Elizabeth Vrba, and many other workshop participants for thoughtful discussion; Bob Brain for allowing me to examine the new Swartkrans cercopithecid sample; the many other museum curators who permitted me to study specimens in their care; and Elizabeth Strasser and David Dean for assistance with Fig. 21.1. This work was financially supported, in part, by a J. S. Guggenheim Memorial Fellowship and by grants from the National Science Foundation and the PSC–CUNY Faculty Research Program.

References

Brain, C. K. (1981). *The hunters or the hunted?* University of Chicago Press, Chicago.
Brown, F. H. and Feibel, C. S. (1986). Revision of lithostratigraphic nomenclature in the Koobi Fora region, Kenya. *J. Geol. Soc. Lond.*, *143*: 297–310.
Brown, F. H., McDougall, I., Davies, T. and Maier, R. (1985). An integrated Plio-Pleistocene chronology for the Turkana Basin. *In* E. Delson (ed.), *Ancestors: the hard evidence*, pp. 82–90. Alan R. Liss, New York.

Delson, E. (1984). Cercopithecid biochronology of the African Plio-Pleistocene: correlation among eastern and southern hominid-bearing localities. *Cour. Forsch. Inst. Senckenberg,* 69: 199–218.

Eisenhart, W. L. (1974). The fossil cercopithecoids of Makapansgat and Sterkfontein. Unpublished B. A. Thesis, Harvard College.

Feibel, C. S., Brown, F. H. and McDougall, I. (1988). Stratigraphic context of fossil hominids from the Omo Group deposits, northern Turkana Basin, Kenya and Ethiopia. *Amer. J. Phys. Anthropol.* (in press).

Grine, F. E. (1982). A new juvenile hominid (Mammalia, Primates) from Member 3, Kromdraai Formation, Transvaal, South Africa. *Ann. Tvl. Mus., 33:* 165–239.

Grine, F. E. (1985). Australopithecine evolution: the deciduous dental evidence. In E. Delson (ed.), *Ancestors: the hard evidence,* pp. 153–167. Alan R. Liss, New York.

Harris, J. M., Brown, F. H., Leakey, M. G., Walker, A. C. and Leakey, R. E. (1988). Pliocene and Pleistocene hominid-bearing sites from West of Lake Turkana, Kenya. *Science, 239:* 27–33.

Partridge, T. C. (1982). The chronological positions of the fossil hominids of southern Africa. In H. de Lumley and M-A. de Lumley (eds.), Prétirage, Ier Cong. Internat. Paléo. Humaine, pp. 617–675 (Vol. II). Cent. Nat. Rech. Sci., Nice.

Partridge, T. C. (1985). Spring flow and tufa accretion at Taung. In P. V. Tobias (ed.), *Hominid evolution: past, present and future,* pp. 171–187. Alan R. Liss, New York.

Partridge, T. C. (1986). Paleoecology of the Pliocene and Lower Pleistocene hominids of southern Africa: how good is the chronological and paleoenvironmental evidence? *S. Afr. J. Sci.,* 82: 80–83.

Pocock, T. N. (1987). Plio-Pleistocene fossil mammalian microfauna of southern Africa—A preliminary report including description of two new fossil muroid genera (Mammalia:Rodentia). *Palaeont. Afr.,* 26: 69–91.

Vogel, J. C. (1985). Further attempts at dating the Taung tufas. In P. V. Tobias (ed.) *Hominid evolution: past, present and future,* pp. 189–194. Alan R. Liss, New York.

Vrba, E. S. (1982). Biostratigraphy and chronology, based particularly on Bovidae, of southern hominid-associated assemblages: Makapansgat, Sterkfontein, Taung, Kromdraai, Swartkrans; also Elandsfontein (Saldanha), Broken Hill (now Kabwe) and Cave of Hearths. In H. de Lumley and M-A. de Lumley (eds.), Prétirage, Ier Cong. Internat. Paléo. Humaine, pp. 707–752 (Vol. II). Cent. Nat. Rech. Sci., Nice.

Vrba, E. S. and Panagos, D. C. (1983). New perspectives on taphonomy, palaeoecology and chronology of the Kromdraai ape-man. In J. A. Coetzee and E. M. Van Zinderen Bakker (eds.), *Palaeoecology of Africa and the surrounding islands,* Vol. 15, pp. 13–26.

Wells, L. H. (1967). Antelopes in the Plesitocene of Southern Africa. In W. W. Bishop and J. D. Clark (eds.), *Background to evolution in Africa,* pp. 99–107. University of Chicago Press, Chicago.

Wells, L. H. (1969). Faunal subdivision of the Quaternary in southern Africa. *S. Afr. Archaeol. Bull.,* 24: 93–95.

"Robust" Hominids and Plio-Pleistocene Paleogeography of the Turkana Basin, Kenya and Ethiopia

22

FRANK H. BROWN AND CRAIG S. FEIBEL

The purpose of this paper is twofold: (1) to provide an up-to-date chronology of Pliocene and Pleistocene sediments in the Turkana Basin from which early hominids have been collected, and (2) to document serious difficulties with some published paleogeographic models of the region (Findlater, 1976; Vondra and Bowen, 1978; de Heinzelin, 1983) and to replace those models with a new model that accords better with stratigraphic observations made over the past six years. The chronology will be treated summarily here, without providing primary data on new age determinations, because revisions that have been made since the last review (Brown et al., 1985) do not seriously affect the age estimates of "robust" hominid specimens. A full review of the chronologic placement of all hominid specimens from the region has been prepared for publication elsewhere (Feibel, et al., 1988). Most of the localities mentioned in the text are shown in Fig. 22.1.

Age Estimates of "Robust" Hominid Specimens

Hominid specimens have been collected from four formations in the Turkana Basin—the Shungura Formation, the Usno Formation, the Koobi Fora Formation, and the Nachukui Formation. The stratigraphy of these has been described elsewhere (see de Heinzelin, 1983; Brown and Feibel, 1986; Harris et al., 1988a), and they are securely linked stratigraphically by chemical analysis of volcanic ash layers. The stratigraphic relations between three of these formations, based on the volcanic ash layers, is shown in summary fashion in Fig. 22.2.

The chronology developed for these hominid-bearing formations is based mainly on conventional K/Ar and ^{40}Ar/^{39}Ar dates on anorthoclase feldspar separated from pumice clasts contained within the volcanic ash layers. Most of the data have been published in McDougall et al. (1980), McDougall (1985), Brown et al. (1985), and McDougall et al. (1985). These data have been supplemented with K/Ar age determinations on two new horizons (Feibel et al., 1988), and by paleomagnetic polarity zonation of the Shungura and Koobi Fora Formations (Brown et al., 1978; Hillhouse et al., 1986). The chronology is consistent with that derived by correlation to calcareous nannoplankton zones in the Gulf of Aden (Sarna-Wojcicki et al., 1985).

There is a discrepancy of about 0.07 Myr between the time scale based on K/Ar and that based on magnetostratigraphy, the reasons for which are not completely understood (see Hillhouse et al., 1986, for discussion). The age estimates provided for the "robust" hominid specimens in this paper are derived by striking a compromise between the two scales. The error bars in Figs. 22.3 and 22.4 represent both analytical uncertainty in the age determinations and uncertainty in the stratigraphic position of the hominid specimens in the section. The hominids lie within the time interval shown by the error bars, but closer placement is not possible at the present time.

It was somewhat difficult to decide which specimens to include in these figures because of lack of unanimity on the taxonomic attribution of specimens. We have tried to include all specimens that have been attributed to or discussed as "robust" taxa in the literature, but inadvertently

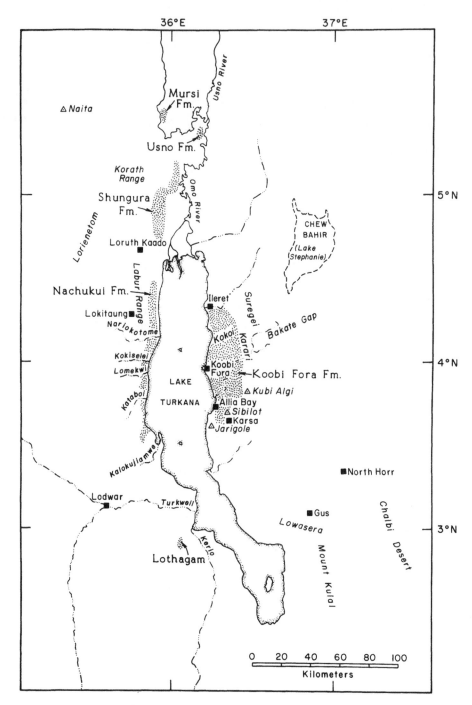

FIGURE 22.1. Map of the Turkana Basin showing the locations of features mentioned in the text. Stippled areas represent principal outcrop areas of various Plio-Pleistocene formations.

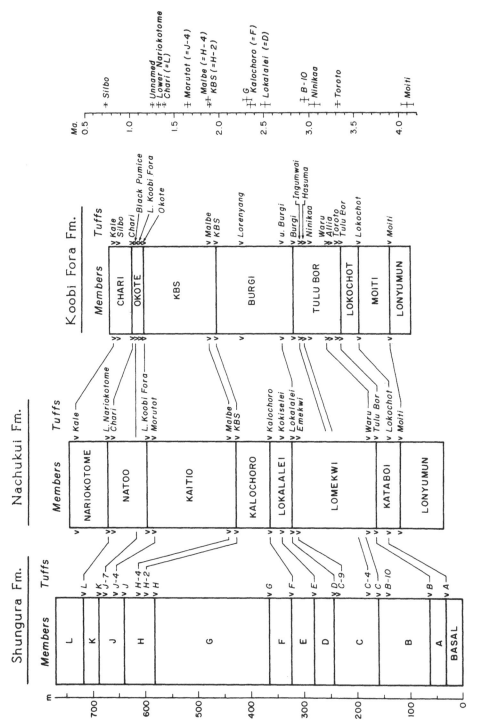

FIGURE 22.2. Stratigraphic sections of the Shungura, Nachukui, and Koobi Fora Formations, showing members, correlations and named tuffs.

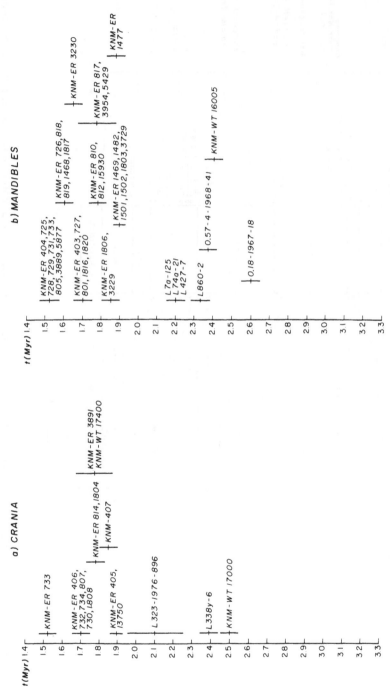

FIGURE 22.3. (a) Temporal distribution of "robust" hominid crania from the northern Turkana Basin. Specimens within groups are listed numerically. No internal stratigraphic ordering within groups is implied. The vertical bars represent the range of possible ages for each group. (b) Temporal distribution of "robust" hominid mandibles from the northern Turkana Basin. Specimens within groups are listed numerically. No internal stratigraphic ordering within the groups is implied. The vertical bars represent the range of possible ages for each group.

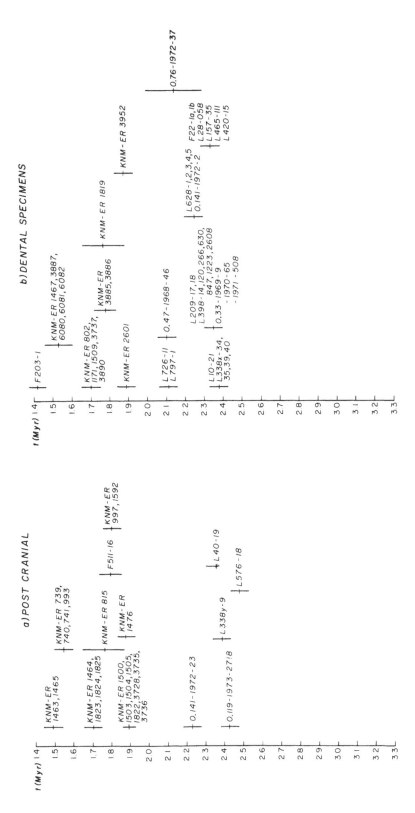

FIGURE 22.4. (a) Temporal distribution of "robust" hominid postcrania from the northern Turkana Basin. Specimens within groups are listed numerically. No internal stratigraphic ordering within groups is implied. The vertical bars represent the range of possible ages for each group. (b) Temporal distribution of "robust" hominid dental specimens from the northern Turkana Basin. Specimens within groups are listed numerically. No internal stratigraphic ordering within groups is implied. The vertical bars represent the range of possible ages for each group.

may have omitted a few. The compilations of McHenry (this volume, chapter 9) and White (this volume, chapter 27) were especially helpful, and it is likely that all of the more complete specimens have been included. In addition, a few specimens may be shown in these figures, which the anthropological community will decide are not "robust" hominids. We claim no expertise in hominid taxonomy, and inclusion of a specimen on these diagrams is not intended as a statement about taxonomic attribution by the authors or their co-workers. The specimens were grouped such that the chronologic placement is as precise as possible. In many cases the age ranges overlap, but the possible range of ages differs. No internal ordering of specimens within groups is indicated on the diagrams, although for some specimens such ordering is possible.

Paleogeography

The maximum utility of paleogeographic models to paleoanthropology in the Turkana Basin will be realized only when they can be constructed for a large number of very brief time intervals, and although it is not yet possible to do this, it is clear that the presently published paleogeographic models are incorrect. This chapter sets out those features of the paleogeography which we now believe to be secure.

The Shungura Formation and the Koobi Fora Formation (Fig. 22.1) were studied independently, and no serious attempt was made to relate deposits between the two regions until the late 1970's (Brown et al., 1978). The first precise stratigraphic correlations between the two regions were also achieved at about the same time (Cerling et al., 1979). The data used to construct paleogeographic models included variations in thickness of distinct lithologic entities, lateral variations in grain size of particular units, heavy mineral analyses to establish source regions of the sediments, paleocurrent determinations to establish direction of flow of streams, sedimentary structures to establish lateral changes in depositional environments, and floral and faunal analyses to suggest reasonable ranges of ecological conditions. Authigenic mineralogy and stable isotope variations related to climatic changes in the region were provided for the Koobi Fora Formation by Cerling (1979) and Abell (1982).

The Current Model

Models developed for sedimentation in the Koobi Fora region and the lower Omo Valley contain within them several implicit elements that have greatly influenced ideas about deposition in the Turkana basin. Workers in both areas thought in terms of a permanent lake located somewhere south of the lower Omo Valley and west of the Koobi Fora region, which existed in a closed basin for the past 4 Myr. The geography was thought to have been very similar to that of the present day in both areas, thus the Shungura formation was postulated to have been deposited by a major permanent meandering river flowing southward along a course similar to the modern Omo River until it reached the lake. Sedimentation in the Koobi Fora region was thought to have been the result of a second meandering stream entering the basin from the east through the topographically low region known as the Bakate Gap (or Ol Bakate). It was believed that lateral changes in deposition in both regions were controlled by fluctuations in the level of the lake, but whether these fluctuations were controlled by climatic changes or tectonic deformations was not specified.

Clearly, the present geography unduly influenced the thinking of all workers. It was not until most local sections at Koobi Fora were placed in proper stratigraphic order that the local paleogeographic story came into question (Brown and Cerling, 1982). When stratigraphic relations between there and exposures north and west of the lake were established, the entire model came into serious difficulty.

Difficulties with the Current Model

New data that have caused us to doubt the veracity of the current working model include (1) revision of the stratigraphy of the Koobi Fora region (Brown and Feibel, 1986); (2) correlations between the Koobi Fora and Shungura Formations based on chemical analysis and K/Ar dating of volcanic ash layers at a number of points in the section (Cerling and Brown, 1982; Brown *et al.*, 1985); (3) documentation of the Nachukui Formation west of the lake (Harris *et al.*, 1988a), and its correlation to deposits east and north of the lake; (4) discovey of related deposits and faunal elements east of the present basin margin; (5) knowledge of the time of formation of various highland features of the present topography, both volcanic and structural; (6) observations of the grain size and lithology of sediments in the present streams draining the region; and (7) critical re-examination and review of the depositional features of the deposits.

Pliocene and Pleistocene strata of the region have been described by de Heinzelin (1983) for the Shungura Formation in the lower Omo Valley, by Brown and Feibel (1986) for the Koobi Fora Formation exposed east of Lake Turkana, and by Harris *et al.* (1988a) for the Nachukui Formation west of the lake. Summary stratigraphic columns, formal stratigraphic terms, and K/Ar chronological data are shown in Fig. 22.1.

A number of problems are inherent in the current working model of deposition in the Turkana Basin over the past 4 Myr. There is no sedimentologic evidence for an extensive and long-lived lake in the region for the periods 4 to 2 Myr ago and 1.7 to 0.2 Myr ago, although there were brief intervals of lacustrine deposition over restricted areas during these times. In addition, the similarity of fluvial deposits between 4 and 2.5 Myr at Koobi Fora and Shungura is so striking that it is extremely unlikely that they represent deposition by different fluvial systems. These similarities include style of sedimentation, heavy mineral assemblages, number of fining-upward cycles, and stratigraphic thickness between correlated tuffs. Basalt clasts are essentially absent in the section at Koobi Fora prior to 1.9 Myr ago, yet the present ephemeral streams carry basalt pebbles and cobbles nearly to the lake. This argues strongly against Plio-Pleistocene drainage configurations similar to those of the present. Such basalt pebble conglomerates as exist within the Koobi Fora Formation are restricted to the region of the Karari Escarpment.

Most of the highland areas within the Koobi Fora region (Jarigole, Sibilot, the Kokoi, the Suregei highlands, the highlands south of Kubi Algi) are in fault contact with Pliocene and/or Pleistocene sediments, and thus their elevation postdates deposition of the sediments. In several localities (particularly in the Allia Bay region), large channels exist in the stratigraphic record in which the flow directions are to the southeast, yet today there is no possibility of their continuation in that direction because they are blocked by volcanic highlands. Near Lowasera the Tulu Bor Tuff was deposited by fluvial processes, yet it lies outside the present margin of the basin. The only reasonable source for this material is from the Turkana basin itself. Note that *Euthecodon* (a crocodilian) and *Lates* (Nile perch) have been reported from the Marsabit Road site (Nyamweru, 1986) and as both are aquatic elements, it is unlikely that they could have reached an area 120 km east of the nearest point on Lake Turkana without a watercourse connecting the two areas. Moreover, there is no record of fossil stingrays in the Turkana Basin prior to c. 2 Myr ago, yet they must have had a route into the basin, which could not have been the case were the basin closed. There is little sedimentologic evidence that the basin that presently contains Chew Bahir (Lake Stephanie) has been in existence for a protracted period of time. On geomorphological evidence that basin appears quite young. If a highland existed in that region, it is unlikely that it would have supported a perennial stream, and it certainly would have cut off postulated connections to drainage networks farther north in the Ethiopian Rift Valley (Findlater, 1976).

Many tuffs exist in fluvial sediments in all three areas that contain pumice clasts up to 1 m in diameter which could have reached the area only by water transport. Recent work by Wolde-Gabriel and Aronson (1987) lends strong support to the hypothesis that the volcanic source areas are located in the Ethiopian highlands, as proposed originally by Brown (1972). A model in which two streams arise in the same distant region and empty into the Turkana Basin is possible but cannot explain the presence of pumice clasts west of the lake. Finally, the sediment volume contained within the basin (c. 4,000 km^3) is only about one tenth the volume of sediment estimated by Cerling (1986) to have been delivered to the basin in the past 4 Myr (c. 40,000 km^3). These difficulties are sufficient to demonstrate the need for a new model of sedimentation in the region.

A New, Alternative Model

Lake Turkana is unlike other rift-related lakes in that it lies on a drainage divide between the Indian Ocean and the Mediterranean Sea. It is not bounded by steep escarpments, except in its southern reaches, and, if it were filled with sediment, water would flow across it to one ocean or the other. The stratigraphic record of the Pliocene and Pleistocene of the Turkana Basin is a record of entrapment of only part of the sediment delivered to it during subsidence.

Cerling (1986) published an analysis of mass-balance constraints on sediment and water in the Turkana Basin and showed that the Omo River delivers so much sediment to the basin that even today the basin will be filled with *compacted* sediment in about 180 kyr. It follows that any depression that existed in the past would have been rapidly infilled as well, even if the basin were open and the river carried most of the sediment through. In addition, the Omo River delivers so much water that when it flows into the basin, a large lake must form if the basin is closed, as is the case today. From the late Pleistocene record it is known that a lake formed in this manner may vary in area between 7,500 km^2 and 25,000 km^2 because of climatic changes. Such a lake, however, would have a relatively short lifetime, probably not exceeding a few hundred thousand years. Note also that there are several low points in the basin through which water may flow were the basin filled today. These include (1) the broad low area between Lorienetom and Naita, where a hydrographic connection with the Nile River is believed to have existed about 10 kyr ago (Butzer 1971); (2) the low area of the eastern drainage divide near North Horr; and (3) the divide northwest of Lodwar between the Turkwell and Lotagipi drainages. Topographic information is too crude to state the elevations of these divides with precision, but both east and west of the lake Plio-Pleistocene sediments reach elevations of 165 m above the present water surface.

In contrast to the model of deposition outlined above, we find no evidence for a long-lived lake in the Turkana Basin; instead, it appears that for long periods of time there was no lake in the region at all. Nor do we find that a stream entering the basin from the east through the Bakate Gap is an essential element of the paleogeography at any time, except for the latest Pleistocene and Holocene, when it appears that the gravel cap of the Koobi Fora ridge was deposited. Rather, sedimentation in the northern half of the Turkana Basin was dominated by the ancestral Omo River, not only in the lower Omo Valley, but also east and west of present Lake Turkana. Stratigraphic data for areas around the southern part of Lake Turkana are scant; thus our sketch maps for that region are largely conjectural. From the stratigraphic sequence at Lothagam (Powers, 1980), it is clear that a perennial river system flowing from south to north existed in that region sometime prior to 4 Myr ago, and while we have assumed that this sytem flowed into the region occupied by the present lake for much of this time, other scenarios are possible.

The first time period for which we can construct a reasonable paleogeographic picture is between about 4 and 4.5 Myr ago (Fig. 22.5). At that time almost the entire region was occupied by a stable lake. The Muruongori Formation at Lothagam represents deposition in this lake, which extended into the Kerio Valley. Volcanic hills near the upper reaches of Lomekwi shed conglomerates into this lake, and a re-entrant of the lake wrapped around these hills northeast of Lodwar, but its western extent is not well known. Sediments of the same age and lithology exist at Loruth Kaado, at the northern end of the Labur Range, and possibly as far north as the slopes of Lorienetom where again similar sediments are found. The northern extent of the lake is constrained by the presence of fluvial and deltaic sediments of the Mursi Formation and by a sandy section that underlies the basalt of the Usno Formation. Along the eastern margin of Lake Turkana, deposits of this lake make up most of the Lonyumun Member of the Koobi Fora Formation, and they are recognizable in discontinuous patches from the Suregei Highland through the Allia Bay region to the area around Gus (north of Mt. Kulal). The eastern extent of this water body is poorly constrained, but its minimum surface area must have been at least 28,000 km^2. Evidence of buried topography in the Koobi Fora region shows that this lake had a minimum depth of 36 m. Molluscan faunas were dominated by the gastropods *Bellamya* and *Cleopatra*. After an initial period of fine clastic sedimentation, these lake deposits are dominated by richly diatomaceous sediments, followed by renewed clastic deposition. The temporal range of this *first* lacustrine interval was most likely on the order of 100 to 200 kyr. At a number of localities in the basin (Mursi, Usno, Kataboi, Karsa, Suregei), nonporphyritic basalts intrude these lake sediments or are erupted as surface lavas about 4 Myr ago.

The second phase of deposition, extending from about 4.1 to 2.0 Myr ago was dominated by sedimentation by a large river in the Shungura Formation and in the Nachukui Formation. Similar deposits are present in the Koobi Fora Formation from 4.1 to about 2.5 Myr ago, but there was

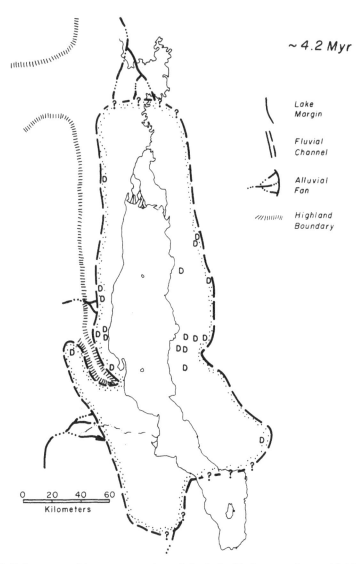

FIGURE 22.5. Paleogeographic reconstruction of the Lake Turkana region at 4.2 Myr. The symbol D represents known localities of diatomite. See text for discussion.

apparently no deposition in the Koobi Fora region from 2.5 to 2.0 Myr ago. During this interval we can recognize the following features. Short-lived lacustrine intervals are recorded in the Koobi Fora Formation and the Nachukui Formation at about 3.6 and 3.2 Myr ago, but these lakes apparently were much smaller than that which existed previously, perhaps having individual surface areas of about 2500 km^2 (Fig. 22.6). From Kalokujiamwe to Kokiselei, the western margin of the basin is well-marked by alluvial conglomerates interfingering with finer-grained sediments of the central part of the basin during this interval. No sediments of this age are exposed between Kokiselei and Loruth Kaado, but at Loruth Kaado fine volcanic pebble conglomerates in this part of the section mark the proximity of the basin margin. The Lokochot Tuff is present in an alluvial fan on the eastern slopes of Lorienetom and also occurs south of Lothagam, where it may be an air fall deposit. The absence of quartz clasts in the conglomerates of this interval at Loruth Kaado provides reasonable evidence that the Labur Range did not exist at this time as it is made up largely of quartz-bearing Cretaceous conglomerates and Precambrian quartz-rich rocks.

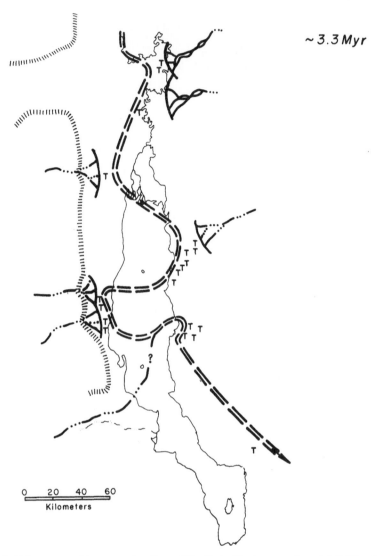

FIGURE 22.6. Paleogeographic reconstruction of the Lake Turkana region at 3.3 Myr. The symbol T represents known proximal channel deposits of the Tulu Bor Tuff. Other symbols as in Fig. 22.5.

Deposits near the basin center consist largely of the fluvial fining-upward sequences that are characteristic of the Shungura Formation. These deposits are well exposed in the lower Omo Valley, and they extend as discontinuous patches as far south as Allia Bay. Only a single outcrop is known southeast of Allia Bay near Lowasera, but this outcrop lies within the present drainage basin of the Chalbi Desert. Channels are large, with measured depths ranging from 15 to 17 m, and widths of about 150 m. *Etheria* (fresh-water oyster) bioherms are found commonly on the bottoms of channels in all three formations, but other molluscs are only sparsely represented. It is during this period that the large channels with flow directions oriented toward the present eastern basin margin are found. There is little evidence on which to establish the position of the eastern basin margin during this time, as the deposits at Koobi Fora are in fault contact with Miocene volcanic rocks at their easternmost extent. It is probably significant, however, that volcanic clast conglomerates are not found in the section during this period at Koobi Fora; the implication is that the basin margin is some distance away. There is almost no evidence on which to reconstruct the course of drainage systems off the Kenyan highlands, dominated by the Turkwell and Kerio Rivers, and the integration of this drainage with the Omo River drainage in Fig. 22.6 is largely conjectural.

A number significant events occurred in the period between 2.5 and 2 Myr ago (Fig. 22.7). It is at this time that Mt. Kulal was built, creating a new highland where none existed previously. (Marsabit and the Huri Hills were built somewhat later, between about 2 and 1 Myr ago.) Deposition apparently ceased in the Koobi Fora region during this interval, perhaps because of gentle upwarping in that area. It is likely that the Labur Range began to be uplifted during this time, because later conglomerates at Loruth Kaado (c 1.6 Myr in age) contain quartz, in contrast to the earlier ones in the same area. At the same time *Etheria* became much less common in the Shungura Formation but much more apparent in channels along the west side of the lake, although the reason for this is obscure.

Whether caused by volcanic damming of the rivercourse, or by upwarp along the eastern basin margin, this phase of fluvial deposition ends about 2 Myr ago. At that time the basin was inundated

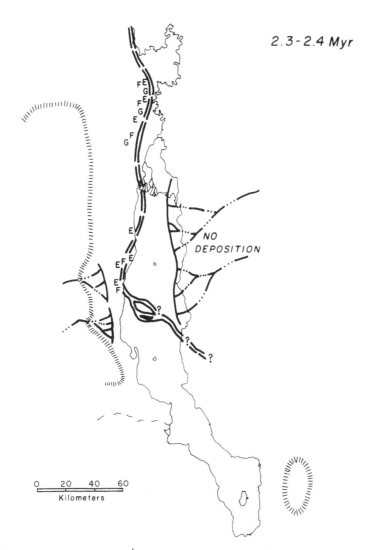

FIGURE 22.7. Paleogeographic reconstruction of the Lake Turkana region at 2.3 to 2.4 Myr. Symbols E, F, and G represent known localities of Tuffs E, F, and G of the Shungura Formation or their correlative units. Other symbols as in Fig. 22.5.

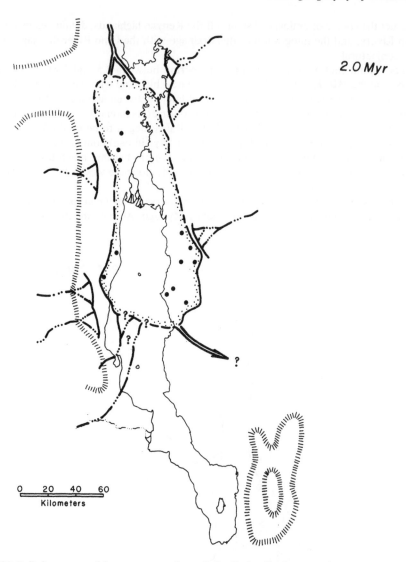

FIGURE 22.8. Paleogeographic reconstruction of the Lake Turkana region at 2.0 Myr. Filled circles represent known locations of deep lacustrine deposits. Other symbols as in Fig. 22.5.

by development of a large stable lake (Fig. 22.8). The transition is marked by an abrupt lithologic change to claystones and siltstones in all areas. The size of this lake is difficult to determine, because the northern, southern and eastern margins are poorly constrained, but a minimum area is on the order of 9,000 km^2. Molluscs, including numerous endemic forms (Williamson, 1982), are reasonably common in this interval, implying that the lake was relatively fresh and presumably it still drained to the east (i.e., the basin trapped sediment, but there was an outlet for water). Further evidence for this direction of drainage is offered by fossil fresh-water stingrays, which entered the basin during this period, perhaps because the lake damped out large oscillations in the flow downstream from its exit (Feibel and Brown, in prep.). It is possible, however, that some sort of hydrographic connection to the Nile drainage also developed at this time. The northern and eastern part of this lake was infilled by deltaic sedimentation from the Omo River. The KBS Tuff (= H-2), for example, was deposited 26 m above the base of the deltaic deposits in the Omo Valley, but it lies only 15 m above the base of these deposits near Koobi Fora Spit, and within lacustrine deposits near the northern boundary of Allia Bay and in the Nachukui Formation. There is no known record of lacustrine (or other) sedimentation during this interval south of Allia Bay, hence the southern limit shown in Fig. 22.8 is insecure.

This lacustrine interval very likely had other important consequences that are pertinent to the later part of the record. If an integrated drainage network existed prior to the formation of the lake, it certainly would have been disrupted. Riverine plant communities would have been destroyed over a stretch of river c 200 km in length, which may explain some of the differences in the fossil flora before and after this interval (Bonnefille and Deschamps, 1983). The basin would have retained most of the sediment delivered to it, which was not the case when the river flowed through, and therefore would have been rapidly infilled.

A complex depositional system followed in the period between 1.8 and 1.3 Myr ago, perhaps initiated because the basin had been filled with sediment shortly before. It was characterized by alternating fluvial and lacustrine conditions recorded by relatively thin (a few meters) intervals of claystone and sandstone. The characteristic floodplain and channel sedimentation of the earlier fluvial phase of deposition never returned. Channels were shallower than those that existed previously (<10 m), but some still contained *Etheria*, marking the persistence of perennial water.

At Koobi Fora there was a clear change in the aquatic communities between 1.9 and 1.6 Myr. Cosmopolitan molluscs abounded for brief periods during the earlier part of this interval, but they decreased in importance later on. The later mollusc beds commonly are capped with algal stromatolites. Still higher in the section algal stromatolities dominate the aquatic community, marking a new feature of the record. The region near Koobi Fora spit was characterized by low energy shorelines during lacustrine intervals, with broad flats where bivalves *(Mutela)* burrowed into a sandy substratum. In contrast, contemporary deposits west of the modern lake show development of high-energy, wave-dominated beaches.

Along the Karari Escarpment the basin margin was clearly marked by volcanic clast conglomerate filled channels, as was the basin margin west of the lake. These channels are distinct from the complex of channels associated with the KBS Tuff, which are filled largely with coarse to medium sand composed dominantly of quartz, plagioclase and K-feldspar. This latter channel complex often has been interpreted as a single erosional interval (called the post-KBS erosion surface), but such description is clearly inaccurate as many erosional and depositional events are recorded within it, and the complex was initiated before deposition of the KBS Tuff (= H-2) in some areas. In the Koobi Fora region this period of deposition was largely complete by the time the Chari Tuff was deposited, and there is again a hiatus in the section in this region lasting from about 1.35 to 0.8 Myr ago.

Deposits of the Shungura Formation from this interval are obviously different from those of the preceding fluvial phase. The pronounced cyclicity is much subdued; sandstones are finer grained, siltstones more abundant, and the overall range of grain size is generally reduced. Volcanic ash layers in this part of the section are also distinct from the earlier ones (G 11 and below), in that they are generally thinner, much finer grained, and have smaller scale cross-stratification.

Pumice-bearing volcanic ash layers continued to be supplied to the region, but their distribution was much more restricted, with a few (fortunate) exceptions. It appears that sedimentation occured only locally at any particular time, rather than over broader areas as was the case before. As a result, correlations between tuffs are more useful locally than regionally, although a few (e.g., Chari (= L), Black Pumice Tuff (= J-7), Lower Koobi Fora and KBS (= H-2), Malbe (= H-4) are still valuable for correlation basinwide.

West of Lake Turkana, lacustrine conditions persisted in the area near Nariokotome until c 1.7 Myr ago. This is difficult to explain if the Omo River again flowed through to the Indian Ocean. Rather, it appears that the Omo River became less and less noticeable as a constant feature of the local geography, and it may have alternated its flow from the Turkana Basin to the Nile system during this time. Possibly a remnant lake existed, fed predominantly by waters from the Turkwell and Kerio Rivers, with some additional water supplied by the Usno River and other, smaller streams draining from the highland regions between Chew Bahir and the Turkana Basin. This arrangement (Fig. 22.9) is far more tentative than the conditions shown on other figures.

A certain element of special pleading seems to be required to explain the continued introduction

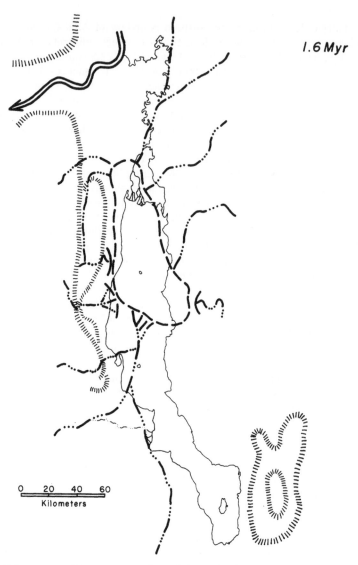

FIGURE 22.9. Paleogeographic reconstruction of the Lake Turkana region at 1.6 Myr. Symbols as in Fig. 22.5.

of large pumice clasts into the tuffs of northern part of the Turkana Basin, for which we believe the Omo River to be essential, and yet to deny the continued presence of that river in the region. In other words, we will have the Omo in when we want it, but otherwise not. This may not be as arbitrary as it first appears, as the area between the Turkana Basin and the broad flats east of the Nile River was probably of very low gradient. During large floods, perhaps caused in part by volcanic eruption and temporary damming of the course of the Omo River in its headquarters, overbank sediments still may have been dispersed into the Turkana Basin (see Brown, 1972), while most of the water and sediment exited to the west. The materials one would expect from such events would be of fine grain size, except for pumice clasts, which, because of their low density, would be included with the overbank materials. Once dispersed into the local drainage systems of the Turkana Basin, the pumice might be concentrated into channels by a variety of mechanisms. Alternatively, the Omo River may have reentered the basin for brief intervals; the details of specific depositional events have yet to be worked out.

Sedimentation continued west of the lake until about 1.2 Myr ago, but alluvial fans extended several kilometers farther eastward after deposition of the Nariokotome Tuff (1.33 Myr) than they did before. Associated with these alluvial fans are well-developed algal stromatolites, and occasional mollusk-rich horizons. There is no certain record of deposition west of the lake from c 1.2 to c 0.8 Myr, while in the lower Omo Valley sedimentation continued until about 1 Myr ago.

The existence of the Williamson's Molluscan Range Zone 10 in Member L of the Shungura Formation (Williamson, 1982; pers. comm.) some 300 kyr earlier than its appearance at Koobi Fora, may imply continued lacustrine conditions somewhere in the northern part of the basin during the interval 1.0–0.7 Myr ago, but there is no documented stratigraphic record for this period. The uppermost strata of the Shungura Formation contain prominent ostracod and mollusk beds that are both thicker and more widespread than any others recorded from that formation.

Sometime after this period of deposition, the Kokoi Horst was elevated to form a prominent highland within the basin at Koobi Fora, and intrabasinal faulting affected earlier strata of all three regions. It is clear that the Shungura Formation was deformed and exposed during the last million years, for it had been eroded prior to deposition of the Kibish Formation. The basin largely took on its modern appearance at this time, with continued elevation of mountains to the west and broad highlands to the east, and eruption of volcanoes internal to the basin (Korath Range, islands of Lake Turkana).

Relevance to "Robust" Australopithecine Evolution

"Robust" australopithecine fossils (attributed to various taxa by various authors, but including *Australopithecus boisei* and *A. aethiopicus*) have been recovered from sediments 2.5 to 1.4 Myr in age in the Turkana Basin. The fossils were collected from floodplain and channel deposits of a large meandering river, deltaic deposits of a large river, alluvial fans associated with ephemeral streams, and marginal deposits of both reasonably fresh and somewhat saline lakes. As discussed above, several fundamental changes occurred in the paleogeography and depositional conditions in the basin during this interval, and there are clear changes in the composition of the flora during this interval as well (Bonnefille and Deschamps, 1983). Fossil wood becomes much less common in strata younger than 1.6 Myr in the Shungura Formation, which either reflects a change in the vegetation, with less woody species present, or a change in the mode of preservation. Faunal changes are also apparent; for example, colobine monkeys and impala are much better represented in the earlier part of the record than at later times, and several workers (e.g., Bonnefille, 1976; Cerling, 1979; Behrensmeyer and Cooke, 1985; Harris *et al.*, 1988a) have argued that, in addition, there were pronounced climatic changes in the region during this interval. Paleogeographic changes alone, however, may have contributed significantly to the floral and faunal changes.

It seems then, that the "robust" australopithecines were quite adaptable creatures, for they continued to exist in the region despite all these changes. There has been much argument about the specific habitat occupied by the "robust" australopithecines (Shipman and Harris, this volume, chapter 23; Vrba, this volume, chapter 25), whether or not it differed from that of early *Homo*, and even whether or not such information is extractable from the fossil record (White, this volume, chapter 27). Since hominids are neither infaunal nor fossorial, it is clear that taphonomic processes intervene between their death and their burial. With our present knowledge, it may be wise to look at the broader environmental contexts of hominid distribution and acknowledge that hominids may have existed within any of the more restricted habitats within those.

Conclusions

Most "robust" hominid specimens from the Turkana Basin are now reasonably well constrained stratigraphically and chronologically. Over the past 4 Myr the Turkana Basin has seen several fundamental changes in the configuration of dominant geographical elements, but none of these changes were sufficient to eradicate hominids from the region.

Acknowledgments

The work reported herein was supported by NSF Grants BNS-8210735, BNS-840637, and BNS-8605687, and by the National Geographic Society, Washington, D.C. A Sienko, T. Davies, and R. Maier of the Australian National University, Canberra provided much help with K/Ar analyses. Logistic support was provided in part by the National Museums of Kenya, for which we are grateful. R. Leakey and M. Leakey are particularly thanked for their continued encouragement. A. Ekdale and B. Boschetto made valuable comments on a preliminary draft of the manuscript, and P. Onstott drafted the figures.

References

Abell, P. (1982). Paleoclimates at Lake Turkana, Kenya, from oxygen isotope ratios of gastropod shells. *Nature, 297:* 321–323.

Behrensmeyer, A. K. and Cooke, H. B. S. (1985). Paleoenvironments, stratigraphy, and taphonomy in the African Pliocene and early Pleistocene. *In* E. Delson (ed.), *Ancestors: the hard evidence*, pp. 60–62. Alan R. Liss, New York.

Bonnefille, R. (1976). Palynological evidence for an important change in the vegetation of the Omo Basin between 2.5 and 2 million years. *In* Y. Coppens, F. C. Howell, G. Ll. Isaac and R. E. F. Leakey (eds.), *Earliest Man and environments in the Lake Rudolf Basin*, pp. 421–431. University of Chicago Press, Chicago.

Bonnefille, R. and Deschamps, R. (1983). Data on fossil flora, in *The Omo Group, Archives of the International Omo Research Expedition.* Musee Royal de l'Afrique Central, Tervuren, Belgique, Annales, serie in 8°, Sciences Geologiques, no. 85, pp. 191–207.

Brown, F. H. (1972). Radiometric dating of sedimentary formations in the lower Omo Valley, southern Ethiopia. *In* W. W. Bishop and J. A. Miller (eds.), *Calibration of hominoid evolution*, pp. 273–287. Scottish Academic Press, Edinburgh.

Brown, F. H. and Cerling, T. E. (1982). Stratigraphical significance of the Tulu Bor Tuff of the Koobi Fora Formation. *Nature, 299:* 212–215.

Brown, F. H. and Feibel, C. S. (1986). Revision of lithostratigraphic nomenclature in the Koobi Fora region, Kenya. *Geol. Soc. Lond., 143*: 297–310.

Brown, F. H., Howell, F. C. and Eck, G. G. (1978a). Observations on problems of correlation of late Cenozoic hominid-bearing formations in the north Lake Turkana Basin. *In* W. W. Bishop (ed.), *Geological background to fossil man*, pp. 473–498. Scottish Academic Press, Edinburgh.

Brown, F. H., Shuey, R. T. and Croes, M. K. (1978b). Magnetostratigraphy of the Shungura and Usno Formations, southwestern Ethiopia. New data and comprehensive reanalysis. *Geophys. Roy. Astron. Soc., 54*: 519–538.

Brown, F. H., McDougall, I., Davies, T. and Maier, R. (1985). An integrated Plio-Pleistocene chronology for the Turkana Basin. *In* E. Delson (ed.), *Ancestors: the hard evidence*, pp. 82–90. Alan R. Liss, New York.

Butzer, K. (1971). *Recent history of an Ethiopian delta.* University of Chicago, Dept. of Geography Research Paper no. 136. University of Chicago Press, Chicago.

Cerling, T. E. (1979). Paleochemistry of Plio-Pleistocene Lake Turkana, Kenya. *Palaeogeog., Palaeoclimat., Palaeoecol., 27:* 247–285.

Cerling, T. E. (1986). A mass-balance approach to basin sedimentation: constraints on the recent history of the Turkana Basin. *Palaeogeog., Palaeoclimat., Palaeoecol., 54*: 63–86.

Cerling, T. E. and Brown, F. H. (1982). Tuffaceous marker horizons in the Koobi Fora region and the Lower Omo Valley. *Nature, 299*: 216–221.

Cerling, T. E., Brown, F. H., Cerling, B. W., Curtis, G. H. and Drake, R. E. (1979). Preliminary correlations between the Koobi Fora and Shungura Formations, East Africa. *Nature, 279*: 118–121.

de Heinzelin, J. (1983). *The Omo Group, Archives of the International Omo Research Expedition.* Musee Royal de l'Afrique Central, Tervuren, Belgique, Annales, serie in 8, Sciences Geologiques, no. 85.

Feibel, C. S. and Brown, F. H. Freshwater Stingrays from the Plio-Pleistocene of the Turkana Basin, Kenya and Ethiopia (in prep).

Feibel, C. S., Brown, F. H. and McDougall, I. (1988). Stratigraphic context of fossil hominids from the Omo Group deposits, northern Turkana Basin, Kenya and Ethiopia. *Amer. J. Phys. Anthropol.* (in press).

Findlater, I. (1976). Isochronous surfaces within the Plio-Pleistocene sediments east of Lake Turkana. In W. W. Bishop (ed.), *Geological background to fossil Man*, pp. 415–420, Scottish Academic Press, Edinburgh.

Harris, J. M., Brown, F. H. and Leakey, M. G. (1988a). Geology and Paleontology of Plio-Pleistocene localities west of Lake Turkana, Kenya. *Contributions in Science, Los Angeles County Museum of Natural History, 399:* 1–123.

Harris, J. M., Brown, F. H., Leakey, M. G., Walker, A. C. and Leakey, R. E. (1988b). Pliocene and Pleistocene hominid-bearing sites from west of Lake Turkana, Kenya. *Science, 239:* 27–33.

Hillhouse, J., Cerling, T. E. and Brown, F. H. (1986). Magnetostratigraphy of the Koobi Fora Formation, Lake Turkana, Kenya. *Jour. Geophys. Res., 91:* 11581–11595.

McDougall, I. (1985). K/Ar and 40Ar/39Ar dating of the hominid-bearing Pliocene-Pleistocene sequence at Koobi Fora, Lake Turkana, northern Kenya. *Geol. Soc. Amer. Bull., 96:* 159–175.

McDougall, I., Maier, R., Sutherland-Hawkes, R. and Gleadow, A. J. W. (1980). K-Ar age estimates for the KBS Tuff, East Turkana, Kenya. *Nature, 284:* 230–234.

McDougall, I., Davies, T., Maier, R. and Rudowski, R. (1985). Age of the Okote Tuff Complex at Koobi Fora, Kenya. *Nature, 316:* 792–794.

Nyamweru, C. K. (1986). Quaternary environments of the Chalbi Basin, Kenya: sedimentary and geomorphological evidence. In L. E. Frostick (ed.), *Sedimentation in the African Rifts*, Geological Society Special Publication no. 25, pp. 297–310.

Powers, D. (1980). Geology of Mio-Pliocene sediments of the Lower Kerio Valley, Kenya. Unpublished Ph.D. Dissertation, Princeton University.

Sarna-Wojcicki, A. M., Meyer, C. E., Roth, P. H. and Brown, F. H. (1985). Ages of tuff beds at East African early hominid sites and sediments in the Gulf of Aden. *Nature, 313:* 306–308.

Vondra, C. F. and Bowen, B. E. (1978). Stratigraphy, sedimentary facies, and paleoenvironments East Lake Turkana, Kenya. In W. W. Bishop (ed.), *Geological background to fossil Man*, pp. 395–414. Scottish Academic Press, Edinburgh.

Wolde-Gabriel, G. and Aronson, J. L. (1987). Chow Bahir rift: a "failed" rift in southern Ethiopia. *Geology, 15:* 430–433.

Williamson, P. G. (1982). Molluscan biostratigraphy of the Koobi Fora hominid-bearing deposits. *Nature, 295:* 140–142.

Habitat Preference and Paleoecology of *Australopithecus boisei* in Eastern Africa

23

PAT SHIPMAN AND JOHN M. HARRIS

In this paper we evaluate the evidence of the ecological niche and habitat preferences of *Australopithecus boisei* in eastern Africa in the light of new sites, new specimens and a new technique for classifying modern and fossil habitats based on bovid faunas. Four sites or general fossiliferous areas in eastern Africa yielded adequate faunal remains to be used here: Olduvai, Koobi Fora, West Turkana, and Omo. Taxonomic attributions of hominids are based in nearly all cases on cranial remains and are conservative, differing little from those given by White (this volume, chapter 27), Day (1986) and others. *Australopithecus boisei* is taken here to include remains sometimes attributed to *A. aethiopicus* (see Kimbel et al., this volume, chapter 16). The habitat indications derived from those localities bearing "robust" australopithecine cranial remains are compared and contrasted with those from the same sites that lack these hominids. In addition, data from these eastern African sites are compared with those from the *A. robustus* and other hominid-bearing sites from South Africa. Finally, the distribution of three early hominid species (*A. boisei, A. robustus* and *Homo*) across the entire sample of East and South African sites is considered, to determine whether any of these species exhibits a distinct habitat preference.

Techniques and Taphonomy

Although the ecology of any species is clearly a crucial aspect of its adaptation and evolution, it is sometimes remarkably difficult to establish the paleoecology of an extinct species with certainty. Various sets of factors may contribute to the difficulty of paleoecological reconstruction. The first set of factors is related to the numerous taphonomic events that may selectively destroy or distort the fossil record and the associations among species. These account for the differences between living animal communities and bone or fossil assemblages (Behrensmeyer and Dechant Boaz, 1980; Shipman, 1981; White, this volume, chapter 27). The second type of factor is related to the behavioral flexibility of species. Animals are not always strongly tied to particular extreme habitats, and even those that are may stray out of their preferred habitats into other areas. The third set of factors is related to depositional variables which bias the fossil record by sampling a disproportionate number of habitats related to water (e.g., lake margins, streams, channels, deltas) and by failing to sample many open-country habitats farther away from water sources. Another type of factor that creates problems with paleoecological reconstruction occurs because of the necessity for classification of habitats into simple, readily identifiable and distinct types, when in reality habitats are often complex and mosaic. Compounding these problems are the effects of time-averaging and migration of ecological zones or habitats across basins in response to climatic and other fluctuations.

Despite these problems, even a crude glimpse of the habitat preferences of our ancestors and near-relatives proves thought provoking. It has been a standard assumption since the writings of Darwin that hominids first evolved in open, savannah habitats. The strength of this assumption is attested to by the many graphic reconstructions of early hominids that almost without exception place them in savannahs (e.g., those in Howell, 1973 or Isaac and McCown, 1976). While there has been some solid evidence in support of this assumption from detailed analyses of particular sites (e.g., Vrba, 1975, 1976, 1984), few if any of these works have attempted to summarize both the East and South African data in any systematic or quantitative way for individual hominid species.

Understanding the inherent limitations of the data about early hominid paleoecology is important. We contend that since hominids are rare species in the Pleistocene, their absence from any particular site does not necessarily indicate that the habitat in question was avoided. Nonetheless, a strong pattern of repeated association between a particular habitat type and a particular hominid species, if found, is likely to be indicative of habitat preference. The approach taken here is to establish the paleoecology of the localities where *A. boisei* or *A. robustus* bones are present, and then to determine whether the localities bearing "robust" australopithecines are distinct from other localities at the same sites.

Here, we analyze the faunas associated with early hominids to deduce habitats by using proportions of bovid tribes with known, different habitat preferences in a technique derived from work by Vrba (1975, 1976, 1980). The rationale behind using bovids is that they are abundant in many African habitats, some tribes have well-developed habitat preferences that can be traced back through the Pleistocene, and their range in body size includes the range of early hominid body sizes. The rough comparability in size will minimize some of the potential biases caused by common taphonomic agents [but see Brain (1981) for a summary of taphonomic differences between bovids and primates]. One question to be resolved is whether the technique used here is capable of reliably classifying modern, known areas (game parks or reserves) into habitat groups. We present data below that demonstrate the power and utility of the approach we use to classify modern habitats.

If analyses of bovid faunas are to be used to reconstruct paleoenvironments, an important question to be answered is: do bovids die where they live? The only good data available on the relationship between numbers of living animals, numbers of bones from dead animals, and habitats are those of Behrensmeyer and Dechant Boaz (1980) and Western (1980) on Amboseli National Park in Kenya. Their data reveal a clear, strong relationship between the expected number of carcasses, based on the abundance of live animals and their life expectancy, and the observed abundance of carcasses or bones of those same species. In five of six Amboseli habitats (swamp, open woodland, plains, lake bed and bush), the census data on six different bovid species correlated significantly with the census data on carcasses of those species ($r > .88$ in each case, 5 df, $p < 0.01$; Fig. 23.1). Thus it may be concluded that bovids generally do die where they live. The exception to this rule is the dense woodland habitat, in which the live and dead census data correlate poorly ($r = 0.66$, 5 df, $p > 0.1$). A possible expanation for the lack of correlation in dense woodland is that this is the habitat in which live animal censuses might be least accurate due to poor visibility (Pennycuick and Western, 1972).

Not only is the abundance of the living bovid population correlated with that of the dead bovid population in most habitats, but most (four out of six) of the slopes of the regression lines for the habitats include 1 within their 95% confidence interval. This means that the living and dead populations are highly correlated with each other and that the proportions of different species stay approximately the same in most habitats. If the migratory wildebeest are removed from the data sets for the swamp and dense woodland (habitats in which wildebeest are present only transiently during their migration) then all six habitats show regression lines that include a slope of 1 within their 95% confidence interval.

If bovids generally die where they live, and are found dead in proportions that closely resemble their proportions in the living community, do they fossilize where they die? As a general rule, the chances of fossilization depend upon the density of the individual bones and upon their chances of becoming rapidly buried in sediments (Shipman, 1981). By restricting the analysis to bovids, the variation in overall body plan and pattern of density within the skeleton is minimized. The major variable at work, then, is depositional setting. It is apparent that proximity to water, in the form of either lakes or rivers, enhances the probability of fossilization. All of the eastern African sites analyzed here are associated either with a lake (Olduvai) or with a large river system (Omo, Koobi Fora, West Turkana). In contrast, the South African early hominid sites, which occur in limestone caves, were not formed by major rivers. The possible significance of this major taphonomic distinction between the eastern and southern sites is discussed further below.

There is another important taphonomic distinction among the sites analyzed here. The Olduvai material comes from excavated and small-scale localities, some of which are archaeological in origin and some of which are paleontological (Leakey, 1971). Most Olduvai assemblages were probably formed over relatively brief periods of time (i.e., less than 10 years) (Potts, 1982) and thus represent truly contemporaneous samples of past animal populations. In contrast, localities from Omo, Koobi Fora and West Turkana mostly represent fluvial deposits; the animals in the fossil assemblages from these sites were derived from larger geographic areas and longer time spans than those preserved in localities at Olduvai. Further, the "localities" from Omo, Koobi Fora and West Turkana, as used here, are aggregate samples from tuff-bounded geologic members rather than samples from individual localities. Thus, the Olduvai remains may be expected to represent more restricted geographic and temporal sampling of the animal population and local habitats than the other assemblages.

Materials and Methods

All localities ($N = 76$) from Olduvai Beds I and II, Omo, Koobi Fora, and West Turkana in eastern Africa, as well as Swartkrans Members 1 and 2, Sterkfontein Members 4 and 5, the Sterkfontein Dumps, and the Sterkfontein West Pit in South Africa were analyzed. The designation of sites as "robust" australopithecine sites was made, with one exception, on the basis of cranial/dental remains to lessen the possibility of incorrect taxonomic identifications. FLKNN level 3 from Olduvai has yielded what may be a partial foot of *A. boisei* (OH 8), based on the morphological similarity of its talus to one from Koobi Fora that is associated only with *A. boisei* cranial remains (B. Wood and A. Walker, pers. comm.), and to one from Swartkrans. This tentative identification of OH 8 is not universally accepted (Susman, pers. comm.; White, this volume, chapter 27), but the exclusion of FLKNN level 3 from the data set would have little impact on the patterning in habitat preferences as a whole.

Some individual sites (Kromdraai, Makapansgat, Chesowanja, Natron, and Peninj) and particular localities or members from Olduvai, Koobi Fora, West Turkana, and Swartkrans were eliminated from the sample, either because they had too few associated bovid remains for the application of the technique used here, or because appropriate data on those remains were not available. Bovid data were derived from Vrba (1976), Brain (1981), Boaz (1977b, 1985), Gentry (1976, 1985) and our own original work. A total of sixty-nine East African and six South African localities were included in the sample (Table 23.1); twenty-five of the East African and two of the South African sites yield specimens of "robust" australopithecines.

The technique used here for reconstructing paleoenvironments at these sites relies upon the proportions of different bovid tribes in the faunas. Vrba (1975, 1976, 1980) pointed out that the Alcelaphini and Antilopini are committed to adaptations for open habitats and that these adaptations apparently arose by the Early Pleistocene. Therefore, she suggested that the combined percentage of these two tribes relative to all other bovids at a fossil site would indicate the openness of the habitat. This approach has been widely adopted, with minor modifications, for assessing the paleoecology of African sites from the Plio-Pleistocene (e.g., Bunn, 1982; Potts, 1982, 1984; Shipman, 1986; Boaz, 1977a, b).

Inspection of Vrba's (1980) tables comparing abundance of different bovid tribes in different habitats supports our interpretation that two other pairs of tribes also show consistent patterns of abundance in different habitats. Tragelaphines (e.g., kudu, bongo, eland) and aepycerotines (impala) show marked preferences for bushy, closed habitats and tend to be found in high frequencies in game parks and reserves where such habitats are dominant. While eland are also found in more open habitats, we feel the overall preference of the tribe for more closed habitats is sufficient to warrant its use here. Similarly, reduncines (e.g., waterbuck, kob) and bovines (buffalo) are usually found near standing water and in generally wetter, often closed habitats. In making these observations, we echo Vrba's comments about the danger of oversimplifying the issue of habitat preferences by caricaturing habitats and species' behavior in this way. However, Vrba (1984: 64–65) points out that most bovid species are restricted to a habitat that is the norm for their tribe.

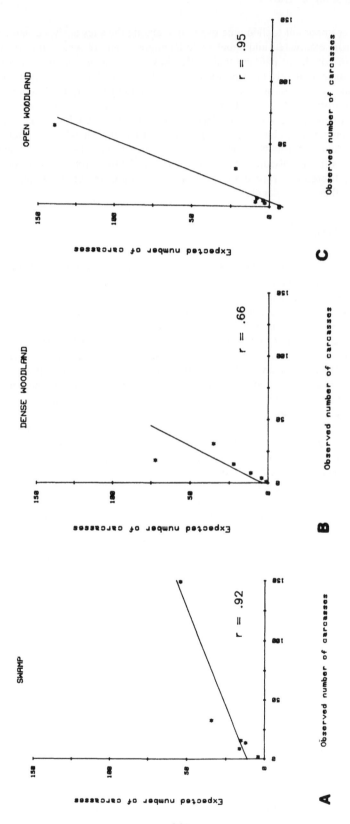

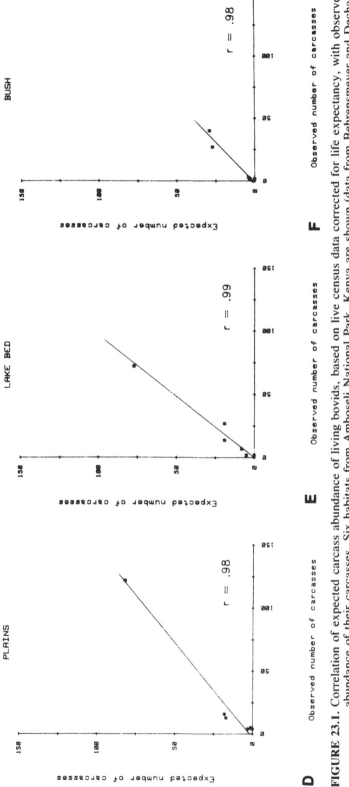

FIGURE 23.1. Correlation of expected carcass abundance of living bovids, based on live census data corrected for life expectancy, with observed abundance of their carcasses. Six habitats from Amboseli National Park, Kenya are shown (data from Behrensmeyer and Dechant Boaz, 1980). (A) Swamp; (B) dense woodland; (C) open woodland; (D) plains; (E) lake bed; (F) bush. See text for discussion.

By using percentage frequency of each of three pair of bovid tribes, we are able to classify habitats more precisely than in Vrba's original approach. However, our approach relies on simplifying the data by considering only six of nine modern bovid tribes. Data on bovid abundances in twenty-nine modern game parks or preserves in central, eastern, western and southern Africa were gathered from the literature (Tables 23.2 and 23.3); sixteen of these were the same as those used by Vrba (1980). The total abundance of all individual bovids in the designated six tribes was calculated for each park and set equal to 100%, and the percentage abundance of each of the pairs—Alcelaphini–Antilopini, Tragelaphini–Aepycerotini, and Reduncini–Bovini—was calculated. Parks were clustered using SPSS/PC+ programs on the basis of these three variables. Both a centroid and a nearest neighbor method were used and produced similar results (a fuller description of the techniques can be found in Norusis, 1986:B70–89).

Table 23.1. Fossil Localities Used in This Study

"Robust" australopithecine sites	Other sites
Olduvai (Fig. 23.4)	
Bed I: FLKNN level 3 (2)	FLKNN level 2 (3)
FLK Zinj (4)	DK (1)
	FLKN levels 1–3 (5)
	FLKN level 4 (6)
	FLKN level 5 (7)
	FLKN level 6 (8)
Bed II: HWK (9)	FLKN II Deinotherium (11)
BK (17)	HWKE levels 1–2 (10)
	HWKE level 3–5 (13)
	HWK EE (12)
	MNK Skull + Main (14)
	FC and FC West (15)
	SHK (16)
Omo Shungura formation members (Fig. 23.5)	
C (2)	B (1)
E (4)	D (3)
F (5)	late G–H (7)
Lower G (6)	J–L (8)
Koobi Fora (Fig. 23.6)	
Upper Burgi 102 (1)	Upper Burgi 12 (6)
Upper Burgi 105 (2)	Upper Burgi 14 (5)
Upper Burgi 123 (3)	Upper Burgi 129 (7)
Upper Burgi 131 (4)	Upper Burgi 130 (8)
	Upper Burgi 115 (9)
	Upper Burgi 116 (10)
	Upper Burgi 104 (11)
	Upper Burgi 100 (12)
KBS 6A (13)	KBS 1A (20)
KBS 102 (14)	KBS 6 (21)
KBS 103 (15)	KBS 8 (22)
KBS 104 (16)	KBS 8A (23)
KBS 105 (17)	KBS 8B (24)
KBS 119 (18)	KBS 9 (25)
KBS 130 (19)	KBS 12 (26)
	KBS 15 (27)
	KBS 101 (28)
	KBS 106 (29)
	KBS 123 (30)
Okote 1 (31)	Okote 103 (35)

Table 23.1. Continued

"Robust" australopithecine sites	Other sites
Okote 1A (32)	Okote 103 (35)
Okote 3 (33)	
Okote 6A (34)	
West Turkana Members (Fig. 23.7)	
Upper Nachukui (4)	Upper Kataboi (1)
Kaitio (7)	Lower Nachukui (2)
	Middle Nachukui (3)
	Lokalalei (5)
	Kalachoro (6)
	Natoo (8)
	Nariokotome (9)
South Africa (Fig. 23.8)	
Swartkrans Member 1 (1)	Sterkfontein Member 4 (3)
Swartkrans Member 2 (2)	Sterkfontein Member 5 (4)
	Sterkfontein West Pit (5)
	Sterkfontein Dumps (6)

*a*Number in parentheses following each site is used to identify point on ternary diagrams (Figs. 23.4–23.8).

Table 23.2. Data for African Wildlife Preserves

Number on Fig. 23.3	Wildlife preserve	Rainfall (mm/yr)	Altitude (m)	Approx. Latitude	Major references
1	Kruger National Park, South Africa	600	300–1000	23°S	Vrba (1980) and references therein
2	Lake Manyara National Park, Tanzania	915	960–1828	4°S	Vrba (1980) and references therein
3	Quiçama, Angola	450	0–265	10°S	Vrba (1980) and references therein
4	Bicuar National Park, Angola	750	1150–1350	15°S	Vrba (1980) and references therein
5	Luando National Park, Angola	1200	1040–1455	11°S	Vrba (1980) and references therein
6	Mupa National Park, Angola (formerly Cuelei)	800	2000–3000	15°S	Vrba (1980) and references therein
7	Mkuzi Game Reserve and Nkwala State Lands, South Africa	700	40–400	28°S	Vrba (1980) and references therein
8	Hluhluwe/Corridor/Umfolozi Game Reserve Complex, South Africa	850	100–550	15°S	Vrba (1980) and references therein
9	Kafue National Park (Ngoma area only), Zambia	1000	970–1470	15°S	Vrba (1980) and references therein
10	Hwange National Park, Zimbabwe (formerly Wankie)	640	938–1152	19°S	Vrba (1980) and references therein

continued

Table 23.2. Continued

Number on Fig. 23.3	Wildlife preserve	Rainfall (mm/yr)	Altitude (m)	Approx. Latitude	Major references
11	Kalahari Gemsbok National Park, South Africa; Kalahari Gemsbok Park, Botswana; Mabua Reserve, Botswana: (all continuous)	150	500–2000	25°S	Vrba (1980) and references therein
12	Nairobi National Park, Kenya	500	1533–1710	1°S	Vrba (1980) and references therein
13	Lake Turkana Game Reserve, Kenya	300	900–2270	4°N	Vrba (1980) and references therein
14	Serengeti Ecological Unit, Tanzania	800	920–1850	2°S	Vrba (1980) and references therein
15	Ngorongoro Crater, Tanzania	750	1500–3000	3°S	Vrba (1980) and references therein
16	Etosha National Park, Namibia	350	1000–1500	18°S	Vrba (1980) and references therein
17	Arli, Burkina Faso	1000	100–500	12°N	Milligan et al. (1982); MacKinnon and MacKinnon (1987); Green (1979); Harroy and Elliot (1971)
18	Deux Bales, Burkina Faso	970	250–310	11°N	Milligan et al. (1982); MacKinnon and MacKinnon (1987); Harroy & Elliot (1971)
19	Po, Burkina Faso	900	200–400	11°N	Harroy & Elliot (1971)
20	Saint-Floris, Central African Republic	1150	400	10°N	Milligan et al. (1982)
21	Bouba Ndjida, Cameroon	1200	350–900	9°N	Afolayan and Ajayi (1980); van Lavieren and Bosch (1977); Esser and van Lavieren (1979); van Lavieren and Esser (1980)
22	Yankari, Nigeria	1000	200–500	8°N	MacKinnon and MacKinnon (1987); Milligan et al. (1982)
23	Fina, Mali	1000	200–500	15°N	MacKinnon and MacKinnon (1987)
24	Pendjari, Benin	1050	100–500	13°N	Green (1979); MacKinnon and MacKinnon (1987)
25	Comoe, Ivory Coast	1200	119–658		Geerling and Bokdam (1973)
26	Kainji, Nigeria	1100	120–339	10°N	Child (1974); Ayeni (1980)
27	W, Niger	730	230–373	12°N	Poche (1973, 1976)
28	Omo National Park,	804	500–2000	5°N	Baba et al. (1982)
29	Timbavati Reserve, South Africa	483	300–450	24°S	Bearder (1977); Hirst (1969)

Table 23.3. Bovid Abundances in 29 Modern Wildlife Areas[a]

Wildlife preserve[b]	Alcelaphini			Antilopini			Aepycerotini	Tragelaphini	
	Alcelaphus	Connachaetes	Damaliscus	Antidorcas	Gazella	Aepyceros	Taurotragus	Tragelaphus
1 Kruger	0	9,360	640	0	0	153,000	424	10,700
2 Manyara	0	0	0	0	0	700	0	25
3 Quiçama	0	0	0	0	0	0	2,500	3,000
4 Bicuar	0	500	0	0	0	150	250	200
5 Luando	0	0	0	0	0	0	200	700
6 Mupa	0	250	0	0	0	0	200	550
7 Mkuzi	0	1,397	0	0	0	9,394	0	533
8 Hluhluwe	0	3,509	0	0	0	4,894	0	3,202
9 Kafue	90	100	0	0	0	2	2	40
10 Hwange	0	2,500	130	0	0	8,000	1,800	3,650
11 Kalahari	14,180	6,378	0	24,040	0	0	6,569	0
12 Nairobi	1,095	253	0	0	845	633	58	20
13 Turkana	0	0	2,023	0	1,059	0	0	0
14 Serengeti	18,000	410,000	27,000	0	190,000	65,000	7,000	2,500
15 Ngorongoro	100	13,530	0	0	5,000	0	400	0
16 Etosha	600	4,000	0	12,000	0	0	500	2,000
17 Arli	100		90					80
18 Deux Bales	453		0					198
19 Po	543		0					108
20 Saint-Floris	907		2,115					—
21 Bouba Ndjida	6,984		144				954	846
22 Yankari	675		—					—
23 Fina	2,975		—					—
24 Pendjari	8,768		599					1,000
25 Comoe	8,000		—					550
26 Kainji	6,436		—					240
27 W	1,020		420					3
28 Omo	—		2,090		646		944	78
29 Timbavati		1,247	—		—	4,659	—	—

continued

Table 23.3. Continued

Wildlife preserve[b]	Reduncini		Bovini	Hippotragini		Neotragini			Cephalophini	
	Kobus	Redunca	Syncerus	Hippotragus	Oryx	Ourebia	Raphicerus	Cephalophus	Sylvicapra
1 Kruger	3,220	1,510	24,200	1,400	0	100	5,000	300	1,000
2 Manyara	15	40	1,500	0	0	0	0	0	0
3 Quiçama	0	1,500	8,000	1,500	0	0	0	2,500	1,000
4 Bicuar	100	50	100	500	0	50	100	0	500
5 Luando	1,750	500	150	3,000	0	250	0	100	1,000
6 Mupa	600	1,000	0	200	0	500	500	0	500
7 Mkuzi	0	69	0	0	0	0	6	1	3
8 Hluhluwe	820	189	2,195	0	0	0	0	0	15
9 Kafue	78	120	85	31	0	50	40	30	40
10 Hwange	1,000	250	13,000	2,500	120	0	3,000	0	2,000
11 Kalahari	0	0	0	0	16,070	0	1,645	0	710
12 Nairobi	90	15	0	0	0	0	2	0	2
13 Turkana	0	0	0	0	1,342	0	0	0	0
14 Serengeti	3,000	2,500	50,000	2,500	2,500	0	0	0	0
15 Ngorongoro	60	60	60	0	0	0	0	0	0
16 Etosha	0	0	0	296	4,500	0	500	0	250
17 Arli	1,185	55	65	1,900	—	1,000	—	—	2,150
18 Deux Bales	176	51	40	1,200	—	650	—	153	498
19 Po	99	188	248	777	—	505	—	—	482
20 Saint-Floris	3,224	—	1,813	504	—	—	—	—	—
21 Bouba Njida	1,386	5,670	1,746	4,356	—	11,736	—	51	5,400
22 Yankari	653	—	675	360	—	—	—	—	—
23 Fina	1,283	—	—	3,938	—	4,613	—	—	—
24 Pendjari	18,915	822	3,480	5,014	—	7,864	—	—	2,466
25 Comoe	7,500	—	500	100	—	17,000	—	6,500	6,500
26 Kainji	4,827	—	118	5,886	—	2,775	—	157	1,256
27 W	6,120	—	4,140	2,850	—	2,100	—	—	—
28 Omo	—	—	350	—	—	—	—	—	—
29 Timbavati	—	33	—	—	—	—	—	—	—

[a]Data from sources in Table 23.2.
[b]See Table 23.2 for data about preserves.

Statistical clustering procedures produce three main groups of wildlife areas, with only two parks failing to cluster closely with the others. Figure 23.2 shows the dendrogram of clustering sites produced by the centroid method. For convenience, the results of the statistical clustering procedures can also be displayed graphically on a three-component or ternary diagram (Fig. 23.3). The cluster in the lefthand corner of Fig. 23.3 represents game parks with a high percentage of Alcelaphini and Antilopini (the AA cluster); that in the upper corner represents areas with a high percentage of Tragelaphini and Aepycerotini (the TA cluster); and that in the righthand corner represents areas with high percentages of Reduncini and Bovini (the RB cluster). Two parks (points 4 and 18 on Fig. 23.2 and 23.3) do not group strongly and unambiguously with one of these three associations.

The arcs drawn onto Fig. 23.3 indicate the approximate location of boundaries between the habitat clusters. It must be emphasized that Fig. 23.3 is only a graphic presentation of statistically defined entities. That is, membership in a cluster is defined by statistical procedures, not by visual inspection of ternary diagrams. Thus, for example, points 23 and 26, which appear to be close to each other on Fig. 23.3, are well-separated in statistical clustering procedures. These points do not cluster in the same group until the seventh of eight iterations using a centroid clustering technique (see Fig. 23.2).

As a test of the extent to which these clusters reflect distinct habitats, they were compared with an independently assigned set of classifications of 28 of the 29 modern parks conducted by MacKinnon and MacKinnon (1987), using a detailed classification of African vegetative types produced by White (1983). White describes 21 African phytochoria, or centers of floral endemism, subdivided into 77 specific vegetation categories. MacKinnon and MacKinnon give data on the

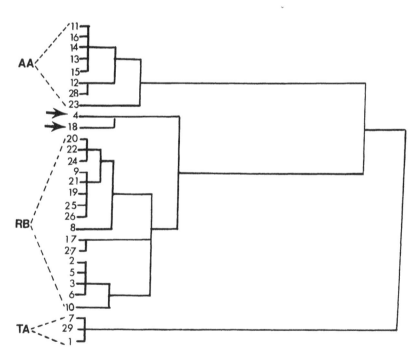

FIGURE 23.2. Dendrogram of twenty-nine modern African wildlife areas clustered using squared Euclidean distances and the centroid method on the basis of the abundance of Alcelaphini + Antilopini (AA), Tragelaphini + Aepycerotini (TA), and Reduncini + Bovini (RB) (see Tables 23.2 and 23.3 for details). Three main groups are formed, each having a high percentage of one pair of tribes; arrows indicate two outliers that do not cluster closely with any group. A similar grouping is obtained if a nearest neighbor (single linkage) clustering technique is used.

29 MODERN AFRICAN HABITATS

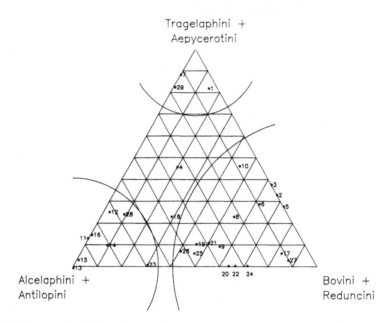

FIGURE 23.3 Data on bovid abundances from twenty-nine modern African wildlife areas plotted on a ternary diagram (see Tables 23.2 and 23.3 for data, and text for explanation). Arcs indicate approximate boundaries of three clusters of parks: the AA cluster (lefthand corner), the TA cluster (upper corner), and the RB cluster (righthand corner). On the basis of primary ecological information about these areas (Table 23.4), the AA cluster is considered to represent open/arid habitats, the TA cluster samples closed/dry habitats, and the RB cluster indicates closed/wet habitats.

parks both by phytochorion and vegetation categories, the latter being presented on the basis of area covered (Table 23.4).

Comparison of their habitat classifications with our cluster analysis reveals a remarkable degree of agreement. If the single wildlife area (Timbavati) not classified by MacKinnon and MacKinnon and the two ambiguous parks are removed from the data set, then 23 (88%) of the remaining 26 parks cluster with the same suite on the basis of phytochoria and vegetation classifications as in our clustering based on bovid abundances. [In the discussion that follows, all statements about habitats or vegetation are taken from MacKinnon and MacKinnon (1987) unless otherwise indicated.]

Two sources of data were used to characterize the habitats indicated by these clusters. First, the MacKinnons' classifications of the parks in terms of phytochorion, vegetation types, and coverage were inspected for consistencies that might characterize the parks that fall within the same cluster. Second, the mean rainfall of each of the three clusters was calculated as a crude way of assessing the wetness of the habitat (using data presented in Table 23.2). Below, an attempt is made to characterize each cluster in terms of habitat; exceptions to these generalizations are presented after the habitat sketches.

All of the parks in the AA cluster had a major portion of their area classified as belonging to either the Somalia–Masai or the Kalahari–Highveld phytochorion. The dominant vegetation types in the Somalia–Masai phytochorion are deciduous bushland and thickets, with the *Acacia–Commiphora* association being common; large regions of this phytochorion are very dry and support semidesert grassland and shrubland. The Kalahari–Highveld is a complex mosaic of woodland, bushland, and scrubland with significant areas of grassland. Thus, in general terms, these phytochoria represent relatively open, often arid habitats. Small areas of two individual parks also

Table 23.4. Phytochorion Affinities and Vegetation of Modern Wildlife Areas[a]

Wildlife area	Phyto-chorion	Vegetation type	Description[b]	Area of animal census (km^2)	Area of vegetation classification (km^2)	Remarks
Kruger	Zambezian	34% Woodland 29d	South Zambezian undifferentiated woodland; arid lowveld and scrub woodland (p. 96)	19,084	19,484	
		44% Woodland 28	Zambezian mopane woodland dominated by *Colophospermum*; scrub woodland (p. 94)			
	Tongaland–Pondoland	23% Woodland 29e	Evergreen and semi-evergreen bushland and thicket; *Aloe* and *Euphorbia* common (p. 200–201)			
Manyara	Somalia–Masai	100% Grassland 59	Edaphic grassland on volcanic soils; seasonally waterlogged; may occur as glades within *Acacia*–*Commiphora* bushland (p. 116)	91	320	Entire area of park apparently not included in animal census
Quiçama	Zambezian	100%	North Zambezian undif-	9,960	9,960	

continued

Table 23.4. Continued

Wildlife area	Phyto-chorion	Vegetation type	Description[b]	Area of animal census (km^2)	Area of vegetation classification (km^2)	Remarks
		Woodland 29c	ferentiated woodland and wooded grassland lacking miombo or mopane; common species include *Afzelia quanzensis, Burkea africana, Dombeya rotundiflora, Pericopsis angolensis, Pseudolachnostylis maprouneifolia, Pterocarpus angolensis,* and *Terminalia sericea* (p. 95)	9,960	9,960	
Bicuar	Zambezian	11% Forest/Grassland 22a	Zambezian mosaic of dry deciduous forest (*Baikiaea plurijuga*) and secondary grassland of *Brachiaria brizantha, Dactyloctenium giganteum, Digitaria milanjiana, Panicum maximum* and *Sporobolus pyramidalis* (p. 101)	7,900	7,900	
		76% Woodland 25	Zambezian miombo woodland dominated by *Brachystegia* alone or with *Julbernardia* or *Isoberlinia* (p. 93)			

356

Luando	Zambezian	13% Woodland 28 100% Woodland 25	Zambezian mopane woodland (p. 94) Zambezian miombo woodland (p. 93)	8,280	8,280	
Mupa	Zambezian	30% Forest/Grassland 22a 30% Woodland 25 40% Woodland 28	Zambezian mosaic of dry deciduous forest and secondary grassland (p. 101) Zambezian miombo woodland (p. 93) Zambezian mopane woodland (p. 94)	4,500	6,600	
Mkuzi	Tongaland–Pondoland	55% Forest/Grassland 16c 45% Woodland 29e	Coastal mosaic of the Zanzibar–Inhambane type; undifferentiated forest with 10-30 m canopy and badly drained grasslands; scattered palms (pp. 199–200, 202) Evergreen and semi-evergreen deciduous bushland and thicket (p. 116)	323	251	Nkwala State Lands not included in vegetation classification
Hluhluwe	Tongaland–Pondoland	6% Forest/Grassland 16c 94% Woodland 29e	Coastal mosaic; Zanzibar-Inhambane type (pp. 199–200, 202) Evergreen and semi-evergreen deciduous bushland and thicket (p. 116)	960	709	Corridor area not included in vegetation classification

continued

Table 23.4. Continued

Wildlife area	Phyto-chorion	Vegetation type	Description[b]	Area of animal census (km²)	Area of vegetation classification (km²)	Remarks
Kafue	Zambezian	2% Forest 6	Zambezian dry evergreen forest; rarely >25 m height; mean annual rainfall 900–1200 mm/yr; dominated by *Berlinia giorgii, Cryptoscpalum pseudotaxus, Daniellia alsteeniana, Entandrophragma delevoyi, Marquesa acuminata, M. macroura, Parinari excelsa* and *Syzygium guineese* (p. 89–90)	91	22,400	Animal census is only Ngoma area but vegetation classification is entire park
		89% Woodland 25	Zambezian miombo woodland (p. 93)			
		9% Grassland 64	Edaphic grassland mosaic with semiaquatic vegetation; seasonally waterlogged river valleys or floodplains; mosaic of permanent swampland, valley grassland, and "bushgroup" grassland (p. 100)			
Hwange	Zambezian	49% Forest/Grassland 22a	Zambezian mosaic of dry deciduous forest and secondary grassland (p. 101)	13,300	14,650	

Kalahari	Kalahari–Highveld	36% Bushland 44	Kalahari deciduous *Acacia* bushland and wooded grassland; widely spaced trees <7 m tall; plentiful shrubs or grass (p. 193)	36,692	35,183
		64% Semi-desert 56	Kalahari/Karoo–Namib transition; mosaic of lightly wooded grassland on dune crests, pure grassland in shallow depressions between dunes, and *Rhigozum trichotomum* shrubby grassland in deeper hollows (p. 193)		
Nairobi	Somalia–Masai	100% Bushland/Grassland 45	Mosaic of East African evergreen bushland and secondary *Acacia* wooded grassland; ecotone between forest and grassland in Nairobi Park is altered by grazing, burning, and browsing to create more grassland in area capable of supporting bushland or forest; *Acacia–Commiphora* bushland (p. 116)	122	117

Mbua Reserve in animal census but not in vegetation classification

continued

Table 23.4. Continued

Wildlife area	Phyto-chorion	Vegetation type	Description[b]	Area of animal census (km²)	Area of vegetation classification (km²)	Remarks
Turkana	Somalia–Masai	73% Bushland 42	*Acacia–Commiphora* deciduous bushland and thicket; low bushes and stunted trees, principally *Acacia reficiens misera* forming a thin cover over ground layer of small shrubs and ephemeral grasses (pp. 113–114)	2,050	1,091	Census from narrow grassy strip close to lake
		27% Bushland/Grassland 45	Mosaic of East African evergreen bushland and secondary *Acacia* wooded grassland (pp. 116, 123–130)			
Serengeti	Somalia–Masai	73% Bushland 42	*Acacia–Commiphora* deciduous bushland and thicket or wooded grassland (pp. 128–129)	25,500	18,463	Part of Serengeti ecosystem included in animal census is area surrounding Ngorongoro; in vegetation classification, this area is lumped with Ngorongoro, not Serengeti
		19% Grassland 59	Edaphic grasslands on volcanic soils (pp. 116, 125–126)			
		8% Bush-	Mosaic of East African			

		land/Grassland 45	evergreen bushland and secondary *Acacia* wooded grassland (p. 116)		
Ngorongoro	Somalia–Masai	48% Bushland 42	*Acacia–Commiphora* deciduous bushland and thicket or wooded grassland (pp. 128–129)	360	6,280	Vegetation classification includes areas surrounding crater; animal census was carried out in the crater only
		28% Grassland 59	Edaphic grassland on volcanic soils (pp. 125–126)			
	Afromontane	24% Montane 19a	Undifferentiated vegetation; montane heath, forest, secondary thicket, grassland and woodland complex (pp. 129–130, 165)			
Etosha	Zambezian	89% Woodland 28	Zambezian mopane woodland (p. 94)	22,270	21,200	
		11% Halophytic 76	Halophytic vegetation in saline soils, usually semi-arid and arid regions; principal vegetation is *Suaeda articulata, Atriplex vestita, Sporobolus spicatus, S. tenellus, S. virginicus,* and *Odyssea paucinervis* surrounded by dwarf shrub zone (p. 267)			

continued

Table 23.4. Continued

Wildlife area	Phyto-chorion	Vegetation type	Description[b]	Area of animal census (km²)	Area of vegetation classification (km²)	Remarks
Arli	Sudanian	100% Woodland 29a	Undifferentiated Sudanian woodland; probably originally *Isoberlinia* dominated but heavily disturbed by cultivation; open woodland; common species: *Afzelia africana, Burkea africana, Anogeissus leiocarpus, Pteleopsis suberosa, Combretum glutinosum, C. nigricans, Pericopsis laxiflora, Lonchocarpus laxiflorus, Terminalia avicennioides, T. laxiflora, Lannea schimperi* and, in rocky areas, *Detarium microcarpum* (p. 106–107)	1,000	2,330	
Deux Bales	Sudanian	50% Woodland 30	Undifferentiated Sudanian woodland with islands of *Isoberlinia*, especially on rocky hills (p. 106)	566	560	
		50% Woodland 29a	Undifferentiated Sudanian woodland (pp. 106–107)			

Po	Sudanian	100% Woodland 29a	Undifferentiated Sudanian woodland (pp. 106–107)	468	1,494	
Saint-Floris	Sudanian	100% Woodland 27	Sudanian woodland with abundant *Isoberlinia* lacking *Brachystegia* or *Julbernardia*; only one species each of *Monotes* and *Uapaca*; woodland rarely more than 15 m tall (p. 106)	1,007	1,300	
Bouba Ndjida	Sudanian	9% Woodland 27	Sudanian woodland with abundant *Isoberlinia* (see above; also, p. 106)	1,800	2,200	
		91% Woodland 29a	Undifferentiated Sudanian woodland (p. 106–107)			
Yankari	Sudanian	100% Woodland 30	Undifferentiated Sudanian woodland with islands of *Isoberlinia* (pp. 106–107)	2,250	2,240	
Fina	Sudanian	100% Woodland 29a	Undifferentiated Sudanian woodland (pp. 106–107)	2,250	1,360	
Pendjari	Sudanian	100% Woodland 29a	Undifferentiated Sudanian woodland (p. 106–107)	2,750	2,755	
Comoe	Sudanian	83% Woodland 27	Sudanian woodland with abundant *Isoberlinia* (p. 106)	12,500	11,500	Protected area recently expanded
	Guinea–Congoland/ Sudanian	17% Forest/ Grassland 11a	Mosaic of Guinea–Congoland rain forest and secondary grassland;			

continued

Table 23.4. *Continued*

Wildlife area	Phytochorion	Vegetation type	Description[b]	Area of animal census (km²)	Area of vegetation classification (km²)	Remarks
	transition		diverse forest with canopy >30 m with small patches grass and thicket about 2 m tall (pp. 84–5)			
Kainji	Sudanian	100% Woodland 27	Sudanian woodland with abundant *Isoberlinia* (see above; also p. 106–107)	3,924	5,341	
W	Sudanian	100% Woodland 29a	Undifferentiated Sudanian woodland (see above; also p. 106–107)	3,000	2,200	
Omo	Somalia–Masai	35% Bushland 42	Somalia–Masai *Acacia–Commiphora* deciduous bushland and thicket; dense bushland 3–5 m tall with scattered emergent trees; Capparidaceae and *Grewia* nearly always present (p. 113–114)	4,015	3,465	

Sudanian	36% Woodland 29b	Ethiopian undifferentiated woodland dominated by *Anogeissus leiocarpus, Combretum collinum, C. hartmannianum, Balanites aegytiaca, Boswellia papyrifera, Commiphora africana, Dalbergia melanoxylon, Erythrina abyssinica, Gardenia terniflora (lutea), Lannea schrimperi, Lonchocarpus laxiflorus, Piliostigma thonningii, Stereospermum kunthianum,* and *Terminalia brownii* (p. 107)
Afromontane	29% Grassland 17	Cultivation and secondary grassland replacing upland montane forest (p. 162)

[a] Areal data are taken from MacKinnon and MacKinnon (1987) using vegetation categories and descriptions from White (1983).
[b] Page numbers are from White, 1983. See also discussion in text of this chapter.

sampled the Afromontane phytochorion, typically covered in high altitude bamboo and evergreen forests, *Hagenia* woodland, or a grassland/alpine complex. Specifically, each of the parks in the open cluster had from about 80 to 100% of its area given over to grassland and/or bushland of various types. T-tests on the data in Table 23.2 show that parks in the AA cluster also had significantly lower rainfall than those in the other clusters combined or than the RB cluster (AA vs TA + RB: student's t = 2.46, p = .01; AA vs RB: student's t = 2.96, p = .007), but not lower than the TA cluster. Thus, the AA cluster is an open habitat that is drier than the RB cluster.

Both parks in the TA cluster had a significant portion of their area in the Tongaland–Pondoland Mosaic phytochorion. This region is characterized as a coplex mosaic of forest, scrub forest, evergreen, and semi-evergreen bushland and thicket in a matrix of secondary grassland and wooded grassland. Within this phytochorion, both parks are dominated by woodland of various types, especially category W29e which covers 23% of Kruger and 45% of Mkuzi. Data from Table 23.2 show that rainfall in the TA cluster is significantly lower than in the RB cluster (student's t = 36, p < .001) but is not different from the AA cluster. Thus, the TA cluster can be characterized as a closed habitat that is as dry as that in the AA cluster.

With one exception, all parks in the RB cluster derive from either the Zambezian or the Sudanian phytochorion. The Zambezian phytochorion is characterized as one of three woodland types: the miombo, high altitude scrub type; the mopane type; or the Zambezian undifferentiated woodland type. The Sudanian phytochorion is described as dry forest and woodland with small areas of swamp forest or riparian forest and very little bushland or thicket. With only one exception, wet habitat parks show 45–100% woodland vegetation within one of these two phytochoria, with all but one park being over 70% woodland. Woodland types 25, 28, and 29a are especially common. Grassland or forest/grassland mosaics occur in a few wet habitat parks in low proportions (6–30%). Thus, although both TA and RB clusters are dominated by woodland, the specific types of woodland (i.e., combinations of species or genera represented) vary. This may be related to the general wetness of the habitats; rainfall in the RB cluster is higher than in either the AA or TA clusters (p <.01 in each case). Thus, the RB cluster can be characterized as a closed/wet habitat.

The exceptions to these generalizations include two parks classified as AA that do not show very high percentages of bushland or grassland and one classified as RB that was cited by MacKinnon and MacKinnon (1987) as entirely grassland. The two AA cluster exceptions were Etosha and Fina. Etosha National Park, according to MacKinnon and MacKinnon (1987), has 89% woodland and only 11% halophytic scrubland. Probably the low rainfall and the limitations of the soil type account for the grouping of Etosha with the more open, arid parks, despite an ecological classification that implies more closed habitat than is usual in the AA group. Fina National Park, in Mali, is entirely woodland and has a high rainfall combined with a high frequency of hartebeest (alcelaphines). Although the bovid and vegetation data produce different clusterings, it is heartening that Fina falls closest to the RB cluster of any of the AA cluster parks. The final exception concerns Lake Manyara National Park in Tanzania, which has many reduncines and bovines and a high rainfall but which is described by the MacKinnons as 100% edaphic grassland on volcanic soils. Our personal experience and data presented in other publications (i.e., Douglas-Hamilton, 1972; Mwalyosi, 1977), suggest that the MacKinnons' classification seriously underestimates the percentage of scrubland and groundwater forest, dominated by *Acacia tortillus* and *Trichikia roka,* in the park. Other sources suggest that perhaps 30–50% of Manyara is covered in closed, wet habitats (scrubland and forest), which would bring the vegetation classification into close agreement with other parks in the RB cluster.

In summary, remarkably few of the parks would be grouped in different clusters on the basis of two independent methods of habitat classification carried out by two different groups of researchers. Given that habitats grade into one another, that many parks sample a variety of different microhabitats, that three tribes of bovids are eliminated in our analyses, and that bovid habitat preferences may be somewhat flexible, these results suggest that the technique used here is a very powerful means of characterizing known habitats and of deducing habitat information for fossil assemblages with a substantial component of bovids.

Clearly, ecologic classification of African habitats is subtle and complex. The reduction of this variability into three habitat clusters risks oversimplification, but in the context of fossil localities, the risk may be offset by a recognition of the limits of resolution of the approach and its possible errors. The variability of habitats *within* clusters, as least as observed in the modern wildlife areas, is considerable. Thus, for example, the AA cluster includes the Kalahari Gemsbok complex of parks, Fina in Mali, and Ngorongoro Crater; similar variability can be documented within the closed/wet cluster and would probably occur in the closed/dry cluster if more than three parks were represented. This means that two fossil localities falling within the same cluster are only crudely "the same" in habitat; their actual habitats may have been as diverse as the modern members of any single habitat cluster.

As a shorthand means of indicating general habitat types, the AA cluster will be referred to here as representing open/arid habitats, the TA cluster as closed/dry habitats, and the RB cluster as representing closed/wet habitats.

Data on the minimum number of individual (MNI) bovids were compiled for all of the fossiliferous sites listed above (Table 23.5) and percentage abundance for each of the three designated pairs were calculated. Fossil sites with a total MNI within the six designated tribes of less than five were eliminated from the sample. The data on 27 of the 29 modern habitats were then used to predefine three clusters or habitat groups [the 2 outliers in Figs. 23.2 and 23.3 (points 4 and 18) were excluded]. Discriminant analysis was then used to classify fossil localities into these preexisting clusters; the probability that each unknown fit into the nearest two predefined groups was calculated using Bayes' rule (Norusis, 1986:B10).

Results

Data on 69 fossil localities from East Africa and six localities from South Africa (Table 23.5) were classified using the discriminant analysis program. Twenty-five (37%) of the total sample of East African localities and two (33%) of the South African localities have yielded "robust" australopithecine remains. The results are shown graphically on ternary diagrams in Figs. 23.4 through 23.8. Table 23.6 shows the modern habitat that is the nearest neighbor to each fossil site that has yielded "robust" australopithecine remains. These results are discussed on a site-by-site basis below.

Olduvai

Minimum number of individual bovids were available for seventeen localities from Beds I and II, of which four (26%) yielded *A. boisei* material. One of the four *A. boisei* sites yielded only a single, surface collected specimen and was therefore regarded as possibly less reliable. Thirteen of the Olduvai localities had a high probability (77 to 100%) of belonging in the open/arid cluster (Fig. 23.4); none belonged in the closed/dry habitat cluster. The remaining four localities were all most likely to belong to the closed/wet habitat cluster, with two having a very high probability (99–100%) and two having a somewhat lower probability (63 and 70%). The locality that yielded only a surface *A. boisei* specimen, HWK, had a 99% probability of falling within the open/arid cluster. In contrast, the three localities that yielded *in situ* remains of "robust" australopithecines–FLKNN level 3 (the controversial OH 8 foot), FLK Zinj (OH 5), and BK (OH 3)—all tend to be classified into the closed/wet habitat group, with probabilities of 100, 63, and 70%, respectively. Figure 23.4 shows that these sites fall along the closed/wet–open axis of the three-component diagram. In summary, Olduvai samples a diverse set of habitats ranging mostly from open/arid to closed/wet; the surface find of *A. boisei* occurs in a clearly open/arid habitat and the *in situ* remains occur in distinctly more closed and often wetter habitats. Thus, the Olduvai data do not unambiguously suggest a strong habitat preference for "robust" australopithecines.

Omo

Data on bovid abundances were available from eight Shungura Members or combinations of members from the Omo succession (B, C, D, E, F, G 1 through 13, late G through H, and J

Table 23.5. Abundances of Bovid Tribes in Fossil Localities (Minimum Number of Individuals)[a]

	Alcel.	Ant.	Trag.	Aepy.	Red.	Bov.	Hippo.	Neo.	Cepha.	Capr.	Ovi.	Total
Olduvai												
FLKNN level 2	2	1	1	0	7	0	2	0	0	0	0	13
FLKNN level 3	1	4	1	0	10	0	0	0	0	0	0	16
DK	13	8	8	0	6	3	4	0	0	0	0	29
FLK Zinj	9	7	2	0	9	1	1	0	0	0	0	29
FLKN levels 1–3	32	23	2	0	1	1	1	0	0	1	0	60
FLKN level 4	7	3	2	0	2	0	0	1	0	0	0	15
FLKN level 5	12	16	3	0	0	0	0	1	0	0	0	32
FLKN level 6	8	5	6	0	0	1	1	0	0	0	0	21
HWK	4	2	0	0	1	0	0	0	0	0	0	7
FLKNII Deinotherium	3	2	1	0	1	0	0	0	0	0	0	7
HWKE levels 1–2	20	5	1	0	3	2	1	0	0	0	0	32
HWKE levels 3–5	9	4	0	0	0	1	1	0	0	0	0	15
HWK EE	8	2	1	0	1	0	0	0	0	0	0	12
MNK Skull + Main	15	1	1	0	2	5	1	0	0	0	0	25
FC and FC West	2	1	0	0	1	2	1	1	0	0	0	8
SHK	14	12	4	0	1	4	2	0	0	0	0	37
BK	30	4	1	1	2	22	3	0	0	0	0	65
Omo												
B	11	2	56	101	173	52	0	0	0	0	0	395
C	14	3	268	171	152	109	1	1	0	0	1	720
D	8	0	90	65	41	19	0	0	0	0	1	224
E	14	0	196	136	127	20	0	1	0	0	0	494
F	101	5	185	227	149	53	0	1	0	0	0	721
Lower G	106	7	1031	979	2347	85	1	14	0	0	1	4571
Late G–H	41	6	33	76	260	5	1	8	0	0	0	430
J–L	92	3	9	54	156	7	0	0	0	0	0	321
Koobi Fora												
Upper Burgi 12	1	1	3	1	5	0	0	0	0	0	0	11
Upper Burgi 14	13	0	7	4	23	0	0	0	0	0	0	47

Upper Burgi 100	10	4	8	11	28	4	0	0	0	0	65
Upper Burgi 102	6	4	2	1	10	2	1	0	0	0	26
Upper Burgi 104	3	1	4	0	7	0	0	0	0	0	15
Upper Burgi 105	22	16	17	2	24	2	1	0	0	0	84
Upper Burgi 115	3	1	8	0	2	0	0	0	0	0	14
Upper Burgi 116	3	2	0	0	6	1	0	0	0	0	12
Upper Burgi 123	0	0	0	0	5	0	0	0	0	0	5
Upper Burgi 129	2	3	3	1	8	3	0	0	0	0	20
Upper Burgi 130	7	6	6	1	51	1	0	0	0	0	72
Upper Burgi 131	3	7	7	14	20	0	1	0	0	0	52
KBS 1a	0	0	2	0	3	1	0	0	0	0	6
KBS 6	4	5	1	1	7	0	0	0	0	0	18
KBS 6A	4	0	5	1	9	2	1	0	0	0	22
KBS 8	5	2	2	3	9	2	1	0	0	0	24
KBS 8A	5	0	1	0	4	2	0	0	0	0	12
KBS 8B	10	1	3	0	8	4	0	0	0	0	26
KBS 9	1	0	2	0	3	0	0	0	0	0	6
KBS 12	6	3	3	2	13	1	0	0	0	0	28
KBS 15	7	2	2	0	5	3	0	0	0	0	19
KBS 101	7	2	1	1	5	3	0	0	0	0	19
KBS 102	11	3	8	6	46	8	4	0	0	0	86
KBS 103	53	16	22	4	60	11	1	1	0	0	168
KBS 104	22	6	22	16	36	7	0	0	0	0	109
KBS 105	15	2	5	1	10	4	1	0	0	0	38
KBS 106	1	1	1	3	9	1	0	0	0	0	16
KBS 119	5	2	4	0	12	1	0	0	0	0	24
KBS 123	10	11	16	12	36	2	0	0	0	0	87
KBS 130	2	1	1	0	4	0	0	0	0	0	8
Okote 1	8	0	7	1	21	0	0	0	0	0	37
Okote 1A	2	2	5	2	17	3	0	0	0	0	31
Okote 3	3	1	4	0	10	0	0	0	0	0	18
Okote 6A	4	1	4	0	0	1	0	0	0	0	10
Okote 103	6	3	7	1	6	6	0	0	0	0	29

continued

Table 23.5. Continued

	Alcel.	Ant.	Trag.	Aepy.	Red.	Bov.	Hippo.	Neo.	Cepha.	Capr.	Ovi.	Total
West Turkana												
Lower Kataboi	0	0	0	0	0	0	0	0	0	0	0	0
Upper Kataboi	3	2	2	4	2	3	0	0	0	0	0	13
Lower Nachukui	34	4	10	41	13	3	1	0	0	0	0	105
Middle Nachukui	4	12	0	7	6	4	0	0	0	0	0	33
Upper Nachukui	7	19	5	9	51	2	1	0	0	0	0	94
Lokalalei	1	1	3	1	11	1	0	0	0	0	0	18
Kalachoro	12	12	4	5	27	4	0	0	0	0	0	64
Kaitio	2	22	27	11	12	10	1	0	0	2	0	84
Natoo	13	7	4	5	13	7	0	0	0	2	0	51
Nariokotome	4	0	2	3	6	2	0	0	0	2	0	19
South Africa												
Swartkrans Member 1	18	12	4	0	6	3	1	0	1	0	3	48
Swartkrans Member 2	30	86	11	0	9	0	9	11	0	0	0	147
Sterkfontein Member 4	17	5	1	0	1	1	8	0	0	0	8	41
Sterkfontein Member 5	13	3	1	0	0	0	1	1	0	0	1	20
Sterkfontein West Pit	14	5	1	0	0	0	1	1	0	0	1	23
Sterkfontein Dumps	18	9	3	0	3	0	4	5	0	0	0	42

[a]Key to abbreviations: Alcel, Alcelaphini; Ant, Antiliopini; Trag, Tragelaphini; Aepy, Aepycerotini; Red, Reduncini; Bov, Bovini; Hippo, Hippotragini; Neo, Neotragini; Cephal, Cephalophini; Capr, Caprini; Ovi, Ovibovini.

Table 23.6. Closest Modern Habitat Equivalents to Fossil Localities Bearing "Robust" Australopithecines Based on Bovid Abundances

Fossil locality	Best modern analog
Olduvai	
FLKNN level 3	Saint-Floris, Central African Republic
FLK Zinj	Kainji, Nigeria
HWK	Serengeti Ecological Unit, Tanzania
BK	Kainji, Nigeria
Omo Shungura Formation Members	
C	Hwange National Park, Zimbabwe
E	Kruger National Park, South Africa
F	Hwange National Park, Zimbabwe
G 1–13	Hwange National Park, Zimbabwe
West Turkana Members	
Upper Nachukui	Hluhluwe/Corridor/Umfolozi Game Reserve Complex, South Africa or Kafue National Park (Ngoma area only), Zambia
Kaitio	Bicuar National Park, Angola
Koobi Fora	
Upper Burgi 102	Bouba Ndjida, Cameroon
Upper Burgi 105	Deux Bales, Burkina Faso
Upper Burgi 123	W, Niger
KBS 6A	Hluhluwe/Corridor/Umfolozi Game Reserve Complex, South Africa
KBS 102	Mupa Regional Nature Park, Angola
KBS 103	Po, Burkina Faso
KBS 104	Hwange National Park, Zimbabwe
KBS 105	Deux Bales, Burkina Faso
KBS 119	Kafue National Park (Ngoma area only), Zambia
KBS 130	Kafue National Park (Ngoma area only), Zambia
Okote 1	Hluhluwe/Corridor/Umfolozi Game Reserve Complex, South Africa
Okote 1A	Luando National Park, Angola
Okote 3	Hluhluwe/Corridor/Umfolozi Game Reserve, South Africa
Okote 6	Bicuar National Park, Angola
South Africa	
Swartkrans Member 1	Ngorongoro Crater or Serengeti Ecological Unit, Tanzania
Swartkrans Member 2	Kalahari Gemsbok National Park, South Africa; Kalahari Gemsbok Park, Botswana; and Mabua Reserve, Botswana (all continuous)

through L) based on published data by Gentry (1985). These differed from the data for the other sites in that MNIs were not presented; only the number of specimens was given. However, it has been assumed previously that each Omo specimen probably represents a different individual due to the fragmentary nature of nearly all of the remains; that assumption is followed here in the absence of indications to the contrary. At least four Members from Omo (C, E, F and lower G) clearly contain "robust" australopithecines (Walker and Leakey, this volume, chapter 15;

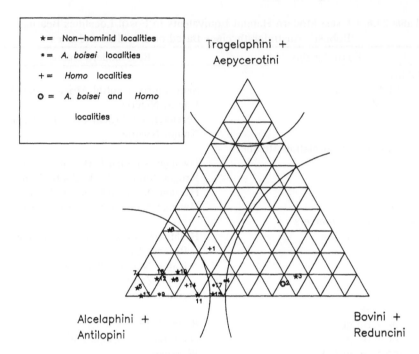

FIGURE 23.4. Localities from Olduvai Beds I and II plotted according to bovid abundances. As in Fig. 23.3, arcs indicate approximate boundaries of the three habitat clusters.

Howell, 1978); it has been argued that D (Grine, 1985) and K (White, this volume, chapter 27)—represented in these data as J through L—also contain "robust" specimens.

Four of the Omo Members group with a very high probability (99 to 100%) in the closed/dry habitat cluster and four group with equally high probability with the closed/wet cluster (Fig. 23.5). No localities fall in the open/arid cluster. Three of the "robust" localities fall into the closed/dry cluster and the fourth falls in the closed/wet cluster. Member D falls within the closed/dry cluster and J through L falls in the closed/wet cluster, thus preserving the even split between closed/dry and closed/wet cluster sites. Thus, the Omo succession samples only closed habitats of varying wetness and the "robust" remains occur in these.

Koobi Fora

As with the Omo, the bovid data from Koobi Fora represent numbers of specimens rather than calculated MNIs. Since one of us (J. H.) personally supervised the collection of these specimens, we are reasonably confident that they represent different individuals. Thirty-five localities from Koobi Fora were analyzed, of which fifteen contain *A. boisei* remains. Thirty-three localities fall into the closed/wet and two into the closed/dry group, all with probabilities greater than 93% (Fig. 23.6). No locality falls within the open/arid group. All but one of the *A. boisei* localities fall within the closed/wet cluster and the remaining locality is in the closed/dry cluster group. Thus, like the Omo, the Koobi Fora localities sample a restricted range of habitats. The *A. boisei* sites from Koobi Fora overwhelmingly represent closed/wet habitats.

West Turkana

As with Koobi Fora, the bovid data from West Turkana represent numbers of specimens rather than calculated MNIs but are judged likely to be insignificantly different from MNIs for the same reasons as at Koobi Fora. Eight localities from West Turkana were analyzed, two of which have yielded "robust" australopithecine remains. Six localities fall within the closed/wet

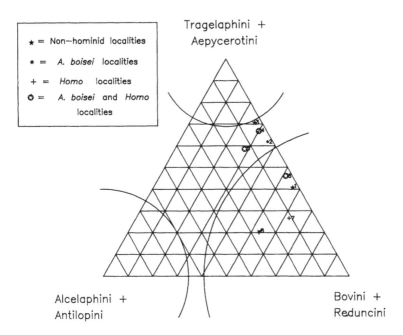

FIGURE 23.5. Localities from Omo Shungura Formation Members B through L, plotted according to bovid abundances. Arcs as in Fig. 23.3. Note that the range of habitats sampled falls along the closed axis from dry to wet.

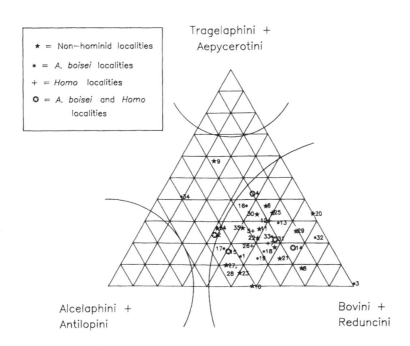

FIGURE 23.6. Localities from Koobi Fora plotted according to bovid abundances. Arcs as in Fig. 23.3. Note that the habitats sampled are nearly all closed/wet.

cluster and one falls within the closed/dry cluster, with very high probability (>99%) (Fig. 23.7). The remaining site is ambiguous, having a 55% probability of grouping with the closed/dry cluster and 45% of grouping with the closed/wet cluster. One of the two *A. boisei* localities is a closed/wet habitat and the other is the ambiguous locality. Again, the habitats sampled are restricted in the West Turkana region to closed/dry and closed/wet habitats and perhaps to a mixed habitat lying along the closed/dry–wet axis. The "robust" remains are distributed in the closed/wet habitat and in the possibly mixed site.

South Africa

Six localities were analyzed, of which two have yielded *A. robustus* remains. Data are calculated MNIs. All six plot with very high probability (>99%) in the open/arid group and, therefore, all sites yielding "robust" australopithecine remains fall within that habitat (Fig. 23.8). Like Koobi Fora, West Turkana, and the Omo, the South African cave sites sample a very limited range of habitats. Unlike the East African sites, the range in South Africa is entirely restricted to quite open habitats, whether *A. robustus* is present or not.

These results differ somewhat from those reported by Vrba (1976), in which she concluded that the Sterkfontein Type Site (Members 4 and 5 together) and the D 16 Sterkfontein dumps sampled relatively more closed habitats whereas the Swartkrans sites sampled more open habitat species. Her conclusions were based on the representation of Alcelaphini + Antilopini relative to all other bovids. Vrba (1976: 60–61) notes that the large numbers of *Makapania* and hippotragines in Sterkfontein, where they are common species, cause the relative paucity of alcelaphines and antilopines, but it is unclear what the rise in *Makapania* and hippotragines indicates in terms of habitat. Vrba considers it unlikely that these were open country species like antilopines and alcelaphines. Because our analysis uses the relative proportions of only six tribes and excludes ovibovines and hippotragines altogether, the paleoecological information their changing abundances might reveal is lost. At the level of habitat classification that our technique provides, there is little significant habitat variability among the South African cave sites.

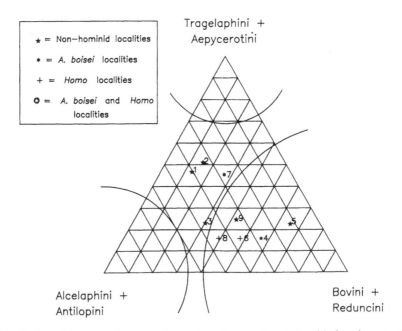

FIGURE 23.7. Localities from West Turkana plotted according to bovid abundances. Arcs as in Fig. 23.3. Note the restricted range of habitats sampled.

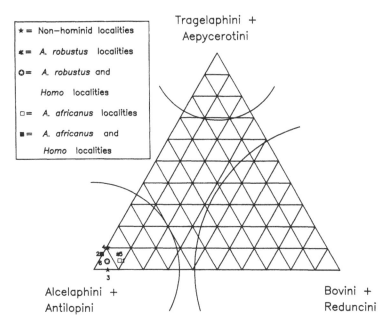

FIGURE 23.8. Localities from the South African cave sites plotted according to bovid abundances. Arcs as in Fig. 23.3. Note that all habitats sampled are strongly open/arid.

"Robust" Australopithecine Habitats

The sites analyzed here show three striking patterns of habitat sampling that correlate closely with the depositional environment. The only lake margin localities with "robust" australopithecines are from Olduvai, and these sample a spectrum from open to wet habitats. The three riverine sites derived from the basin of the proto-Omo (Koobi Fora, West Turkana, and Omo) sample only closed/wet and closed/dry habitat sites. The five cave localities sample only open/arid habitats. These results clearly demonstrate a strong taphonomic bias in habitat representation dictated by the mode of deposition.

Apart from highlighting the influence of depositional environment, however, these results do suggest something about the habitat preferences of the "robust" australopithecines. There is a strong, persistent association of *A. boisei* remains with closed/wet habitats. Of twenty-five *A. boisei* sites from East Africa, seventeen (68%) fall within the closed/wet cluster with a very high degree of probability and two more (8%) fall within that cluster with a somewhat lesser degree of probability (63–70%); four fall in the closed/dry cluster with very high probability (>92%); one falls in the open/arid cluster with a very high degree of probability (>99%); and one is ambiguous but falls along the closed/dry–wet habitat axis. Accepting Members D and K from the Omo as "robust"-bearing sites changes these results insignificantly by increasing by one each the number of closed/wet and closed/dry habitat sites.

The question is whether the association of *A. boisei* with closed/wet habitat sites is meaningful. Given the small number of total sites (seventy-six), do the "robust"-bearing sites represent a significantly different pattern of habitat abundances than in the non-"robust" sites? The answers are complex, in part because of the limited data available.

As an initial attempt to answer this question, the East African sites were classified into clusters according to the discriminant analysis, with any probability value > 50% accepted as indicating membership. Chi-squared tests were run, using Yates' correction for continuity because of small sample size (as was done on all χ^2 tests reported here). On this basis, the *A. boisei* sites are clearly distributed differently across the habitats than the non-"robust" sites ($\chi^2 = 6.0$, 2 *df*,

$p = 0.05$). Inspection of the data reveals a disproportionate occurrence of *A. boisei* in closed/wet habitats. This is an extremely interesting result, but it must be interpreted cautiously because changes in the assignment of a small number of sites may cause the results to fall to insignificance. For example, if the ambiguous fossil sites (probabilities of $< 90\%$) are removed from the data set, the results show a similar, but statistically insignificant trend ($\chi^2 = 4.9$, 2 df, $p = .08$).

Other possible changes include the status of Omo Members D and K; when they are included as *A. boisei*-bearing sites, and ambiguous cases are either retained or removed, the differences between *A. boisei* and non-*A. boisei* sites remain significant (with ambiguous cases retained: $\chi^2 = 7.8$, 2 df, $p = .02$; with ambiguous cases removed: $\chi = 6.5$, 2 df, $p = .04$).

In any case, these results suggest that *A. boisei* probably preferred closed over open habitats and favored wetter rather than drier closed habitats. More data are obviously needed to improve the size of the sample and increase the confidence in these results. However, even as a provisional conclusion pending better data, these results are considerably different from that suggested on the basis of the South African *A. robustus* localities or on previous analyses by Boaz (1977a,b, 1985) for the Omo succession.

Different Techniques for Paleoecological Reconstruction

The East African data invite comparison between the pollen and micromammal data available for Omo and Olduvai with the habitats assigned on the basis of the bovids. At Olduvai, the correspondence between the three lines of evidence is close, with the localities from the lower levels of Bed I (FLKNN, both levels; DK; FLK Zinj) being wetter than the upper levels of Bed I (the FLKN levels).

A comparable situation can be demonstrated for the rather older Omo Shungura sequence. Circumstantial molluscan evidence (Williamson, 1985) suggests the presence of tropical rainforest in Shungura Member B. The Omo pollen data suggests wet, treed habitats in Member C (Bonnefille, 1976), with a significant increase in grasses and drought-tolerant trees in Shungura Member E. The bovid data for Member C classify it in the closed/wet cluster and for E classify it in the closed/dry cluster. Thus, the bovid data suggest habitat changes in the general direction suggested by the pollen but there is no apparent influx of grassland, open habitat bovids to accompany the rise in grass pollens. The micromammals tell a similar story to the pollen. A continued drying trend throughout the Omo succession is suggested by the growing percentage of more xeric micromammals (Wesselman, 1984) from Member C to lower Member G. In Member C, Wesselman's (1984:197) data show that 43% of the micromammals are "forest and forest edge (closed), gallery forest" species; another 34% are "mesic savannah, woodlands," and only 23% are "dry savannah, savannah–woodland (open)." In the lower Member G levels at the Omo, where the frequency of more xeric micromammals is highest, 38% are "arid, semi-arid steppe, scrub" species and 62% are "dry savannah or savannah–woodland (open)." This shift toward dryness and openness is reflected in a general way by the progression of Members B to E from closed/wet to closed/dry habitats and by the slight movement of points representing E through Members J through L toward the open cluster. However, none of the Omo localities ever reaches a position that could be classified as belonging in the open habitat cluster.

There are four possible explanations for these discrepancies, all of which may be true. First, the Omo bovid habitats may be dominated and stabilized by the presence of the large, proto-Omo river that is responsible for the deposition of the sites. Thus, the habitat as reflected by the bovids may be more stable than the habitat in general, farther away from the river where there is obviously open grassland. In fact, Vrba (this volume, chapter 25) suggests that the Omo River may have represented a refugium into which closed habitat species migrated during times of overall climatic perturbation. This would account for the apparent stability of habitats in the Omo, Koobi Fora, and West Turkana localities across the 2.5-Myr climatic disturbance attested to by Vrba's data (this volume, chapter 25) on bovid species turnovers and by Prentice and Denton's data (this volume, chapter 24) on the changes in the Antarctic ice sheets (Fig. 23.9).

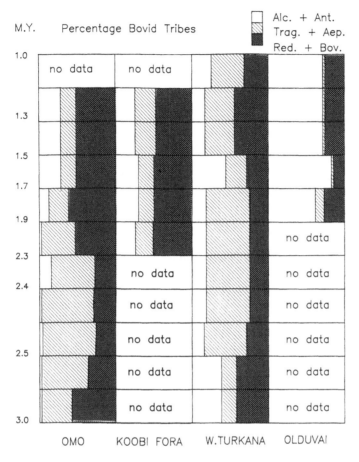

FIGURE 23.9. Changes in bovid abundances through time in Omo Shungura Formation, Koobi Fora, West Turkana and Olduvai. The vertical dimensions are not proportionate to years sampled. Note the differences in bovid abundances and inferred habitats at localities along the same river (Omo, Koobi Fora, West Turkana) at the same point in time. Further, the climatic shift at about 2.5 Myr (Vrba, this volume, chapter 25; Prentice and Denton, this volume chapter 24) is not reflected in a significant shift in habitats sampled at these sites.

In other words, the habitats away from the proto-Omo may have shifted and become connected to other areas via continuous habitat corridors through which immigrant species passed, but the habitats immediately around the Omo apparently did not change much through time.

Second, both the pollen and the micromammals may represent more general and less localized measures of habitat than the bovids. Grass pollen is notorious for traveling long distances (Bonnefille, 1976) and pollen in general gives a somewhat generalized indication of overall habitats. Micromammals travel very limited distances under their own power, but may be collected and carried by birds of prey over distances of several miles. Since many micromammal accumulations are formed by raptors (Mellett, 1974), and since raptors often more readily catch their prey in open habitats, micromammals also may tend to give an overrepresentation of open habitats. Third, the lowest part of the Olduvai succession correlates chronologically with Shungura Member H, and the animals from Shungura Members H through L were less intensively sampled or studied than those from earlier horizons. Given the geographic separation, ecologic disparity, and chronological distance between the sites of Omo and Olduvai, the differences in paleoecological interpretation are not entirely unexpected. Nevertheless, the drying trend inferred from the lower Omo succession may be interpreted as continuing in the Olduvai sequence. Finally, the method of classifying fossil sites into habitat clusters used here may be too crude to reveal the changes attested to by other approaches. It must be remembered that entire Omo Members are treated as single units and that only six of all of the possible bovid tribes are utilized here.

Our technique also produces a discrepancy with assessments by Boaz (1977a,b, 1985) of Omo paleoecology based on fauna, most of which is comprised of bovids. Boaz concluded that the Omo habitats are much more open and drier than our analyses suggest. We agree with Boaz (pers. comm.) that the difference in our results is caused by his classification of impalas and their relatives as Alcelaphini, following Gentry (1976, 1985), whereas in our work they are treated as a separate tribe, the Aepycerotini. Whatever the appropriate taxonomic status of impalas might be, ecologically they are clearly distinct from the Alcelaphini in preferring more closed habitats. Therefore, we would follow Vrba (1975, 1980) and others in treating them as a distinct tribe for the purposes of ecological reconstruction.

"Robust" Australopithecines, Early *Homo*, and Tools

Two persistent major issues in paleoanthropology concern the relationship between different sympatric and contemporaneous species of hominid and the role of tools in their ecologic niches. Plotting the occurrence of remains attributable to *Homo* at all East African localities against those of *A. boisei* shows that the two species co-occur at many sites. Our data do suggest that the two species were sympatric at least in part.

However, statistical tests of the distribution of *A. boisei*, *A. robustus*, and *Homo* across the sites studied here reveals some important distinctions. As pointed out above, *A. boisei* remains are more likely to be recovered from closed/wet habitat localities in East Africa. This conclusion is likely to be valid because all three types of habitats are represented at East African sites during the time intervals from which *A. boisei* remains are known. In contrast, early *Homo* is randomly distributed across the habitat types, whether East African sites are considered separately from South African sites or pooled with them ($\chi^2 = 1.03$ for pooled sites or 0.15 for East Africa alone, 2 df, $p > 0.5$ in each case). Again, all three types of habitats are sampled from areas from which *Homo* has been recovered. *Australopithecus robustus* has an apparent preference for open country sites, but all South African sites in the sample fall into the open habitat group, so this conclusion may not be valid.

These data suggest the following working hypotheses, which should be re-examined critically when more data on more localities are available:

1. *Australopithecus boisei* apparently preferred closed/wetter habitats.
2. Early *Homo* was less restricted in its use of habitats, ranging into open and dry, bushy areas as well as closed/wet habitats.
3. *Australopithecus robustus* may have been an open, arid habitat specialist, but this hypothesis cannot be tested until comparable sites representing closed habitats have been located and analyzed.

While the greater flexibility in use of habitats by *Homo* might be interpreted as evidence that it was the only toolmaker and tool-user among these early hominid species, the data are unclear. Both "robust" and *Homo* species occur about as often in association with Oldowan remains as without any archaeologic industry at all. At Olduvai, the only site where such studies have been performed, two of the *A. boisei* localities (FLK Zinj and BK) are remarkable for the relative abundance of cutmarks on bovid or other large mammal postcrania (pers. obs.). However, the third Olduvai site where *in situ* remains of *A. boisei* have been recovered (FLKNN level 3) is likely to be nonarchaeological (Potts, 1982), and it bears several striking resemblances to modern hyena lair assemblages. Thus, it is not possible to determine from these data whether or not *A. boisei* or *A. robustus* was involved in toolmaking or cutmarking activities.

Acknowledgments

We gratefully acknowledge funding and support from the following sources: the National Science Foundation (BNS 8512001), the Boise Fund, the Wenner-Gren Foundation (4625), the Los Angeles County Museum, the National Geographic Society, and an Institutional Research Grant from the Johns Hopkins University School of Medicine. Work was made possible by the kind permission and assistance of the Governors and staff of the National Museums of Kenya, the Government of Tanzania, and the Office of the President of Kenya. We thank A. Walker, F. Brown, and E. Vrba for especially useful comments, J. Richtsmeier for statistical discussions, and F. Grine for organizing the workshop that led to the writing of this paper. We thank L. Baldwin of the World Wildlife Fund for making MacKinnon and MacKinnon's report available to us.

References

Afolayan, T. A. and Ajayi, S. S. (1980). The influence of seasonality on the distribution of large mammals in Yankari Game Reserve, Nigeria. *Afr. J. Ecol.*, 18: 87–96.
Ayeni, J. S. O. (1980). Management problems of the Kainji National Park, Nigeria. *Afr. J. Ecol.*, 18: 97–112.
Baba, M., Doi, T., Ikeda, H., Iwamoto, G. and Ono, Y. (1982). A census of large mammals in Omo National Park, Ethiopia. *Afr. J. Ecol.*, 20: 207–210.
Bearder, S. (1977). Feeding habits of spotted hyaenas in a woodland habitat. *E. Afr. Wildl. J.*, 15: 263–280.
Behrensmeyer, A. K. and Dechant Boaz, D. E. (1980). The recent bones of Amboseli Park, Kenya, in relation to East African paleoecology. *In* A. K. Behrensmeyer and A. P. Hill (eds.), *Fossils in the making*, pp. 72–93. University of Chicago Press, Chicago.
Boaz, N. T. (1977a). Paleoecology of early Hominidae in Africa. *Kroeber Anthropol. Soc. Papers*, 50: 37–61.
Boaz, N. T. (1977b). Paleoecology of Plio-Pleistocene Hominidae in the Lower Omo Basin, Ethiopia. Ph.D. Dissertation, University of California, Berkeley.
Boaz, N. T. (1985). Early hominid paleoecology in the Omo Basin, Ethiopia. *In* Y. Coppens (ed.), *L'Environnements des Hominides au Plio-Pleistocene*, pp. 279–308. Singer-Polignac, Paris.
Bonnefille, R. (1976). Palynological evidence for an important change in the vegetation of the Omo Basin between 2.5 and 2 million years ago. *In* Y. Coppens, F. C. Howell, G. Ll. Isaac and R. E. F. Leakey (eds.), *Earliest man and environments in the Lake Rudolf Basin*, pp. 421–431. University of Chicago Press, Chicago.
Brain, C. K. (1981). *The hunters or the hunted?* University of Chicago Press, Chicago.
Bunn, H. T. (1982). Meat-eating and human evolution. Ph.D. Dissertation, University of California, Berkeley.
Child, G. S. (1974). An ecological survey of the Borgu Game Reserve, Nigeria. United Nations Development Programme, Food and Agriculture Organization of the United States, Rome.
Day, M. H. (1986). *Guide to fossil man*. University of Chicago Press, Chicago.
Douglas-Hamilton, I. (1972). On the ecology of the African elephant. D. Phil. thesis, Oxford University.
Esser, J. D. and van Lavieren, L. P. (1979). Importance, repartition et tendance evolutive des populations de grands herbivores et de l'autruche dans le Parc National de Waza, Cameroun. *La Terre et La Vie*, 33: 3–26.
Geerling, C. and Bokdam, J. (1973). Fauna of the Comoe National Park, Ivory Coast. *Biol. Cons*, 5: 251–257.
Gentry, A. W. (1976). Bovidae of the Omo Group Deposits. *In* Y. Coppens, F. C. Howell, G.

L1. Isaac and R. E. F. Leakey (eds.), *Earliest man and environments in the Lake Rudolf Basin*, pp. 275–292, University of Chicago Press, Chicago.
Gentry, A. W. (1985). The Bovidae of the Omo Group Deposits, Ethiopia. *Les Faunes Plio-Pleistocene de la Basse Vallee de l'Omo (Ethiopie) 1*: 119–191. CRNS, Paris.
Gentry, A. W. and Gentry, A. (1978a). Fossil Bovidae (Mammalia) of Olduvai Gorge, Tanzania, Part I. *Bull. Brit. Mus. Nat. Hist. (Geol)., 29*: 289–446.
Gentry, A. W. and Gentry, A. (1978b). Fossil Bovidae (Mammalia) of Olduvai Gorge, Tanzania. Part II. *Bull. Brit. Mus. Nat. Hist. (Geol)., 30*: 1–83.
Green, A. (1979). Density estimates of the larger mammals of Arli National Park, Upper Volta. *Mammalia, 43*: 59–70.
Grine, F. E. (1985). Australopithecine evolution: the deciduous dental evidence. *In* E. Delson (ed.), *Ancestors: the hard evidence*, pp. 185–206. Alan R. Liss, New York.
Harroy, J-P. and Elliott, H. (1971). *United Nations List of National Parks and Equivalent Reserves*. Brussels, Hayez.
Hirst, S. M. (1969). Populations in a Transvaal Lowveld Nature Reserve. *Zool. Afr., 4*: 199–230.
Howell, F. C., ed. (1973) *Early Man*. Time-Life Books, New York.
Howell, F. C. (1978) Hominidae. *In* V. Maglio and H. B. S. Cooke (eds.), *Evolution of African mammals*, pp. 154–248. Harvard University Press, Cambridge, Mass.
Isaac, G. Ll. and McCown, E. (eds.) (1976). *Human Origins*. W. A. Benjamin, Menlo Park.
Leakey, M. D. (1971). *Olduvai Gorge, vol. III*. Cambridge University Press, London.
MacKinnon, J. and MacKinnon, S. (1987) *Protected areas systems: Review of the Afro-Tropical realm*. World Conservation Center, Gland, Switzerland.
Mellett, J. (1974). Scatological origins of microvertebrate fossils. *Science, 185*: 349–350.
Milligan, K., Ajayi, S. S. and Hall, J. B. (1982). Density and biomass of the large herbivore community in Kainji National Park, Nigeria. *Afr. J. Ecol., 20*: 1–12.
Mwalyosi, R. B. B. (1977). A count of large mammals in Lake Manyara National Park. *E. Afr. Wildl. J., 15*: 333–335.
Norusis, M. J. (1986). *Advanced Statistics SPSS/PC+ for the IBM PC/XT/AT*. SPSS Inc., Chicago, Illinois.
Pennycuick, C. and Western D. (1972). An investigation of some sources of bias in aerial transect sampling of large mammal populations. *E. Afr. Wildl. J., 10*: 175–192.
Poche, R. (1973). Niger's threatened Park W. *Oryx, 12*: 216–222.
Poche, R. (1976). Notes on primates in Parc National du W du Niger, West Africa. *Mammalia, 40*: 187–198.
Potts, R. (1982). Lower Pleistocene site formation and hominid activities at Olduvai Gorge, Tanzania. Ph.D. Dissertation, Harvard University, Cambridge, Mass.
Potts, R. (1984). Home bases and early hominids. *Amer. Sci., 72*: 338–347.
Shipman, P. (1981). *Life history of a fossil*. Harvard University Press, Cambridge, Mass.
Shipman, P. (1986). Scavenging or hunting in early hominids. *Amer. Anth., 88*: 27–43.
van Lavieren, L. P. and Bosch, L. P. (1977). Evaluations des densites de grandes mammiferes dans le Parc National de Bouba Ndjida, Cameroun. *La terre et la vie, 31*: 3–32.
van Lavieren, L. P. and Esser, J. D. (1980). Numbers, distribution and habitat preference of large mammals in bouba Ndjida National Park, Cameroon. *Afr. J. Ecol., 18*: 141–154.
Vrba, E. S. (1975). Some evidence of chronology and paleoecology of Sterkfontein, Swartkrans, and Kromdraai from the fossil Bovidae. *Nature, 254*: 301–304.
Vrba, E. S. (1976). The fossil Bovidae of Sterkfontein, Swartkrans, and Kromdraai. *Tvl. Mus. Mem., 21*: 1–166.
Vrba, E. S. (1980). The significance of bovid remains as an indicator of environment and predation patterns. *In* A. K. Behrensmeyer and A. P. Hill (eds.), *Fossils in the making*, pp. 247–272. University of Chicago Press, Chicago.
Vrba, E. S. (1984). Ecological and adaptive changes associated with early hominid evolution. *In* E. Delson (ed.) *Ancestors: the hard evidence*, pp. 63–71. Alan R. Liss, New York.
Wesselman, H. (1984). *The Omo micromammals*. S. Karger, Basel.

Western, D. (1980). Linking the ecology of the past and present mammal communities. *In* A. K. Behrensmeyer and A. P. Hill (eds.), *Fossils in the making*, pp. 41–54. University of Chicago Press, Chicago.

White, F. (1983). *The Vegetation of Africa: A Descriptive Memoir to Accompany the Unesco/AETFAT/UNSO Vegetation Map of Africa.* Unesco, Paris.

Williamson, P. (1985). Evidence for an early Plio-Pleistocene rainforest in East Africa. *Nature, 315*: 487–489.

The Deep-Sea Oxygen Isotope Record, The Global Ice Sheet System and Hominid Evolution

24

MICHAEL L. PRENTICE AND GEORGE H. DENTON

Several evolutionary histories of *Australopithecus* and *Homo* species have been presented during the past decade (e.g., Johanson and White, 1979; Tobias, 1980; White et al., 1981; Olson, 1981), and the recent discovery of a late Pliocene "robust" cranium in the Lake Turkana Basin (Walker et al., 1986) has led to improved phylogenetic schemes. One such scheme is shown in Fig. 24.1. Eventually, as the phylogeny of hominids becomes more firmly known, the chronology and underlying causes of their evolutionary history will emerge. Meanwhile, most current hominid phylogenies show more than one splitting event in late Pliocene time between 3.0 and 2.0 Myr (Fig. 24.1) (Grine, this volume, chapter 30; Kimbel et al., this volume, chapter 16; Walker and Leakey, this volume, chapter 15; Wood, this volume, chapter 17).

A series of recent papers has served to develop the hypothesis that widespread environmental change forced by climate change was the initiating cause of late Pliocene evolutionary events in hominid phylogeny, particularly near 2.5 Myr (Brain, 1981; Grine, 1981, 1982, 1985, 1986; Vrba, 1975, 1982, 1985a). It is further possible that late Miocene cooling coincided with the origin of the Hominidae (Brain, 1981), and that a global climatic change near 0.9 Myr caused the extinction of "robust" australopithecines and the spread of *Homo erectus* into Eurasia (Vrba, 1985b, this volume, chapter 25; Klein, this volume, chapter 29).

The background for the first-mentioned hypothesis comes from emerging knowledge of widespread environmental and climatic changes in late Pliocene time that could have driven evolutionary events. In the Transvaal of southern Africa, changes in fossil bovids suggest a fundamental shift from relatively mesic and bush-covered environments to more arid and open environments between 2.5 and 2.0 Myr (Vrba, 1975, 1985b). In East Africa, micromammals (Wesselman, 1985), pollen assemblages (Bonnefille, 1976, 1983), and bovids (Vrba, 1985d) all show expansion of arid and open grasslands close to 2.5 Myr. The Transvaal and East African changes probably represented the same marked spread of open grassland near 2.5 Myr across sub-Saharan Africa in response to global cooling and associated rainfall changes (Vrba, 1985d; Van Zinderen Bakker and Mercer, 1986). More recently, Janecek and Ruddiman (1987) showed from sediment composition changes in the South Atlantic Ocean that the African continent became increasingly arid about 2.5 Myr, a result in accord with terrestrial evidence. Less significant environmental changes preceded the dramatic 2.5-Myr shift (Williamson, 1985; Bonnefille et al., 1987).

Outside of Africa, major late Pliocene environmental change, probably associated with cooling, is recorded in several localities. Nearly three decades ago Zagwijn (1960) recognized pronounced late Pliocene cooling in north-central Europe at the base of the cold Praetiglian Stage, when tundra-like open vegetation first replaced the transitional elements of the Reuverian Stage and warm–temperature elements of the older Brunssumian Stage (Fig. 24.2). During the Praetiglian, arctic conditions were reached for the first time in north-central Europe and the cold–warm cycles of the Quaternary began (Fig. 24.2). In fact, the base of the Praetiglian Stage has long been considered in north-central Europe to mark the beginning of the Quaternary Period. The Reuverian/Praetiglian boundary is near 2.4 to 2.5 Myr (Fig. 24.2) (van der Hammen et al., 1971; Boellstorff, 1978; Zagwijn and Doppert, 1978; Grube et al., 1986). This runs counter to international

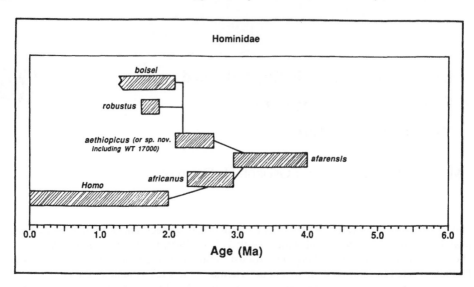

FIGURE 24.1. Generalized phylogeny for Hominidae (after Grine, this volume, chapter 30). The generic name of the species *afarensis, aethiopicus, robustus* and *boisei* has been omitted because of uncertainty as to whether it is *Australopithecus* or *Paranthropus*.

stratigraphic convention which places the base of the Quaternary in the Vrica Section of Calabria, southern Italy, at 1.6 Myr (Berggren *et al.*, 1985).

Coincident with Praetiglian cooling in north-central Europe was the onset of major glaciation in the Northern Hemisphere. Extensive iceberg rafting of glacial sediments began in the North Atlantic at 2.4 Myr (Shackleton *et al.*, 1984), indicating the first extensive ice sheet growth on adjacent land masses. Previous workers (e.g., Poore, 1981) estimated the age of initial widespread ice-rafting in the North Atlantic at 3.0 Myr. Oxygen isotopic records of both benthic and planktonic foraminifers from widespread locations exhibit gradual enrichment starting at 3 Myr followed by significant abrupt enrichment at 2.4 to 2.5 Myr that record some combination of glaciation and declining water temperatures (Shackleton and Opdyke, 1977; Blanc *et al.*, 1983; Hodell *et al.*, 1983; Thunell and Williams, 1983; Shackleton *et al.*, 1984; Prell, 1985a; Keigwin, 1987).

In central China, widespread and thick loess rests stratigraphically on Red Clays, a formation of late Pliocene age. The boundary between the Red Clays and overlying loess is transitional. The initiation of loess deposition, which marks a fundamental change from warm and humid environments to harsh continental steppes, is dated between 2.3 and 2.5 Myr (Figs. 24.2 and 24.3) (Kukla, 1987). Finally, a 357 m long core recovered from the high plain of Bogotá, Colombia, contains evidence for a significant lowering of the "forest line" and high plain temperature at 2.5 Myr from the relatively inflated levels of the preceding 0.5 Myr (Fig. 24.3) (Hooghiemstra, 1984).

Such strong indications of widespread climatic and environmental change in late Pliocene time, particularly at 2.4 to 2.5 Myr, are consistent with hypotheses relating paleoclimate, environment, and hominid evolution. Therefore, we here review several tests to determine if events in hominid evolution were intertwined with paleoclimatic change. We then outline our knowledge of an important component of the Pliocene climate system, namely, the polar ice sheets. Our immediate purpose is to determine if polar regions were cold and ice-covered through the critical late Pliocene interval when *Homo* first appeared on the African continent, and to determine whether polar ice sheet changes conform to Vrba's global climate-forcing hypothesis, which predicts that polar ice sheets either originated or experienced significant size changes coeval with the major late Pliocene evolutionary events. Our long-term goal is to afford the background for atmospheric modeling experiments to determine the influence exerted by the Antarctic Ice Sheet itself on late Pliocene climate and hence hominid evolution (Denton, 1985).

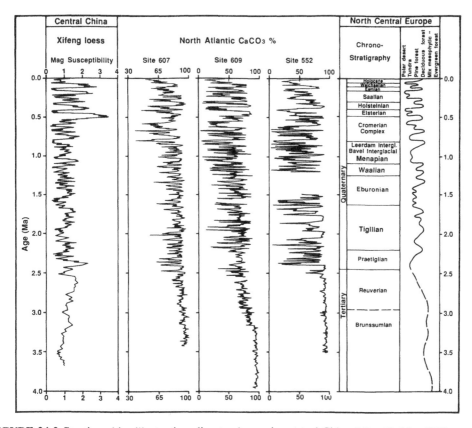

FIGURE 24.2. Stratigraphies illustrating climate change in central China (after Kukla, 1987), north-central Europe (after Grube *et al.*, 1986), and the North Atlantic (after Ruddiman and Raymo, 1988). Figure 24.3 shows site locations. North Atlantic $CaCO_3\%$ is uniformly high until 2.4 Myr when it dropped dramatically as the flux of ice-rafted detritus increased. Magnetic susceptibllility at Xifeng in central China exhibits little variance and moderate values until 2.4 Myr when the variance increases significantly as the susceptibility fluctuates between low susceptibility (cold) loess layers and high susceptibility (warm) soils. Magnetic susceptibility changes may reflect a difference in concentration of fine-grained magnetic minerals. Vegetation changes in north-central Europe based on palynologic studies mimic these trends. The Tertiary/Quaternary boundary in north-central Europe is still placed at the base of the Praetiglian Stage at 2.4 to 2.5 Myr contrary to international stratigraphic convention.

Tests of the Hypothesis

The hypothesis that changes in various parts of the climate system initiated major biotic evolutionary events contains predictions of temporal and spatial relations among climate, environment, and biotic evolution. Determining whether such predictions are met in the geologic record constitutes strong testing of the hypothesis (Vrba, 1985c). Figure 24.4 is a partial blueprint of the climate system network that underlies our problem and is pertinent to test development.

One test is to examine *temporal* patterns of macroevolution as recorded in African fossil records and then to compare these patterns with the chronology of African environmental change. The hypothesis predicts that patterns of speciation, extinction, and migration within a biotic group should occur as nearly synchronous events in the geologic record rather than being randomly

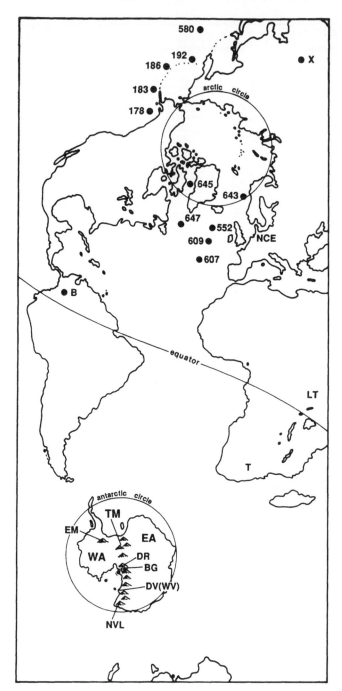

FIGURE 24.3. Location of selected sites discussed in text. On Antarctica EA = East Antarctica, TM = Transantarctic Mountains, BG = Beardmore Glacier, DR = Dominion Range, WA = West Antarctica, DV = Dry Valleys, EM = Ellsworth Mountains, NVL = Northern Victoria Land. In central China, X = Xifeng loess section. In Africa, LT = Lake Turkana, T = Transvaal. NCE = North Central Europe. In South America, B = Bogotá, Colombia. Numbers refer to DSDP drill sites.

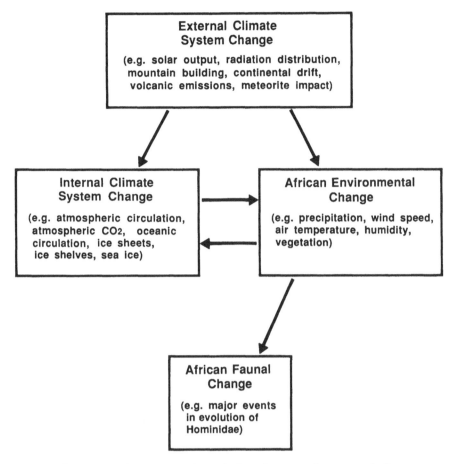

FIGURE 24.4. Schematic diagram showing relations between the external climate system, the internal climate system, the African environment, and African evolution.

distributed through time. Further, turnover events of diverse biotic groups should occur as nearly synchronous pulses that correlate with climatically induced environmental changes determined independently from the geologic record (Vrba, 1985c). Specifically, the important late Pliocene cladogenetic events in hominid evolution should be embedded in widespread turnover pulses coeval with environmental change. Of particular importance is whether hominid evolutionary events coincide with the turnover of coevolving antelopes, because antelopes are narrowly habit specific and, like early hominids, were mobile and herbivorous savannah mammals. Therefore, the hypothesis predicts that the evolutionary response to vicariance caused by physical environmental change should be synchronous in both biotic groups (Vrba, 1985c).

A second test involves the *spatial* distribution of environmental change and biotic turnover. Because it depends on vicariance caused by changes in the physical environment, the hypothesis predicts that there should be centers of evolutionary activity where physical environmental fluctuations are marked, but not so extreme as to make the environment uninhabitable (Vrba, 1985c). During times of extreme polar climates, the tropics should be centers of evolutionary turnover pulses. Specifically, the huge continental land mass of Africa, because it straddles the equator and extends into both Northern and Southern Hemispheres, should be uniquely important for macroevolutionary events in late Pliocene time if both polar regions became ice covered. Even within Africa, certain restricted geographical areas should be particularly favorable for speciation (Vrba, 1985c).

A third test, which comprises the primary focus of this chapter, involves the geological record outside of Africa. The hypothesis predicts that global climate change, not a local African event, initiated late Pliocene environmental and evolutionary shifts (Vrba, 1985b). The first step in this test is to compare the temporal pattern of African environmental shifts with paleoclimatic records elsewhere. Appropriate records include those of internal climate system variables such as ice sheets, sea ice, sea level, ocean temperatures, vegetation, rainfall (e.g., monsoons) and atmospheric composition (eg., CO_2) (Fig. 24.4). Temporal correspondence would initiate further tests aimed at isolating the cause(s) of the climatic change and determining whether this change could produce environmental shifts and associated vicariance of severity sufficient to initiate turnover pulses in Africa. Such an exercise requires external climate system records such as those of mountain range uplift, insolation and volcanism. All these records provide the boundary conditions for sensitivity tests with atmospheric circulation models to determine triggering causes and localized climate responses.

Pliocene Record of the Global Ice Sheet System

We now outline the Pliocene evolution of the global ice sheet system as one aspect of the third test (above). We first examine low- to mid-latitude isotopic data and progress to high-latitude marine and continental records.

Deep-Sea Oxygen Isotopic Record

Oxygen isotope data on deep-sea foraminifers afford the only continuous, well-dated, and high resolution record of global ice volume and ocean temperature variation. Such records are central to our purpose. Because no high resolution and high quality $\delta^{18}O$ record from a single site spans the entire Pliocene and Pleistocene, we have spliced together the best $\delta^{18}O$ records available to form composite isotopic records for this long time interval. Figure 24.5 presents the resulting composite isotope curves for tropical shallow-dwelling planktonic foraminifers and deep-water benthic foraminifers (Prentice, 1988). The composites follow the Berggren et al. (1985) time scale. We rely on these curves for our analysis of ice sheet and ocean temperature changes.

Because the planktonic composite was designed for equatorial western Pacific surface waters, the modern value of the composite equals the average of Holocene $\delta^{18}O$ on *Globigerinoides sacculifer* from equatorial western Pacific core V28-238 (Shackleton and Opdyke, 1973). These waters were selected as the base for the composite because both theoretical (e.g., Newell and Chiu, 1981) and empirical (Prell, 1985b; Shackleton, 1987) data suggest that they may be more thermally stable than any other open-ocean water mass.

Construction of the planktonic composite involved two primary steps. The first was to project high-quality planktonic $\delta^{18}O$ data from low-latitude, non-upwelling regions to the equatorial western Pacific. This was accomplished using $\delta^{18}O$ gradients between the equatorial western Pacific and the sites of the selected data. The resulting $\delta^{18}O$ values, in effect normalized to the equatorial western Pacific, were then averaged over selected short time intervals to produce target values for the composite. These target values represent average tropical (non-upwelling) $\delta^{18}O$ in the equatorial western Pacific.

The second step in construction of the planktonic composite was to select the premier shallow-dweller $\delta^{18}O$ records from low-latitude, non-upwelling regions. Four records were chosen to span the last 6 Myr, and these were then projected to the composite target values in the equatorial western Pacific. For each input $\delta^{18}O$ record, three to four composite target values were typically used to calculate three to four $\delta^{18}O$ gradients between that record and the composite. We used the average of all gradients calculated for each input $\delta^{18}O$ record to project that input record to the composite. The half-range in the gradients for each input record averages 0.15‰.

The benthic composite reflects the global average $\delta^{18}O$ of *Cibicides wuellerstorfi* over the water depth interval of 2 to 4 km (Fig. 24.5). This species was selected as the benthic composite target because its oxygen isotopic disequilibrium is less variable than that of other common benthic foraminifers (Graham et al., 1981). The selected depth range permits evaluation of deep water $\delta^{18}O$ variation, which provides a strong contrast to the tropical surface water record. The

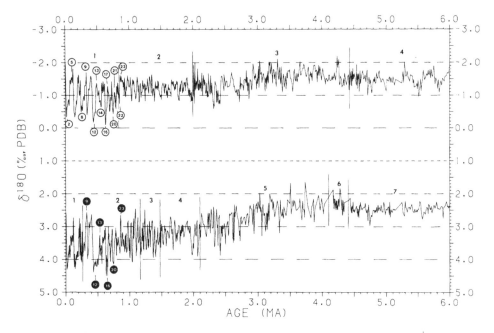

FIGURE 24.5. Composite oxygen isotope curves for tropical shallow-dwelling planktonic foraminifers in the equatorial western Pacific (upper) and benthic foraminifers of the genus *Cibicides* in deep water of 2- to 4-km depth (lower). Vertical lines separate different records spliced into the composites. Numbers within clear and dark circles refer to oxygen isotopic stages. By convention, odd-numbered stages reflect interglacial periods of relative minimum ice volume and seawater $\delta^{18}O$. During even-numbered stages, ice volume and ocean $\delta^{18}O$ reached significant relative maxima. Larger numbers without circles identify the references for this data. With P = planktonic and B = benthic: P1 = Prell *et al.* (1986); B1 = Pisias *et al.* (1984); B2,B4 = Shackleton and Hall (1984); P2 = Shackleton and Opdyke (1976); B3 = Ruddiman *et al.* (1986); P3 = Prell (1985a); B5 = Keigwin (1987); P4,B6,B7 = Prentice (1988).

standard deviation of *C. wuellerstorfi* $\delta^{18}O$ from this depth interval in the modern ocean is only 0.2‰.

Construction of the benthic composite parallels that of the planktonic composite. The primary targets for the last 6 Myr are provided by seven different estimates of the global average $\delta^{18}O$ of *C. wuellerstorfi* from 2 to 4 km paleodepth. Each of these targets is, in fact, the average of data from as many as 36 globally distributed sites at these paleodepths. Several independent estimates were made of the $\delta^{18}O$ gradient between each input record and the various composite targets. Each input record was projected to the composite by using the average of all $\delta^{18}O$ gradients calculated for that record. The half-range in the gradients for each input record averages 0.1‰.

Paleoclimatic interpretation of $\delta^{18}O$ curves, including the composite curves in Fig. 24.5, is not straightforward. Difficulties arise from the fact that foraminiferal $\delta^{18}O$ varies not only with water $\delta^{18}O$, which is primarily a function of ice volume, but also with water temperature. Various strategies have been employed to control this ambiguity. One scheme has been to assume that deep/bottom water temperature change has been negligible and to attribute benthic $\delta^{18}O$ variations entirely to ice volume (e.g., Ninkovitch and Shackleton, 1975). However, this scheme has been largely discredited by overwhelming evidence that deep-water temperature has changed significantly through glacial–interglacial cycles (Dodge *et al.*, 1983; Matthews, 1986; Chappell and Shackleton, 1986).

An alternative strategy, which we employ here, is to assume that average tropical sea-surface temperature (SST) outside upwelling regions has been constant (Matthews and Poore, 1980;

Prentice and Matthews, 1988). This permits us to interpret the planktonic $\delta^{18}O$ composite completely in terms of ice volume. Because we consider deep-water temperature change to have been significant, we suggest that benthic $\delta^{18}O$ cannot confirm or negate interpretation of tropical planktonic $\delta^{18}O$ in terms of ice volume. What benthic $\delta^{18}O$ brings to the quantitative study of ice volume and climate is a chance to examine the contrast between tropical surface water and deep water $\delta^{18}O$. This contrast is independent of ice volume and primarily reflects important deep-water temperature changes.

We acknowledge that ice volume estimates based on our interpretation of the tropical planktonic $\delta^{18}O$ composite can err for two primary reasons. The first is that average tropical SST outside upwelling regions may indeed have varied slightly. We suggest that variation was limited to 2°C. Tropical SST 2°C cooler than today is one way to reconcile the sea level record derived from New Guinea coral reef terraces (Chappell and Shackleton, 1986) with the planktonic composite during isotope stage 3. A 2°C deviation from modern tropical SST would mean that 0.4‰ of planktonic composite change reflects temperature rather than ice volume change.

The second principal reason that our ice volume estimates may be in error relates to partial dissolution of the planktonic foraminifers used to generate the isotopic data. Prentice (1988) demonstrated that partial dissolution of a monospecific planktonic foraminifer population from the Walvis Ridge resulted in enrichment of their $\delta^{18}O$ by up to 0.5‰. The effect of dissolution on the planktonic composite would be to alter the projection of input records to the equatorial western Pacific. Most likely, this would make the composite erroneously heavy and so would lead to excessive ice volume estimates. We appraise this effect at less than 0.2‰ over the last 6 Myr (Fig. 24.5).

Following our interpretation scheme, the planktonic composite indicates important changes in the pattern of global ice volume variability since 6 Myr (Fig. 24.5). From 6 until 4.3 Myr, global ice volume fluctuated about a constant mean with an average amplitude of about 0.5‰ (Prentice, 1988; Keigwin, 1988). From 4.3 until about 2.8 Myr, high frequency ice volume fluctuations occurred about a smaller mean ice volume. At about 2.8 Myr, global ice volume began a cyclic but steady increase that culminated with an abrupt increase about 2.4 Myr to a maximum volume significantly larger than had been attained in the preceding 4 Myr. Mean ice volume then decreased steadily until 2.1 Myr when it stabilized at a level greater than it had achieved prior to 2.8 Myr (Prell, 1985a). From 2.1 to 0.9 Myr, the ice budget varied about a constant mean primarily at 40 kyr periodicity (Pisias and Moore, 1981; Ruddiman et al., 1986). This pattern was broken at 0.9 Myr when mean global ice volume increased abruptly to a new maximum at about 0.7 Myr (Shackleton and Opdyke, 1976; Prell, 1982). Since this time, large amplitude, 100 kyr fluctuations in the ice sheet system have dominated earth history (Imbrie et al., 1984).

North Atlantic and Pacific Ice-Rafting Records of Arctic Glaciation

Hydraulic piston coring in the North Atlantic Ocean, Norwegian and Labrador Seas, and Baffin Bay indicates that moderate-to-large scale glaciation in the Northern Hemisphere commenced about 2.4 Myr (Fig. 24.6b) (Shackleton et al., 1984; Ruddiman and Raymo, 1988). It is at this time that the first major flood of ice-rafted detritus swept across the North Atlantic Ocean (Fig. 24.2). However, small quantities of ice-rafted sediment appeared in Baffin Bay (Site 645) as long ago as 3.5 Myr (Arthur et al., 1986) and on the Voring Plateau in the Norwegian Sea (Site 643) as early as 3.0 Myr (Fig. 24.2) (Jansen et al., 1986). The implication is that small-scale glaciation occurred in the Baffin Bay region and the Scandinavian mountains prior to 2.4 Myr (Fig. 24.6b). These Arctic ice masses were extensive enough by 2.4 Myr that the discharged icebergs could reach drill sites as far south as 41°N (Figs. 24.2 and 24.3).

Rotary-drilled DSDP sites distributed throughout the North Pacific down to 51°N indicate that ice-rafting first became widespread there about 2.5 Myr (Fig. 24.2) (Rea and Schrader, 1985). Ice-rafted detritus older than 2.5 Myr is rare but has been found in small quantities at Sites 178 and 186. Krissek and Carney (1986) reported the first unequivocal appearance of ice-rafted detritus in hydraulically piston-cored sediment from Site 580 at 2.47 to 2.6 Myr (Fig. 24.2).

Antarctic Continental Record

The continental record of Pliocene Antarctic glaciation is fragmentary and poorly dated relative to deep-sea oxygen isotopic and ice-rafting records. However, the data permit five different Pliocene configurations to be postulated for the Antarctic Ice Sheet. Four configurations are based directly on deposits and erosional features in the Transantarctic Mountains (Denton et al., 1988b). Each was probably achieved on several occasions in the Pliocene. The last configuration is inferred indirectly from retransported marine microfossils in glacial sediments in the Transantarctic Mountains (Webb et al., 1984).

The last-mentioned configuration (Fig. 24.6d) is characterized by massive deglaciation in both West and East Antarctica. The evidence is marine microfossils that occur in glacial sediments in the Transantarctic Mountains. Webb et al. (1984) proposed that these microfossils were eroded from marine basins within the East Antarctic craton and transported into the Transantarctic Mountains by the East Antarctic Ice Sheet. This implies massive ice recession in interior East Antarctica to allow deposition of marine sediments there. The marine-based relatively unstable West Antarctic Ice Sheet should also have collapsed. The age and paleoecology of the recycled diatoms imply one or more Pliocene deglaciations under climatic conditions much warmer than at present (Harwood, 1986).

Such a configuration for the Antarctic Ice Sheet disagrees with our interpretation of the deep-sea oxygen isotope record. Our planktonic isotope composite does not show $\delta^{18}O$ negative enough to reflect such substantial loss of East Antarctic ice in the Pliocene (Fig. 22.5). This would be true even if the average $\delta^{18}O$ of the remnant ice was as negative as the $\delta^{18}O$ of the ice accumulating at late Pleistocene glacial maximums (Lorius et al., 1985), an unlikely situation given the much lower elevation of the remnant ice surface, the reduced distance to moisture sources, and warmer temperatures of the deglaciated Antarctic region.

Drastic Pliocene deglaciation of Antarctica is supported by the latest "global eustatic curve" based on seismic stratigraphy, which shows highstands of up to 60 m around 3 and 4.5 Myr (Haq et al., 1987). However, we question this interpretation primarily because the underlying data as well as the method for constructing global curves from regional data are unavailable. Further, we suspect that two of the regional sections averaged into the global curve, the San Joaquin (California) and North Sea Basins (Vail et al., 1977), have quite complicated subsidence histories that make it difficult to separate local from global effects.

The Peleus configuration (Fig. 24.6c) shows an ice sheet confined to East Antarctica with small ice caps on West Antarctic highlands. This reconstruction is based on two distinctive classes of glacial deposits within the Transantarctic Mountains; both were deposited by large temperature outlet glaciers of the East Antarctic Ice Sheet (Denton et al., 1988b). One class consists of thick wedges of glacial sediment with numerous beds of waterlaid material that are plastered alongside and at the confluence of major troughs that transect the mountains. Southern beech *(Nothofagus)* twigs and pollen occur within one such deposit at about 86°S in the Dominion Range of the Beardmore Glacier area (Fig. 24.3) (Prentice et al., 1986; Carlquist, 1987). The other class consists of discontinuous sheets of basal till with upper margins that define an ice edge, the best documented example being the Peleus till sheet in Wright Valley (Prentice et al., 1987). Diatoms indicative of a former warm-water fjord occur within Peleus till as well as in marine sediments beneath Peleus till (Burckle et al., 1986).

A variety of age control data, including marine diatoms, $^{87}Sr/^{86}Sr$ of carbonates, and isotopic dates on volcanics indicate that some deposits on which the Peleus configuration is based are Early Pliocene in age. From the character of its deposits, the height of its equilibrium line, and the associated wood twigs (Mercer, 1988), we infer that the periphery of the Peleus ice sheet existed under a maritime subpolar climate with surface air temperatures at the ice margin substantially warmer than today. This ice sheet had an extensive melting margin, unlike the present polar Antarctic Ice Sheet.

We also hypothesize that the Antarctic Ice Sheet expanded to an overriding configuration when its margin was influenced by a maritime subpolar climate cooler than that characterizing a Peleus configuration (Fig. 24.6b, dashed line). The evidence includes high-elevation trimlines

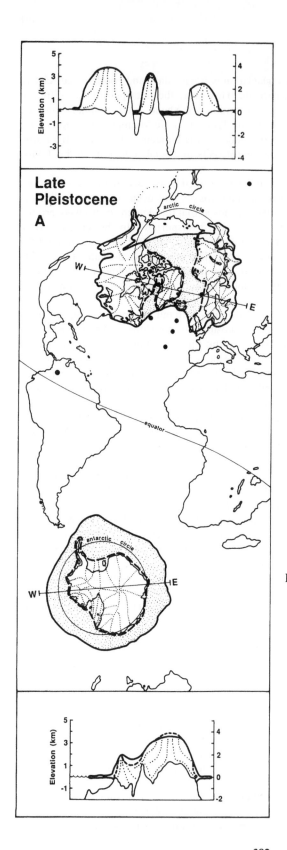

FIGURE 24.6. Schematic maps and sections of the global ice sheet system since 6 Myr. Flow lines are sketched for grounded portions of the cryosphere. The maximum yearly extent of floating ice (sea ice or ice shelf) is dotted. Large dots depict the locations of important records identified in Fig. 24.3. (A) In sections, solid lines depict 18 kyr numerical reconstructions. Arctic and Antarctic reconstructions for 18 kyr are from Denton and Hughes (1981) and Denton *et al.* (1988b), respectively. Dashed lines show the modern configuration. (B) The robust

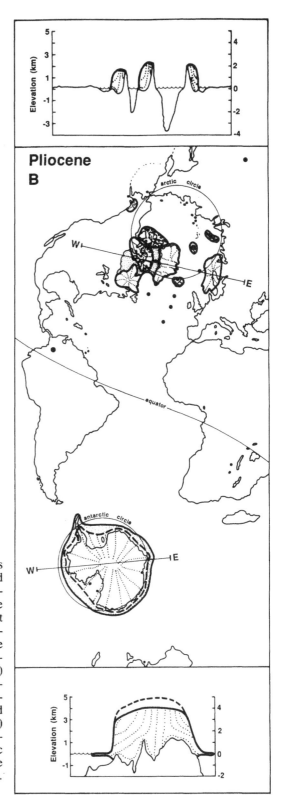

and overriding configurations for Antarctic ice (solid and dashed lines in section, respectively). The floating ice corresponds to the robust configuration and is conjectural. The Arctic cryosphere schematically reflects an icecap phase since 3 Myr. (C) The Peleus ice sheet configuration. (D) Conjectural Antarctic ice configuration based on the Webb *et al.* (1984) interpretation of marine microfossils in Transantarctic Mountain glacial drift. [Figure 24.6 (c) and (d) continue on pp. 394–395.]

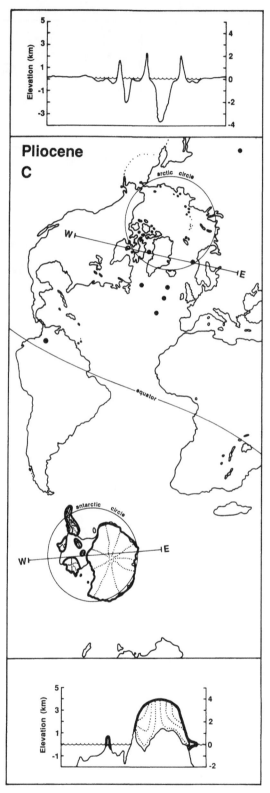

FIGURE 24.6(c)

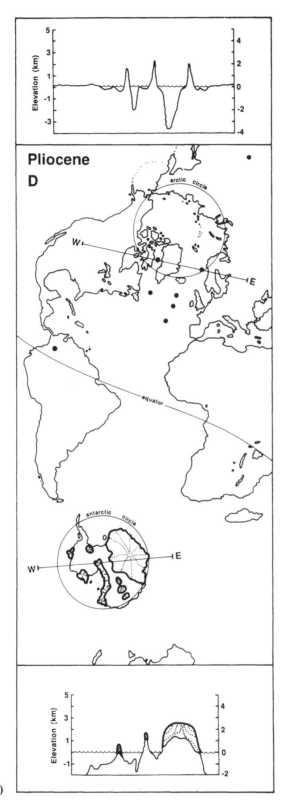

FIGURE 24.6(d)

at the extremities of the continent in Northern Victoria Land (Denton et al., 1987a) and the Ellsworth Mountains (Denton et al. 1988a) (Fig. 24.3). In contrast, striated bedrock surfaces and patches of basal till with far-traveled erratics characterize high elevation terrain farther south. Such striated surfaces occur up to 4200 m and till patches occur up to 4115 m in the Beardmore Glacier region (Prentice et al., 1986). The observations that the erratics could not have been derived locally and that the striations show a uniform direction regardless of irregular local topography indicate continental as opposed to alpine glaciation. In the Dry Valleys region, glacially molded sandstone bedrock occurs to 2885 m (Denton et al., 1984); striated bedrock occurs at least as high as 3215 m; and the higher peaks show no trimlines (Denton et al., 1988b).

Figure 24.6b (solid line) shows a robust Antarctic Ice Sheet that was dominantly cold-based where it passed across the Transantarctic Mountains. We infer that the ice sheet attained this configuration when a maritime polar climate prevailed along its margin. Evidence for the robust configuration includes boulder-belt moraines deposited alongside cold-based ice up to 700 m above the present margin of the Beardmore Glacier in the Dominion Range (Prentice et al., 1987). The primary evidence in the Dry Valleys region is selective minor erosion of bedrock, colluvium, and glacial drift up to at least 2000 m (Denton et al., 1984). We speculate that the dome(s) of the robust configuration was lower in elevation than that of the overriding configuration because accumulation on the robust ice surface under a polar climate should have been less than under the warmer subpolar climate governing the overriding configuration. Age control data place one occurrence of this configuration in the Pliocene, subsequent to the aforementioned configurations.

We also infer that the Antarctic Ice Sheet shrank to near its present dimensions during several intervals throughout the Pliocene (Fig. 24.6a). The evidence for our inference is the areal distribution of Pliocene polar desert pavements in the Transantarctic Mountains. This suggests that a continental polar climate colder than that of the robust configuration dominated the ice sheet periphery.

Hypothesis for the Pliocene Global Ice Sheet System and Its Impact on Hominidae

Figure 24.7 illustrates our leading hypothesis for the coevolution of important climate system components and Hominidae through the Pliocene. Figure 24.7a exhibits an hypothesis for glacioeustatic sea level derived from the planktonic $\delta^{18}O$ composite in Fig. 24.5. The hypothesis was derived by assuming that our composite reflects only ice volume and then applying the isotope-to-sea-level calibration derived from late Pleistocene coral terraces on Barbados (Fairbanks and Matthews, 1978). According to this calibration, a 0.11‰ change in ocean $\delta^{18}O$ equals a 10 m change in sea level. We have already outlined the potential problems of assuming constant tropical SST and negligible dissolution. A smaller uncertainty surrounds our isotope-to-sea-level calibration, which assumes that the isotopic composition of the global ice mass is constant at about -41‰. In detail this cannot be true because the oxygen isotopic composition of ice sheets will change over the course of a glacial cycle (Mix and Ruddiman, 1984) and thus over the last 6 Myr. However, plausible variation in the $\delta^{18}O$ of the global ice sheet system would change the sea-level side of the calibration by at most 3 m. In other words, a 0.11‰ change in ocean $\delta^{18}O$ will reflect between 7 and 13 m of sea-level change. We estimate that this effect may result in a slight overestimation (< 15 m) of sea-level lowering during glacial maxima.

Figure 24.7b shows the results of subtracting the planktonic isotope composite from the benthic isotope composite. Assuming that the planktonic curve represents only ice volume, then the difference reflects deep-water temperature change. Figure 24.7c shows our chronologic arrangement of the Antarctic ice reconstructions in Fig. 24.6 based on available age control.

We interpret glacioeustatic sea level fluctuations prior to 2.8 Myr as being almost entirely due to the Antarctic Ice Sheet. We think that the ice sheet attained each of the aforementioned ice sheet configurations, except the massive deglaciation scenario of Webb et al. (1984), repeatedly

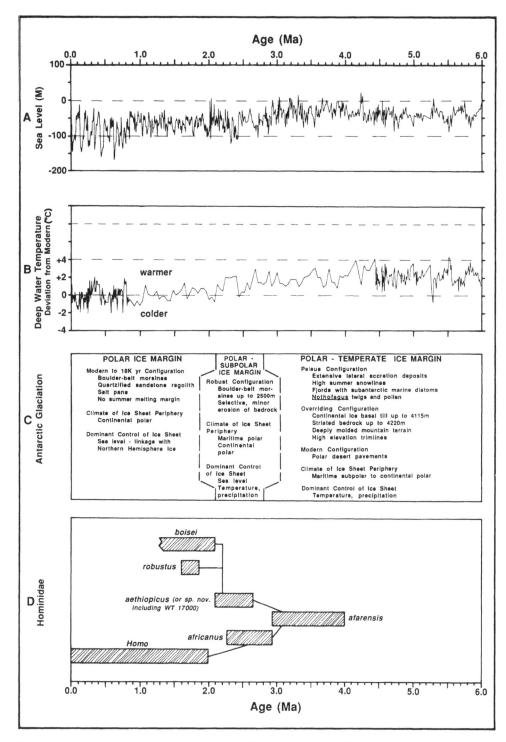

FIGURE 24.7. Our preferred hypothesis for (A) glacioeustasy, (B) deep-water temperature, and (C) Antarctic glaciation for the last 6 Myr. Hominid phylogeny (D) is from Grine (this volume, chapter 30). The generic name of the species *afarensis, aethiopicus, robustus,* and *boisei* has been omitted because of uncertainty as to whether it is *Australopithecus* or *Paranthropus*. See text for discussion.

through this interval. Hence, the ice sheet underwent volume changes that shoud have been sufficient to drive sea level between the proposed extremes depicted in Fig. 24.7a. Such massive volume changes resulted from climate shifts between maritime subpolar and continental polar superimposed on an average subpolar climate.

An average subpolar climatic regime at the Antarctic ice margin prior to 2.8 Myr is supported by our interpretation that global deep-water temperature in this interval was commonly 2°C warmer than present (Fig. 24.7b). Because this deep water likely upwelled at the divergence south of the polar front, as it does today, an increase in its temperature should have resulted in an increase in heat transfer to the Antarctic environment. The relatively warm deep-water upwelling around the Antarctic may have provided the increased moisture necessary to build the ice sheet to an overriding configuration (Prentice and Matthews, in prep.).

Our hypothesis does not yet include significant deglaciation in East Antarctica of the scale depicted in Fig. 24.6d. This is because of the discrepancy between our interpretation of the deep-sea oxygen isotope record and the prevalent interpretation of marine microfossils in till at high Transantarctic Mountain elevations. Figure 24.7b suggests that Antarctic temperatures may not have been warm enough to initiate massive deglaciation in East Antarctica. In fact, we associate our warmest inferred deep-water temperatures with more Antarctic ice than exists today, suggesting ice expansion from increased accumulation.

We interpret glacioeustatic sea level from 2.8 Myr until about 2.4 Myr as primarily reflecting Antarctic Ice Sheet fluctuations with a minor but increasing component attributable to Arctic ice caps. Around 2.4 Myr, the global ice sheet system shifted from a unipolar to a bipolar distribution. We attribute the significant increase in global ice volume at 2.4 Myr to both Antarctic and Arctic ice, the latter having the dominant effect. Because of the general decline in mean deep-water temperature through this interval (Fig. 24.7b), we suggest that the character of Antarctic glaciation became increasingly polar. The dimensions of the Antarctic Ice Sheet probably did not exceed those of the robust configuration. Such a configuration could have been achieved because water temperatures around 2.4 Myr were still warm enough to provide significant precipitation, while marine ice appendages reached a new outer limit due to the additional lowering of sea level driven by growth of Arctic ice.

We suspect that our hypothesized changes in the Antarctic cryosphere prior to and at 2.4 Myr were of a magnitude sufficient to influence Southern Hemisphere climate at mid- and low latitudes. That hemispheric climate can be seriously affected by a large polar ice mass was amply demonstrated by the atmospheric modeling experiments of Manabe and Broccoli (1985). Their experiments with and without the huge Northern Hemisphere ice sheets of the last glacial maximum indicate important changes in Northern Hemisphere climate. Additionally, Barron (1985) modeled the impact of significant Antarctic cryosphere expansion and found significant decreases in Southern Hemisphere temperature. We speculate that the our postulated changes in the Antarctic cryosphere, by themselves, could have altered: (1) the hemispheric radiation balance and therefore surface air temperature; (2) the area and intensity of the polar anticyclone and, hence, the strength of the circumpolar westerlies; and (3) the boundary in Africa between the descending dry air of the subtropical anticyclone and moist air in the ascending limb of the Hadley cell. Our speculations can be tested by sensitivity tests with global atmospheric circulation models that specify our Antarctic cryosphere reconstructions as boundary conditions. Of course, variations in other components of the climate system (e.g., CO_2), some of which may have initiated Antarctic cryosphere changes, could have amplified or even nullified the low latitude affects of the cryosphere changes.

Since 2.1 Myr the style of Antarctic glaciation has been uniformly polar. The Antarctic Ice Sheet has undergone only minor changes, driven principally by sea-level fluctuations induced by waxing and waning Northern Hemisphere ice sheets (Fig. 24.6a) (Denton et al., 1988b). Hence, the Pleistocene and latest Pliocene global ice sheet system was interlocked by sea level, with the dynamism portrayed in Fig. 24.7a evidenced in the Northern but not the Southern Hemisphere. Since 2.1 Myr, Southern Ocean SST has probably been low enough to permit significant expansion of sea ice around Antarctica. It is even possible that a ring of floating shelf ice surrounded Antarctica during late Quaternary glacial maximums (Denton and Hughes, 1986).

Conclusions

Paleoclimate inferred from deep-sea oxygen isotopic records and ice sheets varied significantly during Pliocene intervals of hominid speciation shown in Fig. 24.7d. Hence, our analysis is consistent with contentions that climatic change was the principal initiating factor of Pliocene hominid evolution (Brain, 1981; Grine, 1985; Vrba, 1985a). This analysis shows not only an important climatic event at 2.4 Myr involving a coupled bipolar ice sheet system, but also emphasizes a prior interval of progressively cooling high-latitude climate that began about 2.8 Myr under a unipolar ice sheet system. If important events in hominid evolution are linked to paleoclimate, they should not be randomly distributed through the entire 3.5 Myr span of the Pliocene. Rather, they should be concentrated in the window of climatic deterioration that began at 2.8 Myr and culminated at 2.4 Myr. Rigorous testing of these hypotheses must await a refined chronology of paleoclimate and evolution, as well as substantial advances in our understanding of Pliocene climate system components and their cumulative effects on the African environment.

One of the most pronounced changes in the global ice budget to have occurred in the last 6 Myr is that between 0.9 and 0.7 Myr (Figs. 24.5, 24.7a). Hence, this is a critical time to test the climate-forcing hypothesis by examining the fossil record for important evolutionary events. Unfortunately, the African record is not yet well represented in this important climatic interval. But the "end Villafranchian" turnover pulse in Eurasia certainly occurred close to 0.9 Myr (Azzaroli, 1983). It is also possible that the extinction of "robust" australopithecines in Africa, as well as the spread of *Homo erectus* from Africa to Eurasia, also date to this time (Klein, this volume, chapter 29; Vrba, this volume, chapter 25).

Acknowledgments

Our Antarctic research is funded by the Division of Polar Programs of the National Science Foundation. MLPs isotopic research was funded substantially by the Marine Geology Program of the National Science Foundation. Dick Kelly drafted the figures and Nancy Kealiher processed the text. We thank F. E. Grine and E. S. Vrba for reviewing this manuscript.

References

Arthur, M. and Leg 105 shipboard party (1986). High-latitude paleoceanography. *Nature, 320*: 17–18.

Azzaroli, A. (1983). Quaternary mammals and the "end-Villafranchian" dispersal event—a turning point in the history of Eurasia. *Palaeogeog. Palaeoclimatol. Palaeoecol., 44*: 117–139.

Barron, E. J. (1985). Explanations of the Tertiary global cooling trend. *Palaeogeog. Palaeoclimatol. Palaeoecol., 50*: 45–61.

Berggren, W. A., Kent, D. V. and van Couvering, J. A. (1985). Neogene geochronology and chronotratigraphy. *In* N. J. Snelling (ed.), *The chronology of the geological record. Geol. Soc. Lond. Mem., 10*: 211–260.

Blanc, P. L., Fontugne, M. R. and Duplessy, J. C. (1983). The time-transgressive initiation of boreal ice-caps: continental and oceanic evidence reconciled. *Palaeogeog. Palaeoclimatol. Palaeoecol., 42*: 211–224.

Boellstorff, J. (1978). A need for redefinition of North American Pleistocene stages. *Trans. Gulf Coast Assoc. Geol. Soc., 28*: 65–74.

Bonnefille, R. (1976). Palynological evidence for an important change in the vegetation of the Omo Basin between 2.5 and 2.0 million years ago. *In* Y. Coppens, F. C. Howell, G. L. Isaac and R. E. F. Leakey (eds.), *Earliest Man and environments in the Lake Rudolf Basin*, pp. 421–431. University of Chicago Press, Chicago.

Bonnefille, R. (1983). Evidence for a cooler and drier climate in Ethiopia uplands towards 2.5 Myr ago. *Nature, 303*: 487–491.

Bonnefille, R., Vincens, A. and Buchet, G. (1987). Palynology, stratigraphy and palaeoenvironment of a Pliocene hominid site (2.9–3.3 M.Y.) at Hadar, Ethiopia. *Palaeogeog. Palaeoclimatol. Palaeoecol., 60*: 249–281.

Brain, C. K. (1981). The evolution of man in Africa: was it a consequence of Cainozoic cooling? *Ann. Geol. Soc. S. Afr., 84*: 1–19.

Burckle, L. H., Prentice, M. L. and Denton, G. H. (1986). Neogene Antarctic glacial history: new evidence from marine diatoms in continental deposits. *EOS Trans., 67*: 295.

Carlquist, S. (1987). Pliocene *Nothofagus* wood from the Transantarctic Mountains. *Aliso, 11*: 571–583.

Chappell, J. and Shackleton, N. J. (1986). Oxygen isotopes and sea level. *Nature, 324*: 137–140.

CLIMAP project members (1981). Seasonal reconstructions of the earth's surface at the last glacial maximum. *Geol. Soc. Amer. Map Chart Ser., MC–36.*

Denton, G. H. (1985). Did the Antarctic ice sheet influence Late Cenozoic climate and evolution in the southern hemisphere? *S. Afr. J. Sci., 81*: 224–229.

Denton, G. H. and Hughes, T. J. (1981). *The last great ice sheets*. Wiley (Interscience), New York.

Denton, G. H. and Hughes, T. J. (1986). Potential influence of floating ice shelves on the climate of an ice age. *S. Afr. J. Sci., 82*: 509–513.

Denton, G. H., Bockheim, J. G., Wilson, S. C. and Schluchter, C. (1987). Late Cenozoic history of Rennick Glacier and Talos Dome, Northern Victoria Land, Antarctica. *In* E. Stump (ed.), *Geological investigations in northern Victoria Land*. Antarctic Research Series 46, pp. 339–375. Amer. Geophy. Union, Washington, D.C.

Denton, G. H., Bockheim, J. G., Rutford, R. H. and Andersen, B. G. (1988a). Glacial history of the Ellsworth Mountains. *In* G. F. Webers, J. F. Splettstoesser and C. Craddock (eds.), *Geology of the Ellsworth Mountains*, Geological Society of American Memoir. in press.

Denton, G. H., Prentice, M. L. and Burckle, L. H. (1988b). Cenozoic history of the Antarctic Ice Sheet. *In* R. Tingey (ed.), *The geology of Antarctica*. Oxford University Press, London, in press.

Dodge, R. E., Fairbanks, R. G., and Maurrasse, F. (1983). Pleistocene sea levels from raised coral reefs of Haiti. *Science, 219*: 1423–1425.

Fairbanks, R. G. and Matthews, R. K. (1978). The marine oxygen isotope record in Pleistocene coral, Barbados, West Indies. *Quat. Res., 10*: 181–196.

Graham, D., Corliss, B., Bender, M. and Keigwin, L. D. (1981). Carbon and oxygen isotopic disequilibria of Recent deep-sea benthic foraminifera. *Marine Micropaleont., 6*: 483–497.

Grine, F. (1981). Trophic differences between "gracile" and "robust" australopithecines: a scanning electron microscope analysis of occlusal events. *S. Afr. J. Sci., 77*: 203–230.

Grine, F. (1982). A new juvenile hominid (Mammalia; Primates) from Member 3, Kromdraai Formation, Transvaal, South Africa. *Ann. Tvl. Mus., 33*: 165–239.

Grine, F. (1985). Was interspecific competition a motive force in early hominid evolution? *In* E. S. Vrba (ed.), *Species and speciation*, pp. 143–152. *Monogr. Tvl. Mus.*, Pretoria.

Grine, F. (1986). Ecological causality and the pattern of Plio-Pleistocene hominid evolution in Africa. *S. Afr. J. Sci., 80*: 87–89.

Grube, F., Christensen, S. and Vollmer, T. (1986). Glaciations in north west Germany. *Quat. Sci. Rev., 5*: 347–358.

Haq, B. U., Hardenbol, J. and Vail, P. R., (1987). Chronology of fluctuating sea levels since the Triassic. *Science, 235*: 1156–1167.

Harwood, D. M. (1986). Recycled siliceous microfossils from the Sirius Formation. *Ant. J. U.S., 21:* 101–103.

Hodell, D. A., Kennett, J. P., and Leonard, K. A. (1983). Climatically induced changes in vertical water mass structure of the Vema Channel during the Pliocene: evidence from Deep Sea Drilling Project Holes 516A, 517, and 518. *In* P. F. Barker, R. L. Carlson, D. A. Johnson et al. (eds.), *Initial reports of the Deep Sea Drilling Project*, vol. 72, pp. 907–919. U.S. Gov. Printing Office, Washington, D.C.

Hooghiemstra, H. (1984). Vegetational and climatic history of the high plain of Bogota, Colombia: A continuous record of the last 3.5 million years. *Dissert. Botan., 79*, J. Cramer Verlag.

Imbrie, J., Hays, J. D., Martinson, D. G., McIntyre, A., Mix, A. C., Morley, J. J., Pisias, N. G., Prell, W. L. and Shackleton, N. J. (1984). The orbital theory of Pleistocene climate: support from revised chronology of the marine $\delta^{18}O$ record. *In* A. L Berger, J. Imbrie, J. Hays, G. Kukla and B. Saltzman (eds.), *Milankovitch and Climate, Part 1*. Reidel, pp. 269–305.

Janecek, T. R. and Ruddiman, W. F. (1987). Plio-Pleistocene sedimentation in the south equatorial Atlantic divergence: XII INQUA Research, Ottawa, Canada. Program and Abstracts: 193.

Jansen, E., Kringstad, L. and Sejrup, H. P. (1986). Evolution of climate and ocean circulation in the Norwegian Sea and the North Atlantic, Miocene to Quaternary: ODP Leg 104, DSDP Leg 94. *Second International Conference on Paleoceanography*. Abstracts with program.

Johanson, D. C. and White, T. D. (1979). A systematic assessment of early African hominids. *Science, 203*: 321–330.

Keigwin, L. D., Jr. (1987). Pliocene stable-isotope record of Deep Sea Drilling Project Site 606: Sequential events of $\delta^{18}O$ enrichment beginning at 3.1 Ma. *In* W. F. Ruddiman, R. B. Kidd et al. (eds.), *Initial reports of the Deep Sea Drilling Project. v. 94*, pp. 911–920. U.S. Gov. Printing Office, Washington, D.C.

Keigwin, L. D., Jr. (1988). Toward a high-resolution chronology for latest Miocene paleoceanographic events. *Paleoceanography, 2*: 639–660.

Krissek, L. A. and Carney, T. (1986). The late Pliocene and Pleistocene records of ice-rafting in the northwest Pacific and the Norwegian Sea. *Second International Conference on Paleoceanography*. Abstracts with program.

Kukla, G. (1987). Loess stratigraphy in central China. *Quat. Sci. Rev., 6*: 191–219.

Lorius, C., Jouzel, J., Ritz, C., Merlivat, L., Barkov, N. I., Korotkevich, Y. S. and Kotlyakov, V. M. (1985). A 150,000-year climatic record from Antarctic ice. *Nature, 316*: 591–596.

Manabe, S. and Broccoli, A. J. (1985). The influence of continental ice sheets on the climate of an ice age. *J. Geophys. Res., 90*: 2167–2190.

Matthews, R. K. (1986). The $\delta^{18}O$ signal of deep-sea planktonic foraminifera at low latitudes as an ice-volume indicator. *S. Afr. J. Sci., 82*: 521–522.

Matthews, R. K. and Poore, R. Z. (1980). Tertiary $\delta^{18}O$ record and glacio-eustatic sea-level fluctuations. *Geology, 8*: 501–504.

Mercer, J. H. (1988). The Antarctic Ice Sheet during the Neogene. *Palaeoecol. Afr.*, in press.

Mix, A. C. and Ruddiman, W. F. (1984). Oxygen-isotope analyses and Pleistocene ice volumes. *Quat. Res., 21*: 1–21.

Newell, R. E. and Chiu, L. S. (1981). Climatic changes and variability: a geophysical problem. *In* A. Berger (ed.), *Climatic variations and variability: facts and theories*, pp. 21–61. D. Reidel, Dordrecht.

Ninkovitch, D. and Shackleton, N. J. (1975). Distribution, stratigraphic position and age of ash layer "L" in the Panama Basin region. *Earth Planet. Sci. Let., 27*: 20–34.

Olson, T. R. (1981). Basicranial morphology of the extent hominids and Pliocene hominids: The new material from the Hadar Formation, Ethiopia and its significance in early human evolution and taxonomy. *In* C. B. Stringer (ed.), *Aspects of human evolution*, pp. 99–128: Taylor and Francis, London.

Pisias, N. G. and Moore, T. C., Jr. (1981). The evolution of Pleistocene climate: a time series approach. *Earth Planet. Sci. Let., 52*: 450–458.

Pisias, N. G., Martinson, D. G., Moore, T. C., Jr., Shackleton, N. J., Prell, W. L., Hays, J. and Boden, G. (1984). High resolution stratigraphic correlation of benthic oxygen isotopic records spanning the last 300,000 years. *Marine Geol., 56*: 119–136.

Poore, R. Z. (1981). Temporal and spatial distribution of ice-rafted mineral grains in Pliocene sediments of the North Atlantic: Implications for late Cenozoic climatic history. *Soc. Econ. Paleontol. Mineral. Spec. Pub., 32*: 505–515.

Prell, W. L. (1982). Oxygen and carbon isotope stratigraphy for the Quaternary of Hole 502B: Evidence for two modes of isotopic variability. *In* W. L. Prell, J. V. Gardner et al. (eds.), *Initial reports of the deep sea drilling project, v. 68*, pp. 455–464. U.S. Gov. Printing Office, Washington, D.C.

Prell, W. L. (1985a). Pliocene stable isotope and carbonate stratigaphy (Holes 572C and 573A): Paleoceanographic data bearing on the question of Pliocene glaciation. *In* L. Mayer, F. Theyer, et al. (eds.), *Initial reports of the deep sea drilling project, v. 85*, pp. 723–734. U.S. Gov. Printing Office, Washington, D.C.

Prell, W. L. (1985b). The stability of low-latitude sea-surface temperatures: An evaluation of the CLIMAP reconstruction with emphasis on the positive SST anomalies. *Depart. of Energy (DOE/ER/60167-1)*, p. 60.

Prell, W. L., Imbrie, J., Martinson, D. G., Morley, J. J., Pisias, N. G., Shackleton, N. J. and Streeter, H. F. (1986). Graphic correlation of oxygen isotope stratigraphy application to the late Quaternary. *Paleocean., 1*: 137–162.

Prentice, M. L. (1985). Peleus glaciation of Wright Valley, Antarctica. *S. Afr. J. Sci, 81*: 241–243.

Prentice, M. L. (1988). The deep sea oxygen isotopic record: Significance for Tertiary global ice volume history, with emphasis on the latest Miocene/early Pliocene. PhD Thesis, Brown University, Providence, RI.

Prentice, M. L. and Matthews, R. K. (1988). Cenozoic ice volume history: Development of a comoposite oxygen isotope record. *Geology*, (in press).

Prentice, M. L. and Matthews, R. K. (in prep.). Tertiary ice sheet dynamics: the snow-gun hypothesis.

Prentice, M. L., Denton, G. H., Lowell, T. L. and Conway, H. (1986). Pre-late Quaternary glaciation of the Beardmore Glacier region, Antarctica. *Ant. J. U.S., 21*: 95–98.

Prentice, M. L., Denton, G. H., Burckle, L. H., and Hodell, D. A. (1987). Evidence from Wright Valley for the response of the Antarctic Ice Sheet to climate warming. *Antarc. J. U.S.*, in press.

Rea, D. K. and Schrader, H. (1985). Late Pliocene onset of glaciation: ice-rafting and diatom stratrigraphy of North Pacific DSDP cores. *Palaeogeog., Palaeoclim., Palaeoecol., 49*: 313–325.

Ruddiman, W. F. and Raymo, M. E. (1988). Northern hemisphere climate regimes during the last 3 myr: Possible tectonic connections. *Phil. Trans. R. Soc.*, in press.

Ruddiman, W. F., Raymo, M. and McIntyre, A. (1986). Matuyama 41,000 year cycles: North Atlantic ocean and northern hemisphere ice sheets. *Earth Planet. Sci. Let., 80*: 117–129.

Shackleton, N. J. (1987). Oxygen isotopes, ice volume, and sea level. *Quat. Sci. Rev., 6*: 183–190.

Shackleton, N. J. and Hall, M. A. (1984). Oxygen and carbon isotope stratigraphy of Deep Sea Drilling Project Hole 552A: Plio-Pleistocene glacial history. *In* D. G. Roberts, D. G Schnitker, et al. (eds.), *Initial reports of the deep sea drilling project, v. 81.*, pp. 599–609, U.S. Gov. Printing Office, Washington, D.C.

Shackleton, N. J. and Opdyke, N. D. (1973). Oxygen isotope and palaeomagnetic stratigraphy of equatorial Pacific core V28-238: Oxygen isotope temperatures and ice volumes on a 10^5 year and 10^6 year scale. *Quat. Res., 3*: 39–55.

Shackleton, N. J. and Opdyke, N. D. (1976). Oxygen-isotope and paleomagnetic stratigraphy of Pacific core V28-239 late Pliocene to latest Pleistocene. *In* R. M. Cline and J. Hays (eds.), *Investigation of late Quaternary paleoceanography and paleoclimatology. Geol. Soc. Amer. Mem., 145*, pp. 449–464.

Shackleton, N. J. and Opdyke, N. D. (1977). Oxygen isotope and paleomagnetic evidence for early Northern Hemisphere glaciation. *Nature, 270*: 216–219.

Shackleton, N. J., Backman, J., Zimmerman, H., Kent, D. V., Hall, M. A., Roberts, D. G., Schnitker, D., Baldauf, J. G., Desprairies, A., Homrighausen, R., Huddlestun, P., Kenne, J. B., Kaltenback, A. J., Krumsiek, K. A. O., Morton, A. C., Murray, J. W. and Westberg-Smith, J. (1984). Oxygen isotope calibration of the onset of ice-rafting and history of glaciation in the North Atlantic region. *Nature, 307*: 620–623.

Thunell, R. C. and Williams, D. F. (1983). The stepwise development of Pliocene-Pleistocene paleoclimatic and paleoceanographic conditions in the Mediterranean: oxygen isotopic studies of DSDP Sites 125 and 132. *In* Meulenkamp, J. E. (ed.), *Reconstruction of Marine Paleoenvironments, Utrecht Micropal. Bull., 30*: 111–127.

Tobias, P. V. (1980). *"Australopithecus afarensis"* and *A. africanus:* critique and alternative hypothesis. *Palaeont. Afr.,* 23: 1–17.

Vail, P. R., Mitchum, R. M., Jr. and Thompson, S., III (1977). Seismic stratigraphy and global changes of sea level, Part 4: Global cycles of relative changes of sea level. *In* C. E. Payton (ed.), *Seismic Stratigraphy—applications to hydrocarbon exploration. Amer. Assoc. Pet. Geol. Mem.,* 26, pp. 83–97. Tulsa.

van der Hammen, T., Wijmstra, T. A. and Zagwijn, W. H. (1971). The floral record of the late Cenozoic of Europe. *In* K. K. Turekian (ed.), *The late Cenozoic glacial ages,* pp. 391–424: Yale University Press, New Haven.

Van Zinderen Bakker, E. M. and Mercer, J. H. (1986). Major late Cainozoic climatic events and palaeoenvironmental changes in Africa viewed in a world wide context. *Palaeogeog. Palaeoclimatol. Palaeoecol.,* 56: 217–235.

Vrba, E. S. (1975). Some evidence of chronology and paleoecology of Sterkfontein, Swartkrans and Kromdraai from the fossil Bovidae. *Nature,* 254: 301–304.

Vrba, E. S. (1982). Southern African hominid-associated assemblages: biostratigraphy, chronology and evolutionary pulses. Address to 1982 Congress International de Paleontologie Humaine, Union Internat. des Sciences Prehistoriques et Protohistoriques, Nice.

Vrba, E. S. (1985a). Ecological and adaptive changes associated with early hominid evolution. *In* E. Delson (ed.), *Ancestors: the hard evidence,* pp. 63–71. Alan R. Liss, New York.

Vrba, E. S. (1985b). Early hominids in southern Africa: Updated observations on chronological and ecological background. *In* P. V. Tobias (ed.), *Hominid evolution: past, present and future,* pp. 195–200. Alan R. Liss, New York.

Vrba, E. S. (1985c). Environment and evolution: alternative causes of the temporal distribution of evolutionary events. *S. Afr. J. Sci.,* 81: 229–236.

Vrba, E. S. (1985d). African Bovidae: evolutionary events since the Miocene. *S. Afr. J. Sci.,* 81: 263–266.

Vrba, E. S. (1988). The environmental context of the evolution of early hominids and their culture. *In* R. Bonnichsen and M. H. Sorg (eds.), *Bone modification.* Center for the Study of Early Man, in press.

Walker, A., Leakey, R. E., Harris, J. M. and Brown, F. H. (1986). 2.5 Myr *Australopithecus boisei* from west of Lake Turkana, Kenya. *Nature,* 322: 517–522.

Webb, P. N., Harwood, D. M., McKelvey, B. C., Mercer, J. H. and Stott, L. D. (1984). Cenozoic marine sedimentation and ice volume variation on the East Antarctic craton. *Geology,* 12: 287–291.

Wesselman, H. B. (1985). Fossil micromammals as indicators of climatic change about 2.4 Myr ago in the Omo Valley, Ethiopia. *S. Afr. J. Science,* 81: 260–261.

White, T. D., Johanson, D. C. and Kimbel, W. H. (1981). *Australopithecus africanus:* its phyletic position reconsidered. *S. Afr. J. Sci.,* 77: 445–470.

Williamson, P. G. (1985). Evidence of an early Plio-Pleistocene rainforest expansion in East-Africa. *Nature,* 315: 487–489.

Zagwijn, W. H. (1960). Aspects of the Piocene and Early Pleistocene vegetation in the Netherlands. *Mededel. Geol. Sticht.,* Ser. C, 3, 1, 5, 1.

Zagwijn, W. H. and Doppert, J. W. Chr. (1978). Upper Cenozoic of the southern North Sea Basin: Palaeoclimatic and palaeogeographic evolution. *Geol. En Mijnbouw,* 57: 577–588.

Late Pliocene Climatic Events and Hominid Evolution

25

ELISABETH S. VRBA

The Late Pliocene was a time of both global climatic changes and critical events in hominid evolution. Recent advances have led to new and sharpened perceptions of the nature and chronology of climatic shifts in many parts of the world, of biotic changes in Africa in particular, and of hominid evolutionary events during this time. In this contribution I analyse some hypotheses and predictions that link the new environmental and hominid data.

The Hominid Record

One previous set of phylogenetic hypotheses placed *Australopithecus africanus* as the direct ancestor of the bifurcation to *Homo*, on the one hand, and to the "robust" clade containing *Australopithecus robustus* and *A. boisei*, on the other (e.g., Tobias, 1985; Skelton *et al.*, 1985). These hypotheses, as well as mine that this particular bifurcation was initiated by climatic cooling close to 2.5 Myr ago (Vrba, 1985a: fig. 3), are contradicted by the morphology of the cranium KNM-WT 17000 that was recently discovered west of Lake Turkana in a stratum dated 2.5 Myr. In terms of current consensus among the discoverers (Walker *et al.*, 1986; Walker and Leakey, this volume, chapter 15) and others (Grine, this volume, chapter 30), KNM-WT 17000 is close to the ancestry of *A. boisei*, its character combination is incompatible with descent from *A. africanus*, and KNM-WT 17000–*A. boisei* and *A. africanus* belong on two lineages that were separate before 2.5 Myr.

Thus, future phylogenetic hypotheses will center on at least two lineages already present during the time of Makapansgat and Sterkfontein *A. africanus* (~3.0 to 2.5 Myr), on their branching patterns relative to each other and to *A. afarensis*, and on their subsequent evolutionary histories. Two such alternatives are represented in Fig. 25.1. Cladistic taxonomies based on any of the phylogenetic alternatives will necessitate revision of previously accepted hominid generic names, and until the choice between alternatives becomes more clear-cut, I prefer to omit generic names. Thus, in Fig. 25.1 and throughout the text I use *afarensis, africanus, boisei, aethiopicus, robustus* and *crassidens* for species previously assigned to *Australopithecus* or *Paranthropus*.

There are four kinds of statements implicit in Fig. 25.1 that differ in the number of substatements (assumptions and hypotheses) they contain. The first and simplest concerns the earliest and latest records of hominid species (solid rectangles in Fig. 25.1). Such a statement contains assumptions about which phenotypes belong to the same species and about chronology, but none concerning the relationships among species. In Fig. 25.1 I use Delson's (1984) suggestion that Taung rather than Sterkfontein Member 4 contains the latest evidence of *africanus*.

Statements of the second kind are cladograms, requiring hypotheses that pairs of taxa share a closer common ancestry with each other than either does with any other taxon. The cladistic arrangement underlying Fig. 25.1a treats "robust" australopithecines as a monophyletic group (Grine, this volume, chapter 30). In contrast, Fig. 25.1b belongs to the set of hypotheses that posit a polyphyletic origin for "robust" australopithecines, with *africanus* ancestral to *robustus* but not to *boisei* (Walker and Leakey this volume, chapter 15; Wood, this volume, chapter 17).

Statements of the third kind are phylogenetic trees, which add hypotheses of ancestry to the

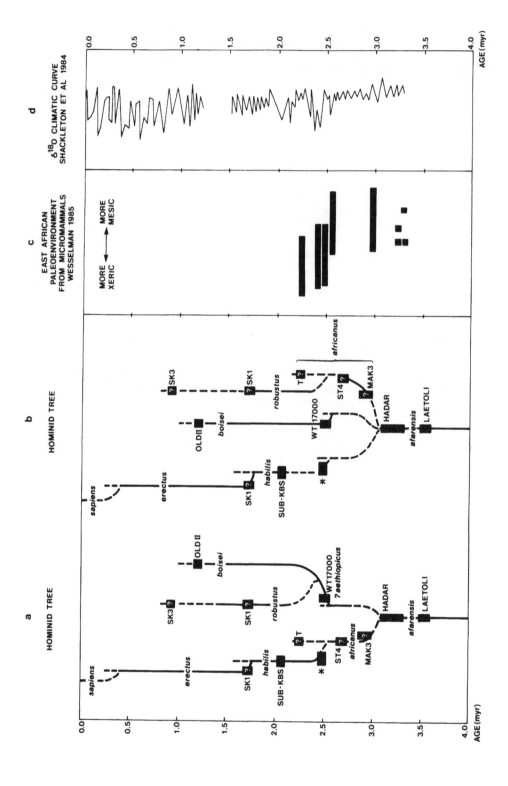

cladistic ones. Those of the fourth and most complex kind add further assumptions of how the splitting events relate to absolute time: they are phylogenetic trees plotted against time, or "time trees". In the time trees in Fig. 25.1, each species is assumed to have originated by what is technically a branching event, even though only one species may have continued from (while the parent, or sister species disappeared close to) the split, and the assumption is made that at least some of the first and last records (e.g., those for *africanus, habilis,* and the KNM-WT 17000–*boisei* lineage) fall close to times of species' origins and extinctions, respectively.

The range of plausible phylogenetic hypotheses has been thrown wide open, with the schemes represented in Fig. 25.1 merely two among several. The common denominator between the rival phylogenies is the fossil record. Many of my arguments will refer directly to these simplest statements in Fig. 25.1a and 25.1b.

Major Paleoclimatic Changes

Physical changes that act as evolutionary causes can come from two sources: tectonic changes with relatively localized signatures in the record; and global climatic changes that leave imprints in many parts of the world. An example of the first kind is the development of the East African Rift System. Starting in the Late Eocene there were three major episodes of warping and uplift, accompanied by volcanism and faulting in East Africa. Baker *et al.* (1982) place the last of these in the Late Pliocene. Such episodes formed barriers and changed drainage patterns and climate [see Brown and Feibel (this volume, chapter 22) for evidence in the hominid sequences], and must have influenced the local biota strongly (Denys *et al.*, 1986).

Superimposed on the effects of the regional tectonism were the more rapid and regular global climatic cycles (the so-called astronomical or Milankovitch cycles). Some of the major long-term alterations in the mean, amplitude and dominant periodicity of this cyclic pattern during the Plio-Pleistocene are reviewed in Prentice and Denton (this volume, chapter 24). Here I refer briefly to results related to hominid evolution.

There is Pleistocene evidence that these astronomically caused cycles were accompanied not only by large-scale expansion and retreat of ice at the poles (Hays *et al.*, 1976; Shackleton *et al.*, 1984; Denton, 1985), but also by major climatic and vegetational changes in the terrestrial tropics (Rind and Peteet, 1985). Thus, African evidence ranging from the Eastern low latitudes (Livingstone, 1975; Flenley, 1979; Hamilton, 1982) all the way to the Southern Cape (Vogel, 1985) shows that at 18 kyr BP (the last glacial maximum) it was substantially cooler (5°–6°C) and more arid, with more reduced forests and more widespread grasslands, than today. During the Holocene, warmer and wetter conditions with higher wood–grass ratios than at present are reported from many areas (Vogel, 1984). [We should not expect the same temperature–wetness–vegetation correlations at all past times, as they varied with fluctuations in major circulation features such as monsoonal cycles (Kutzbach and Otto-Bliesner, 1982; Street-Perrot and Roberts, 1983).] The

◁—————

FIGURE 25.1. Hominid evolution and paleoenvironmental change since the Late Pliocene a.b: Two hypothetical hominid trees each of which is supported by several workers (see Grine's summary, this volume, chapter 30). Solid rectangles represent chronologic estimates for some earliest and latest records of hominid taxa (mainly after Brown *et al., 1985* for East Africa and Vrba, 1982, for South Africa). The ?'s denote estimates based on biochronology. The symbol * refers to the hypothesis that at least some of the earliest known stone tools represent the behavioral phenotype of *Homo habilis*. SK 1–3 Swartkrans Members 1 through 3; ST 4 = Sterkfontein Member 4; MAK 3 = Makapansgat Member 3; T = Taung; WT = Turkana; OLD II = Olduvai Bed II level BK; SUB-KBS = stratum below KBS Tuff at Koobi Fora. c: Paleoecological analysis of micromammals from Shungura Members B, C, F, G. Adapted from Wesselman (1985: fig. 1). d; deep sea oxygen isotope data from a North Atlantic site, adapted from Shackleton *et al.*, 1984; ice volume increase to the left.

known environmental differences that accompanied the Late Quaternary oxygen isotope cycles were massive relative to the habitat specificities of most species.

Were there Pliocene changes of comparable magnitude? For the Late Pliocene some continuous, well-dated records are available, and they agree on a major global cooling event at ~2.4 to 2.5 Myr. Shackleton et al.'s (1984) deep-sea $\delta^{18}O$ record (Fig. 25.1d) shows evidence near 2.4 Myr of the first widespread North Polar glaciation and a second intensification of cold cyclic extremes after 1.0 Myr. Similarly, Thunell and Williams (1983) report "mean cooling shifts" at ~2.5 Myr and 0.9 Myr in $\delta^{18}O$ values from several areas in the world's oceans.

The 2.5-Myr event is also recorded in land sequences of different continents. The floral record shows the first onset of Arctic conditions in north-central Europe close to this time (during the Praetiglian following the warm Reuverian stage) (Van der Hammen et al., 1971). For South America, Hooghiemstra (1986) reports palynological changes from the high plain of Bogota, Colombia, where between ~3.2 and 2.5 Myr and again between ~0.96 and 0.0 Myr the plain was predominantly covered with Upper Andean Forest reflecting two relatively warm periods. At ~2.5 Myr a change occurred to open vegetation types, indicating a predominantly cold period lasting until ~0.96 Myr. This massive response in continental areas is as expected by the 18kyr analogy: Shackleton et al. (1984) describe the 2.4-Myr excursion (Fig. 25.1d) as toward a truly glacial interval with ice volumes comparable to those at the Quaternary maxima. I will discuss below to what extent the African evidence agrees with this.

The Pliocene climatic record pre-2.5 Myr is of poorer quality (review in Mercer, 1985). Evidence on its larger features, going back in time from 2.5 Myr, is as follows:

(1) A very warm period (predominantly warmer than today), the onset of which as a global feature is not clear, is indicated ~3.2 to 2.5 Myr in Colombia (Hooghiemstra, 1986); in Antarctica tentatively somewhere between 3.1 and 2.6 Myr (Prentice and Denton this volume, chapter 24); and ~3.4 to 2.6 Myr, with particularly warm excursions near 3.1 and 2.9 Myr, in Shackleton et al.'s (1984) curve (Fig. 25.1).

(2) Some reports of cooling just previous to or within (depending on which data set is referred to) this warm phase indicate glaciation in Iceland starting at 3.1 Myr (referenced in Shackleton et al., 1984), ice-sheet surging in Antarctica at ~3.2 Myr (Harwood, 1985), and a cold period of open vegetation in Colombia ~3.5 to 3.2 Myr (Hooghiemstra, 1986).

(3) An unusually warm period in the early Pliocene placed by Burckle (1985) between ~4.5 and 3.5 Myr. Once this climatic picture for the earlier Pliocene becomes clearer, it will be of great interest to students of African environments as a context for early hominid evolution.

General Evolutionary and Ecological Hypotheses

How do environmental changes relate to biotic evolution? Here I will discuss a few general hypotheses and predictions, and in the next section several specific ones concerning hominids, that are testable by the African record. Some key terms will be used as follows (Vrba, 1987a):

BIOME—a major ecological community characterized by distinctive life forms and principal taxa, although the particular species may vary within the same biome at different times and in different areas.

DISTRIBUTION DRIFT—changes in species' geographic distributions, including shrinkage/fragmentation (vicariance) and expansion/coalescence (mobilism). "Habitat drift" is used analogously for species' habitats.

BIOTIDAL AREA—an area is biotidal over a time interval if distinct biomes replace each other during that time (Fig. 25.2). A markedly biotidal area (or stratigraphic column) has "biome tides" sweeping across it at each excursion of the astronomical climatic cycles.

REFUGIUM—an area that, over a given time interval, escapes major climatic changes typical of a region as a whole, and in which a biome persists that has disappeared elsewhere (Fig. 25.2).

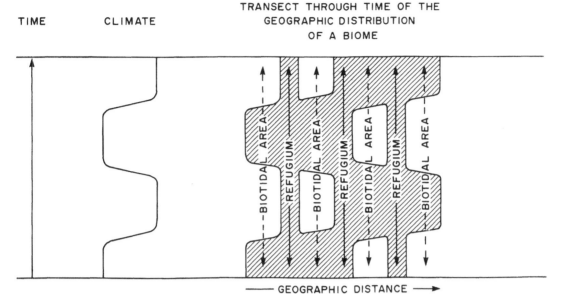

FIGURE 25.2. Transect through time of the geographical distribution of a biome (hatched). Refugia and biotidal areas leave contrasting patterns in stratigraphic columns in response to climatic changes.

LINEAGE TURNOVER—includes speciation, extinction and distribution drift all of which "turn over" the composition of species in particular areas.

TURNOVER-PULSE—concentration against time of turnover events in many biotic groups in response to a widespread enviromental change. Thus, if a high number of first and last species' records in diverse lineages occur within 200kyr or less, preceded and postdated by ~1.0 Myr of significantly less turnover, a turnover-pulse is recognized (Vrba, 1985b).

GLOBAL CLIMATIC CHANGE—alterations of climatic signals in many parts of the world during a particular time period, such as those that can result from astronomical forcing. My use of the term recognizes that not all localities are affected, that climatic signals differ among those that are affected, and that particular climatic results may be out of phase.

INITIATING CAUSES OF SPECIATION AND EXTINCTION—I distinguish between physical causes from (1) global climatic changes or from (2) regional topographic and tectonic changes; and (3) biotic causes, such as interspecific competition and predation that can act independently of physical factors to cause turnover. To label a factor "an initiating cause on earth" does not imply that other extraterrestrial or subsequent terrestrial causes are not operating as well in the larger causal chain.

Hypothesis 1: Refugial Versus Biotidal Areas

Distribution drift, the most common turnover-response of lineages to environmental changes, consistently affects two basic kinds of geographic areas differently over a given time period: in a refugium the biome persists, while in a biotidal area it does not.

This is related to Haffer's (1982) "refuge theory." He emphasises refugia and predictions for extant distributions. My hypothesis focuses on both refugia and biotidal areas and on predictions of their present and paleontological signatures.

Predictions of Hypothesis 1 include the following. A biotidal record in a particular stratigraphic column (Fig. 25.2) is characterized by disappearances and new appearances of dominant (abundant) higher taxa at times of climatic change; while some resistant peripheral taxa persist. The same suite of dominant taxa (the same biome) may return repeatedly to the same area as similar habitats recur. However, if a major event of extinctions and speciations intervened between two strata, they will contain different dominant taxa even if they represent a similar climate. In contrast, a refugial record is characterized by persistence of dominant taxa (although new species within these taxa may be added). Climatic change in the larger region is recorded in a refugium only close to its ecotonal limits, by the new appearances (or disappearances) of peripheral taxa (few organisms in few species) that represent occasional intrusive elements from the alternative biome. Such intrusive taxa may include wind-borne (e.g., pollen) or bird-transported (e.g., microfauna in owl pellets) elements and mobile organisms that typically transgress ecotones (e.g., some ungulates).

A modern African analogue is the Okavango Delta. It is a species-rich, riverine-deltaic ecosystem within the "Middle Kalahari" semidesert region of Botswana (Cooke, 1976). The dominant, resident large-bovid taxa are impalas *(Aepyceros)*, buffaloes *(Syncerus)*, several species of reduncines, tragelaphines (several species of *Tragelaphus*), alcelaphines (blue wildebeests, Connochaetes; and tsessebes, *Damaliscus*), and hippotragines (roan and sable antelopes, *Hippotragus*). Arid-adapted elements, such as *Oryx*, springbuck *(Antidorcas)*, and hartebeest *(Alcelaphus)* cross the ecotone bordering the semidesert seasonally and in relatively small numbers (Patterson, 1976; Astle and Graham, 1976). There is evidence that the Kalahari as a whole experienced alternating wet and dry phases in the past (Cooke, 1976). Thus, the larger region surrounding the Delta has been a biotidal area. If we find a suitable fossil sequence in the Okavango refugium, we may expect to find the onset of the current arid phase to be marked only by the relatively sudden addition of small numbers of arid-adapted specimens to the persistently abundant resident taxa.

Hypothesis 2: The Turnover-Pulse Hypothesis

Physical environmental change is required to initiate most speciations, extinctions and distribution drift. Thus, most lineage turnover has occurred in pulses, near-synchronous across diverse groups of organisms and associated with particular physical changes (Vrba, 1985b).

Predictions include that, given sufficient resolution, first and last records of species across diverse taxa should be associated nonrandomly with each other aginst the time scale, and with independent evidence of physical environmental change. An alternative hypothesis is this: Numerous different, mostly local, initiating causes, including a significant proportion of purely biotic causes such as interspecific competition, contribute to turnover. The predictions include that most speciations and extinctions should not associate with each other or across diverse taxa and should form a largely random frequency distribution against time (Vrba, 1985b).

Hypothesis 3: Climatic Initiating Cause

This hypothesis states that a particular environmental cause of turnover was global, or at least widespread, climatic change.

Predictions include that the turnover record has a global or widespread distribution that matches the climatic record consistently. An alternative hypothesis about a physical initiating cause is that it was tectonic (such as uplift, drift, or volcanism). This predicts a turnover signature that is appropriately geographically restricted.

Specific Hypotheses on Hominid Environments and Evolution

Dating estimates for the major Late Pliocene climatic change, recorded in marine and land areas world wide, are close to 2.5 to 2.4 Myr. I shall refer to this change as the 2.5-Myr event. I now evaluate some hypotheses that relate directly to hominid environments and evolution; and, in doing so, I will also refer to the foregoing three hypotheses.

Hypothesis 4: The 2.5-Myr Event in East Africa

This hypothesis states that the 2.5-Myr global cooling event initiated the spread of more arid and vegetationally open environments in Eastern Africa.

Figure 25.3 presents the stratigraphic ranges of 61 sub-Saharan species of Bovidae. The left part of the graph, containing species either confined to the East African record or known from both East and South Africa, provides a test of this hypothesis. It is based on the cited literature plus personal study of all the bovid collections except that from West Turkana. The most useful data set for testing this hypothesis is currently that from the Shungura Formation, Ethiopia, because it provides provenance records correct to subunits within radiometrically dated members that sample fairly closely across the critical time period. Figure 25.3 contains minor departures from Gentry's (1985) analysis of the Omo bovids. For instance, I differ in not assigning isolated bovine dentitions from Members B and C to *Pelorovis,* Member E through G bovine horn core fragments to *Syncerus* [previously assigned by Gentry (1985) to *Hemibos*], or a Member B dentition to *Antidorcas recki.* Also I do not see remains of *Tragelaphus gaudryi* in Members B through C as implied in Table 1 of Gentry (1985) [although on p. 127 of Gentry (1985) the range for the species is cited as from Member E onward, which accords with my observations].

The time 2.6 to 2.4 Myr includes first records of possible Eurasian immigrants (marked I) and locally evolved species, as well as last species records. Based on phylogenetic analyses at least some of the last and first records reflect terminal extinction and true speciation (lineage splitting) respectively, rather than unbranching "phyletic extinction and speciation" (see Vrba, 1985b,c).

I used χ^2 analyses to test the null hypothesis that the first records (disregarding dashed lines) in Fig. 25.3 are randomly distributed in time, as follows: (a) with the time scale divided into 200-kyr-long intervals, (i.e., 1.0–1.2 Myr, etc.) including the 2.4 to 2.6-Myr record, (b) excluding this record, (c) with the time scale divided into 400-kyr-long intervals (to satisfy the χ^2 requirement that each expected frequency should be ≥ 5) including the 2.2- to 2.6-Myr record, and (d) excluding this record. The results are (a) $\chi^2_{10} = 43.56$ ($\chi^2_{10(.01)} = 23.21$); (b) $\chi_9 = 16.50$ ($\chi^2_{9(.05)} = 16.92$); (c) $\chi^2_4 = 12.46$ ($\chi^2_{4(.02)} = 11.67$); and (d) $\chi^2_2 = 2.74$ ($\chi^2_{2(.05)} = 5.99$). There results reject the null hypothesis and indicate that first records are significantly overrepresented in the 2.4 to 2.6-Myr interval. The outcome for (b) tentatively suggests additional significant aggregations of first records from 1.6 to 1.8 Myr and 1.8 to 2.0 Myr; but the more reliable results for (c) and (d) pinpoint only the 2.4- to 2.6-Myr record as significantly nonrandom.

Thus, for the relevant period and area, the bovid data on their own support Hypothesis 2: they show a turnover-pulse between 2.6 and 2.4 Myr. It must have been initiated by physical change that was widespread at least in the East African region. Also, the habitat associations of most of the first recorded species (inferred from the ecology of extant close relatives) tell us about the nature of the environmental stimulus. The appearances of new species of *Gazella* (Fig. 25.3, 21), *Antilope* (25), *Antidorcas* (29), *Oryx* (30), *Megalotragus* (31) and *Pelorovis* (32), albeit in small numbers of individuals, indicate the spread of open grassland and increased aridity in the larger region. The arrival of Eurasian immigrants is consistent with the equatorward spread, at times of global cooling, of previously temperate habitats.

Other East African data agree with these bovid results on the timing and nature of this climatic change. Wesselman's (1984,1985) analyses of Omo (Shungura) micromammals indicate the presence near 3 Myr of equatorial high forests close to the ancient Omo River; continuing mesic conditions with some signs of more open vegetation towards the top of Member C (unit 8); and a shift toward arid, open conditions between 2.5 to 2.4 Myr by Member F times (see Fig. 25.1c). A shift in pollen taxa between the top of Shungura Member C (units 7 through 9) and Member E (unit 4) shows grassland extension at the expense of wooded savannah between 2.5 and 2.4 Myr, probably related to a decrease in temperature and rainfall in the larger Omo area (Bonnefille, 1976). These results from the Omo sequences specify more precisely the dating and nature of a major climatic change that was already noticed by Coppens (1975) as occurring between what he termed Omo 1 and 2 faunas. A pollen record from Gadeb, Ethiopia, shows a 2.5 to 2.4 Myr shift in vegetation belts of the same magnitude as that at 18kyr (Bonnefille, 1983). The near-contemporaneous responses of taxa as distantly related as plants, micromammals, and antelopes confirms the occurrence of a turnover pulse.

BIOCHRONOLOGY OF AFRICAN BOVIDAE

EASTERN AND EASTERN-SOUTHERN SPECIES — SOUTHERN ENDEMICS

I hypothesize that some fluvial–deltaic records from East Africa, such as the Omo (Shungura) strata, represent refugia. That is, the predictions for such records are those given for Hypothesis 1: the abundances of dominant higher taxa should persist (although new species within these taxa may be added). Climatic change in the larger region may be recorded only by the appearances and disappearances of peripheral taxa that represent intrusive elements from the alternative biome. The percentages of bovid tribes across the 3.0- to 1.0-Myr period (Shipman and Harris, this volume, chapter 23) for the Omo and West Turkana sequences, show no significant change across 2.5 Myr. Thus, on the one hand, we have conservative abundance of dominant resident bovid tribes across 2.5 Myr (Shipman and Harris, this volume, chapter 23; fig. 23.9), and on the other the first appearances of peripheral bovid species (Fig. 25.3) plus turnover in pollen (Bonnefille, 1976) and micromammal taxa (Wesselman, 1985) near 2.5 Myr. This combination is precisely as predicted if the Omo-Turkana area represented a fluvial–deltaic refugium in the midst of widespread climatic change (Hypothesis 1). It is interesting that the biome represented in these assemblages seems to have persisted until today, albeit with changes in geographic location and in species composition. For instance, the close correspondence between the dominant bovid genera of the Pliocene Shungura sequence (Gentry, 1985) and those of the modern Okavango Delta (Astle and Graham, 1976) suggests the same, long-surviving biome.

Could this East African change have been a purely local phenomenon, or was it part of the global 2.5-Myr event? I think that the evidence for the latter is very good for four reasons. First, the magnitude of the first Arctic glaciation at this time (Shackleton *et al.*, 1984), together with the 18kyr evidence cited above leads one to expect an impact in continental areas, including tropical Africa. Second, the nature of the change observed in East Africa—grassland spread, aridification, and arrival of Eurasian immigrants—is precisely as predicted for global cooling from the 18-kyr analogy. Third, the consistent timing and nature of changes in land areas as far-flung as South America (Hooghiemstra, 1986), Europe (Van der Hammen *et al.*, 1971), and central Africa strongly suggests a common global cause. Finally, some data directly link African climate to evidence from deep-sea cores. Ruddiman and Janacek (1988) report an increase in terriginous sediment in marine cores off West Africa, indicating opening of African vegetation and an increase in wind-borne dust, at 2.5 Myr.

Thus, I accept the East African climatic change as part of the global 2.5-Myr event and the local biotic turnover as part of a global turnover-pulse (see Hypotheses 2,3,4). There are still puzzling aspects of this evidence, and close study of the 2.5-Myr event should give insights into both "macroclimatic" and macroevolutionary processes. For instance, there is a discrepancy between the dating estimates for the deep-sea change (~2.4 Myr for the major glacial event, after the Gauss–Matuyama transition that is dated elsewhere at 2.48 Myr, with a minor Arctic ice-rafting episode near 2.5 Myr; Shackleton *et al.*, 1984) and those for the onset of East African biotic changes [e.g., the earliest bovid species arrive below Tuff D, before 2.52 ± 0.5 Myr by Brown *et al.*'s (1985) estimates]. One possibility is that future revision of errors in the two sets

◁—————————————————————————————————————

FIGURE 25.3. Biostratigraphy of sub-Saharan Bovidae from 3.3 to 1.0 Myr. Each line represents a species, except 2, 24, 30, 48 which may each refer to more than one species. Tribes and genera represented: Bovini: *Syncerus* (2), *Simatherium* (9), *Hemibos* (23), *Pelorovis* (32); Tragelaphini: *Tragelaphus* (4, 17, 18, 20, 22, 34, 47, 57), *Taurotragus* (51, 61); Reduncini: *Kobus* (3, 5, 13, 19, 28, 33), *Menelikia* (16, 26, 27), *Redunca* (52); Hippotragini: *Praedamalis* (6), *Oryx* (30), *Wellsiana* (46), *Hippotragus* (50); Neotragini: *Madogua* (12), *Oreotragus* (53), *Raphicerus* (6); Antilopini: *Antidorcas* (15, 29, 59), *Gazella* (11, 21, 36, 48), *Antilope* (25); Ovibovini: gen. indet. (24, refers to Shungura record only), *Makapania* (49); Peleini: *Pelea* (8, 58); Aepycerotini: *Aepyceros* (1); Alcelaphini: *?Damalops* (7), *Parmularius* (10, 14, 39, 40, 43, 45), *Megalotragus* (31), *Sigmoceros* (35), *Damaliscus* (37, 38, 42, 56), *Connochaetes* (41), *Rabaticeras* (44, 55); Tribe indet.: gen. indet. (54). L = living; I = immigrants from Eurasia. See Vrba (1985c) for references for chronology and taxonomy, plus Gentry (1985), Harris (1985), Vrba (1987 b,c). c; Shungura records are corrected to units within members.

of dates will align them more closely. It is noteworthy that the Gauss–Matuyama transition occurs below Tuff D and that the isotopic ages from tuffs in the Koobi Fora and Shungura formations generally tend to be 50 to 100 kyr older than expected from the magnetostratigraphy (Hillhouse et al., 1986). If a similar lag between the recorded and actual polarity boundary was present in deep-sea sediments, the $\delta^{18}O$ change would appear to be later than it really is. A more interesting contribution to this time gap may result because the terrestrial response to astronomical forcing really did precede the sea temperature change.

Hypothesis 5: The 2.5-Myr Event in Southern Africa

This hypothesis states that the global 2.5-Myr event initiated climatic and biotic changes in Southern Africa.

New bovid fossils from ongoing excavations at Swartkrans and Sterkfontein have not yet been published. My present discussion, although consistent with preliminary observations of these new materials, will refer to the old assemblages (Vrba, 1976, 1982, 1985a) using the abbreviations in Fig. 25.1.

The southern hominid record has not yet been radiometrically dated. One set of faunal estimates (e.g., Vrba, 1982; Delson, 1984) shows a gap in the Transvaal record, roughly 2.7 to 2.0 Myr (Fig. 25.3). This rules out the direct tracing of environmental change in successive strata such as is possible, for instance, in the Shungura sequence. But we can compare the assemblages that flank this gap—ST 4 (preceded by MAK 3) before and SK 1 (close to ST 5) after the gap. Because the *in situ* assemblage from Kromdraai B East Member 3 has not been chronologically informative (Vrba, 1981), it is omitted in most of this discussion and in Fig. 25.1.

There is a major change in the dominant bovid fauna between ST4 and SK 1/ST 5 (Vrba 1974,1975,1980,1985 a,b,c). One way to express this is in terms of percentage of minimum numbers of Alcelaphini-plus-Antilopini (AA) of all bovids present per assemblage, which is close to 50% for ST 4 (estimated to be even higher for MAK 3) and to 80% for SK 1 and ST 5. My earlier specific proposal, that higher AA percentages during SK 1/ST 5 times show less wood cover than during ST 4 time, involves statistical and ecologic arguments (Vrba, 1975,1980; Greenacre and Vrba, 1984) that are not necessary for considering Hypothesis 5.

Another way to look at the difference across the ~2.7 to 2.0 Myr gap in the Transvaal successions (Fig. 25.3) is simply to note that *the dominant taxa up to tribal level change decisively*. If dominant taxa at time A disappear thereafter to be replaced by new dominants at time C, there is no rational alternative explanation than that significant physical change intervened at some time between A and C. Thus, within the interval 2.7 to 2.0 Myr (if present date estimates are correct, see below) we have major biotic turnover in both East and South Africa, major physical change in both areas (which, in the Eastern case, is deduced to be part of a global climatic change), and both indicate the same direction of environmental change. Parsimonious argument must favor the hypothesis that a physical change that accounts for the bulk of the southern turnover (Fig. 25.3) fell near 2.5 Myr as in East Africa. An alternative of a temporally separate cause for the southern pattern, while the globally induced eastern change was not felt in the South, is clearly less parsimonious.

One contrasting possibility is that the entire ST 4 assemblage accumulated between 2.5 to 2.0 Myr, and that the biotic turnover between it and SK 1 belongs closer to 2.0 Myr than to 2.5 Myr. A second is that the ST 4 assemblage (as traditionally construed) is an aggregated one that spans a long time period including both pre- and post-2.5-Myr elements. Several lines of evidence support the second possibility, such as the presence of *Equus* (known from ~2.3 Myr in East Africa) and some current hominid evaluations (Clarke, this volume, chapter 18; Kimbel and White, this volume, chapter 11). I find the hypothesis of a time-aggregated ST 4 sample, with the dominant, archaic bovid taxa in the earlier fraction (because they are also found in MAK 3 but never again after ST 4) and taxa like *Equus* in the later fraction, a very viable one.

In sum, there is strong paleontological support for the notion of major environmental change and biotic turnover at some time(s) between ST 4 and SK 1. If at least a part of ST 4 accumulated pre-2.5 Myr, there is also support, less direct and weaker, for Hypothesis 5 that the global 2.5-Myr event left an imprint on the southern hominid sequences. One can also note that the Transvaal localities, whatever particular climatic or chronologic interpretations one may favor, obviously represent a biotidal area unlike the eastern refugial ones (see Hypothesis 1, Fig. 25.2).

Hypothesis 6: Environmental Comparison between ST 4 and SK 1/ST 5

This hypothesis states that the ST 4 population(s) of *africanus* lived in a more wooded and mesic environment than did the SK 1 *robustus* and *Homo* or the ST 5 *Homo* populations.

Support for the last hypothesis would not necessarily make this one true. If, between two temporally distant assemblages A and C, a major turnover pulse intervened at time B, then we would still expect their dominant taxa to differ even if they are situated at similar climatic extremes. Could this be so in the case of ST 4 and SK 1? I shall next reevaluate my previous conclusion that these bovid assemblages reflect different environments (Vrba, 1975,1980a,1985a).

There is probably agreement that, unlike some East African hominid contexts, "the Sterkfontein Valley environment apparently belonged in the vegetationally open part of the habitat spectrum during all known deposition phases" (Vrba, 1980a: 264). The previous argument that ST 4 and SK 1 environments nevertheless differed within that characterization relies mainly on the two most dominant ST 4 taxa, on their morphologies and on their ecological and abundance relations to the tribes A&A. As posited above, especially the dominant taxa are significant as signals of climatic change in biotidal sequences.

The two taxa are *Makapania broomi* (minimum number of eight out of the bovid total of 43), tribe Ovibovini whose last Southern African record is ST 4; and what I previously called ?Hippotragini and recently cf. *Hippotragus cookei* (Vrba, 1975, 1987c) also represented by eight individuals. *Hippotragus* is the genus to which the extant sable and roan antelopes belong. *Makapania* is dominant and cf.*H.cookei* is also present in the MAK 3 assemblage. Neither is known from anywhere after ST 4. Hippotragini and Ovibovini together account for 42% of ST 4 bovids and are the chief cause of the relatively low A&A percentage at ST 4. The third-most dominant species is a small alcelaphine, *Parmularius* sp. (seven individuals), that is probably also present in the MAK 3 and Hadar assemblages.

Could the *Makapania* and *Hippotragus* species have been as open-adapted as we infer Pliocene alcelaphines to have been, thus indicating similar environments for ST 4 and SK 1 (and ST 5)? This is particularly relevant in relation to *Makapania,* as the tribe Ovibovini features among the immigrants to East Africa as grasslands spread near 2.5 Myr (Fig. 25.3: 24). On the other hand, *M. broomi* has less hypsodont teeth, with less reduced premolars than contemporary alcelaphines of similar size. Greater differences from alcelaphines and from other tough-forage feeders are seen in cf. *H. cookei*, whose masticatory morphology is distinctly reminiscent of modern soft-vegetation feeders (Vrba, 1976). Thus, the morphology of the dominant bovids suggests ecological differences between ST 4 and SK 1/ST 5. This support for Hypothesis 6 remains tentative, and the problem of hominid-associated environmental change in South Africa is badly in need of additional approaches.

Shipman and Harris's (this volume, chapter 23) analysis relates to Hypothesis 6. Using the bovid tribal pairs Tragelaphini–Aepycerotini (TA), Reduncini–Bovini (RB) and Alcelaphini-Antilopini (AA) they classified 29 African game parks into three categories representing closed/dry, closed/wet and open/arid habitats, respectively. The hominid localities were then classified, using their TA, RB, and AA proportions, into one of these three clusters as predefined by the modern data. By this method ST 4, SK 1, and other Transvaal strata group closely together among open/arid areas.

I have reservations about Shipman and Harris's (this volume, chapter 23) grouping of TA versus RB (see Greenacre and Vrba, 1984: figs. 4 and 7) and about their clustering procedure which forces intermediate assemblages into one of three uniform categories. But my strongest doubt is one that Shipman and Harris themselves acknowledge: because they used only six tribes, excluding all other antelopes present even where these comprise a substantial proportion of the fossil assemblages, their results may misrepresent actual environmental changes in the Transvaal strata. In the case of ST 4 their method has a disastrous effect. The two dominant bovid taxa, Ovibovini *(Makapania)* and Hippotragini, which comprise 42% of all bovids and, apart from primates, represent the most dominant of all ST 4 large mammals, are excluded. This procedure leaves the criteria for ecological placement of ST 4 to one dominant extinct alcelaphine species (*Parmularius* sp. that probably also occurs in Hadar and MAK 3) plus a motley assortment of about a dozen minor taxa. If a similar procedure is applied to modern ecosystems, whether in the analysis of Shipman and Harris (this volume, chapter 23) or in a multidimensional one such as Greenacre and Vrba (1984, from which most of Shipman and Harris' census data came), it can play havoc with any presumed "ecologic indications." For example, the modern reserve Kafue, which has moderate to high wood cover, a high rainfall and riverine floodplains (Dowsett, 1966), is closed/wet which is indeed how Shipman and Harris (this volume, chapter 23) classify it. Now let us exclude just one dominant taxon, Reduncini comprising 30% of all bovids (Vrba, 1980a: table 14.1) and rerun their analysis using TA, B, and AA groupings. As a result Kafue "jumps" from the closed/wet to the open/arid cluster simply because the AA% in their tribal sample has now artificially (by exclusion of reduncines) jumped from 37 to 60%. I suggest that Shipman and Harris's (this volume, chapter 23) placement of ST 4 as "open/arid" has occurred by an analogous effect.

The application of modern ecological analogues to ancient faunas is highly problematic, especially if a turnover-pulse intervened, as inferred between MAK 3 and ST 4 times, on the one hand, and subsequent sediments like SK 1, on the other. It involves assumptions about the habitat specificities of extinct taxa that are phylogenetically (and perhaps ecologically) far removed from extant relatives. Thus, all paleoecological comparisons between the archaic MAK 3/ST 4 and the essentially modern SK 1/ST 5 assemblages that use modern faunal analogues are plagued by this difficulty of risky assumptions. It is as much a problem for Shipman and Harris's approach as it is for others, such as my use of the A&A percentage criterion (Vrba, 1980a,1985a). Essentially, Shipman and Harris (this volume, chapter 23) have to assume that *Makapania* and cf.*Hippotragus cookei* are ecologically similar to alcelaphines and antilopines and can, therefore, safely be excluded without distorting their conclusions in the manner illustrated by the Kafue example. In contrast, my criterion had to assume that the two dominant taxa are ecologically different from A&A, and can therefore safely be included in the category "other bovids." To decide which of these assumptions is more likely to be true, they can be tested by functional morphological studies of *Makapania* and cf. *H. cookei*. This examination supports the assumption of ecological difference from alcelaphines and antilopines: the cf.*Hippotragus* species is markedly brachyodont, with shallow mandibles and other features that strongly suggest an habitual soft-vegetation feeder. In general, it is surely preferable to include dominant taxa, and to include study of their anatomies, rather than to exclude them in analyses of this kind.

In sum, I conclude that the currently available bovid evidence supports Hypothesis 6: The ST 4 (and MAK 3) populations of *africanus* lived in more wood-covered and mesic environments than did either the SK 1 *Homo* and *robustus* or the ST 5 *Homo* populations, although all these assemblages seem to belong within the vegetationally more open part of the habitat spectrum. This conclusion is tentative. It should and will be tested by additional data from the South African hominid sites.

The last three Hypotheses (numbers 7, 8, and 9) have an hierarchical pattern of interdependence, with each requiring that the previous, but not the subsequent, be true: 7 is the most general and does not necessitate the truth of 8 and 9; 8 is more specific and requires 7 but not 9; and 9 is the most specific and assumes that both 7 and 8 are true.

Hypothesis 7: Hominid Evolution in Response to the 2.5-Myr Event

There are earliest records near 2.5 Myr of new hominid phenotypes (a new "robust" morphology assigned by some to *aethiopicus,* and stone tools at least some of which I take to be a product of the *Homo habilis* behavioral phenotype) that represent evolutionary responses to the large-scale environmental changes at that time.

This hypothesis leaves open whether the hominid lineages evolved locally in the East African environments represented by the fossil localities, or whether they evolved as they immigrated to the area from elsewhere. One alternative hypothesis is that the earliest records near 2.5 Myr reflect *only* immigration of phenotypes already evolved elsewhere prior to that time. An extreme alternative is that the 2.5 Myr records reflect neither immigration nor a spurt of evolution at that time (distinct from preceding and succeeding gradual evolution), but that their synchronous appearances near 2.5 Myr are an artefactual association.

I have argued above that some of the predictions of Hypothesis 7 enjoy support; for instance, the records of new hominid phenotypes at 2.5 Myr are embedded in a turnover-pulse that indicates synchronous evolutionary responses in other diverse taxa. The basic requirement for this hypothesis to be true is that at least some of the advanced phenotypic characters of *aethiopicus* and *habilis* should not be recorded prior to 2.6 Myr in East Africa or anywhere else.

It seems to me that the evidence available to date is consistent with Hypothesis 7, and that the alternative of artefactual association is contradicted. However, the evidence is also consistent with the notion that the hominid records near 2.5 Myr reflect immigration without notable evolutionary change. It seems likely that a response to climatic change of immigration alone should occur more quickly than one of speciation. The bovid record (Fig. 25.3) agrees with this: three of the earliest appearances between 2.4 and 2.6 Myr are putative Eurasian immigrants. It is notable that the time of appearance of *aethiopicus*, particularly mandible Omo 18-1967-18 from Shungura C (unit 8), is very close to those of the bovid immigrants. Future hominid finds previous to 2.6 Myr may throw light on this possibility. The current lack in the pre-2.6 Myr record of fossils ascribable to either *aethiopicus/boisei* or to *Homo habilis*, plus the nature of the earliest hominid phenotypes near 2.5 Myr in relation to the known particulars of the climatic change (see below) marginally favors evolution over pure immigration.

Hypothesis 7 leaves open whether lineage splitting or simply rapid phyletic evolution occurred. If either of the hominid trees in Fig. 25.1 is correct, and if the Taung hominid really postdates 2.4 Myr (Delson, 1984), then splitting in at least one lineage is indicated.

This hypothesis also leaves open whether or not the hominid phenotypes include adaptation to cooler, more open and arid environments. Theoretically, such a climatic change could initiate vicariance, and speciation via stochastic evolution in small refugial populations in mesic-adapted lineages without adaptation to less mesic conditions. This question is addressed by the next hypothesis.

Hypothesis 8: Hominid Adaptation Near 2.5 Myr to Open, Xeric Environments

Some of the phenotypic evidence, first recorded near 2.5 Myr, of "robust" australopithecines and of *Homo habilis* represents adaptation for use of resources characteristic of open xeric environments. Such adaptations were absent in *afarensis* and *africanus*.

This cannot be tested easily by broad contextual evidence, such as whether or not the hominid fossils are found associated with permanent water, grazing antelopes, etc. It is best tested by examination of the fossils themselves, of their functional anatomy, of the details of occlusal wear on the teeth, of isotopic signatures in bones and teeth as indicators of diet, and of any behavioral evidence they may have left in the record.

Hypothesis 8 leaves open the question of ecological specialization vs generalization and merely posits that the hominids that appeared near 2.5 Myr possessed new capabilities (with or without losing ancestral ones) and characters that evolved by selection for using resources that are more prevalent in relatively open and arid than in alternative environments.

How does the earliest evidence of the *Homo habilis* phenotype relate to this? I here assume that *H. habilis* was a producer of stone tools that are known from about 2.5 Myr. This assumption does not conflict with the proposal that *robustus* was also capable of stone culture (Susman, this volume, chapter 10). It merely requires that *H. habilis* was responsible for at least some of the 2.5-Myr stone tools. If so, then those tools provide the earliest currently known evidence of the *H. habilis* phenotype. It is not obvious how the evolution of stone tool manufacture and use should be functionally linked particularly to open, arid environments, although Hatley and Kappelman (1980) argued for such a link. They demonstrated that a high below-ground plant biomass is characteristic of xeric open areas, and argued that digging out of such foods, first by hand and later by digging sticks and other tools must have evolved as an important feeding strategy of early hominids when the African savannah became more open and arid (see also Brain, this volume, chapter 20).

The proposal of a functional link between "robust" morphology and food resources prevalent in open, arid environments, as suggested tentatively by Vrba (1975) and as argued for in detail by Grine (1981), can be supported more convincingly. It seems very likely, based on the functional morphologies of other mammals, that the "robust" cranial and dental morphology at least partly reflects selection for feeding on tough forage: greatly thickened tooth enamel (Grine and Martin, this volume, chapter 1), expansion of posterior and reduction of anterior teeth, pronounced and expanded areas of origin and insertion of masticatory muscles, and other anatomical features discussed in this volume, are well-known to recur in parallel in many mammal groups in association with feeding on tough forage (for bovids, see for example Caithness, in prep.). Some scanning electron microscopic analyses of early hominid teeth support this proposal of "robust" adaptation (e.g., Grine 1981; Kay and Grine, this volume, chapter 26).

It remains to be seen whether or not the hominid phenotypes that appear near 2.5 Myr were relatively more bipedal and less arboreal than *afarensis* and *africanus*. Several studies are consistent with this, although consecutive postcranial evidence across the Late Pliocene is not currently available. For instance, comparison of the findings of Stern and Susman (1983) on *afarensis* postcranial fossils [see also Vrba (1979) on the Sts 7 *africanus* scapula], on the one hand, with Susman's (this volume, chapter 10) analysis of Swartkrans *robustus* postcrania, on the other, suggests such a difference in locomotor adaptation.

Alternative proposals for the evolution of phenotypes that involve adaptation to aspects of mesic, well-wooded environments, or that involve exaptation are clearly also possible. (They might sound plausible in the case of the origin of stone tool culture; but in the case of "robust" masticatory morphology, it seems to me, they would be a bit strained.) Nevertheless, one can note that the available evidence does not contradict Hypothesis 8, but is consistent with it.

Hypothesis 9: Breadth of Resource Use in Early Hominids

The 2.5-Myr climatic event initiated changes in the breadth of resource-use of early hominids: (a) At least some phenotypic characters in the *Homo* lineage evolved towards greater ecological generalization relative to all other hominids. (b) In at least some respects the "robust" lineage(s) became more ecologically specialized than other hominids' and the specialization represents adaptation to resources prevalent in open, arid environments.

The term "resource" has a wide meaning: it refers to any component of the environment, both physical and biotic, that can be utilized by an organism or the organisms in a species. Thus, the concepts of generalist and specialist make sense only when specified in relation to particular categories of resources (Vrba, 1987a). I will refer to three such categories in relation to early hominids.

1. Breadth of biome association. A stenobiomic organism or species is virtually restricted to a particular biome, implying in the terrestrial case a particular vegetational physiognomy, or wood–grass ratio. A eurybiomic organism or species can exist and use resources in more than one biome, implying in the terrestrial case a wide vegetational range in terms of wood–grass ratio.

2. Breadth of water dependence. Specialists of this kind are dependent on permanent water, while generalists are less so.

3. Breadth of feeding behavior. In contrasting two species, the one whose organisms can feed on a wider range of food items, within more food categories, is relatively euryphagous; while the other is relatively stenophagous.

Each species (and each organism within it) commonly has a mosaic of generalizations and specializations, although one or the other category usually predominates. With respect to breadth of biome association and water dependence, the sample of fossil localities currently available is insufficient for a characterization of each of (and comparison between) the post-2.5-Myr hominid lineages. To do this adequately one needs, within the known distribution range of each species (e.g., the relevant East African area for *boisei*), a sample of sites that represents at least two distinct biomes as indicated by independent evidence. One could then test whether the species *is restricted* to only one biome or whether it ranges across the known habitat spectrum.

One can argue that for *Homo* we have samples between 2.5 and 1.5 Myr that at least partly fulfill these requirements. Thus, *H. habilis* is known from margins of major water bodies (lake and major river and delta margins in East Africa), associated with substantial wood cover fringing at least some of these water margins, *and* from an area without such major water bodies in predominant open grassland (ST 5). The same is true for early *H. erectus* from East Africa and the Transvaal (SK 1), respectively.

However, these requirements are met neither in the case of South African *robustus* (here I include Swartkrans *crassidens* in the same species as Kromdraai *robustus*) nor in the case of East African *boisei*. I have argued that Swartkrans *robustus* occurred in predominantly open grassland (Vrba, 1975); and we do not know whether or not a permanent river was present at the time. In contrast, the new *in situ* Kromdraai Member 3 excavations indicated that this population of *robustus* was associated with permanent water, but in this case information on vegetation cover is lacking (see Vrba (1981), which replaced earlier paleoecological conclusions (Vrba, 1975,1976) based on the old stratigraphically mixed assemblage). Thus, South African *robustus* is known from what may be only one kind of habitat: open grassland in the vicinity of a river.

It has been known for some decades that East African *boisei* was associated with the margins of major water bodies, and the currently available sample of sites indicates that this is basically the only broad context from which the species is known (Shipman and Harris, this volume, chapter 23). It is potentially a single biome: woodlands along water margins.

Thus, at the broad level ecological variety is lacking both within the total East African sample of assemblages covering the relevant time period, and within the South African one. Therefore we do not know whether or not *boisei* and *robustus* also occurred in ecosystems other than the ones from which they are known. That is, the evidence tells us only about *occurrence* of each species in one kind of biome, but it cannot tell us about *preference* among biomes.

Shipman and Harris (this volume, chapter 23) investigate a more subtle contrast *within* the the water-marginal biome represented by their samples from Omo, Koobi Fora, West Turkana, and Olduvai; namely the contrast between relatively more closed/wet, closed/dry, and open/arid habitats, defined by the bovid tribal pairs RB%, TA%, and AA%, respectively. They interpret their results to "suggest that *A. boisei* probably preferred closed over open habitats and favored wetter rather than drier closed habitats." But the currently available evidence, on closer inspection, clearly does not support habitat preference by *boisei* even at this subtle level. Because this statement contradicts the major conclusion of Shipman and Harris (this volume, chapter 23), it is appropriate that I detail my reasons (I ask the reader who does not wish to follow these details to skip the next paragraph).

The data for Omo, Koobi Fora and West Turkana cannot indicate preference of closed/wet over open/arid habitats (because no open/arid assemblages are included), nor do they indicate *preference* of closed/wet over closed/dry habitats (in each case the occurrences of *boisei* across these two kinds of habitats do not depart significantly from random). That leaves only the 17 Olduvai localities, including four *boisei* localities, to support the notion of *boisei* habitat preference. Shipman and Harris (this volume, chapter 23) judge 13 of these (including one *boisei* locality, HWK II) as open/arid and 4 (including three *boisei* localities, FLKNN I level 3, FLK I Zinj, BK II) as closed/wet. However, errors seem to have entered into Shipman and Harris's (this volume, chapter 23: table 23.5) frequencies for FLK I Zinj and BK II, if these data derive from Gentry and Gentry (1978: tables 3 and 9, respectively). These errors result in a significant bias towards closed/wet indications. For FLK I Zinj level 22, compare 57% AA and 36% RB, of total AA plus RB plus TA, in Shipman and Harris (this volume, chapter 23: table 23.5), with 67% AA and 24% RB derived from Gentry and Gentry (1978: table 3). In either case it is difficult to see why the lower RB% should overwhelm the higher AA% to force this locality into the closed/wet rather than the open/dry category. For BK II, Gentry and Gentry (1978: table 9) give specimen frequencies that comprise 63% AA versus 36% RB, of total AA plus RB plus TA, and note (p.55) that the Alcelaphini percentage has been "brought low by the distorting effects" of a herd of the bovine *Pelorovis oldowayensis*. As these authors give only a specimen total of more than 390 individuals but no minimum number for alcelaphines, it is difficult to see how Shipman and Harris (this volume, chapter 23: table 23.5) arrived at their minimum number of 30 alcelaphine individuals, giving them 57% AA relative to 40% RB. Again, even in the case of these latter figures it is difficult to see why the lower RB% should have overwhelmed the higher AA% in

Shipman and Harris's (this volume, chapter 23) analysis to push BK II into their closed/wet rather than the open/arid category. The bias towards closed/wet indications in Shipman and Harris's (this volume, chapter 23) analysis of BK II resulting from an apparently erroneously low alcelaphine % relative to high bovine %, is compounded by the nature of the BK II bovine occurrence. Two-thirds of the bovine individuals belong to *Pelorovis,* a genus that throughout its Plio-Pleistocene duration shows a striking association with independent evidence of relatively cold and/or open vegetational habitats (Vrba, 1987b). Together, these arguments regarding the Olduvai sample imply the following: Of 15 open/arid localities, three contain *boisei,* and of two closed/wet localities, one contains *boisei.* Thus, the distribution of *boisei* across Olduvai habitats does not depart significantly from random. The same conclusion holds for the entire East African sample of relevant assemblages.

In sum, the evidence does not support a notion that *boisei preferred* some habitats over others, within the biome represented by the the margins of major water bodies from which the species is known. On this point I disagree with Shipman and Harris (this volume, chapter 23). Nevertheless, they demonstrated convincingly that *boisei* was *present* in at least three kinds of water-marginal habitats. Thus, the known occurrence of *boisei* contradicts an hypothesis of extreme stenobiomy in open environments and/or in areas without permanent water. It is compatible with the degree of stenobiomy observed for instance in wildebeests (gnus, *Connochaetes taurinus*) or hartebeests (*Alcelaphus buselaphus*), that attain peak abundance in very open areas although they are also found in woodlands and water-marginal areas. It is not clear whether *boisei* was a rare inhabitant or infrequent visitor of such habitats (which might be judged consistent with the low frequency of recovered specimens), or an habitual resident restricted to this biome or ranging across more than one biome. Thus, the known occurrence of *boisei* is also compatible with the proposal that the species was eurybiomic across higher-to-lower wood cover and more-or-less well-watered areas. Similarly, *robustus* may have been stenobiomic with respect to the predominant grassland biome in which it is inferred to occur, or it may have ranged across other environments that are simply not represented in our sample of localities.

The occurrences of post-2.5-Myr *Homo* suggest a degree of eurybiomy because they seem to span open grassland to water-marginal woodland.

Let us next consider Hypothesis 9 in relation to breadth of feeding behavior. Did members of early *Homo* include both plants and meat in their diets, while "robust" australopithecines were exclusive vegetarians that included a larger proportion of tough plant foods in their diet than did previous hominids?

Keeley and Toth (1981) found wear traces on stone artefacts from Koobi Fora dated ~1.5 Myr that variously resemble traces induced experimentally by cutting animal tissue and plant materials. Potts and Shipman (1981) identified cutmarks on fossil bones from Olduvai Beds I and II. If this evidence pertains to feeding behavior of early *Homo,* as seems likely, then it supports the notion of a dietary shift from ancestral herbivory to hominine omnivory.

At least for South African hominids there is evidence that *africanus* and *robustus* differed in dietary habits. Thus, Kay and Grine (this volume, chapter 26) report that the wear pattern of "robust" dentitions resemble those of living primates that eat hard plant items while *africanus* wear differs in the direction of extant taxa that subsist more on fleshy fruits and foliage. Walker (1981) examined the microwear on molar teeth from nearly 20 *boisei* from East Africa and suggested that it is consistent with bulk eating of fruits without much preparation.

Thus, the available evidence is consistent with a notion of hominine omnivory in contrast to "robust" herbivory, and to this extent it supports Hypothesis 9 in regard to feeding behavior. It appears that South African "robust" australopithecines differed as predicted from *africanus*, and that they may have also differed in plant intake from East African *boisei*. I expect that future studies using isotopic signatures in fossils will test more conclusively for differences in diet and dietary breadth among the the early hominids.

General Remarks on Paleoclimate and Hominid Evolution

The available evidence implicates climatic changes to colder, more arid and more vegetationally open environments, either one at a time or in combination, in critical hominid evolutionary events not only in the Late Pliocene but also at other times (Vrba, 1988). For instance, during the end of the Miocene, when temperatures plunged to a low point unprecedented during the entire previous Tertiary, a number of bovid tribes appeared for the first time (Vrba, 1985c). Thus, global climatic change at this time may have coincided with hominid origins (Brain, 1981), and also with a spectacular proliferation of bovid taxa that still characterize the African savannah today. This has led to the suggestion that the Hominidae were probably "founder members" of the biota of the the extensive African savannah (Vrba, 1985a).

During the Pleistocene, changes to open and arid conditions may have triggered the origin of *Homo erectus* and of his characteristic tool kit, and later the beginnings of Middle and Late Stone Ages during the cultural evolution of *Homo sapiens* (Vrba, 1988).

There is one other Pleistocene period that should be mentioned here because it might have had a particular relevance to the history of "robust" lineages. This is the time around 0.9 Myr. Unfortunately the relevant African fossil assemblages are few in number and have mostly been dated only biochronologically. If Vogel's (1985) relative thermolumine-scence dates for SK 1 through 4 are close to the truth, then the *robustus* (= *crassidens*) occurrence in SK 3 (Brain, this volume, chapter 20) at about 0.9 Myr is the last record of any "robust" australopithecine species. East African *boisei* is last recorded at about 1.2 Myr, but East African assemblages that are reliably dated to the period 1.2 to 0.9 Myr are so sparse that we cannot conclude real absence of *boisei* based on its absence in the fossil record (Klein, this volume, chapter 29; White, this volume, chapter 27). The period around 0.9 Myr may also coincide with the earliest massive geographic expansion of any hominid species, namely of *Homo erectus*. Clark (1980: 44) describes this as follows:

> Man appears to have remained in the open savannas until about one million years ago when there was a rapid radiation out into a great variety of other habitats: arid steppe, deciduous woodland, high-altitude grassland, and forest grassland ecotone. For instance, it was about 1.0 Myr BP that *Homo erectus* occupied the high plateau grasslands of Ethiopia on both sides of the Rift . . . From this time, too, appears the first good evidence for the occupation of Europe and western Asia.

More recent researches confirm that all of the Asian *H. erectus* samples may be less than 1.0 Myr in age (Pope, 1983; Rightmire, 1985).

The global paleoclimatic record shows a shift toward colder extremes in astronomical cycles (e.g., Thunell and Williams, 1983; Shackleton *et al.*, 1984) (Fig. 25.2d) and toward more open vegetation types in at least some terrestrial areas (e.g., Hooghiemstra, 1986) close to 0.9 Myr. Land mammals across Eurasia undergo a dramatic turnover pulse, involving waves of extinction, speciation and distribution drift. Azzaroli (1983:117) calls this "end-Villafranchian dispersal event", which he places between 1.0 and 0.9 Myr, "a turning point in the history of Eurasia" and notes that it was marked by "a practically total rejuvenation of the fauna." I suggest that the global cooling near 0.9 Myr very likely caused both the Eurasian mammalian turnover event and the dispersal and exodus from Africa of *H. erectus*. More tentatively, it seems possible that the same climatic change was responsible for the final disappearance of "robust" australopithecines. This seems a simpler hypothesis than to invoke competitive exclusion of "robust" australopithecines, by *H. erectus* populations that were undergoing "niche expansion" and behavioral advances (Klein, this volume, chapter 29). While there is no convincing evidence that interspecific competition, on its own in the absence of physical change, ever initiated extinction in the history of life, there is massive evidence that severe climatic changes have done so.

The hominid record is species-poor in comparison with those of some other Plio-Pleistocene mammal clades. Yet, it hints at an interesting difference in the macroevolutionary responses of hominid lineages. Grine (1981, this volume, chapter 30) has suggested that "robust" australopithecines are relatively more speciose than other hominid lineages over comparable time periods, and even a conservative, "lumping" evaluation of a monophyletic "robust" clade (Fig. 25.1a) necessitates at least one splitting event and two terminal extinctions, while no events in either category are required in the tree of *Homo*. This would be as predicted if "robust" australopithecines were more narrowly habitat-specific with respect to at least one crucial resource category, and therefore more frequently susceptible to strong directional selection pressures and vicariance during their evolutionary history. As Grine (pers. comm.) has suggested previously, this evolutionary contrast among hominid lineages may exemplify the Effect Hypothesis (Vrba, 1980b, 1983).

Conclusions

I have analysed some hypotheses and predictions that link new hominid interpretations to new global and African climatic data. There is evidence from deep-sea and several continental records of a major global cooling event near 2.5 Myr. Furthermore, there is sound evidence that the change to colder, more vegetationally open and arid environments, recorded in East Africa at this time, is part of the global 2.5-Myr event. The pattern of biotic responses in some of the East African stratigraphic sequences, notably the Shungura sequence, is as expected if these riverine–deltaic–lacustrine areas were refugia, surrounded by widespread changes in the larger region, during the Late Pliocene.

It seems likely that the global 2.5-Myr event initiated climatic and biotic changes in South Africa as well. There is also support for the more specific hypothesis that ST 4 (and MAK 3) *A. africanus* lived in more wood-covered and mesic environments than did SK 1 *Homo* and "robust" ape-men and ST 5 *Homo*. I have outlined reservations concerning the analysis of Shipman and Harris (this volume, chapter 23), who suggest that these Transvaal hominid environments were all the same.

The evidence is consistent with the notion that the earliest records near 2.5 Myr of new hominid phenotypes (a new "robust" morphology assigned by some to the species *aethiopicus* and stone tools at least some of which I take to represent the *H. habilis* behavioral phenotype) reflect evolutionary responses to the 2.5-Myr climatic change. But it is also consistent with the rival notion that these earliest records reflect only immigration into the relevant East African area of phenotypes that had already evolved elsewhere prior to that time.

Do the phenotypes of *Homo habilis* and "robusts" represent adaptations for use of resources characteristic of open, xeric environments? Some have argued that the origin of stone tools is linked adaptively to such environments; but the connection is not an obvious one. A functional link between "robust" cranial and dental morphology and resources prevalent in open, arid environments (Grine, 1981; Kay and Grine, this volume, chapter 26) can be argued more convincingly.

I have hypothesized that in some respects the *Homo* lineage evolved toward greater ecological generalization, while in contrast the "robust" lineage(s) became more specialized on resources prevalent in more open environments. With respect to breadth of biome association and water dependence, the sample of fossil localities currently available is insufficient to characterize any one of the "robust" species adequately, and therefore to test this hypothesis. Thus, the East African fossil localities relevant to the times of *boisei* and *aethiopicus* sample what is basically only one biome—the vegetational margins of major water bodies. Similarly, the post-2.5-Myr Transvaal assemblages potentially sample only one biome—predominant grassland in the vicinity of a river. The known occurrence of *boisei* clearly contradicts an hypothesis of extreme stenobiomy in open, waterless environments. But it seems to me (*contra* Shipman and Harris, this volume, chapter 23) that the known record of *boisei* (or of any other "robust" species) is insufficient to argue for particular *preferences* among alternative habitats. First, at the broad level of the biome, there is essentially no alternative to the water-marginal biome represented in the relevant East African sample of sites. Second, at the level of more detailed distinctions between closed/wet, closed dry, and open/arid habitats within the water-marginal biome (Shipman and Harris, this volume, chapter 23), the known distribution of *boisei* across these habitats does not depart from random. Only in the case of post-2.5-Myr *Homo* can one make a case for breadth of biome association: the records span open grassland to water-marginal woodland, which suggests that *H. habilis* was a biome generalist.

The evidence is consistent with the proposal that members of early *Homo* included both plants and meat in their diet, whereas "robust" australopithecines were exclusive vegetarians who ate tough plant foods at least some of the time (the East and South African "robust" forms may have differed in their vegetarian diets). If true then, in this respect, "robust" australopithecines were indeed more specialized than *Homo*.

The climatic and hominid evidence implicates climatic changes to colder, more arid and more open environments (either one at a time or in combination) in critical hominid evolutionary events not only in the Late Pliocene but also at other times. Of particular note in the present context is the major global event near 0.9 Myr recorded in several marine and terrestrial sequences, which may have caused both the massive geographical expansion of *Homo erectus* and the extinction of "robust" australopithecines. Such an explanation, invoking climatic change, is simpler than the alternative one that *H. erectus* competitively excluded *robustus* and *boisei*. While there is no convincing evidence that interspecific competition, on its own in the absence of physical change, ever initiated extinction (or speciation) in paleontologic history, there is massive evidence that severe climatic changes have done so.

Acknowledgments

I thank Fred Grine for inviting me to the stimulating workshop on The Evolutionary History of the "Robust" Australopithecines. Following the editor's request to discuss each other's contributions, Pat Shipman, John Harris and I started a debate on habitat indicators and preferences. I thank these authors for our open and lively discussions which, although they did not remove all disagreements, were certainly enjoyable.

References

Astle, W. L. and Graham, A. (1976). Ecological investigations of the UNDP in the Okavango Delta. *In* F. Youngman (ed.), *Proceedings of the symposium on the Okavango Delta and its future utilisation*, pp. 81–91. Okavango Wildlife Society, Johannesburg.

Azzaroli, A. (1983). Quaternary mammals and the "End-Villafranchian" dispersal event—a turning point in the history of Eurasia. *Palaeogeog., Palaeoclimatol., Palaeoecol., 44*:117–139.

Baker, B. H., Mohr, P. A. and Williams, L. A. J. (1982). Geology of the Eastern Rift System of Africa. *In* A. M. Quennell (ed.), *Rift Valleys Afro-Arabian*, pp. 242–276. Hutchinson Ross, Stroudsburg, Pennsylvania.

Bonnefille, R. (1976). Palynological evidence for an important change in the vegetation of the Omo Basin between 2.5 and 2 million years ago. *In* Y. Coppens, F. C. Howell, G. L. Isaac and R. E. F. Leakey (eds.), *Earliest Man and environments in the Lake Rudolph Basin*, pp. 421–432. University of Chicago Press, Chicago.

Bonnefille, R. (1983). Evidence for a cooler and drier climate in the Ethiopian uplands towards 2.5 Myr ago. *Nature, 303*:487–491.

Brain, C. K. (1981). The evolution of Man in Africa: was it a consequence of Cainozoic cooling? *Ann. Geol. Soc. S. Afr., 84*:1–19.

Brown, F. H., McDougall, I., Davies, T. and Maier, R. (1985). An integrated Plio-Pleistocene chronology for the Turkana basin. *In* E. Delson (ed.), *Ancestors: the hard evidence*, pp. 82–90. Alan R. Liss, New York.

Burckle, L. H. (1985). Diatom evidence for Neogene palaeoclimate and palaeoceanographic events in the world ocean. *S. Afr. J. Sci., 81*: 249.

Caithness, N. (in prep.). Cranial morphology in relation to ecology and evolution in African Bovidae. Unpublished PhD Thesis, University of the Witwatersrand, Johannesburg.

Clark, J. D. (1980). Early human occupation of African Savanna environments. *In* D. R. Harris (ed.), *Human ecology in savanna environments*, pp. 41–71. Academic Press, London.

Cooke, H. J. (1976). The palaeogeography of the Middle Kalahari of Northern Botswana and adjacent areas. *In* F. Youngman (ed.), *Proceedings of the symposium on the Okavango Delta and its future utilization*, pp. 21–28. Okavango Wildlife Society, Johannesburg.

Coppens, Y. (1975). Evolution des Hominides et de leur environnement au cours du Plio-Pleistocene dans la basse vallee de l'Omo en Ethiopie. *C. R. Acad. Sci. Paris, 281*:1693–1696.

Delson, E. (1984). Cercopithecid biochronology of the African Plio-Pleistocene: correlation among eastern and southern hominid-bearing localities. *Cour. Forsch. Inst. Senckenberg, 69*:199–218.

Denton, G. H. (1985). Did the Antarctic ice sheet influence Late Cainozoic climate and evolution in the southern hemisphere? *S. Afr. J. Sci., 81*:224–229.

Denys, C., Chorowicz, J. and Tiercelin, J. J. (1986). Tectonic and environmental control on rodent diversity in the Plio-Pleistocene sediments of the African Rift System. *In* L. E. Frostick, R. W. Renault, I. Reid and Tiercelin, J. J. (eds.), *Geological Society special publication no. 25*, Blackwell Scientific Publications, London.

Dowsett, R. J. (1966). Wet season game populations and biomass in the Ngoma area of the Kafue National Park. *The Puku, 4*:135–145.

Flenley, J. R. (1979). *The equatorial rain forest, a geological history.* Butterworths, London.

Gentry, A. W. (1985). The bovidae of the Omo Group deposits, Ethiopia. *In: Les faunes Plio-Pleistocene de la Basse Vallee de l'Omo (Ethiopia), Tome 1. Perissodactyles—Artiodactyles (Bovidae) 1*, pp. 119–191. Editions du CNRS, Paris.

Gentry, A. W. and Gentry, A. (1978). Fossil Bovidae (Mammalia) of Olduvai Gorge, Tanzania. Parts I and II. *Bull. Br. Mus. Nat. Hist. (Geol.), 29*:289–446; *30*:1–83.

Greenacre, M. J. and Vrba, E. S. (1984). A correspondence analysis of biological census data. *Ecology, 65*:984–997.

Grine, F. E. (1981). Trophic differences between "gracile" and "robust" australopithecines: a scanning electron microscope analysis of occlusal events. *S. Afr. J. Sci., 77*:203–230.

Haffer, J. (1982). General aspects of the refuge theory. *In* G. T. Prance (ed.), *Biological diversification in the tropics*, pp. 6–24. Columbia University Press, New York.

Hamilton, A. C. (1982). *Environmental history of East Africa.* Academic Press, New York.

Harris, J. M. (1985). Fossil ungulate faunas from Koobi fora. *In* Y. Coppens (ed.), *L'environnment des hominides au Plio-Pleistocene*, pp. 151–163. Masson, Paris.

Harwood, D. M. (1985). Late Neogene climatic fluctuations in the southern high-latitudes: implications of a warm Pliocene and deglaciated Antarctic continent. *S. Afr. J. Sci., 81*:239–241.

Hatley, T. and Kappelman, J. (1980). Bears, pigs and Plio-Pleistocene hominids: a case for the exploitation of belowground food resources. *Hum. Ecol., 8*:371–387.

Hays, J. D., Imbrie, J. and Shackelton, N. J. (1976). Variations in the earth's orbit: pacemaker of the ice ages. *Science, 194*:1121–1132.

Hillhouse, J. W., Cerling, T. E. and Brown, F. H. (1986). Magnetostratigraphy of the Koobi Flora Formation, Lake Turkana, Kenya. *J. Geophys. Res., 91*:11, 581–11, 595.

Hooghiemstra, H. (1986). A high-resolution palynological record of 3.5 million years of Northern Andean climatic history: the correlation of 26 "glacial cycles" with terrestrial, marine and astronomical data. *Zbl. Geol. Palaont., 1*:1363–1366.

Keeley, L. H. and Toth, N. (1981). Microwear polishes on early stone tools from Koobi Fora, Kenya. *Nature, 293*:464–465.

Kutzbach, J. and Otto-Bliesner, B. L. (1982). The sensitivity of the African-Asian monsoonal climate to orbital parameter changes for 9000 yrs BP in a low resolution general circulation model. *J. Atmos. Sci., 21*: 1177–1188.

Livingstone, D. A. (1975). Late Quaternary climatic change in Africa. *Ann. Rev. Ecol. Syst., 6*:249–280.

Mercer, J. H. (1985). When did open-marine conditions last prevail in the Wilkes and Pensacola Basins, East Antarctica, and when was the Sirius Formation emplaced? *S. Afr. J. Sci., 81*:243–245.

Patterson, L. (1976). An introduction to the ecology and zoo-geography of the Okavango Delta. *In* F. Youngman (ed.), *Proceedings of the symposium on the Okavango Delta and its future utilization*, pp. 55–60. Okavango Wildlife Society, Johannesburg.

Pope, G. G. (1983). Evidence on the age of the Asian Hominidae. *Proc. Nat. Acad. Sci., 80*:4988–4992.

Potts, R. and Shipman, P. (1981). Cutmarks made by stone tools on bones from Olduvai Gorge, Tanzania. *Nature, 291*:577–560.

Rightmire, G. P. (1985). The tempo of change in the evolution of Mid-Pleistocene *Homo*. In E. Delson (ed.), *Ancestors: the hard evidence*, pp. 255–264. Alan R. Liss, New York.

Rind, D. and Peteet, D. (1985). Terrestrial conditions at the last glacial maximum and CLIMAP sea-surface temperature estimates: are they consistent? *Quat. Res., 24*:1–22.

Ruddiman, W. F. and Janacek, T. (1988). Plio-Pleistocene biogenic and terriginous fluxes at equatorial Atlantic sites 662-3 and 664. *Ocean Drilling Project Initial Report 108B;* in press.

Shackleton, N. J., Backman, J., Zimmerman, H., Kent, D. V., Hall, M., Roberts, D. G., Schnitker, D., Baldauf, J. G., Desprairies, A., Homrighausen, R., Huddlestun, P., Keene, J. B., Kaltenbach, A. J., Krumsieck, K. A. O., Morton, A. C., Murray, J. W. and Westberg-Smith, J. (1984). Oxygen isotope calibration of the onset of ice-rafting and history of glaciation in the North Atlantic region. *Nature, 307*:620–623.

Skelton, R. R., McHenry, H. M., and Drawhorn, G. M. (1985). Phylogenetic analysis of early hominids. *Curr. Anthropol., 27*:21–43.

Stern, J. T. and Susman, R. L. (1983). The locomotor anatomy of *Australopithecus afarensis*. *Amer. J. Phys. Anthropol., 60*:279–317.

Street-Perrott, F. A. and Roberts, N. (1983). Fluctuations in closed lakes as an indicator of past atmospheric circulation patterns. In F. A. Street-Perrott and N. Roberts (eds.), *Variations in the global water budget*, pp. 331–345. D. Reidel, Dordrecht, The Netherlands.

Thunell, R. C. and Williams, D. F. (1983). The stepwise development of Pliocene-Pleistocene paleoclimate and paleoceanographic conditions in the Mediterranean: oxygen isotope studies of DSDP sites 125 and 132. In J. E. Meulenkamp (ed.), *Reconstruction of marine paleoenvironments, Utrecht Micropal. Bull., 30*:111–127.

Tobias, P. V. (1985). Punctuational and phyletic evolution in the hominids. In E. S. Vrba, (ed.), *Species and speciation*, pp. 131–141. Transvaal Museum, Pretoria.

Van der Hammen, T., Wijmstra, T. A. and Zagwijn, W. H. (1971). The floral record of the Late Cenozoic of Europe. In K. K. Turekian, (ed.), *The Late Cenozoic glacial ages*, pp. 391–424. Yale Univ. Press, New Haven.

Vogel, J. C. (1984). Ed. *Late Cainozoic palaeoclimates of the Southern Hemisphere*. A. A. Balkema, Rotterdam.

Vogel, J. C. (1985). Southern Africa at 18000 yr B.P. *S. Afr. J. Sci., 81*:250–251.

Vrba, E. S. (1974). Chronological and ecological implications of the fossil Bovidae at the Sterkfontein australopithecine site. *Nature, 256*:19–23.

Vrba, E. S. (1975). Some evidence of chronology and palaeoecology of Sterkfontein, Swartkrans and Kromdraai from the fossil Bovidae. *Nature, 254*:301–304.

Vrba, E. S. (1976). The fossil Bovidae of Sterkfontein, Swartkrans and Kromdraai. *Tvl. Mus. Mem., 27*:1–166.

Vrba, E. S. (1979). A new study of the scapula of *Australopithecus africanus* from Sterkfontein. *Amer. J. Phys. Anthropol., 51*:117–129.

Vrba, E. S. (1980a). The significance of bovid remains as indicators of environment and predation patterns. In A. K. Behrensmeyer and A. P. Hill (eds.), *Fossils in the making*, pp. 247–271. University of Chicago Press, Chicago.

Vrba, E. S. (1980b). Evolution, species and fossils: how does life evolve? *S. Afr. J. Sci., 76*:61–84.

Vrba, E. S. (1981). The Kromdraai australopithecine site revisited in 1980: recent investigations and results. *Ann. Tvl. Mus., 33*(3):18–60.

Vrba, E. S. (1982). Biostratigraphy and chronology, based particularly on Bovidae, of southern African hominid-associated assemblages: Makapansgat, Sterkfontein, Taung, Kromdraai, Swartkrans; also Elandsfontein (Saldanha), Broken Hill (now Kabwe) and Cave of Hearths. In H. de Lumley and M. A. de Lumley (eds.), Proceedings of *Congress international de paleontologie humaine*, Vol. 2, pp. 707–752. Union Internationale des Sciences Prehistoriques et Protohistoriques, Nice.

Vrba, E. S. (1983). Macroevolutionary trends: new perspectives on the roles of adaptation and incidental effect. *Science, 221*:387–389.

Vrba, E. S. (1985a). Ecological and adaptive changes associated with early hominid evolution. *In* E. Delson (ed.), *Ancestors: the hard evidence*, pp. 63–71. Alan R. Liss, New York.

Vrba, E. S. (1985b). Environment and evolution: alternative causes of the temporal distribution of evolutionary events. *S. Afr. J. Sci.*, *81*:229–236.

Vrba, E. S. (1985c). African Bovidae: evolutionary events since the Miocene. *S. Afr. J. Sci.*, *81*:263–266.

Vrba, E. S. (1987a). Ecology in relation to speciation rates: some case histories of Miocene-Recent mammal clades. *Evol. Ecol.*, *1*, in press.

Vrba, E. S. (1987b). A revision of the Bovini (Bovidae) and a preliminary revised checklist of Bovidae from Makapansgat. *Palaeont. Afr.*, *26*:33–46.

Vrba, E. S. (1987c). New species and a new genus of Hippotragini (Bovidae from Makapansgat Limeworks. *Palaeont. Afr.*, *26*:47–58.

Vrba, E. S. (1988). The environmental context of the evolution of early hominids and their culture. *In* R. Bonnichsen and M. H. Sorg (eds.), *Bone modification*, Center for the Study of Early Man, Orono, Maine, in press.

Walker, A. (1981). Diet and teeth. *Phil. Trans. R. Soc. Lond.*, B, *292*:57–64.

Walker, A., Leakey, R. E., Harris, J. M. and Brown, F. H. (1986). 2.5-Myr *Australopithecus boisei* from west of Lake Turkana, Kenya. *Nature*, *322*:517–522.

Wesselman, H. B. (1984). The Omo micromammals: systematics and palaeoecology of early man sites from Ethiopia. *Contrib. Vert. Evol.*, *7*:1–219.

Wesselman, H. B. (1985). Fossil micromammals as indicators of climatic change about 2.4 Myr ago in the Omo Valley, Ethiopia. *S. Afr. J. Sci.*, *81*:260–261.

Tooth Morphology, Wear and Diet in *Australopithecus* and *Paranthropus* from Southern Africa

26

RICHARD F. KAY AND FREDERICK E. GRINE

For over 30 years, the dietary proclivities of the australopithecines have been the subject of considerable study and debate. The longevity of these debates is understandable because the question of diet is central to the construction of viable models of early hominid ecology and evolution. Not only do dietary factors directly affect the structures of the masticatory system, the anatomical parts that comprise the bulk of the fossil record, but they also bear upon musculoskeletal developments associated with food acquisition.

From his initial description of *Australopithecus africanus*, Dart (1925, 1949, 1957, 1958) championed the idea that it was a predatory and carnivorous hominid. The australopithecine specimens from the sites of Kromdraai and Swartkrans, attributed by Broom (1938, 1949) to the genus *Paranthropus*, were considered by Robinson (1954a,b) to have subsisted on an essentially herbivorous diet. According to Robinson (1954b; 1961), the omnivorous–herbivorous distinction between *Australopithecus* and *Paranthropus* was attested to by differences in the sizes of their teeth and masticatory muscles, the configuration of their facial skeletons, and by differences in the incidences of antemortem enamel chipping on their postcanine dentitions. Robinson (1954a,b, 1961) argued that the inferred trophic differences between *Australopithecus* and *Paranthropus* were related to their having occupied different "adaptive zones" rather than different aspects of the same "adaptive zone," and that these differences attested to their long evolutionary separation as members of distinct lineages.

Studies by Tobias (1967) and Wallace (1973) found no appreciable difference between *Australopithecus* and *Paranthropus* specimens in tooth chipping, and Tobias (1967) postulated that since the enamel chips were all "rather large" they may have been produced during bouts of bone chewing. With regard to the differences noted by Robinson (1954b, 1963, 1967) in the crania and dentitions of *Australopithecus* and *Paranthropus*, Tobias (1967), Brace (1969) and Wolpoff (1973, 1974, 1978) have argued that they are minimal in number, magnitude and significance, and that those that do exist are related primarily to presumed differences in the body sizes of these forms. Wolpoff (1974: 128), for example, concluded that the morphological patterns displayed by both give the "uniform interpretation of adaptation to a very heavily masticated herbivorous diet." Many workers have come to regard *Paranthropus* as simply a "robust" version of *Australopithecus*.

The questions raised about whether the morphological differences between *Australopithecus* and *Paranthropus* are not simply the scaling effects of concomitant body size differences, rather than evidence of distinct functional adaptations, have been the subject of debate for the past decade. Indeed, highly correlated size related changes may themselves reflect major adaptive differences along a size gradient of individuals or taxa (Smith, 1980; Jungers, 1984; Fleagle, 1985). Much of the recent work on the problem of allometric scaling in early hominid evolution has focused on the relative sizes of the incisors, canines and cheek-teeth in these taxa (Pilbeam and Gould, 1974; Kay, 1975, 1985; Wood and Stack, 1980; Wood *et al.*, 1983; Wood, 1984; Jungers and Grine, 1986).

Robinson (1954a,b) considered the disparity in the sizes of the canines and cheek-teeth in

Australopithecus and *Paranthropus* to provide perhaps the most compelling evidence for their dietary distinction. The relatively small canine size seen in *Paranthropus* represents a departure from the scaling effects seen in other primates, and the comparative diminution of these teeth may have been related to a functional shift from shearing to apical crushing activity (Wallace, 1978; Grine, 1981, 1984).

In addition to their relatively small canines, *Paranthropus* incisors are small in comparison to the sizes of their cheek-teeth, and Hylander (1975) observed that relative incisor size tends to be correlated with the size of dietary items in extant anthropoids. By analogy to Hylander's (1975) data, the relatively smaller incisors of *Paranthropus* may have been related to their having fed principally upon smaller food items than were ingested by *Australopithecus*. Alternatively, or perhaps concomitantly, incisal preparation and/or manipulation of food items may have differed between these early hominid taxa. Szalay (1972) suggested that the small, orthognathically implanted incisors of *Paranthropus* were related to bone-cracking activity, although Grine (1981) noted the considerable dissimilarity between the dentition of *Paranthropus* and those of canids, felids and hyaenids in which bone-cracking activities are observed.

It has been argued that postcanine tooth size in *Australopithecus* and *Paranthropus* scaled with positive allometry to estimates of average body weight—although it would appear that body weights themselves were estimated from tooth dimensions in at least some of these studies—and this has been taken to signify functional equivalence in *Australopithecus* and *Paranthropus* cheek-tooth size (Pilbeam and Gould, 1974; Wood and Stack, 1980; Peters, 1981). That is, *Paranthropus* required larger postcanine teeth simply to process more food per chewing cycle in order to maintain a larger body size, but the foods that were masticated would not necessarily have differed from those eaten by *Australopithecus*.

Contrary to these arguments of functional (i.e., metabolic) equivalence, however, tooth and body size regressions for a variety of mammals, including primates, are equivalent to or lower than the isometric slope of constant proportionality (Kay, 1975, 1985; Creighton, 1980; Gingerich and Smith, 1985). Larger animals do not appear to require relatively larger teeth simply to process greater quantities of the same sorts of foods eaten by their smaller relatives in order to maintain functional, or metabolic equivalence. What must be of concern in attempts to relate tooth size to metabolic rate is the frequency, or rate of masticatory cycles by which foods of given caloric content are processed (Calder, 1984). It is perhaps because tooth size is related to energy capacity rather than to energy rate that postcanine tooth area in mammals does not appear to parallel basal metabolic rate.

Moreover, the most recent body weight estimates that have been calculated from independent data sets by Jungers (this volume, chapter 7) and McHenry (this volume, chapter 9) are very similar in suggesting that the so-called "robust" australopithecines (*Paranthropus robustus* and *P. boisei*) were not notably larger than the "gracile" species of *Australopithecus*. Indeed, the body weight estimates provided by Jungers and McHenry would seem to make *Paranthropus* postcanine tooth size remarkably large.

It also has been argued that because the apparent lines of action of the masticatory muscles of *Paranthropus* were similar to those of modern humans, the manner in which bite forces can be applied to foods are also similar, and that because the postcanine occlusal area of *Paranthropus* is some four to five times that of modern humans, *Paranthropus* would have had to have generated four to five times the bite force of modern humans just to maintain equivalent occlusal pressures (Walker, 1981). From this, Walker has suggested that *Paranthropus* did not necessarily chew anything that could not have been masticated also by modern humans; rather, *Paranthropus* was simply processing greater quantities of these sorts of foods. Occlusal masticatory pressures, however, are not necessarily (if ever) applied evenly across the entire occlusal area, and Hylander (this volume, chapter 3) has demonstrated that the medial fibers of the masseter muscles play a significant role in mastication; a factor that cannot be overlooked when comparing the flaring, phaenozygous arches of *Paranthropus* to the delicate temporal arches of modern humans.

While analyses of cheek-tooth size scaling and masticatory biomechanics have provided invaluable insights as to the possible functional significance of the trophic structures of the Plio-Pleistocene hominids, extrapolation by analogy must be treated with a certain amount of circumspection. *Australopithecus* and *Paranthropus* were unique taxa in their own rights. Because no living primate possesses the unique dental proportions and craniofacial morphology displayed by *Paranthropus*, interpretations of its dietary habits that are based upon the odontometrics of extant primate species will be of a rather *ad hoc* nature (Kay and Covert, 1984). The interpretations that result from such studies, however, may be tested to a degree by examination of the tooth surfaces of these extinct taxa, for it is the occlusal surfaces that made intimate contact with the food items comprising their diets.

Tooth Wear and Diet

Microscopic dental wear patterns have been demonstrated to be related to dietary habits in extant animals (Walker, 1976; Walker *et al.*, 1978; Rensberger, 1978; Walker, 1981; Teaford and Walker, 1984; Teaford, 1985, 1986), and the quantification of such patterns, pioneered largely by Gordon (1982), has proved useful in distinguishing among living primate species with different diets. Despite the potentially valuable information that may be gained through such an examination of tooth wear, there have been surprisingly few attempts to apply such observations in the reconstruction of early hominid diets (Grine, 1977, 1981, 1984; Ryan, 1979; Walker, 1980, 1981; Puech *et al.*, 1983). These studies, moreover, have relied solely upon qualitative visual impressions in the assessment of wear patterns. Only very recently has quantification of microwear data for early hominid species been undertaken (Grine, 1986, 1987a,b).

Previous studies have had some success in atomizing microwear fabrics into individual constituent elements (e.g., "scratches," "pits," "gouges," etc.) although there continues to be disagreement over the means by which such wear feature categories should be defined (Gordon, 1982; Teaford and Walker, 1984; Grine, 1986). Because microwear fabrics commonly comprise numerous, overlapping features (Fig. 26.1), the definition and mensuration of individual elements are time consuming and difficult to replicate precisely. As a solution to these problems, Kay (1987) recently introduced numerical Fourier transformation as an analytical tool to characterize and distinguish different patterns of tooth wear. Quite different power spectra were obtained for extant primate species whose wear fabrics are dominated by "scratches" and "pits." This method and other quantitative techniques are employed here in the analysis of microwear details on australopithecine molars.

Materials and Methods

Permanent maxillary second molars (M^2s) of five specimens of *A. africanus* from Member 4 of the Sterkfontein Formation (Sts 28, Sts 30, Sts 31, Sts 52 and TM 1511) and five specimens

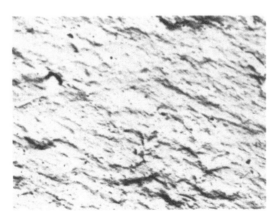

FIGURE 26.1. Dental microwear on facet 9 of an *Aotus trivirgatus* M_2 (United States National Museum 443734) illustrating the difficulty of identifying individual wear features in a single fabric. Field width = 400 μm.

of *Paranthropus* from Member 1 ("Hanging Remnant") of the Swartkrans Formation and Member 3 of the Kromdraai B East Formation (SK 13, SK 16, SK 42, SK 48 and TM 1517) were used in this study. These specimens were drawn at random from the larger sample that has been quantitatively analyzed by Grine (1986), who gives details about the specimens. Although it has been argued that the *Paranthropus* fossils from Kromdraai and Swartkrans should be accorded separate specific status as *P. robustus* and *P. crassidens*, respectively (Howell, 1978; Grine, 1982, 1985), analyses of microwear details on both the permanent and deciduous molars have revealed no appreciable differences between specimens from these two sites (Grine, 1981, 1987b). Thus, in this, as in past studies, the Kromdraai and Swartkrans specimens are combined in the *Paranthropus* sample.

Enamel wear on the Phase II facet 9 of Kay and Hiiemae (1974), which is located on the distolingual aspect of the protocone, was examined by scanning electron microscopy. Each facet was micrographed at magnifications of $100\times$ and $200\times$ at what was judged to be the center of the facet. When diagenetic or preparation damage marred the center of the facet, micrographs were recorded nearer the edge. In all instances, the entire enamel facet was examined to ensure that the micrographed field was representative of the surface as a whole.

Following the quantitative methods of Gordon (1982), Teaford and Walker (1984), and Grine (1986, 1987a), three groups of variables were recorded directly from the micrographs: *(i)* the numbers of scratches and pits per mm^2, *(ii)* the orientation of wear scratches, and *(iii)* the dimensions of the wear features. In addition, following the methods established by Kay (1987), the micrographs were analyzed by image processing techniques.

Wear Feature Incidences

Micrographs at $100\times$ were used to count the number of wear features. The contact micrograph was placed under a clear acetate sheet and a rectangular field equivalent to $0.5\ mm^2$ was ruled at its center. Each feature within the field was scored on the acetate sheet in order to ensure that large or elongate features were not counted more than once. Scratches and pits were scored independently following the definitions of Gordon (1982), although the "scratch" and "gouge" categories recognized by her were subsumed here under the designation "scratch." Subsequent measurements of the features indicated that virtually all of the pits recognized in the present study possessed length–width ratios that fell between 4:1 and 1:1.

Scratch Orientation

Micrographs at $200\times$ were used to determine the orientation of the wear scratches on each facet, with the micrograph being placed under a clear acetate sheet upon which the scratches were traced with ink. The angle at which the preferred orientation of the scratch diverged from the principal buccolingual (BL) axis of the crown, as recorded from the CRT (cathode ray tube), was measured. Because of the problems associated with the assignment of a straight line to oftentimes curvilinear and/or truncated scratches, the recorded angles are probably accurate to within only about ± 5°. For this reason, the scratch orientations were scored as falling within one of 18 blocks of 10° each, with the BL axis regarded as the 0–180° axis. Scratches that diverged to the distal (D) side, i.e., those with a distobuccal–mesiolingual orientation, were assigned values between 0° and 90°, while those that diverged to the mesial (M) side and which displayed a mesiobuccal–distolingual orientation were assigned values between 90° and 180° (Fig. 26.2).

Wear Feature Dimensions

The dimensions of the wear features on each facet were recorded from the $200\times$ micrographs, but because some of the diameters (e.g., scratch width) are so small on $200\times$ contact micrographs as to be virtually immeasurable, each of these micrographs was enlarged to a scale of $500\times$. Scratch length was measured on the enlargements using a Graf-Pen sonic digitizer, and scratch width, pit length, and pit width were measured using a dial-equipped caliper with tapered ends. The longest axis of a pit was defined as its length, and the width was measured as the greatest

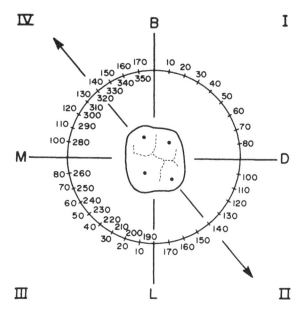

FIGURE 26.2. Diagram of hominid left M^2 indicating the method used to determine the preferred orientation of microwear scratches. The principal buccolingual (BL) crown axis is at 0° to 180°, and the mesiodistal (MD) axis is at 90° to 270°. The two axes define four quadrants, labeled I through IV. Scratches falling within quadrants I and III were assigned values between 0° and 90°; those in quadrants II and IV were assigned values greater than 90° and less than 180°.

diameter perpendicular to its length. The data for each of these parameters were recorded separately. The scratch lengths constitute conservative values in as much as truncated scratches were measured (Teaford and Walker, 1984).

Image Processing

The negative of each 200× micrograph was cropped to eliminate artifacts at the edges of the image, and scanned on a computer-controlled digital microdensitometer (Perkin-Elmer Corp.), which yields a 512^2 pixel matrix of optical density values. The matrix was analyzed on a Digital Equipment Corp. (DEC) PDP-11/45 computer using the SPIDER software system (Frank *et al.*, 1981) to obtain a power spectrum by computation of a numerical Fourier transform. The power spectrum thus obtained, which was stored in a new 512^2 matrix of optical intensities, is equivalent to a diffraction pattern that would be obtained if one passed a beam of coherent (laser) light through the original micrograph placed on the front focal plane of a curved lens and collected the resultant optical diffraction pattern on the back focal plane of a converging lens.

In order to facilitate quantitative comparisons of the power spectra obtained from the micrographs, the integrated optical intensity (IOI) was calculated radially from the center to the outer edge beginning at an arbitrary axis for each power spectrum. This procedure was repeated with the power spectrum rotated at 5° intervals through 180°, and the resulting array of 36 IOI values was aligned such that the largest IOI value corresponded to 90°. (The IOI is not calculated through 360° because the power spectrum of a microwear field is symmetrical such that when the image is bisected by any line through its center, one half of the power spectrum is the reverse mirror image of the other. Thus, only half of the original image need be studied.)

When the IOI values are plotted against degrees of rotation through 180° with the points connected, the resultant curve is often somewhat asymmetrical, resembling a bell-shaped curve with a shoulder at about 30° to 45° to one side of the the major peak. The possible significance of this shoulder, which is produced by a secondary (fainter) set of striations, is unclear at present, although we suspect that it may be related to contralateral tooth gliding activity. In this study, the primary wear features, which result in the form of the major peak, are of principal interest; thus, the shoulder is regarded as "noise" and its effects eliminated by analysis of one-half (90°) of the 180° scatterplot from which it was absent. In this instance, the IOI values from 0° to 90° were summed for each micrograph, and the individual values were expressed as percentages of the summed total. The resultant plots of percentage IOI versus degrees of rotation of the power spectrum are analyzed here.

Further details concerning these analytical protocols are presented in Grine (1986) and Kay (1987).

Results

The facet 9 surfaces of *Paranthropus* M²s tend to display higher densities of microwear features ($\overline{X}$ = 461.6 per mm²; SD = 116.2) than *Australopithecus* homologues ($\overline{X}$ = 331.6 per mm²; SD = 63.3), although the difference between them is not statistically significant. The respective values for the larger samples recorded by Grine (1986) are 437.6 per mm² and 290.4 per mm², and the means for the larger samples from which the present specimens derive are significantly different (t = 3.39; $p<0.005$).

The percentage incidences of scratches and pits derived from the total number of features recorded for each sample, as well as those calculated from individual percentage values (Table 26.1) reveal significantly higher incidences of pitting on *Paranthropus* molars. Thus, while pits comprise some 50% of features on *Paranthropus* molars, scratches constitute approximately 71% of features on *Australopithecus* teeth (Fig. 26.3). The corresponding percentage incidences obtained for the larger sample from which these specimens derive are very similar to those recorded here, and they also differ significantly (Grine, 1986).

The diameters of the wear features recorded for the present *Australopithecus* and *Paranthropus* samples are recorded in Table 26.2. While scratches tend to be longer on *Australopithecus* molars, the difference between the two sample means is not statistically significant. On the other hand, the scratches on *Australopithecus* crowns are significantly narrower, and the pits are significantly smaller than on *Paranthropus* homologues. As might be expected, the averages for the pooled or combined features indicate a distinct tendency for the wear fabrics on *Paranthropus* teeth to be composed of shorter and broader features.

While the measurement of sufficient numbers of scratches and pits per field of examination will permit a reasonable assessment of the separate dimensions of these feature categories, the pooling of scratch and pit diameters is justifiable only if the relative proportions of measured pits and scratches are equivalent to the observed proportions of these features. Since not every feature within a given field can be measured, even by the most meticulous observer, this situation

Table 26.1. Comparison of Scratch and Pit Frequencies on M² Facet 9

	Scratch	Pit	χ^2	p
Percentages from sample total				
Australopithecus	70.7	29.3	10.41	<0.005
Paranthropus	48.3	51.7		
Averages of individual percentages				
Australopithecus	70.8	29.2	8.31	<0.005
Paranthropus	50.9	49.1		

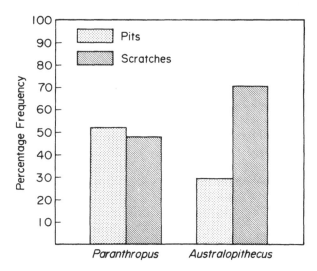

FIGURE 26.3. Histogram of the relative frequencies of scratches and pits on facet 9 of *Australopithecus* and *Paranthropus* M²s. Frequencies from averages of individual percentage values comprising each sample.

will be unlikely to be realized in any study. It certainly is not the situation in the present study, where no attempt was made to measure features in relation to their relative abundances. As discussed by Grine (1986), this problem can be overcome by computing "frequency adjusted" feature dimensions where the mean values for a given diameter (e.g., width) are multiplied by the observed percentage incidences of the two feature categories for the same individual facet.

The "frequency adjusted" diameters of the wear features for the present *Australopithecus* and *Paranthropus* samples are recorded in Table 26.3. Compared to the unadjusted diameters (Table 26.2), the "frequency adjusted" value for pooled feature length is slightly lower for *Paranthropus* and somewhat larger for *Australopithecus*, while the mean for pooled feature width is slightly higher for *Paranthropus* and slightly lower for *Australopithecus*. The unadjusted (Table 26.2) and "frequency adjusted" (Table 26.3) means for pooled feature length and width do not differ significantly within either sample. As was observed for the unadjusted values, the "frequency adjusted" means for pooled feature length do not differ significantly between *Australopithecus* and *Paranthropus*, but the *Paranthropus* sample mean for pooled feature width is significantly larger than that for the *Australopithecus* sample.

The values recorded here for wear feature dimensions (Table 26.2 and 26.3) are very similar to the corresponding means recorded by Grine (1986) for the larger samples from which the present samples derive.

The angles at which wear scratches on facet 9 diverge from the principal BL crown axis in the *Australopithecus* and *Paranthropus* samples are depicted graphically in Fig. 26.4. Scratches are predominantly of a distobuccal–mesiolingual orientation in both samples, with 78.1% of scratches on *Paranthropus* molars and 73.4% on *Australopithecus* teeth running from straight BL (0°) to 89° distal. The *Australopithecus* sample distribution, which appears more leptokurtic would appear also to display greater orientational homogeneity than the *Paranthropus* distribution (Fig. 26.4). A Kolmogorov–Smirnov test, which will be conservative for grouped data of this sort (Sokal and Rohlf, 1981), reveals a significant difference between these distributions (where Observed $D = 0.265$ and Expected $D = 0.175$) at the $p < 0.05$ level, as was found for the distributions of the larger samples from which the present data derive (Grine, 1986, 1987).

A G-test applied to the distributions shown in Fig. 26.4 reveals that they differ significantly ($G_{adj} = 31.5457$; $p < 0.001$), although a Watson and Williams test, which can be legitimately

Table 26.2. Sample Statistics on Microwear Feature Dimensions (μm) for *Australopithecus* and *Paranthropus*

Sample[a]	$\bar{X}^b$	SD	t	p
Scratch length				
Australopithecus	114.74	27.52	1.39	NS
Paranthropus	85.10	39.04		
Scratch width				
Australopithecus	0.98	0.07	3.87	<0.005
Paranthropus	1.94	0.55		
Pit length				
Australopithecus	8.41	1.52	4.54	<0.002
Paranthropus	12.18	1.07		
Pit width				
Australopithecus	5.00	0.91	4.99	<0.002
Paranthropus	8.54	1.30		
Pooled feature length				
Australopithecus	75.03	24.41	1.65	NS
Paranthropus	50.77	21.90		
Pooled feature width				
Australopithecus	2.60	0.88	4.91	<0.002
Paranthropus	5.21	0.80		

[a]$N = 5$ for each sample.
[b]means computed from individual specimen averages.

applied to angular data such as these (Watson and Williams, 1956; Mardia, 1972), reveals that there is no significant difference in angular variance. Since the angular deviation is similar in the two samples, the results of the Kolmogorov–Smirnov and G-test indicate that the *Australopithecus* and *Paranthropus* distributions differ only in mean direction.

Thus, the protoconal Phase II surfaces of *Australopithecus* and *Paranthropus* M^2s differ significantly in the incidences of occlusal pitting, the sizes, and especially the widths of the wear features, and in the mean orientation assumed by the wear scratches. *Paranthropus* displays, on average, significantly more and larger occlusal pits on this wear surface than does *Australopithecus*.

Examples of *Australopithecus* and *Paranthropus* Phase II surfaces and the corresponding power spectra for these wear facets are illustrated in Fig. 26.5. The power spectrum for the *Aus-*

Table 26.3. Sample Statistics on "Frequency Adjusted" Microwear Feature Dimensions (μm) for *Australopithecus* and *Paranthropus*

Sample[a]	$\bar{X}$	SD	t	p
Pooled feature length				
Australopithecus	86.77	34.06	2.18	NS
Paranthropus	48.89	18.83		
Pooled feature width				
Australopithecus	2.24	0.82	5.03	<0.002
Paranthropus	5.36	1.12		

[a]$N = 5$ for each sample.

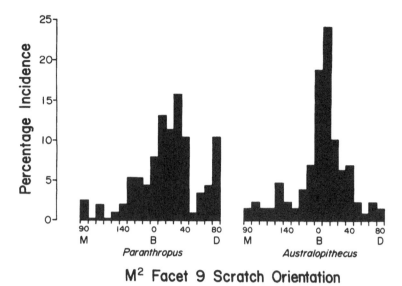

FIGURE 26.4. Histogram of occlusal scratch orientation on facet 9 of *Australopithecus* and *Paranthropus* M^2s. M, mesial; B, buccolingual crown axis; D, distal. Data for specimens comprising the present samples ($N = 5$ for each).

tralopithecus specimen (Sts 28) (Fig. 26.5b) reveals two bands of light concentrated at right angles to the predominant long axes of the wear scratches (the orientational preferences are separated by approximately 40°), while the *Paranthropus* (SK 42) spectrum (Fig. 26.5d) is somewhat more diffuse with less light concentrated in a narrow interval.

The bivariate plots of integrated light intensity versus angle of measurement for these two specimens and for the others comprising the present samples are depicted in Figs. 26.6 and 26.7. It is readily apparent that the *Paranthropus* sample displays markedly less internal variability and that the bivariate plots for the *Paranthropus* specimens tend to be considerably flatter than those for *Australopithecus* specimens. Were these plots extended to a full 180° by mirror-imaging, those for the *Paranthropus* specimens would be seen to be notably more platykurtic than those of four of the five *Australopithecus* specimens. The obvious exception is the Sts 31 tooth, for which the light intensity profile is comparatively flat (Fig. 26.6), indicating a somewhat more diffuse power spectrum with less light concentrated in a narrow interval than other *Australopithecus* molars. While the relatively diffuse power spectrum and platykurtic bivariate plot obtained for Sts 31 are similar to those for *Paranthropus* specimens, it is evident that the pattern evinced by Sts 31 relates primarily to the diffuse array of scratch orientations on this facet since comparatively few pits comprise this wear fabric (Fig. 26.8). The diffuse power spectra and flat plots obtained for the *Paranthropus* specimens, on the other hand, appear to be related primarily to the high frequencies of occlusal pitting displayed by these surfaces.

Figure 26.9 summarizes radial light intensity means vs degrees of rotation for the *Australopithecus* and *Paranthropus* samples (Table 26.4). *Paranthropus* displays a more platykurtic plot owing primarily to the higher incidence of pitting, with higher light intensity values between 0° and 45° and lower values between 55° and 90°, and especially between 70° and 90°. Wilcoxon two-sample and Kruskal–Wallis tests of first-order differences between the *Australopithecus* and *Paranthropus* sample means (Table 26.4) yield a Z of 1.661 ($p = 0.0967$) and a X^2 of 2.81 ($p = 0.0936$), respectively. The probabilities of the two sample means being different are 0.0967 and 0.0936 by these respective tests; i.e., differences of this magnitude in samples of these sizes would be expected to occur at random about 9.0% of the time.

Similar results for these data are obtained by application of randomization tests for absolute and squared differences of means (Fig. 26.10). Of the 126 possible specimen combinations, eight (6.35%) produce greater absolute differences and nine (7.14%) yield greater squared differences

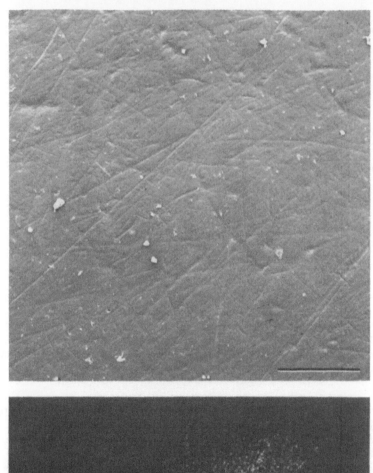

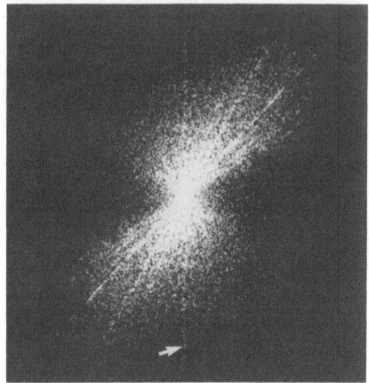

FIGURE 26.5. Examples of scanning electron micrographs of occlusal wear fabrics and the power spectra produced by Fourier transformation. a,b: *Australopithecus* (Sts 28). c,d: *Paranthropus* (SK 42). Scale bars = 100 μm. Solid vertical line on b (arrow) is an artefact of the Fourier transform.

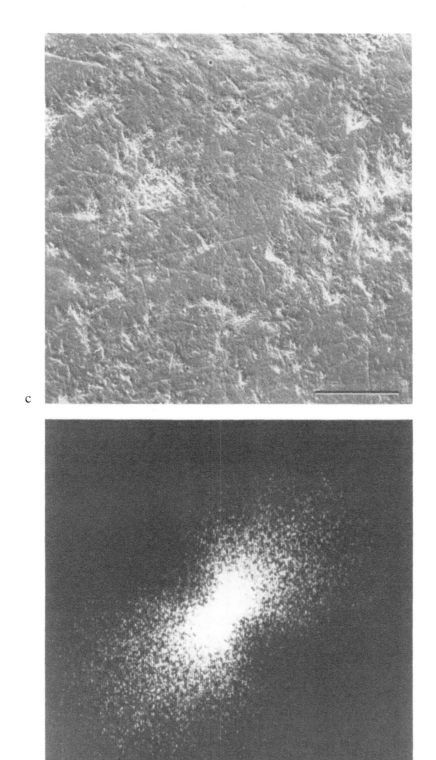

Figure 26.5. *Continued.*

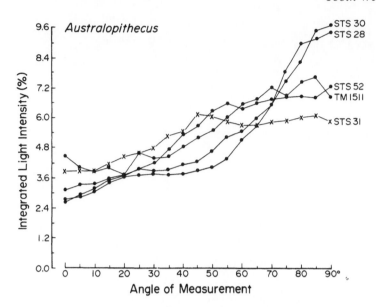

FIGURE 26.6. Bivariate plots of integrated light intensity (measured radially from the centers of the power spectra) versus angle of measurement (in degrees) for five *Australopithecus* specimens. Note the comparatively flat profile shown by Sts 31.

than obtained for the present sample configurations. The specimen combinations that yielded larger absolute and squared differences involved the removal of Sts 31 from the *Australopithecus* sample and its replacement by one of the *Paranthropus* specimens. It is worth noting, however, that randomization tests are very conservative and the observed mean differences stand in close proximity to the 0.05 probability limits (Fig. 26.10).

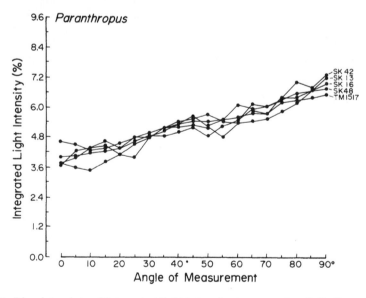

FIGURE 26.7. Bivariate plots of integrated light intensity (measured radially from the centers of the power spectra) vs angle of measurement (in degrees) for five *Paranthropus* specimens.

FIGURE 26.8. Scanning electron micrograph of occlusal wear fabric of facet 9 of Sts 31. Scale bar = 100 μm.

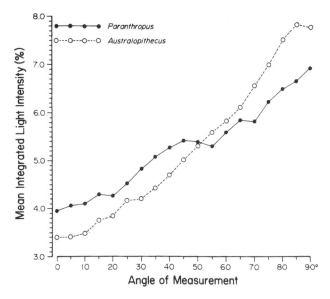

FIGURE 26.9. Bivariate plots of mean integrated light intensity (measured radially from the centers of the power spectra) versus angle of measurement (in degrees) for *Australopithecus* and *Paranthropus* samples. The data for these plots are recorded in Table 26.4.

Table 26.4. Sample Statistics of Adjusted Radial Light Intensity for Power Spectra of *Australopithecus* and *Paranthropus* Facet 9 Micrographs[a]

Degrees	Australopithecus		Paranthropus	
	$\bar{X}$	SD	$\bar{X}$	SD
0	3.407	0.708	3.950	0.346
5	3.411	0.489	4.059	0.301
10	3.483	0.358	4.099	0.326
15	3.749	0.315	4.289	0.274
20	3.853	0.307	4.275	0.171
25	4.167	0.363	4.532	0.291
30	4.206	0.379	4.831	0.070
35	4.429	0.561	5.082	0.134
40	4.704	0.660	5.266	0.145
45	5.027	0.852	5.424	0.160
50	5.312	0.843	5.291	0.287
55	5.597	0.751	5.296	0.278
60	5.840	0.538	5.598	0.271
65	6.121	0.457	5.852	0.222
70	6.572	0.441	5.819	0.200
75	6.979	0.669	6.244	0.208
80	7.524	1.011	6.489	0.303
85	7.846	1.310	6.662	0.147
90	7.775	1.431	6.942	0.290

[a] $N = 5$ for each sample.

Discussion

The demonstration of differences in the microwear patterns of *Australopithecus* and *Paranthropus* molars suggests a dietary difference between these two early hominid taxa.

As noted elsewhere (Grine, 1984), microwear patterns may not be indicative of diet *per se* because similar wear details can be produced by different agents. Enamel scratches, for example, may be engendered by intrinsic dietary items such as opaline phytoliths (Baker *et al.*, 1959; Walker *et al.*, 1978) or by extraneous material such as exogenous "grit" (Covert and Kay, 1981; Gordon and Walker, 1983). Similar diets may also produce different wear patterns if the food preparation techniques employed by two species differ.

Nevertheless, there does seem to be a reasonable relationship between microwear and some broad dietary parameters in extant mammals, including primates (Walker *et al.*, 1978; Gordon, 1982; Teaford and Walker, 1984; Teaford, 1985, 1986; Kay, 1987). Thus, it appears possible to distinguish between browsing and/or grazing habits, and/or between taxa that feed on hard food objects and those that do not.

The distinctive power spectra and resultant bivariate arrays obtained for *Australopithecus* and *Paranthropus* enamel surfaces are similar to, albeit less pronounced than the spectral differences recorded for *Ateles geoffroyi* and *Chiropotes satanas* respectively by Kay (1987). Behavioral and experimental studies (van Roosmalen, 1984) show that while both *Ateles* and *Chiropotes* are predominantly frugivorous, *A. geoffroyi*, which evinces fewer wear pits, feeds chiefly on mature fleshy fruits, swallowing the pulp and seeds whole without much effort at breaking the seeds. *Chiropotes satanas*, by contrast, feeds primarily upon seeds that are encased within a tough or hard exocarp; it splits open the exocarp with its stout canines and incisors and crushes the seeds between its cheek-teeth.

The percentage incidence of occlusal pitting on *Paranthropus* molars is significantly higher than that shown by *Australopithecus*, and Teaford and Walker (1984) have shown that pitting frequencies are higher in extant primates (e.g., *Cercocebus albigena*, *Cebus apella* and *Pongo*

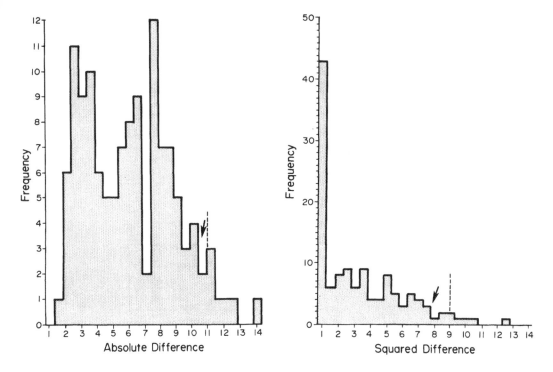

FIGURE 26.10. Frequency distributions of absolute and squared differences between means for all possible $N=5$ samples using *Australopithecus* and *Paranthropus* specimens. The arrow points to the observed sample combinations being tested; the broken line marks the 0.05 limits to the right.

pygmaeus) which feed on hard food items such as the seeds of the date palm, palm nuts and bark (Chalmers, 1968; Rodman, 1977; Izawa and Mizuno, 1977), than in those taxa (e.g., *Pan troglodytes, Colobus guereza, Alouatta* and *Gorilla*) whose diets do not include such items (Hladik, 1977; Wrangham, 1977; Oates, 1977; Milton, 1980; Fossey and Harcourt, 1977). The incidences of occlusal pitting recorded by Grine (1986, 1987) for the larger *Australopithecus* and *Paranthropus* samples are compared with frequencies for homologous facets of extant primate species in Fig. 26.11. The data for the extant primate samples have been modified by courtesy of M. Teaford (pers. comm.) from those presented by Teaford and Walker (1984) so that pits are defined by a length–width ratio of 4:1 instead of 10:1. The comparability of the *Paranthropus* pitting incidence with those of living primates that masticate hard food objects stands in sharp contrast to the similarities of the frequencies for *Australopithecus*.

Data for the widths of pooled microwear features on protoconal Phase II facets of extant primates have been recorded by Teaford and Walker (1984) and Teaford (1985, 1986). While the data presented by Teaford and Walker (1984) were calculated using questionable procedures (where species means were obtained from the total number of features in each sample rather than from the averages for the individuals comprising each species sample) (Grine, 1986), these data have been recalculated by courtesy of M. Teaford (pers. comm.) (Fig. 26.12). As would be expected from the pitting frequencies, the means for pooled feature breadth for the early hominid samples recorded by Grine (1986, 1987) suggest that the diet of *Paranthropus* consisted of comparatively hard objects.

Our analysis reveals that while the majority of the scratches are aligned along a distobuccal–mesiolingual axis in both early hominid taxa, indicating similarity in the general direction of the masticatory power stroke (and the absence of any supposed "orthal retraction"), *Paranthropus* molars tend to show more heterogeniety of scratch orientation, although the differences are not

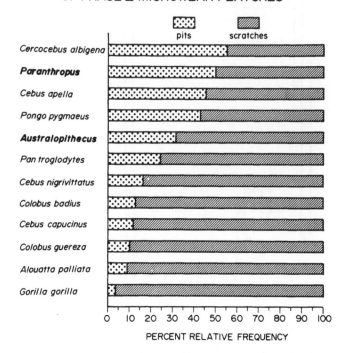

FIGURE 26.11. Histograms of percentage frequency of microwear pitting recorded for M^2 facet 9 of *Australopithecus, Paranthropus,* and extant primate taxa. Extant species frequencies determined by use of length–width ratio of 4:1 to 1:1 to define pits. Data for extant species supplied courtesy of M. F. Teaford (pers. comm.).

statistically significant, and a difference in mean orientation. The tendency to display greater heterogeneity of wear scratch orientation may be related to an increased amount of eccentric jaw movement at the terminal portion of Phase I activity and/or during Phase II activity. Increased eccentricity would seem likely to be related to differences in the mechanical properties of the food items that were processed by the two taxa, where the sudden structural failure of hard, brittle items as the molars approached centric occlusion would act to impart such directional change.

Thus, the differences between the microwear patterns displayed by *Australopithecus* and *Paranthropus* molars are strongly suggestive of concomitant differences in their dietary habits. The overall design of the masticatory apparatus of these early hominids suggests herbivory, and the wear fabrics evinced by *Australopithecus* specimens are similar to those of extant primates whose diets consist primarily (although not exclusively) of fleshy fruits and foliage. Thus, *Australopithecus* falls between *Pan* and *Pongo* in pitting frequencies and between *Pan* and *Alouatta*, on the one hand, and *Cebus nigrivittatus* and *Colobus* species, on the other, with regard to microwear feature breadths. *Paranthropus* microwear, on the other hand, is comparable to that shown by living forms that process more hard food items. These quantitative results for the permanent molars support previous interpretations based upon qualitative analyses of deciduous molar wear in *Australopithecus* and *Paranthropus* (Grine, 1977, 1981, 1984). Parenthetically, the relative pitting incidence for *Sivapithecus* calculated using a 4:1 as opposed to a 10:1 length–width ratio (courtesy of M. F. Teaford, pers. comm.), if interpolated into Fig. 26.11 would fall midway between *Pongo* and *Australopithecus,* suggesting that the diet of *Sivapithecus* may have comprised hard objects as well as fruits.

It would appear, therefore, that the diets of *Australopithecus* and *Paranthropus* were quali-

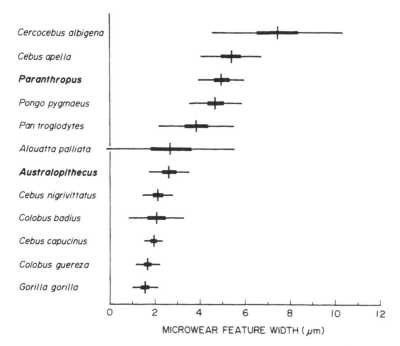

FIGURE 26.12. Comparison of widths of pooled microwear features on M^2 facet 9 for *Australopithecus*, *Paranthropus*, and extant primate taxa. These data are not frequency adjusted. Data for extant species supplied courtesy of M. F. Teaford (pers. comm.). Vertical lines = means; horizontal lines = 1 *SD* on either side of mean; horizontal bars = 1 *SE* on either side of mean.

tatively dissimilar. *Paranthropus* does not appear to have been masticating simply greater quantities of the same foods that were chewed by *Australopithecus*. These apparent dietary differences, and the substantial differences in their trophic structures do not appear to be related solely to estimated differences in their body sizes. There is scant justification for regarding *Paranthropus* as simply a "robust" version of *Australopithecus*.

Conclusions

There is longstanding debate in hominid paleontology over the dietary habits of *Australopithecus* and *Paranthropus*, the so-called "gracile" and "robust" australopithecines. Opinions have ranged from the classic dietary hypothesis formulated by Robinson (1954b), which held that their diets differed in the proportions of meat and vegetables consumed, to more recent formulations which maintain that the "robust" forms simply processed greater quantities of the sorts of items consumed by the "gracile" forms in order to maintain their presumably larger body sizes (Pilbeam and Gould, 1974; Walker, 1981).

Microscopic dental wear patterns, and more particularly the quantitative analyses of these wear fabrics have been shown to be useful in distinguishing among extant primate species with different dietary habits (Gordon, 1982; Teaford and Walker, 1984; Teaford, 1985, 1986). The quantitative analyses of microwear on Phase II protoconal facets of *Australopithecus* and *Paranthropus* M^2s presented here indicate that these taxa differed in their diets, with the implication that differences in their craniofacial and dental morphologies may be accounted for likewise. *Paranthropus* wear fabrics comprise significantly greater incidences of pits, significantly broader wear features, and significantly greater heterogeneity of wear scratch orientation; and these differences are confirmed by the power spectra obtained for these samples through the Fourier transformation of optical diffraction patterns.

Wear on the permanent (and deciduous) molars of *Paranthropus* resembles that of extant primates that eat hard food items, whereas the *Australopithecus* pattern differs in the direction of living species that subsist more on leaves and fleshy fruits.

Acknowledgments

We thank C. K. Brain and E. S. Vrba (Transvaal Museum) and P. V. Tobias and A. R. Hughes (University of the Witwatersrand) for permission to cast and examine the early hominid specimens in their care, E. A. Peterson for her painstaking care and expertise in the manufacture of the casts, G. Shidlovsky (Brookhaven National Laboratory) for access to and assistance with the AMR 1400 SEM, and J. Gwinnett for the use of coating facilities. We are grateful to R. R. Sokal and M. Burgman for their invaluable statistical advice and assistance, and we thank D. Kopf and M. Lamvik for technical assistence with image processing. This work was supported by a National Science Foundation Equipment Grant (NSF PCM 8306638) to M. Lamvik and by a grant from the L. S. B. Leakey Foundation to F. E. G.

References

Baker, G., Jones, L. H. P. and Wardrop, I. D. (1959). Causes of wear in sheep's teeth. *Nature, 184*: 1583–1584.
Brace, C. L. (1969). The australopithecine range of variation. *Amer. J. Phys. Anthropol., 31*: 255.
Broom, R. (1938). The Pleistocene anthropoid apes of South Africa. *Nature, 142*: 377–379.
Broom, R. (1949). Another type of fossil ape-man *(Paranthropus crassidens)*. *Nature, 163*: 57.
Calder, W. A. (1984). *Size, function and life history*. Harvard University Press, Cambridge, Mass.
Chalmers, N. R. (1968). Group composition, ecology and daily activities of free living mangabeys in Uganda. *Folia Primatol., 8*: 247–262.
Covert, H. H. and Kay, R. F. (1981). Dental microwear and diet: implications for determining the feeding behaviors of extinct primates, with a comment on the dietary pattern of *Sivapithecus*. *Amer. J. Phys. Anthropol., 55*: 331–336.
Creighton, G. K. (1980). Static allometry of mammalian teeth and the correlation of tooth size and body size in contemporary mammals. *J. Zool. (Lond.), 191*: 435–443.
Dart, R. A. (1925). *Australopithecus africanus:* the man-ape of South Africa. *Nature, 115*: 195–199.
Dart, R. A. (1949). The predatory implemental technique of *Australopithecus*. *Amer. J. Phys. Anthropol., 7*: 1–16.
Dart, R. A. (1957). The osteodontokeratic culture of *Australopithecus prometheus*. *Mem. Tvl. Mus., 10*: 1–105.
Dart, R. A. (1958). The minimal bone-breccia content of Makapansgat and the australopithecine predatory habit. *Amer. Anthropol., 60*: 923–931.
Fleagle, J. G. (1985). Size and adaptation in primates. *In* W. L. Jungers (ed.), *Size and scaling in primate biology*, pp. 1–19. Plenum, New York.
Fossey, D. and Harcourt, A. H. (1977). Feeding ecology of the free-ranging mountain gorilla *(Gorilla gorilla beringei)*. *In* T. H. Clutton-Brock (ed.), *Primate ecology*, pp. 415–447. Academic Press, New York.
Frank, J., Shimkin, B. and Dowse, H. (1981). SPIDER—a modular software system for electron image processing. *Ultramicroscopy, 6*: 343–358.
Gingerich, P. D. and Smith, B. H. (1985). Allometric scaling in the dentition of primates and insectivores. *In* W. L. Jungers (ed.), *Size and scaling in primate biology*, pp. 257–272. Plenum, New York.

Gordon, K. D. (1982). A study of microwear on chimpanzee molars: implications for dental microwear analysis. *Amer. J. Phys. Anthropol., 59*: 195–215.

Gordon, K. D. and Walker, A. C. (1983). Playing 'possum: a microwear experiment. *Amer. J. Phys. Anthropol., 60*: 109–112.

Grine, F. E. (1977). Analysis of early hominid deciduous molar wear by scanning electron microscopy: a preliminary report. *Proc. Electron Microsc. Soc. Sthn. Afr., 7*: 157–158.

Grine, F. E. (1981). Trophic differences between "gracile" and "robust" australopithecines: a scanning electron microscope analysis of occlusal events. *S. Afr. J. Sci., 77*: 203–230.

Grine, F. E. (1982). A new juvenile hominid (Mammalia: Primates) from Member 3, Kromdraai Formation, Transvaal, South Africa. *Ann. Tvl. Mus., 33*: 165–239.

Grine, F. E. (1984). Deciduous molar microwear of South African australopithecines. In D. J. Chivers, B. A. Wood and A. Bilsborough (eds.), *Food acquisition and processing in primates*, pp. 525–534. Plenum, New York.

Grine, F. E. (1985). Australopithecine evolution: the deciduous dental evidence. In E. Delson (ed.), *Ancestors: the hard evidence*, pp. 153–167. Alan R. Liss, New York.

Grine, F. E. (1986). Dental evidence for dietary differences in *Australopithecus* and *Paranthropus*: a quantitative analysis of permanent molar microwear. *J. Hum. Evol., 15*: 783–822.

Grine, F. E. (1987a). Quantitative analysis of occlusal microwear in *Australopithecus* and *Paranthropus*. *Scanning Microsc., 1*: 647–656.

Grine, F. E. (1987b). L'alimentation des Australopitheques d'Africa du Sud, d'apres l'etude des microtraces d'usure sur les dents. *L'Anthropologie, 91*: 467–482.

Hladik, C. M. (1977). Chimpanzees of Gabon and chimpanzees of Gombe: some comparative data on the diet. In T. H. Clutton-Brock (ed.), *Primate ecology*, pp. 481–501. Academic Press, New York.

Howell, F. C. (1978). Hominidae. In V. J. Maglio and H. B. S. Cooke (eds.), *Evolution of African mammals*, pp. 154–248. Harvard University Press, Cambridge, Mass.

Hylander, W. L. (1975). Incisor size and diet in anthropoids with special reference to Cercopithecidae. *Science, 189*: 1095–1098.

Izawa, K. and Mizuno, A. (1977). Palm-fruit cracking behavior of wild black-capped capuchin (*Cebus apella*). *Primates, 18*: 773–792.

Jungers, W. L. (1984). Aspects of size and scaling in primate biology with special reference to the locomotor skeleton. *Yrbk. Phys. Anthropol., 27*: 73–97.

Jungers, W. L. and Grine, F. E. (1986). Dental trends in the australopithecines: the allometry of mandibular molar dimensions. In B. A. Wood, L. B. Martin and P. Andrews (eds.), *Major topics in primate and human evolution*, pp. 203–219. Cambridge University Press, Cambridge.

Kay, R. F. (1975). Allometry and early hominids. *Science, 189*: 63.

Kay, R. F. (1985). Dental evidence for the diet of *Australopithecus*. *Ann. Rev. Anthropol., 14*: 315–341.

Kay, R. F. (1987). Analysis of primate dental microwear using image processing techniques. *Scanning Microsc., 1*: 657–662.

Kay, R. F. and Covert, H. H. (1984). Anatomy and behavior of extinct primates. In D. J. Chivers, B. A. Wood and A. Bilsborough (eds.), *Food acquisition and processing in primates*, pp. 467–508. Plenum, New York.

Kay, R. F. and Hiiemae, K. M. (1974). Jaw movement and tooth use in recent and fossil primates. *Amer. J. Phys. Anthropol., 40*: 227–256.

Mardia, K. V. (1972). *Statistics of directional data*. Academic Press, New York.

Milton, K. (1980). *The foraging strategy of howler monkeys*. Columbia University Press, New York.

Oates, J. F. (1977). The guereza and its food. In T. H. Clutton-Brock (ed.), *Primate ecology*, pp. 275–322. Academic, New York.

Peters, C. R. (1981). Robust vs. gracile early hominid masticatory capabilities: the advantages of the megadonts. In L. L. Mai, E. Shanklin and R. W. Sussman (eds.), *The perception of human evolution*, pp. 161–181. University of California Press, Los Angeles.

Pilbeam, D. R. and Gould, S. J. (1974). Size and scaling in human evolution. *Science, 186*: 892–901.

Puech, P. F., Albertini, H. and Serratrice, C. (1983). Tooth microwear and dietary patterns in early hominids from Laetoli, Hadar and Olduvai. *J. Hum. Evol., 12*: 721–729.

Rensberger, J. M. (1978). Scanning electron microscopy of wear and occlusal events in some small herbivores. *In* P. M. Butler and K. A. Joysey (eds.), *Development, function and evolution of teeth*, pp. 415–438. Academic Press, New York.

Robinson, J. T. (1954a). The genera and species of the Australopithecinae. *Amer. J. Phys. Anthropol., 12*: 181–200.

Robinson, J. T. (1954b). Prehominid dentition and hominid evolution. *Evolution, 8*: 324–334.

Robinson, J. T. (1961). The australopithecines and their bearing on the origin of man and of stone tool making. *S. Afr. J. Sci., 57*: 3–16.

Robinson, J. T. (1963). Adaptive radiation in the australopithecines and the origin of man. *In* F. C. Howell and F. Bourliere (eds.), *African ecology and human evolution*, pp. 385–416. Aldine, Chicago.

Robinson, J. T. (1967). Variation and taxonomy of early hominids. *In* T. Dobzhansky, M. K. Hecht and W. C. Steere (eds.), *Evolutionary biology*, vol. 1, pp. 69–100. Meredith, New York.

Rodman, P. S. (1977). Feeding behaviour of orang-utans of the Kutai Nature Reserve, East Kalimantan. *In* T. H. Clutton-Brock (ed.), *Primate ecology*, pp. 383–413. Academic Press, New York.

Ryan, A. S. (1979). Scanning electron microscopy of tooth wear on the anterior teeth of *Australopithecus afarensis*. *Amer. J. Phys. Anthropol., 50*: 478.

Smith, R. J. (1980). Rethinking allometry. *J. Theor. Biol., 87*: 97–111.

Sokal, R. R. and Rohlf, F. J. (1981). *Biometry: the principles and practice of statistics in biological research*. 2nd ed. W. H. Freeman, New York.

Szalay, F. S. (1972). Hunting-scavenging protohominids: a model for hominid origins. *Man, 10*: 420–429.

Teaford, M. F. (1985). Molar microwear and diet in the genus *Cebus*. *Amer. J. Phys. Anthropol., 66*: 363–370.

Teaford, M. F. (1986). Dental microwear and diet in two species of *Colobus*. *In* J. G. Else and P. C. Lee (eds.), *Primate ecology and conservation*, pp. 63–66. Cambridge University Press, Cambridge.

Teaford, M. F. and Walker, A. C. (1984). Quantitative differences in dental microwear between primate species with different diets and a comment on the presumed diet of *Sivapithecus*. *Amer. J. Phys. Anthropol., 64*: 191–200.

Tobias, P. V. (1967). *The cranium and maxillary dentition of* Australopithecus (Zinjanthropus) boisei. Olduvai Gorge, vol. 2. Cambridge University Press, Cambridge.

van Roosmalen, M. G. M. (1984). Subcategorizing foods in primates. *In* D. J. Chivers, B. A. Wood and A. Bilsborough (eds.), *Food acquisition and processing in primates*, pp. 167–176. Plenum, New York.

Walker, A. C. (1980). Functional anatomy and taphonomy. *In* A. K. Behrensmeyer and A. P. Hill (eds.), *Fossils in the making*, pp. 182–196. University of Chicago Press, Chicago.

Walker, A. C. (1981). Dietary hypotheses and human evolution. *Phil. Trans. Roy. Soc. (Lond.), B292*: 57–64.

Walker, A. C., Hoeck, H. N. and Perez, L. (1978). Microwear of mammalian teeth as an indicator of diet. *Science, 201*: 808–810.

Walker, P. L. (1976). Wear striations on the incisors of cercopithecoid monkeys as an index of diet and habitat preference. *Amer. J. Phys. Anthropol., 45*: 299–308.

Wallace, J. A. (1973). Tooth chipping in the australopithecines. *Nature, 244*: 117–118.

Wallace, J. A. (1978). Evolutionary trends in the early hominid dentition. *In* C. Jolly (ed.), *Early hominids of Africa*, pp. 285–310. Duckworth, London.

Watson, G. S. and Williams, E. J. (1956). On the construction of significance tests on the circle and the sphere. *Biometrika, 43*: 344–352.

Wolpoff, M. H. (1973). Posterior tooth size, body size and diet in South African gracile australopithecines. *Amer. J. Phys. Anthropol., 39*: 375–394.

Wolpoff, M. H. (1974). The evidence for two australopithecine lineages in South Africa. *Yrbk. Phys. Anthropol., 17*: 113-139.

Wolpoff, M. H. (1978). Some aspects of canine size in the australopithecines. *J. Hum. Evol., 7*: 115-126.

Wood, B. A. (1984). Interpreting the dental peculiarities of the "robust" australopithecines. *In* D. J. Chivers, B. A. Wood and A. Bilsborough (eds.), *Food acquisition and processing in primates*, pp. 535-544. Plenum, New York.

Wood, B. A. and Stack, C. G. (1980). Does allometry explain the difference between "gracile" and "robust" australopithecines? *Amer. J. Phys. Anthropol., 52*: 55-62.

Wood, B. A., Abbott, S. A., and Graham, S. H. (1983). Analysis of the dental morphology of Plio-Pleistocene hominids. II. Mandibular molars—study of cusp area, fissure pattern and cross-sectional shape of the crown. *J. Anat., 137*: 287-314.

Wrangham, R. W. (1977). Feeding behaviour of chimpanzees in Gombe National Park, Tanzania. *In* T. H. Clutton-Brock (ed.), *Primate ecology*, pp. 503-538. Academic Press, New York.

The Comparative Biology of "Robust"[1] *Australopithecus:* Clues from Context

27

TIM D. WHITE

Robert Broom's 1938 announcement of the initial Kromdraai "robust" *Australopithecus* specimen represents the first attempt to understand aspects of this organism's ecology. Broom applied information based on the context of the specimen and wrote: "The apes lived on the plains and among rocky krantzes . . . the larger bones in the caves seem all to have been introduced by carnivorous animals. . . ." (1938: 379).

In the years that have passed since Broom's pioneering work, the samples of fossils representing *Australopithecus robustus, Australopithecus boisei* and *Australopithecus aethiopicus* have increased dramatically. Many new contextual data are available from sites in southern and eastern Africa. This paper is about the application of contextual data to an understanding of the biology of "robust" *Australopithecus* species.

Background

Attributes of any mammal's biology may be partitioned into five basic but overlapping categories: anatomy, ecology, demography, behavior, and phylogeny. Figure 27.1 illustrates the relationships between skeletal and contextual data and various aspects of the five categories in extinct hominids. Some aspects will be understood only if further discoveries are made. Many aspects will remain in the realm of informed speculation.

Materials and Methods

This analysis will be confined to fossils of *A. robustus* and *A. boisei*. The available sample of *A. aethiopicus* (see Kimbel *et al.,* this volume, chapter 16) is presently too small for meaningful investigation based on context. The binomen *A. boisei* is applied to the collections of "robust" *Australopithecus* fossils from the eastern African sites of Olduvai Gorge (Beds I and II), Peninj, the Omo Shungura Formation (Members E through K; "robust" specimens from Omo Shungura Members C and D probably belong to *A. aethiopicus* and are not included in the analysis), West Turkana, Koobi Fora, and Chesowanja. The name *A. robustus* is applied to the Kromdraai and Swartkrans ("Hanging Remnant" only) collections from southern Africa. The early *Homo* sample contains *H. habilis* and early *H. erectus* and only includes specimens derived from strata that yield the "robust" *Australopithecus* samples.

The data set analyzed includes southern African hominid specimens listed by Brain (1981a) [newer specimens introduced by Grine (this volume, chapter 14) and Susman (this volume, chapter 10) do not figure in the analysis]. Table 27.1 lists specimens from eastern African sites, with approximate individual age [very roughly divided into adult (AD) and immature (IM)], geological age, depositional context, and classification. Data in the table are compiled from Brown and Feibel (1986), Brown *et al.* (1985b), Day (1986), de Heinzelin (1983), Findlater (1978), Gowlett

---

[1]The author does not agree with the use of quotation marks around the word robust.

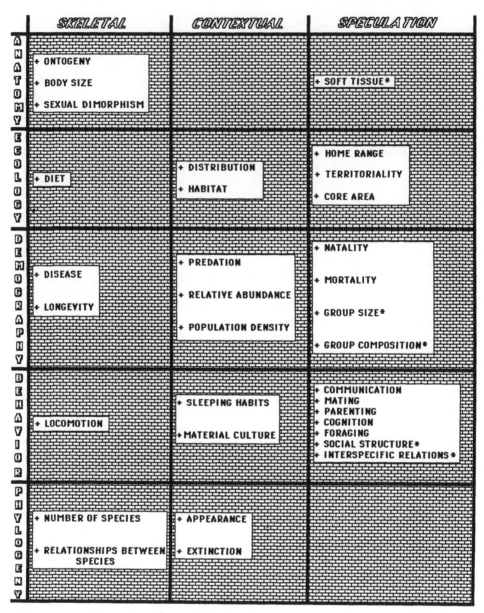

FIGURE 27.1. An illustration of how skeletal and contextual data from the fossil record (top) might be used to understand five broad, overlapping categories of "robust" *Australopithecus* biology (side). Aspects listed under speculation with asterisks might be susceptible to understanding if unique preservational circumstances not yet encountered are discovered. See text for details.

et al. (1981), Howell and Coppens (1974), Leakey (1971), Leakey and Leakey (1978), and Feibel (pers. comm.).

Specimens listed in Table 27.1 include at least fragmentary cranial and/or dental elements. I was able to sort many intact premolars and molars into *Homo* (H) or *A. boisei* (R). Work in progress by Suwa (this volume, chapter 13) and others will certainly make it possible to identify many additional teeth listed as indeterminate. Taxonomic attributions are conservative and very similar to those of White (1977) and White *et al.* (1981). Identifications of the available specimens differ only slightly from those of Wood and his colleagues for the Olduvai and Koobi Fora material

(Wood and Abbott, 1983; Wood et al., 1983; Chamberlain and Wood, 1985; Beynon and Wood, 1986; Wood and Uytterschaut, 1987). Hominids from Omo Shungura Members E through K have been placed in *Homo, A. boisei,* or indeterminate. The presence of *A. africanus* at Omo was not convincingly demonstrated by Howell (1978) or Coppens (1980) as noted by Howell et al. (1987).

Isolated hominid postcranial elements are not listed in Table 27.1. The only postcranial elements attributable with confidence to "robust" *Australopithecus* (diagnosed by associated cranial remains) are the carpals and tarsals of Omo 323–896 (but see Grausz et al., this volume, chapter 8). Although the Kromdraai TM 1517 limb and tarsal elements were found in the same cubic foot of breccia as the *A. robustus* holotype skull fragments (Broom and Schepers, 1946), proximity in a South African karst cave depository is an uncertain guide to individual association. These remains and most isolated postcranial elements from the Swartkrans hanging remnant are probably those of *A. robustus* simply because of the rarity of other hominid taxa in this unit. For postcrania in Member 1 "Lower Bank", Members 2 and 3 (Brain, this volume, chapter 20), this probability is lessened because of smaller samples, taphonomic differences between the units, and somewhat higher relative *Homo* abundance (Grine, this volume, chapter 14)

Species level attribution of isolated postcranial elements in eastern Africa has been practiced since their initial recovery in the 1960's. Even without associated cranial or dental evidence, specimens like the Omo L 40-19 ulna or the Koobi Fora KNM-ER 739 humerus were attributed to *Australopithecus* (Leakey, 1971), and sometimes to *A. boisei* (Howell and Wood, 1974). Some speculation held that "The Rudolf australopithecines, in fact, may have been close to the 'knuck-lewalker' condition, not unlike the extant African apes" (Leakey, 1971: 245). Such attributions and speculations may relate to preconceptions about "robust" hominid "primitiveness." From almost the beginning, many assessments of "robust" *Australopithecus* seem to have been based on the mistaken but pervasive notion that sagittal crests somehow signal great phylogenetic and functional distance.

Taxonomic attributions of early hominid postcrania unassociated with cranial remains have recently been called into question by discoveries of partial and complete skeletons of *Homo habilis* (Johanson et al., 1987) and *Homo erectus* (Brown et al., 1985a). It is probable that future discoveries will allow for the differential diagnosis of some early hominid postcranial elements. Until the discovery of whole or partial skeletons that include cranial remains of *A. robustus* or *A. boisei*, however, it is premature to attribute isolated postcrania from Late Pliocene or early Pleistocene deposits to genus. The relevant available sample of 51 isolated hominid postcranial specimens is therefore not included in Table 27.1 but is divided roughly as follows: 3 fragmentary lower legs, 2 femora, 18 femoral fragments, 3 tibial fragments, 1 fibular fragment, 1 partial foot, 2 tali, 3 metatarsals, 2 partial arms, 2 humeri, 3 humeral fragments, 3 radial fragments, 2 ulnae, 2 ulnar fragments, 1 partial hand, 2 phalanges, and one clavicle belonging to *A. boisei* or early *Homo*.

An appreciation of the relevant current sample sizes of hominid fossils identifiable to genus may be gained from Table 27.1. The available collection dating to between c. 2.4 and c. 1.2 Myr from eastern Africa consists of nearly 300 cranial/mandibular/dental specimens. Of these, about 100 belong to *A. boisei* and about 85 to *Homo*, while the remaining specimens are indeterminate. For Swartkrans, analogous numbers are nearly 200 *A. robustus* and 6 *Homo* when no specimens are left unassigned and no isolated specimen is joined to another isolated piece. Given the depositional background and recovery methods for the specimens from eastern African sites, specimens listed in the table represent separate individuals in nearly every case. In southern Africa, Brain (1981a) estimates a minimum number of 85 *A. robustus* individuals from Swartkrans compared to three *Homo* individuals from the same "pink breccia," hanging remnant unit.

From Death to Recovery: Bias in the Contextual Database

That the fossil record usually represents a very tiny (Walker and Leakey, 1978) and very distorted reflection of ancient plant and animal communities has become increasingly appreciated with the rise of "taphonomic" studies over the last few decades (Dodson, 1971; Behrensmeyer and Hill, 1980). As Western (1980: 49) points out: "An analysis of the biases associated with a collection of fossils should be central to the process of interpreting past environments."

Consider an African environment such as the Turkana Basin two million years ago. Populations representing different hominid species lived and died there. For the Plio-Pleistocene observer each species exhibited anatomy, ecological relationships, demography and behavior. Our challenge is to use the available fossil database to accurately reconstruct the biology of these extinct species on the basis of fossil samples that have been sifted through and distorted by a series of successive situational and preservational filters (Fig. 27.2). Each filter has the potential to bias the final samples.

Eastern African strata that have produced the samples of "robust" *Australopithecus* and *Homo* were deposited in waterside locations. As Western (1985) points out, the sedimentary basins were often saline and/or alkaline, conditions that favored grasslands and thus grazers even though bushlands and browsers might have existed slightly away from the lake margins.

What was going on in drier, upland environments analogous to the mid-Pliocene situation at Laetoli is unknown in eastern Africa for the time period under consideration (c 2.4 to c 1.2 Myr). The record is therefore skewed toward a representation of the more watered, axial areas of basins associated with the Eastern Rift. Further skewing of bone assemblages from these environments is expected to have resulted from water-dependent animals being attracted to water sources and suffering proportionally large amounts of predation and deposition here.

Carnivores in both southern and eastern Africa also impacted bone assemblages. The apparent loss of very young individuals to complete consumption and the predominance of cranial elements reflect, in part, this taphonomic filter (Fig. 27.2). Behrensmeyer *et al.* (1979) have shown that there is size-biasing against smaller species in a modern bone assemblage from Amboseli due to the vulnerability of bones from smaller mammals to damage by carnivores, weathering, and trampling. Predators have had dramatic effects on the accumulation and destruction of hominid skeletal remains in southern Africa (Brain, 1981a). Local predators and their preferences may have also been responsible for biasing the eastern African hominid record.

Many hominid fossils from the Omo Shungura and Koobi Fora Formations are derived from sedimentary rocks deposited under fluviatile conditions. Table 27.1 shows that over 70% of *all* hominid specimens whose sedimentary envelope could be established were derived from sand or gravel bodies deposited by moving water. It is appropriate to consider how such conditions can influence the nature of the fossils embedded in the rock units. Dechant-Boaz summarizes relevant information on the geomorphology of modern riverine systems, concluding that "the fluvial system can rarely guarantee permanent burial, particularly in the perennial river or its floodplain. Vertical and lateral erosion constantly move alluvium downstream to new, usually temporary sites of deposition" (1982: 90).

A river's channel migrates laterally at widely varying rates ranging from 2 to 2,000 feet per year (Wolman and Leopold, 1970). The entire floodplain can be reworked in centuries (Schumm, 1977). The consequent mixing of time-successive and ecologically disparate mammalian bone assemblages can be dramatic. Behrensmeyer (1982: fig. 4) illustrates the effects of this migration and concludes:

> The time-averaged samples available to the vertebrate paleontologist will combine morphological characteristics from populations spanning thousands of years. Likewise, the fauna represented in a single-unit assemblage will contain species that were "contemporaneous" only at this level of resolution. Morphologists and ecologists working in recent systems and aware of the variation that can occur over years or decades are typically skeptical about the potential for deriving useful information from such a fossil record, although some respond to the idea that a positive effect of time-averaging may be to damp out short-term "noise" and reveal important longer term trends (pp. 224–225).

The operation of taphonomic filters associated with fluviatile systems has biased the associations of early hominids (Fig. 27.2). In addition, the destructive capability of these fluviatile systems has resulted in selective destruction, alteration and transport of different body elements. Even teeth and jaws are subject to these effects. For example, Omo specimen L 860-2 displays the *morphology-altering effects of fluviatile abrasion* in its paper-thin mandibular cortex and heavily abraded tooth crowns. Taxonomic attribution of this *A. boisei* mandible to *A. africanus* (Howell,

Tim D. White 453

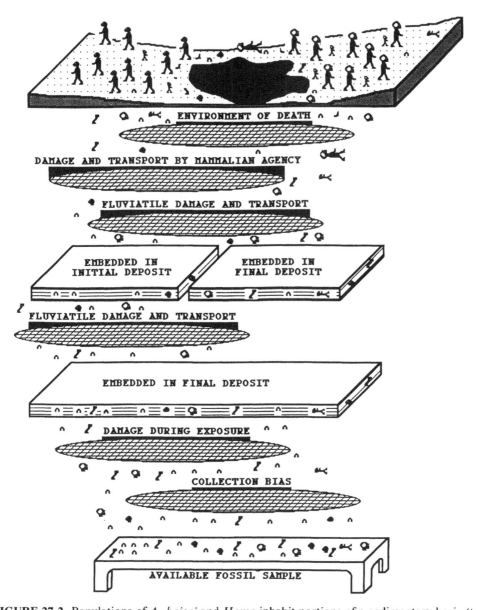

FIGURE 27.2. Populations of *A. boisei* and *Homo* inhabit portions of a sedimentary basin (top). Over time (top to bottom) skeletal parts of individuals are preserved, destroyed, embedded, recycled, exposed and collected. As remains pass through successive taphonomic screens or filters they are mixed and differentially lost. Each filter has the potential to bias the sample passing through and/or around it. The ultimate fossil collections available for study are therefore usually highly skewed with respect to the representation of elements, ages, sexes, and death environments. See text for discussion.

1978) or *incertae sedis* (Chamberlain and Wood, 1985) did not mention the reducing effects of this abrasion on the specimen, a phenomenon common in the Omo vertebrate assemblages.

After final embedding, fossils can be damaged as they approach and reach the surface (Fig. 27.2). Mineral expansion can shatter entire crania that were embedded intact. Discovery of specimens that have weathered out of their context intact is an extremely rare event. Larger, less fragile elements will be least susceptible to exposure-related damage. Leakey *et al.* (1978) note the importance of this in the disproportionate number of *A. boisei* mandibles from Koobi Fora. These massive mandibular corpora are relatively abundant in the Koobi Fora assemblage, probably because they are resistant to carnivore destruction when fresh, resistant to fluviatile damage in transport, and resistant to exposure damage after fossilization. Such specimens were commonly recovered during the first few years of collection at Koobi Fora. They tended to behave, geomorphologically, as cobbles in stabilized erosional (desert pavement) surfaces that had formed in place. Once collected, such surfaces no longer bore hominid specimens and the recovery rate dropped dramatically in the mid-1970's.

Finally, specimens become available for study only after a collector discovers them. The southern African fossils are embedded in cemented breccia that fills limestone cavities. They are usually discovered by systematic examination of the broken breccia surfaces. In eastern Africa, on the other hand, most hominid fossils are recovered by collectors searching stable or active erosional outcrops of fossiliferous sedimentary strata. The potential biases associated with such surface collection are rarely recognized but nevertheless significant. Depending on collector skill and survey methodology, major biases may be introduced to the hominid samples recovered. For example, Boaz and Beherensmeyer (1976) attribute the preponderance of isolated teeth in the Omo deposits to the fluviatile nature of the sediments and contrast this situation with Koobi Fora. Different depositional environments undoubtedly contributed to this bias but many of the Omo teeth were recovered from excavations, while the Koobi Fora collections are almost exclusively from surface exposures. Perhaps even more importantly, Omo collectors were working on an isolated tooth search image while Kamoya Kimeu's team of collectors moved rapidly across the Koobi Fora outcrops with a search image that was initially oriented toward hominid cranio-dento-mandibular specimens and then incorporated postcrania after about 1970. With the wealth of more complete fossils available, the recovery of isolated teeth at Koobi Fora was not encouraged. A different collection strategy would undoubtedly help correct the bias against isolated teeth at Koobi Fora.

When attempting to establish the distribution, sympatry, environment, habitat, predation, relative abundance, population density, sleeping habits, material culture, appearance and extinction of "robust" *Australopithecus,* it will be necessary to consider the accuracy of our assessments in light of the taphonomic factors detailed above. The biases identified here and illustrated in Fig. 27.2 render suspect many conclusions that have been drawn about the biology of these organisms.

Plio-Pleistocene Hominid Distribution and Sympatry

Australopithecus robustus and *A. boisei* are known from a total of eight African areas: two sites in the Transvaal and six fossil fields in the Eastern Rift Valley of northern Tanzania, Kenya and southern Ethiopia. Thus, the total known range of these organisms is approximately 2200 miles. This geographic distribution pattern is merely a reflection of geological circumstance and intensity of search. The actual distribution of species in the "robust" *Australopithecus* clade(s) will never be known, but will certainly be expanded when appropriate sites are discovered elsewhere in Africa.

Grine (1985b:145) states: "Sympatry of fossil species can be safely assumed . . . only when specimens occur cheek-by-jowl in the same stratigraphic horizon and it can be demonstrated that they have not undergone any significant taphonomic transportation." Within their known ranges of time and space, *A. robustus* and *A. boisei* overlap with *Homo,* and several sites have provided evidence for this *broad* sympatry. At Swartkrans both *A. robustus* and early *Homo* are found in the pink breccia of the "Hanging Remnant" but carnivore transport of bones to the depository is probable (Grine, 1985b).

In eastern Africa the evidence for *tight* sympatry is even more limited. At Olduvai, *A. boisei* and *Homo* overlap in time throughout Bed I and Lower Bed II but none of the sites yields unequivocal skeletal evidence of both taxa in the same microstratigraphic unit. This pattern of temporal overlap is repeated for the upper Burgi, KBS and Okote members at Koobi Fora. Only in the case of KNM-ER 1805 (a cf. *Homo* skull) and KNM-ER 1806 (an *A. boisei* mandible) have both taxa been recovered from the same microstratigraphic unit—a channel deposit of richly fossiliferous, tuffaceous sands at site FxJj-38NW. From the Omo Shungura, at excavated sites L 398 and Omo 33 of Member F-0, isolated teeth of cf. *Homo* and *A. boisei* have been recovered together in a highly fossiliferous sandy tuffite. Localities L 28 and Omo 47 have also yielded isolated teeth of both taxa from channel gravels.

In the Omo, as at Koobi Fora and Olduvai, *Homo* and *A. boisei* are found in the same basin and overlap completely in time. In the few eastern African cases where fossils representing the two taxa have been found embedded in the same depositional unit, that unit is a fluviatile channel—a context known to be capable of combining animals whose deaths occurred many miles and hundreds or thousands of years apart (Behrensmeyer, 1982; Dechant-Boaz, 1982). From the contextual data we can currently conclude that *Homo* and robust *Australopithecus* species were sympatric *only* at these limited scales of resolution.

Plio-Pleistocene Hominid Environment

An enormous amount of effort has been expended since 1960 in attempts to characterize the habitats of early hominids in eastern Africa. Because of the biases discussed above, it is unlikely that the habitats associated with available hominid remains are the only habitats (or even the primary habitats) occupied by the hominids. It cannot even be ascertained whether these discovery contexts are *representative* of the habitats that hominids might have occupied. Non-basin, non-waterside habitats are not even sampled in eastern Africa.

However, given the abundance of "robust" *Australopithecus* and early *Homo* taxa at Swartkrans, Olduvai, Koobi Fora and the Omo, it is useful to examine the broad environmental context of each site of deposition for clues about early hominid environment.

The Swartkrans depository is complex (Brain, this volume, chapter 20) but the *A. robustus*-bearing infill is rich in vertebrate remains useful in establishing the site's Lower Pleistocene environment. Brain (1981a: 266) describes the Swartkrans Member I breccia as sampling an environment ". . . on the open highveld, which appears to have been largely a grassland in Swartkrans Member I times." Patches of wooded cover (Brain, 1985) and proximity to aquatic resources necessary for hippo and otter (Brain, this volume, chapter 20) are indicated.

The *A. boisei*-bearing Beds I and II at Olduvai have yielded sedimentological, macrobotanical, paleontological and palynological data allowing a broad reconstruction of the habitats that hominids were buried in. For the hominids between the Bed I lavas and Tuff ID the climate was moister than at any later time recorded by the stratigraphy (Hay, 1976). Much of the shift to a drier climate occurred about 1.7 Myr (Bonnefille and Vincens, 1985). Bed I hominid fossils and archaeological materials are concentrated in lake-margin settings where a perennial supply of fresh water was available on the southeastern margin of a 7 to 25 km diameter, shallow, closed lake whose level fluctuated. Fauna here is dominated by swamp-dwelling animals but a semiarid environment somewhat moister than in the same area today is indicated. Above Tuff ID the climate became appreciably drier, but *Homo* and *A. boisei* persisted. They continued to persist above the Lemuta Member as the proportions of plains bovids increased and the lake turned more saline, following the trends seen in upper Bed I (Hay, 1976).

The Omo Shungura Formation has been documented in detail by de Heinzelin (1983). During the period represented by Members E through K, *A. boisei* and *Homo* specimens are found in Lower Omo Valley deposits related to ". . . a large perennial river bordered by forest . . . which flowed southward into a lake . . . The area surrounding the river was rather dry, receiving between 40 and 80 cm of rainfall each year. The floral composition of the riverine forest and surrounding region changed in time, but there is no clear trend toward either much wetter or much drier conditions than those which presently prevail in the area" (Brown, 1981:161–162). The lake was much larger than the Olduvai lake but still capable of great fluctuation in level. For example, lake levels have risen 60 to 80 meters above the modern level at least three times in the last 10,000 years, pushing the Omo River's delta far to the north (Butzer, 1971). The bulk of the hominid remains recovered in the Omo Shungura Formation come from deposits accumulated during the latter part of a 1.5 Myr fluviatile phase. Subsidence controlled the accumulation of successive floodplains of the Pliocene Omo River. After a critical assessment of the evidence available for habitat reconstruction, de Heinzelin summarizes as follows:

> Altogether, surprisingly little can be deduced from the fossil mammal assemblages in terms of paleoenvironment, except in a very broad way: a tropical African fauna, ancestral to the living one, lived in biota similar to the present ones in the same basin, the proportions of the represented biota shifted from more closed woodland to more open grassland with innumerable interactions betwixt and between (1983: 215).

Finally, strata of the Koobi Fora Formation east of Lake Turkana have yielded abundant hominid remains in four of the five major sedimentary environments defined by Findlater (1978): Alluvial Valley Plain (AVP), Alluvial Coastal Plain (ACP), Alluvial Delta Plain (ADP) and Lacustrine High Energy (LHE). In contrast to the Omo situation, deposition at Koobi Fora during the upper Burgi, KBS and Okote Members was a complex process related to fluctuating lake levels and generally westwardly prograding alluvial deltaic and alluvial coastal plains (Findlater, 1978: fig. 2.8). Satisfactory stratigraphic control over the hominids collected from these complex Koobi Fora deposits has taken almost two decades to complete. Harris (1985) suggests more humid conditions in the region in the past, noting an overall trend toward more arid conditions through the sequence. He describes the environment of hominid burial as ". . . a pattern of well vegetated lake margins, perennial and ephemeral rivers fringed with gallery forest, and with wooded savanna predominating away from the lake. The extent and disposition of these habitats must have fluctuated in accordance with prevailing climactic conditions but all appear to have been present throughout the succession" (1985:160).

In summary, the major sites of burial of early hominids between 2.4 and 1.2 Myr range from the grassland environment at Swartkrans to the dense riverine forest indicated in the Lower Omo Valley. Proximity of standing or flowing fresh water is indicated for all of these environments. In addition, the remains of the hominids came to be buried within easy reach of a variety of habitats in which they could have ranged. Western's (1985:403) remarks regarding the eastern African sites are appropriate in this regard: ". . . I am impressed that many different environments are represented in each fossil location. I do not think there is any region where we can say that grasslands are exclusive. We are therefore dealing with habitat mosaics".

Analogous modern African environments are characterized by substantial seasonality and this was likely to have been true during the time in question. O'Connell *et al.* (1988) studied the effect of modern seasonality in similar environments in the Eyasi Basin and have identified dramatic consequences for early hominid subsistence strategies. Western (1975) has pointed out the marked effects of water availability on the structure and dyamics of large mammals in the modern Amboseli ecosystem. Given these findings, the caution of Mann (1981) is especially appropriate in assessing early hominid habitats:

> Only a small number of habitats offer optimal conditions for deposition of bony remains. Therefore, most fossil samples may derive from those environments most conducive to long-term preservation, even though these environments may have been exploited only marginally by hominids or occupied for only brief intervals during the year (p. 27).

Is it possible, as Behrensmeyer (1985) has suggested, to go beyond the knowledge that early hominids lived and were buried in savanna-mosaic habitats? To answer this we must return to the comparative contextual database for the early hominid fossils.

The Comparative Ecology of *Homo* and "Robust" *Australopithecus*

Because the Swartkrans hominids are sampled from a single depository and Shungura Formation hominids from predominantly fluviatile deposits, Olduvai and Koobi Fora provide the best opportunities to make meaningful comparisons between the depositional environments of *A. boisei* and *Homo*.

Speculation about the comparative ecology of Koobi Fora hominids began with Behrensmeyer's doctoral dissertation (1975a). She concluded that the stratigraphy and dating at Koobi Fora were ". . . reasonably well established . . ." (p. 475) and went on to consider two terrestrial faunas in the Koobi Fora assemblages. One was an "open habitat" or "grassland" fauna, including alcelaphine bovids, high-crowned suids and horses. The other was a "closed habitat" or "bush" fauna, characterized by reduncine and tragelaphine bovids and low-crowned suids. She found an overlap of these faunas in the deposits, but a tendency for delta margin sediments to preserve a greater proportion of closed habitat forms and channel sediments to preserve more open habitat forms. She then surveyed the available sample of 50 *"Australopithecus"* specimens (the genus name has often been employed by Koobi Fora researchers to identify the species *A. boisei*), 34 *"Homo"* specimens, and 10 unassigned hominid specimens. Behrensmeyer (1975a) attributed the large number of *A. boisei* mandibles to the survivability of these elements. Differentiation of *Homo* and *A. boisei* with regard to lake margin and fluvial deposits is asserted in a footnote.

Behrensmeyer (1975b) compared the Olduvai, Omo and Koobi Fora hominid samples and argued that taphonomic sorting accounts for the different pattern of element representation at these sites. Assessing the 120 hominid specimens available in 1973, Behrensmeyer found "gracile" and "robust" forms in both fluvial and lake margin deposits, with "robust" forms more common in the former. Behrensmeyer noted that taphonomic factors or differential mortality could explain the perceived patterning and advocated further research.

Behrensmeyer worked with the same hominid sample in 1978, concluding that hominids were about equally distributed in fluvial (channel plus floodplain) and lake margin environments, but that *Homo* was relatively under-represented in fluvial environments. She went on to contend that the hominids were buried ". . . in the general vicinity of areas inhabited (at least temporarily) by the original populations . . ." (p. 174) and that the relative abundance of *A. boisei* in Upper Member deposits was due to the increase in fluvial sediments in these beds. Behrensmeyer (1978) went on to conclude that the most reasonable explanation for the patterning observed is an ecological one whereby: "More australopithecines lived (or died) in fluvial habitats, probably gallery forest and bush near rivers, than *Homo*. Both inhabited lake margins," (p. 187). Despite *A. boisei* abundance in fluvial deposits, Behrensmeyer placed this form in a closed (bush) rather than open (grassland) environment because of an unexpectedly high frequency of low crowned suids at the hominid localities.

Behrensmeyer's recent (1985) suggestions about the use of subhabitats by Koobi Fora hominids rely on comparisons of two "contemporaneous" (within 100,000 years!) deltas (Ileret and Area 103) and a more fluvial inland (Karari) environment. Minor differences in mammalian abundances were found between the three areas. Indeed, the most significant faunal contrasts were between the levee-floodplain and channel deposits *within* the Ileret area! Behrensmeyer's 1985 study notes the need for a new assessment of the relationship between depositional environment and hominid taxa at Koobi Fora. It goes on to suggest, however:

> From the paleoecology study of the KOLM [a tuffaceous unit], it appears that *Australopithecus* was more commonly associated with faunas indicating more vegetated environments. The fact that it has not been found in area 103, while both *Homo* and numerous smaller primates occur in a large fossil sample, is surely suggestive of (at least) partial ecological exclusion (1985: 322).

Finally, summarizing a decade of speculation on habitat preference for the Koobi Fora hominids, Behrensmeyer concludes that ". . . clear evidence for hominid utilization of different sub-habitats within the 'savanna-mosaic' remains elusive" (1985: 322).

Any conclusions regarding differential habitat preference of the hominids from sites like Olduvai and Koobi Fora are impacted by three major limiting factors. First, the taxonomic attribution of the hominid skeletal parts must be accurate. Second, we must be able to identify and differentiate ancient habitats on the basis of geological and paleontological data. Finally, the association of hominids with various geological and paleontological habitat indicators must not represent fortuitous death site associations but should, instead, correspond to actual hominid habitat preference during life.

The consistent occurrence of hippos, crocodiles, alcelaphine bovids, and horses in the same channel deposits at sites like Koobi Fora is not surprising from a geomorphological perspective in which floodplain deposits and their contents are recycled into laterally migrating channels. These co-occurrences, however, do little to inspire confidence in our ability to provide sharp, precise habitat placements for the hominids embedded in such deposits.

The failure of previous investigators to convincingly demonstrate habitat preferences for early hominids is not unexpected given the limitations described above. Recognizing these limitations I have used the data presented in Table 27.1 to examine the depositional contexts (not necessarily habitat preferences) of early hominids from eastern Africa. My data sets differ from Behrensmeyer's primarily in tighter chronological control for the hominids. In addition, postcranial elements do not figure in my analysis. I have been forced to reluctantly rely on Behrensmeyer's and Findlater's interpretations of the specific depositional context for each Koobi Fora hominid (Leakey and Leakey, 1978) because they are the only currently available interpretations. These

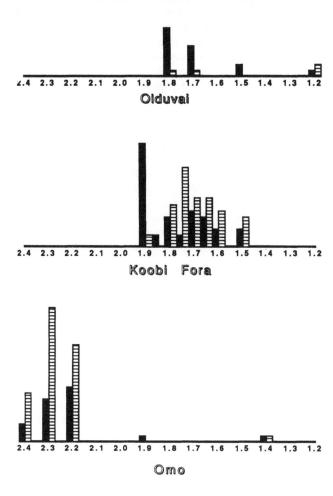

■ = 1 Individual

FIGURE 27.3. Contemporary *Homo* (solid bars) and *A. boisei* (striped bars) from eastern Africa plotted against time at the three major sites. Data from Table 27.1. See text for details.

interpretations will be tested as work progresses. Figure 27.3 plots the available identified hominids from Olduvai, Koobi Fora, and Omo against time. The Omo collections mostly predate the Olduvai and Koobi Fora samples by about 0.25 Myr. "Robust" *Australopithecus* consistently outnumbers *Homo,* but this effect could be related to the fluvial nature of the deposits. "Robust" forms also consistently outnumber *Homo* in the Koobi Fora collections, *except* in the upper Burgi Member (c 1.9 Myr) where *Homo* is very abundant. At Olduvai, *A. boisei* is also relatively rare in a sample notable for its small size.

Figure 27.4 plots the identifiable Koobi Fora and Olduvai hominids by geological macroenvironment, relying on Findlater (1978) and Hay (1976). At Koobi Fora, *A. boisei* mandibles are more abundant than *Homo* mandibles in most environments. The only additional patterning evident in these data is a slight tendency for *A. boisei* to be relatively more abundant in the coastal settings.

Figure 27.5 displays the identifiable Koobi Fora sample against geological microenvironment. Mandibles, as expected taphonomically, dominate in the fluvial microenvironments. Note that Behrensmeyer's compilations, unlike those of Fig. 27.5, exclude channel-derived specimens in coastal environments from the fluvial microenvironment. Nevertheless, it is evident that the bulk of the Koobi Fora hominids come from channels—depositional microenvironments dominated by running water and capable of mixing bones from diverse original embedding sites and habitats.

Figure 27.6 plots identified hominid fossils from Koobi Fora, Olduvai and Omo against the

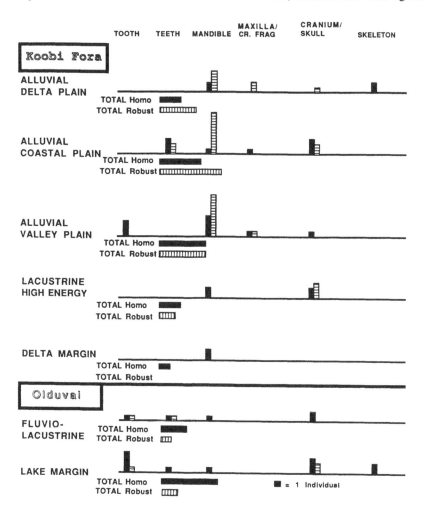

FIGURE 27.4. Hominids from Koobi Fora and Olduvai plotted by element and depositional macroenvironment. Data from Table 27.1. See text for details.

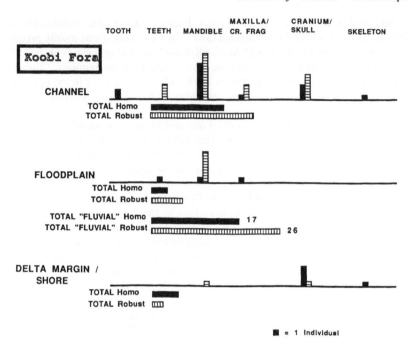

FIGURE 27.5. Hominids from Koobi Fora and Olduvai plotted by element and depositional microenvironment. Data from Table 27.1. See text for details.

matrix in which they were found. The Omo hominid sample comes from depositional conditions of the highest energy and is strongly skewed toward dense, strong skeletal elements. Koobi Fora hominids come from mostly sand and gravel matrices but compared to the Omo, relatively more hominids have been found embedded in lower-energy microenvironments. Olduvai hominids derive from such lower-energy microenvironments at a greater frequency than is the case at Omo or Koobi Fora but this may relate to the fact that many of them have been recovered in the course of archeological excavations.

Behrensmeyer's studies of Koobi Fora hominid habitat preference (1975a,b, 1976, 1978, 1985) relied on now-outdated hominid classification and chronostratigraphic placement and included postcrania of dubious affinity. These studies, however, consistently noted the relatively high abundance of *A. boisei* in fluvial environments. Thus, according to Leakey *et al.* (1978): "specimens identified as *Homo* sp. appear to be significantly more common in lake margin than fluvial deposits, whereas specimens of *A. boisei* are equally common in fluvial and lake margin environments" (87). These studies, however, did not assess the data within the chronological framework of the Koobi Fora Formation. Perhaps the most significant difference between depositional environments of Plio-Pleistocene hominids has remained obscured because of this drawback. Figure 27.7 shows data on the geographic placement of identifiable Koobi Fora hominid fossils according to Findlater (1978). Fossils from coastal (Findlater's alluvial coastal plain, alluvial delta plain, lacustrine high energy; Behrensmeyer's lake margin) and more inland (Findlater's alluvial valley plain and Behrensmeyer's floodplain and channel) environments are plotted by Member. Mean age for the Upper Burgi (Lower Member) hominids is about 1.9 Myr while that of the KBS Member sample is about 1.75 Myr and the Okote Member hominids range from 1.5 to 1.65 Myr. Differences between the Upper Member data sets stem from Behrensmeyer's use of the term "floodplain" for some sites that Findlater places in the alluvial coastal plain.

The results shown in Fig. 27.7 reflect, in part, the geologic situation described by Burggraf *et al.* (1981:133): "The sediments of the Lower Member of the Koobi Fora Formation indicate dominant lacustrine, beach, and deltaic conditions. . . . Between 1.5 and 1.8 m.y. B.P., however, a dramatic change began which initiated widespread deposition of fluvial sediments in the eastern portion of the basin." The Upper Burgi Member's paucity of hominids from inland sediments is

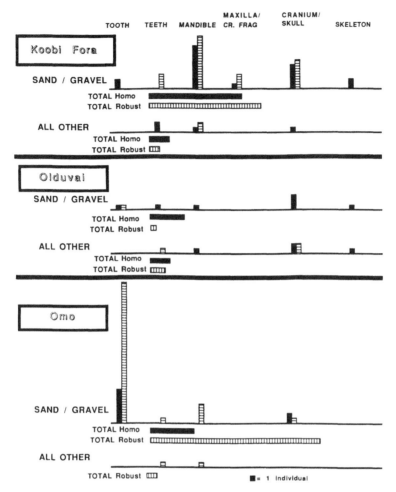

FIGURE 27.6. Hominids from Koobi Fora, Olduvai and Omo plotted by element and sedimentary matrix. Data from Table 27.1. See text for details.

explained by this geographical picture. The small inland hominid sample of the KBS Member shows *A. boisei* and *Homo* in roughly equal proportions, but *A. boisei* comes to dominate in the Okote Member assemblages (Fig. 27.7).

It is in coastal settings, however, that the most provocative results of the time-geographic analysis are evident. Behrensmeyer stressed the equal numbers of *A. boisei* and *Homo* in lake margin settings, but her analysis unfortunately pooled hominids from the Upper and Lower Members. Figure 27.7 avoids this pitfall by dividing the Koobi Fora hominids by Member. It shows that *Homo* is four times more common than *A. boisei* in the Upper Burgi Member coastal settings. In the subsequent KBS Member, however, the relative abundance is reversed, with *A. boisei* two to three times more common than *Homo in the same coastal depositional environment*. The taxa are nearly equally abundant in the Okote Member. The dramatic numbers of *Homo* in 1.9 Myr Koobi Fora sediments seen in Fig. 27.3 thus take on added significance when considered geographically.

Of the 13 identifiable Upper Burgi hominids, six are from area 131 and seven are from areas 123/127. An additional four identifiable Upper Burgi specimens are known, but depositional data are not available in Findlater (1978). All of these, however, are *Homo,* with one being from area 130 and three from area 105/116 (collected from cf. lake margin settings). Hence, the 8:2 ratio of *Homo* to *A. boisei* in Upper Burgi lake margin settings in Fig. 27.7 represents a highly con-

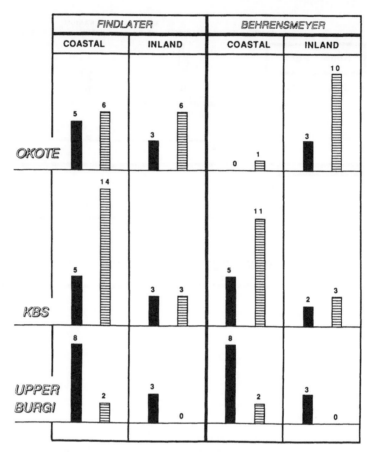

FIGURE 27.7. Paleogeographic distribution of Koobi Fora hominid remains. *Homo* represented by solid bars, *A. boisei* by striped bars. Data from Table 27.1 and Findlater (1978). See text for details.

servative estimate. *Homo* also dominates in lake margin settings in the small Olduvai sample (5:2, *Homo: A. boisei,* Fig. 27.3). In dramatic contrast to the Olduvai situation, however, Koobi Fora lake margin settings have yielded virtually no stone artifacts, causing considerable consternation among archeologists (Isaac and Harris, 1978). Archeology aside, the pattern of reversing relative abundance in lake margin settings of *Homo* and *A. boisei* through time at Koobi Fora is a dramatic one. Is this pattern merely an artifact of sampling? Could it represent a decreasing susceptibility of *Homo* individuals to predation in lake margin environments, perhaps related to the *H. habilis* to *H. erectus* shift? Could the pattern indicate changing population sizes through time or relative shifts in habitat preference? Choosing between these and other alternatives represents a research challenge.

Predation and Sleeping Sites

Carnivore-induced modifications of hominid bones in the context of eastern African sites represent an impossible-to-unravel combination of carnivore involvement in the killing of hominids (predation), and carnivore feeding on already dead hominid carcasses (scavenging). Toth and White have not yet studied the Koobi Fora hominid collections for evidence of postmortem alteration, but their analysis of the Olduvai hominids, included in Table 27.1 shows that there are eleven available, relevant hominids with bony surfaces capable of showing evidence of carnivore chewing. Of these, seven show evidence of carnivore alteration. Such damage has already been noted in the case of OH 7 and OH 8 but my analysis of the OH 5 *A. boisei* cranium shows that even this fairly intact fossil was chewed in both mastoid and zygomatic regions by a medium-sized carnivore (jackal or immature hyena). These findings do not bear directly on the issue of predation on early hominids, but the extensive carnivore involvement already documented for the fauna on the *H. habilis* and FLK Zinj horizons (Bunn, 1981) in Bed I has clearly impacted the disposition of hominid remains on these surfaces.

Brain (1981) lists a wide variety of predators that played a potential role in accumulating the hominids at Swartkrans, noting that "It is to be expected that each of the three bone accumulations has resulted from the independent activity of cats and hyaenas" (1981a: 273). Mann (1975) noted that the demographics of the Swartkrans sample may have been biased by carnivore preference for younger hominid individuals, and White's (1978) analysis suggested that Swartkrans *A. robustus* individuals with enamel hypoplasia had increased susceptibility to predation.

Brain (1981a) identifies hominids SK 54, SK 3978 and SK 3155b as demonstrating "clear" carnivore alteration and another twelve specimens as being "very probably" carnivore damaged. Toth and White have recently analyzed the same sample. They found "certain" evidence of carnivore alteration (punctures, gnawing) on SK 54, SK 82, SK 859 and SK 3155b, "probable" evidence on SK 438 and SK 1585 and "possible" damage on SK 50 and SK 2831. The latter workers were more conservative in not including overall damage patterning as evidence of involvement. No matter how the Swartkrans collection is approached, however, a dominant role was played by predators in accumulating the *A. robustus* fossils. It seems evident that at least the "robust" *Australopithecus* populations living at and/or around Swartkrans were (happily for us) susceptible to considerable predation. The low relative abundance of *Homo* could reflect a real scarcity on the local landscape or a lower (perhaps behaviorally moderated) susceptibility to predation and/or accumulation. Teasing these alternatives apart is a challenge.

The morphologically homogenous makeup of the Swartkrans *A. robustus* sample (Kimbel and White this volume, chapter 11; Grine, this volume, chapter 14), the faunal and geologic contexts of the site, and the evidence of carnivore accumulation all lead to the conclusion that this site, because of the care with which it has been excavated and analyzed, offers perhaps the best opportunity now available for achieving a fine-grained understanding of "robust" *Australopithecus* habitats and habits. Brain's (1981a) suggestion that the site represents a sleeping locus on the highveld for these creatures is of interest in this regard.

Demography

Taphonomic biases associated with the African Plio-Pleistocene hominids make it difficult or impossible to accurately reconstruct most demographic variables of the once living populations. For example, even if we can decide how to accurately gauge ontogenetic rates and hence ages of death for the Swartkrans *A. robustus* hominids, the age profile of that sample has undoubtedly been skewed by differential predation and destruction of body parts.

For eastern Africa, in an incredible argument about early hominid demography, Boaz (1985) used standards of modern human dental attrition to estimate the individual age of 173 Omo teeth spanning several taxa sampled across over 2 Myr. Using the resulting curve to predict average maximum lifespan for *Australopithecus* at 60 years, Boaz then inferred sexual maturation for this genus to have been only slightly elevated from the chimpanzee condition!

One demographic variable that has received a surprising amount of attention is population density. Both *Homo* and *A. boisei* are extremely rare elements of the eastern African faunas. This is probably because compared to more common mammals they were relatively low in abundance, had long lives, and were intelligent. Boaz (1979) "calculated" overall hominid population density based on proportional representation of remains in a set of excavated Omo fossil assemblages at between 0.001 and 2.48 individuals per square kilometer. Martin (1981) used inferred diet and body size, and Blumenberg (1981) used body weight and rainfall estimates to speculate about the population densities of early hominid species. None of their results inspire confidence.

Material Culture

At least two hominid lineages existed in *broad* sympatry over much of the late Pliocene and early Pleistocene in Africa. The acheological record demonstrates that at least one of these hominids was involved with the manufacture of stone tools. Speculative attribution of this material culture to the genus *Homo* has traditionally been accomplished as follows:

> The repeated association of *H. habilis* with the Oldowan industry, the scarcity of *A. boisei* and the apparent absence of any other recognisable hominid taxon make it reasonable to assume that *H. habilis* was responsible for making the Oldowan tools. In fact, it is possible that the abundance of Oldowan artifacts at Olduvai compared to their relative scarcity at East Rudolf may be due to the greater numbers of *Homo habilis* in the Olduvai area. (M. D. Leakey, 1978:5).

A review of recent archeological literature reveals a tendency to ignore the issue of which hominid made and/or used the tools by referring to "the hominids" (Potts, 1983, 1986; Shipman, 1983; Bunn, 1983). Toth (1985) notes the coincident appearance of *Homo* and stone tools in the eastern African record but cautiously stops short of insisting on a causal relationship and/or denying *A. boisei* the status of toolmaker or tool user. He terms the issue "unresolved."

Table 27.1 shows that hominids and stone tools co-occur in the same geological stratum at the same place at 14 Olduvai sites, three Koobi Fora sites, possibly two Chesowanja sites, and one Omo locality. Although all of these represent associations in the temporal and spatial sense, *none* of them can be demonstrated to be functionally associated. Many of the co-occurrences, particularly those like OH 3 or L 860-2 are from fluviatile deposits, and probably represent entirely fortuitous associations. As noted above, even the "association" of the carnivore-gnawed OH 5 *A. boisei* cranium in low-energy depositional circumstances at FLK Zinj tells us nothing about the relationship of this taxon to the artifacts or hominid butchery activities documented at this locality. At Olduvai, in the six cases of low-energy deposits where hominids co-occur with stone tools, OH 5 is the only *A. boisei* representative, but the five *Homo* individuals are more fragmentary, being represented by a molar (OH 44), a mandible scrap (OH 4), two crania (OH 16 and 24) and the *H. habilis* holotype fragments (OH 7). At Koobi Fora the only low-energy co-occurence is with *A. boisei* (KNM-ER 3230). Even the discovery of an intact *A. boisei* skeleton with Oldowan chopper in hand would fail to resolve the issue.

The present impossibility of attributing the material culture so abundant at Olduvai and Koobi Fora was best summarized by Glynn Isaac:

> There exist three possibilities as to the authorship of the artifacts: (1) the early forms of *Homo* made them, (2) the robust australopithecines made them, and (3) both species made them. At present I see no way of distinguishing among these possibilities, though one tends to favor the first and possibly the third. (1984:10).

Indeed, it will probably be anatomical (lateralization of hand and upper limb osteologic features not yet available) rather than contextual (archeological) data that ultimately might allow a choice between these alternatives.

Contextual Data Bearing on Phylogeny

Considerable attention has been devoted to the issues of "robust" *Australopithecus* origins, evolution and extinctions. The contextual database has little to offer in the way of illuminating these issues because of the tiny site and sample sizes currently available. The *Homo* clade probably originated between 3 and 2.6 Myr. The latest sampled populations of species in the "robust" *Australopithecus* clade(s) lived about 1.2 Myr ago (see Table 27.1; Klein, this volume, chapter 29).

The data plotted in Fig. 27.3 show no tendency in the known hominid samples for the abundance of *Homo* to increase relative to "robust" *Australopithecus* through time. This is hardly surprising given the small available database. The hominid-bearing Koobi Fora, Omo and Olduvai successions effectively "top out" at c 1.0 to 1.3 Myr. Subsequent sites in eastern Africa are much smaller, more widely-scattered, and virtually undated. While many have pointed to the "disappearance" of *A. boisei* at c 1.2 Myr, few have noted that *H. erectus* also effectively disappears at roughly the same time, becoming paleontologically "invisible" in Africa until the Middle Pleistocene except for a few isolated occurrences. It is probable that the apparent extinction of "robust" *Australopithecus* at c 1.2 Myr is merely an artifact of the fossil record and that future discoveries will show these creatures to have persisted until Middle Pleistocene times. Only further fieldwork will tell.

Recent years have witnessed a tendency on the part of anthropologists and others to link contextual evidence of global, continental, or local climatic shifts to apparent "events" in the history of fossil hominid taxa. An example is Boaz (1985), who suggests that the appearance and disappearance of *A. boisei* at Omo is related to dry periods of open vegetation. Vrba (1985), following Brain (1981b), has suggested a link between a global temperature minimum at c 2.6 Myr and the differentiation of the "robust" *Australopithecus* clade. Grine (1981) comments on the "approximately concurrent" first appearance of *A. boisei* and a shift to drier, more open environments at Koobi Fora and Omo (Member E). Bonnefille and Vincens (1985) reviewed the paynological data collected from a variety of localities in eastern Africa, and they noted dramatic cooling and drying just before 2.48 Myr, several grassland extensions in the Turkana basin at c 1.9-2.4 Myr (Omo E4, F1, G; Upper Burgi) and an arid episode atop Bed I at Olduvai (before Tuff IF).

As more environmental "episodes" are established, of course, the more probable it becomes that one of them will correlate with some hominid evolutionary "event", and the less likely it becomes that correlation is evidence of causation. Recent confirmation (KNM-WT 17000; Walker and Leakey, this volume, chapter 15) of "robust" *Australopithecus* at c 2.6 Myr in the Turkana Basin (Arambourg and Coppens, 1967; White, 1977; Grine, 1985a; Kimbel *et al.*, this volume, chapter 16) is a dramatic caution to those inclined to accept correlation of environmental change as evidence of evolutionary causality. For further critical consideration of similar issues, see Hill (1987). Changes in physical environment clearly played major roles in mammalian evolution in Africa. It remains a challenge to future researchers to determine what roles these changes might have played in the evolution of our relatives and ancestors.

Conclusion

Contextual data provide little insight into the lifeways of "robust" *Australopithecus* species. Despite what appears to be a large fossil database (Table 27.1), the evidence currently available is plagued by a number of taphonomically induced biases. Broad sympatry with *Homo* can be established in savannah-mosaic environments in eastern and southern Africa, but other environments remain paleontologically unsampled. Relative abundance data in different depositional environments hint at differential mortality or habitat preference of the contemporary hominids, but a satisfactory choice between these alternatives is not yet possible. Demographic characteristics of early hominid species cannot yet be defined from contextual data. Cultural capabilities of these forms are also impossible to gauge from the available record. The origin of the clade(s) is open to debate, while the disappearance of terminal species members may reflect real extinction at 1.2 Myr or an artifact of a still fragmentary fossil record.

From this summary it might be concluded that the contextual database should be abandoned in favor of anatomical analysis as a vehicle for understanding "robust" *Australopithecus* biology. The characterization of an extinct group of organisms through speculation, assumption, and faulty inference is effective only in the construction of myths about the past. Perhaps myths about "robust" *Australopithecus* built on such foundations have prevented us from acknowledging our ignorance about these creatures. My summary may appear to be unjustifiably negative, but we will only eliminate our ignorance if we are willing to recognize it. Fieldwork aimed at recovering vastly increased samples of early hominids is the research most urgently needed to eliminate this ignorance. In the course of that fieldwork close attention to context may yet yield substantial knowledge about "robust" *Australopithecus* biology.

Table 27.1. Database of Eastern African Plio-Pleistocene Hominids Dated to between c 2.4 and c 1.2 Myr[a,b]

Specimen No.	Element	Individual age	Depositional macroenvironment	Depositional microenvironment
KNM-CH 1 + 302	f cranium	AD	—	—
KNM-CH 304	ff cranium	AD	—	—
KNM-ER 0164	ff skeleton	AD	alluvial delta plain	delta margin distributory
KNM-ER 0403	f mandible	AD	alluvial delta plain	large distributory channel
KNM-ER 0404	f mandible	AD	alluvial coastal plain	interdistributory channel
KNM-ER 0405	f maxilla	AD	alluvial valley plain	channel
KNM-ER 0406	cranium	AD	lacustrine high energy	small distributory channel
KNM-ER 0407	f cranium	AD	lacustrine high energy	shoreline or interdistributory
KNM-ER 0417	ff parietal	AD	—	—
KNM-ER 0725	f mandible	AD	alluvial valley plain	floodplain
KNM-ER 0726	f mandible	AD	alluvial valley plain	channel
KNM-ER 0727	ff mandible	AD	alluvial coastal plain	distributory or interdistributory channel
KNM-ER 0728	f mandible	AD	alluvial valley plain	floodplain with $CaCO_3$
KNM-ER 0729	mandible	AD	alluvial coastal plain	floodplain, channel edge
KNM-ER 0730	ff skull	AD	alluvial coastal plain	delta margin
KNM-ER 0731	ff mandible	AD	alluvial valley plain	—
KNM-ER 0732	f cranium	AD	lacustrine high energy	dist. or interdist. channel; emergent deltaic
KNM-ER 0733	ff skull	AD	alluvial coastal plain	small channel on floodplain
KNM-ER 0734	f parietal	AD	alluvial coastal plain	delta margin
KNM-ER 0801	ff skull	AD	alluvial coastal plain	distributory channel
KNM-ER 0802	teeth	AD	alluvial coastal plain	distributory channel
KNM-ER 0803	ff skeleton	AD	alluvial coastal plain	levee or floodplain
KNM-ER 0805	ff mandible	AD	alluvial valley plain	floodplain or small channel
KNM-ER 0806	teeth	AD	alluvial coastal plain	—
KNM-ER 0807	ff maxilla	AD	alluvial coastal plain	floodplain or levee

Matrix	Carnivore damage	Stone tools	Stratigraphic unit	Geol. age	Taxon
—	—	possibly Oldowan	—	c 1.4	R
—	—	possibly Oldowan	—	c 1.4	R
—	—	0[b]	KBS Mbr	c 1.8	?
sand	—	0	KBS Mbr	c 1.7	R
sand	—	0	Oko Mbr	c 1.5	R
fine sand	—	0	Kbs Mbr	c 1.75	R
medium-coarse sand	—	0	KBS Mbr	c 1.75	R
sand	—	0	KBS Mbr	c 1.8	R
—	—	0	Bur Mbr	c 1.9	?
$CaCO_3$ + sand	—	0	Oko Mbr	c 1.6	R
—	—	0	Oko Mbr	c 1.6	R
gravel and sand	—	0	KBS Mbr	c 1.7	R
—	—	0	Oko Mbr	c 1.6	R
tuffaceous sand	—	0	Oko Mbr	1.65	R
silty clay soil	—	0	KBS Mbr	1.7	H
—	—	0	Oko Mbr	c 1.6	H
silty sand	—	0	KBS Mbr	c 1.75	R
tuffaceous sands	—	0	Oko Mbr	c 1.65	R
—	—	0	KBS Mbr	1.65	?
—	—	0	KBS Mbr	c 1.7	R
—	—	0	KBS Mbr	c 1.7	R
—	—	0[b]	Oko Mbr	1.65	?
—	—	0	Oko Mbr	c 1.6	R
—	—	0	Oko Mbr	1.65	H
—	—	0	Oko Mbr	1.65	H

continued

Table 27.1. *Continued*

Specimen No.	Element	Individual age	Depositional macroenvironment	Depositional microenvironment
KNM-ER 0808	teeth	IM	alluvial coastal plain	—
KNM-ER 0809	teeth	AD	alluvial coastal plain	—
KNM-ER 0810	f skull	AD	alluvial delta plain	channel
KNM-ER 0811	ff parietal	AD	—	—
KNM-ER 0812	ff mandible	IM	alluvial delta plain	channel
KNM-ER 0814	f frontal	AD	alluvial delta plain	channel
KNM-ER 0816	teeth	AD	alluvial delta plain	channel
KNM-ER 0817	ff mandible	AD	alluvial valley plain	—
KNM-ER 0818	f mandible	AD	alluvial coastal plain	—
KNM-ER 0819	ff mandible	AD	alluvial coastal plain	floodplain
KNM-ER 0820	mandible	IM	alluvial coastal plain	floodplain
KNM-ER 0992	mandible	AD	alluvial valley plain	channel
KNM-ER 1171	teeth	IM	alluvial coastal plain	distributory channel
KNM-ER 1462	molar	AD	—	—
KNM-ER 1466	f parietal	AD	alluvial valley plain	channel
KNM-ER 1467	molar	AD	—	—
KNM-ER 1468	f mandible	AD	alluvial valley plain	channel or floodplain
KNM-ER 1469	f mandible	AD	alluvial delta plain	distributory channel
KNM-ER 1470	cranium	AD	alluvial coastal plain (or L.H.E.)	distributory channel
KNM-ER 1474	ff parietal	AD	alluvial coastal plain (or L.H.E.)	delta margin
KNM-ER 1477	mandible	IM	alluvial valley plain	channel
KNM-ER 1478	ff cranium	AD	alluvial valley plain	top channel or floodplain
KNM-ER 1479	teeth	?	alluvial valley plain	top channel or floodplain
KNM-ER 1480	molar	AD	alluvial valley plain	channel
KNM-ER 1482	f mandible	AD	alluvial coastal plain (or L.H.E.)	delta margin
KNM-ER 1483	f mandible	AD	alluvial delta plain	channel or levee
KNM-ER 1501	f mandible	AD	delta margin	distributory channel
KNM-ER 1502	ff mandible	AD	delta margin	distributory channel
KNM-ER 1506	ff mandible	AD	—	distributory channel
KNM-ER 1507	f mandible	IM	alluvial valley plain	channel
KNM-ER 1508	molar	AD	alluvial valley plain	channel

Matrix	Carnivore damage	Stone tools	Stratigraphic unit	Geol. age	Taxon
tuff	—	0	Oko Mbr	1.65	H
tuff	—	0	Oko Mbr	1.65	H
—	—	0	KBS Mbr	c 1.8	R
—	—	0	cf. KBS	c 1.8	?
—	—	0	KBS Mbr	c 1.8	R
—	—	0	KBS Mbr	c 1.8	R
—	—	0	KBS Mbr	c 1.8	?
—	—	0	KBS	1.7	H
tuff	—	0	Oko Mbr	1.65	R
—	—	0	Oko Mbr	c 1.65	R
fine sand	—	0	Oko Mbr	1.65	H
—	—	0	Oko Mbr	c 1.5	H
—	—	0	KBS Mbr	c 1.75	R
—	—	0	Bur Mbr	c 1.9	H
silty sand	—	0	Oko Mbr	c 1.5	H
—	—	0	Oko Mbr	c 1.5	R
—	—	0	Oko Mbr	c 1.65	R
sand	—	0	Bur Mbr	c 1.9	R
fine sand	—	0	Bur Mbr	c 1.9	H
sands and silts	—	0	Bur Mbr	c 1.9	?
sand	—	0	KBS Mbr	c 1.75	R
—	—	0	KBS Mbr	c 1.75	?
—	—	0	KBS Mbr	c 1.75	?
—	—	0	KBS Mbr	c 1.75	H
—	—	0	Bur Mbr	c 1.9	R
sand	—	0	Bur Mbr	c 1.9 H	
sand	—	0	Bur Mbr	c 1.9	H
—	—	0	Bur Mbr	c 1.9	H
—	—	0	KBS Mbr	c 1.7	?
—	—	0	Bur Mbr	c 1.9	H
—	—	0	Bur Mbr	c 1.9	H

continued

Table 27.1. *Continued*

Specimen No.	Element	Individual age	Depositional macroenvironment	Depositional microenvironment
KNM-ER 1509	teeth	AD	—	—
KNM-ER 1515	incisor	AD	—	—
KNM-ER 1590	ff cranium	IM	high energy lake margin	delta margin
KNM-ER 1593	ff skull	AD	lacustrine high energy	delta margin
KNM-ER 1648	ff parietal	?AD	—	—
KNM-ER 1800	ff par. + occ.	?AD	lacustrine high energy (or A.C.P.)	delta margin
KNM-ER 1801	f mandible	AD	lacustrine high energy (or A.C.P.)	lacustrine
KNM-ER 1802	f mandible	AD	lacustrine high energy (or A.C.P.)	delta margin/channel
KNM-ER 1803	ff mandible	AD	alluvial delta plain	shallow offshore delta margin
KNM-ER 1804	f maxilla	AD	alluvial delta plain	channel
KNM-ER 1805	skull	AD	alluvial valley plain	channel
KNM-ER 1806	mandible	AD	alluvial valley plain	channel
KNM-ER 1808	f skeleton	AD	alluvial delta plain	distributory channel or transgressive sand
KNM-ER 1811	ff mandible	AD	ailuvial delta plain	distributory channel
KNM-ER 1812	ff skeleton	AD	alluvial delta plain	delta margin
KNM-ER 1813	cranium	AD	alluvial or coastal plain	delta margin
KNM-ER 1814	teeth	AD	alluvial valley plain	floodplain
KNM-ER 1816	ff mandible	IM	alluvial coastal plain	distributory channel
KNM-ER 1817	ff mandible	AD	alluvial coastal plain	floodplain
KNM-ER 1818	incisor	AD	—	—
KNM-ER 1819	molar	AD	—	—
KNM-ER 1820	f mandible	IM	alluvial delta plain	—
KNM-ER 1821	ff parietal	?AD	—	—
KNM-ER 2592	ff parietal	?AD	—	—
KNM-ER 2593	f molar	?	—	—
KNM-ER 2595	ff parietal	?AD	—	—
KNM-ER 2597	molar	AD	—	—
KNM-ER 2598	ff occipital	AD	—	—
KNM-ER 2599	f premolar	AD	—	—
KNM-ER 2600	f molar	AD	—	—
KNM-ER 2601	f molar	IM	—	—
KNM-ER 2607	f molar	?	alluvial valley plain	channel
KNM-ER 3229	f mandible	AD	—	—
KNM-ER 3230	mandible	AD	alluvial valley plain	floodplain
KNM-ER 3729	f mandible	AD	—	—

Matrix	Carnivore damage	Stone tools	Stratigraphic unit	Geol. age	Taxon
—	—	0	KBS Mbr	c 1.7	R
—	—	0	KBS Mbr	1.7	?
—	—	0	KBS Mbr	c 1.8	H
—	—	0	KBS Mbr	c 1.8	H
—	—	0	KBS Mbr	c 1.75	?
—	—	0	Bur Mbr	c 1.9	?
—	—	0	Bur Mbr	c 1.9	H
sand	—	0	Bur Mbr	c 1.9	H
—	—	0	Bur Mbr	c 1.9	?
—	—	0	KBS Mbr	c 1.8	R
tuffaceous sand	—	? KBS industry	KBS Mbr	1.7	H
tuffaceous sands	—	? KBS industry	KBS Mbr	1.7	R
sand	—	0	KBS Mbr	c 1.7	H
—	—	0	KBS Mbr	1.8	H
—	—	0	Bur Mbr	c 1.9	H
silty fine sand	—	0	Bur Mbr	c 1.9	H
—	—	0	Bur Mbr	c 1.9	H
—	—	0	KBS Mbr	c 1.7	R
—	—	0	Oko Mbr	1.6	R
—	—	0	KBS Mbr	c 1.7	?
—	—	0	KBS Mbr	c 1.75	R
—	—	0	KBS Mbr	c 1.75	R
—	—	0	Bur Mbr	c 1.9	?
—	—	0	Oko Mbr	c 1.6	?
—	—	0	Oko Mbr	1.65	?
—	—	0	Oko Mbr	c 1.6	?
—	—	0	KBS Mbr	c 1.85	H
—	—	0	Bur Mbr	c 1.9	H
—	—	0	KBS Mbr	c 1.85	H
—	—	0	Bur Mbr	c 1.9	H
—	—	0	Bur Mbr	c 1.9	?
—	—	0	KBS Mbr	c 1.75	?
—	—	0	KBS Mbr.	c 1.75	R
silt	—	Karari industry	Oko Mbr	1.65	R
—	—	0	KBS Mbr	c 1.75	R

continued

Table 27.1. *Continued*

Specimen No.	Element	Individual age	Depositional macroenvironment	Depositional microenvironment
KNM-ER 3731	f mandible	AD	—	—
KNM-ER 3732	f cranium	AD	—	channel
KNM-ER 3733	cranium	AD	—	—
KNM-ER 3734	f mandible	AD	—	channel
KNM-ER 3735	ff skeleton	AD	—	—
KNM-ER 3737	f teeth	AD	—	distributory channel
KNM-ER 3883	f cranium	AD	—	—
KNM-ER 3885	premolar	AD	—	—
KNM-ER 3886	d molar	IM	—	—
KNM-ER 3887	f molar	AD	—	—
KNM-ER 3889	ff mandible	AD	—	—
KNM-ER 3890	molar	AD	—	—
KNM-ER 3891	ff cranium	AD	—	—
KNM-ER 3892	ff frontal	?	—	—
KNM-ER 3950	ff mandible	AD	—	—
KNM-ER 3952	teeth	AD	—	—
KNM-ER 3953	molar	AD	—	—
KNM-ER 3954	f mandible	AD	—	—
KNM-ER 5429	f mandible	AD	—	—
KNM-ER 5877	ff mandible	AD	—	—
KNM-ER 5879	ff parietal	AD	—	—
KNM-ER 5884	incisor	AD	—	—
KNM-ER 6080	molar	IM	—	—
KNM-ER 6081	f molar	IM	—	—
KNM-ER 6082	f premolar	IM	—	—
KNM-ER 7330	f maxilla	AD	—	—
KNM-WT 15000	skeleton	IM/AD	lake margin	alluvial plain
L F203-1	molar	AD	—	—
L f22-1A	teeth	AD	—	—
L F22-2	incisor	AD	—	—
L P933-1	molar	IM	—	—
L P996-17	ff cranium	AD	—	—
L SH1.1-17	molar	IM	—	—
L 007-279	molar	AD	—	—
L 007A-125	mandible	AD	floodplain	—
L 010-21	f molar	AD	—	mixed
L 026-1G	molar	IM	—	?
L 026-59	f molar	IM	—	?
L 028-028	f premolar	?	—	—
L 028-030	molar	?	—	—
L 028-031	f molar	?	—	—
L 028-058	molar	AD	—	—
L 028-125	canine	AD	—	—
L 028-126	f premolar	AD	—	—
L 074a-21	f mandible	AD	—	—
L 157-35	molar	AD	—	—
L 209-17	f molar	AD	—	—
L 209-18	f molar	AD	—	—
L 209-30	f d molar	IM	—	—
L 238-35	molar	AD	—	—

Matrix	Carnivore damage	Stone tools	Stratigraphic unit	Geol. age	Taxon
—	—	0	Bur Mbr	c 1.9	?
sand	—	0	Bur Mbr	c 1.9	H
sand	—	0	KBS Mbr	c 1.7	H
sand	—	0	Bur Mbr	c 1.9	H
sand	—	0	Bur Mbr	c 1.9	H
—	—	0	KBS Mbr	c 1.75	R
—	—	0	Oko Mbr	c 1.5	H
—	—	0	KBS Mbr	1.8	R
—	—	0	KBS Mbr	1.8	R
—	—	0	Oko Mbr	c 1.5	R
—	—	0	Oko Mbr	c 1.5	R
—	—	0	KBS Mbr	c 1.75	R
—	—	0	KBS Mbr	c 1.75	H
—	—	0	Oko Mbr	c 1.6	H
—	—	0	KBS Mbr	1.8	H
—	—	0	KBS Mbr	c 1.75	R
—	—	0	KBS Mbr	1.8	H
—	—	0	KBS Mbr	c 1.75	R
—	—	0	KBS Mbr	c 1.7	R
—	—	0	Oko Mbr	c 1.5	R
—	—	0	Bur Mbr	c 1.9	?
—	—	0	KBS Mbr	c 1.75	?
—	—	0	Oko Mbr	1.6	R
—	—	0	Oko Mbr	1.65	R
—	—	0	Oko Mbr	1.65	R
—	—	0	KBS Mbr	?	H
tuffaceous silt	—	0	—	c 1.6	H
—	—	0	K 4^	1.4	R
clay	—	0	F 1^	2.3	R
clay	—	0	F 1^	2.3	?
—	—	0	F 3^	2.3	H
—	—	0	upper K	1.4	H
—	—	0	G 8–9	2.2	H
—	—	0	G 5	2.2	H
clayey silt	—	0	G 5	2.2	R
—	—	0	E 2^	2.4	R
—	—	0	E 2^	2.4	H
—	—	0	E 2^	2.4	?
gravels	—	0	F 1^	2.3	H
gravels	—	0	F 1^	2.3	?
gravels	—	0	F 1^	2.3	?
gravels	—	0	F 1^	2.3	R
gravels	—	0	F 1^	2.3	?
gravels	—	0	F 1^	2.3	?
sand	—	0	G 3–5	2.2	R
—	—	0	F 1	2.3	R
—	—	0	F 0/1	2.3	R
—	—	0	F 0/1	2.3	R
—	—	0	F 0/1	2.3	?
—	—	0	F 3^	2.3	R

continued

Table 27.1. *Continued*

Specimen No.	Element	Individual age	Depositional macroenvironment	Depositional microenvironment
L 338x-32	f molar	?	—	—
L 338x-33	f premolar	AD	—	—
L 338x-34	molar	AD	—	—
L 338x-35	premolar	AD	—	—
L 338x-39	molar	AD	—	—
L 338x-40	molar	AD	—	—
L 338x-89	f premolar	AD	—	—
L 338y-6	f cranium	IM	alluvial	—
L 398-0014	d molar	IM	—	—
L 398-0120	premolar	IM	—	—
L 398-0264	f molar	AD	—	—
L 398-0266	f molar	AD	—	—
L 398-0573	molar	AD	—	—
L 398-0630	molar	AD	—	—
L 398-0847	f molar	AD	—	—
L 398-1215	f molar	AD	—	—
L 398-1223	premolar	IM	—	—
L 398-1699	canine	AD	—	—
L 398-2230	f premolar	AD	—	—
L 398-2282	f molar	AD	—	—
L 398-2608	molar	AD	—	—
L 398-2626	f premolar	AD	—	—
L 398-2627	f molar	?	—	—
L 420-15	premolar	?	—	—
L 427-7	f mandible	IM	—	—
L 465-111	premolar	AD	—	—
L 465-112	f molar	?	—	—
L 465-113	f molar	?	—	—
L 628-001	premolar	AD	—	—
L 628-002	molar	AD	—	—
L 628-003	molar	AD	—	—
L 628-004	premolar	AD	—	—
L 628-005	premolar	?	—	—
L 628-007	f molar	AD	—	—
L 628-008	f molar	AD	—	—
L 628-009	molar	IM	—	—
L 628-010	molar	AD	—	—
L 628-224	f premolar	AD	—	—
L 726-11	premolar	?	—	—
L 797-1	f premolar	AD	—	—
L 860-2	f mandible	AD	—	channel
L 894-1	ff cranium	AD	—	—
O 029.1-43	premolar	AD	—	channel
O 033-0009	molar	AD	—	—
O 033-0010	f molar	AD	—	—
O 033-0031	f premolar	AD	—	—
O 033-0062	f premolar	AD	—	—
O 033-0063	molar	AD	—	—
O 033-0064	f molar	AD	—	—
O 033-0065	f molar	AD	—	—

Matrix	Carnivore damage	Stone tools	Stratigraphic unit	Geol. age	Taxon
coarse sand	—	0	E 3	2.4	?
coarse sand	—	0	E 3	2.4	?
coarse sand	—	0	E 3	2.4	R
coarse sand	—	0	E 3	2.4	R
coarse sand	—	0	E 3	2.4	R
coarse sand	—	0	E 3	2.4	R
coarse sand	—	0	E 3	2.4	?
sands and silt	—	0	E 3	2.4	R
sandy tuffite	—	0	F 0	2.3	R
sandy tuffite	—	0	F 0	2.3	R
sandy tuffite	—	0	F 0	2.3	?
sandy tuffite	—	0	F 0	2.3	R
sandy tuffite	—	0	F 0	2.3	H
sandy tuffite	—	0	F 0	2.3	R
sandy tuffite	—	0	F 0	2.3	R
sandy tuffite	0	0	F 0	2.3	?
sandy tuffite	—	0	F 0	2.3	R
sandy tuffite	—	0	F 0	2.3	H
sandy tuffite	—	0	F 0	2.3	?
sandy tuffite	—	0	F 0	2.3	?
sandy tuffite	—	0	F 0	2.3	R
sandy tuffite	—	0	F 0	2.3	?
sandy tuffite	—	0	F 0	2.3	?
—	—	0	F 3^	2.3	R
sand and gravel	—	0	G 4	2.2	R
—	—	0	F 1^	2.3	R
—	—	0	F 1^	2.3	?
—	—	0	F 1^	2.3	?
gravel and sand	—	0	G 3	2.2	R
gravel and sand	—	0	G 3	2.2	R
gravel and sand	—	0	G 3	2.2	R
gravel and sand	—	0	G 3	2.2	R
gravel and sand	—	0	G 3	2.2	R
gravel and sand	—	0	G 3	2.2	?
gravel and sand	—	0	G 3	2.2	?
gravel and sand	—	0	G 3	2.2	?
gravel and sand	—	0	G 3	2.2	?
gravel and sand	—	0	G 3	2.2	?
sand	—	0	G 5	2.2	R
sand	—	0	G 5^	2.2	R
conglomerate	—	cf. Oldowan	F 1	2.3	R
silty sand	—	0	G 28	1.9	H
—	—	0	lower G	2.2	H
silty tuffite	—	0	F 0	2.3	R
silty tuffite	—	0	F 0	2.3	?
silty tuffite	—	0	F 0	2.3	?
silty tuffite	—	0	F 0	2.3	?
silty tuffite	—	0	F 0	2.3	?
silty tuffite	—	0	F 0	2.3	?
silty tuffite	—	0	F 0	2.3	R

continued

Table 27.1. *Continued*

Specimen No.	Element	Individual age	Depositional macroenvironment	Depositional microenvironment
O 033-0506	premolar	AD	—	—
O 033-0507	premolar	AD	—	—
O 033-0508	premolar	AD	—	—
O 033-0509	f molar	AD	—	—
O 033-0740	f premolar	AD	—	—
O 033-3282	premolar	AD	—	—
O 033-3325	f molar	AD	—	—
O 033-3721	f molar	AD	—	—
O 033-5496	premolar	AD	—	—
O 033-5497	f molar	AD	—	—
O 033-6172	molar	AD	—	—
O 033E-3495	f premolar	?	—	—
O 035-4022	f molar	AD	—	—
O 035-4023	molar	AD	—	—
O 035-4024	f molar	AD	—	—
O 044-1410	f premolar	AD	—	channel
O 044-2466	f mandible	AD	—	channel
O 047-0020	f molar	AD	—	channel
O 047-0045	molar	AD	—	channel
O 047-0046	molar	AD	—	channel
O 047-0047	f molar	?	—	channel
O 047-1500	molar	AD	—	channel
O 057.4-041	f mandible	AD	—	—
O 057.4-042	f molar	?	—	—
O 057.4-147	f premolar	AD	—	—
O 057.4-148	f premolar	?	—	—
O 057.5-123	f molar	AD	—	—
O 057.5-319	f premolar	?	—	—
O 057.5-320	f molar	?	—	—
O 057.5-371	f molar	?	—	—
O 057.5-372	f molar	?	—	—
O 057.6-244	f molar	?	—	—
O 075.14A	ff skull	AD	—	—
O 0751-1255	premolar	AD	—	—
O 075S-15	molar	AD	—	—
O 075S-16	molar	AD	—	—
O 076-13	molar	?	—	—
O 076-37	f molar	AD	—	—
O 076R-11	f molar	AD	—	—
O 076R-12	f molar	AD	—	—
O 123-5495	premolar	AD	—	—
O 136-1,2,3	teeth	AD	—	channel
O 141-001	f molar	AD	—	—
O 141-002	f molar	AD	—	—
O 141-069	f molar	AD	—	—
O 141-123	incisor	AD	—	—
O 166-781	molar	AD	—	—
O 177-4525	premolar	IM	—	—
O 195-1630	molar	AD	—	—
O 222-2744	ff mandible	IM	—	—
O 323-896	ff skeleton	AD	—	—

Matrix	Carnivore damage	Stone tools	Stratigraphic unit	Geol. age	Taxon
silty tuffite	—	0	F 0	2.3	R
silty tuffite	—	0	F 0	2.3	R
silty tuffite	—	0	F 0	2.3	R
silty tuffite	—	0	F 0	2.3	?
silty tuffite	—	0	F 0	2.3	H
silty tuffite	—	0	F 0	2.3	?
silty tuffite	—	0	F 0	2.3	?
silty tuffite	—	0	F 0	2.3	?
silty tuffite	—	0	F 0	2.3	H
silty tuffite	—	0	F 0	2.3	?
silty tuffite	—	0	F 0	2.3	R
—	—	0	mid E	2.4	?
—	—	0	G 1	2.2	?
—	—	0	G 1	2.2	?
—	—	0	G 1	2.2	?
—	—	0	E 1	2.4	?
—	—	0	E 1	2.4	R
gravel	—	0	G 8^	2.2	?
gravel	—	0	G 8^	2.2	H
gravel	—	0	G 8^	2.2	R
gravel	—	0	G 8^	2.2	?
gravel	—	0	G 8^	2.2	R
—	—	0	E 4	2.4	R
—	—	0	E 4	2.4	?
—	—	0	E 4	2.4	?
—	—	0	E 4	2.4	?
—	—	0	E 5	2.4	?
—	—	0	E 5	2.4	?
—	—	0	E 5	2.4	?
—	—	0	E 5	2.4	?
—	—	0	E 5	2.4	?
—	—	0	E 5	2.4	?
silts and sand	—	0	G 13	2.2	H
sand	—	0	G 3-8	2.2	H
—	—	0	G 1–13	2.2	?
—	—	0	G 1–13	2.2	H
—	—	0	lower G	2.2	?
—	—	0	lower G	2.2	R
gravel	—	0	F 0	2.3	?
gravel	—	0	F 0	2.3	?
—	—	0	F 3	2.3	H
sand	—	0	G 1	2.2	R
gravel	—	0	G 3	2.2	R
gravel	—	0	G 3	2.2	?
gravel	—	0	G 3	2.2	?
gravel	—	0	G 3	2.2	?
—	—	0	E 1	2.4	H
—	—	0	upper E	2.4	H
—	—	0	G 7-8	2.2	H
—	—	0	G 7^	2.2	H
—	—	0	G 6–7	2.2	R

continued

Table 27.1. *Continued*

Specimen No.	Element	Individual age	Depositional macroenvironment	Depositional microenvironment
OH 03	d̄ molar	IM	fluvial–lacustrine	channel
OH 04	ff mandible	AD	lake margin	—
OH 05	cranium	AD	lake margin	weak paleosol/small channel
OH 06	ff cranium	IM	lake margin	—
OH 07	ff skeleton	AD	lake margin	—
OH 09	f cranium	AD	fluvial–lacustrine	—
OH 13	f skull	AD	fluvial–lacustrine	—
OH 14	ff cranium	—	fluvial–lacustrine	—
OH 15	teeth	AD	fluvial–lacustrine	—
OH 16	ff skull	AD	lake margin	—
OH 17	f d molar	IM	lake margin	—
OH 21	molar	IM	?lake margin	—
OH 24	cranium	AD	lake margin	—
OH 26	molar	AD	?lake margin	—
OH 27	molar	AD	?lake margin	—
OH 30	ff cranium	IM	cf. lake margin	—
OH 31	f molar	AD	?lake margin	—
OH 32	f molar	IM	—	—
OH 33	f par + f occ	—	?lake margin	—
OH 37	ff mandible	AD	fluvial–lacustrine	channel
OH 38	teeth	AD	fluvial–lacustrine	—
OH 39	teeth	IM	lake margin	lacustrine
OH 40	f molar	?	fluvial–lacustrine	channel
OH 41	molar	AD	fluvial–lacustrine	channel
OH 42	premolar	AD	—	—
OH 44	molar	IM	lake margin	—
OH 45	molar	IM	lake margin	—
OH 46	f molar	AD	lake margin	—
OH 52	f temporal	AD	—	—
OH 54	f molar	AD	—	—
OH 55	molar	IM	—	—
OH 56	f parietal	—	—	—
OH 57	premolar	IM	—	—
OH 58	f premolar	IM	—	—
OH 60	molar	AD	—	—
OH 62	ff skeleton	AD	lake margin	small channel
WN 64-160	mandible	AD	—	channel

a Data assembly is described in the text. Symbols: KNM-CH = Chesowanja, KNM-ER = Koobi Fora, KNM-WT = West Turkana, L = American Omo, O = French Omo, OH = Olduvai, WN = Peninj; ff = very fragmented, f = fragmented; d = deciduous; R = *A. boisei;* H = *Homo.* OH 3 refers to deciduous molar only; KNM-ER 1170 = 801 and 814 = 810 in this table.

b 0 = Absent; —, no information; ^, specimen may derive from adjacent unit.

Matrix	Carnivore damage	Stone tools	Stratigraphic unit	Geol. age	Taxon
coarse sand	—	"developed" Oldowan	>IID	c 1.2	R
consolidated tuff	0	inferred Oldowan	<IB	1.8	H
silty clay	+ +	Oldowan	<IC	1.8	R
coarse sand	+	inferred Oldowan	Mid I	c 1.8	H
clay	+ +	Oldowan	<IC	1.8	H
tuffaceous medium sand	?	0	>IID	c 1.2	H
sandy tuffaceous clay	+	Oldowan	<IIB	c 1.5	H
fine to medium sand	+ +	inferred Oldowan	<IIB	c 1.5	?
clayey sandstone	—	0	<IIB	c 1.5	H
claystone	+	inferred Oldowan	<IIA	1.7	H
—	—	0	<IIA	c 1.7	?
—	—	0	?	?	H
tuffaceous clay	0	inferred Oldowan	lower I	c 1.8	H
—	—	0	? II	?	R
—	—	0	I or II	?	H
clay	0	0	lower II	c 1.7	R
clay	—	0	upper I	1.7	?
—	—	0	? II	?	?
—	?	0	II	?	?
medium sand	—	0	Bed II	?	H
siliceous earthy clay	—	?	>IID	c 1.2	R
algal concretion	—	0	I	c 1.8	H
sandstone	—	0	II	?	?
sandstone	—	"developed" Oldowan	II	?	H
—	—	0	I or II	?	H
—	—	inferred Oldowan	I	c 1.7	H
—	—	inferred Oldowan	I	c 1.7	H
—	—	inferred Oldowan	I	1.7	?
—	+ +	0	lower I	c 1.8	H
—	—	0	I	c 1.7	?
—	—	0	I	c 1.7	H
CaCO$_3$	0	—	<IB	1.8	?
—	—	—	lower II	1.8	H
—	—	—	I	c 1.8	H
—	—	—	I	c 1.7	H
fine to medium sand	0	—	<IC	1.8	H
sand	—	0	—	1.4	R

Acknowledgments

This paper is dedicated to the late Maundo Muluila. Thanks are extended to Fred Grine for inviting me to participate in this workshop and for providing the flexibility of topic that allowed this analysis. Part of the research reported on here was funded by the Harry Frank Guggenheim Foundation. Results on carnivore damage reported for the Swartkrans and Olduvai hominid collections represent collaborative research conducted by the author and Nicholas Toth with whose permission they have been cited. Gen Suwa and Bill Kimbel assisted in the hominid identifications in the table, but I take full responsibility for these identifications. Craig Feibel and Frank Brown provided valuable assistance in chronostratigraphic placement of the Koobi Fora hominids. Fred Grine, Eric Delson, Pat Shipman and Richard Klein provided helpful criticisms of versions of this paper. Thanks go to C. K. Brain and E. S. Vrba for assistance in studying the Swartkrans and Kromdraai collections, and to F. Masao for access to Tanzania's Olduvai hominid collections.

References

Arambourg, C. and Coppens, Y. (1967). Sur la découverte dans le Pléistocéne Inferieur de la Vallée de l'Omo (Éthiopie) d'une mandibule d'australopithécien. *C. R. Acad. Sci.*, Paris, *265*:589–590.

Behrensmeyer, A. K. (1975a). The taphonomy and paleoecology of the Plio-Pleistocene vertebrate assemblages east of Lake Rudolf, Kenya. *Bull. Mus. Comp. Zool., 146*:473–578.

Behrensmeyer, A. K. (1975b). Taphonomy and paleoecology in the hominid fossil record. *Yrbk. Phys. Anthropol., 19*:36–50.

Behrensmeyer, A. K. (1976). Fossil assemblages in relation to sedimentary environments in the East Rudolf succession. *In* Y. Coppens, F. C. Howell, G. L. Isaac and R. E. F. Leakey (eds.), *Earliest man and environments in the Lake Rudolf Basin*, pp. 383–401. University of Chicago Press, Chicago.

Behrensmeyer, A. K. (1978). The habitat of Plio-Pleistocene hominids in East Africa: taphonomic and micro-stratigraphic evidence. *In* C. Jolly (ed.), *Early hominids of Africa*, pp. 165–189. Duckworth, London.

Behrensmeyer, A.K. (1981). Vertebrate paleoecology in a recent East African ecosystem. *In* J., Gray, A. T. Boucot and W. B., Berry (eds.), *Communities of the past*, pp. 591–615. Hutchinson Ross, Stroudsburg, Pa.

Behrensmeyer, A. K. (1982). Time resolution in fluvial vertebrate assemblages. *Paleobiol., 8*:211–227.

Behrensmeyer, A. K. (1985). Taphonomy and the paleoecological reconstruction of hominid habitats in the Koobi Fora Formation. *L'Environment des hominidés au Plio-Pléistocène*, pp. 309–323. Masson, Paris.

Behrensmeyer, A. K. and Hill, A. P. (1980). *Fossils in the making*. University of Chicago Press, Chicago.

Behrensmeyer, A. K., Western, D. and Dechant-Boaz, D. (1979). New perspectives in vertebrate paleoecology from a recent bone assemblage. *Paleobiol., 5*:12–21.

Beynon, A. D. and Wood, B. A. (1986). Variations in enamel thickness and structure in East African hominids. *Amer. J. Phys. Anthropol., 70*:177–193.

Blumenberg, B. (1981). Observations on the paleoecology, population structure and body weight of some Tertiary hominoids. *J. Hum. Evol., 10*:543–564.

Boaz, N. T. (1979). Early hominid population densities: new estimates. *Science, 206*:592–595.

Boaz, N. T. (1985). Early hominid paleoecology in the Omo Basin, Ethiopia. *L'environment des hominidés au Plio-Pléistocène*, pp. 279–308. Masson, Paris.

Boaz, N. T. and Behrensmeyer, A. K. (1976). Hominid taphonomy: transport of human skeletal parts in an artificial fluviatile environment. *Amer. J. Phys. Anthropol., 45*:53–60.

Bonnefille, R. and Vincens, A. (1985). Apport de la palynologie à l'environnement des hominidés d'Afrique Orientale. In: *L'Environnement des hominidés au Plio-Pléistocène*, pp. 237–273. Masson, Paris.
Brain, C. K. (1972). An attempt to reconstruct the behavior of australopithecines: the evidence for interpersonal violence. *Zool. Afr.*, 7:379–401.
Brain, C. K. (1981a). *The hunters or the hunted?* University of Chicago Press, Chicago.
Brain, C. K. (1981b). The evolution of man in Africa: was it a consequence of Cainozoic cooling? 17th Alex du Toit Memorial Lecture. *Geol. Soc. S. Afr.* (suppl.), 64:1–19.
Brain, C. K. (1985). New insights into early hominid environments from the Swartkrans Cave. *L'environnement des hominidés au Plio-Pléistocène*, pp. 325–343. Masson, Paris.
Broom, R. (1938). The Pleistocene anthropoid apes of South Africa. *Nature*, 142:377–379.
Broom, R. and Schepers, G. W. H. (1946). The South African fossil ape-men, the Australopithecinae. *Tvl. Museum Memoir*, 2:1–153.
Brown, F. H. (1981). Environments in the Lower Omo Basin from one to four million years ago. *In* G. Rapp and C. Vondra (eds.), *Hominid sites: their geologic setting*, pp. 149–163. Westview Press, Boulder.
Brown, F. H. and Feibel, C. S. (1986). Revision of lithostratigraphic nomenclature in the Koobi Fora region, Kenya. *J. Geol. Soc. Lond.*, 143:297–310.
Brown, F. H., Harris, J., Leakey, R. E. and Walker, A. C. (1985a). Early *Homo erectus* skeleton from West Lake Turkana, Kenya. *Nature*, 316:788–792.
Brown, F. H., McDougall, I., Davies, T. and Maier, R. (1985b). An integrated Plio-Pleistocene chronology for the Turkana Basin. *In* E. Delson (ed.), *Ancestors: the hard evidence*, pp. 82–90. Alan R. Liss, New York.
Bunn, H. T. (1981). Archaeological evidence for meat eating by Plio-Pleistocene hominids from Koobi Fora and Olduvai Gorge. *Nature*, 291:574–577.
Bunn, H. T. (1983). Evidence on the diet and subsistence patterns of Plio-Pleistocene hominids at Koobi Fora, Kenya and Olduvai Gorge, Tanzania. *In* J. Clutton-Brock and C. Grigson (eds.), *Animals and archaeology: Hunters and their prey*. Br. Archaeolog. Rep., Oxford, 163:21–30.
Burggraf, D. R., White, H., Frank, H. J. and Vondra, C. F. (1981). Hominid habitats in the Rift Valley: Part 2. *In* G. Rapp and C. F. Vondra (eds.), *Hominid sites: their geological settings*, pp. 115–147. Westview Press, Boulder.
Butzer, K. W. (1971). Recent history of an Ethiopian delta. Department of Geography Research Paper, 136, University of Chicago, Chicago.
Chamberlain, A. T. and Wood, B. A. (1985). A reappraisal of variation in hominid mandibular corpus dimensions. *Amer. J. Phys. Anthropol.*, 66:399–405.
Coppens, Y. (1980). The difference between *Australopithecus* and *Homo;* preliminary conclusions from the Omo Research Expedition's studies. *In* L. Königsson (ed.), *Current argument on early man*, pp. 207–225. Pergamon Press, Oxford.
Day, M. H. (1986). *Guide to fossil man*. (4th ed.) Cassell, London.
de Heinzelin, J. (ed.). (1983). The Omo Group. *Musée Royale de l'Afrique Centrale Annales, Science Geologique*, Serie no. 8:85:1–365.
Dechant-Boaz, D. (1982). Modern riverine taphonomy: Its relevance to the interpretation of Plio-Pleistocene Hominid paleoecology in the Omo basin. Ph.D. Dissertation, University of California, Berkeley. Univ. Microfilms Int. no. 8312762.
Dodson, P. (1971). Sedimentology and taphonomy of the Oldman Formation (Campanian) Dinosaur Provincial Park, Alberta, Canada. *Palaeogeog., Palaeoclimatol., Palaeoecol.*, 10:21–74.
Findlater, I. (1978). Stratigraphy. *In* M. G. Leakey and R. E. Leakey (eds.), *Koobi Fora Research Project Volume 1*, pp. 14–31. Clarendon Press, Oxford.
Gowlett, J. G., Harris, J. W. K., Walton, D. and Wood, B. A. (1981). Early archaeological sites, hominid remains and traces of fire from Chesowanja, Kenya. *Nature*, 294:125–129.
Grine, F. E. (1981). Trophic differences between 'gracile' and 'robust' australopithecines: A scanning electron microscope analysis of occlusal events. *S. Afr. J. Sci.*, 77:203–230.

Grine, F. E. (1985a). Australopithecine evolution: the deciduous dental evidence. *In* E. Delson (ed.), *Ancestors: the hard evidence*, pp. 153–167. Alan R. Liss, New York.

Grine, F. E. (1985b). Was interspecific competition a motive force in early hominid evolution? *In* E. S. Vrba (ed.), *Species and speciation*, pp. 143–152. Transvaal Museum (Monograph 4), Pretoria.

Harris, J. M. (1985). Fossil ungulate faunas from Koobi Fora. *L'environnement des Hominidés au Plio-Pléistocène*, pp. 151–163. Masson, Paris.

Hay, R. L. (1976). *Geology of the Olduvai Gorge*. University of California Press, Berkeley.

Hill, A. P. (1987). Causes of perceived faunal change in the later Neogene of East Africa. *J. Hum. Evol.*, 16:583–596.

Howell, F. C. (1978). Hominidae. *In* V. J. Maglio and H. B. S. Cooke (eds.), *Evolution of African mammals*, pp. 154–248. Harvard University Press, Cambridge.

Howell, F. C. and Coppens, Y. (1974). Inventory of remains of Hominidae from Pliocene/Pleistocene Formations of the Lower Omo Basin, Ethiopia (1967–1972). *Amer. J. Phys. Anthropol.*, 40:1–16.

Howell, F. C. and Wood, B. A. (1974). Early hominid ulna from the Omo Basin, Ethiopia. *Nature*, 249:174–176.

Howell, F. C., Haesaerts, P. and de Heinzelin, J. (1987). Depositional environments, archeological occurrences and hominids from Members E and F of the Shungura Formation (Omo Basin, Ethiopia). *J. Hum. Evol.*, 16:665–700.

Isaac, G. L. (1984). The archaeology of human origins: studies of the Lower Pleistocene in East Africa 1971–1981. *In* F. Wendorf and A. Close (eds.), *Advances in world archaeology*, Vol. 3, pp. 1–87. Academic Press, Orlando.

Isaac, G. L. and Harris, J. W. K. (1978) Archaeology. *In* M. G. Leakey and R. E. Leakey (eds.), *Koobi Fora Research Project Volume 1*, pp. 64–85. Clarendon Press, Oxford.

Johanson, D. C., Masao, F., Eck, G. G., White, T. D., Walter, R. C., Kimbel, W. H., Asfaw, A., Manega, P., Ndessokia, P. and Suwa, G. (1987). New partial skeleton of *Homo habilis* from Olduvai Gorge, Tanzania. *Nature*, 327:205–209.

Leakey, M. D. (1971). *Olduvai Gorge Volume 3*. Cambridge University Press, London.

Leakey, M. D. (1978). Olduvai fossil hominids: their stratigraphic positions and associations. *In* C. J. Jolly (ed.), *Early hominids of Africa*, pp. 3–16. Duckworth, London.

Leakey, M. G. and Leakey, R. E. F. (eds.) (1978). *Koobi Fora Research Project Volume 1*. Clarendon, Oxford.

Leakey, M. G., Leakey, R. E. F. and Behrensmeyer, A. K. (1978). The hominid catalog. *In* M. G. Leakey and R. E. F. Leakey (eds.), *Koobi Fora Research Project Volume 1*, pp. 86–187. Clarendon, Oxford.

Leakey, R. E. F. (1971). Further evidence of Lower Pleistocene hominids from East Rudolf, Northern Kenya. *Nature*, 231:241–245.

Leakey, R. E. F. and Walker, A. C. (1985). Further remains from the Plio-Pleistocene of Koobi Fora, Kenya. *Amer. J. Phys. Anthropol.*, 67:135–163.

Mann, A. E. (1975). Some paleodemographic aspects of the South African australopithecines. *Univ. Penna. Publ. Anthropol.*, No. 1. Philadelphia.

Mann, A. (1981). Diet and human evolution. *In* R. Harding and G. Teleki (eds.), *Omnivorous primates*, pp. 10–36. Columbia University Press, New York.

Martin, R. A. (1981). On extinct hominid population densities. *J. Hum. Evol.*, 10:427–428.

O'Connell, J. F., Hawkes, K. and Blurton-Jones, N. (1988). Hazda scavenging: Implications for Plio-Pleistocene hominid subsistence, in press.

Potts, R. B. (1983). Foraging for faunal resources by early hominids at Olduvai Gorge, Tanzania. *In* J. Clutton-Brock and C. Grigson (eds.), *Animals and archaeology: hunters and their prey*. Br. Archaeol. Repts., Oxford, 163:51–62.

Potts, R. (1986). Temporal span of bone accumulations at Olduvai Gorge and implications for early hominid foraging behavior. *Paleobiology*, 12:25–31.

Schumm, S. A. (1977). *The fluvial system*. Wiley, New York.

Shipman, P. (1983). Early hominid lifestyle: hunting and gathering or foraging and scavenging. *In* J. Clutton-Brock and C. Grigson (eds.), *Animals and archaeology: hunters and their prey*. Br. Archaeol. Repts, Oxford 163:31–49.

Toth, N. (1985). The Oldowan reassessed: a close look at early stone artifacts. *J. Archaeol. Sci.*, *12*:101–120.

Vrba, E. S. (1985). Palaeoecology of early Hominidae, with special reference to Sterkfontein, Swartkrans and Kromdraai. *L'environnement des Hominidés au Plio-Pléistocène*, pp. 151–163. Masson, Paris.

Walker, A. and Leakey, R. E. F. (1978). The hominids of east Turkana. *Sci. Amer.*, *239*:54–66.

Western, D. (1975). Water availability and its influence on the structure and dynamics of a savannah large mammal community. *E. Afr. Wild. J.*, *13*:265–286.

Western, D. (1980). Linking the ecology of past and present mammal communities. *In* A. K. Behrensmeyer and A. Hill (eds.), *Fossils in the making*, pp. 41–54. University of Chicago Press, Chicago.

Western, J. (1985). Discussions. *L'environnement des Hominidés au Plio-Pléistocène*, pp. 402–405. Masson, Paris.

White, T. D. (1977). The anterior mandibular corpus of early African Hominidae: The functional significance of size and shape. Ph.D. Dissertation, The University of Michigan, Ann Arbor.

White, T. D. (1978). Early hominid enamel hypoplasia. *Amer. J. Phys. Anthropol.*, *49*:79–84.

White, T. D., Johanson, D. C. and Kimbel, W. H. (1981). *Australopithecus africanus:* its phyletic position reconsidered. *S. Afr. J. Sci.*, *77*:445–470.

Wolman, M. G. and Leopold, L. B. (1970). Floodplains. *In* G. H. Dury (ed.), *Rivers and river terraces*, pp. 166–196. MacMillan, New York.

Wood, B. A. and Abbott, S. A. (1983). Analysis of the dental morphology of Plio-Pleistocene hominids. I. Mandibular molars: Crown area measurements and morphological traits. *J. Anat.*, *136*:197–219.

Wood, B. A. and Uytterschaut, H. (1987). Analysis of the dental morphology of Plio-Pleistocene hominids. III. Mandibular premolar crowns. *J. Anat.*, *154*:121–156.

Wood, B. A., Abbott, S. A. and Graham, S. H. (1983). Analysis of the dental morphology of Plio-Pleistocene hominids. II. Mandibular molars—study of cusp areas, fissure pattern and cross-sectional shape of the crown. *J. Anat.*, *137*:287–314.

Divergence between Early Hominid Lineages: The Roles of Competition and Culture

28

MILFORD H. WOLPOFF

While a "single species hypothesis" for the hominids was originally proposed by Weidenreich (1943: 246; 1946: 3), it was Mayr (1950) who first applied the competitive exclusion principle of Gause to hominid taxonomy. I believe that I coined the phrase "single species hypothesis" (Wolpoff, 1968), as a description of the most likely consequence of competition between culture-bearing hominid species.

The subsequent discovery of contemporary hominid species in the late Pliocene and early Pleistocene of eastern Africa disproved any general application of this hypothesis (Leakey and Walker, 1976; Wolpoff, 1976; Brace, 1980; Cartmill et al., 1986). However, this did not invalidate the principle of competitive exclusion (cf. Winterhalder, 1980, 1981). This paper explores the related roles of competition and culture in the evolution of Pliocene and early Pleistocene hominids of southern and eastern Africa and in the appearance of the diverging lineages in at least one of these regions.

Competition, Culture and Taxonomy

The idea that only one hominid species would be expected at a given time rested on the contention that contemporary culture-bearing hominid species were necessarily in competition over limiting resources (Wolpoff, 1968, 1971; Walker, 1984; Gould, 1987). This contention was made because of the claim that cultural adaptation makes a unique contribution to the human ecological niche (Swedlund, 1974); it was argued that culture in the hominids came to be both the adaptive *mechanism* and the *milieu* to which this adaptation oriented genetic change. Therefore, cultural adaptation to a new realized niche was expected to expand the fundamental niche. The competitive exclusion principle became the link between culture and competition, providing basis for the argument that there was likely to be only one human species at a time.

If this perception is extended into the past, however (Weiss, 1972), the narrower the human fundamental niche was the more difficult the divergence of realized niches became, and as Robinson put it (1972:214): "presumably culture did not come into existence instantaneously and fully fashioned." At some earlier stage of human evolution, niche divergence resulting from competition (cf. Brown, 1958; Schaffer, 1968) would have had much more of a biological effect and be recognizable as character displacement (Brown and Wilson, 1956) in some adaptively important aspect of morphology. Even so, the idea that culture played a critical role in the earliest hominids had come to be an ubiquitous assumption among human paleontologists (Mayr, 1982), myself included. For instance Robinson (1963b) postulated a fundamental role for cultural adaptation in his dietary hypothesis. Thus, the vast majority of authors who recognized closely related sympatric hominid species in the Pliocene and/or early Pleistocene explicitly or implicitly accepted the idea that the competitive exclusion principle accounted for the differences between them and looked for its consequences in the evidence for character displacement (Tobias, 1967) at the eastern African sites where they were sympatric (Brown, 1958; Cachel, 1975; Leakey and Walker, 1976; Lucas et al., 1985; Mayr, 1963; Peters, 1979, 1981; Pilbeam and Vaišnys, 1975; Pilbeam and Zwell, 1973; Rak, 1983; Schaffer, 1968; Shaklee and Shaklee, 1975; Walker, 1981;

Washburn and Ciochon, 1978; Wolpoff, 1980a). Yet, there remained the question of how important the cultural adaptation was (Durham, 1976). In fact, while clarifying many problems, the eastern African discoveries of the last two decades led to renewed questioning of whether an assumption of culture was valid for both lineages (for instance, see Leakey and Walker, 1976).

While the magnitude of morphological divergence and differing adaptive patterns provided a convincing demonstration of two hominid lineages in eastern Africa, this information did not resolve the southern African situation, where the two-lineage hypothesis began. Instead, the eastern African model was used to buttress *both* interpretations of the southern African specimens. On the one hand, in support of Robinson's dietary hypothesis, the adaptive differences between the eastern African species were regarded as more dramatic expressions of the differences Robinson (1963a) had already recognized between two southern African varieties. These largely involved the same morphological comparisons he had used, focusing on features that reflect dietary specializations and behavioral complexity—the masticatory system and brain size. At their extreme these comparisons resulted in the interpretation of two adaptively distinct contemporary taxa in Swartkrans Member 1 (Clarke, 1977, 1985) and (in my opinion even less credibly) in Sterkfontein Member 4 (Clarke, this volume, chapter 18). On the other hand, the single species interpretation of the southern African data focused on the low magnitude or even *lack* of equivalent adaptive distinctions between the southern African varieties, in contrast to the eastern African varieties. The comparisons are most instructive where the varieties are thought to have been contemporary such as at the Swartkrans cave in Members 1 and 2. Are members of a distinct "*Homo* lineage" really coincident with "robust" australopithecines here, or are these the small (and perhaps therefore the most Sterkfontein-like) end of a normal sample range (Clark, 1981) that included a few senile females?

In my view, these problems are potentially resolvable in the context of the competitive exclusion principle because they concern closely related taxa that were likely to have been sympatric at least for some times and at some places. Indeed, the natural history of that other human, *Australopithecus boisei*, is probably inexplicable and is almost certainly confusing outside of this context (McEachron, 1984). But to examine the consequences of competition between hominid populations without taking the potential effects of culture into account is unrealistic. Therefore, in order to ascertain the possible roles of culture and competition, I will examine the adaptive relationships of *A. boisei* with the other hominids in eastern Africa, where there is known time depth and recognizable morphological divergence, first focusing on the dental remains because they are numerous, well dated, and a direct reflection of dietary adaptations that may involve limiting resources.

Relation of the Eastern African Clades

To identify the adaptive situation possibly relating the two hominid lineages in eastern Africa, the dental remains of four teeth were analyzed. These four (M_2, P_4, M^1, and P^4) were chosen because a large sample of each was available for the three time spans described below and since they fully represented the new dental remains from the west Turkana deposits. Since there are few complete specimens, and virtually no jaw associations, the individuals represented by each of these four teeth are largely different from each other. For each tooth, three samples were constructed. Except for the most recently discovered materials (measured by A. Walker and used with his permission), I measured all of the specimens used in this analysis. These samples differ from those used in my earlier analyses (Wolpoff, 1980a,b) in that their use takes into account more recent theories of chronology and stratigraphic considerations (Brown and Feibel, this volume, chapter 22). Table 28.1 indicates the composition of each sample.

The early span begins in the late Miocene with the earliest hominid dental remains and extends to approximately 2.6 Myr. This span was defined so as to just predate the earliest *A. boisei* material and largely, if not exclusively, incorporate specimens attributed to *A. afarensis*. I have defined the lower boundary for the late span at approximately 1.9 Myr. The date is defined by the base of Olduvai Gorge and can be clearly ascertained in the circum-Turkana sequences. Using it avoids the problems of Olduvai Bed II dates and their relation to the Turkana hominid

Table 28.1. East African Samples

	Early span (5.5 Myr to 2.6 Myr)	Middle span (2.6 Myr to 1.9 Myr)	Late span (1.9 Myr to 1.4 Myr)
East Turkana (Koobi Fora Formation)	Lokochot Member and below	Burgi Member Tulu Bor Member	Okote Member KBS Member
West Turkana (Nachukui Formation)	Kataboi Member and below	Kalochoro Member Lokalalei Member Lomekwi Member	Natoo Member Kaitio Member
Omo Basin (Usno and Shungura Formations)	Usno Sands Basal Members A–B	Members C–G	Member H and above
Other sites	Laetoli Hadar Tabarin		Olduvai Beds I–II Peninj Chesowanja

dates (which would be especially problematic if the earliest appearance of *Homo erectus* had been used instead). The late span extends to as recently as the latest *A. boisei* remains of about 1.4 Myr. The mixture of a *Homo habilis* and *H. erectus* sample in this latest span is conservative for the hypotheses examined here, since the mixture minimizes the observed divergence with the *A. boisei* lineage in an analysis hinging on the prediction of a marked magnitude of divergence. In my view, morphological links and temporal relationships at single sites clearly indicate that these *H. erectus* specimens are the descendants of *H. habilis*.

Using these three time spans, frequency distributions of the occlusal areas were constructed and plotted (Fig. 28.1) without regard to taxonomic identity. Thus, besides including newer specimens such as those from west Turkana, this analysis differs from previous analyses of my data by Gingerich and Schoeninger (1977) and Hunt and Vitzthum (1986) in that both of these studies identified and plotted some or all of the dental data by taxon and each was restricted to only one of the circum-Turkana sites.

For all four teeth the early span distribution range is the narrowest and the mean is lower than that of the middle span sample. In my opinion the specimens in this earliest span are all (or should be all) allocated to *A. afarensis*. The distributions are unimodal for all of the teeth.

The middle span distributions are uniformly broader in range and higher in mean than the early span. This is more evident for the two mandibular teeth, perhaps because they have much larger sample sizes than the two maxillary teeth. The distributions are unimodal in their form. However, at least two hominid lineages are known to be present. The samples include specimens attributed to the "*A. boisei* lineage" that can be recognized unambiguously on the basis of cranial and mandibular characteristics. *A. boisei* and its immediate ancestor (whatever its taxonomy) are shown in Table 28.2, which enumerates all of the specimens preserving at least one of the teeth examined. At the same time specimens such as KNM-ER 1801, KNM-ER 1482, KNM-ER 3734, and Omo L 18s-32 represent a different lineage and are at the very least predecessors of *Homo habilis* if not actually that species. The latest of these, L 894-1 and KNM-ER 1590, almost certainly represent *H. habilis*. This knowledge of what varieties of hominids are in the middle span sample reflects 20/20 hindsight as applied to the morphology as well as metrics of the more complete specimens. However, there is nothing about the distributional characteristics of the sample, except perhaps for the extent of its range, that would clearly reflect the presence of

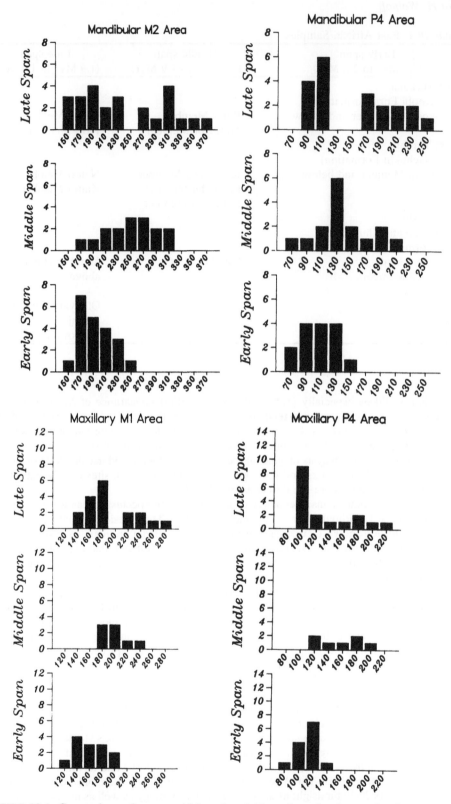

FIGURE 28.1. Comparison of eastern African hominid tooth occlusal area distributions in mm² for three time spans of the Plio-Pleistocene, as defined in Table 28.1 and discussed in the text.

Table 28.2. *Australopithecus boisei* Specimens in the Middle Span Preserving at Least One of the Teeth Analyzed

Circum-Turkana Sites
West Turkana
KNM-WT 16005
KNM-WT 17000
Omo Basin
Omo 33-508
Omo 47-46
Omo 75-14
Omo 136-1
Omo 323
F 22-1
L 7a-125
L 74a-21
L 157-35
L 238-35
L 338x-40
L 628-1
L 860-2

more than one taxon, and even an argument from the sample range is not totally convincing when comparisons are made with other higher primates (Wolpoff, 1978, 1980a).

The samples of hominids that are not *A. boisei* have larger tooth areas for all four teeth (Table 28.3), compared with the early span sample of *A. afarensis*. On the average they are about 13% bigger. Contra Hunt and Vitzthum (1986), larger teeth do not necessarily make these specimens *A. africanus*. This taxonomic issue will have to be examined in the craniomandibular portion of the sample to ascertain whether it can be resolved. A reading of what has been published thus far, for instance with regard to the identity of the Omo 338 juvenile vault and the edentulous Omo 18-1967-18 mandible, does not provide comfort or suggest that the taxonomic disputes are likely to be clarified soon. Whatever the specific resolution, and whether or not specimens on the "*Homo* lineage" have been correctly divided from those on the "*A. boisei* lineage," the fact is that *there is postcanine expansion in Homo*. This assessment is conservative, since if some *Homo* teeth were mistakenly identified as *A. boisei*, the expansion in *Homo* was even greater, while if some *A. boisei* teeth were mistakenly left in the sample attributed to *Homo*, the subsequent expansion in later *A. boisei* was even greater. The expanding postcanine area of *Homo* seems to reflect a real evolutionary trend in eastern Africa, especially since the sample compositions for each of the teeth are mostly independent of each other. Therefore, if this lineage is correctly identified as the clade ancestral to *H. erectus*, these data suggest that postcanine *expansion* characterizes our lineage during the Pliocene.

The latest span is bimodal for the distributions of each of the four teeth. The division is clearest in the two mandibular samples, in my opinion a consequence of their larger sample sizes. However, the gap is greatest for the P_4, where its magnitude is approximately 50 mm^2. Since the sample size for this tooth is the second largest of the four, it follows that there is no clear relation between the size of the sample and the magnitude of the gap between the separate distributions in it.

The *A. boisei* sample in this latest span has larger postcanine teeth than the middle span *A. boisei* sample *no matter how the isolated teeth in this middle span sample are divided*. At the same time, there is significant reduction in postcanine size for the *Homo* clade—a reduction that would be of even greater magnitude if the *H. erectus* sample was isolated from the (in my opinion earlier) *H. habilis* sample in this latest span. The comparisons of mean characteristics for the samples based on Table 28.2 identifications are reported in Table 28.3.

Table 28.3. Sample Characteristics[a]

	M_2	P_4	M^1	P^4
Early span				
Mean area	*191.6*	*106.6*	*159.7*	*112.9*
sigma	28.6	19.7	21.5	14.8
N	21	15	13	13
Non-*A. boisei* middle span				
Mean area	*219.0*	*115.1*	*185.2*	*128.0*
sigma	28.8	23.1	13.3	
N	8	9	6	3
Middle span *A. boisei*[a]				
Mean area	*281.9*	*168.4*	*229.9*	*179.0*
sigma	25.1	28.8		
N	8	7	2	4
Late span *A. boisei*				
Mean area	*298.1*	*201.9*	*239.6*	*180.0*
Sigma	39.5	25.8	23.6	30.0
N	12	10	6	6
Late span *Homo*				
Mean area	*181.7*	*103.5*	*165.2*	*101.0*
Sigma	27.2	12.0	16.1	7.2
N	13	10	12	11
Late span trophic ratio	*1.3*	*1.4*	*1.2*	*1.3*

[a]The sample identified in Table 28.2.

How divergent the postcanine tooth areas of the later span hominids became is suggested by the trophic ratios of their tooth sizes. The idea of the trophic ratio was developed by Hutchinson (1959), who argued that similar competing species with the same food requirements would be expected to maintain trophic related structures at a ratio of 1:1.28, presumably to avoid the consequences of this competition. The difference reflected in this ratio is a consequence of competition avoidance, and thus reflects the competitive exclusion principle as weakly defined. Further developed by Pearson and Mury (1979) this was considered an optimum ratio, while Robison (1975) regarded it as a minimum ratio, as do I (cf. Grine 1981, 1985, 1986). It is likely that trophic ratios are more complexly determined and are dependent on a number of relative variabilities of the two competing species (Wilson, 1975; Maiorana, 1978; Weins and Rotenberry, 1981), on the density ratios of sympatric and allopatric populations of the species (Fagerstrom, 1978; Grant, 1972), and on the extent to which there is trophic competition over limiting resources (Brooks and Wiley 1986). Nevertheless, these ratios may give some insight into the results of competition.

Grine (1985, 1986) gives trophic ratios based on postcanine tooth diameters for what he regards as *A. boisei* and early *Homo* and finds these to approximate the 1:1.28 value. Yet, he does not consider the magnitude of this ratio in his comparisons as the consequence of competition because of two problems: (1) when the South African *A. robustus* (=Swartkrans) sample is compared with early *Homo*, the trophic ratio does not attain this magnitude, and (2) the expected relationship between variability and trophic ratios (according to Maiorana, 1978) does not characterize these two comparisons (*A. robustus* with early *Homo* and *A. boisei* with early *Homo*). Grine (1985:149) instead concludes that "the morphological differences evinced by synchronic and possibly sympatric species of *Australopithecus* and *Homo* may reflect autecological adaptations."

I question whether this conclusion is necessary and whether the problems raised by Grine may not have a simpler explanation. The *A. robustus* comparisons may introduce a red herring since *this* taxon is not sympatric with *Homo* from eastern Africa and in my opinion it is unclear exactly *what* it is sympatric within Swartkrans Member 1. If there are *Homo* specimens along with *A. robustus* in Swartkrans Member 1, these taxa are less divergent than *A. boisei* and early *Homo* in the middle span of eastern Africa. This is indicated by a comparison of the dispersion and variation for the unimodal samples from Swartkrans Member 1 and the eastern African middle and late span hominids (Fig. 28.2). In these comparisons, the Swartkrans sample shows less variation and an even greater approach toward a unimodal distribution form (certainly a greater central tendency and more specimens at the mode) than the middle span sample from eastern Africa, which is the comparison of the two eastern African clades that is earliest and least diverged. The contrast with the late span sample, known to include *Homo erectus* remains, is markedly greater.

The simple explanation of the differences between southern and eastern Africa is that the distributions are not the same because the samples are not the same. In Swartkrans Member 1 what Grine (1981; this volume, chapter 14) regards as *A. robustus* is not the same species and may not even be the same clade as *A. boisei* of eastern Africa (a point also suggested by the late date and small size of the Swartkrans teeth, even when they are compared with the middle span *A. boisei* sample). Further, what Grine (1985) and Clarke (1977, 1985) regard as early *Homo* may not be *H. habilis* or *H. erectus*. At the least, assuming the southern and eastern African samples are validly comparable, the comparison would suggest that if there are two lineages in Swartkrans Member 1, they are either not involved in the same pattern of competition that they are in eastern Africa (perhaps one or both use different resources, or perhaps they are diurnally or seasonally allopatric), or Swartkrans is temporally comparable to the middle span in east Africa and therefore reflects an earlier phase of the competition insofar as the morphologic consequences are less dramatic [see also Wolpoff (1980b)]. Ironically, while the important discoveries from eastern Africa have clarified many of the problems of early hominid evolution raised by analysis of the earlier discovered southern African hominids, they may not have resolved the problems in the specific analysis of the Transvaal hominids.

The eastern African situation is easier to analyze, especially when hominids from the region are isolated from the southern African sample and divided into different time spans. Divergence of two lineages recognizable morphologically by the middle span has distinct adaptive consequences by the later span, for instance, in terms of trophic ratios (Table 28.3) for the postcanine teeth of sufficient sample size that, if anything, exceed the 1:1.28 prediction. These data are compatible with the hypothesis that there is (probably seasonally specific) competition between the hominid lineages over food as a limiting resource and that it eventually has predictable evolutionary consequences in eastern Africa. At the same time, the question of exactly what is represented in the Swartkrans Member 1 sample becomes even more difficult to understand because clearcut predictions of the competitive exclusion principle seem so obviously met in eastern Africa.

The pattern of postcanine tooth variation in eastern Africa is clearly one of expansion for the *A. boisei* lineage, in parallel with the earlier postcanine size expansion that characterized populations of the *Homo* lineage. Competition between contemporary populations of these two clades over (almost certainly seasonal) food resources that were potentially limiting factors is one implicated cause. However, there is much more involved in the evolution of *A. boisei*. The postcanine teeth are not the only aspect of morphology that seems to expand.

Unless KNM-WT 17000 is a very unusual male in the species it samples, the fact that the 410 cc cranial capacity of this specimen is the smallest male capacity of any hominid is significant. Cranial capacities of *A. boisei* specimens in the middle and later spans for both sexes are markedly greater. For instance, the two male capacities known for the middle span are 510 cc for KNM-ER 406 and 530 cc for OH 5. These may even underestimate the magnitude of capacity expansion, since in my opinion (albeit based on 20/20 hindsight after the discovery of KNM-ER 406) the OH 5 reconstruction is too short and the capacity of the specimen correctly reconstructed could be dramatically greater than 530 cc. Whatever the case, less certain evidence from even later *A. boisei* crania (in the late span sample) indicates further continued increases in cranial capacity. Walker (1971) has suggested an enlarged capacity for the Chesowanja female and Wolpoff (1977) has indicated the same interpretation for the KNM-ER 733 male vault. In all, I believe there is strong evidence for continued cranial capacity expansion in the *A. boisei* lineage. Based on my interpretation of the data, the rate of capacity expansion appears to exceed expansion of the postcanine tooth areas and therefore both cannot be allometric reflections of any possible (but unknown) increase in body size, if in fact either is. Because cranial size scales to body size with

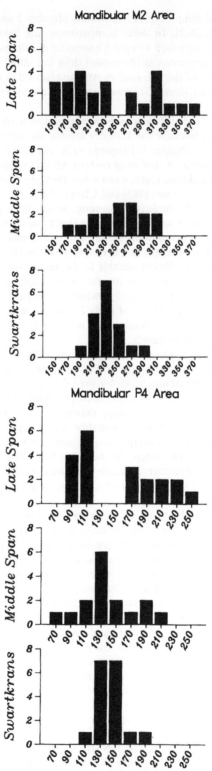

FIGURE 28.2. Comparison of eastern African middle and late span hominids and the Swartkrans Member 1 sample for tooth area variation in mm^2.

a much lower allometric exponent than postcanine tooth size scaling (Wolpoff, 1985), even if the dental changes reflect allometry, the changes in cranial capacity are to at least some extent independent of body size and thus are a real evolutionary trend within the lineage.

Somewhat different changes may characterize the canines in this lineage. As Walker and Leakey (this volume, chapter 15) point out, there is nothing particularly unusual about the canine socket size of KNM-WT 17000. However, in terms of the *sample*, reduction in canine size can be shown between the middle and late span *A. boisei* mandibular canine dimensions. In the middle span mandibular sample the mean canine breadth for three specimens is 10.1 mm, while 11 specimens in the late span *A. boisei* sample average 9.2 mm. In the maxillary sample there is probably also reduction, but an accurate inference cannot be made since a single middle span *A. boisei* canine and the broken KNM-WT 17000 socket are all that is preserved from the earlier time.

Dissimilarity between middle and later span canines may also extend to the analysis of canine function for at least some of the specimens. At the moment this is a supposition that can only be made from the form of the cranial musculature, and it will remain unconfirmed until canines unambiguously associated with early *A. boisei* male specimens such as KNM-WT 17000 are actually found. There are dramatic differences between the nuchal regions of KNM-WT 17000 superior and of all later *A. boisei* crania. The angulation and marked development of the KNM-WT 17000 superior nuchal line in terms of its projection and height above the Frankfurt Horizontal and the extent of the occiput involved in supporting the nuchal musculature combine with the lack of evidence for *Pan*-like incisor hypertrophy (an alternative structure requiring large nuchal muscles) to suggest to me that this expanded musculature was present in order to provide horizontal forces for powerful and possibly projecting canines. No canines known for later *A. boisei* specimens are worn in a manner that could reflect cutting against the premolar, but it will be interesting to discover what the canines of early males similar to KNM-WT 17000 are like, and whether they were used to cut as were some of the canines of *A. afarensis*.

What may be most interesting is that evolutionary trends involving expanding brain size and reduction of the canine cutting complex in a biped effect just those features that Darwin, and many following him (reviewed by Jolly, 1973; Wolpoff, 1980a), regarded as reflecting a cultural adaptation. In the eastern African hominids of the Pliocene and early Pleistocene these evolutionary trends are known for both lineages (as is the third trend of expanding postcanine tooth size). Moreover, these are not the only homoplasies recognized that may reflect behavioral changes; characteristics of the endocasts (Holloway, 1975), features of the cranial base, and the delayed eruption times of the molars (Dean, 1986), provide additional evidence for the appearance of similar or shared behaviors in the two lineages.

Recognition of a number of seemingly independent evolutionary trends shared by two very closely related lineages creates problems in phylogenetic analysis since homoplasies are generally thought to be minimized in the evolutionary process. What appear to be numerous homoplasies could, of course, reflect an inappropriate division of the lineages—the single species hypothesis resurrected. However, in any guise, the hypothesis fails at the same point it failed before, with the unlikelihood that contemporary morphologies as dissimilar as *A. boisei* and *H. erectus* are to be found in the same species. And if a single species is an unreasonable interpretation of the late span hominid sample, it is probably an unreasonable interpretation of the middle span samples as well.

The solution to the problems stemming from the discovery of numerous seemingly independent homoplasies in different lineages is probably elsewhere, specifically in the assumption that the homoplasies are unrelated. Darwin's thinking, in fact, suggests just how these parallel trends may have a common origin—through the emergence of an adaptation in both hominid clades that involved using culture as a way of dealing with the world. In each hominid, morphological adaptations adjusted to the requirements imposed by the cultural adaptation. Under these circumstances many more homoplasies might appear than would normally be expected in similarly adapting closely related species, since cultural developments would most likely spread from one species to the other, regardless of which population they originated in, and thereby adaptive innovations and consequent changes in selection initiated by one would immediately spread to the other. Common evolutionary responses to these rapidly shared innovations in the two culture-bearing lineages could be posited as the main common factor underlying the parallel trends. This would suggest that the verb "to ape" has picked on the wrong hominoid taxon.

Comments

The eastern African comparisons reviewed here seem to reflect the consequences of competition between two early hominid varieties that were preadapted to sharing information through their cultural adaptations. The form of their relationships brings the possibility of a new understanding to the analysis of adaptations in both by more accurately determining their interactions. The essential importance of culture in hominid evolution is brought into focus. A particular innovation or adaptation may have appeared twice, more or less independently, but it will forever remain unclear as to what extent the members of one species copied the behavior of the other. It is likely that, as Robinson (1963a,b) originally suggested, cultural adaptation in the Pliocene hominids did not provide so broad a fundamental niche that competing culture-bearing hominids could not coexist. I am not the first to speculate that the incorporation of stone tools in the adaptations of both lineages may have initially provided the basis for subsequent adaptive divergence by broadening the fundamental niches enough for the realized niches to not overlap in the utilization of limiting resources.

Subsequent niche expansion in the lineage leading to the genus *Homo* was evidently based on continued changes in adaptability stemming from important developments in both social organization and technology. Additional elements may have included the utilization of more food resources allowed by a division of labor in food acquisition, regular structured distribution of the gathered food resources throughout the group (perhaps with a home base), more effective use of the environment resulting from more knowledge about it (for instance regarding seasonal availability and digestibility), and more efficient food preparation technology. Niche expansion in the other lineage was affected through continued increase in the size of the masticatory system, along with continued brain size expansion that almost certainly reflected changes in behavioral complexity and the accurate transmission of knowledge about the environment from generation to generation. These probably combined to also affect an increased range of usable food resources. To what extent other cultural innovations were shared by members of the *A. boisei* lineage, or perhaps were even initiated by them, will always be unclear because of the enormous potential for copying. In any event, large grinding teeth are hardly an adaptive speciality; they *expand* rather than restrict the range of usable food resources.

Through the continued expansion of brain size and posterior tooth size, the *A. boisei* lineage continued the earlier trends in australopithecine evolution. It is our own line that embarked on a novel evolutionary pathway that ultimately proved to be the more successful. And if the case can be made that the increasing extent of niche overlap ultimately contributed to the extinction of our less fortunate cousins, one can also wonder about the extent to which both copying from and competition with these other hominids influenced the evolutionary pathway taken by our own ancestors. The Pliocene australopithecines were already very successful hominoid forms, possibly the most intelligent organisms on the planet, with an eclectic diet (Mann, 1981) and changes in their reproductive strategies resulting in fairly well developed colonizing abilities (Lovejoy, 1981). With their spread across the drier and more open environments of sub-Saharan Africa, a reasonably stable adaptive pattern was attained. One wonders what would have happened if there had remained only one species of this higher primate. Had there been no competition between the closely related hominid species of the early Pleistocene, would the broad adaptive patterns, complex social organization, and enhanced technological abilities of *H. erectus* ever have evolved? Or, would the already successful pattern of reproductive, behavioral, and dietary adaptability been retained in the subsequent evolutionary development of what would have remained a very successful, intelligent, grasslands-dwelling ape? Perhaps it is best to leave this "future history" of a different world for another Heinlein to develop, but one cannot help speculating about the true and complete role of the "robust" australopithecines in human evolution.

Acknowledgments

I am deeply indebted to Fred Grine for inviting me to participate in the "robust" australopithecine workshop and for the gift of his editorial skills in preparing this contribution. I am grateful for the subsistence grant allowing me to attend the workshop, and for additional travel expenses paid from the University of Michigan Faculty Assistance Fund. Participating in this attempt to resolve issues about the natural history of the "robust" australopithecines provided new meaning to that Chinese curse, "May you live in interesting times."

References

Brace, C. L. (1980). Reply to Washburn. *Amer. Anthropol., 82*:394.
Brooks, D. R. and Wiley, E. O. (1986). *Evolution as entropy. Toward a unified theory of biology.* University of Chicago Press, Chicago.
Brown, W. L. (1958). Some zoological concepts applied to the problems of evolution in the hominid lineage. *Amer. Sci., 46*:151–158.
Brown, W. L. and Wilson, E. O. (1956). Character displacement. *Syst. Zool., 5*:49–54.
Cachel, S. (1975). A new view of speciation in *Australopithecus*. In R. Tuttle (ed.), *Paleoanthropology, morphology, and paleoecology*, pp. 183–201. Mouton, The Hague.
Cartmill, M., Pilbeam, D. and Isaac, G. LI (1986). One hundred years of paleoanthropology. *Amer. Sci., 74*:410–420.
Clark, G. A. (1981). Multivariate analysis of *"Telanthropus capensis"*: implications for hominid sympatry in South Africa. *Quaternaria, 22*:39–63.
Clarke, R. J. (1977). A juvenile cranium and some adult teeth of early *Homo* from Swartkrans. *S. Afr. J. Sci., 73*:46–49.
Clarke, R. J. (1985). *Australopithecus* and early *Homo* in southern Africa. In E. Delson (ed.), *Ancestors: the hard evidence.* pp. 171–177. Alan R. Liss, New York.
Dean, M. C. (1986). *Homo* and *Paranthropus:* similarities in the cranial base and developing dentition. In B. Wood, L. Martin and P. Andrews (eds.), *Major trends in primate and human evolution*, pp. 249–265. Cambridge University Press, Cambridge.
Durham, W. H. (1976). The adaptive significance of cultural behavior. *Hum. Ecol., 4*:89–121.
Fagerstrom, J. A. (1978). Paleobiologic application of character displacement and limiting similarity. *Syst. Zool., 27*:463–468.
Gingerich, P. D. and Schoeninger, M. (1977). The fossil record and primate phylogeny. *J. Hum. Evol., 6*:483–505.
Gould, S. J. (1987). Bushes all the way down. *Nat. Hist., 87*:12–19.
Grant, P. R. (1972). Convergent and divergent character displacement. *Biol. J. Linn. Soc., 4*:39–68.
Grine, F. E. (1981). Trophic differences between "gracile" and "robust" australopithecines: a scanning electron microscope analysis of occlusal events. *S. Afr. J. Sci., 77*:203–230.
Grine, F. E. (1985). Was interspecific competition a motive force in early hominid evolution? In E. Vrba (ed.), *Species and speciation*, pp. 143–152. Transvaal Museum Monograph, Pretoria.
Grine, F. E. (1986). Ecological causality and the pattern of Plio-Pleistocene evolution in Africa. *S. Afr. J. Sci., 82*:87–89.
Holloway, R. L. (1975). The role of human social behavior in the evolution of the brain. Forty-third James Arthur lecture on the evolution of the human brain. *American Museum of Natural History*, New York.
Hunt, K. and Vitzthum, V. (1986). Dental metric assessment of the Omo fossils: implications for the phyletic position of *Australopithecus africanus*. *Amer. J. Phys. Anthropol., 71*:141–155.
Hutchinson, G. E. (1959). Homage to Santa Rosalia, or why are there so many kinds of animals. *Amer. Nat., 93*:145–159.
Jolly, C. J. (1973). Changing views of hominid origins. *Yrbk. Phys. Anthropol., 16*:1–17.

Leakey, R. E. F. and Walker, A. C. (1976). *Australopithecus, Homo erectus*, and the single species hypothesis. *Nature, 261*:572–574.
Lovejoy, C. O. (1981). The origin of man. *Science, 211*:341–350.
Lucas, P. W., Corlett, R. T. and Luke, D. A. (1985). Plio-Pleistocene hominid diets: an approach combining masticatory and ecological analysis. *J. Hum. Evol., 14*:187–202.
McEachron, D. L. (1984). Hypothesis and explanation in human evolution. *J. Soc. Biol. Struct., 7*:9–15.
Maiorana, V. (1978). An explanation of ecological and developmental constraints. *Nature, 273*:375–376.
Mann, A. E. (1981). Diet and human evolution. *In* R. S. O. Harding and G. Teleki (eds.), *Omnivorous primates. Gathering and hunting in human evolution*, pp. 10–36. Columbia University Press, New York.
Mayr, E. (1950). Taxonomic categories in fossil hominids. *Cold Spring Harbor Symp. Quant. Biol., 15*:108–118.
Mayr, E. (1963). *Animal species and evolution*. Belknap, Cambridge.
Mayr, E. (1982). Reflections on human paleontology. *In* F. Spencer (ed.), *A history of American physical anthropology 1930–1980*, pp. 231–237. Academic Press, New York.
Pearson, D. L. and Mury, E. J. (1979) Character divergence and convergence among tiger beetles Coleoptera: Cicindelidae. *Ecology, 60*:557–566.
Peters, C. R. (1979). Toward an ecological model of African Plio-Pleistocene hominid adaptations. *Amer. Anthropol., 81*:261–78.
Peters, C. R. (1981). Robust vs. gracile early hominid masticatory capabilities: the advantages of the megadonts. *In* L. L. Mai, E. Shanklin and R. W. Sussman (eds.), *The perception of human evolution*, pp. 161–181. University of California Press, Los Angeles.
Pilbeam, D. R. and Vaišnys, J. R. (1975). Hypothesis testing in paleoanthropology. *In* R. Tuttle (ed.), *Paleoanthropology, morphology, and paleoecology*, pp. 3–13. Mouton, The Hague.
Pilbeam, D. R., and Zwell, M. (1973). The single species hypothesis, sexual dimorphism, and variability in early hominids. *Yrbk. Phys. Anthropol., 16*:69–79.
Rak, Y. (1983). *The australopithecine face*. Academic Press, New York.
Robinson, J. T. (1963a). Adaptive radiation in the australopithecines and the origin of man. *In* F. C. Howell and F. Bourliere (eds.), *African ecology and human evolution*, pp. 385–416. Aldine, New York.
Robinson, J. T. (1963b). Australopithecines, culture, and phylogeny. *Amer. J. Phys. Anthropol., 21*:595–605.
Robinson, J. T. (1972). *Early hominid posture and locomotion*. University of Chicago Press, Chicago.
Robison, R. A. (1975). Species diversity among agnostid trilobites. *Fossils and Strata, 4*:219–226.
Schaffer, W. M. (1968). Character displacement and the evolution of the Hominidae. *Amer. Nat., 102*:559–571.
Shaklee, A. B. and Shaklee, R. B. (1975). Ecological models in relation to early hominid adaptations. *Amer. Anthropol., 77*:611–615.
Swedlund, A. C. (1974). The use of ecological hypotheses in australopithecine taxonomy. *Amer. Anthropol., 76*:515–529.
Tobias, P. V. (1967). *Olduvai Gorge, vol. II. The cranium and maxillary dentition of* Australopithecus (Zinjanthropus) boisei. Cambridge University Press, London.
Walker, A. C. (1971). Late australopithecine form Baringo district, Kenya. *Nature, 230*:513–514.
Walker, A. C. (1981). Diet and teeth: dietary hypotheses and human evolution. *Phil. Trans. Roy. Soc., Lond., B, 292*:57–64.
Walker, A. C. (1984). Extinction in human evolution. *In* M. Niteki (ed.), *Extinctions*, pp. 119–152. University of Chicago Press, Chicago.
Washburn, S. L. and Ciochon, R. L. (1976). The single species hypothesis. *Amer. Anthropol., 78*:96–98.
Weidenreich, F. (1943). *The skull of Sinanthropus pekinensis: A comparative study of a primitive hominid skull. Paleonto. Sin.*, n.s. D, No. 10 (whole series No. 127).

Weidenreich, F. (1946). *Apes, giants, and man.* University of Chicago, Chicago.
Weiss, K. M. (1972). A generalized model for competition between hominid populations. *J. Hum. Evol., 1*:451–456.
Wiens, J. A. and Rotenberry, J. T. (1981) Morphological size ratios and competition in ecological communities. *Amer. Nat., 117*:592–599.
Wilson, D. S. (1975). The adequacy of body size as a niche difference. *Amer. Nat., 109*:769–784.
Winterhalder, B. (1980). Hominid paleoecology: the competitive exclusion principle and determinants of niche relationships. *Yrbk. Phys. Anthropol., 23*:43–63.
Winterhalder, B. (1981). Hominid paleoecology and competitive exclusion: limits to similarity, niche differentiation, and the effects of cultural behavior. *Yrbk. Phys. Anthropol., 24*:101–121.
Wolpoff, M. H. (1968). "Telanthropus" and the single species hypothesis. *Amer. Anthropol., 70*:447–493.
Wolpoff, M. H. (1971). Competitive exclusion among lower Pleistocene hominids: the single species hypothesis. *Man, 6*:601–614.
Wolpoff, M. H. (1976). Some aspects of the evolution of early hominid sexual dimorphism. *Curr. Anthropol., 17*:579–606.
Wolpoff, M. H. (1977). A reexamination of the ER 733 Cranium. *Z. Morphol. Anthropol., 68*:8–13.
Wolpoff, M. H. (1978). Analogies and interpretations in paleoanthropology. *In* C. J. Jolly (ed.), *Early man in Africa*, pp. 461–503. Duckworth, London.
Wolpoff, M. H. (1980a). *Paleoanthropology.* Knopf, New York.
Wolpoff, M. H. (1980b). Morphological dating of the Swartkrans australopithecines. *In* R. E. Leakey and B. A. Ogot (eds.), *Proceedings of the 8th Panafrican congress of prehistory and Quaternary studies*, pp. 69–170. The Louis Memorial Institute for African Prehistory, Nairobi.
Wolpoff, M. H. (1985). Tooth size–body size scaling in a human population: theory and practice of an allometric analysis. *In* Jungers, W. L. (ed.), *Size and scaling in primate biology*, pp. 273–318. Plenum, New York.

The Causes of "Robust" Australopithecine Extinction

29

RICHARD G. KLEIN

Extinction is easy to explain in the abstract. It is what happens when a species cannot cope with changed natural selective pressures and mortality begins to consistently outpace recruitment. Depending on the gap, the path to extinction may be long or short, but the history of life shows that most species eventually take it. Again, speaking abstractly, the geologic record reveals more than enough natural selective (environmental) change to explain the frequency of extinction.

A problem arises only when a specific extinction or set of contemporaneous extinctions must be explained. Thus, after decades of study, there is still no consensus on why the dinosaurs disappeared at the end of the Cretaceous or why 75% of the large mammals of North America vanished at the end of the Pleistocene. Broadly speaking, investigators recognize two potential causes—a change in the physical environment, especially a change in climate, or a change in the biotic environment, especially the appearance of a new competitor or predator. To achieve a consensus, alternative explanatory events must be ruled out, either because their timing was wrong (they did not coincide closely enough with the extinction event) or because their impact on the species would have been too limited (e.g., the reconstructed ecology of the species or its prior history suggests it could have adapted to the changed circumstances). Perhaps more than any other case, the North American end-Pleistocene extinctions show how difficult it can be to eliminate competing hypotheses. Because the North American extinctions occurred so recently, they are relatively well documented and dated, yet proponents of a climatic explanation (Pleistocene to Holocene climatic change) remain deadlocked with proponents of a biotic one (the first appearance of people in North America) (Martin and Klein, 1984). Some investigators also favor a combination of climatic and biotic change.

The extinction of the "robust" australopithecines occurred long before the North American end-Pleistocene episode, and its timing and context are much more poorly documented. Not surprisingly then, no single compelling explanation can be offered. My purpose here is simply to present what is known about its timing and then to discuss the major alternative explanations from which a compelling choice may one day emerge. I assume here that the eastern and southern African "robust" australopithecines (*Australopithecus boisei* and *A. robustus*) may be regarded as ecological vicars, whose somewhat distinctive morphologies mean no more than those of other closely related eastern and southern African species pairs (such as the wildebeests, *Connochaetes taurinus* and *C. gnou* or the bastard hartebeests, *Damaliscus lunatus* and *D. dorcas*.) In this case, their behavior and ecology were probably similar enough that the same event(s) could explain their extinction.

The Timing of Extinction

The youngest known "robust" australopithecine fossils in eastern Africa may be isolated teeth from Upper Bed II at Olduvai Gorge. Their age is estimated at about 1.2 Myr (Leakey, 1980; Leakey and Hay, 1982), based on the assumption that the disconformity separating Bed II from Bed III reflects major Rift Valley faulting dated by K/Ar elsewhere to about 1.2 Myr. "Robust" australopithecines that are probably only slightly older, perhaps 1.4 to 1.3 Myr, have been found in the Okote Member of the Koobi Fora Formation (Walker and Leakey, 1978; Walker, 1981) and in Member K of the Shungura Formation in the lower Omo Valley (Coppens, 1980; Suwa, this volume, chapter 13; dates from Brown et al., 1985; Brown and Feibel, 1986). The "robust" australopithecines from the Chemoigut Formation, Chesowanja (Baringo) may also postdate 1.5 Myr, but it is just as likely that they are older (Bishop et al., 1975; Gowlett et al., 1981).

In southern Africa, the youngest "robust" australopithecine fossils are probably those from Swartkrans Member 3. When only a single "robust" australopithecine fossil was known from this unit, the possibility existed that it had been reworked from older units (either Member 1 or Member 2) (Brain, 1982, 1985). However, Member 3 has now provided 25 skeletal parts from at least six "robust" australopithecine individuals, and some are undoubtedly *in situ* (Brain, this volume, chapter 20; Grine, this volume, chapter 14). Their geologic age is difficult to gauge (Brain, this volume, chapter 20), but on faunal and artifactual grounds, Member 3 does not appear to be readily distinguishable from Member 1, which is bracketed between roughly 1.8 and 1.5 Myr by faunal correlation to eastern Africa (Vrba, 1982). On the other hand, thermoluminescence determinations by Vogel (1985) suggest that Member 3 may be only half as old as Member 1, in which case its "robust" australopithecine fossils could postdate 1 Myr. They could thus be substantially younger than any known "robust" australopithecines from eastern Africa.

Depending on how the Swartkrans data are interpreted, the "robust" australopithecines may have disappeared earlier in eastern than in southern Africa, or at roughly the same time—1.4 to 1.2 Myr—in both places. A figure in the 1.4 to 1.2 Myr range is commonly cited by both specialists and popular authors, but the issue is more complicated. As pointed out by White (this volume, chapter 27), the fossiliferous portions of the Plio-Pleistocene sequences in the Turkana Basin essentially "top out" at 1.4 to 1.2 Myr and, thus, would be very unlikely to provide younger "robust" australopithecines, even if they existed afterward. Excepting Olduvai Gorge and perhaps Melka Kunturé (Chavaillon, 1982; Geraads, 1985) and the Middle Awash (Kalb et al., 1982; Clark et al., 1984), there are no fossiliferous sites in eastern Africa that clearly monitor the immediately succeeding interval, and deposits postdating 1.2 Myr at Olduvai, Melka Kunturé, and the Middle Awash have not provided large fossil samples. It is therefore possible that "robust" australopithecines persisted much later than 1.2 Myr, independent of how the dating at Swartkrans is ultimately resolved.

How much later is impossible to pinpoint. In southern, eastern, and northern Africa, there are scattered Lower and Middle Pleistocene sites that have provided animal fossils, artifacts, and/or remains of archaic *Homo*, but no "robust" australopithecine fossils. (Figure 29.1 shows the approximate locations of what are arguably the most important sites.) Two problems arise in using these sites to date "robust" australopithecine extinction: (1) most of them are very imprecisely dated, and (2) the northern and southern African sites lie outside the known geographic range of the "robust" australopithecines. Even sites within this range may not sample local environments in which the "robust" australopithecines could live, and the bone accumulator(s) at some sites may not have had access to "robust" australopithecine remains. Also, at most sites, the absence of "robust" australopithecines could be a result of the small size of the fossil sample although, despite small sample size, early *Homo* is occasionally represented, especially at Melka Kunturé.

If we assume that "robust" australopithecines could inhabit any savannah environment in Africa, then an upper limit on their existence can perhaps be inferred from their absence in the relatively large faunal samples from Ternifine in Algeria (Geraads et al., 1986) and Elandsfontein in South Africa (Hendey, 1974; Klein, 1983). Both sites have provided Acheulean artifacts and remains of early *Homo*, accompanied by faunal remains that suggest a late Lower or early Middle Pleistocene age, roughly equivalent to Olduvai Bed IV. Chronometric estimates must be inexact, but dates somewhere between 700 and 500 kyr seem likely for both sites. Ternifine may be somewhat older, and its human fossils are commonly assigned to *Homo erectus*, while those from Elandsfontein are placed in early *Homo sapiens*. Together with other, smaller and equally poorly dated occurrences elsewhere in Africa, Ternifine and Elandsfontein can be used to argue that the "robust" australopithecines had probably disappeared by 700 to 500 kyr ago. A firmer estimate will depend on major new discoveries that may be many years in coming.

Richard G. Klein 501

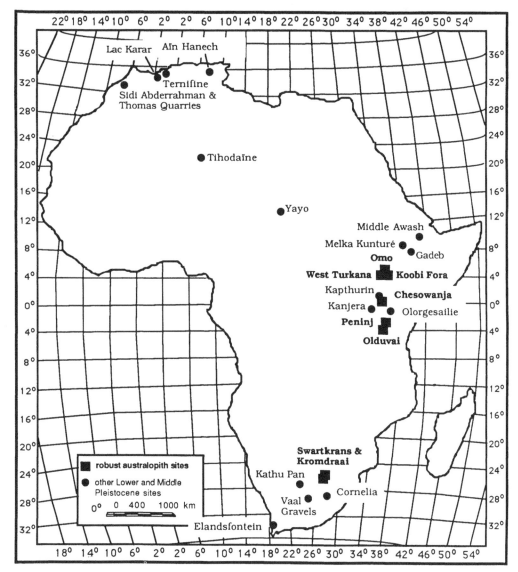

FIGURE 29.1. Approximate locations of "robust" australopithecine sites and other Lower and Middle Pleistocene sites that may be relevant to dating "robust" australopithecine extinction.

Climatic Explanations for "Robust" Australopithecine Extinction

Global climatic fluctuations between colder and warmer characterized the entire time span of the "robust" australopithecines, from at least 2.5 Myr until 1.2 Myr or later. Superimposed on these fluctuations were two major climatic events (Shackleton and Opdyke, 1977; Thunell and Williams, 1983; Shackleton *et al.*, 1984) that could have affected the "robust" australopithecines. The first was a sharp downturn in mean global temperatures about 2.5 Myr that may be linked to their emergence (Vrba, 1985; this volume, chapter 25), assuming they were not already present. The second was a further, broadly comparable steplike downturn that occurred about 0.9 Myr and that was probably accompanied by a major change in the nature of climatic fluctuations (Prell *et al.*, 1982; Roberts, 1984; Prentice and Denton, this volume chapter 24), such that the colder intervals ("glaciations") became longer and colder.

In Africa, mammal species were probably less affected by long-term fluctuations in mean temperature than by accompanying fluctuations in precipitation. By analogy to the relatively well known Last Glaciation, in Africa (and most other places), glaciations (especially glacial maxima) were generally drier than interglaciations (Butzer, 1978, 1984; Roberts, 1984; Deacon and Lancaster, 1988). In eastern Africa, Upper Pleistocene prehistory is not well known, but in both southern and northern Africa, where it has been better studied, archeological occupations dating to the middle of the Last Glaciation (between ≥ 40 and 20 kyr BP) are remarkably rare, probably because widespread hyperaridity had dramatically reduced human populations (Guillien and Laplace, 1978; Deacon and Thackeray, 1984; Klein, 1984). The people who lived in Africa during the Last Glaciation were nearly or even fully modern *H. sapiens*, and if extreme aridity affected them so severely, its impact on the "robust" australopithecines would surely have been even more severe. At least hypothetically then, "robust" australopithecine extinction may have resulted from the onset of particularly arid glacial intervals (or glacial maxima) beginning around 0.9 Myr ago.

This hypothesis will be difficult to test until the date of "robust" australopithecine extinction is better fixed, but its plausibility would clearly be enhanced by evidence that other African savannah dwellers disappeared following global climatic change about 0.9 Myr, or at least that this time coincided with an increase in faunal turnover rates. Measuring turnover rates during the "robust" australopithecine time span is complicated by severe sampling and dating problems, coupled with taxonomic and phylogenetic uncertainties. However, so far there is no evidence for unusual turnover during the long, late Lower Pleistocene to early Middle Pleistocene interval centering on 0.9 Myr. The famous "faunal break" in Olduvai Lower Bed II, which may record a major extinction/speciation episode in non-hominid taxa, clearly occurred much earlier, before 1.6 Myr ago. More generally, turnover as measured by major first and last appearances (for example, the first for *Equus, Phacochoerus, Lycaon,* and *Crocuta* and the last for *Parapapio, Dinopithecus, Gorgopithecus, Deinotherium, Ancylotherium, Notochoerus, Menelikia, Chasmoporthetes, Dinofelis,* and *Machairodus*) probably peaked sometime between 2.5 and 1.6 Myr. Circumstantially this peak could be linked to the global climatic change of 2.5 Myr, but if so, it remains unclear that a second peak followed the broadly comparable or even more dramatic change of 0.9 Myr.

Biotic Explanations for "Robust" Australopithecine Extinction

In the abstract, climatic and biotic explanations for extinction may not be separable, since climatic change may have driven most past biotic change, including the evolution or dispersal of potential competitors and predators. However, I am concerned here only with the proximate effects of biotic change, not with its ultimate causation. With regard to "robust" australopithecine extinction, at least two possible biotic causes must be considered: (1) changes in the nature of the African carnivore guild, and (2) on-going evolution in the genus *Homo*.

The large faunal samples available from early Pliocene and mid-Pleistocene sites in southern Africa indicate that carnivore species diversity declined significantly between roughly 5.5 and 0.5 Myr, while herbivore (prey) diversity remained more or less constant (Klein, 1977). Carnivore diversity probably declined simultaneously elsewhere in Africa, but sampling problems make this difficult to prove. In both southern and eastern Africa, the probable change in carnivore diversity was accompanied by a basic change in large carnivore taxa (Walker, 1984). In particular, sabertooth and "false" sabertooth cats (various Machairodontinae and their feline mimic *Dinofelis*) disappeared, along with several hyena species. The vanished hyenas included some that may have depended largely on meat and bones the sabertooths left at carcasses and some, especially *Chasmoporthetes,* that were probably cursorial forms that ran down their prey in the manner of hunting dogs or cheetahs. The niches of the sabertooths, "false" sabertooths, and the various hyenas were probably incorporated into those of their still extant large carnivore successors, especially species of *Panthera* (lion and leopard), *Acinonyx* (cheetah), *Crocuta* (spotted hyena), *Canis* (various jackals), and *Lycaon* (hunting dog). The decline of the sabertooths and hyenas may have been prompted by the evolution or immigration of these successors, and the replacement may have been gradual, since temporal overlap between the archaic and extant carnivore taxa is reasonably well documented. This is especially true for the sabertooths/"false" sabertooths and modern large cats (*Panthera* species), which coexisted for perhaps two Myr. The latest known African sabertooth is probably that from Elandsfontein (Hendey, 1974), with an estimated age between 700 and 500 kyr.

Sabertooths, "false" sabertooths, and cursorial ("hunting") hyenas are all represented at Swartkrans, where Brain (1981) suggests that either "false" sabertooths or cursorial hyenas were the major bone accumulators in Member 1. Both probably inhabited caves, and either could have produced the pattern of "robust" australopithecine skeletal part representation in Member 1, where limb bones seem to have been crunched beyond recognition and only skulls were left more or less intact. If either "false" sabertooths or cursorial hyenas were responsible, then the remarkable abundance of "robust" australopithecines in Member 1 (approximately 27% of all individual animals) suggests that "robust" australopithecines were among their favored prey. It follows that interference competition may have shielded "robust" australopithecines from other, potentially more devastating predators and that when the "false" sabertooths or cursorial hyenas vanished, other predators may literally have hunted the "robust" australopithecines to extinction (Walker, 1984). This hypothesis would be strengthened if it were possible to cite an historic analogy, wherein a prey species became extinct when one of its major predators was finally displaced by a long-time competitor, and if it could be shown that "robust" australopithecine and sabertooth/"false" sabertooth or cursorial hyena numbers declined together. Currently, however, it appears that most sabertooths/"false" sabertooths and cursorial hyenas disappeared before 1.6 Myr, long before the "robust" australopithecines.

The argument that *Homo* was responsible for "robust" australopithecine extinction is less intricate, but no less circumstantial. The "robust" australopithecines certainly survived several important events in the early evolution of *Homo*, including the emergence of *H. erectus* at roughly 1.6 Myr (Walker *et al.*, 1982) and the appearance of the Acheulean Industrial Tradition, at roughly 1.4 Myr (Isaac, 1984). However, future research may suggest that these events precipitated a progressive ebb in "robust" australopithecine numbers. In this context, it may be germane that the emergence of *H. erectus* and the Acheulean coincided closely with a significant reduction in terrestrial cercopithecoid diversity (Delson, 1984). The eventual demise of the "robust" australopithecines may have been caused by continuing behavioral advances and niche expansion in *H. erectus* after 1.4 to 1.2 Myr. Arguably, these advances are signaled by an enhanced ability to acquire meat and by the mastery of fire, and more certainly, by enlargement of the hominid ecumene, including the first colonization of temperate environments (Clark and Kurashina, 1979; Clark and Harris, 1985; Isaac, 1984). If *H. erectus* was responsible for "robust" australopithecine extinction, the mechanism was probably not direct hunting or violence. More likely, it was competitive exclusion, perhaps through the growing ability of *H. erectus* to exploit shared (?plant) food resources more efficiently.

Summary

The date of "robust" australopithecine extinction is not firmly established, but a time between 1.2 Myr and perhaps 0.7 Myr seems most probable. Among possible causes, three stand out: (1) a likely increase in the aridity associated with glacial intervals in Africa after 0.9 Myr; (2) the decline and disappearance of sabertooth/"false" sabertooth cats or cursorial ("hunting") hyenas, one or both of which may have preferentially preyed on "robust" australopithecines and thus shielded them from other, more devastating predators; and (3) on-going biological and behavioral evolution in *Homo*, whose expanding ecological niche perhaps ultimately absorbed that of the "robust" australopithecines, eliminating them through competitive exclusion. A denser, more precisely dated fossil record will be necessary to choose among these alternatives.

Acknowledgments

I thank F. E. Grine for inviting me to participate in the "robust" australopithecine workshop and F. E. Grine, F. C. Howell, E. Trinkaus, R. H. Tuttle, T. P. Volman, and C. A. Wolf for helpful comments on a draft of this paper.

References

Bishop, W. W., Pickford, M. and Hill, A. (1975). New evidence regarding the Quaternary geology, archaeology and hominids of Chesowanja, Kenya. *Nature*, 258: 204–208.

Brain, C. K. (1981). *The hunters or the hunted? An introduction to African cave taphonomy.* University of Chicago Press, Chicago.

Brain, C. K. (1982). The Swartkrans site: stratigraphy of the fossil hominids and a reconstruction of the environment of early *Homo*. In H. deLumley (ed.), *L'Homo erectus et la place de l'homme de Tautavel parmi les hominidés fossiles*, pp. 676–706. Centre National de la Recherche Scientifique, Nice.

Brain, C. K. (1985). Cultural and taphonomic comparisons of hominids from Swartkrans and Sterkfontein. In E. Delson (ed.), *Ancestors: the hard evidence*, pp. 72–75. Alan R. Liss, New York.

Brown, F. H. and Feibel, C. S. (1986). Revision of lithostratigraphic nomenclature in the Koobi Fora region, Kenya. *J. Geol. Soc. Lond.*, 143: 297–310.

Brown, F. H., McDougall, I., Davies. T. and Maier, R. (1985). An integrated Plio-Pleistocene chronology for the Turkana Basin. In E. Delson (ed.), *Ancestors: the hard evidence*, pp. 82–90. Alan R. Liss, New York.

Butzer, K. W. (1978). Climate patterns in an un-glaciated continent. *Geog. Mag.*, 51: 201–208.

Butzer, K. W. (1984). Late Quaternary environments in South Africa. In J. C. Vogel (ed.), *Late Cainozoic palaeoclimates of the Southern Hemisphere*, pp. 235–264. A. A. Balkema, Rotterdam.

Chavaillon, J. (1982). Position chronologique des hominidés fossiles d'Éthiopie. In H. de Lumley (ed.), *L'Homo erectus et la place de l'homme de Tautavel parmi les hominidés fossiles*, pp. 766–797. Centre National de la Recherche Scientifique, Nice.

Clark, J. D. and Harris, J. W. K. (1985). Fire and its roles in early hominid lifeways. *Afr. Archaeol. Rev.*, 3: 3–27.

Clark, J. D. and Kurashina, H. (1979). Hominid occupation of the east-central highlands of Ethiopia in the Plio-Pleistocene. *Nature*, 282: 33–39.

Clark, J. D., Asfaw, B., Assefa, G., Harris, J. W. K., Kurashina, H., Walter, R. C., White, T. D. and Williams, M. A. J. (1984). Palaeoanthropological discoveries in the Middle Awash Valley, Ethiopia. *Nature*, 307: 423–428.

Coppens, Y. (1980). The differences between *Australopithecus* and *Homo:* preliminary conclusions from the Omo Research Expedition's studies. In L.-K. Königsson, (ed.), *Current argument on early man*, pp. 207–225. Pergamon Press, Oxford.

Deacon, H. J. and Thackeray, J. F. (1984). Late Pleistocene environmental changes and implications for the archaeological record in southern Africa. In J. C. Vogel (ed.), *Late Cainozoic Palaeoclimates of the Southern Hemisphere*, pp. 375–390. A. A. Balkema, Rotterdam.

Deacon, J. and Lancaster, N. J. (1988). *Late Quaternary palaeoenvironments of southern Africa.* Oxford University Press, Oxford.

Delson, E. (1984). Cercopithecid biochronology of the African Plio-Pleistocene: correlation among eastern and southern hominid-bearing localities. *Cour. Forsch. Inst. Senckenberg*, 69: 199–218.

Geraads, D. (1985). La faune des gisements de Melka Kunturé (Éthiopie). *L'Environnement des hominidés au Plio-Pléistocène*, pp. 165–174. Paris, Masson.

Geraads, D., Hublin, J.-J., Jaeger, J.-J., Tong, H., Sen, S. and Tourbeau, P. (1986). The Pleistocene hominid site of Ternifine, Algeria: new results on the environment, age, and human industries. *Quat. Res.*, 25: 380–386.

Gowlett, J. A. J., Harris, J. W. K., Walton, D. and Wood, B. A. (1981). Early archaeological sites, hominid remains and traces of fire from Chesowanja, Kenya. *Nature*, 294: 125–129.

Guillien, Y. and Laplace, G. (1978). Les climates et les hommes en Europe et en Afrique septentrionale de 28000 B.P. à 10000 B. P. *Bull. Assoc. Fr. Quat.*, 4: 187–193.

Hendey, Q. B. (1974). The Late Cenozoic Carnivora of the south-western Cape Province. *Ann. S. Afr. Mus.*, 63: 1–369.

Isaac, G. Ll. (1984). The archaeology of human origins: studies of the Lower Pleistocene in East Africa: 1971–1981. In F. Wendorf and A. E. Close (eds.), *Advances in world archaeology*, vol. 3 pp. 1–87. Academic Press, New York.

Kalb, J. E., Jolly, C. J., Mebrate, A., Tebedge, S., Smart, C., Oswald, E. B., Cramer, D., Whitehead, P., Wood, C. B., Conroy, G. C., Adefris, T., Sperling, L. and Kana, B. (1982). Fossil mammals and artefacts from the Middle Awash Valley, Ethiopia. *Nature, 298*: 25–29.

Klein, R. G. (1977). The ecology of early man in southern Africa. *Science, 197*: 115–126.

Klein, R. G. (1983). Palaeoenvironmental implications of Quaternary large mammals in the fynbos region. *S. Afr. Nat. Sci. Program. Rep., 75*: 116–138.

Klein, R. G. (1984). Mammalian extinctions and stone age people in Africa. In P. S. Martin and R. G. Klein (eds.), *Quaternary extinctions: a prehistoric revolution*, pp. 553–573. University of Arizona Press, Tucson.

Leakey, M. D. (1980). Early man, environment and tools. In L.-K. Königsson (ed.), *Current argument on early man*, pp. 114–133. Pergamon Press, Oxford.

Leakey, M. D. and Hay, R. L. (1982). The chronological position of the fossil hominids of Tanzania. In H. de Lumley (ed.), *L'Homo erectus et la place de l'homme de Tautavel parmi les hominidés fossiles*, pp. 753–765. Centre National de la Recherche Scientifique, Nice.

Martin, P. S. and Klein, R. G. (eds.) (1984). *Quaternary extinctions: a prehistoric revolution*. University of Arizona Press, Tucson.

Prell, W. L., Gardner, J. V. et al. (1982). Oxygen and carbon isotope stratigraphy for the Quaternary of Hole 502B: evidence for two modes of isotopic variability. *Init. Repts. DSDP, 68*: 455–464.

Roberts, N. (1984). Pleistocene environments in time and space. In R. Foley (ed.), *Hominid evolution and community ecology*, pp. 25–53. Academic Press, London.

Shackleton, N. J. and Opdyke, N. D. (1977). Oxygen-isotope and palaeomagnetic evidence for early Northern Hemisphere glaciation. *Nature, 270*: 216–219.

Shackleton, N. J., Backman, J., Zimmerman, H., Kent, D. V., Hall, A., Roberts, D. G., Schnitker, D., Baldauf, J. G., Desprairies, A., Homrighausen, R., Huddlestun, P., Keene, J. B., Kaltenback, A. J., Krumsiek, K. A. O., Morton, A. C., Murray, J. W. and Westberg-Smith, J. (1984). Oxygen isotope calibration of the onset of ice-rafting and history of glaciation in the North Atlantic region. *Nature, 307*: 620–623.

Thunell, R. C. and Williams, D. F. (1983). The stepwise development of Pliocene-Pleistocene paleoclimatic and paleoceanographic conditions in the Mediterranean: oxygen isotopic studies of DSDP sites 125 and 132. *Utrecht Microbot. Bull., 30*: 111–127.

Vogel, J. C. (1985). Further attempts at dating the Taung tufas. In P. V. Tobias (ed.), *Hominid evolution: past, present and future*, pp. 189–194. Alan R. Liss, New York.

Vrba, E. S. (1982). Biostratigraphy and chronology, based particularly on Bovidae of southern hominid-associated assemblages: Makapansgat, Sterkfontein, Taung, Kromdraai, Swartkrans; also Elandsfontein (Saldanha), Broken Hill (now Kabwe) and Cave of Hearths. In H. de Lumley (ed.), *L'Homo erectus et la place de l'homme de Tautavel parmi les hominidés fossiles*, pp. 706–752. Centre National de la Recherche Scientifique, Nice.

Vrba, E. S. (1985). Ecological and adaptive changes associated with early hominid evolution. In E. Delson (ed.), *Ancestors: the hard evidence*, pp. 63–71. Alan R. Liss, New York.

Walker, A. C. (1981). The Koobi Fora hominids and their bearing on the origins of the genus *Homo*. In B. A. Sigmon and J. S. Cybulski (eds.), *Homo erectus—papers in honor of Davidson Black*, pp. 193–215. University of Toronto Press, Toronto.

Walker, A. C. (1984). Extinction in hominid evolution. In M. H. Nitecki (ed.), *Extinctions*, pp. 119–152. University of Chicago Press, Chicago.

Walker, A. C. and Leakey, R. E. F. (1978). The hominids of East Turkana. *Sci. Amer., 239*: 54–66.

Walker, A. C., Zimmerman, M. R. and Leakey, R. E. F. (1982). A possible case of hypervitaminosis A in *Homo erectus*. *Nature, 296*: 248–250.

Summary Comments VI

IV

Evolutionary History of the "Robust" Australopithecines: A Summary and Historical Perspective

30

FREDERICK E. GRINE

This volume was compiled in order to provide a vehicle for the presentation of recent research pertaining to the Plio-Pleistocene hominids commonly referred to as "robust" australopithecines. While these creatures are fascinating in their own right, they also provide for an enhanced appreciation of the questions associated with the evolution of our own lineage. The "robust" australopithecines were, after all, probably our closest cousins.

The papers that comprise this volume address a number of issues relating to "robust" australopithecine variability and taxonomy, the developmental and adaptive aspects of "robust" australopithecine morphology, locomotor behavior and body size, paleoecologic parameters such as habitat preference and diet, the timing and causative factors underlying the origin(s) of the "robust" australopithecines, their phylogenetic relationships, possible ecological interactions with early members of the genus *Homo,* and the timing and possible causes of "robust" australopithecine extinction.

While the questions themselves are not novel, the papers that comprise this volume indicate that considerable new evidence has been brought to bear upon these issues in recent years. Previous perceptions about "robust" australopithecine natural history certainly have been altered and, as might be expected, the pictures have become more, rather than less complex. While it is neither possible nor desirable to summarize all that has been written (even that which has been written here) about these early hominids, certain issues emerge that should perhaps be viewed from an historic perspective.

What Is a "Robust" Australopithecine?

The first early hominid specimen to be recovered from ancient African sediments—the partial juvenile skull from the Buxton Limeworks at Taung—was described by Dart (1925), who designated it the type of a new genus and species, *Australopithecus africanus*. Dart's claims for the evolutionary status of *Australopithecus* did not go unchallenged, and despite the early recognition of the fundamentally hominid features of its craniofacial skeleton and dentition (Broom, 1925a,b, 1929; Sollas, 1926; Dart, 1929, 1934; Romer, 1930), many workers refused to accept that *Australopithecus* was in any way relevant to human evolution. Much of the early history of interpretation of African fossils revolved around the struggles by Dart and Broom to convince the paleoanthropological community that these were the remains of human cousins if not direct forebears. When Broom (1936) obtained the first hominid specimen from the site of Sterkfontein, he was convinced of its close affinities to the Taung specimen and thus named it *Australopithecus transvaalensis*. Two years later, Broom (1938) transferred the Sterkfontein species to a new genus, *Plesianthropus,* based upon an incomplete mandibular symphysis which he considered to differ from that of the Taung specimen.

In 1938 the first hominid specimen from the site of Kromdraai was obtained. Broom (1938) was impressed by the facial and dental differences between it and the Sterkfontein fossils such that he made it the type of a new taxon, *Paranthropus robustus*. (Given 50 years hindsight, Broom's choice of a nomen is undoubtedly the most prophetically appropriate to have been employed in the field of hominid paleontology.) In 1948 the first hominid fossil was recovered from the site of Swartkrans; this was attributed by Broom (1949) to a second species of *Paranthropus*, *P. crassidens*. He noted that the teeth were morphologically similar to, albeit larger than those of the Kromdraai specimen. A year before the Swartkrans find, the first hominid specimen was recovered from the limeworks dumps at Makapansgat; this was designated by Dart (1948) as a new species of *Australopithecus*, *A. prometheus*.

Gregory and Hellman (1939) established the subfamily Australopithecinae to accommodate the fossils from Taung, Sterkfontein, and Kromdraai and to delineate them from "true" humans (including fossils), the Homininae. The term "australopithecine," a vernacular form of the taxonomic rank name, carries with it the implicit recognition of a subfamilial distinction between the Australopithecinae and Homininae, although most authors who use the term australopithecine would probably deny that such a distinction is warranted. That is, most would recognize an australopithecine *grade* but not an australopithecine *clade*.

The level of taxonomic distinction that should be accorded the fossils from the various South African sites was considered by Simpson (1945), who opined that *Paranthropus* was, at most, deserving of subgeneric rank within *Australopithecus*. Mayr (1950) argued that both *Australopithecus* and *Paranthropus* represented junior synonyms of *Homo*, although he later wrote that "the tremendous evolution of the brain since *Australopithecus* permitted man to enter a niche so completely different that generic separation is fully justified . . . whether or not one wants to admit a second genus *Paranthropus*, is largely a matter of taste" (Mayr, 1963).

Broom (1950), on the other hand, recognized some four genera and five species from the South African sites. Arguing the "case of the splitter of the South African ape-men" he placed these taxa into three distinct subfamilies: Australopithecinae *(Australopithecus africanus* and *Plesianthropus transvaalensis)*, Paranthropinae *(Paranthropus robustus* and *Paranthropus crassidens)*, and Archanthropinae *[(Australopithecus) prometheus]*. Robinson (1954a) maintained the generic separation of *Australopithecus* and *Paranthropus*, but he recognized only subspecific level distinction between the various sites; all were grouped within the subfamily Australopithecinae by him.

In 1959 a large-toothed cranium was recovered from Bed I, Olduvai Gorge; this was designated by Leakey (1959) as the type of a novel taxon, *Zinjanthropus boisei*. Robinson (1960) was quick to recognize the close morphological resemblance of this specimen to those from Kromdraai and Swartkrans; thus, he attributed the Olduvai specimen to *Paranthropus boisei*. Craniodental fossils that are undoubtedly referrable to this species have been recovered from a variety of sites in Tanzania [e.g., the Humba Formation at Peninj near Lake Natron (Leakey and Leakey, 1964; Isaac, 1965)], Kenya [e.g., the Chemoigut Formation near Chesowanja (Carney *et al.*, 1971; Walker, 1972; Hooker and Miller, 1979) and the Koobi Fora Formation east of Lake Turkana (Leakey and Leakey, 1978)], and southern Ethiopia [e.g., Shungura Formation Members E, F, G and L (Howell, 1978)].

Workers who had argued that *Paranthropus* should be regarded as a junior synonym of *Australopithecus* prior to the discoveries in East Africa were joined by others following the 1959 find (Campbell, 1963; Le Gros Clark, 1964; Pilbeam and Simons, 1965) and Tobias's (1967) comparative analysis of the specimen (Brace, 1967, 1969, 1973; Campbell, 1972, 1974; Pilbeam and Gould, 1974; Wolpoff, 1974). The studies by Tobias (1967), Wolpoff (1974) and others that have stressed a fundamentally similar grade of organization among "australopithecines," have resulted in *A. africanus*, *P. robustus* and *P. boisei* being regarded by most students of hominid evolution as scaled variants on a single theme. That is, they are simply *"gracile"* and *"robust"* versions of the australopithecine grade of organization.

The use of the terms "gracile" and "robust" with reference to Plio-Pleistocene hominids has become widespread in the paleoanthropological literature, although their use has often been misleading and confused. These terms are most often taken to connote size—for example, Tobias (1973a) writes of "a robust or large ape-man"—and as such are generally synonymous with *Australopithecus* ("gracile") and *Paranthropus* ("robust" and/or "hyper-robust"). However, the "gracile" australopithecines are not invariably smaller than the "robust" ones in body size

(McHenry, this volume, chapter 9; Jungers, this volume, chapter 7), nor do the "robust" forms invariably possess larger teeth or more robust mandibles. Nevertheless, in their most common usage, the terms "gracile" and "robust" are employed as synonyms for *Australopithecus* and *Paranthropus*, respectively. Their use in this context, however, can lead to confusion. For example, in reference to KNM-ER 732, an undoubted female cranium of *P. boisei*, Leakey (1971) wrote that "the concept of the gracile australopithecine being ancestral to *Homo* in the lower Pleistocene requires careful re-examination on the basis that at East Rudolf the gracile variety may prove to be the female of the robust form." Some workers (e.g., Swedlund, 1974) have used the term "gracile" to refer to both *Australopithecus* and members of the genus *Homo*.

Because of the confusion that they can cause, and because of their implicit (albeit possibly erroneous) body size connotations, the terms "gracile" and "robust" should be dropped from the paleoanthropological vocabulary. This was suggested at least seven years ago, although it was noted at the time that "because these terms are so ensconced in the literature and therefore so familiar, it seems doubtful that they will fall into disuse in the near future" (Grine, 1981: 204).

How Many "Robust" Australopithecine Taxa?

This question relates to species identity (How many species can be identified in the available fossil record?) and phylogeny (Are the putative ancestors of "robust" australopithecines also themselves "robust?"), although for the moment only the former will be considered. This is essentially a problem of alpha taxonomy; in this context, "robust" australopithecine is taken to be a synonym for *Paranthropus*.

As noted above, Broom (1950) held that the specimens from Kromdraai and Swartkrans warranted specific separation as *P. robustus* and *P. crassidens* respectively, while Robinson (1954a) maintained that the observed morphological differences between them attested to only subspecific separation. Most workers have either followed Robinson or have disregarded the differences between the Kromdraai and Swartkrans samples. However, Howell (1978) iterated features suggestive of a species-level distinction, and other studies (Grine, 1982; 1985a,b, this volume, chapter 14; Jungers and Grine, 1986) have cited a number of subtle dental differences that lend support to the taxonomic distinctiveness of the Kromdraai and Swartkrans samples.

Numerous specimens from Olduvai Beds I and II, the Humba and Chemoigut Formations, Members E, F, G and L of the Shungura Formation, and the Burgi, KBS and Chari Members of the Koobi Fora Formation have been been confidently referred to as *P.* (or *Australopithecus*) *boisei*. While some workers (e.g., Rak, 1983) have argued that these fossils represent a species that was distinct from that represented by southern African remains, others envision a closer relationship between these materials. Thus, Tobias (1975) suggested that the eastern and southern African specimens represent "a superspecies—*A. robustus* comprising two semispecies, *A. robustus* and *A. boisei*;" Clarke (1977) and Bilsborough (1978) postulated an even closer relationship by viewing them as subspecies, while White *et al.* (1981) employed the designation *A. robustus/boisei* to signify a very close relationship.

Although many specimens have been universally accepted as being attributable to *P. boisei*, the allocation of some fossils has been questioned. For example, Rak and Howell (1978) have argued that the L 338y-6 partial juvenile cranium from Member E (Unit E-3) of the Shungura Formation represents this taxon, whereas Holloway (1981) has opined that it is a "gracile australopithecine, more closely related to *A. africanus* or *A. afarensis*" in view of its lack of an occipital/marginal sinus, the branching pattern of its middle meningeal vessel, its low endocranial volume, and the rounded form of its cerebellar lobes (see also Saban, 1983; Holloway, this volume, chapter 5; Walker and Leakey, this volume, chapter 15). At the same time, isolated deciduous teeth from Member D of the Shungura Formation that have been attributed to *P. boisei* by Grine (1985a) together with other attributions from sub-G members have been thrown into question by Suwa (this volume, chapter 13).

Perhaps most significantly, the KNM-WT 17000 cranium from the Lomekwi Member of the Nachukui Formation (equivalent of Shungura Member C, Unit C-9) has been attributed to *P.*

boisei (actually *A. boisei*) by Walker *et al.* (1986) and Walker and Leakey (this volume, chapter 15), although they noted that differences between it and other *P. boisei* crania may warrant specific distinction. In this context, they opined that a mandible (KNM-WT 16005) from the Lokaleli Member (equivalent to middle Member D) is a female specimen of the same species as represented by the male KNM-WT 17000 cranium and that comparison of the former with the edentulous Omo 18-1967-18 mandible from Member C of the Shungura Formation leads to the conclusion that all three represent the same taxon.

The edentulous Omo 18-1967-18 mandible from Unit C-8 is stratigraphically slightly older than Tuff D at 2.52 Myr (Brown *et al.*, 1985), and thus nearly equivalent to the age of the KNM-WT 17000 cranium. This mandible was described by Arambourg and Coppens (1967), who ascribed it to a provisional taxon, *Paraustralopithecus aethiopicus*. The name was accordingly not available (Article 15 ICZN) until the publication of a paper the following year in which they provided a nearly verbatim diagnosis of the taxon and a nonprovisional citation of the nomen. Arambourg and Coppens (1968) cited several features of the corpus of Omo 18-1967-18 that appeared to differ from jaws of *P. robustus* and the *single known mandible* of *P. boisei* from the Humba Formation in Tanzania. Coppens (1970), following Arambourg and Coppens (1967, 1968) noted the similarities of the Omo 18-1967-18 mandible to *P. robustus* and *P. boisei* homologues although the Omo specimen was regarded as being more primitive, whereas Howell and Coppens (1976) noted closer affinities with *A. africanus* specimens from southern Africa. This apparently equivocal placement of the Omo mandible found advocacy with Johanson and White (1979: fig. 7), who placed it *temporally* between *A. africanus* and later members of their perceived "robust" lineage. Wood and Chamberlain (1986) have attributed the Omo 18-1967-18 mandible to *A. boisei*.

While considerable retrospective emphasis has been placed on the supposedly diagnostic features of the Omo 18-1967-18 mandible (Kimbel *et al.*, this volume, chapter 16), the allocation of the KNM-WT 17000 cranium to the same species as the Omo 18-1967-18 mandible, *Paraustralopithecus* (= *Australopithecus* or *Paranthropus* depending upon preference) *aethiopicus*, hangs by an extremely tenuous thread of logic. Indeed, the bottom line espoused by some colleagues for the attribution of KNM-WT 17000 to *P. aethiopicus* is remarkably reminiscent of the single species hypothesis: "if KNM-WT 17000 is not *A. aethiopicus*, then a new species would have to be erected for it. This would imply the existence of two essentially sympatric and synchronic, primitive hominid taxa during the time period represented by Members C and (lower) D of the Shungura Formation" (Kimbel *et al.*, this volume, chapter 16).

Despite the fact that the attribution of KNM-WT 17000 to the same taxon as the Omo 18-1967-18 mandible is rather spurious, there can be little doubt that KNM-WT 17000 is *not* attributable to *P. boisei*. Kimbel *et al.* (this volume, chapter 16) have clearly spelled out numerous morphological features in which the KNM-WT 17000 cranium differs from acknowledged *P. boisei* crania (e.g., OH 5, KNM-ER 406) and from *P. robustus* and *P. crassidens* specimens, and in which it resembles the more primitive *A. afarensis* condition. Features discussed also by Holloway (this volume, chapter 5) and Walker and Leakey (this volume, chapter 15) leave little doubt that KNM-WT 17000 represents a novel species. Suwa (this volume, chapter 13), who has offered dental evidence that specimens from below Member G of the Shungura Formation differ from those definitely attributable to *P. boisei*, provides support for the presence of a distinct species of "robust" australopithecine in eastern Africa between c 2.6 and 2.3 Myr BP. It is noteworthy that in a variety of facial and basicranial features (e.g., the presence of a "zygomaxillary step"; the form of the entoglenoid process), the KNM-WT 17000 cranium resembles those from Kromdraai and Swartkrans (Kimbel *et al.*, this volume, chapter 16) as opposed to those attributed to *P. boisei*.

In addition, Dean (this volume, chapter 2) has suggested that mandibular metrics may indicate yet another (third) species in the currently recognized *P. boisei* hypodigm. Although this opinion is difficult to support at present, it highlights the fact that morphometrical variability within the *P. boisei* hypodigm warrants reexamination. At present, the existence of *at least* four species of "robust" australopithecine must be afforded serious consideration: *P. robustus, P. crassidens, P. "aethiopicus"* and *P. boisei*. The possibility of more should not be dismissed out of hand.

While it is becoming increasingly evident that there were a number of "robust" australopithecine species, their relationships have been called into question by recent discoveries.

Frederick E. Grine

Do "Robust" Australopithecines Form a Clade?

As noted above, a very close relationship between the Kromdraai and Swartkrans forms is universally agreed; indeed, almost all workers view them as belonging to a single species. Similarly, it is generally agreed that the earlier and later eastern African "robust" forms (i.e., as represented by KNM-WT 17000 and KNM-ER 406, respectively) are very closely related. They are regarded either as members of a single species (Walker *et al.*, 1986; Walker and Leakey, this volume, chapter 15) or belonging to two species in a single lineage (Kimbel *et al.*, this volume, chapter 16). In a word, the closeness of relationship of "robust" australopithecines *within* the separate eastern and southern geographic regions is a point of consensus. There is, however, less agreement about the relationship *between* the eastern and southern African fossils (Walker *et al.*, 1986; Walker and Leakey, this volume, chapter 15; Kimbel *et al.*, this volume, chapter 16; Wood, this volume, chapter 17).

Prior to the discovery of KNM-WT 17000 there was seemingly no doubt that *P. robustus* (or both *P. robustus* and *P. crassidens*) and *P. boisei* were very closely related, and many workers regarded them as warranting only semi- or subspecies distinction. Wood (this volume, chapter 17), however, has provided a timely reminder that evolutionary convergence—especially in functional complexes such as the masticatory system—may lead to false phylogenetic conclusions. Although Wood (this volume, chapter 17) observes that acceptance of the "robust" australopithecines as a monophyletic group is the most parsimonious hypothesis, he notes that there is no really compelling evidence to reject a diphyletic hypothesis out of hand. Thus, the southern African "robust" australopithecines may have evolved from *A. africanus* (or an *A. africanus*-like ancestor), while the eastern "robust" forms may have evolved from *A. afarensis* (or an *A. afarensis*-like ancestor), in which case their morphologic resemblances would have come about through functional convergence.

As noted by Wood (this volume, chapter 17), many of the features that appear to unite the southern and eastern "robust" forms are directly related to mastication (e.g., the relative hypertrophy of the postcanine dentition, the relative anterior placement of the zygomatic process, "dished" midfacial architecture, and perhaps the high hafting of the face onto the neurocranium with the attendant frontal trigone). Other morphological similarities may be indirectly related to mastication by being developmental concomitants of craniodental features that are intimately associated with chewing (e.g., the presence of a naso-alveolar clivus that grades smoothly into the nasal cavity floor, the insertion of the vomer at the anterior nasal spine, nearly coincident nasion and glabella, and coronally orientated petrous bones).

Nevertheless, there are a number of characters or character states that are shared by acknowledged *P. robustus* and *P. boisei* specimens that are difficult to envisage as necessarily being associated with adaptations for enhanced masticatory capability (e.g., the vertically inclined and deep tympanic plate with its strong vaginal process, the presence of an occipital/marginal sinus, the unique branching pattern of the middle meningeal vessels, and other aspects of endocranial form). Indeed, the KNM-WT 17000 cranium with its obvious facial, dental and ectocranial structures related to masticatory capability differs from later "robust" australopithecines in eastern and southern Africa in a number of endocranial characters [e.g., as noted by Walker and Leakey (this volume, chapter 15) and Holloway (this volume, chapter 5), it lacks the characteristic "robust" middle meningeal vessel pattern, and the cerebellar lobes display greater lateral and posterior protrusion].

The fact that some morphological features shared by later "robust" australopithecines from both eastern and southern Africa are not developed in KNM-WT 17000 suggests that they are not necessary concomitants of masticatory adaptation. Since these features are unlikely to be functionally convergent, the likelihood that they are synapomorphies is greatly increased. Moreover, developmental similarities such as those related to the timing of tooth calcification and emergence (Dean, 1985, 1986, this volume, chapter 2; Smith, 1986) and the formation of very thick tooth enamel (Grine and Martin, 1988; also, this volume, chapter 1), certainly appear to bear testament to common ancestry.

These characters appear to support Wood's (this volume, chapter 17) conclusion that while a case for convergent evolution may be stronger than many would have suspected, "robust" australopithecine monophyly continues to be the strongest hypothesis. Given this, there are valid systematic grounds for the use of the taxonomic nomen *Paranthropus;* indeed, it seems to be the application of the name *Australopithecus* that requires reevaluation.

What Are the Systematic Relationships of the "Robust" Australopithecines?

Questions about phylogeny seem to lead to the most vigorous debates in the field of hominid paleontology. Early on, Robinson (1954b, 1961) argued that the substantial morphological differences between *Australopithecus* and *Paranthropus* attested to their having belonged to divergent evolutionary lineages that had been separated for a considerable period of time. He considered that *Australopithecus*, specifically *A. africanus*, was a tool-making omnivore that had attained a "hominine" grade (Robinson, 1961, 1963), and he later (1965, 1967, 1972) proposed that because *A. africanus* was an early member of the *Homo* lineage, the name *Australopithecus* should be abandoned as a junior synonym of *Homo*. Thus, according to Robinson (1965, 1967, 1968, 1972), *Paranthropus* and *Homo* shared a common ancestor; he also opined that *Paranthropus* exhibited the primitive pattern of craniofacial morphology from which *Homo* evolved (see also Robinson and Steudel, 1973).

Robinson's suggestions regarding the presence of separate *Paranthropus* and *Homo* lineages, with the latter containing fossils generally attributed to *A. africanus*, were met with widespread scepticism. For example, Tobias (1967, 1968) argued that the "gracile" and "robust" australopithecines were "very much more closely related" than Robinson (and Broom) had averred. Tobias indicated that *A. africanus*, the *A. boisei* to *A. robustus* lineage, and *Homo* evolved from an unidentified "hypothetical ancestral australopithecine," and while he held that "we cannot exclude the chance of crossing between *A. africanus* and members of the *A. boisei* to *A. robustus* line" (1967: 244), he also envisaged *A. africanus* as exhibiting greater morphological and behavioral similarities to humans. Later, Tobias (1978a,b, 1980a,b, this volume, chapter 19) argued that the "robust" australopithecine (*A. robustus* and *A. boisei*) and *Homo* lineages diverged from a common ancestor in the form of *A. africanus*, and that the Hadar and Laetoli fossils represented subspecific variants of this taxon.

Studies by Dean (1986) and Skelton *et al.* (1986) provided some *apparent* support for Tobias's hypothesis regarding *A. africanus* as the last common ancestor of the "robust" australopithecine and human lineages. It must be stressed, however, that the study by Skelton *et al.* (1986) employed *only* data gleaned from the paper by White *et al.* (1981) in which the "robust" affinities of *A. africanus* were strongly emphasized. Skelton *et al.* (1986) recognized correctly that the *actual* last common ancestor could not be identified in the known *A. africanus* fossil sample inasmuch as these specimens lack a number of hypothesized "robust" australopithecine and *Homo* synapomorphies (e.g., coronally orientated petrous pyramids). Tobias (this volume, chapter 19) has responded that the Taung specimen itself may have been a member of the 'derived' *A. africanus* population that was the last common ancestor of these two lineages; but this newest claim for the Taung skull is difficult to reconcile with his earlier (1973b, 1978c) opinion that it was a late-surviving "robust" australopithecine.

Studies by Johanson and White (1979), White *et al.* (1981), and Rak (1983) suggested the divergence of the *Homo* and "robust" australopithecine *(A. africanus–A. robustus–A. boisei)* lineages from *A. afarensis*. This argument was criticized by Tobias (1978b, 1980a), who maintained that the *A. afarensis* specimens from Hadar and Laetoli could be accommodated within the hypodigm of *A. africanus*, albeit as separate subspecies—the Hadar fossils as *A. africanus aethiopicus* and the Laetoli remains as *A. africanus tanzaniensis*. Olson (1981, 1985), on the other hand, argued that the Laetoli and some of the Hadar fossils (e.g., those from site 333) belonged to one species, *Paranthropus africanus*, while the remainder of the Hadar sample represented another taxon, *Homo aethiopicus*. (The use of the trivial nomen *aethiopicus* by both Tobias and Olson has provided for added confusion because this trivial name was used over a decade earlier by Arambourg and Coppens.)

Olson's proposal that the Hadar and Laetoli samples represent two species has been disputed

(Grine, 1985a; Kimbel et al., 1985; White, 1985), and although this position has been championed by others [see Tobias (this volume, chapter 19) and Falk (this volume, chapter 4) for references], such studies have been inconsistent in their delineation of species within the relevant fossil samples, and they commonly have been based on fallacious characters and criteria. Indeed, during the course of the workshop, Olson stated that he no longer subscribed to the position that more than a single hominid species is represented by the Hadar and Laetoli fossils.

However, Olson's (1981, 1985) analyses of the Hadar and Laetoli fossils resulted in a taxonomic scheme that was consistent with his earlier (1978) support of Robinson's (1963, 1965, 1967, 1968, 1972) conclusion that the *Paranthropus* and *Homo* lineages had a remote ancestry. Olson's (1981, 1985) proposal that the Laetoli and at least some of the Hadar fossils showed "robust" australopithecine affinities found support in studies of occipital/marginal sinus distributions (Falk and Conroy, 1983). The notion that a phylogenetic relationship was discernible between the Hadar and "robust" australopithecine fossils has received some support from the cladistic studies by Wood and Chamberlain (1986) (cf. Chamberlain and Wood, 1987).

Finally, the KNM-WT 17000 cranium (Walker et al., 1986), which displays numerous synapomorphies with *P. robustus* and *P. boisei*, in conjunction with more primitive morphological conditions (shared with *A. afarensis*), is strongly suggestive of a phylogenetic link between *A. afarensis* and the "robust" australopithecines to the exclusion of *A. africanus*.

While it is abundantly evident that the phylogenetic relationships of the "robust" australopithecines still have yet to be satisfactorily resolved (Walker and Leakey, this volume, chapter 15; Kimbel et al., this volume, chapter 16; Grine, this volume, chapter 14), it would seem that *Paranthropus* can trace its origins to the late Pliocene at circa 2.6 Myr ago, and that *Australopithecus africanus* likely belonged to a separate lineage. It would appear that John Robinson was correct after all in his interpretation of the distinctiveness of *Paranthropus*.

Questions about "Robust" Australopithecine Adaptations

Although there is strong evidence to suggest that *Paranthropus* evolved in the late Pliocene, about 2.6 Myr ago, from an *A. afarensis*-like ancestor, there is considerably less agreement about the factors that may have accounted for the divergent morphological path taken by this lineage. Indeed, some 35 years after the publication of Robinson's (1954b) classic "dietary hypothesis" the dietary proclivities of *Paranthropus* species are still being debated. Studies of microwear on the deciduous and permanent teeth of the southern African specimens (Grine 1981, 1984a, 1986a, 1987a,b; Kay and Grine, this volume, chapter 26) indicate that *Paranthropus* ate hard food items. *A. africanus* wear is reminiscent of that of extant primates whose diets of leaves and/or fleshy fruits do not regularly include hard items. Walker (1981), on the other hand, has reported that *P. boisei* microwear lacks the characteristic pitting of hard object feeders. Although comparable quantitative microwear data for *P. boisei* have not been published yet, Walker indicated at the workshop that a preliminary study yielded sample statistics comparable to those that have been generated for the common chimpanzee.

It is, of course, possible that taphonomic factors alone may account for potential differences in the microwear patterns displayed by southern and eastern African species of *Paranthropus;* but it is equally possible that their dietary habits differed.

Vrba (1975, 1981, this volume, chapter 25) has provided compelling evidence that the fossil accumulations at the *A. africanus* sites were made under more wooded/closed conditions than those at the sites of Kromdraai and Swartkrans, although she noted elements of the Kromdraai assemblage that suggested well-watered conditions. Moreover, Brain (this volume, chapter 20) has suggested that conditions during the deposition of the Swartkrans Member 1 "Lower Bank" may have been somewhat wetter than during the accumulation of the *Paranthropus*-rich Member 1 "Hanging Remnant." The potential that the Sterkfontein Member 4 breccia may span a considerable period of time, or that it may contain heterochronic elements has been raised by several workers (e.g., Kimbel and White, this volume, chapter 11; Vrba, this volume, chapter 25; Delson, this volume, chapter 21). Thus, while it would seem that the *Paranthropus*-bearing breccias in South Africa attest to depositional environments that were more xeric than those prevalent during the accumulation of the *A. africanus* remains, it is increasingly likely that the mesic–xeric model put forward by Grine (1981) to explain the trophic adaptations of the "gracile" and "robust" australopithecines is overly simplistic. The habitat preferences of *Paranthropus* species in eastern Africa have yet to be resolved adequately (Shipman and Harris, this volume, chapter 23).

Nevertheless, there is evidence that *Paranthropus* emerges in the African fossil record coincident with major changes in the pattern of Antarctic glaciation at approximately 2.6 Myr, and that other faunal changes at this time attest to a significant climatic effect upon the African continent (Vrba, this volume, chapter 25; Prentice and Denton, this volume, chapter 24). Similarly, although the exact timing of the extinction of *Paranthropus* species has yet to be resolved adequately (Brain, this volume, chapter 20; White, this volume, chapter 27; Klein, this volume, chapter 29), it is possible that a second climatic perturbation of considerable magnitude at c 900 kyr may have been related to the demise of this lineage (Klein, this volume, chapter 29; Vrba, this volume, chapter 25). While Grine (1981, 1984b, 1985c, 1986b) has argued that ecological factors brought about by environmental changes are adequate to explain the evolutionary appearance of *Paranthropus* species, and that the masticatory structures of these species attest to novel food resource exploitation, ecologic interaction with early members of the genus *Homo* cannot be ruled out. While there is no evidence for ecologic competition (e.g., character displacement) in southern Africa between *P. crassidens* (or *P. robustus* depending upon preference) and *Homo erectus* (Grine 1984b, 1985c), Wolpoff (this volume, chapter 28) has argued that there is evidence for such divergence in the eastern African record between *P. boisei* and *H. erectus*. Indeed, Klein (this volume, chapter 29) has suggested that *Paranthropus* extinction may coincide with the spread of *H. erectus* out of Africa, and that this might attest to changes in resource exploitation by *H. erectus* that had profound effects upon its "robust" cousins.

References

Arambourg, C. and Coppens, Y. (1967). Sur la decouverte, dans le pleistocene inferieur de la vallee de l'Omo (Ethiopie), d'une mandibule d'Australopithecien. *C. R. Acad. Sci. Paris*, *265*: 589–590.
Arambourg, C. and Coppens, Y. (1968). Decouverte d'un Australopithecien nouveau dans les gisements de l'Omo (Ethiopie). *S. Afr. J. Sci.*, *64*: 58–59.
Bilsborough, A. (1978). Some aspects of mosaic evolution in hominids. *In* D. J. Chivers and K. A. Joysey (eds.), *Recent advances in primatology*, vol. 3, *Evolution*, pp. 335–350. Academic Press, New York.
Brace, C. L. (1967). *The stages of human evolution*. Prentice-Hall, Englewood Cliffs, N. J.
Brace, C. L. (1969). The australopithecine range of variation. *Amer. J. Phys. Anthropol.*, *31*: 255.
Brace, C. L. (1973). Sexual dimorphism in human evolution. *Yrbk. Phys. Anthropol.*, *16*: 50–68.
Broom, R. (1925a). On the newly discovered South African man-ape. *Nat. Hist.*, *25*: 409–418.
Broom, R. (1925b). Some notes on the Taungs skull. *Nature*, *115*: 569–571.
Broom, R. (1929). A note on the milk dentition of *Australopithecus*. *Proc. Zool. Soc. Lond.*, *1928*: 85–88.
Broom, R. (1936). A new fossil anthropoid skull from South Africa. *Nature*, *138*: 486–488.
Broom, R. (1938). The Pleistocene anthropoid apes of South Africa. *Nature*, *142*: 377–379.
Broom, R. (1949). Another new type of fossil ape-man *(Paranthropus crassidens)*. *Nature*, *163*: 57.
Broom, R. (1950). The genera and species of the South African fossil ape-men. *Amer. J. Phys. Anthropol.*, *8*: 1–13.
Brown, F. H., McDougall, I., Davies, T. and Maier, R. (1985). An integrated Plio-Pleistocene chronology for the Turkana Basin. *In* E. Delson (ed.), *Ancestors: the hard evidence*, pp. 82–90. Alan R. Liss, New York.
Campbell, B. G. (1963). Quantitative taxonomy and human evolution. *In* S. L. Washburn (ed.), *Classification and human evolution*, pp. 50–74. Aldine, Chicago.

Campbell, B. G. (1972). Man for all seasons. *In* B. G. Campbell (ed.), *Sexual selection and the descent of Man*, pp. 40–58. Aldine, Chicago.
Campbell, B. G. (1974). A new taxonomy of fossil man. *Yrbk. Phys. Anthrop.*, 17: 195–201.
Carney, J., Hill, A., Miller, J. A. and Walker, A. (1971). Late australopithecine from Baringo district, Kenya. *Nature*, 230: 509–514.
Chamberlain, A. and Wood, B. A. (1987). Early hominid phylogeny. *In* F. E. Grine, J. G. Fleagle and L. B. Martin (eds.), *Primate phylogeny*, pp. 119–133. Academic Press, London.
Clark, W. E. Le Gros (1964). *The fossil evidence for human evolution*. 2nd edition. University of Chicago Press, Chicago.
Clarke, R. J. (1977). The cranium of Swartkrans hominid, SK 847, and its relevance to human origins. Unpublished Ph.D. Thesis, University of the Witwatersrand, Johannesburg.
Coppens, Y. (1970). Les restes d'hominides des series inferieurs et moyennes des formations plio-villafranchiennes de l'Omo en Ethiopie. *C. R. Acad. Sci. Paris*, 271: 2286–2287.
Dart, R. A. (1925). *Australopithecus africanus*: the man-ape of South Africa. *Nature*, 115: 195–199.
Dart, R. A. (1929). A note on the Taungs skull. *S. Afr. J. Sci.*, 26: 648–658.
Dart, R. A. (1934). The dentition of *Australopithecus africanus*. *Folia Anat. Japon.*, 12: 207–221.
Dart, R. A. (1948). The Makapansgat proto-human, *Australopithecus prometheus*. *Amer. J. Phys. Anthropol.*, 6: 259–283.
Dean, M. C. (1985). The eruption pattern of the permanent incisors and first permanent molars in *Australopithecus (Paranthropus) robustus*. *Amer. J. Phys. Anthropol.*, 67: 251–259.
Dean, M. C. (1986). *Homo* and *Paranthropus*: similarities in the cranial base and developing dentition. *In* B. A. Wood, L. B. Martin and P. Andrews (eds.), *Major topics in primate and human evolution*, pp. 249–265. Cambridge University Press, Cambridge.
Falk, D. and Conroy, G. (1983). The cranial venous sinus system in *Australopithecus afarensis*. *Nature*, 306: 779–781.
Gregory, W. K. and Hellman, M. (1939). The dentition of the extinct South African man-ape *Australopithecus (Plesianthropus) transvaalensis* Broom. A comparative and phylogenetic study. *Ann. Tvl. Mus.*, 19: 339–373.
Grine, F. E. (1981). Trophic differences between "gracile" and "robust" australopithecines: a scanning electron microscope analysis of occlusal events. *S. Afr. J. Sci.*, 77: 203–230.
Grine, F. E. (1982). A new juvenile hominid (Mammalia; Primates) from Member 3, Kromdraai Formation, Transvaal, South Africa. *Ann. Tvl. Mus.*, 33: 165–239.
Grine, F. E. (1984a). Deciduous molar microwear of South African australopithecines. *In* D. J. Chivers, B. A. Wood and A. Bilsborough (eds.), *Food acquisition and processing in primates*, pp. 525–534. Plenum, New York.
Grine, F. E. (1984b). Ecological models of early hominid evolution: the doubtful role of competition. *Amer. J. Phys. Anthropol.*, 63: 167.
Grine, F. E. (1985a). Australopithecine evolution: the deciduous dental evidence. *In* E. Delson (ed.), *Ancestors: the hard evidence*, pp. 153–167. Alan R. Liss, New York.
Grine, F. E. (1985b). Dental trends in the australopithecines. *J. Anat.*, 140: 513.
Grine, F. E. (1985c). Was interspecific competition a motive force in early hominid evolution? *In* E. S. Vrba (ed.), *Species and speciation*, pp. 143–152. Tvl. Mus. Monogr. 4, Transvaal Museum, Pretoria.
Grine, F. E. (1986a). Dental evidence for dietary differences in *Australopithecus* and *Paranthropus*: a quantitative analysis of permanent molar microwear. *J. Hum. Evol.*, 15: 783–822.
Grine, F. E. (1986b). Ecological causality and the pattern of Plio-Pleistocene hominid evolution in Africa. *S. Afr. J. Sci.*, 80: 87–89.
Grine, F. E. (1987a). L'alimentation des Australopitheques d'Africa du Sud, d'apres l'etude des microtraces d'usure sur les dents. *L'Anthropoligie*, 91: 467–482.
Grine, F. E. (1987b). Quantitative analysis of occlusal microwear in *Australopithecus* and *Paranthropus*. *Scanning Microsc.*, 1: 615–630.
Grine, F. E. and Martin, L. B. (1988). New evidence for the distinctiveness of *Paranthropus*. *Amer. J. Phys. Anthropol.*, 75: 217–218.

Holloway, R. L. (1981). The endocast of the Omo 338y-6 juvenile hominid: gracile or robust *Australopithecus? Amer. J. Phys. Anthropol., 54*: 109–118.

Hooker, P. J. and Miller, J. A. (1979). K-Ar dating of the Pleistocene fossil hominid site at Chesowanja, North Kenya. *Nature, 282*: 710–712.

Howell, F. C. (1978). Hominidae. *In* V. J. Maglio and H. B. S. Cooke (eds.), *Evolution of African mammals,* pp. 154–248. Harvard University Press, Cambridge, Mass.

Howell, F. C. and Coppens, Y. (1976). An overview of Hominidae from the Omo Succession, Ethiopia. *In* Y. Coppens, F. C. Howell, G. L. Isaac and R. E. F. Leakey (eds.), *Earliest Man and environments in the Lake Rudolf Basin,* pp. 522–531. University of Chicago Press, Chicago.

Isaac, G. Ll. (1965). The stratigraphy of the Peninj beds and the provenance of the Natron australopithecine mandible. *Quaternaria, 7*: 101–130.

Johanson, D. C. and White, T. D. (1979). A systematic assessment of early African hominids. *Science, 203*: 321–330.

Jungers, W. L. and Grine, F. E. (1986). Dental trends in the australopithecines: the allometry of mandibular molar dimensions. *In* B. A. Wood, L. B. Martin and P. Andrews (eds.), *Major topics in primate and human evolution,* pp. 203–219. Cambridge University Press, Cambridge.

Kimbel, W. H., White, T. D. and Johanson, D. C. (1985). Craniodental morphology of the hominids from Hadar and Laetoli: evidence of *"Paranthropus"* and *Homo* in the mid-Pliocene of eastern Africa? *In* E. Delson (ed.), *Ancestors: the hard evidence,* pp. 120–137. Alan R. Liss, New York.

Leakey, L. S. B. (1959). A new fossil skull from Olduvai. *Nature, 184*: 491–493.

Leakey, L. S. B. and Leakey, M. D. (1964). Recent discoveries of fossil hominids in Tanganyika: at Olduvai and near Lake Natron. *Nature, 202*: 5–7.

Leakey, M. G. and Leakey, R. E. F. (1978). *Koobi Fora research project. Vol. 1. The fossil hominids and an introduction to their context 1968–1974.* Clarendon Press, Oxford.

Leakey, R. E. F. (1971). Further evidence of lower Pleistocene hominids from East Rudolf, North Kenya. *Nature, 231*: 241–245.

Mayr, E. (1950). Taxonomic categories in fossil hominids. *Cold Spring Harbor Symp. Quant. Biol., 15*: 108–118.

Mayr, E. (1963). The taxonomic evaluation of fossil hominids. *In* S. L. Washburn (ed.), *Classification and human evolution,* pp. 332–346. Aldine, Chicago.

Olson, T. R. (1978). Hominid phylogenetics and the existence of *Homo* in Member I of the Swartkrans Formation, South Africa. *J. Hum. Evol., 7*: 159–178.

Olson, T. R. (1981). Basicranial morphology of the extant hominoids and Pliocene hominids: the new material from the Hadar Formation Ethiopia and its significance in early hominid evolution and taxonomy. *In* C. B. Stringer (ed.), *Aspects of human evolution,* pp. 99–128. Taylor and Francis, London.

Olson, T. R. (1985). Cranial morphology and systematics of the Hadar Formation hominids and *"Australopithecus" africanus. In* E. Delson (ed.), *Ancestors: the hard evidence,* pp. 102–119. Alan R. Liss, New York.

Pilbeam, D. R. and Gould, S. J. (1974). Size and scaling in human evolution. *Science, 186*: 892–901.

Pilbeam, D. R. and Simons, E. L. (1965). Some problems of hominid classification. *Amer. Sci, 53*: 237–259.

Rak, Y. (1983). *The australopithecine face.* Academic Press, New York.

Rak, Y. and Howell, F. C. (1978). Cranium of a juvenile *Australopithecus boisei* from the lower Omo Basin, Ethiopia. *Amer. J. phys. Anthropol., 48*: 345–366.

Robinson, J. T. (1954a). The genera and species of the Australopithecinae. *Amer. J. Phys. Anthropol., 12*: 181–200.

Robinson, J. T. (1954b). Prehominid dentition and hominid evolution. *Evolution, 8*: 324–334.

Robinson, J. T. (1960). The affinities of the new Olduvai australopithecine. *Nature, 186*: 456–458.

Robinson, J. T. (1961). The australopithecines and their bearing on the origin of man and stone-tool making. *S. Afr. J. Sci., 57*: 3–13.

Robinson, J. T. (1963). Adaptive radiation in the australopithecines and the origin of man. *In* F.

C. Howell and F. Bourliere (eds.), *African ecology and human evolution*, pp. 385–416. Aldine, Chicago.

Robinson, J. T. (1965). Homo 'habilis' and the australopithecines. *Nature*, 205: 121–124.

Robinson, J. T. (1967). Variation and the taxonomy of the early hominids. *In* T. Dobzhansky, M. K. Hecht and W. Steere (eds.), *Evolutionary biology*, vol. 1, pp. 69–100. Appleton-Century-Crofts, New York.

Robinson, J. T. (1968). The origin and adaptive radiation of the australopithecines. *In* G. Kurth (ed.), *Evolution und Hominization*, 2nd edition, pp. 150–175. Gustav Fischer, Stuttgart.

Robinson, J. T. (1972). *Early hominid posture and locomotion*. University of Chicago Press, Chicago.

Robinson, J. T. and Steudel, K. (1973). Multivariate discriminant analysis of dental data bearing on early hominid affinities. *J. Hum. Evol.*, 2: 509–527.

Romer, A. S. (1930). *Australopithecus* not a chimpanzee. *Science*, 71: 482–483.

Saban, R. (1983). Les veines meningees moyennes des australopitheques. *Bull. Mem. Soc. d'Anthrop. Paris*, 10: 313–324.

Simpson, G. G. (1945). The principles of classification and a classification of mammals. *Bull. Amer. Mus. Nat. Hist.*, 85: 1–350.

Skelton, R. R., McHenry, H. M., and Drawhorn, G. M. (1986). Phylogenetic analysis of early hominids. *Curr. Anthropol.*, 27 21–43.

Smith, B. H. (1986). Dental development in *Australopithecus* and early *Homo*. *Nature*, 323: 327–330.

Sollas, W. J. (1926). A sagittal section of the skull of *Australopithecus africanus*. *Quart. J. Geol. Soc. Lond.*, 82: 1–11.

Swedlund, A. C. (1974). The use of ecological hypotheses in australopithecine taxonomy. *Amer. Anthropol.*, 76: 515–529.

Tobias, P. V. (1967). *The cranium and maxillary dentition of* Australopithecus (Zinjanthropus) boisei. Olduvai Gorge, Vol. 2. Cambridge University Press, Cambridge.

Tobias, P. V. (1968). The taxonomy and phylogeny of the australopithecines. *In* B. Chiarelli (ed.), *Taxonomy and phylogeny of Old World Primates with references to the origin of Man*, pp. 277–315. Rosenberg and Sellier, Torino.

Tobias, P. V. (1973a). Darwin's prediction of the African emergence of the genus *Homo*. *Proc. Darwin Centenary Symp.*, Rome 1971, pp. 63–85. Acad. Nas. Lincei, Rome.

Tobias, P. V. (1973b). Implications of the new age estimates of the early South African hominids. *Nature*, 246: 79–83.

Tobias, P. V. (1975). New African evidence on the dating and the phylogeny of the Plio-Pleistocene Hominidae. *Trans. R. Soc. New Zealand*, 13: 289–296.

Tobias, P. V. (1978a). The place of *Australopithecus africanus* in hominid evolution. *In* D. J. Chivers and K. A. Joysey (eds.), *Recent advances in primatology*, vol. 3, *Evolution*, pp. 373–394. Academic Press, New York.

Tobias, P. V. (1978b). Position et role des australopithecines dans la phylogenese humaine, avec etude particuliere de *Homo habilis* et des theories controversees avancees a propos des premiers hominides fossiles de Hadar et de Laetolil. *Les origines humaines et les Epoques de l'Intelligence*, pp. 38–75. Foundation Singer-Polignac/Masson, Paris.

Tobias, P. V. (1978c). South African australopithecines in time and hominid phylogeny, with special reference to dating and affinities of the Taung skull. *In* C. J. Jolly (ed.), *Early hominids of Africa*, pp. 45–84. Duckworth, London.

Tobias, P. V. (1980a). "*Australopithecus afarensis*" and *A. africanus*: critique and an alternative hypothesis. *Palaeont. Afr.*, 23: 1–17.

Tobias, P. V. (1980b). A survey and synthesis of the African hominids of the late Tertiary and early Quaternary Periods. *In* L. K. Konigsson (ed.), *Current argument on early Man*, pp. 86–113. Pergamon, Oxford.

Vrba, E. S. (1975). Some evidence of chronology and palaeoecology of Sterkfontein, Swartkrans and Kromdraai from the fossil Bovidae. *Nature*, 254: 301–304.

Vrba, E. S. (1981). The Kromdraai australopithecine site revisited in 1980: recent investigations and results. *Ann. Tvl. Mus.*, 33: 17–60.

Walker, A. C. (1972). Chesowanja australopithecine. *Nature*, 238: 108–109.

Walker, A. C. (1981). Dietary hypotheses and human evolution. *Phil. Trans. Roy. Soc. Lond., B292*: 57–64.

Walker, A. C., Leakey, R. E. F., Harris, J. M. and Brown, F. H. (1986). 2.5-Myr *Australopithecus boisei* from west of Lake Turkana, Kenya. *Nature, 322*: 517–522.

White, T. D. (1985). The hominids of Hadar and Laetoli: an element-by-element comparison of dental samples. *In* E. Delson (ed.), *Ancestors: the hard evidence*, pp. 138–152. Alan R. Liss, New York.

White, T. D., Johanson, D. C. and Kimbel, W. H. (1981). *Australopithecus africanus:* its phyletic position reconsidered. *S. Afr. J. Sci., 77*: 445–470.

Wolpoff, M. H. (1974). The evidence for two australopithecine lineages in South Africa. *Yrbk. Phys. Anthropol., 17*: 113–139.

Wood, B. A. and Chamberlain, A. (1986). *Australopithecus:* grade or clade? *In* B. A. Wood, L. B. Martin and P. Andrews (eds.), *Major topics in primate and human evolution*, pp. 220–248. Cambridge University Press, Cambridge.

Specimen Index*

AL 58-22, 266
AL 162-28, 99–103, 266
AL 277-1, 183
AL 288-1, 100, 117, 122, 130–132, 133, 137, 140, 157, 266
AL 333-3, 140
AL 333-19, 157
AL 333-45, 91–92, 100, 266
AL 333-57, 157
AL 333-62, 157
AL 333-63, 157
AL 333-86, 48
AL 333-93, 157
AL 333-105, 100
AL 333-115, 152
AL 333W-4, 157
AL 333W-12, 266
AL 333W-32, 266
AL 333W-60, 266
KB 5223, 5–7, 22–24, 27, 30, 32, 33, 37, 38, 234
KNM-CH 1, 255, 491
KNM-CH 304, 253
KNM-ER 405, 197
KNM-ER 406, 49, 51, 91, 97, 101, 184–187, 193, 195–197, 248, 255, 491, 512, 513
KNM-ER 407, 50, 51, 97, 186–187, 255, 304
KNM-ER 729, 49, 50
KNM-ER 731, 141
KNM-ER 732, 50, 51, 97, 185–186, 197–198, 248, 511
KNM-ER 733, 44, 253, 491
KNM-ER 736, 137, 144
KNM-ER 738, 118, 140, 141, 144
KNM-ER 739, 451
KNM-ER 741, 118
KNM-ER 802, 208

KNM-ER 812, 45, 47, 49, 50
KNM-ER 818, 50, 183, 212
KNM-ER 820, 49, 50, 141
KNM-ER 992, 144
KNM-ER 993, 140
KNM-ER 1171. 211–212
KNM-ER 1463, 140
KNM-ER 1464, 127
KNM-ER 1465, 141
KNM-ER 1468, 50
KNM-ER 1469, 50
KNM-ER 1470, 91, 100, 299
KNM-ER 1472, 140, 141
KNM-ER 1475, 141
KNM-ER 1477, 45, 47, 47, 50
KNM-ER 1480, 141
KNM-ER 1481. 140, 141
KNM-ER 1482, 212, 487
KNM-ER 1500, 118, 127–132
KNM-ER 1503, 118, 141
KNM-ER 1505, 118
KNM-ER 1507, 49, 50
KNM-ER 1509, 212
KNM-ER 1590, 487
KNM-ER 1801, 487
KNM-ER 1802, 208, 211
KNM-ER 1804, 45
KNM-ER 1805, 91, 93–94, 100, 101
KNM-ER 1808, 141
KNM-ER 1813, 93, 100, 101, 299
KNM-ER 1820, 45, 47–51
KNM-ER 1823, 127
KNM-ER 1824, 127
KNM-ER 1825, 127
KNM-ER 3228, 141
KNM-ER 3229, 212
KNM-ER 3230, 50, 170, 249, 253, 254, 464
KNM-ER 3734, 487
KNM-ER 3735, 141

---

*This index includes only those specimens that are illustrated and/or referred to in the text. Fossil hominid specimens are listed also in Table 7.2 (pp. 119–120), Table 9.1 (p. 134), 9.4 (p. 139), Table 10.1 (p. 151), Table 12.1 (p. 194), Table 13.6 (p. 207), Table 13.10 (p. 213), Table 13.11 (p. 214), Table 14.1 (p. 226), Table 18.1 (p. 286), Fig. 22.3 (p. 328), Fig. 22.4 (p. 329), Table 27.1 (pp. 466–479), and Table 28.2 (p. 489).

KNM-ER 5429, 212
KNM-ER 13750, 97–101, 247, 249
KNM-ER 15930, 128, 248–249, 253, 254
KNM-WT 15000, 140, 141, 144, 168
KNM-WT 16005, 182, 219, 248, 249, 251, 512
KNM-WT 17000, 39, 49, 51, 85, 91–92, 97–102, 107, 108, 171, 185–186, 189, 193, 198, 218, 241–242, 248–256, 259–268, 278, 290, 291, 301, 384, 405, 407, 465, 491, 493, 511–515
KNM-WT 17400, 97, 100, 101, 247–248, 251, 255
LH 2, 48, 50
LH 3, 48, 208
LH 6, 48
LH 21, 48, 89, 90, 93
MLD 2, 48, 50, 290
MLD 6, 290
MLD 37, 101, 186, 304
MLD 38, 101, 186, 304
OH 3, 367, 464
OH 4, 464
OH 5, 49, 97–103, 184–186, 188, 193, 195–197, 247, 248, 251, 255, 304, 367, 463, 464, 491, 512
OH 7, 100, 152, 153, 167–170, 463, 464
OH 8, 127, 132, 152, 153, 169, 345, 367, 463
OH 13, 100
OH 16, 464
OH 24, 100, 304, 464
OH 35, 118, 132
OH 44, 464
OH 62, 132, 136, 141
Omo 7, 50
Omo 18-31, 206, 218
Omo 18-1967-18, 61, 127, 218, 248–249, 251, 260–261, 263, 266, 417, 489, 512
Omo 33-507, 215
Omo 33-508, 213–217
Omo 57.4-41, 213, 219
Omo 323, 141
Omo 323-896, 451
Omo 338, 489
Omo L 10-21, 5, 6, 8, 11, 22–26, 29, 30, 32, 37
Omo L 40-19, 451
Omo L 55-33, 213, 218
Omo L 60-2, 216–217
Omo L 394-1, 299
Omo L 398-120, 206
Omo L 398-847, 5, 6, 8, 11, 22–26, 29, 30, 32, 36–38
Omo L 398-1223, 215
Omo L 420-15, 215
Omo L 427-7, 206, 215–217, 219
Omo L 465-111, 206
Omo L 628-4, 213–217
Omo L 860-2, 206, 215, 219, 452, 464
Omo L 894-1, 199, 487
Omo L 7a-125, 183, 204, 213, 218–219
Omo L 74a-21, 215–217, 219
Omo L 18s-32, 487
Omo L 338x-40, 213–217
Omo L 338y-6, 98, 99, 251, 253, 511
Omo W 978, 218
Peninj mandible, 212, 249
SK 13, 430, 438
SK 15, 168, 294, 313
SK 16, 430, 438
SK 18, 168
SK 23, 291
SK 34, 212
SK 42, 430, 435, 436, 438
SK 43, 168
SK 46, 183–184
SK 48, 430, 438
SK 50, 141, 463
SK 54, 463
SK 61, 47, 49, 50
SK 62, 49, 50
SK 63, 49, 50
SK 64, 49, 50
SK 82, 141, 463
SK 84, 152, 155, 163, 168
SK 85, 163, 168
SK 88, 212
SK 97, 141
SK 438, 49, 50, 463
SK 826, 208, 212
SK 847, 294
SK 859, 463
SK 1585, 97–103, 463
SK 1587, 212
SK 2831, 463
SK 3155, 141, 463
SK 3978, 45, 49, 50, 463
SK 45690, 152
SKW 1261, 164, 168
SKW 2954, 159, 161, 163
SKW 3646, 159–162
SKW 3774, 159, 161
SKW 14147, 163, 168
SKW 34805, 159–161
SKX 162, 46, 224–225, 291
SKX 163, 224–225
SKX 240, 225
SKX 242, 225
SKX 247, 163, 168

Specimen Index

SKX 265, 224–225, 290, 291
SKX 308, 46, 224–225
SKX 310, 225
SKX 312, 225
SKX 344, 164, 168
SKX 1084, 160, 162, 168
SKX 1788, 46, 224–225
SKX 3062, 164
SKX 3342, 159–162
SKX 3498, 164–165
SKX 3602, 156, 158, 160
SKX 3646, 163
SKX 3699, 160, 168
SKX 4446, 224–225
SKX 5013, 224–225
SKX 5016, 152, 154, 169
SKX 5017, 152, 153, 164
SKX 5018, 155, 157, 164, 168
SKX 5019, 153, 156
SKX 5020, 152, 153, 155, 163, 168–169
SKX 5021, 153, 156
SKX 5022, 153, 156
SKX 8761, 156, 159
SKX 8963, 152, 154, 169
SKX 9449, 153, 156
SKX 12814, 156, 160
SKX 13476, 156
SKX 19576, 157, 166
SKX 21841, 5, 6, 8, 22, 23, 30, 32, 36–38
SKX 22511, 157, 166
SKX 22741, 157, 164, 166, 168
SKX 27431, 157, 164, 166, 168
SKX 27504, 154, 166, 168
SKX 30220, 166
SKX 31117, 162, 164, 168
SKX 33355, 156, 166
SKX 33380, 165, 166, 168
SKX 35439, 156, 166
SKX 35822, 166
SKX 36712, 156, 166
Sts 5, 89, 99–101, 183–187, 285, 290, 293, 304
Sts 7, 418
Sts 13, 185, 288
Sts 14, 117, 133, 136–138, 161, 288
Sts 17, 185–186, 287, 290
Sts 19, 99, 186–187, 304
Sts 24, 48
Sts 25, 186
Sts 28, 429, 435, 436, 438
Sts 30, 429, 438
Sts 31, 429, 435, 438, 439
Sts 34, 140
Sts 36, 290
Sts 52, 50, 184–186, 287, 290, 429, 438
Sts 53, 185–186
Sts 60, 99, 100
Sts 70, 100
Sts 71, 92, 184–185, 287, 289, 290
Stw 13, 186
Stw 27, 156, 158
Stw 28, 157, 166
Stw 53, 299
Stw 65, 161
Stw 68, 162
Stw 252, 91–92, 285–292
Stw 284, 5–7, 22, 23, 29, 30, 32, 37, 38
Stw 402, 5–7, 22, 23, 29, 30, 32, 36–38
Taung, 48, 50, 51, 89, 101–103, 290, 291, 299, 300, 302–305, 407, 509, 514
Ternifine mandibles, 144
TM 1511, 185–186, 429, 438
TM 1517, 3, 141, 156, 159, 161, 168–169, 186, 430, 438, 451
TM 1601, 212

Subject Index

Allometry
  body size, 61, 62, 115–125, 133–148, 427–429, 491–493, 510–511
  encephalization. *See* Cranial capacity
  tooth size, 122, 133–148, 212, 240–241, 270, 274–275, 291–292, 427–429, 491–494, 510–511
Antarctica geology, 391, 396
Australopithecus
  afarensis
    cranial features, 85, 86, 97, 98, 103, 185–186, 198
    dentition, 39, 122, 176–182, 204–214, 229, 487–492
    mandible, 61–77, 182–183
    postcrania, 130–132, 141, 144–145, 157
    taxonomic considerations of, 92, 93, 120, 122, 175–176, 188–189, 263–266, 270, 295, 300–301
  africanus
    cranial features, 103, 183–186, 197, 304
    dentition, 7, 23, 25, 27–30, 32, 33, 122, 176–182, 204–213, 229, 427–447, 488–492
    dietary reconstruction, 427–447
    mandible, 182–183
    postcrania, 118, 130, 141, 144–145, 161
    taxonomic considerations of, 93, 100, 188–189, 259–260, 263–265, 285, 293–305
  see also Paranthropus

Basicranium, 107–112, 186–187, 189, 247, 253, 255, 256, 296–299, 304, 493
Bipedalism, 94, 101–103
Bite force. *See* Masticatory apparatus
Body size. *See* Allometry
Bovidae. *See* Paleoenvironmental reconstruction
Brain, 109–110, 255; *see also* Cranial capacity

Carpals. *See* Individual bones
Cercopithecidae, mandibles of, 55, 62–72, 75, 77, 81–83

Cerebral rubricon, 168
Chimpanzees. *See Pan*
Competitive exclusion, 485–497, 502–503
Cope's law, 133
Cranial blood flow. *See* Endocranium
Cranial capacity, 91, 94, 97–105, 110, 133–148, 251, 256, 491–494
Cranial features, 247–248, 253, 255, 256–261, 266, 285–292, 296–299, 493
  masticatory effects on, 195–198; *see also* Dentition; Facial morphology; Mandible
Culture. *See* Tools; Fire
Cuneiforms, 162, 164

Dating; *see also* Individual sites
  biostratigraphy and, 232, 313–14, 317–325
  global climate and, 232, 314
  paleomagnetism, 232, 325
  radiometric, 233, 314, 320, 325
  thermoluminescence, 232, 314, 320
Demography, 43, 107–108, 463, 465
Dentine formation, 43
Dentition; *see also* Allometry; Dentine formation; Dietary reconstruction; Enamel
  canine, 63, 64, 185, 227, 493
  eruption sequences, 47, 48, 185, 256, 493
  growth and development, 51
  incisors, 227, 233–234, 236
    crown formation in, 45, 47
  microwear, 427–447, 515
  molars, 3–5, 228–229, 234–238, 272–276, 429–444, 488
  odontometrics, 176–182, 286, 298–299
  premolars, 74, 168, 199–222, 228, 234–238, 272–277, 488
    molarization of, 44
  root formation, 46
Dietary reconstruction, 75–77, 149, 169–170, 418, 420, 422–423, 427–447, 515

Ecology comparisons among early hominid taxa, 456–462

Effect hypothesis, 422
Enamel
  formation, 33, 34, 39, 43, 44, 46, 49, 51
  hunter-schreger bands, 4, 39
  incremental lines, 4, 33–35, 37, 39, 43–45
  perikymata, 4, 37, 45
  prisms
    decussation of, 4
    developmental rates of, 4
    packing patterns in, 4, 34, 43
  thickness, 3–42, 45
Endocranium, 85–96, 97–105, 108–110, 251–252, 255, 256, 296–299, 493
  lunate sulcus, 99, 101, 102
  venus sinuses, 85–96, 251, 256
    occipital marginal, 85–96, 98–103
  vertebral plexus, 87–88, 90, 94
Erection, 269
Extant great apes, 47, 152, 157; see also Gorilla, Homo, Hylobates, Pan, Pongo
Extinction, 383, 465, 499–505, 516

Facial morphology, 183–186, 188–189, 193–198, 224, 247, 253–256, 260–261, 266, 285–287, 291, 296–299
Femora, 133–141, 144–145
Fire, 315
Foot, 149–173
Footprints (Laetoli), 93
Foramen magnum, 107–110
Foramina, emissary, 85–96

Garusi, 2, 3
Geology. See Individual sites
Gibbons. See Hylobates
Global climatic events, 376–377, 383–403, 411–420
  and effects on biota, 383–403, 405–426, 465, 501–502, 516
Gorilla
  cranial features, 98, 108, 188, 304
  dentition, 5, 12–21, 28, 180–182, 194–195, 197–198
  mandibles, 63–77, 182–183, 194–195, 197–198
  postcrania, 116–117, 135–139, 144, 157, 182

Habitat preference, 343–381, 456–462, 465
Hands, 149–173; see also Individual bones
Homo
  early
    cranial features, 103
  dentition, 32, 33, 49, 204–213, 488–492
  postcrania, 140–141, 166
  taxonomic considerations of, 464–465
  tool use, 149, 167–171
  erectus
    cranial features, 108, 304
    migration to Eurasia, 383, 399, 516
    postcrania, 141, 144–145, 150, 168
  habilis
    cranial features, 304
    postcrania, 132, 144–145, 170
    taxonomic considerations of, 100, 293–305
  sapiens
    cranial features, 85
    dentition, 5, 11–21, 25, 27–29, 39, 44, 47
    postcrania, 116–117, 127, 130–132, 135–137, 144–145, 154–166
Humerus, 159–161
Hylobates
  dentition, 33
  mandibles, 63–77
  postcrania, 116–117

Koobi fora. See Turkana Basin
Kromdraai, dating, 300, 302, 303, 317, 321–323

Limbs
  lower, 149–173
  upper, 149–173
Locomotion, 169–170; see also Bipedalism
Lunate sulcus. See Endocranium

Makapansgat, dating, 302, 317, 320
Mandible(s), 55–83, 127, 182–183, 188–189, 213, 224, 248–249, 252–253, 260, 266, 452–454
Masticatory apparatus; see also Dentition; Mandible(s); Facial morphology; Cranium
  stress in, 39, 55–8, 265, 428–429
  variation in, 193–198, 218–219, 255–256
Maxilla, 57, 185, 196, 224, 249, 288
Megadontia. See Allometry
Metacarpals, 152, 155, 161–163, 168
Metatarsals, 152, 153, 163, 165, 166, 168
Miopithecus, 62
Muscles
  masseter, 57, 73–75, 185, 195–196
  temporalis, 73–74

Nomenclature considerations, 509–511

Olduvai Gorge, paleoenvironmental reconstruction, 367, 375, 455
Omo. *See* Turkana Basin
Orangutan. *See Pongo*

Paleoenvironmental reconstruction, 410–423, 515–516; *see also* Individual sites
 and faunal remains, 343–381, 414–421
 and icesheet systems, 383–403
 and oxygen isotope analysis, 383–403
 and palynology, 377
Pan
 cranial features, 98–103, 108, 304
 dentition, 5, 12–21, 28, 180–182, 493
 mandibles, 63–77, 182–183
 postcrania, 116–117, 130–132, 135–137, 144, 153–166
Paranthropus
 aethiopicus
  taxonomic status of, 189, 251, 260, 262–263, 512, 514
 crassidens
  taxonomic status of, 233–242, 259, 269, 430, 511
 see also Australopithecus
Patellae, 160, 162, 168
Pelvis, 141, 169
Phalanges
 manual, 152–157, 164–166, 168
 pedal, 152, 168
Plantar aponeurosis, 152, 153
Pongo
 cranial features, 188, 304
 dentition, 5, 13–21, 28, 180–182
 mandibles, 63–77
 postcrania, 116–117, 135–139, 144, 157
Postcrania, 127–132, 451
 and body proportions, 130–132, 133, 136
 diaphyses of, 133, 136
 joint size of, 116–122, 133
 taxonomic considerations and, 139–141, 451
Predation, 502–503

Radii, 156, 158, 160, 168
Reproductive strategies, 494

Scanning electron microscopy
 dental microwear, 427–447
 enamel structure, 9, 43, 45

Sexual dimorphism
 craniodental, 48, 51, 175–192, 251–252, 254, 263, 289
 postcranial, 92, 118–120, 136–138, 144–145
Siamang. *See Hylobates*
Sivapithecus, 28, 30, 33, 39, 442
Sterkfontein
 dating, 290, 300, 302–303, 317, 320, 323
 paleoenvironmental reconstruction, 374
Swartkrans
 dating, 229–233, 300, 302, 313–314, 317–319, 322–323, 500
 environmental reconstruction, 314, 374, 455–456
 fossil provenience, 149–152, 223–224, 313–314
 geology, 223, 311–313
 postcrania, 110, 149–173
Sympatry, 300, 378, 454–455, 465, 485–497

Taphonomy, 343–345, 377, 451–454, 462–463, 465, 515
Tarsals. *See* Individual bones
Taung dating, 300, 302–303, 317–321, 323
Taxonomic considerations, 100, 171, 175–192, 264, 270–272, 279–282, 384, 405–407, 486–494, 511–515; *see also* Specific taxa
Teeth. *See* Dentition
Tibio-fibular articulation, 127
Tools, 100, 149–173, 313–315, 378, 416–418, 464, 465, 494, 500
Triquetrals, 164–165
Turkana Basin
 dating, 319, 325–330
 environmental reconstruction, 330–339, 367, 371–376, 455–456
 stratigraphic correlations, 327
 temporal distribution of hominid specimens, 328–329

Ulnae, 156, 159, 160

Vertebrae, 161–162

Zygomatic arch, 74

For Product Safety Concerns and Information please contact our
EU representative GPSR@taylorandfrancis.com Taylor & Francis
Verlag GmbH, Kaufingerstraße 24, 80331 München, Germany